Lecture Notes in Networks and Systems

Volume 596

The series "Lecture Notes in Networks and Systems" publishes the latest developments in Networks and Systems—quickly, informally and with high quality. Original research reported in proceedings and post-proceedings represents the core of LNNS.

Volumes published in LNNS embrace all aspects and subfields of, as well as new challenges in, Networks and Systems.

The series contains proceedings and edited volumes in systems and networks, spanning the areas of Cyber-Physical Systems, Autonomous Systems, Sensor Networks, Control Systems, Energy Systems, Automotive Systems, Biological Systems, Vehicular Networking and Connected Vehicles, Aerospace Systems, Automation, Manufacturing, Smart Grids, Nonlinear Systems, Power Systems, Robotics, Social Systems, Economic Systems and other. Of particular value to both the contributors and the readership are the short publication timeframe and the world-wide distribution and exposure which enable both a wide and rapid dissemination of research output.

The series covers the theory, applications, and perspectives on the state of the art and future developments relevant to systems and networks, decision making, control, complex processes and related areas, as embedded in the fields of interdisciplinary and applied sciences, engineering, computer science, physics, economics, social, and life sciences, as well as the paradigms and methodologies behind them.

Indexed by SCOPUS, INSPEC, WTI Frankfurt eG, zbMATH, SCImago.

All books published in the series are submitted for consideration in Web of Science.

For proposals from Asia please contact Aninda Bose (aninda.bose@springer.com).

Radek Silhavy · Petr Silhavy · Zdenka Prokopova
Editors

Software Engineering Application in Systems Design

Proceedings of 6th Computational Methods in Systems and Software 2022, Volume 1

Springer

Editors
Radek Silhavy
Faculty of Applied Informatics
Tomas Bata University in Zlín
Zlín, Czech Republic

Petr Silhavy
Faculty of Applied Informatics
Tomas Bata University in Zlín
Zlín, Czech Republic

Zdenka Prokopova
Faculty of Applied Informatics
Tomas Bata University in Zlín
Zlín, Czech Republic

ISSN 2367-3370 ISSN 2367-3389 (electronic)
Lecture Notes in Networks and Systems
ISBN 978-3-031-21434-9 ISBN 978-3-031-21435-6 (eBook)
https://doi.org/10.1007/978-3-031-21435-6

This Springer imprint is published by the registered company Springer Nature Switzerland AG
The registered company address is: Gewerbestrasse 11, 6330 Cham, Switzerland

Preface

This book contains the refereed proceedings of the Computational Methods in Systems and Software 2022 (CoMeSySo 2022) conference, which took place in October 2022.

Papers in the book Software Engineering Application in Systems Design describe novel techniques and methodologies applied to real-world situations. Furthermore, the fundamental study provides unique software engineering and informatics methods in systems and software systems.

The CoMeSySo 2022 conference aims to provide an international venue to discuss the most recent high-quality research achievements in all areas linked to intelligent systems, software, and networks.

The theoretical foundations and applications of software engineering, computational techniques in engineering and science, and artificial intelligence are all discussed. The articles include subjects such as software engineering, cybernetics and automation control theory, econometrics, mathematical statistics, and artificial intelligence.

In 2022, CoMeSySo has received 280 entries (all parts), 165 of which were accepted for publication.

The editors trust that the subsequent proceedings will interest and benefit readers in their research work.

September 2022

Radek Silhavy
Petr Silhavy
Zdenka Prokopova

Program Committee

Program Committee Chairs

Petr Silhavy, Ph.D.	Department of Computers and Communication Systems, Faculty of Applied Informatics, Tomas Bata University in Zlin, Czech Republic
Radek Silhavy, Ph.D.	Department of Computers and Communication Systems, Faculty of Applied Informatics, Tomas Bata University in Zlin, Czech Republic
Zdenka Prokopova, Ph.D.	Associate Professor at Department of Computers and Communication Systems, Tomas Bata University in Zlin, Czech Republic
Krzysztof Okarma, Ph.D., D.Sc.	Faculty of Electrical Engineering, West Pomeranian University of Technology, Szczecin, Poland
Prof. Roman Prokop, Ph.D.	Professor at Department of Mathematics, Tomas Bata University in Zlin, Czech Republic
Prof. Viacheslav Zelentsov	Doctor of Engineering Sciences, Chief Researcher of St. Petersburg Institute for Informatics and Automation of Russian Academy of Sciences (SPIIRAS), Russian Federation
Prof. Lipo Wang	School of Electrical and Electronic Engineering, Nanyang Technological University, Singapore
Silvie Belaskova	St. Anne's University Hospital Brno, International Clinical Research Center, Czech Republic
Roman Tsarev, Ph.D.	Associate Professor, Department of Information Technology, International Academy of Science and Technologies, Moscow, Russia
Stefano Cirillo	Department of Computer Science, University of Salerno, Fisciano (SA), Italy

International Program Committee Members

Pasi Luukka, D.Sc.	President of North European Society for Adaptive and Intelligent Systems and School of Business and School of Engineering Sciences Lappeenranta University of Technology, Finland
Izabela Jonek-Kowalska, Ph.D.	Faculty of Organization and Management, The Silesian University of Technology, Poland
Maciej Majewski, Ph.D., D.Sc.	Department of Engineering of Technical and Informatic Systems, Koszalin University of Technology, Koszalin, Poland
Alena Vagaska, Ph.D.	Department of Mathematics, Informatics and Cybernetics, Faculty of Manufacturing Technologies, Technical University of Kosice, Slovak Republic
Boguslaw Cyganek, Ph.D., D.Sc.	Department of Computer Science, University of Science and Technology, Krakow, Poland
Piotr Lech. Ph.D.	Faculty of Electrical Engineering, West Pomeranian University of Technology, Szczecin, Poland
Monika Bakosova, Ph.D.	Associate Professor, Institute of Information Engineering, Automation and Mathematics, Slovak University of Technology, Bratislava, Slovak Republic
Pavel Vaclavek, Ph.D.	Associate Professor, Faculty of Electrical Engineering and Communication, Brno University of Technology, Brno, Czech Republic
Miroslaw Ochodek, Ph.D.	Faculty of Computing, Poznan University of Technology, Poznan, Poland
Olga Brovkina, Ph.D.	Global Change Research Centre Academy of Science of the Czech Republic, Brno, Czech Republic
Elarbi Badidi, Ph.D.	College of Information Technology, United Arab Emirates University, Al Ain, United Arab Emirates
Gopal Sakarkar	Shri. Ramdeobaba College of Engineering and Management, Republic of India
V. V. Krishna Maddinala	Assistant Professor, GD Rungta College of Engineering and Technology, Republic of India
Anand N. Khobragade	Scientist, Maharashtra Remote Sensing Applications Centre, Republic of India
Abdallah Handoura	Assistant Professor, Computer and Communication Laboratory, Telecom Bretagne—France

Organizing Committee Chair

Radek Silhavy, Ph.D. Tomas Bata University in Zlin, Faculty of Applied Informatics, email: comesyso@openpublish.eu

Conference Organizer (Production)

Silhavy s.r.o.
Web: https://comesyso.openpublish.eu
Email: comesyso@openpublish.eu

Conference website, Call for Papers

https://comesyso.openpublish.eu

Contents

Determining the Level of Awareness of Electronic Waste Impact on the Environment Among Students in at University of Technology: A Case Study

Ramadile Moletsane(✉)

Vaal University of Technology, Private Bag x 021, Vanderbijlpark, South Africa
ramadilem@vut.ac.za

Abstract. Information and Communication Technology (ICT) has transformed the way we work, live and interact. ICT is everywhere and grows in an unprecedented manner. Most people embrace the disruptive change brought by it. Sadly, though people are not aware of detrimental effects of ICT equipment when it has reached its end-of-life. ICT equipment that has reached end-of-life is often referred to as Electronic waste (e-waste). E-waste when managed in a way not friendly to the environment has negative impacts to both environment and health. The objective of the study was twofold. Firstly, to determine the level of awareness of detrimental effects of e-waste generated by ICT equipment on the environment among Information Technology (IT) Diploma first year students at University of Technology. Secondly, based on the findings then suggests the solution. Lack of awareness about detrimental effects of e-waste generated by ICT equipment in Higher Education Institutions (HEIs). In this qualitative study semi structured interviews were used to gather the data. Qualitative content analysis was used to analyse data. ATLAS.ti software was used to assist in the process of analysing the data. The students in IT diploma classroom were found to have low levels of awareness of the negative impacts of electronic waste generated by ICT equipment on the environment. Syllabus on the impact of retired ICT equipment on the environment should be part of the IT Diploma curriculum. HEIs should start implementing initiatives that promote the wellbeing of our environment.

Keywords: Awareness · Environment and health · Electronic waste · Information communication technology · Higher education institutions

1 Introduction

E-waste is one of today's challenges facing humanity [1]. The main problem of e-waste is the amount of toxic components it comprises with and growth rate each year [2, 3]. Various studies have shown that toxins released into the air or water, as a result of e-waste, cause disorders e.g. skin diseases and under-development of the brain in children [4, 5]. The worldwide ICT market is growing and expected to reach 3.68 trillion United

R. Silhavy et al. (Eds.): CoMeSySo 2022, LNNS 596, pp. 1–10, 2023.
https://doi.org/10.1007/978-3-031-21435-6_1

States dollars in value by end 2021 [6] The concepts central to this study namely ICT and E-waste will be discussed in subsequent Sections.

According to Ratheeswari [7] ICT refers to an array of technologies that provide access to information through telecommunication. These technologies are not limited to the internet, cell phones, and wireless networks. ICT is similar to Information Technology except that if focuses to communication technology [7]. ICT is focused on storage, retrieval, transmission, manipulation of digital data. Impotently, it covers how these different uses of ICT work with one another [8]. ICT is a field of work and study that "includes technologies such as desktop and laptop computers, software, peripherals, and connections to the Internet that are intended to fulfil information processing and communications functions" [9]. Another definition for ICT comes from UNESCO, which states ICT is "a scientific, technological and engineering discipline and management technique used in handling information, its application and association with social, economic and cultural matters" [10]. Thus, ICT uses the newest technologies to handle information. ICT has changed the way we live, work and interact with one an- other in unimaginable ways. ICT has impacted social factors such as education, profession including economic factors like income and resources [11]. ICT have both negative and positive impacts on our everyday life. Its impacts are not limited to the following spheres: education, healthcare, social development, etc.

Not long ago, literature have reported the opportunities and benefits of ICT for improving pedagogy, be at schools or higher education [12]. Every individual who is related to education is directly influenced by ICT including students, teachers and administrators [13]. Teachers are using ICT to make pedagogy easier and accessible every-where and anytime. ICT "alters thought patterns, enriches existing educational models and provides new training models" [14]. E-learning, is an example of the methods of education or training created by ICT. According to Mouyabi [15] e-learning grows at an unprecedented rate that forces institutions and organisations to consider it as cost and time saving training model. Research has indicated that the suitable utilisation of ICTs can catalyse the paradigmatic move in both content and instructional method that is at the core of education change in the 21st Century [16]. ICT has the potential to improve students' 21st century learning skills [17]. However, Livingstone [18] noted that "a simple increase in ICT provision does not guarantee enhanced educational performance". ICT also penetrated healthcare sector.

ICT can play a critical role in improving health care of communities and individuals around the world. With the potential of ICT hospitals are able to store medical data that can be available anytime and anywhere. Patients are able to receive their medical data at hand. ICT foster collaboration between doctors and prevent medical omissions. ICT helps healthcare service providers to collect, store, retrieve and transfer information electronically. According to Gulavani and Kulkarni [19] by making right patient and medical information readily available to nurses, physicians and technicians at the right time should improve health. ICT through the use of information systems like clinical decision support systems (CDSS) makes this activity easy and possible. CDSS are "are computer systems designed to impact clinician decision making about individual patients at the point in time that these decisions are made" [20]. Hypoglycemia events in critically ill people are greatly decreased with CDSS that measures blood glucose [21]. Castillo and Kelemen [22] noted that CDSS can help in the decrease of clinical mistakes and decrease in unfriendly medication occasions, guarantee thorough treatment of patient sicknesses and conditions, encourage the adherence to procedures. "to err is human" therefore CDSS coupled with other medical systems are a major approach to improve patients' safety [20]. In the United States of America around 98 000 people lives are lost annually due to medical errors [23]. The discussions in the previous texts are some of the benefits offered by ICT to our daily lives. The list is endless and every corner of our lives is somehow touched by this digital innovation. ICT have both negative and positive impacts on our everyday life. In the previous texts the positive impacts of ICT were discussed. With that said, research has noted the inherent impediments of ICT [24, 25]. The worth noting detrimental effect of ICT in our life related to this study is when these ICT devices have reached their end-of-life stage. The next Sections discusses E-waste and its impact on health and the environment.

There is no universally accepted definition of e-waste [26]. According to Balakrishnan, Anand [27] e-waste is any device dependent on electric or electromagnetic fields to function correctly. The retired ICT equipment are often referred to e-waste or WEEE [28]. E-waste as any electrical powered device that has reach end-of-life stage [29]. E-waste can be classified into three categories: household appliances, ICT equipment and consumer equipment [3]. However, a wider definition encompasses 10 categories as defined by the European Union directives on Waste Electrical and Electronic Equipment (WEEE) [30, 31]. They put forward the following categories: Large household appliances, Small household appliances, Consumer equipment, IT and telecommunications equipment, Consumer equipment, lighting equipment, Electric and electronic tools, but excluding large scale stationary industrial tools, Toys, leisure and sports equipment, automated dispensers, monitoring and control tools, medical appliances with exception of all implants and infected products [32]. The focus of this study is IT and telecommunications equipment category. In the context of this study e-waste is defined as any electrical or electronic ICT equipment that is no longer needed by the owner or has regardless functional or non-functional.

ICT equipment when has reached its end-of-life stage present a huge threat on the environment and health. These ICT devices get obsolete because of reasons like short life span [33]. According to Lobos and Babbitt [34] short life span is caused by planned obsolescence—a deliberate action by product manufacturers of making products obsolete before their technical functional obsolescence [34, 35]. The problem with obsolete units is that they are often improperly disposed of, thereby ending up in landfills (onto or into land), polluting the environment– water, air and air [36]. E-waste chemicals can persist in the environment for longer time because e-waste toxic metals are not "biodegradable" [37, 38]. When e-waste is disposed onto landfills the toxic heavy metals have a potential to leach and contaminate soil and underground water [39]. Toxic water and soil then become unusable for humans, animals and plants [40]. Another problem is recycling methods used to extract valuable materials from e-waste. E-waste also contain valuable metals like gold, silver and aluminium [28]. The annual value of this market is estimated over 52 billion USD, which is far greater than the gross domestic products of some countries [41].

There are two types of recycling namely formal and informal recycling. Formal recycling takes place under controlled environments and workers have personal protective equipment [42]. Informal recycling uses harmful methods to extract these valuable metals from e-waste and workers often are without personal protective equipment [43]. Informal recycling typically exposes the workers and the surrounding communities to the exposure of harmful fumes of e-waste and chemicals [44]. Typically, informal recycling is home to the developing countries because is not expensive compared to formal recycling. Formal recycling entails expensive infrastructure that is primary afforded by the developed countries [45].

Literature has noted soaring statistics on e-waste around the world [41]. Each year e-waste escalates at the rate of 3–5% and it outpace the growth rate of any solid waste segment by three times [46]. Moletsane [47] noted that "e-waste repre- sents one of the pressing challenges faced by today's world". In the years 2016 and 2017 about 45.7 million metric tonnes (Mt) and 49.8Mt of e-waste respectively was reported to be generated around the world [48]. It is estimated that in 2021 e-waste volume will surpass 60Mt around the globe [49]. The examples of these toxic heavy metals are the following: cadmium, chromium, lead and cobalt [51]. According to lit- erature people are not aware of e-waste detrimental effects on both the environment and health [36, 51]. The focus of this study is on the disposal of e-waste generated by ICT equipment in higher education institutions. According to research higher education in- stitutions are not aware of the detrimental effects of e-waste when disposed in a way that is not friendly to the environment [51, 52]. The poor management and subsequent negative effects of e-waste are manifestations of inadequate knowledge and awareness about e-waste as a problem [53, 54]. Therefore, it is important to make people aware of the negative impacts e-waste have on the environment when managed in a manner un- friendly to the environment.

The study is structured as follows: Section 2 discusses the methodology of the study. In Section 3 the findings of the study are discussed followed by the findings of the study in Section 4. The study concludes in Section 5.

2 Methods

2.1 Study Design

A qualitative cross sectional study was conducted during the months of February and March 2020 to gather empirical data from students. The purpose was to determine the first year IT Diploma's students' level of awareness of electronic waste detrimental effects on the environment.

2.2 Participants Characteristics and Their Selection

The participants were students from a first year classroom studying for IT Diploma at the University of Technology. The participants were both females and males with ages ranging from 17 to 18 years. 10 students out of a total of 25 voluntarily took part in the study. Participants were studying at one of the University of Technology found in Gauteng province. About 80% (8 participants) of the participants originated from different parts of the country and the remaining 20% (2 participants) were international students. The sampling strategy was purposeful sampling [56].

2.3 Gathering the Data

Semi structured interviews were used as data gathering strategy. The interview guide contained the questions related to the objective of the study. All the interviews were conducted in English and lasted until saturation [57]. All the interviews were conducted in campus at the time and venue suitable to the participants. All the interviews were recorded using the smartphone and field notes were taken. Each interview lasted on average between 30 and 45 minutes. Out of 10 participants only 40% (4 participants) withdraw from the study citing that studies workload is the reason for their withdrawal.

2.4 Data Analysis

The data was analysed using qualitative content analysis assisted with Atlas.ti software for qualitative data analysis as per suggestion from [58] and [59]. In qualitative content analysis, data are presented in words and according to themes, which enables the researcher to portray results detailed in, for example, text passages, and seek to answer questions about what [58]. Krippendorff [60] defines content analysis as "a research technique for making replicable and valid inferences from texts (or other meaningful matter) to the contexts of their use.

2.5 Ethics for the Study

The participants were all people and hence ethical considerations were critical for the study. None of the participants was coerced to participate. The participation in this study was voluntary and participants were freely to exit the study at any time. None of the interviews took place without the participants' signed consent.

3 Findings

The following themes were generated after analysis of the data. Disposal, institutional management of electronic waste, electronic waste and knowledge of electronic waste impact on the environment.

3.1 Theme 1: Disposal

Some of the participants' response corresponding to this theme were:

All the participants dispose their electronic waste by throwing them away. P1 said that "*I discard most of the e-waste in the dustbins*". The problem with mixing e-waste with municipality waste is that often the waste end up in or onto the landfills [61]. When e-waste is on into or onto the landfills leaches [37]. Leaching eventually contaminates groundwater, soil meant for cultivation [37]. Throwing away e-waste is the indication of low level of awareness on how thrown away e-waste could affect the environment [62]. P4 said "*I usually burn it in our backyard*". P2 said that "*We normally store it in hose and during spring cleaning at home we discard into municipality bean or burn it in the veld*". Burning e-waste releases fumes that are harmful to health in the air [63]. Under this theme only P5 showed somehow a friendly way to the environment for e-waste disposal. P5 said "*I give to family or friends cell phones that needed no longer, I am a fashionable person, I change with time...*" When you donate you encourage reuse and indirectly reducing the production of IT equipment that eventually will be obsolete [67].

3.2 Theme 2: Institutional Management of e-waste

Some of the participants' comments that go with the theme were:

None felt that the institution is caring for the environment when coming to the disposal of electronic waste. P5 said "Wherever you go around the campus you would see the e-waste components such as computer screens lying around without any attention". P6 said that "I have not seen any advertisement or awareness program about e-waste se I came here". P5 said "What do you mean by that, e-waste management?". On the other hand, P3 said "Aaag I don't know how is managed because no one has ever said anything about e-waste to us as students". According to Suryawanshi and Narkhede [64] HEIs procurement practices are the cause or challenges to manage e-waste in a way that is friendly to the environment. Research has noted that HEIs lack awareness of how to dispose e-waste in a way that is friendly to the environment [51, 65]. Kitila and Woldemikael [54] noted that HEIs are "engrossed" with e-waste from ICT equip- ment such as computers, printers and cartridges. It is shocking to HEIs to harbour e-waste at their door steps. HEIs are obligated to promote and engage in activities that promote sustainable development because of inherited expertise from students and staff [66, 67].

3.3 Theme 3: E-waste Impact on the Environment

Some of the participants' comments that go with the theme were:

P6 said "Cartridges leftovers when inhaled will affect the lungs". P3 said that "Oh yes, the loose cables may trip others and cause them to fall or incur injury". Both participants have a limited level of awareness on how e-waste affects wellbeing. According to Perkins, Brune Drisse [3] to minimise or eliminate the e-waste negative effects to health we must dispose it in a friendly manner to the environment.

3.4 Theme 4: Knowledge of E-waste Terminology

Some of the participants' responses regarding the theme were:

P1 said "*E-waste, I don't know what it means*". P2 said that "*I never heard about the term e-waste*". Not knowing what e-waste is, suggests a lack of awareness about e-waste and its detrimental effects on health and the environment. On the other hand, P3 said e-waste "*is a scrap of computers*". Though correct, e-waste is more than just junk of ICT equipment. A working ICT equipment can be classified as e-waste so long it no longer satisfies the needs of the owner [68].

4 Discussion

The participants showed a lack of awareness on negative impacts of e-waste on the environment. Most participants reported unfriendly methods of disposing e-waste like throwing e-waste into municipal general waste. Literature noted that institutions of higher learning lack awareness of e-waste detrimental effects on the environment and health [69, 70]. With that said, participants need to be made aware of the negative impacts of e-waste on the environment and health. According to Miner, Rampedi [36] people need to be made aware through educational interventions on disposal methods friendly to the environment like recycling, reduce and reuse [45]. E-waste is one of the 21st century challenges facing humanity. The main concern is the rate at which e-waste increasing and lack of awareness on how to manage it in a way friendly to the environment.

It is therefore suggested that education is key to create awareness of e-waste among higher education institutions. This could be achieved by including e-waste management content in syllabi and around the campus campaigns that promote management of e-waste in a friendly way to the environment.

References

1. Rama, M., Turaga, R., Bhaskar, K., Sinha, S., Hinchliffe, D., Hemkhaus, M., Arora, R., Chatterjee, S., Khetriwal, D.S., Radulovic, V., Singhal, P., Sharma, H.: E-Waste Management in India: issues and strategies. Vikalpa J. Decision Makers **44**(3), 127–162 (2019)
2. Singh, N., Duan, H., Tang, Y.: Toxicity evaluation of E-waste plastics and potential repercussions for human health. Environ. Int. **137**(2020), 1–8 (2020)
3. Perkins, D.N., Brune Drisse, M.-N., Nxele, T., Sly, P.D: E-Waste: a global hazard. Ann. Global Health **80**(4), 286–295 (2014)
4. Janagam, D., Jeyamani, M.: E-Waste–a major threat to environment and health. Indian J. Sci. Technol. **4**(3), 313–317 (2011)
5. Decharat, S., Kiddee, P.: Health problems among workers who recycle electronic waste in Southern Thailand. Public Health Res. Perspect **11**(1), 34–43 (2020)
6. Ratheeswari, K.: Information communication technology in education. J. Appl. Adv. Res. **3**(1), 45–47 (2018)
7. Rahman, M.: Impacts of information and communication technology to health and environment impacts of information and communication technology to health and environment. Patuakhali Sci. Tech. Univ. (2011)
8. Evroro, E., Okumoku-Evroro, O.: Imperative of information and communication technology in tertiary education. Asian J. Educ. e-Learn. **2**(5), 419–422 (2014)
9. Roztocki, N., Soja, P., Weistroffer, H.R.: The role of information and communication technologies in socioeconomic development: towards a multi-dimensional framework. Inform. Technol. Develop. **25**(2), 171–183 (2019)
10. Sangrà, A., González-Sanmamed, M.: The role of information and communication technologies in improving teaching and learning processes in primary and secondary schools. ALTJ **18**(3), 207–220 (2010)
11. Khanday, K.A.: Role of ICT (information & communication technology) in 21st century's teacher education. Int. J. Adv. Multidisci. Sci. Res. **2**(1), 10–14 (2019)
12. Talebiana, S., Mohammadia, H.M., Rezvanfara, A.: Information and communication technology (ICT) in higher education: advantages, disadvantages, conveniences and limitations of applying e-learning to agricultural students in Iran. Proc. Soc. Behav. Sci. **152**(2014), 300–305 (2014)

13. Mouyabi, S.M.: Higher education in the wake of new ICT: Reaping benefits or creating more problems through e-learning? South African J. Higher Educ. **25**(6), 1178–1189 (2011)
14. Noor-Ul-Amin, S.: An effective use of ICT for education and learning by drawing on world-wide knowledge, research, and experience: ICT as a change agent for education. Scholarly J. Educ. **2**(4) 38–45 (2013)
15. Pheeraphan, N.: Enhancement of the 21st century skills for Thai higher education by integration of ICT in classroom. Procedia. Soc. Behav. Sci. **103**(203), 365–373 (2013)
16. Livingstone, S.: Critical reflections on the benefits of ICT in education. Oxf. Rev. Educ. **38**(1), 9–24 (2012)
17. Berner, E.S., La Lande, T.J.: Overview of clinical decision support systems. In Berner, E.S. (eds.) Health informatics series clinical decision support systems: Theory and Practice. vol. 16, pp. 3–22. Springer New York (2007)
18. Eslami, S., de Keizer, N.F., Dongelmans, D.A., de Jonge, E., Schultz, M.J., Abu-Hanna, A.: Effects of two different levels of computerized decision support on blood glucose regulation in critically ill patients. Int. J. Med. Inform **81**(1), 53–60 (2012)
19. Maheshwari, R.K., Rani, B., Singh, U.: Dribble of E-Waste: It's impact on human health & management strategies for sustainable development. Res. J. Chem. Environ. Sci. **1**(2), 3–16 (2013). Castillo, R.S., Kelemen, A.: Considerations for a successful clinical decision support system. CIN: Comput. Inform. Nursing **31**(7), 319–326 (2013)
20. Ehteshami, A., Rezaei, P., Tavakoli, N., Kasaei, M.: The role of health information technology in reducing preventable medical errors and improving patient safety. Int. J. Health Syst. Disaster Manage. **1**(4), 195–199 (2013)
21. Gašparović, S.: Impact of ICT on some Segments of Everyday Life of High School Population of the City of Zagreb. In International Conference on Traffic and Transport Engineering, pp. 155–164. University of Zagreb, Serbia (2016)
22. Rouleau, G., Gagnon, M.-P., Côté, J.: Impacts of information and communication technologies on nursing care: an overview of systematic reviews (protocol). Syst. Rev. **4**(1), 1–8 (2015)
23. Xianbing, L., Masaru, T., Yasuhiro, M.: Electrical and electronic waste management in China: progress and barriers to overcome. J. waste Manage. Res. **24**(1), 92–101 (2006)
24. Balakrishnan, R.B., Anand, K.P., Chiya, A.B.: Electrical and electronic waste: a global problem. J. Waste Manage. Res. **25**(4), 307–317 (2007)
25. Ding, Y., Zhang, S., Liu, B., Zheng, H., Chang, C.-C., Ekberg, C.: Recovery of precious metals from electronic waste and spent catalysts: a review. Resour. Conserv. Recycl. **141**(2019), 284–298 (2019)
26. Sinha-Khetriwal, D., Kraeuchib, P., Markus, S.: A comparison of electronic waste recycling in Switzerland and in India. Environ. Impact Assess. Rev. **25**(2005), 492–504 (2005)
27. European Union, https://www.informea.org/en/legislation/directive-200296ec-european-parliament-and-council-waste-electrical-and-electronic
28. Forti, V., Baldé, C.P., Kuehr, R.: E-waste statistics: guidelines on classification. Report. Indicators. United Nations University, Germany (2018)
29. Needhidasan, S., Samuel, M., Chidambaram, R.: Electronic waste—an emerging threat to the environment of urban India. J. Environ. Health Sci. Eng. **12**(36), 1–9 (2014)
30. Widmer, R., Oswald-Krapf, H., Sinha-Khetriwal, A.: Global perspectives on the e-waste. Environ. Impact Assess. Rev. **25**(5), 436–458 (2005)
31. Lobos, A., Babbitt, C.W.: Integrating emotional attachment and sustainability in electronic products design. Challenges **4**(1), 19–33 (2013)
32. Needhidasan, S., Samuel, M., Chidambaram, R.: Electronic waste: an emerging threat to the environment of urban India. J Environ Health Sci Eng **12**(36), 1–9 (2014)
33. Minter, A.: The burning truth behind an E-waste dump in Africa: ending the toxic smoke rising from an iconic dump in Ghana will take more than curbing Western waste. In: Orisakwe, O.E.,

Frazzoli, C., Ilo, C.E., Oritsemuelebi, B.: Public health burden of E-waste in Africa. J. Health Pollut. **9**(22), 1–12(2019). https://www.smithsonianmag.com/science-nature/burning-truth-behind-e-waste-dump-africa-180957597. Last Accessed 25 Dec 2018
34. Heacock, M., Kelly, C.B., Asante, K.A., Birnbaum, L.S., Bergman, Å.L., Bruné, M.-N., Buka, I., Carpenter, D.O., Chen, A., Huo, X., Kamel, M., Landrigan, P.J., Magalini, F., Diaz-Barriga, F., Neira, M., Omar, M., Pascale, A., Ruchirawat, M., Sly, L., Sly, P.D., Van den Berg, M., Suk, W.A.: E-Waste and harm to vulnerable populations: a growing global problem. Environ. Health Perspect. **124**(5), 550–555 (2016)
35. Dharini, K., Cynthia, J.B., Kamalambikai, B., Celestina, J.P.A.S., Muthu, D.: Hazardous E-waste and its impact on soil structure. IOP Conf. Ser. Earth Environ. Sci. **80**(1), 1–5 (2017)
36. Ignatuschtschenko, E.: E-waste management in China: bridging the formal and informal sectors. J. Chinese Govern. **2**(4), 385–410 (2017)
37. Hoeltl, A., Brandtweiner, R., Müller, H.: Approach to solving the E-waste problem: case study Ghana. Int. J. Sustain. Dev. Plan. **12**(6), 1050–1060 (2017)
38. Annamalai, J.: Occupational health hazards related to informal recycling of E-waste in India: an overview. Indian J. Occupat. Environ. Med. **19**(1), 61–65 (2015)
39. Ceballos, D.M., Dong, Z.: The formal electronic recycling industry: challenges and opportunities in occupational and environmental health research. Environ. Int. **95**(2016), 157–166 (2016)
40. Mishra, S., Shamanna, B.R., Kannan, S.: Exploring the Awareness Regarding E-waste and its Health Hazards among the Informal Handlers in Musheerabad Area of Hyderabad. Indian J. Occupat. Environ. Med. **21**(3), 143–148 (2017)
41. Grant, R.: E-waste challenges in cape town: opportunity for the green economy? Urbani izziv **30**(2019), 1–23 (2019)
42. Joon, V., Shahrawat, R., Kapahi, M.: The emerging environmental and public health problem of electronic waste in India. J. Health Pollut. **7**(15), 1–7 (2017)
43. Kumar, A., Holuszko, M.: Electronic waste and existing processing routes: a Canadian perspective. Resources **5**(35), 1–19 (2016)
44. Orisakwe, O., Frazzoli, C., Elochukwu, C., Oritsemuelebi, B.: Public health burden of E-waste in Africa. J. Health. Pollut. **9**(22), 1–12 (2019)
45. Julander, A., et al.: Formal recycling of e-waste leads to increased exposure to toxic metals: an occupational exposure study from Sweden. Environ. Int. **73**(2014), 243–251 (2014)
46. Miner, K.J., Rampedi, I.T., Ifegbesan, A.P., Machete, F.: Survey on household awareness and willingness to participate in E-waste management in Jos, Plateau State. Nigeria. Sustain. **12**(1047), 1–16 (2020)
47. Kitila, A.W.: Electronic waste management in educational institutions of Ambo Town, Ethiopia, East Africa. Int. J. Sci. Basic Appl. Res. **24**(4), 319–331 (2015)
48. Thomson, S., Belle, J.: Antecedents of green IT adoption in South Africa higher education institutions. Electron. J. Inf. Syst. Evaluation **18**(2), 172–186 (2015)
49. Subhaprada, C.S., Kalyani, P.: Study on awareness of e-waste management among medical students. Int. J. Community Med. Publ. Health **42**(2), 506–510 (2017)
50. Halady Indrani, R.: Does awareness to climate change lead to behavioral change? Int. J. Climate Change Strateg. Manage. **2**(1), 6–22 (2010)
51. Mertens, D.M.: Inclusive evaluation: implications of transformative theory for evaluation. Am. J. Eval. **20**(1), 1–14 (1999)
52. Bednar, P.M., Welch, C.: Critical systemic thinking as a foundation for information systems research practice. J. Inf. Commun. Ethics Soc. **10**(3), 144–155 (2012)
53. Renault, E.: Critical theory and processual social ontology. J. Soc. Ontol. **2**(1), 17–32 (2016)
54. Kitila, A., Woldemikael, S.: Waste electrical and electronic equipment management in the educational institutions and governmental sector offices of Addis Ababa. Ethiopia. Waste Manage. **85**(2019), 30–41 (2019)

55. Surmiak, A.: 'Should we maintain or break confidentiality? the choices made by social researchers in the context of law violation and harm. J. Acad. Ethics **18**(2019), 229–247 (2019)
56. Palinkas, L.A., Horwitz, S.M., Green, C.A., Wisdom, J.P., Duan, N., Hoagwood, K.: Purposeful sampling for qualitative data collection and analysis in mixed method implementation research. Adm Policy Ment Health **42**(5), 533–544 (2015)
57. Leedy, P.D.: Practical Research: planning and design (Prentice Hall, 1992 (1992)
58. Bengtsson, M.: How to plan and perform a qualitative study using content analysis. Nursing Plus Open **2**(2016), 8–14 (2016)
59. Saldaña, J.: The coding manual for qualitative researchers. SAGE Publications Inc., London (2009)
60. Krippendorff, K.: Content analysis: an introduction to its methodology. Sage Publications Inc., CA (2004)
61. Bhutta, S., Omar, A., Yang, X.: Electronic waste: a growing concern in today's environment. Econ. Res. Int. **2011**(474230), 1–8 (2011)
62. Suryawanshia, K., Narkhedeb, S.: Green ICT for sustainable development: a higher education perspective. Proc. Comput. Sci. **70**(2015), 701–707 (2015)
63. Bailey, J., Pena, M., Tudor, T.: Strategies for improving recycling at a higher education institution: a case study of the university of the West Indies, Cave Hill Campus. Barbados Open Waste Manage. **8**(2015), 1–11 (2015)
64. Grant, K., et al.: Health consequences of exposure to e-waste: a systematic review. Lancet Glob Health **1**(e350–61), 350–361 (2013)
65. Kiddee, P., Naidu, R., Wong, M.H.: Electronic waste management approaches: an overview. Waste Manage. **33**(5), 1237–1250 (2013)
66. Panda, R.: E-waste management: a step towards green computing. Int. J. Environ. Eng. Manage. **4**(5), 417–424 (2013)
67. Kapukha, R.N., Mbuguah, S., Kilwake, J.: Challenges facing disposal of E-waste in Kenyan public Universities. J. Eng. (IOSRJEN) **9**(4), 20–25 (2019)
68. Chugh, R., Wibowo, S., Grandhi, S.: Environmentally sustainable information and communication technology usage: awareness and practices of Indian information and communication technology professionals. J. Clean. Prod. **131**(2016), 435–446 (2016)
69. Gaidajis, G., Angelakoglou, K., Aktsoglou, D.: E-waste: Environmental problems and current management. J. Eng. Sci. and Technol. Rev. **3**(1), 193–199 (2010)
70. Edumadze, J.K.E., Tenkorang, E.Y., Armah, F.A., Luginaah, I., Edumadze, G.E.: Electronic waste is a mess: awareness and proenvironmental behavior among university Students in Ghana. Appl. Environ. Educ. Commun. **12**(4), 224–234 (2013)
71. Azodo, A.P., Ogban, P.U., Okpor, J.: Knowledge and awareness implication on e-waste management among Nigerian collegiate. J. Appl. Sci. Environ. Manage. **21**(6), 1035–1040 (2017)

Assessment of the Awareness Level of E-waste and Factors That Influence Informal Recycling: A Case Study of E-waste Pickers in Local Municipality—Gauteng Province

Ramadile Moletsane(✉)

Vaal University of Technology, Private Bag x 021, Vanderbijlpark, South Africa
ramadilem@vut.ac.za

Abstract. Electronic waste (e-waste) is an example of 21st challenges facing humanity. The problem of e-waste is a growing volume at unprecedented pace and toxicity of components when managed in a manner not friendly to the environment. Improper management of e-waste includes informal recycling prevalent in developing countries due to low levels of economic development, absence of regulation to mention the least. **Objective**: To create awareness about informal recycling negative impact on the environment and suggest solution. **Methods**: In this qualitative study design data was gathered through semistructured interviews. The purposive sampling was used to identify the participants. Qualitative content analysis was used to analyse the data. The data analysis strategy was assisted with ATLAS.ti software for qualitative data analysis. **Findings**: Poverty, lack of formal education and unemployment are some of the factors that encourages informal recycling. Most of the workers do not have protective equipment. **Discussion**: E-waste pickers have limited awareness of detrimental effects of informal recycling on the environment and health. Most [e-waste pickers] find themselves in this labour intensive and hazardous working environment due to low levels of economic development. **Conclusion**: E-waste is one of the 21st century challenges facing mankind. Informal recycling is not friendly to the environment and is harmful to health. People must be made aware about the dangers of improper management of e-waste like informal recycling to the environment and health. Awareness of detrimental effects of e-waste on the environment and health could be achieved through education.

Keywords: Awareness · Electronic waste · Environment and health · Electronic waste pickers · Recycling

1 Introduction

E-waste is one of E-waste contains complex, harmful substances that are expensive to manage in a friendly way to the environment. Informal recycling generates a large volume of unwanted waste that eventually end up into or onto landfills thereby posing a threat to the environment [1]. E-waste disposed on landfills increases the chances of

R. Silhavy et al. (Eds.): CoMeSySo 2022, LNNS 596, pp. 11–20, 2023.
https://doi.org/10.1007/978-3-031-21435-6_2

health risks like "cancer and developmental and neurological disorders" [2]. E-waste is composed of heavy metals that are resistant to decompose under natural conditions [3]. Under natural conditions heavy metals have ability to accumulate and become toxic [4]. Accumulated e-waste heavy metals like cadmium lead and chromium can contaminate soil for irrigation, plants and underground water [5]. E-waste is one of the 21st century challenges facing humanity due to its growing volumes at an unprecedented rate and high levels of components toxicity [2]. In 2016 around 44.70 million metric tonnes of e-waste was generated around the world and expected to reach 52.20 million metric tonnes in 2021[6].

There is no universally accepted definition of e-waste [7]. E-waste is defined as "broad and growing range of electronic devices ranging from large household devices such as refrigerators, air conditions, cell phones, personal stereos, and consumer electronics to computers which have been discarded by their users" [8]. E-waste can be classified as "any electrical powered appliance that has reached its end-of-life" [9]. On the other hand, e-waste is defined as "... Various forms of electric and electronic equipment that have ceased to be of any value to their owners" [10]. In the context of this paper adopted definition of e-waste is provided by [10].

E-waste consists of both valuable and toxic substances [11]. Toxic materials include cadmium, lead and chromium while valuable materials are like copper, gold and aluminium [2]. E-waste is a lucrative business [12]. According to a report by the United Nations states that an estimated US$ 55 billion worth of valuable materials like gold, copper and silver were discarded in the United States only, which over and above the 2016 GDP of most countries around the world [13]. E-waste is a golden opportunity due to its valuable components like gold, silver and copper [14]. E-waste market share is estimated to reach US$ 102.67 billion by the year 2027 [15]. Over 120 countries around the world have annual gross domestic products (GDP) lower than the value of global e-waste [16]. The lucrative substances from e-waste are extracted trough either formal or informal recycling [17]. The distinction is that formal recycling is performed under safe and controlled environment to recover valuable materials from e-waste components [18]. On the other hand, informal recycling is unregulated, unsafe recovery of valuable substances from e-waste [19]. Informal recycling is bad because it exposes workers, surrounding communities and environment to the detrimental effects of e-waste toxic substances [20]. Informal recycling sites are characterised with high concentrations of both metallic and natural contaminations in air, soil and water [21]. Hu- mans could be exposed to these harmful substances through inhalation, indigestion or dermal absorption [22]. Inhalation exposure is caused by toxic fumes in the air during open burning [23]. Indigestion exposure as a result of soil or water contamination with toxicants, for example food chain contamination as a result of improper disposal of e-waste [24]. Dermal exposure caused by direct contact (i.e. skin contact) with toxicants during recycling processes [25]. Health implications of such exposure to e-waste toxins are not limited to abnormal thyroid function, neurobehavioral disturbances, damage to liver, irreversible effects on the nervous system [26]. Literature noted that people are uninformed about the negative effects of e-waste informal recycling on the environment and health [27, 28]. With that said the purpose of this study is to create awareness of awareness of detrimental effects of e-waste informal recycling on the environment.

When e-waste is not managed in a way friendly to the environment it becomes harmful to both health and the environment. Therefore, it is crucial to make sure that people are informed or made aware of detrimental effects of e-waste on the environment. This study is structured as follows: Sect. 2 discusses the problem of the study. Followed by the discussion of research methodology in Sect. 3. Section 4 and Sect. 5 discuss the findings and provides the discussion of the study respectively.

2 Methods

2.1 Background and Characteristics of the Participants

The participants were e-waste pickers around the local municipality in Gauteng province, South Africa. All the participants never completed a matric certificate. Most exit at primary levels or secondary levels education in standard (grade 8). The main indigenous languages spoken include Sesotho, IsiZulu, Setswana and IsiXhosa. The participants were man and women working at the dumping site. This site accommodates general municipality waste including waste from electronic and electrical devices like old computers, fridges, and televisions. At the site people swift through the waste searching for anything that can be of value especially electronic and electric devices. They also collect e-waste from household rubbish bins in the suburbs and central business centres (CBD's). For most of them the day starts around 4:00 AM. Their main activities also include dismantling of e-waste and burning cables for copper wires. Whatever salvaged is then taken to scrap yard owners and dispatched to formal recyclers in the province. They believe that "early bird always catches the fattest". They sold what is salvaged from these unwanted devices to local scrap yards' owners for cash. They are mainly after e-waste devices plastics, cathode ray tubes or screens and copper wires and other valuable materials from the circuit boards.

2.2 Selection of the Participants and the Data Gathering

Purposeful sampling was used to select the study participants. Homogenous purposive sampling simplifies analysis and minimizes variation [33]. I gathered the data using semi-structured interviews. Semi-structured interviews are suitable for inquiries where "you know what you want to find out about" [34]. I developed the data gathering guide in advance. The guide contained predetermined open ended question as per suggestion by Jamshed [35]. The interview guide was supplemented by follow up questions, probes and comments as per suggestion from [36]. The main indigenous languages spoken in this area were Sesotho, IsiZulu, IsiXhosa and Setswana. All the participants chose to conduct interviews using Sesotho language.

2.3 Ethics for the Study

The data was gathered from people as participants and therefore it was important to consider ethics. All the participants consent to partake in this study. None of the participants was coerced to take part in the study and participation was voluntarily. Each participant in this study gave consent before the start of each one-on-one interview. An information letter to all people involved in the study was written and informed about the purpose of the study prior interviews. Participants were assured confidentiality of the study and informed about the purpose of the study.

2.4 Data Analysis for the Study

Consent was sought from the participants to record the interview in detail [37]. During the interviews a smartphone was used to record the interviews. The field notes were written to capture information about the setting of the interview and the non-verbal communication that were not recorded as per suggestion from Poland [38].

Each interview was analysed immediately by reviewing field notes and the transcripts as soon as they were available. All the interviews were conducted in Sesotho which was the preferred language by the participants. In the consent forms it was also indicated that the respondents can either use English or Sesotho. "communicating in the respondent's language is of paramount importance" for three reasons: it allows participants to "fully express themselves"; it creates "good rapport"; and permits the interviewer to interpret the interviewee's responses or statements with "cultural understanding" [39]. Local language is a powerful way to acceptance and an indicator of an individual to "willingness to enter into the world of the interviewees" [40].

As a Sesotho native speaker, I transcribed the interviews into Sesotho to improve the quality of the transcripts. I then translated the Sesotho transcripts into English. To ensure the accuracy of the transcripts a method of back translation was used as per suggestion by Smith, et al. [41]. Three transcripts were back-translated by the translator who was fluent in both English and Sesotho from English to Sesotho. I then checked all the data by listening to the recordings from the smartphone and comparing them to the transcripts. The interview data from the smartphone was immediately after the interview copied to the removable memory stick and locked in a cabinet in my office. The original data from smartphone was then deleted from the smartphone. The cabinet keys were also kept safe on a separate location. The interviewee has opportunity to access the recordings on request for duration of 5 years after the study before data being permanently erased. All the interview data were transcribed verbatim. While transcribing data verbatim was time consuming, it was critical in avoiding "bias introduced through selective data extraction" by the researcher [37]. The data was analysed using qualitative content analysis assisted with ATLAS.ti software for qualitative data analysis. The data gathered generated the following themes: harmful effects of e-waste informal recycling, Knowledge of the term e-waste, informal recycling workers; and poverty, education and unemployment. The findings are discussed in the subsequent text.

3 Findings

The following themes were generated after analysis of the data. Harmful effects of e-waste informal recycling, Knowledge of the term e-waste, Informal recycling workers, Poverty, education and unemployment.

3.1 Theme 1: Harmful Effects of e-waste Informal Recycling

Some of the participants' response corresponding to this theme were:

According to [42] informal recycling is bad for the environment and health. Informal recycling is characterised by workers without personal protective equipment, low income levels, labour intensive, marginal technology, unregulated and unregistered, low socio-economic status, hazardous working conditions [43]. The culprit is the toxicity of components found in electronic and electrical equipment [11]. According to International Labour Organization [44] e-waste is "hazardous, complex and costly to treat". P3 said that "All I know is that the smell from burning cables is not good for eyes, and makes us cough a lot". E-waste dismantling releases Phthalate esters (PAEs) in the environment [45]. PAEs are easily released into air because they are not chemically bounded to the products on which they are applied [46]. They [PAEs] are synthetic chemicals that are found in products like electronic products, food products and floorings including cables and wires [47]. P5 said "before joining here, my face was smooth but now is full of goose bumps". The participants seemed to be aware of dangers of working with e-waste as the result of their health status since joining as e-waste pickers. E-waste is detrimental to health and to the environment [48].

When asked why they continue this kind of job given health impacts, this was their reply. P3 said that "there are no jobs and you cannot eat good health". On the other hand, P5 said "The main reason is unemployment and kids to send to school". P6 said "If I stop my woman will leave me. She will go for another man who can provide". It is evident that poverty and unemployment could compromise health and destroy families.

3.2 Theme 2: Knowledge of the Term E-waste

Some of the participants' comments that go with the theme were:

P2 said e-waste "is what we are working with, no one can claim it". P5 said that "e-waste is anything thrown to this dumping site and from suburb residents waste bins". P4 said "I think these old televisions, computers and fridges are e-waste". It is clear that participants have a sense of what is e-waste though limited. For electric or electronic device be classified as e-waste it does not necessarily have to be thrown away. E-waste encompasses any electric or electronic equipment that is no longer wanted by the owner. Functionality of the device is not decisive. According to Heacock, et al. [49] e- waste is defined as "any electrical or electronic equipment and its parts that have been discarded without intention of reuse by the owner". When a piece of equipment no longer satisfies the needs of the initial user or owner, it does not necessarily mean that it is in a poor condition or has become redundant. In this context the participants showed a limited knowledge of what e-waste is

3.3 Theme 3: Informal Recycling Workers

Some of the participants' comments that go with the theme were:

P1 said that "our children help during school holidays sometimes on Saturdays ". P2 said "having our children helping makes them know that you have to work for every cent you earn. This will teach them discipline and become real women and men". It is evident that participants prioritise money over health. Possibly this could be attributed to poverty or lack of awareness about dangers of e-waste or informal recycling [17]. P4 said "we are not employed, there is no work out there. Therefore, it is best for our children to help so we can put more on the table". It is evident that P2 and P4 put money first to health. Their sentiments could be attributed to poverty or lack of awareness on the detrimental effects of e-waste on children. A study conducted by Pascale, et al. [50] reported that a higher blood to those from non-e-waste informal recycling environments. The higher blood lead levels were due to lead exposure from burning cables for copper wires [50]. According to Mittal [51] high lead blood levels have long term health implications especially for children. Lead affects almost every part of the body and causes health problems like increased blood pressure and anemia [52]. P5 said that "children help with dismantling and they are quick and energetic compared to us. That is the reason why I sometimes bring them to help". P6 "I bring my grandchildren, they are 13 years and 15 years old". Exploitation of venerable populations like women and children is rife in informal recycling [53]. Women working informal recycling centres experience "spontaneous abortions, stillbirths and premature births" associated with exposure to e-waste harmful fumes [11]. On the issue of safety P5 said "we don't have proper personal protective equipment. I only have this rubber shoes and rarely wear hand gloves". P3 said that "safety gear makes me work slow, especially hand gloves. Safety boots are ok for me". Literature has noted that informal recycling is characterised by lack of safety gears [42].

3.4 Theme 4: Poverty, Education and Unemployment

Some of the participants' responses regarding the theme were:

Informal recycling is a source of income for individuals that are socio-economically marginalised [54]. P3 said that "we only collect valuables from e-waste because these are hard times, there is no job out there. People are retrenched day in and day out". P4 said that "e-waste recycling is a quick cash, and there are no jobs. South African youth unemployment figures reached a soaring 41% of a total of 20.4 million young South Africans [55]. The pandemic COVID-19 and a downgrade to junk status pours more salt into a gaping wound that is unemployment and economy [56, 57]. Companies are closing their doors and retrench thousands of people [58]. P5 said ". . . is the only way now to make living since I did not have opportunity to complete matric and university or college". When asked what were the barriers to get formal qualification or a profession, P5 said "I am from the farms, there education is not priority you have to work the whole day". P7 said "I did not go to school due to apartheid, we never had opportunities that is why I will end up here". It is clear that background of P7 and P5 had negatively impacted on their constitutional rights to get a decent education. With that said, not only educated people lead decent careers, for these participants' current socio-economic factors played a detrimental role. E-waste market value in 2014 generated US$1.66 billion and is expected to grow at compound annual growth rate of 1.8 percent from 2019 to 2025 to reach a whopping US$ 45.78 billion in the year 2025 [60]. In 2016 around 44.70 million metric tonnes of e-waste was generated around the world and expected to reach 52.20 million metric tonnes in 2021[6]. However, … urges that informal recycling "provides employment and a livelihood for impoverished, marginalised and vulnerable individuals or social groups" [61]. Added that despite detrimental effects of e-waste, it does provide employment and allow those involved to survive often in the regions with high employment rates [43].

4 Discussion

Informal recycling is rife in the developing countries like Serbia, India and Ghana [44]. This is due to low levels of economic development in those countries [43]. Over 90% of e-waste is managed by the informal sector under unsafe, unregulated conditions. The workers place their health in harm's way by not wearing the protective gears. Informal recycling often plays a role in countries where there is no e-waste regulation or no enforcement of such. Unemployment, lack of education and low level of economic development are some of the reasons that compel people to compromise their health for income. Again, limited knowledge about the detrimental effects of e-waste further compromise the e-waste pickers wellbeing. E-waste is one of the 21st century challenges facing mankind. Informal recycling is not friendly to the environment and harmful to health. This paper showed that people need to be informed about the negative impacts of e-waste on their wellbeing and to the environment. This could be achieved through education.

References

1. Rama, M., Turaga, R., Bhaskar, K., Sinha, S., Hinchliffe, D., Hemkhaus, M., Arora, R., Chatterjee, S., Khetriwal, D.S., Radulovic, V., Singhal, P., Sharma, H.: E-waste management in India: issues and strategies. Vikalpa: J. Decision Makers **44**(3), 127–162 (2019)
2. International labour organization: from waste to jobs: decent work challenges and opportunities in the management of e-waste in India. Int. Labour Office, Geneva (2019)
3. Needhidasan, S., Samuel., M., Chidambaram., R.: Electronic waste: an emerging threat to the environment of urban India. J. Environ. Health Sci. Eng. **12**(36), 1–9 (2014)
4. Gangwar, R., Choudhari, A., Chauhan, A., Kumar, A., Singh, Tripathi, A.: Assessment of air pollution caused by illegal e-waste burning to evaluate the human health risk. Environ. Int. J. **125**(2019), 191–199 (2019)
5. Pinto, V.N.: E-waste hazard: the impending challenge. Indian J. Occupat. Environ. Med. **12**(2), 65–70 (2008)
6. Adesokan, M.D., Adie., G.U., Osibanjo, O.: Soil Pollution by toxic metals near E-waste recycling operations in Ibadan, Nigeria. J. Health Pollu. **6**(11), 26–33 (2016)
7. Rahman, M.: Impacts of information and communication technology to health and environment. Patuakhali Sci. Technol. Univ. (2011)
8. Rani, B., Singh, U., Maheshwari, R., Chauhan, A.K.: Perils of electronic waste issues and management strategies. J. Adv. Sci. Res. **3**(1), 17–21 (2012)
9. Puckett, J., Byster, L., Westervelt, S., Gutierrez, R., Davis, S., Hussain, A., Dutta, M.: Exporting harm the high-tech trashing of Asia (2002)
10. Sinha-Khetriwal, A.: The management of electronic waste: a comparative study on India and Switzerland. M.S University of St. Gallen, St. Gallen, Switzerland (2002)
11. Widmer, R., Oswald-Krapf, H., Sinha-Khetriwal, A.: Global perspectives on the e-waste. Environ. Impact Assess. Rev. **25**(2005), 436–458 (2005)

12. Grant, K., et al.: Health consequences of exposure to e-waste: a systematic review. Lancet Global Health **1**(6), 350–361 (2013)
13. Gomathi, N., Rupesh Kumar, P.L.: Study on business opportunities extracted from e-waste a review. Int. J. Eng. Technol. **7**(2.33), 1106–1109 (2018)
14. United Nations University: E-waste rises 8% by weight in 2 years as incomes rise, Prices Fall (2017)
15. Sivaramanan, S.: E-waste management, disposal and its impacts on the environment. Universal J. Environ. Res. Technol. **3**(5), 531–537 (2013)
16. Allied Market Research, https://www.alliedmarketresearch.com/e-waste-management-market
17. Ryder, G., Houlin, H.Z.: The world's e-waste is a huge problem. It's also a golden opportunity. Available: https://www.weforum.org/agenda/2019/01/how-a-circular-approach-can-turn-e-waste-into-a-golden-opportunity/
18. Perkins, N., Brune Drisse, M-N., Nxele., T Sly., PD.: E-Waste: a global hazard. Ann. Global Health **80**(4), 286–295 (2014)
19. Castillo, R.S., Kelemen, A.: Obstacles to preserving precaution and equity in global hazardous waste regulation: an analysis of contested knowledge in the Basel Convention. Int. Environ. Agreements. Comput. Inform. Nursing **31**(7), 319–326 (2013)
20. Zhang, K., Schnoor, J.L., Zeng, E.Y.: E-Waste recycling: where does it go from here? Environ. Sci. Technol. **46**(20), 10861–10867 (2012)
21. Ohajinwa, C.M., van Bodegom, P.M., Vijver, M.G., Peijnenburg, J.G.: Health risks awareness of electronic waste workers in the informal sector in Nigeria. Int. J. Environ. Res. Public Health **14**(911), 1–16 (2017)
22. Cesaro, A., Belgiorno, V., Gorrasi, G., Viscusi,, G., Vaccari, M., Vinti, G., Dias, M.I., Hursthouse, A., Salhofer, S.: A relative risk assessment of the open burning of WEEE. Environ. Sci. Pollut. Res. **26**(1), 11042–11052 (2019)
23. Arain, L.A., Neitzel, L.R.: A review of biomarkers used for assessing human exposure to metals from E-waste. Int. J. Environ. Res. Public Health **16**(10), 1–24 (2019)
24. Machete, F.: Environmental health risks associated with e-waste exposure in Badplaas, Carolina and Elukwatini landfills, Republic of South Africa. Afr. J. Sci. Technol. Innov. Develop. **9**(6), 679–684 (2017)
25. Orisakwe, O.E., Frazzoli, C., Ilo, C.E., Oritsemuelebi, B.: Public health burden of E-waste in Africa. J. Health Pollut. **9**(22), 190610 (2019)
26. Sinha-Khetriwal, D., Kraeuchib, P., Markus, S.: A comparison of electronic waste recycling in Switzerland and in India. Environ. Impact Assess. Rev. **25**(2005), 492–504 (2005). Wu, C-C., Bao, L-J., Tao, S., Zeng, E.Y.: Dermal Uptake from Airborne organics as an important route of human exposure to e-waste combustion fumes. Environ. Sci. Technol **50**(13) 6599–6605 (2016)
27. Joon, V., Shahrawat, R., Kapahi, M.: The emerging environmental and public health problem of electronic waste in India. J. Health Pollut. **7**(15), 1–7 (2017)
28. Davis, J-M.: Defining an equitable Israeli-Palestinian E-waste economy. Doctor of philosophy, geography, humanities and social sciences, school of graduate studies memorial University of Newfoundland, St. John's, Newfoundland (2017)
29. Hector, M.C.: Toxic Trade: E-Waste Disposal and Environmental Governance in West Africa. Master of arts (international studies). Faculty of arts and social sciences, Stellenbosch, Cape Town, South Africa (2017)
30. Bohley, M.K., Harris, C.: An analysis of E-waste: when do electronics die? J. Indiana Acad. Soc. Sci. **17**(1), 16–24 (2014)
31. Global electronic waste monitor.: global E-waste status and trends. https://www.itu.int/en/ITU-D/Climate-Change/Documents/GEM%202017/Global-E-waste%20Monitor%202017%20-%20Chapter%206.pdf

32. Azodo, A.P., Ogban, P.U., Okpor, J.: Knowledge and awareness implication on E-waste management among Nigerian Collegiate. J. Appl. Sci. Environ. Manage. **21**(6), 1035–1040 (2017)
33. Ozdemir, P.D., Aydın, C., Evci, K.E.D.: Electronic waste awareness among students of engineering department. Cukurova Med. J. **44**(1), 101–109 (2019)
34. Palinkas, L.A., Horwitz, S.M., Green, C.A., Wisdom, J.P., Duan, N., Hoagwood, K.: Purposeful sampling for qualitative data collection and analysis in mixed method implementation research. Adm. Policy Ment. Health **42**(5), 533–544 (2015)
35. Fylan, F.: Semi-structured interviewing. A handbook of research methods for clinical and health psychology. Oxford University Press, New York (2005)
36. Jamshed, S.: Qualitative research method-interviewing and observation. J. Basic Clinical Pharm. **5**(4), 87–88 (2014)
37. DeJonckheere, M., Vaughn, L.M.: Semistructured interviewing in primary care research: a balance of relationship and rigour. Family Med. Community Health **7**(2), 1–8 (2019)
38. O'Keeffe, J., Buytaert, W., Mijic, A., Brozovic´, N., Sinha, R.: The use of semi-structured interviews for the characterisation of farmer irrigation practices. Hydrol. Earth Syst. Sci. **20**(5), 1911–1924 (2016)
39. Poland, B.: Transcription quality as an aspect of rigor in qualitative research. Qual. Inq. **1**(3), 290–310 (1995)
40. Tsang, E.W.K.: Mind your identity when conducting cross-national research. Organization Stud. **19**(3), 511–515 (1998)
41. Rubin, H.J., Rubin, I.S.: Qualitative interviewing: the art of hearing data. Thousand Oaks: Sage, United Kingdom (1995)
42. Smith, P.B., Bond, M.H., Kâğıtçıbaşı, Ç.: Understanding social psychology across cultures: living and working in a changing world. Sage, United Kingdom (2006)
43. Annamalai, J.: Occupational health hazards related to informal recycling of E-waste in India: an overview. Indian J. Occupat. Environ. Med. **19**(1), 61–65 (2015)
44. Wilson, D., Velis, C., Cheeseman, C.R.: Role of informal sector recycling in waste management in developing countries. Habitat Int. **30**(2006), 797–808 (2006)
45. International Labour Organization.: The global impact of e-waste Addressing the challenge. Geneva, Switzerland2012, Available: https://www.ilo.org/sector/Resources/publications/WCMS_196105/lang--en/index.html
46. Wang, B., Wang, H., Zhou, W., Chen, Y., Zhou, Y., Jiang, Q.: Urinary Excretion of phthalate metabolites in school children of china: implication for cumulative risk assessment of Phthalate Exposure. Environ. Sci. Technol. **49**(2), 1120–1129 (2015)
47. Guo, Y., Zhang, Z., Liu, L., Li, Y., Ren, N., Kannan, K.: Occurrence and profiles of phthalates in foodstuffs from china and their implications for human exposure. J. Agric. Food Chem. **60**(27), 6913–6919 (2012)
48. Shevchenko, T., Laitala, K., Danko, Y.: Understanding consumer E-waste recycling behavior: introducing a new economic incentive to increase the collection Rates. Sustainability **11**(2656), 1–20 (2019)
49. Heacock, M., Kelly, C.B., Carpenter, D.O., Sly, P.D., Stein, R.T.: E-Waste and harm to vulnerable populations: a growing global problem. Environ. Health Perspect. **124**(5), 550–555 (2015)
50. Pascale, A., et al.: E-Waste informal recycling: an emerging source of lead exposure in South America. Ann. Glob. Health **82**(1), 197–201 (2016)
51. Parween, A., Khan, M.M., Upadhyay, T., Tripathi, R.V.: Prevalence of elevated blood lead level in children of India. Nat. Environ. Pollut. Technol. **17**(3), 703–710 (2018)
52. Wani, A.L., Ara, A., Usmani, J.A.: Lead Toxicity Rev. Interdiscipl. Toxicol. **8**(2), 55–64 (2015)

53. Schmidt, C.W.: Unfair trade: e-waste in Africa. Environ. Health Perspect. **114**(4), 232–235 (2006)
54. Jaffe, K., et al.: Informal recycling, income generation and risk: Health and social harms among people who use drugs. Int. J. Drug Policy **60**(2018), 40–46 (2018)
55. M. Githahu, M.: South Africa's youth unemployment rate rises above 40%. Cape Argus, ed. South Africa (2020)
56. Stoddard, E.: South Africa: Fitch downgrade pours more salt into the gaping wound that is the economy. Daily Maverick, South Africa (2020)
57. Staff Writer. https://businesstech.co.za/news/business/372434/over-9000-planned-job-cuts-have-been-announced-for-south-africa-in-2020-these-are-the-companies-affected/
58. Research and Markets. https://www.prnewswire.com/news-releases/analysis-of-the-worlds-45-78bn-e-waste-management-market-2019-2025--300974594.html
59. Medina, M.: Scavenger cooperatives in Asia and Latin America. Resour. Conserv. Recycl. **31**(1), 51–69 (2000)
60. Mishra, S., Shamanna, B.R., Kannan, S.: Exploring the awareness regarding E-waste and its health hazards among the informal handlers in Musheerabad area of Hyderabad. Indian J. Occupat. Environ. Med. **21**(3), 143–148 (2017)
61. Tran, C.D., Salhofer, S.P.: Processes in informal end-processing of e-waste generated from personal computers in Vietnam. J. Mater. Cycles Waste Manage. **20**(2), 1154–1178 (2017). https://doi.org/10.1007/s10163-017-0678-1

Novel Simulation Modelling of Energy-Efficient Routing Scheme Using Scheduling Approach in Internet-of-Things

Gauri Sameer Rapate[1](✉) and N. C. Naveen[2]

[1] Department of Computer Science and Engineering, PES University, Bengaluru, India
gauri.rapate@gmail.com

[2] Department of Computer Science and Engineering, JSS Academy of Technical Education, VTU, Bengaluru, India

Abstract. The role of routing protocol for massive number of interconnected nodes in Internet-of-Things (IoT) is one of the pivotal research topics in present times. Review of existing approaches towards routing schemes shows not much availability of benchmarked and efficient routing scheme with evident research problems. Therefore, this manuscript contributes towards presenting a novel simulation-based model where scheduling approach considering routing message and temporal attributes of various routing operation is considered to develop the proposed routing scheme. Different from any existing simulators, this model offers more flexibility to testify the data transmission event over various test cases along with implementation of a unique routing approach. The outcome of simulation shows better energy efficiency, lower delay, and faster transmission of proposed scheme compared to existing routing scheme in IoT.

Keywords: Internet-of-Things · Energy efficiency · Routing protocol · Scheduling · Routing message · Temporal

1 Introduction

The Internet-of-Things (IoT) consists of massive number of connected devices called as *things* connected via internet adopting one or multiple technological advancements [1, 2]. However, these massive number of nodes are quite challenging to be connected to each other in order to carry out seamless data transmission according to the targeted application in IoT [3]. Basically, the routing protocols in IoT is classified into three types based on its usage i.e., (i) based on network organization, (ii) based on route discover, and (iii) based on protocol operation [4]. The network organization type further consists of flat routing, hierarchical routing, and location-based routing. The route discovery-based scheme consists of reactive, proactive, and hybrid. The protocol operation-based routing scheme consists of negotiation-based, multipath based, query based, Quality of Service (QoS) based, and coherent based [5]. The impending challenges of conventional routing schemes are security issues, inadequate fault tolerance, intermittent connectivity, diversified standards of networking, presence of heterogeneous IoT devices, and deployment

R. Silhavy et al. (Eds.): CoMeSySo 2022, LNNS 596, pp. 21–30, 2023.
https://doi.org/10.1007/978-3-031-21435-6_3

of massive number of nodes [6]. At present, there are various existing routing schemes in IoT reported in literature [7–10], however, existing literatures are found to possess further problems viz. (i) less emphasis towards the energy efficiency of an IoT nodes working on an large environment, (ii) simulation-based studies are highly symptomatic and address only few set of transmission problems which is not applicable when test-environment changes, (iii) hardware-based implementation scheme are quite narrowed down that is again not scalable to practical world of implementation in large scale much. Therefore, this paper presents a novel computational model where a simulation-based framework is designed toward framing up an energy-efficient optimized routing scheme in IoT. Section 2 discusses about the existing research work followed by problem identification in Sect. 3. Section 4 discusses about proposed methodology followed by elaborated discussion of algorithm implementation in Sect. 5. Comparative analysis of accomplished result is discussed under Sect. 6 followed by conclusion in Sect. 7.

2 Related Work

This section discusses about the existing simulation modelling carried out in the direction of improving routing operation in IoT. The work of Xu et al. [11] have presented an optimization scheme of routing using Software Defined Network (SDN) for realizing the data transmission in IoT. Similar optimization-based scheme was also reported in work of Chen et al. [12] considering IoT integrated with social networks. Improvisation of conventional RPL routing scheme in IoT has been presented by Djamaa et al. [13] for addressing unwanted retransmission. Another RPL-based routing improvised scheme is presented by Mahyoub et al. [14] and Taghizadeh et al. [15] emphasizing over lossy network in IoT. Bhandari et al. [16] have presented another RPL-based Model that targets to construct a multiple topology in IoT. The work carried out by Ding et al. [17] have presented a relay-based routing scheme in SDN-based IoT system. A clustering-based scheme of data transmission in IoT has been presented by Xu et al. [18] with more emphasis on energy. Wang et al. [19] have presented a routing scheme considering vehicle position in IoT considering cache and distance attribute. Homaei et al. [16] have developed an RPL based routing using learning automata targeting towards better QoS performance in IoT. Game theory has been used for improving the routing scheme as reported in work of Zhang et al. [20] along with solution towards mitigating selfish node in IoT. Fu et al. [21] have presented a routing scheme based on position of nodes in industrial IoT application using greedy policy. Usage of genetic algorithm is witnessed in work of Lin et al. [22] which bridges the gap between convergence and diversity. Liu et al. [23] have presented a relaxation-based scheme using Lagrangian method in order to serve for orchestration method in IoT. Yang et al. [24] have developed a data transmission scheme specific to unmanned aerial vehicle. Apart from there, are various other schemes e.g., cross-layer based scheme (Abbasi et al. [25]), graph-based RPL algorithm (Kim et al. [26]), improvising adhoc routing scheme on demand for mobile IoT (Sobral et al. [27]), and Q-Learning based routing scheme (Zhou et al. [28]). All the above-mentioned routing scheme has been claimed for various beneficial factors along with associated loopholes meeting the operational demands of IoT. The next section highlights the research problem.

3 Problem Description

The research problems identified are as follows: (i) existing routing schemes are mainly deployed on conventional RPL protocol without addressing the impending challenges in an IoT [29], (ii) there are less inclusion of any scheduling approach towards modelling routing scheme in IoT which has more capability to control resource dissipation, (iii) irrespective of various optimization-based scheme, existing models doesn't offer any benchmarked energy-efficiency routing scheme in IoT considering challenging traffic scenario, and (iv) existing data transmission scheme has not much emphasized on temporal factors associated with routing operation, which could be used for developing preemptive operation of data transmission in IoT. The next section discusses about proposed solution towards addressing the above-mentioned research problems.

4 Proposed Methodology

The development of the proposed system is carried out considering analytical methodology where an architecture is designed for the complete simulation modelling for IoT. The prime intention of developing this simulation model is to offer novel and flexible features to assess different forms of environment of IoT followed by implementing a unique data transmission scheme.

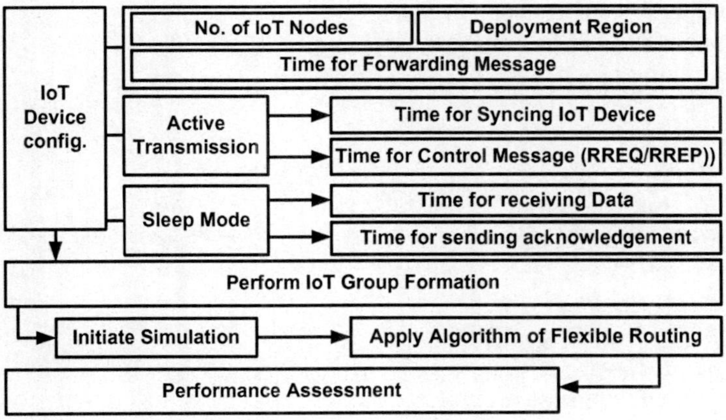

Fig. 1. Proposed architecture

Figure 1 highlights the proposed architecture which initiates by configuring the IoT devices with respect to multiple set of operations. The input parameters also involve temporal parameters associated with node's active and sleeping mode thereby a unique scheduling approach is presented. The novelty of the proposed architecture also resides in group formation where number of groups to be formed is set by the user. After the simulation is initiated, the proposed architecture implements a novel algorithm that perform data transmission followed by performance assessment of the proposed simulation model. The next section discusses about algorithm implementation.

5 Algorithm Implementation

The prime agenda of the proposed system is to mechanize a novel simulation model which facilitates a unique form of data transmission scheme. The proposed simulation model formulates a unique form of frame which possess 4 exclusive fields i.e., (i) syncing field for each IoT devices, (ii) field for control message for route request and reply, (iii) data to be forwarded, and (iv) acknowledgement field. The first two fields are reserved under active transmission period while the last two fields are reserved for sleep period. The proposed system uses random access for active transmission period while it uses scheduled access for sleep period. The initial set of operation presents a discovery of a neighboring node followed by grouping of IoT devices. Each IoT devices will be allocated with a set of time slots called as a routine, which will be interchanged among different IoT nodes. This process makes sure that IoT nodes should go to a sleep mode the moment it is done with its transmission in order to save its energy. Whereas the active period of transmission is maintained as low as possible for an individual IoT nodes while it distributes the task based on the available neighboring IoT nodes. The advantage of this scheduling system is that data packets are well processed from all the IoT nodes and it assists to have a full control of processing within the gateway nodes further. The proposed system is designed in such a way that it can carry out a packet forwarding scheme in energy efficient manner in highly customized mode. The algorithmic steps of the proposed simulation model for performing flexible routing is as follows:

Algorithm for Flexible Routing in IoT
Input: n (number of IoT Devices)
Output: R (routing)
Start
1. **For** i = 1: n
2. $A_t \rightarrow f(\eta_f)$
3. Alloc $T_d \rightarrow n(A_t)$
4. $N_{emer} \rightarrow T_d$*emer_score
5. **For** $j = 1{:}\ g$
6. **For** $k = 1{:}D_{emer}$
7. $Buf \leftarrow f_1(g_t(D_{emer}(k))$
8. $[T_f\ R_f] \rightarrow (q_1\ q_2)$
9. $R(buf) \leftarrow$ message
10. **End**
11. **End**
12. **End**

End

The discussion of steps of the algorithm is as follows: The algorithm considers all the IoT nodes n to be subjected for all the steps of further processing (Line-1). The algorithm constructs a function f(x) considering all the number of frames n_f as its input argument (Line-2). This operation leads to generation of an arbitrary identity of an IoT node A_t. The advantage of considering this step is to offer faster selection of an IoT nodes prior to performing scheduling towards allocation of an energy. This makes the algorithm preliminary operation faster. The next process of the proposed system is to perform an allocation of time for forwarding the message T_d which is based on number of IoT devices n that is selected on the basis of A_t obtained from prior step (Line-3). The proposed simulation model also facilitates processing to assess different level of emergency services in an IoT environment. A variable *emer_score* is formed in order to represent the state of emergency of the data packet to be forwarded. The values of this variable can be customized on the basis of the application used in IoT system. The primary task for this purpose is to compute the number of emergency messages N_{emer} considering T_d value (Line-4). It is to be noted that emergency score involved in the forwarding of the message significantly govern the time duration for forwarding the message T_d for the randomly selected IoT nodes. This is a significant novelty in the design of the simulation model as existing simulators in IoT doesn't facilitate this operation. The operation of permitting the emergency data packet to be forwarded first compared to the other regular data packet is carried out in following manner viz. (i) The algorithm considers all the groups g of the IoT nodes, where group represents a set of sensing devices that performs similar set of operation. It will mean that there are different number of groups $g = \{g_1, g_2, \ldots.., g_m\}$, where m represent maximum number of groups and $m < \ < n$ (Line-5), (ii) the next step is to obtain the identity of all member nodes to find out if they are really a part of a target group followed by assessing the presence of data in each member node within a group, (iii) the third step is to consider degree of emergence D_{emer} by the user (the difference between *emer-score* and D_{emer} is that former is direct value given by user while later is the value transformed in the form of percentage) as shown in Line-6, (iv) the next processing of this algorithm is to generate a temporary matrix of buffer *buf* (Line-7). This is carried out using another explicit function $f_1(x)$ which consider that specific target group g_t with specified value of D_{emer} along with consideration of the data packet (Line-7). Storing this information in a buffer *buf* matrix assists in faster accessibility of different forms of the IoT device in single exploration process. This information is stored within the routing table or memory that is shared within each node in one cluster, and (v) the final step of processing includes flagging the received data R_f and transmission event T_f that considers q_1 and q_2 to represents configured value of receiving flag and size of the data packet respectively (Line-8). Finally, the message is set forward to the buffer space *buf* with respect to the receiver node R (Line-9). It should be noted that proposed algorithm implementation considers three distinct cases of emergency in the form of higher, lower, and medium score set by the user. This completes the total algorithm operation.

The significant contributions of this algorithm are as follows: (i) the algorithm runs on each IoT device using distributed memory system that ensures lesser computational effort and thereby increases device lifetime, (ii) the algorithm offers a specific set of operational time for sleep mode and active transmission mode of operation which potential preserves device resources in IoT, (iii) it offers supportability towards emergency message to reach the destination node in contrast to messages originated from normal traffic and hence it can be deployed for emergency services took in IoT application. The next section discusses about result analysis.

6 Results Discussion

This section presents the briefing of the outcome obtained by executing the proposed simulation model discussed in prior section. The input of following simulation parameters has been considered to assess the outcome viz. Size of buffer = 5 packets, number of IoT devices = 100–1000, data packet size = 30 bytes, receive power = 15 J, idle power = 15 J, length of frame = 1 s. The complete development of proposed simulator is carried out using MATLAB and compared with existing hierarchical routing scheme [30] over similar test environment.

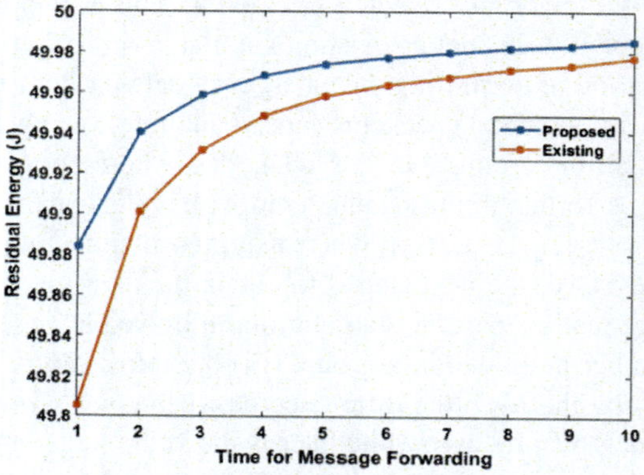

Fig. 2. Comparative analysis for residual energy

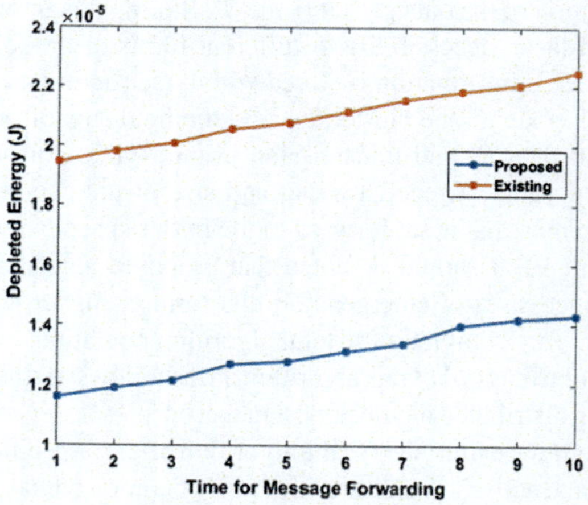

Fig. 3. Comparative analysis for depleted energy

Figures 2 and 3 highlights that proposed system offers better energy conservation compared to existing hierarchical routing protocol in IoT. The prime reason behind this is: proposed scheme makes use of a novel scheduling scheme considering message as well as various temporal factors, which causes better resource availability in reduced set of time, that doesn't exist in conventional scheme. Apart from this, proposed system also offers better resource and transmission scheduling causing reduction in energy consumption (Fig. 3) and retention of more residual energy (Fig. 2).

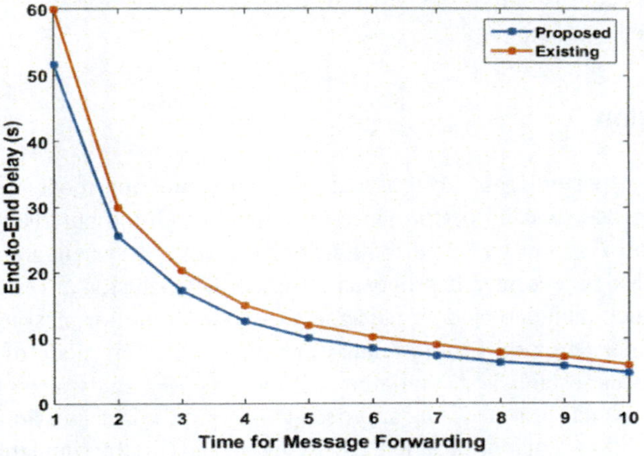

Fig. 4. Comparative analysis for end-to-end delay

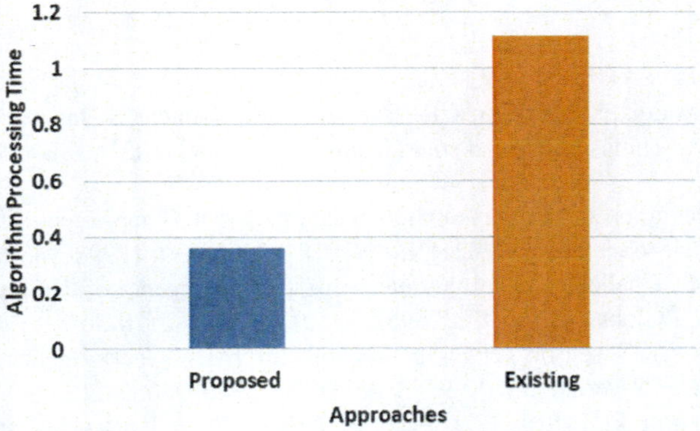

Fig. 5. Comparative analysis for algorithm processing time

Figure 4 exhibits that proposed system offers reduced delay as compared to existing scheme. However, a closer look into this graphical outcome shows very less difference between both. This can be justified as follows: Existing scheme performs data forwarding operation without considering time or considering any dynamic states of IoT nodes. It just forms a cluster, aggregates the data by cluster head and forwards the data. In this process, enough resource is consumed by cluster head as well as it consumes time too. Apart from this, all the nodes are operationally active causing declination of resources. However, owing to proposed scheduling scheme in routing, proposed system offers better synchronization of data frame which facilitates the IoT device to toggle between active and sleep state based on their accomplished task. Hence, proposed scheme offers better delay along with better energy preservation as compared to existing hierarchical scheme. This is also one of the reasons that proposed system takes considerably less amount of time in comparison to existing hierarchical routing scheme in IoT. From Figs. 4 and 5, it can be seen that proposed system offers better response time in contrast to existing scheme and hence it is one of the cost-effective computational frameworks embedded within proposed simulator.

7 Conclusion

The proposed paper presents a discussion of novel simulation model which not only facilitates towards assessing performance of various condition but also offers a novel routing scheme. The novelty / contributions of the proposed system are as follows: (i) the simulator design presented in proposed system offers inclusion of various IoT device parameters which includes random placement of nodes, time for message forwarding, and various other temporal factors related to active and sleep mode of transmission, unlike any existing established simulators, (ii) the proposed study also present a novel routing scheme that uses scheduling approach to balance the transmission demands and energy consumption demands in an IoT environment, and (iii) the simulation design also facilitates user to imply different network case studies as well as compare with existing routing schemes of IoT.

References

1. Sunitha, G., Avanija, J., Madhavi, K.R., Bhushan, S.B., Goundar, S.: Innovations in the industrial Internet of Things (IIoT) and smart factory. IGI Global (2021). ISBN: 9781799833772, 1799833771
2. Tomar, P.: Integration and implementation of the Internet of Things through cloud computing, engineering science reference (2021). ISBN: 9781799869832, 1799869830
3. Velayutham, S.: Challenges and opportunities for the convergence of IoT, big data, and cloud computing. IGI Global. ISBN: 9781799831136, 1799831132 (2021)
4. Yakoi, P.: Wireless sensor networks protocols in IoT. A performance evaluation and comparison, GRIN Verilag. ISBN: 9783346338662, 3346338665 (2021)
5. Buratti, C., Storm, E., Feltrin, L., Clavier, L.: Chapter 7—IoT protocols, architectures, and applications. Science direct, inclusive radio communications for 5G and beyond, pp. 187–220 (2021)
6. Kumar, C., Prakash, S.: Chapter 6—routing protocols: key security issues and challenges in IoT, ad hoc, and sensor networks. ScienceDirect, security and privacy issues in IoT devices and sensor networks. Adv. Ubiq. Sens. Appl. Healthcare, 105–132 (2021)
7. Raoof, A., Matrawy, A., Lung, C-H.: Routing attacks and mitigation methods for RPL-based Internet of Things. IEEE Commun. Surveys Tutorials **21**(2), 1582–1606, Secondquarter 2019. https://doi.org/10.1109/COMST.2018.2885894
8. Balghusoon, A.O., Mahfoudh, S.: Routing protocols for wireless nanosensor networks and internet of nano things: a comprehensive survey. IEEE Access **8**, 200724–200748 (2020). https://doi.org/10.1109/ACCESS.2020.3035646
9. Arshad, S., Azam, M.A., Rehmani, M.H., Loo, J.: Recent advances in information-centric networking-based Internet of Things (ICN-IoT). IEEE Internet Things J. **6**(2), 2128–2158 (2019). https://doi.org/10.1109/JIOT.2018.2873343

10. Nauman, A., Qadri, Y.A., Amjad, M., Zikria, Y.B., Afzal, M.K., Kim, S.W.: Multimedia Internet of Things: a comprehensive survey. IEEE Access **8**, 8202–8250 (2020). https://doi.org/10.1109/ACCESS.2020.2964280
11. Xu, S., Wang, X., Yang, G., Ren, J., Wang, S.: Routing optimization for cloud services in SDN-based Internet of Things with TCAM capacity constraint. J. Commun. Netw. **22**(2), 145–158 (2020). https://doi.org/10.1109/JCN.2020.000006
12. Chen, G., Tang, J., Coon, J.P.: Optimal routing for multihop social-based D2D communications in the Internet of Things. IEEE Internet Things J. **5**(3), 1880–1889 (2018). https://doi.org/10.1109/JIOT.2018.2817024
13. Djamaa, B., Senouci, M.R., Bessas, H., Dahmane, B., Mellouk, A.: Efficient and stateless P2P routing mechanisms for the Internet of Things. IEEE Internet Things J **8**(14), 11400–11414 (2021). https://doi.org/10.1109/JIOT.2021.3053339
14. Mahyoub, M., Hasan Mahmoud, M.S., Abu-Amara, M., Sheltami, T.R., An efficient RPL-based mechanism for node-to-node communications in IoT. IEEE Internet Things J. **8**(9), 7152–7169 (2021). https://doi.org/10.1109/JIOT.2020.3038696
15. Taghizadeh, S., Elbiaze, H., Bobarshad, H.: EM-RPL: enhanced RPL for multigateway Internet-of-Things environments. IEEE Internet Things J **8**(10), 8474–8487 (2021). https://doi.org/10.1109/JIOT.2020.3047079
16. Bhandari, K.S., Ra, I., Cho, G.: Multi-topology based QoS-differentiation in RPL for Internet of Things applications. IEEE Access **8**, 96686–96705 (2020). https://doi.org/10.1109/ACCESS.2020.2995794
17. Ding, Z., Shen, L., Chen, H., Yan, F., Ansari, N.: Energy-efficient relay-selection-based dynamic routing algorithm for IoT-oriented software-defined WSNs. IEEE Internet Things J. **7**(9), 9050–9065 (2020). https://doi.org/10.1109/JIOT.2020.3002233
18. Xu, Y., Yue, Z., Lv, L.: Clustering routing algorithm and simulation of Internet of Things perception layer based on energy balance. IEEE Access **7**, 145667–145676 (2019). https://doi.org/10.1109/ACCESS.2019.2944669
19. Wang, L.-L., Gui, J.-S., Deng, X.-H., Zeng, F., Kuang, Z.-F.: Routing algorithm based on vehicle position analysis for internet of vehicles. IEEE Internet Things J. **7**(12), 11701–11712 (2020). https://doi.org/10.1109/JIOT.2020.2999469
20. Zhang, W., Wang, X., Han, G., Peng, Y., Guizani, M.: SFPAG-R: A reliable routing algorithm based on sealed first-price auction games for industrial Internet of Things networks. IEEE Trans. Veh. Technol. **70**(5), 5016–5027 (2021). https://doi.org/10.1109/TVT.2021.3074398
21. Fu, J., Cui, B., Wang, N., Liu, X.: A distributed position-based routing algorithm in 3-D wireless industrial Internet of Things. IEEE Trans. Industr. Inf. **15**(10), 5664–5673 (2019). https://doi.org/10.1109/TII.2019.2908439
22. Lin, N., Shi, Y., Zhang, T., Wang, X.: An Effective order-aware hybrid genetic algorithm for capacitated vehicle routing problems in Internet of Things. IEEE Access **7**, 86102–86114 (2019). https://doi.org/10.1109/ACCESS.2019.2925831
23. Y. Liu et al., "A Lagrangian-Relaxation-Based Approach for Service Function Chain Dynamic Orchestration for the Internet of Things," in IEEE Internet of Things Journal, vol. 8, no. 23, pp. 17071–17089, 1 Dec.1, 2021, doi: https://doi.org/10.1109/JIOT.2021.3075563
24. Yang, Z., Liu, H., Chen, Y., Zhu, X., Ning, Y., Zhu, W.: UEE-RPL: a UAV-based energy efficient routing for Internet of Things. IEEE Trans. Green Commun. Netw. **5**(3), 1333–1344 (2021). https://doi.org/10.1109/TGCN.2021.3085897
25. Abbasi, U.F., Haider, N., Awang, A., Khan, K.S.: Cross-layer MAC/routing protocol for reliable communication in Internet of health Things. IEEE Open J. Commun. Soc. **2**, 199–216 (2021). https://doi.org/10.1109/OJCOMS.2020.3047888
26. Kim, Y., Paek, J.: NG-RPL for efficient P2P routing in low-power multihop wireless networks. IEEE Access **8**, 182591–182599 (2020). https://doi.org/10.1109/ACCESS.2020.3028771

27. Sobral, J.V.V., Rodrigues, J.J.P.C., Rabêlo, R.A.L., Saleem, K., Kozlov, S.A.: Improving the performance of loading routing protocol in mobile IoT scenarios. IEEE Access **7**, 107032–107046 (2019). https://doi.org/10.1109/ACCESS.2019.2932718
28. Zhou, Y., Cao, T., Xiang, W.: Anypath routing protocol design via Q-learning for underwater sensor networks. IEEE Internet Things J **8**(10), 8173–8190 (2021). https://doi.org/10.1109/JIOT.2020.3042901
29. Subash K., Arockiam, L.: A survey on issues and challenges in RPL based routing for IoT. Ann. . Romanian Soc. Cell Biol. **25**(5) (2021)
30. Poluru, R.K., Naseera, S.: A literature review on routing strategy in the Internet of Things. J. Eng. Sci. Technol. Rev. **10**(5), 50–60 (2017)

BMA Algorithm for Smartphone Purchase Decisions

Dam Tri Cuong(✉)

Industrial University of Ho Chi Minh City, Ho Chi Minh City, Vietnam
damtricuong@iuh.edu.vn

Abstract. Nowadays, individuals are turning out to be increasingly more dependent on technology, particularly when connecting with others. Smartphone has one of the high technology connection tools that permitted people to link with others. Smartphone has carried many advantages to society. A few advantages of the smartphone such as calling and sending messages, catching pictures, getting to the web, socializing, downloading applications, storing data, recreation, etc. The prior reviews of research in the literature suggested factors that affected smartphone purchase decisions such as product features, social influence, product price, and brand name. However, these researches firmed only on the single selection model among models chosen. This paper used the BMA algorithm to choose the optimal models for smartphone purchase decisions. The discovery conducted with the BMA algorithm unveiled that there were the optimal three models for interpreting the relationship between explanatory variables and consumers' smartphone purchase decisions.

Keywords: BMA algorithm · Purchase decision · Smartphone

1 Introduction

Nowadays, individuals are turning out to be increasingly more dependent on technology, particularly when connecting with others. Smartphone has one of the high technology connection tools that permitted people to link with others. Smartphone has carried many advantages to society. A few advantages of the smartphone such as calling and sending messages, catching pictures, getting to the web, socializing, downloading applications, storing data, recreation, etc. [1, 2]. Likewise, the smartphone has progressively joined the everyday activities of individuals because it helps their lives more straightforward to get to the web 24 h per day, seven days every week, and 365 days any place [3]. In the same view above, Lim et al. [4] dedicated that a smartphone was a mobile gadget that was not only just making and getting telephone calls and messages because the fundamental attribute of a cell phone can be access to the web. It can likewise get to the digital environments, like pictures, music, recordings, etc. Likewise, a smartphone can use small personal computer programs through applications or apps. On the other hand, cell phone utilization these days is becoming more dynamic, and the difficulties for cell phone organizations are to analyze potential clients' purchasing choices to confirm selling growth and get competing [5].

R. Silhavy et al. (Eds.): CoMeSySo 2022, LNNS 596, pp. 31–39, 2023.
https://doi.org/10.1007/978-3-031-21435-6_4

At the Vietnamese smartphone market, IDC has just released a report on the Vietnamese smartphone market in 2021. Smartphone sales in the Vietnam market reached 15.9 million units with a growth of 11.9% in 2021. The company forecasts that growth in 2022 will reach 22%, which is almost double the last year. Because of the policy of stopping the import of phones that only support 2G, the Vietnamese smartphone market will have a stronger growth impulse in 2022. Besides, according to data, the 5 biggest smartphone brands in the market in 2021 are still familiar names: Samsung, Oppo, Apple, Xiaomi, and Vivo. In which, the two leading positions have not changed for at least the past 7 years. Moreover, in 2021, the phone brands with impressive growth are Xiaomi (up 70.9% compared to 2020), Apple (66.2%), Vivo (29.4%), Samsung (27.1%), and Oppo (2.3%) [6]. On the other hand, the more a lot of smartphones are available in the marketplace, the more difficult to choose a smartphone. It was difficult for users to select the one for suitable their needs [7].

Together with the tendency growth of smartphones, some researchers are also interested in this subject. For instance, Abdullah et al. [8] showed that the product features variable has a significant affecting on the smartphone buying decision. Haris and Mustaffa [5] debated that social influence and product features influenced smartphone buying decisions. Zamrie et al. [1] revealed the factors (brand name, social influence, and product features) that impacted the smartphone buying decision. Tanveer at el. [2] disclosed that the price of smartphones influenced smartphone purchase decisions. However, these studies could be different from the real Vietnam market. Furthermore, there are still limited research themes in smartphone buying decisions by the BMA algorithm in Vietnam. Thus, this paper focuses on the BMA algorithm for smartphone purchase decisions.

2 Literature Review

2.1 Purchase Decision

The consumer behavioral model by Kotler and Keller [9] recommended that consumers' buy decision goes through 5 stages, earlier and later the real purchase, including needs identification, information review, the alternatives evaluation, buy decision, and post-buy behavior. A buying decision was the process of thinking that will motivate the purchaser or shopper to recognize their needs that result in a selection of a specific item [8]. Some important factors influence consumers' smartphone purchase decisions. Based on the previous research, the critical factors that affected consumers' smartphone purchase decisions such as product features [5], product price [2], brand name, and social influence [1]. So, these factors will be presented in the next items.

2.2 Product Features

Product features can be described as the product attributes or traits that can meet users through the product's possessing, using, and application [10]. Product features contain hardware and software. The smartphone's hardware includes the body of the cell phone, size, weight, color, and design. In contrast, the smartphone's software includes the general terms for programs, procedures, and documentation. Besides, the smartphone's software is the working programs, storage memory, or applications that run the smartphone [4]. In this new time, the cell phones feature currently the remote network, the underlying web program, application establishment, full programmability, the file control system, mixed media show and capture, high-solving displays, a few gigabytes of capacity and placement, and moving sensors. Also, the activity system and camera are the highlights that most smartphone clients focus on. The most famous activity system for cell phones was Google Android and Apple iPhone [11]. Each working system has its exclusive attributes and foundations [12]. On the other hand, various individuals select various features of smartphones that can satisfy their different needs and desires [13]. The former study confirmed that the product features factor affected significantly smartphone purchase decisions [8, 14].

2.3 Social Influence

Social influence was characterized as the manner individuals impact the beliefs, feelings, and behavior of one person [15]. Social influence was also implicated in how people gain proficiency with the abilities, knowledge, and views apt for using goods and services [16]. Social influences imply one individual causes one more to roll out an alteration on his/her sentiments, attitudes, considerations, and behavior, deliberately or unintentionally [17]. People are rarely aware of social influences as they are so widespread in society. Likewise, customers might be prone to the social impact through cognizances, observations, or expectations of cell phone choices made by others [18]. Furthermore, social influence was considered the solid impacted factor on smartphone purchase decisions [13].

2.4 Product Price

Product price is defined as the sum of money expected, needed, or considered when paying for a product [19]. There were two kinds of product prices. High-priced products were marketplace products where the image was the important factor. Low-priced products that products were bought with utilitarian value, the user depending on the price's perceived value. Shoppers regularly search for products or substitutes with low prices to get the best worth [3, 19]. Price also is how much cash a customer is ready to pay in return for items and services that they believe are valued. The money's value shifts for various individuals. Some could think it is significant at a high expense yet others could figure it doesn't worth the worth of cash [9]. Isabella et al. [20] stated that if the discount rate of products was low, it should be introduced in its worth; while on the off chance that the discount rate of products was high, it should be introduced in percentages, so clients will have a higher buy decision. Besides, the product price was the crucial factor that influenced smartphone purchase decisions [2, 21].

2.5 Brand Name

A brand name can be a name, term, or symbol, and is formed to distinguish the firm from its rivals [1]. The brand name can also be the firm's most valuable asset and is an exclusive name that shows the product to the market. The brand name is not only a name and symbol, it's also an element of the association between the organization and clients [10, 13]. Shabrin et al. [22] debated that the brand name makes an impression on the client's insights about the quality of the product or service it's given. A concentration on the impact of a brand name shows individuals trust the understood brand for its quality, and execution and accept that brand displays an individual way of life and social status. A brand is a name, term, plan, image, or any element that distinguishes one vendor's goods or services as unmistakable from those of different merchants [4]. Mudondo [23] confirmed that individuals likewise favor branded items on account of the impression of their function, superior, service, and the societal position related to the brand. The prior research confirmed that the brand name affected smartphone purchase decisions [1, 22].

3 Methodology

3.1 Sample Analysis

The population sample was gathered for quantitative research by the non-probability method through the online survey by Google form. According to Hair et al. [24], in multiple regression analysis, the minimum sample size is 50 and preferably 100 responses. The minimum ratio of observations per variable is 5:1. In this study, the respondents were collected from 320 clients in Ho Chi Minh City, Vietnam.

3.2 Measurement Scales

A five-point Likert scale was applied to measure the factors (with 1 = Totally agree to 5 = Totally disagree). The scale contains five items of product features (PF) from [17], five items of social influence (SI) and five items of product price from [22], four items of brand name (BN) from [3], five items of smartphone purchase decision from [22].

3.3 Analytical Approach

The RStudio program applied the BMA approach to select the optimal models. With BMA not only choose the best model among rational selected models but could select many models to interpret the dependable variable in linear regression analysis [25]. The BMA approach was also one of the methods for solving model unsureness, which enabled the assessment of the strength of results to alternative terms by measuring posterior distributions over coefficients and models. In this study, with independent variables (i.e. product feature - PF, social influence - SI, product price - PP, brand name - BN) and dependent variable (smartphone purchase decision).

4 Results and Discussion

4.1 Descriptive Statistics

As illustrated in Table 1, regarding gender, the sample included 145 male customers accounted for 47.7% and 159 female customers accounted for 52.3%. Regarding age, the sample comprised 191 clients (below 25) estimated at 62.8%, 82 clients (from 25 to 35) estimated at 27.0%, 25 clients (from 36 to 46) estimated at 8.2%, and 6 clients (above 45) estimated at 2.0%. So, the customer with age (from below 25 to 35) account for the majority (89.8%). Regarding income/month, the sample comprised 147 clients (below 5 million VND/month) accounted for 48.4%, 83 clients (from 5 to 10 million VND/month) accounted for 27.3%, 50 clients (below 10 to 20 million VND/month) accounted for 16.4%, 24 clients (above 20 million VND/month) accounted for 7.9%. Thus, the clients with income/month (from below 5 to 10 million VND/month) account for the majority (75.7%).

Table 1. Sample characteristics

Characteristics	Classifications	Frequency	Percent
Gender	Male	145	47.7
	Female	159	52.3
	Total	304	100
Age	< 25	191	62.8
	25–35	82	27.0
	36–45	25	8.2
	> 45	6	2.0
	Total	304	100
Income/month	< 5 mil VND	147	48.4
	5–10 mil VND	83	27.3
	> 10–20 mil VND	50	16.4
	> 20 mil VND	24	7.9
	Total	304	100

4.2 BMA (Bayesian Model Averaging) Algorithm

The chosen models' results from the BMA algorithm as shown in Table 2.

Table 2 revealed the BMA algorithm of independent variables for smartphone purchase decisions. As depicted in Table 2, the BMA algorithm pointed to the best 3 models that were selected. The posterior affect probabilities (p!) of two predictor variables (i.e. product price - PP, and brand name - BN) were 100%. This finding depicted that these two predictor variables had in all the chosen models. Therefore, these two factors were the most crucial factors affecting smartphone purchase decisions. In addition, the posterior affect probability of product features (PF) was 85.0%. This result revealed that product features (PF) were 85.0 times the occurrence of all the choice models. Likewise, the posterior affect probability of social influence (SI) was 95.7%. This result revealed that social influence (SI) was 95.7 times the occurrence of all the choice models. Thus,

Table 2. 3 models were selected

	p! = 0	EV	SD	Model 1	Model 2	Model 3
Intercept	100.0	0.023	0.149	0.011	0.065	0.115
PF	85.0	0.118	0.065	0.136		0.178
SI	95.7	0.163	0.059	0.164	0.203	
PP	100.0	0.263	0.052	0.254	0.296	0.323
BN	100.0	0.460	0.046	0.452	0.492	0.479
nVar				4	3	3
r2				0.729	0.721	0.718
BIC				−373.753	−370.384	−367.871
Post prob				0.808	0.150	0.043

product features (PF) and social influence (SI) also could be considered important factors influencing smartphone purchase decisions.

Moreover, Table 2 also showed that model 1 was the best model of the best three models; model 2 was the second-best model after model 1, and model 3 was the third-best model after model 2.

Model 1 recommended four factors (nVar = 4) including product features (PF), social influence (SI), product price (PP), and brand name (BN). These four independent variables described the 72.9% variance of smartphone purchase decisions. Comparing the three models, model 1 had the smallest BIC indicator (BIC = −373.753). As well, comparing the three models, the posterior probability (post prob) of model 1 that occurred in this study was 80.8%, and also had the largest index.

Model 2 comprised three factors (nVar = 3) that were social influence (SI), product price (PP), and brand name (BN). These three independent variables represented the 72.1% variance of smartphone purchase decisions. Comparing the three models, model 2 had a low BIC indicator (BIC = −370.384), but the posterior probability was pretty small (15.0%).

Model 3 contained three factors (nVar = 3) that were product features (PF), product price (PP), and brand name (BN). These three independent variables explained the 71.8% variance of smartphone purchase decisions. Comparing the three models, model 3 had a low BIC indicator (BIC = −367.871), but the posterior probability was the smallest (only 4.3%).

Compared to other research (e.g. see [1, 5, 8]) as shown in the introduction section, these studies just disclosed the single model among alternative models. For instance, Abdullah et al. [8] debated that the optimal model was the model that contained only one independent variable (product feature) and the dependent variable (smartphone buying decision). Haris & Mustaffa [5] debated that the best model included two predictor factors (social influence and product features) and the response factor (smartphone buying decisions). Zamrie et al. [1] revealed the best model including three factors (brand name, social influence, and product features) and the dependent (smartphone buying decision).

These traditional methods usually ignored the uncertainty in the model's selection. Thus, the outcome of the estimated model could be biased and prone to an imprecise inference. The BMA approach was also one method for resolving model doubtfulness, which empowered the assessment of the strength of results to alternative quantities by estimating posterior distributions over coefficients and models [25]. Specifically, as explained in Table 2, the BMA algorithm found the three best models. Model 1 included four predictor factors (product features - PF, social influence - SI, product price - PP, and brand name - BN) and the dependent variable (smartphone purchase decisions). Model 2 consisted of three independent variables (social influence - SI, product price - PP, and brand name - BN) and the dependent variable (smartphone purchase decisions). Model 3 consisted of three independent variables (social influence - SI, product price - PP, and brand name - BN) and the dependent variable (smartphone purchase decisions). Moreover, this research revealed that the two factors (product price - PP and brand name - BN) were these factors affected significantly consumers' smartphone purchase decisions in all models.

5 Conclusions

This research applied the BMA algorithm for smartphone purchase decisions. The results pointed to the best 3 models. Model 1 included four predictor factors (product features - PF, social influence - SI, product price - PP, and brand name - BN) and the dependent variable (smartphone purchase decisions). Model 2 consisted of three independent variables (social influence - SI, product price - PP, and brand name - BN) and the dependent variable (smartphone purchase decisions). Model 3 consisted of three independent variables (social influence - SI, product price - PP, and brand name - BN) and the dependent variable (smartphone purchase decisions). The findings disclosed that these results fit in the practice because, in real-life, we did not have only one selection, we probably could have replacement choices. The Bayesian method could help us with chances for thoughtful and assessing the model's unsureness. Further, the research results also confirmed the two key overriding factors (product price - PP, and brand name - BN) that influenced meaningful smartphone decision decisions and these factors had in all models. Therefore, these findings also can support the smartphone shop managers in more focus these factors than the two factors (product features - PF and social influence - SI) to enhance consumers' smartphone purchase decisions. Furthermore, smartphone shop managers also should focus on product features (PF) and social influence (SI) as these factors happened representatively 85% and 95.7% in selected models. Thus, managers can combine two key factors (product price - PP, and brand name - BN) with product features (PF) or social influence (SI) to have substitute models to increase smartphone purchase decisions.

References

1. Zamrie, N.A., Azmee, L.I., Zainon, A., Azmi, M.R., Halim, A.I.A., Ahmad, A.H.: Factors influencing the selection of smartphones among University Students: an insight from Universiti Kuala Lumpur, Business School, Malaysia. Asian J. Manage. Entrepreneurship Soc. Sci. **01**, 1–16 (2021)

2. Tanveer, M., Kaur, H., Thomas, G., Mahmood, H., Paruthi, M., Yu, Z.: Mobile phone buying decisions among young adults: an empirical study of influencing factors. Sustainability. **13**, 1–18 (2021). https://doi.org/10.3390/su131910705
3. Suki, N.: Students' demand for smartphones: Structural relationships of product features, brand name, product price and social influence. Campus-Wide Inf. Syst. **30**, 236–248 (2013). https://doi.org/10.1108/CWIS-03-2013-0013
4. Lim Lay-Yee, K., Kok-Siew, H., Chan Yin-Fah, B.: Factors affecting smartphone purchase decision among malaysian generation Y. ISSN Int. J. Asian Soc. Sci. **3**, 2426–2440 (2013)
5. Haris, N.S., Mustaffa, A.H.: Factors affecting consumer buying decision towards choosing a smartphone among young adults. Int. J. Account. Business Manage. **8**, 51–63 (2020)
6. Thao, T.: Vietnamese smartphone market: Xiaomi, Apple grow strongly, https://ictnews.vietnamnet.vn/cuoc-song-so/thi-truong-smartphone-viet-xiaomi-apple-tang-truong-manh-406526.html#:~:text=IDC vừa phát hành báo,gần gấp đôi năm ngoái. Last Accessed 15 Mar 2022
7. Soriton, L., Tumiwa, J.: Factors driving consumer purchase decision in selecting smartphone (study: students of faculty of economics and business). J. EMBA. **4**, 558–567 (2016)
8. Abdullah, N.S.M., Daim, M.I.I., Nickson, N., Masnib, R.: Factors that influence consumer purchase decision in choosing mobile phones among teenagers. Malaysia J. Business Econom. **8**, 109–124 (2021)
9. Kotler, P., Keller, K.L.: Marketing management. Pearson Education Inc, Upper Saddle River, New Jersey (2016)
10. Kotler, P., Armstrong, G.: Principles of marketing. Pearson Education Limited, Harlow, England (2016)
11. Oulasvirta, A., Wahlström, M., Anders Ericsson, K.: What does it mean to be good at using a mobile device? An investigation of three levels of experience and skill. Int. J. Hum. Comput. Stud. **69**, 155–169 (2011). https://doi.org/10.1016/j.ijhcs.2010.11.003
12. Chang, Y.F., Chen, C.S.: Smart phone-the choice of client platform for mobile commerce. Comput. Stand. Interf. **27**, 329–336 (2005). https://doi.org/10.1016/j.csi.2004.10.001
13. Kumar, R., Fernandez, D.: Factors influencing the purchase decision of smartphone users in sultanate of Oman. J. Market. Consumer Res. **72**, 41–52 (2020). https://doi.org/10.7176/jmcr/72-04
14. Malviya, S., Saluja, M.S., Thakur, A.S.: A study on the factors influencing consumer' s purchase decision towards smartphones in indore. Int. J. Adv. Res. Comput. Sci. Manage. Stud. **1**, 14–21 (2013)
15. Mason, W.A., Conrey, F.R., Smith, E.R.: Situating social influence processes: dynamic, multidirectional flows of influence within social networks. Pers. Soc. Psychol. Rev. **11**, 279–300 (2007). https://doi.org/10.1177/1088868307301032
16. Nelson, M.R., McLeod, L.E.: Adolescent brand consciousness and product placements: Awareness, liking and perceived effects on self and others. Int. J. Consum. Stud. **29**, 515–528 (2005). https://doi.org/10.1111/j.1470-6431.2005.00429.x
17. Chow, M., Chen, L., Yeow, J., Wong, P.: Conceptual paper: factors affecting the demand of smartphone among young adult. Int. J. Soc. Sci. Econom. Art. **2**, 44–49 (2012)
18. Ting, D., Lim, S., Patanmacia, T., Low, C., Ker, G.: Dependency on smartphone and the impact on purchase behaviour. Young Consumers. **12**, 193–203 (2011). https://doi.org/10.1108/17473611111163250
19. Swani, K., Yoo, B.: Interactions between price and price deal. J. Prod. Brand Manage. **19**, 143–152 (2010). https://doi.org/10.1108/10610421011033494
20. Isabella, G., Pozzani, A.I., Chen, V.A., Gomes, M.B.P.: Influence of discount price announcements on consumer's behavior. RAE Revista de Admin. de Empresas. **52**, 657–671 (2012). https://doi.org/10.1590/S0034-75902012000600007

21. Sata, M.: Factors affecting consumer buying behavior of mobile phone devices. Mediterr. J. Soc. Sci. **4**, 103–112 (2013). https://doi.org/10.5901/mjss.2013.v4n12p103
22. Shabrin, N., Khandaker, S., Kashem, A.S. Bin, Hie, C.K., Susila, T.: Factors affecting smartphone purchase decisions of generation-Y. J. Contempor. Issues Business Govern. **23**, 47–65 (2017)
23. Mudondo, C.D.: Determinants of generation-y brand preferences in the mobile phone market in Southern Zimbabwe. Res. J. J. Commerce. **2**, 1–12 (2014)
24. Hair, J.F., Black, W. c., Babin, B, J., Anderson, R.E.: Multivariate data analysis. Pearson Education Limited, London, U.K (2014)
25. Raftery, A.E., Madigan, D., Hoeting, J.A.: Bayesian model averaging for linear regression models. J. Am. Stat. Assoc. **92**, 179–191 (1997). https://doi.org/10.1080/01621459.1997.10473615

The Fast Algorithm for Demodulation of Differential Amplitude Phase Shift Keyed Signals

Oleg Chernoyarov[1,2,3](✉), Alexey Glushkov[4], Kaung Myat San[1], Vladimir Litvinenko[5], and Yuliya Litvinenko[5]

[1] National Research University "MPEI", Krasnokazarmennaya Str. 14, 111250 Moscow, Russia
chernoyarovov@mpei.ru

[2] National Research Tomsk State University, Lenin Avenue 36, 634050 Tomsk, Russia

[3] Maikop State Technological University, Pervomaiskaya Str. 191, 385000 Maikop, Russia

[4] Zhukovsky-Gagarin Air Force Academy, Marshal Nedelin Street 133, 394052 Voronezh, Russia

[5] Voronezh State Technical University, Moscow Avenue 14, 394026 Voronezh, Russia

Abstract. A digital algorithm for the demodulation of the differential amplitude phase shift keyed (DAPSK) signals is considered. The transmitted information signal is represented by a known multipath amplitude phase shift keyed (Gray-APSK) signal constellation, in which the information symbol is determined by a relative amplitude change and a phase shift of the received and preceding symbols. Demodulation is implemented based on the digital phase and amplitude detectors. Coherent processing of the signal phase shift is provided while there is no phase locking in the reference generator. An algorithm for a relative estimate of the change in the amplitudes of the adjacent symbols is proposed, which makes it possible to abandon the rigid level calibration of the received signals. #COMESYSO1120.

Keywords: Differential amplitude and phase shift keying · Signal demodulation · Coherent processing · Demodulation error probability · Noise immunity · Statistical simulation

1 Introduction

The amplitude phase shift keyed (APSK) and quadrature amplitude modulation (QAM) signals provide a high-speed transmission of a discrete information in radio engineering [1–5] and optical [6, 7] communication systems, but they require a demodulator phase locking and a normalization of the amplitudes of the received symbols. The structure of these signals is described by a constellation, the points of which display the values of their amplitudes and phases. Multipath constellations (Gray-APSK) [2, 8] are widespread, allowing separate amplitude and phase shift keying.

Differential phase shift keying (DPSK) [1, 9], under the condition of a slow change in the phase of adjacent symbols, eliminates the need for demodulator phase locking, thus

R. Silhavy et al. (Eds.): CoMeSySo 2022, LNNS 596, pp. 40–53, 2023.
https://doi.org/10.1007/978-3-031-21435-6_5

greatly simplifying the receiving equipment design. Introduction of additional symbol amplitude keying allows increasing the number of positions [10]. However, with the amplitude and differential phase shift keyed (ADPSK) signals, the calibration of the received signal amplitudes is required. To avoid that, differential amplitude and phase shift keying (DAPSK) can be applied.

DAPSK signals are widely used in digital television [11], data transmission systems [12, 13], and optical communication systems [6]. A DAPSK signal is represented by a multipath ("mutibeam") constellation, in which the path (beam) angle determines the phase shift between the received and previous symbols, and the amplitude of the current symbol is determined by the amplitude of the previous symbol and transmitted information binary code [11–14], while the amplitude decision is made by comparing the levels of the received and preceding symbols.

Below, a digital DAPSK signal demodulator is introduced, which requires a minimum number of arithmetic operations over signal period, provides high noise immunity and can operate in the conditions of fading and phase instability in the radio path. It is demonstrated that this demodulator can be implemented based on a digital phase detector [15] complimented by an amplitude detector in accordance with [16].

2 Differential Phase Shift Keyed Signal

The element of the M-position DAPSK signal can be represented as.

$$s(t) = S_k \cos(2\pi f_0 t + \psi_0 + \psi_k),$$

where S_k is the amplitude, $f_0 = 1/T_0$ the carrier frequency, T_0 is the carrier period, ψ_k is the signal phase of the kth symbol and ψ_0 is the indefinite initial phase of a signal.

The constellation has a multipath shape (Gray-APSK-like [8]), as one can see in Fig. 1, the dots designating the signal positions. In the polar coordinates, these positions are determined by possible signal amplitude values U_k, $k = \overline{0, M-1}$, where M_1 is a number of the signal levels, and by the phase shift φ between the received and the previous symbols. Thus, the total signal positions is equal to $M = M_1 M_2$, where M_2 is the number of the phase shift levels. In Fig. 1a, one can see that $M_1 = 2$, $M_2 = 8$, $M = 16$ while, in Fig. 1b, – $M_1 = 4$, $M_2 = 8$, $M = 32$.

For the introduced DAPSK signal processing algorithm, it is presupposed that the possible values U_k of the symbol amplitudes S_k satisfy the condition [13]

$$U_k = U_0 a^n, \tag{1}$$

where $a > 1$ is the constant, $n = 0, 1, \ldots$ is the number of the constellation "circle", and, specifically, $a = 3$ in Fig. 1a, and $a = 1.4$ in Fig. 1b [12, 13].

3 Forming a Differential Phase Shift Keyed Signal

The multi-level kth symbol allows transmitting the binary information code with the number of bits $\log_2 M$ and the decimal equivalent d. The lower $\log_2 M_2$ bits of this code (3

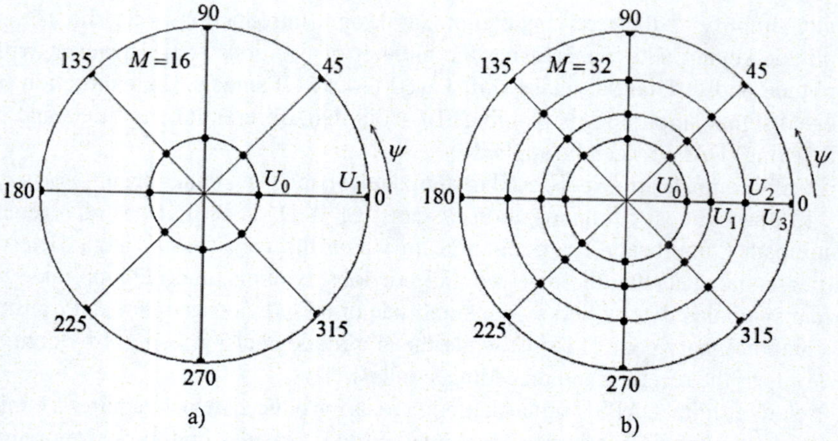

Fig. 1. The DAPSK signal constellations: (**a**) $M = 16$, (**b**) $M = 32$.

bits for the constellations presented in Fig. 1) with a decimal equivalent $d_2 = \overline{0, M_2 - 1}$ determine the phase shift

$$\phi_k = \psi_k - \psi_{k-1} = 2\pi d_2 / M_2. \tag{2}$$

Higher bits of the code d corresponding to the decimal equivalent $d_1 = \overline{0, M_1 - 1}$, so that

$$d = d_1 M_2 + d_2, \tag{3}$$

determine the amplitude $S_k = U_0 a^n$ of the current symbol taking into account the amplitude $S_{k-1} = U_0 a^m$ of the previous symbol in accordance with the rule

$$n = (m + d_1) \bmod M_1. \tag{4}$$

Here y mod x is the operator dividing the number y by the modulo x. For example, for the signal corresponding to the constellation shown in Fig. 1a, the adjacent amplitudes S_k and S_{k-1} are the same, if $d_1 = 0$, and different otherwise. For the signal corresponding to the constellation shown in Fig. 1b, the modulating rule is presented in Table 1.

For the algorithm (4) of the symbol amplitude modulation, the transformation like.

$$G(n - m) = \begin{cases} n - m & if\ n \geq m, \\ n - m + M_1 & if\ n < m \end{cases} \tag{5}$$

results in the expression

$$G(n - m) = d_1. \tag{6}$$

Based on (6) the DAPSK signal amplitude demodulation algorithm is implemented.

Table 1. The modulating rule.

		Information signal d_1			
n		0	1	2	3
m	0	0	1	2	3
	1	1	2	3	0
	2	2	3	0	1
	3	3	0	1	2

4 Digital Demodulation of the Differential Phase Shift Keyed Signal

Digital demodulator of the DAPSK signals can be built based on a digital phase detector [15, 16] and an amplitude detector [17], similar approaches have been realized in [18–20]. The block diagram of the digital demodulator of the DAPSK signal is shown in Fig. 2a. When there are no interferences, the input information signal $s(t)$ is fed to the input of an analog-to-digital converter (ADC), which produces four samples s_{i1}, s_{i2}, s_{i3}, s_{i4} at the ith period T_0 in response to the pulses of the clock generator (CG). And, in this case, the sampling interval is $T_0/4$ and the sampling frequency is $4f_0$. These samples fill the multi-bit four cell shifter (MS4) from which even samples move to the subtractor SUB_0, while odd samples move to the subtractor SUB_1. At the outputs of the subtractors SUB_0 and SUB_1, there are generated the sampling differences $x_{i0} = s_{i3} - s_{i1}$ and $x_{i1} = s_{i4} - s_{i2}$ shifted in time by a quarter of the signal period, that is by $T_0/4$, or by 90° degrees by phase, and then they pass to the identical quadrature processing units QPU_0 and QPU_1.

The block diagram of the quadrature processing unit (QPU) is shown in Fig. 2b. It is a cascade connection of n sample accumulating units. Each of them consists of a summator SUM_m and a multi-bit 2^m cell shifter MR_m ($m = \overline{1, n}$). The sampling differences x_i (which are x_{i0} or x_{i1}) are received at the input of the QPU, and the value $x_i + x_{i-1}$ is calculated at the output of the first summator (the value x_{i-1} has been saved earlier in the MR_1).

At the output of the second summator, the value $x_i + x_{i-1} + x_{i-2} + x_{i-3}$ is calculated (the value $x_{i-2} + x_{i-3}$ has been previously saved in the MR_2), etc. As a result, at the output of the SUM_n, one obtains the responses of the QPU_0 and QPU_1 in the form.

$$y_{0i} = \sum_{j=0}^{N-1} \left[s_{(i-j)1} - s_{(i-j)3} \right],\ y_{1i} = \sum_{j=0}^{N-1} \left[s_{(i-j)0} - s_{(i-j)2} \right],$$

where the number N of the sample accumulation periods is equal to 2^n.

In the normalization unit (NU), the binary codes of the responses y_{0i} and y_{1i} are multiplied by 2 (i.e., shifted by 1 bit to the left) until one of them reaches its maximum modulo value. Thus, at the end of the kth symbol reception (it is presupposed that $k = i/N$ is an integer), one gets.

$$y_{0i} = 2NS_k \cos(\psi_k + \psi_0),\ y_{1i} = 2NS_k \sin(\psi_k + \psi_0)$$

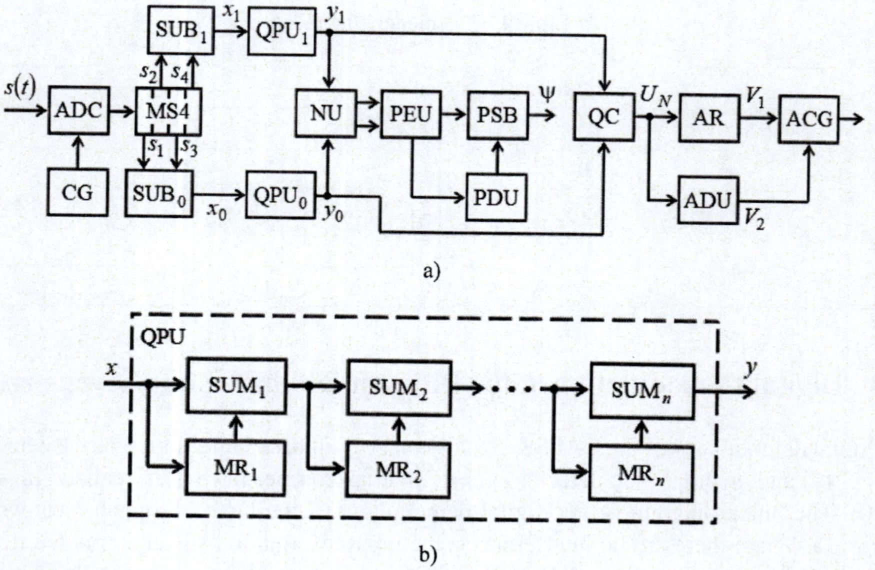

Fig. 2. The block diagram of the digital phase detector, (**a**) and the quadrature processing unit (**b**).

In the phase estimation unit (PEU), the value

$$\psi_i = \begin{cases} -\arctan(y_{1i}/y_{0i}) & \textit{if } y_{0i} \geq 0, \\ -\pi - \arctan(y_{1i}/y_{0i}) & \textit{if } y_{0i} < 0 \end{cases}$$

is determined that, if $k = i/N$ is an integer, is equal to the initial phase of the received symbol $\psi_k + \psi_0$. Under these conditions, the phase shift between the received and previous symbols is formed in the phase subtractor (PSB), and it is equal to

$$\phi_i = \psi_i - \psi_{i-N} \tag{7}$$

with the value ψ_{i-N} passing from the output of the phase delay unit (PDU) for N periods. Based on the value of ϕ_i and at the integer $k = i/N$, the received phase shift code d_2 is formed.

The accumulated amplitude of the received symbol is determined by the expression and with the integer $k = i/N$, the value (8) is equal to $2NS_k$. In the DAPSK signal demodulator presented in Fig. 2a, this operation is implemented by a quadrature converter (QC).

$$U_{Ni} = \sqrt{y_{0i}^2 + y_{1i}^2} \tag{8}$$

If, at the demodulator input, there is a Gaussian interference with independent samples, zero mean and the dispersion σ_n^2, the voltage signal-to-noise ratio (SNR) h_0 for the signal with the minimum amplitude is [17, 18]

$$h_0 = \sqrt{N}U_0 / \sigma_n. \tag{9}$$

With a relative demodulation of the amplitudes of the adjacent DAPSK signal symbols [11–13], the decisions are made in the amplitude code generator (ACG) based on the two amplitude values: $V_1 = U_i$ from the output of the amplitude register (AR) and $V_2 = U_{i-N}$ from the output of the amplitude delay unit (ADU) for N periods (Fig. 2a).

In [13], an approximate demodulation algorithm involves calculating the ratio.

$$p_k = V_1 / V_2 = U_{Nk} / U_{Nk-1},$$

where U_{Nk} is determined according to (8). Then its comparison with a set of threshold values depending on the parameter a (1) is implemented.

To demodulate the amplitude code, it is proposed to use the logarithmic measure

$$g_k = \log_a\left(U_{Nk} / U_{Nk-1}\right). \tag{10}$$

which, in the absence of interference and taking into account (1), is equal to

$g_k = n - m.$

Then, taking into account (5) and (6), the amplitude decoding algorithm in the form

$$G(g_k) = G\left[\log_a\left(U_{Nk} / U_{Nk-1}\right)\right] = G\left[\log_a\left(V_1 / V_2\right)\right] = d_1 \tag{11}$$

is obtained where $G(g_k)$ takes the integer values 0, 1, 2, ..., $(M_1 - 1)$ with a unit interval.

In the presence of interferences at the demodulator input, the values of g_k (10) are random in the neighborhood of $\pm d_1 \pm 0.5$, and, with the deviations greater than 0.5, an erroneous demodulation of the amplitude code occurs. In Fig. 3, there is demonstrated the histogram $W(x)$ of the values of $g_k = x$ obtained as a result of a statistical simulation at $M_1 = 4$, $M_2 = 8$ for the DAPSK signal with $M = 32$ at SNR $h_0 = 24$ dB, the number of wells being 256 and the sample size of 10^6 symbols. The function (11) transforms the values g_k into u_k in accordance with the algorithm

$$u_k = \begin{cases} g_k & \text{if } g_k > -0.5, \\ g_k + 4 & \text{if } g_k \le -0.5. \end{cases} \tag{12}$$

The corresponding histogram $W(x)$ of the values of $u_k = x$ is shown in Fig. 4. Here, by the dotted line, the approximation of the first "lobe" of the histogram by a Gaussian probability distribution is demonstrated.

As one can see, the log measure g_k (10) transforms the ratio of different amplitudes of the received symbols into a uniform range of $g_k = x$ values with the integer decision-making centers $d = \pm d_1$ (in the example: −3, −2,... 3) in the intervals of ± 0.5: the decision is made on the received amplitude code d, if the condition

$$d - 0.5 < g_k \le d + 0.5$$

holds. There the threshold levels depend only on the amplitude codes, as one can see that this is an algorithm for rounding numbers. In accordance with (12), the negative values of g_k are transformed into domains with positive centers, and, as a result, the received amplitude code d_1 taking the possible values 0, 1, 2, 3 is generated (Fig. 4).

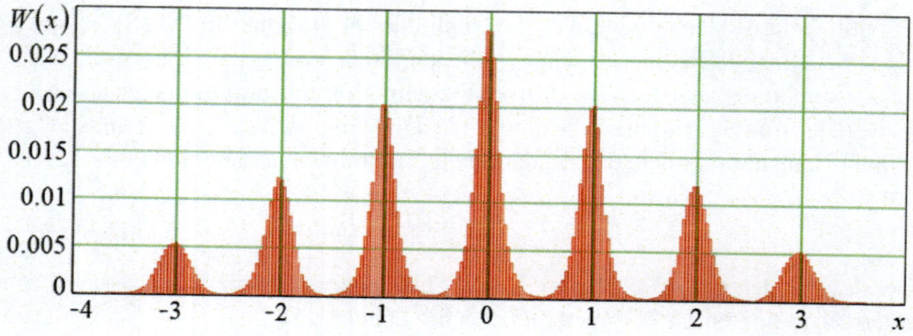

Fig. 3. The histogram of the values of $g_k = x$.

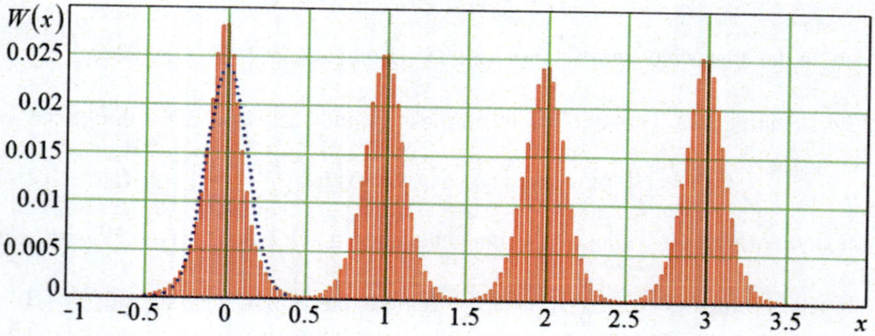

Fig. 4. The histogram of the values of $u_k = x$.

The advantage of the introduced demodulation algorithm is the independence of decision-making thresholds from the signal levels, that is, there is no need to normalize the amplitude characteristics of the reception path. The proposed transformation of the ratio of the symbol amplitudes (11), (12) is simply implemented based on a storage device.

5 The Noise Immunity of the Demodulation Algorithms

Now it is time to estimate the noise immunity of the proposed DAPSK signal demodulation algorithm in the presence of Gaussian noise with independent samples using the statistical simulation method when SNR is equal to h_0 for the symbol with a minimum amplitude (9). Erroneous demodulation of a DAPSK symbol occurs when the phase shift φ between the received and previous symbols is erroneously estimated with the probability P_P or when the amplitude ratio between these symbols is erroneously determined with the probability P_A.

The probability of erroneous coherent demodulation of an M-level differential phase shift keyed (MDPSK) signal can be obtained from the expression [1]

$$P_\phi = 2Q\left(\sqrt{2}h\sin\left(\pi \big/ M_2\sqrt{2}\right)\right)$$

where $Q(x) = \int_x^\infty \exp(-t^2/2)\,dt \Big/ \sqrt{2\pi}$.

For the symbol with the minimum amplitude U_0, the SNR is determined from (9). In this regard, the corresponding dependences $P_0 = P_\phi(h = h_0)$ are shown in Fig. 5, at $a = 1.4$ and $M_2 = 4$ or $M_2 = 16$. The symbols next in level that have the amplitude $U_1 = aU_0$ provide the error probability $P_1 = P_\phi(h = ah_0)$, the corresponding dependencies at $a = 1.4$ are shown in Fig. 5. As one can see, $P_0 >> P_1$, that is, the probability of an erroneous phase demodulation is determined only by the symbols with a minimum amplitude and it is equal to

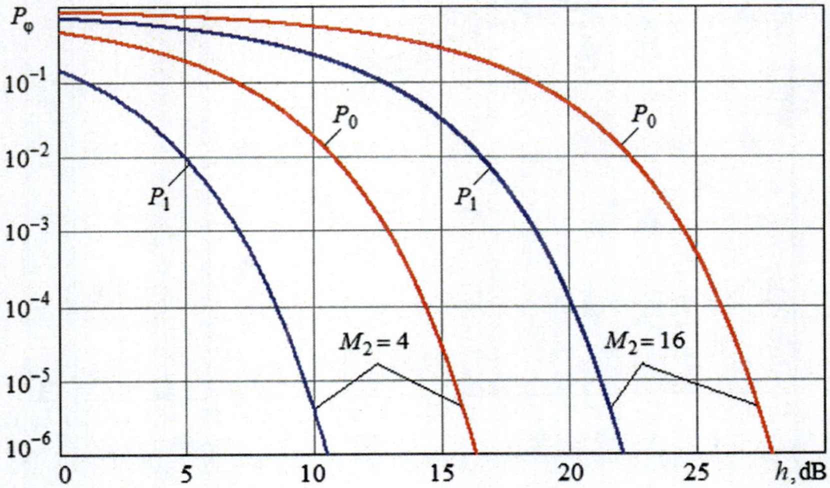

Fig. 5. The probability of the erroneous phase demodulation.

$$P_P = P_\phi(h_0) \big/ M_2. \tag{13}$$

The probability of an erroneous differential amplitude demodulation is determined by the values h_0 and a.

A statistical simulation model of a demodulator has been developed in which the codes for modulating the amplitude $d_1 = \overline{0, M_1 - 1}$ and the phase $d_1 = \overline{0, M_2 - 1}$ generating the M-positional ($M = M_1 M_2$) code of the transmitted symbol d (3) are randomly and equiprobably chosen.

In Fig. 6, for the DAPSK signal with the constellation shown in Fig. 1a the theoretical and experimental (derived from a simulation) dependencies of the erroneous demodulation probabilities are demonstrated. It is presupposed that $M_1 = 2$, $M_2 = 8$ ($M = 16$) and $a = 3$. The squares stand for the experimental values of the erroneous symbol demodulation probability P_{err}; the circles are for the experimental values of the erroneous phase demodulation probability P_P; the rhombuses - for the experimental values of the erroneous amplitude demodulation probability P_{A1} provided that there are no phase demodulation errors; and the triangles denote the experimental values of the unconditional (when no phase error condition is imposed) amplitude error probability P_{A2}. The solid line describes the theoretical dependence of the erroneous phase demodulation probability calculated according to (13). And, for comparison, by the filled rhombuses and triangles respectively, there are shown the experimental values of the amplitude error probabilities P'_{A1} and P'_{A2} for the ADPSK signal while the optimal demodulation of the normalized symbol amplitudes is implemented.

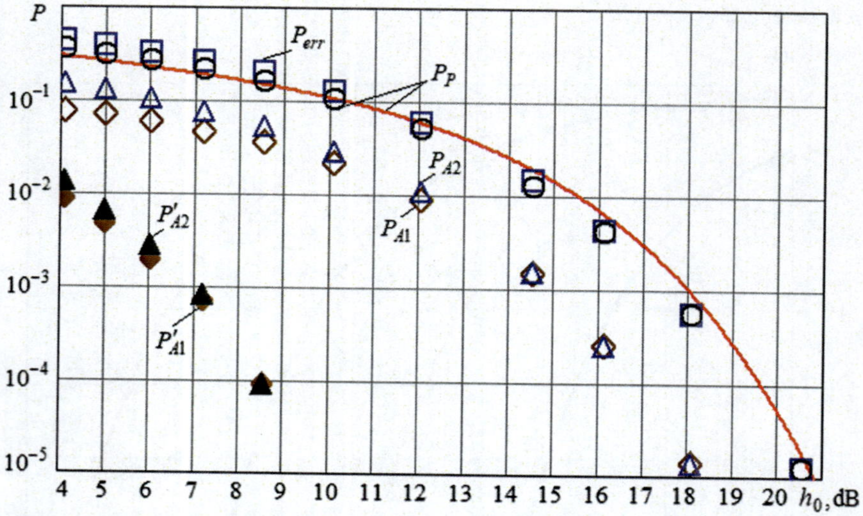

Fig. 6. The probability of the DAPSK signal error demodulation at $M = 16$, $M_1 = 2$, $M_2 = 8$ and $a = 3$.

It follows from Fig. 6 that the differential demodulation of the amplitudes of the received symbols significantly (by 7–8 dB) reduces the noise immunity of the signal level estimation unit. However, in this case also, the probability of an erroneous demodulation of the amplitudes is much lower than the probability of an error in determining the phase shift. It is so due to the high value of a ($a = 3$) and as a result $P_{err} \approx P_{\phi}$. When M_1, the number of the symbol levels, is increasing, these differences are also increasing, as well as the signal peak factor [21].

In Fig. 7, one can find the dependences of the same probabilities $P_{err}, P_P, P_{A1}, P_{A2}$ as the ones demonstrated in Fig. 6, and with the same notations, but for the DAPSK signal at $M = 32$, $M_1 = 4$, $M_2 = 8$ and $a = 2$. Under these conditions, the probabilities of the amplitude and phase errors are approximately the same (balanced), while the overall signal noise immunity is approximately the same when $M_1 = 2$, $a = 3$ and $M_1 = 4$, $a = 2$, but, in the latter case, the information transfer rate is twice as high.

In Fig. 8, at $M = 32$, $M_1 = 4$, $M_2 = 8$ and $a = 1.4$, the experimental dependences of the erroneous demodulation probabilities $P_{err} \approx P_{A1} \approx P_{A2}$ are represented by the squares, while the experimental and theoretical dependences of the erroneous phase demodulation probability P_P is still designated by the circles and the solid line calculated according to (13), respectively. As it can be seen, for low a and $M_1 = 4$, $M_2 = 8$, the probabilities of an erroneous differential demodulation of the amplitudes are many times higher than the probabilities of the phase errors that indicates the inconsistency of the signal parameters. At $a = 1.4$ [12, 13], the peak factor of the signal decreases, but the noise immunity of the symbol amplitude estimation decreases too.

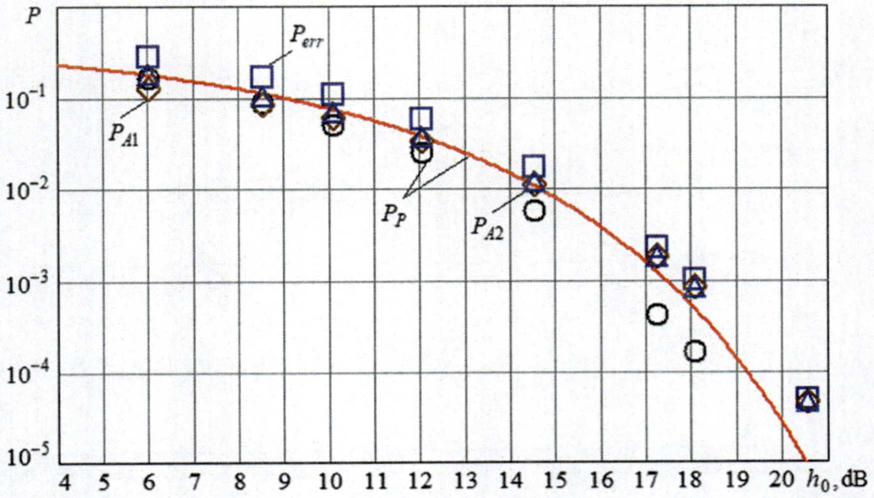

Fig. 7. The probability of the DAPSK signal error demodulation at $M = 32$, $M_1 = 4$, $M_2 = 8$ and $a = 2$.

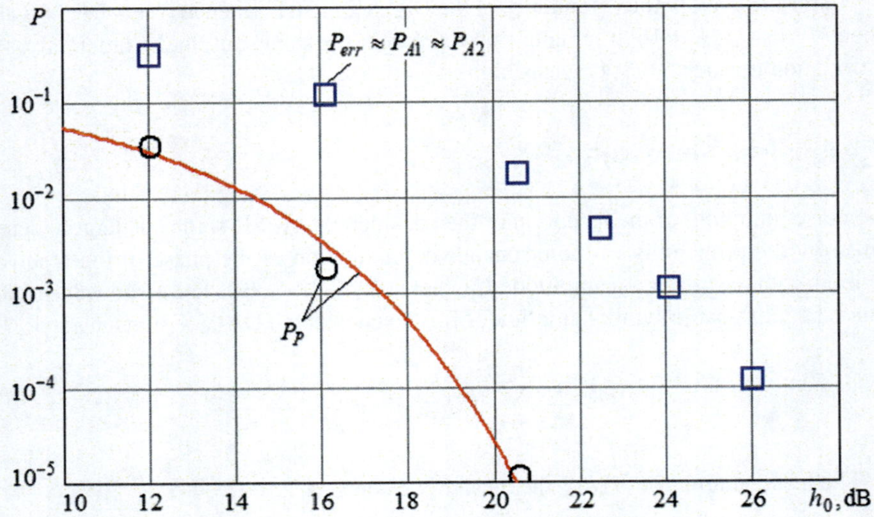

Fig. 8. The probability of the DAPSK signal error demodulation at $M = 32$, $M_1 = 4$, $M_2 = 8$ and $a = 1.4$.

With the same $a = 1.4$ and $M = 64$, $M_1 = 4$, $M_2 = 16$, the DAPSK signal demonstrates its best properties. In this case, the experimental and theoretical results of erroneous demodulation similar to those shown in Fig. 8 are presented in Fig. 9 where one can see that the probabilities of the erroneous differential estimates of the amplitudes and phases are close enough, that is, the signal is balanced.

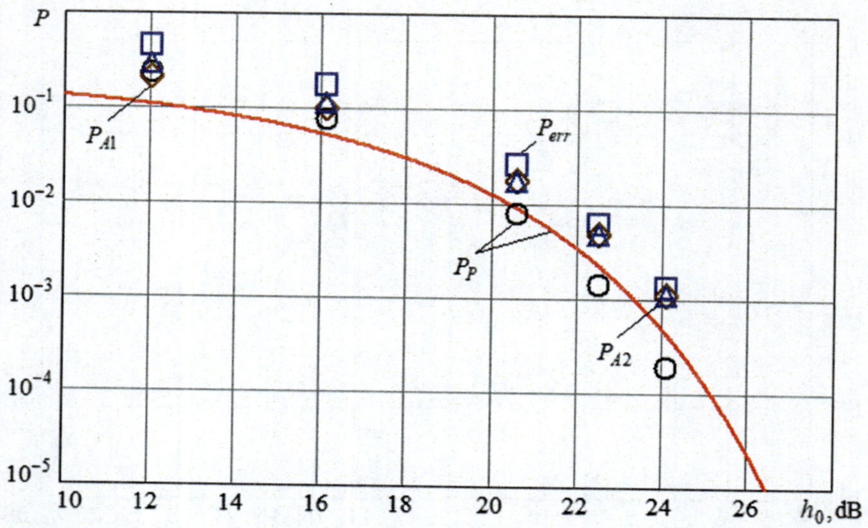

Fig. 9. The probability of the DAPSK signal error demodulation at $M = 64$, $M_1 = 4$, $M_2 = 16$ and $a = 1.4$.

As for the overall noise immunity of the DAPSK signals when $M_1 = 4$, $M_2 = 8$ and when $M_1 = 4$, $M_2 = 16$, it remains approximately the same, but, in the latter case, the information transfer rate is twice as high.

6 Signal-to-Noise Ratio

The noise immunity of the reception of the considered DAPSK signals is largely determined by the probability of the erroneous demodulation of the phase of the received symbols with the minimum amplitude U_0, that is, by SNR h_0 (9). The mean value of the squared equiprobable symbol amplitudes for the considered DAPSK signals is

$$U_{mean}^2 = \frac{1}{M_1} \sum_{n=0}^{M_1-1} \left(U_0 a^n\right)^2 = \frac{U_0^2}{M_1} \frac{a^{2M_1} - 1}{a^2 - 1}$$

Then, accounting for (9), for the mean SNR one gets

$$h_{mean}^2 = U_{mean}^2 h_0^2 \Big/ U_0^2 = h_0^2 \left(a^{2M_1} - 1 \right) \Big/ M_1 \left(a^2 - 1 \right). \tag{14}$$

When comparing the noise immunity of the signals with the different number of positions, the SNR

$$h_B^2 = h_{mean}^2 \Big/ \log_2 M \tag{15}$$

is used, reduced to one bit of the transmitted information [1].

As an example, the estimates for SNR under the error probability $P_{err} = 10^{-4}$ are given in Table 2. Here, one can see the values for h_0 obtained during a statistical simulation. They can be found in Figs. 6–9, while the values of h_{mean} and h_B are the results of the recalculation carried out according to (14), (15).

Table 2. The signal-to-noise ratio estimates.

№	a	M_1	M_2	M	h_0, dB	h_{mean}, dB	h_{B}, dB
1	2	2	8	16	19.6	23.6	17.6
2	2	4	8	16	20	33.3	26.3
3	1.4	4	8	32	26	31.5	24.5
4	1.4	4	16	64	26	31.5	23.8

The first and the second signals in Table 2 are sufficiently balanced in terms of the phase and amplitude demodulation errors (see Figs. 6 and 7). Their h_0 values are approximately the same, but, for the first signal, the value of M_1 is half that of the second, that is, the peak factor [21] is lower and, accordingly, h_{mean} and h_B are also lower, but the second signal has a higher data transfer rate.

The third signal in Table 2 has a much higher probability of amplitude errors than phase errors (see also Fig. 8) due to a decreasing a (it means the reduction in the difference in symbol amplitudes). Thus, the balance is disturbed, leading to a decrease in a noise immunity compared to the second signal at the same transmission rate. In the case of the fourth signal, due to the increase in M_2, an error balance is established, which, at the same SNR, makes it possible to double the information transfer rate.

It follows from these results that, when developing the communication systems using the DAPSK signals, it is necessary to choose the parameters a, M_1 and M_2 accurately to be able to equalize the probabilities of an erroneous amplitude and phase demodulation.

7 Conclusion

A digital DAPSK signal demodulation algorithm has been introduced, making it possible to make decisions in the absence of a phase locking and an accurate calibration of the received signal amplitudes, thus greatly simplifying the demodulator design and allowing it to operate in the conditions of fading and phase instability in the radio path. It is proposed to make a decision on the demodulation of amplitudes by the value of the logarithm of the ratio of the adjacent symbol amplitudes. It is established that in this case it is possible to set the fixed decision-making thresholds that do not depend on signal and noise levels.

Statistical simulation of the operation of the demodulator has been carried out for varied signal parameters. The theoretical and experimental estimates of the probabilities of an erroneous demodulation of the relative values of the amplitude and phase of the adjacent symbols are obtained, and they are in good agreement with each other. Thus, the demodulator efficiency can be determined on a case-by-case basis. It is shown that it is necessary to ensure the balance of the phase and amplitude demodulation errors by choosing the number of positions of the signal amplitude and phase as well as the parameter a.

It is demonstrated that the proposed demodulation algorithm can be effectively implemented based on FPGAs.

Acknowledgements. This work was financially supported by the Ministry of Education and Science of the Russian Federation (research project No. FSWF-2020–0022).

References

1. Sklar, B.: Digital communications: fundamentals and applications, 2nd edn. Prentice Hall, Hoboken (2017)
2. Liu, Z., Xie, Q., Peng, K., Yang, Z.: APSK constellation with gray mapping. IEEE Commun. Lett. **15**(7), 1271–1273 (2011)
3. Ferrand, P., Maso, M., Bioglio, V.: High-rate regular APSK constellations. IEEE Trans. Commun. **67**(2), 2015–2023 (2019)
4. Bao, J., Xu, D., Zhang, X.: A rapid De-mapping method for 64-APSK. In: 6th International Conference on communications and broadband networking. ICCBN, pp. 27–31, Association for Computing Machinery, Singapore (2018)
5. Yao, E., Yang, S., Jiang, W.: A simplified soft decision demapping algorithm of 16-APSK signals in AWGN channels. In: 2nd International conference on networks security, wireless communications and trusted computing, NSWCTC, pp. 103–106, IEEE, China (2010)
6. Elfiqi, A.E., Morra, A.E., Khallaf, H.S., Shalaby, H.M.H., Hranilovic, S.: Two-level MPPM-MDPSK modulation for free-space optical channels. In: 2018 IEEE Global Communications Conference, GLOBECOM, pp. 1–6, IEEE, United Arab Emirates (2018)
7. Prat, J., Velasquez, J.C., Tabares, J.: Direct PSK-ASK modulation for coherent udWDM. In: 2019 21st International conference on transparent optical networks, ICTON, pp. 1–4, IEEE, France (2019)
8. Yan, K., Peng, K., Yang, F., Song, J.: Non-uniform APSK optimization for BICM systems. Tsinghua Sci. Technol. **20**(1), 175–181 (2015)
9. Chernoyarov, O.V., Glushkov, A.N., Litvinenko, V.P., Litvinenko, Y.V., Matveev, B.V.: Fast digital algorithms for the coherent demodulation of the phase-shift keyed signals. In: 11th International IEEE scientific and technical conference "dynamics of systems, mechanisms and machines. Dynamics, pp. 1–5, IEEE, Russia (2017)
10. Takushima, Y., Choi, H.Y., Chung, Y.C.: Transmission of 108-Gb/s PDM 16ADPSK signal on 25-GHz grid using non-coherent receivers. Opt. Express **17**(10), 13458–13466 (2009)
11. Rohling, H., Engels, V.: Differential amplitude phase shift keying (DAPSK)-a new modulation method for DTVB. IBC 95 International Broadcasting Convention, pp. 102–108, IET, Netherlands (1995)
12. Sayhood, K.H., Nan, W.L.: Differential amplitude phase shift keying (DAPSK): a new modulation method for a turbo code in digital radio broadcasting. Eng. J. Univ. Qatar **16**, 73–86 (2003)
13. Liang, D., Ng, S.X., Hanzo, L.: Near-capacity turbo coded soft-decision aided DAPSK/Star-QAM for amplify-and-forward based cooperative communications. IEEE Trans. Commun. **61**(2), 1080–1087 (2013)

14. Wang, L., Hanzo, L.: Low-complexity near-optimum multiple-symbol differential detection of DAPSK based on iterative amplitude/phase processing. IEEE Trans. Veh. Technol. **61**(1), 894–900 (2012)
15. Chernoyarov, O.V., Glushkov, A.N., Litvinenko, V.P., Litvinenko, Y.V., Matveev, B.V., Demina, T.I.: Digital phase detector. Patent RU **2723445**, C2 (2021)
16. Chernoyarov, O.V., Glushkov, A.N., Golikov, A.A., Litvinenko, V.P., Mironov, V.A.: On digital phase detector. J. Phys: Conf. Ser. **2094**(1), 1–7 (2021)
17. Bukharin, S.V., Glushkov, A.N., Kostrov, N.A., Litvinenko, V.P.: Fast digital algorithms for detecting narrow-band signals. Telecommun. Radio Eng. **64**(2), 211–218 (2005)
18. Chernoyarov, O.V., Golpaiegani, L.A., Glushkov, A.N., Lintvinenko, V.P., Matveev, B.V.: Digital binary phase-shift keyed signal detector. Int. J. Eng. Trans. A **32**(3), 510–518 (2019)
19. Chernoyarov, O.V., Glushkov, A.N., Litvinenko, V.P., Matveev, B.V., Makarov, A.A.: Algorithms and devices for noncoherent digital radio signal processing. Eng. Lett. **28**(3), 1238–1248 (2020)
20. San, K.M., Glushkov, A., Chernoyarov, O., Litvinenko, V., Litvinenko, Y., Chernoiarova, E. Non-coherent digital demodulation of ADBPSK and ADQPSK signals. In: 2021 6th International Conference on Frontiers of Signal Processing, ICFSP, pp. 1–5, IEEE, France (2021)
21. Palicot, J., Louët, Y.: Power ratio definitions and analysis in single carrier modulations. In: 13th European Signal Processing Conference, EUSIPCO, pp. 53–56, IEEE, Turkey (2005)
22. Genç, F., Yengel, E., Savaşçihabeş, A., Ertuğ, Ö.: On the optimum ring ratio determination for 16-DAPSK modulation in OFDM systems. In: 2014 22nd Signal processing and communications applications conference, SIU, pp. 742–745, IEEE, Turkey (2014)
23. Rohling, H., May, T., Brüninghaus, R., Grünheid, R.: Broad-band OFDM radio transmission for multimedia applications. Proc. IEEE **87**(10), 1778–1789 (1999)

On the Digital Algorithms for the DASK Signal Noncoherent Demodulation

Oleg Chernoyarov[1,2,3](✉), Alexey Glushkov[4], Kaung Myat San[1], Vladimir Litvinenko[5], and Alexandra Salnikova[1]

[1] National Research University "MPEI", Krasnokazarmennaya Str. 14, 111250 Moscow, Russia
chernoyarovov@mpei.ru

[2] National Research Tomsk State University, Lenin Avenue 36, 634050 Tomsk, Russia

[3] Maikop State Technological University, Pervomaiskaya Str. 191, 385000 Maikop, Russia

[4] Zhukovsky-Gagarin Air Force Academy, Marshal Nedelin Street 133, 394052 Voronezh, Russia

[5] Voronezh State Technical University, Moscow Avenue 14, 394026 Voronezh, Russia

Abstract. The digital algorithms for a noncoherent demodulation of the differential multi-level amplitude shift keyed (DASK) signals are considered, in which the transmitted information determines the ratio of the amplitudes of the received and previous symbols, and the sign is set according to the phase shift of these symbols. Such demodulation does not require neither the calibration of the received signal levels, nor the phase locking of the reference generator. The demodulator response to the DASK signals is studied. The estimation of a demodulation noise immunity is carried out, and the recommendations on the signal parameters are provided. #COMESYSO1120.

Keywords: Differential amplitude shift keying · Signal demodulation · Noncoherent processing · Demodulation error probability · Noise immunity · Statistical simulation

1 Introduction

The amplitude-shift keyed (ASK) signals are used in various radio engineering applications [1–3], while, in the simplest case, a binary on-off keying (OOK) is applied. The multi-level amplitude shift keying is abbreviated as MASK but *M* here is not only "multi-level" in short, it also stands for the particular number of levels. MASK provides an increased speed of information transfer compared to OOK. The best results are obtained when applying the combined ASK and binary phase shift keying (BPSK) that allows displaying the received symbol amplitude sign. Such signals can be considered as a special case of the amplitude and phase shift keyed (APSK) signals [1].

Due to the ease of their implementation, the OOK signals are used to establish an off-chip communication in the modules of electronic devices (in the RF-interconnect (RF-I) systems [4]) as well as a data transmission on a chip [5]. The OOK and ASK signals are used in the communication systems operating over the short distances in single-chip

R. Silhavy et al. (Eds.): CoMeSySo 2022, LNNS 596, pp. 54–67, 2023.
https://doi.org/10.1007/978-3-031-21435-6_6

radio frequency (RF) transceivers [6], for example, TR1000 made by RF Monolithics (RFM), in radio frequency identification (RFID) systems [7, 8] and in optical information transmission systems [9–12].

Most often, a MASK signal noncoherent demodulation is applied, while only the positive values of the symbol amplitudes are used. The latter are compared with the specified thresholds to form the received code, and this operation requires a calibration of the incoming signal levels. If the sign of the symbol amplitude is to be distinguished, thus doubling the information transfer rate, then a coherent signal demodulation appears necessary, and this can significantly complicate the equipment needed for the demodulation.

The transition to the differential amplitude coding methods carried out by combining a comparison of the amplitudes of the received and previous symbols and a noncoherent signal processing eliminates the need to calibrate the levels of the received signals [13, 14].

Below, a digital DAPSK signal demodulator is introduced. It requires a minimum number of arithmetic operations over signal period, provides high noise immunity and can operate in the conditions of fading and phase instability in the radio path. It is demonstrated that this demodulator can be implemented based on the noncoherent narrow-band signal detector [15, 16].

2 A Differential Amplitude Shift Keyed Signal

A DASK signal is a sequence of harmonic symbols with the number k of the form

$$s(t) = S_k \cos(2\pi f_0 t + \psi) \tag{1}$$

where S_k is the symbol amplitude, while $f_0 = 1/T_0$, T_0 and ψ are the carrier frequency, period and initial phase. The symbol duration T_E is equal to

$$T = NT_0,\ N = 2^n \tag{2}$$

where n is the integer. An example of the time diagram of the sequence constituted of 4 symbols at $N = 4$ is shown in Fig. 1.

As it is suggested in [13], the amplitude of the k-th symbol that is S_k takes one of the values:

$$U_m = U_0 a^m \tag{3}$$

where $a > 1$, $m = \overline{0, M-1}$ is the integer, M is the number of the positions of the symbol amplitudes.

The signal positions are represented by the constellations [2, 13], one can see an example of it for the APSK signal at $M = 4$ and $a = 1.4$ in Fig. 2.

The amplitude $S_k = U_m = U_0 a^m$ of the current symbol of the DASK signal is determined by both the amplitude of the previous symbol $S_{k-1} = U_l = U_0 a^l$ and the transmitted multi-position information signal $d = \overline{0, M-1}$ in accordance with the rule.

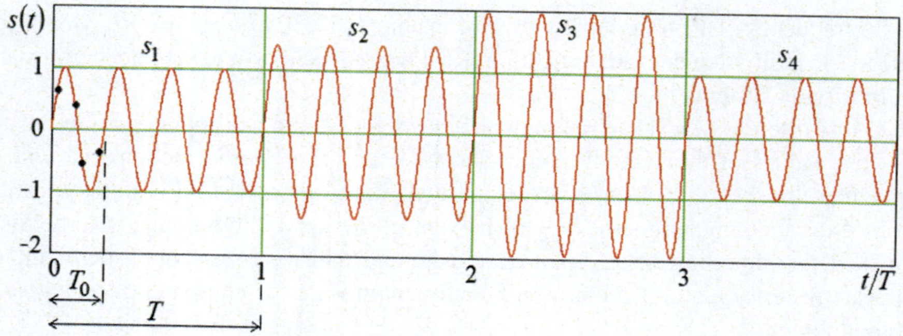

Fig. 1. DASK signal

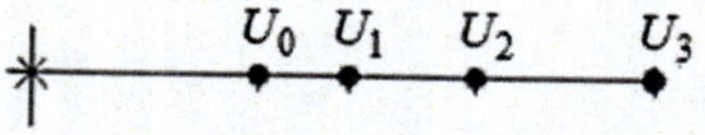

Fig. 2. A constellation of the ASK signal

$$m = (l + d) \bmod M \tag{4}$$

where $y \bmod M$ is the operator for calculating the number y by the modulo M. The expression (4) presents a DASK signal modulation algorithm. In (4), the value d is the decimal equivalent of the transmitted binary code. For example, at $M = 4$, this code takes the values 00, 01, 10 and 11 and, accordingly, the values of d are 0, 1, 2 or 3.

For the demodulation of the amplitude code in [13], the ratio of the amplitudes of the received symbol and the previous one is used that is S_k / S_{k-1}, but it results in a cumbersome demodulation algorithm. Thus, it is proposed to use the logarithmic measure of the amplitude ratio

$$g_k = \log_a(S_k / S_{k-1}) = \log_a(U_m / U_l) \tag{5}$$

for which, in the absence of interferences, one gets:

$$g_k = \log_a\left(a^m / a^l\right) = m - l$$

Accounting for (4),

$$g_k = \begin{cases} d & \text{if } m \geq l, \\ d - M & \text{if } m < l. \end{cases} \tag{6}$$

From that the DASK signal demodulation algorithm follows for the case when the interferences are absent:

$$d = \begin{cases} g_k & \text{if } g_k \geq 0, \\ g_k + M & \text{if } g_k < 0. \end{cases} \tag{7}$$

The magnitudes of g_k take the integer values from $-(M-1)$ to $(M-1)$, while the values of d are from the range between 0 and $(M-1)$, respectively.

In the presence of interferences, the amplitudes of the symbols S_k acquire random deviations from U_m (3) and the values of g_k undergo fluctuations close to the integer values (6). Then the DASK signal demodulation algorithm in the presence of interference takes the form

$$d = \begin{cases} round(g_k)\ if\ g_k \geq 0, \\ round(g_k + M)\ if\ g_k < -0.5, \end{cases} \tag{8}$$

where *round* (x) is the operator performing rounding a number to the nearest integer value, thus, if $|g_k| < 0.5$, then $d = 0$, respectively. The demodulation result d does not necessarily correspond to its binary equivalent; it is advisable to use the Gray code [1, 2].

3 The Demodulation Algorithm

The block diagram of the DASK signal demodulator is built based on the APSK signal demodulator [15] and is shown in Fig. 3. The digital implementation of this device follows from [16–18].

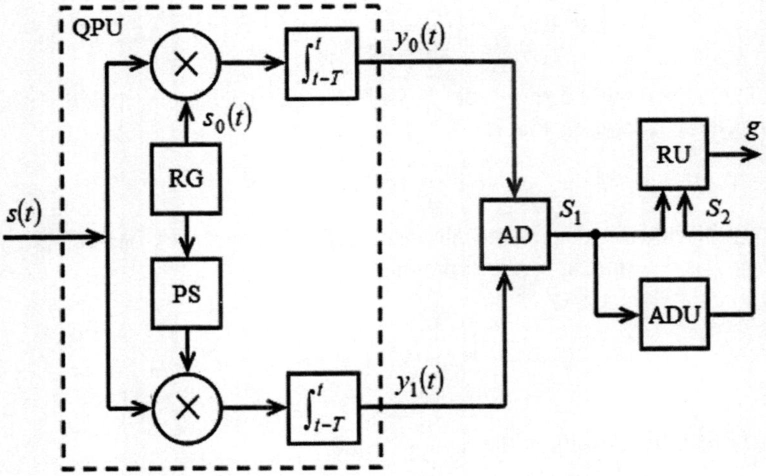

Fig. 3. The block diagram of the DASK signal demodulator

Signal processing is carried out by means of the quadrature algorithm [1]. It is performed in the quadrature processing unit (QPU) and uses 90° phase shifter (PS) together with the multipliers and integrators. In fact, during the interval of the duration of the symbol T (2), in accordance with the input signal $s(t)$ and the signal $s_0(t) = S_0 \cos(2\pi f_0 t + \psi_0)$ arriving from the reference generator (RG), the QPU responses $y_0(t)$

and $y_1(t)$ are calculated in the form

$$\begin{Bmatrix} y_0(t) \\ y_1(t) \end{Bmatrix} = S_m S_0 \int_{t-T}^{t} \cos(2\pi f_0 t + \psi) \begin{Bmatrix} \cos(2\pi f_0 t) \\ \sin(2\pi f_0 t) \end{Bmatrix} dt. \quad (9)$$

The accumulated symbol amplitude S is determined by the amplitude detector (AD) by means of the QPU responses (9):

$$S = \sqrt{y_0^2 + y_1^2}.$$

Its values are fed into the amplitude delay unit (ADU) for the symbol duration interval. At the end of the symbol reception, the accumulated amplitudes S_1 and S_2 of the current and preceding symbols are sent to the resolver unit (RU) where the value of g_k (5) is then determined and the decision on the received information code is made.

4 The Noise Immunity of the Amplitude Ratio Demodulation

According to [19, 20], the probability density $w(x)$ of the values of the response X at the AD output when receiving the signal with the amplitude S against Gaussian noise with the dispersion σ^2 and independent samples corresponds to the Rice distribution of the form.

$$w(x, S) = \frac{x}{\sigma^2} \exp\left(-\frac{x^2 + S^2}{2\sigma^2}\right) I_0\left(\frac{xS}{\sigma^2}\right), \quad x \geq 0,$$

where $I_0(x)$ is the modified zero-order Bessel function. Then the probability density of the ratio of these responses, that is

$$Y = X_k \big/ X_{k-1} \quad (10)$$

(it may be noticed that in the absence of the interferences one refers to $Y = U_0 a^m \big/ U_0 a^l$) is determined by the expression [21]

$$w_1(y, l, m) = \int_0^\infty x\, w\left(x, U_0 a^l\right) w\left(x, U_0 a^m\right) \mathrm{d}x, \quad (11)$$

while the probability density of the random values

$$Z = g_k = \log_a Y \quad (12)$$

is correspondingly equal to

$$w_2(z, l, m) = a^z w_1\left(a^z, l, m\right) \ln a, \quad (13)$$

The signal-to-noise ratio for the signal (1) with the minimum amplitude can be found using the expression [20].

$$h^2 = U_0^2 \big/ 2\sigma^2.$$

The examples of the dependences $w_2(z, l, m)$ (13) at $a = 1.4$ and $h^2 = 50$ (17 dB) are shown in Fig. 4 by solid lines. As it can be seen, the maxima of the probability density (13) practically coincide with the values $r = m - l.$. In accordance with the demodulation algorithm (6), for $r \geq 0$, the r value is equal to the code of the demodulated symbol d, and for $r < 0$ one gets $d = r + M$, respectively. Then, in accordance with (8), the decision formation thresholds are equal to $d \pm 0.5$. The number of the combinations of values of the pairs l and m is equal to M^2, the smaller they are, the higher is the probability of an erroneous demodulation of the corresponding symbol.

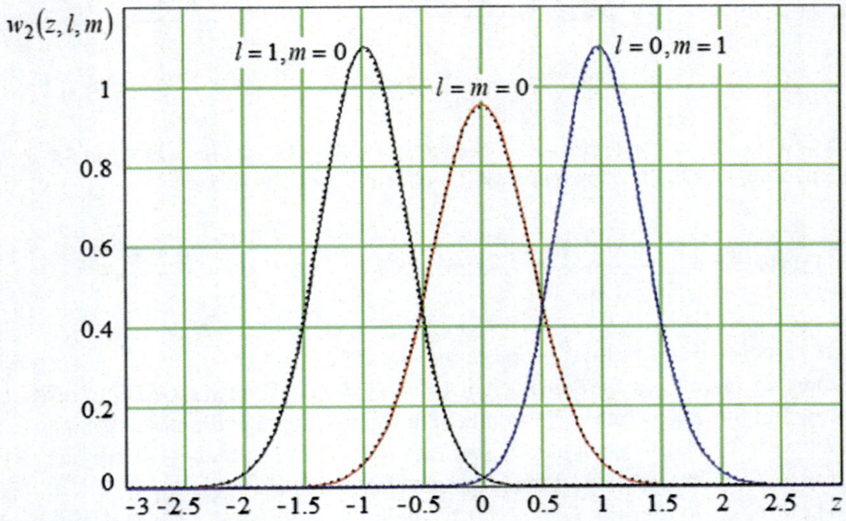

Fig. 4. The probability density of the logarithmic ratio of the signals at the AD output

The probability p_A of the erroneous demodulation of the ratio of the information symbol amplitudes at the specified l and m is determined by the expression

$$p_A(h, l, m) = 1 - \int_{m-l-0.5}^{m-l+0.5} w_2(z, l, m)\, dz. \quad (14)$$

5 Gaussian Approximation

As it follows from (14), the calculations of the probability p_A for the probability density of the symbol amplitudes $w(x, S)$ (10) are quite laborious and do not allow one to obtain analytical expressions. But, it is well known [19] that for $h >> 1$ the Rice distribution is approximated by a Gaussian probability distribution with the probability density of the form

$$w(x, S) = \exp\left[-(x - S)^2 \Big/ 2\sigma^2\right] \Big/ \sigma\sqrt{2\pi}. \quad (15)$$

Then the probability density (11) of the ratio Y of the amplitudes of the adjacent symbols after substituting (15) is equal to

$$w_1(y,l,m) = \frac{1}{2\pi\sigma^2}\int_0^\infty w\exp\left[-\frac{\left(x-U_0a^l\right)^2}{2\sigma^2}\right]\exp\left[-\frac{(yx-U_0a^m)^2}{2\sigma^2}\right]\mathrm{d}x. \quad (16)$$

After the algebraic transformations one gets

$$w_1(y,l,m) = \frac{1}{2\pi\sigma^2}\int_0^\infty x\exp\left[-\frac{(x-A(y,l,m))^2 - A^2(y,l,m) + B(y,l,m)}{2\sigma^2/\left(1+y^2\right)}\right]\mathrm{d}x, \quad (17)$$

where $A(y,l,m) = U_0\left(a^l + ya^m\right)/\left(1+y^2\right)$, $B(y,l,m) = U_0^2\left(a^{2l}+a^{2m}\right)/\left(1+y^2\right)$

The integral (17) is approximately transformed into the form

$$w_1(y,l,m) = \sqrt{\frac{h^2}{\pi}}\frac{a^l + ya^m}{\left(1+y^2\right)^{3/2}}\exp\left[-h^2\left(a^{2l}+a^{2m}\right) - \frac{\left(a^l+ya^m\right)^2}{1+y^2}\right], \quad (18)$$

where signal-to-noise ratio h^2 is determined by (14).

In Fig. 5, one can see the dependencies of $w_1(y,l,m)$.. Here the solid line corresponds to (16), and the dotted line corresponds to the approximate expression (18), at $a = 1.4$ and $h^2 = 500$ (27 dB). As it can be seen, for $h > 22$ the formula (18) provides a high accuracy of a representation of the Gaussian approximation.

The probability density $w_2(z,l,m)$ of the random values Z (12) is determined by (13) after substituting (18) as follows

$$w_2(z,l,m) = a^z\sqrt{\frac{h^2}{\pi}}\frac{a^l + a^{z+m}}{\left(1+a^{2z}\right)^{3/2}}\exp\left[-h^2\left(a^{2l}+a^{2m}\right) - \frac{\left(a^l+a^{z+m}\right)^2}{1+a^{2z}}\right]\ln a. \quad (19)$$

The corresponding dependences are shown by the dotted lines in Fig. 4 and agree well with the results of the calculations based on the Rice distribution (solid lines).

The functions $w_2(z,l,m)$ depend on the pairs of values l and m, which define the information symbol $d = \overline{0, M-1}$ according to (7). For example, when $M = 4$, the values of d are 0, 1, 2 or 3, and in the Table 1 their interrelations with l and m are demonstrated.

As one can see, each value of d corresponds to 4 pairs of l and m with their own probability densities $w_2(z,l,m)$, the examples of which are shown in Fig. 6 at $a = 1.4$, $h^2 = 500$ (27 dB) and $l = m$. Similar curves also take place for other values of l and m.

It follows that as l and m increase, the curves become narrower, since the amplitude of the corresponding symbol increases, thus leading to a decrease in the probability of erroneous decisions $p_A(h,l,m)$ (14).

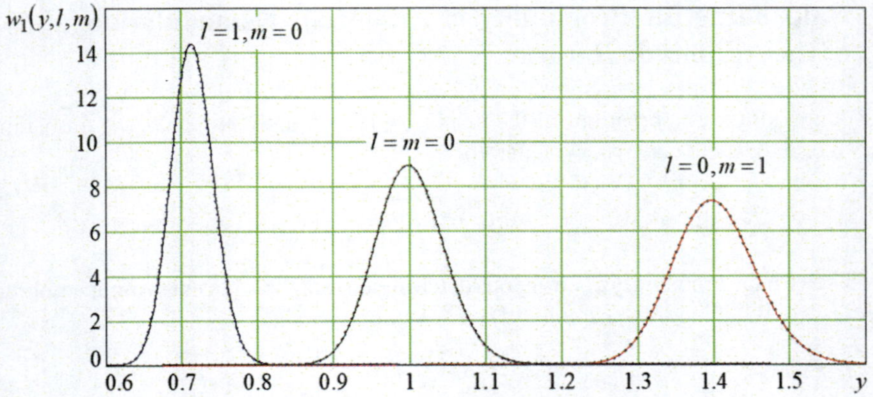

Fig. 5. The probability density of the ratio of the signals at the AD output and its approximation based on the Gaussian distribution

Table 1. The modulating rule

l	m			
	0	1	2	3
0	0	1	2	3
1	3	0	1	2
2	2	3	0	1
3	1	2	3	0

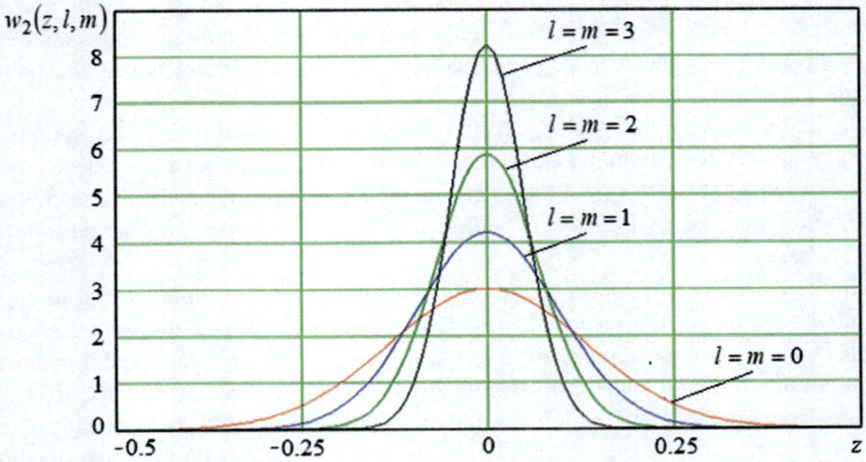

Fig. 6. The probability density of the logarithmic ratio of the signals at the AD output when using the Gaussian approximation of the QPU responses

6 Calculating the Probability of Erroneous Demodulation of the Amplitude Ratio

Assuming all the combinations of the pairs (l, m) are equally probable, the probability p_{lm} of each of them occurring is equal to

$$p_{lm} = 1\big/M^2$$

Then for the probability of erroneous determination of the adjacent symbol amplitude ratio $p_{errA}(h)$, taking into account (14) and (19), one obtains

$$p_{errA}(h) = \sum_{l=0}^{M-1} \sum_{m=0}^{M-1} p_{lm} p_A(h, l, m). \tag{20}$$

The dependencies (20) are shown in Fig. 7 for the different values of a and M by solid lines. By squares, circles and triangles, the results of the statistical simulation are marked obtained at $M = 2$, 4 and 8, respectively. As it can be seen, the probability $p_{errA}(h)$ decreases with an increasing a due to an increase in the ratio of the , of the adjacent symbols, and it decreases with an increase in the number of positions of the amplitude M due to a decrease in the probability of appearance of the symbols with small amplitudes. The results of the calculation performed using the approximate formula (20) are in good agreement with the simulation, but for the high $M \geq 8$, the discrepancies are observed in the region of low error probabilities.

The dashed line in Fig. 7 shows a limit curve at $M = 1$ demonstrating a reception of the harmonic signal symbols at a constant amplitude. In this case, the probability $p_{errA}(h)$ characterizes the intensity of exceeding of a specified threshold a by the ratio of the adjacent amplitudes (in the example it is 1.4) due to interferences in the communication channel. This can be used while estimating the quality of a communication channel by the received signal with a constant amplitude (in the case of the frequency or phase shift keying).

It should be taken into account that the signal-to-noise ratio h (14) corresponds to the symbol with the minimum amplitude U_0. At $M \geq 2$, , the higher amplitude symbols appear, resulting in an increase in the average symbol power (i.e. the mean signal-to-noise ratio) and the signal peak factor. In addition, as M increases, the information transfer rate increases too.

In Fig. 8, the dependencies of the error probability $p_{errA}(h)$ on the signal-to-noise ratio h are shown for the case of the two-level ASK ($M = 2$) and at different relative levels of the adjacent symbols a (solid lines), while the circles, triangles and rhombuses represent the results of the statistical simulation for $a = 1.4$, 1.8 and 3, respectively. It follows from Fig. 8 that the error probability decreases rapidly with an increasing a and the theoretical values of the probability $p_{errA}(h)$ are consistent with the corresponding experimental data.

The dashed line in Fig. 8 shows the dependence of the error probability p_{err2} on h in the case of the ideal demodulation of the two-level ASK signal at $a = 3$ (here the symbol amplitudes are equal to U_0 and $3U_0$) taking the form.

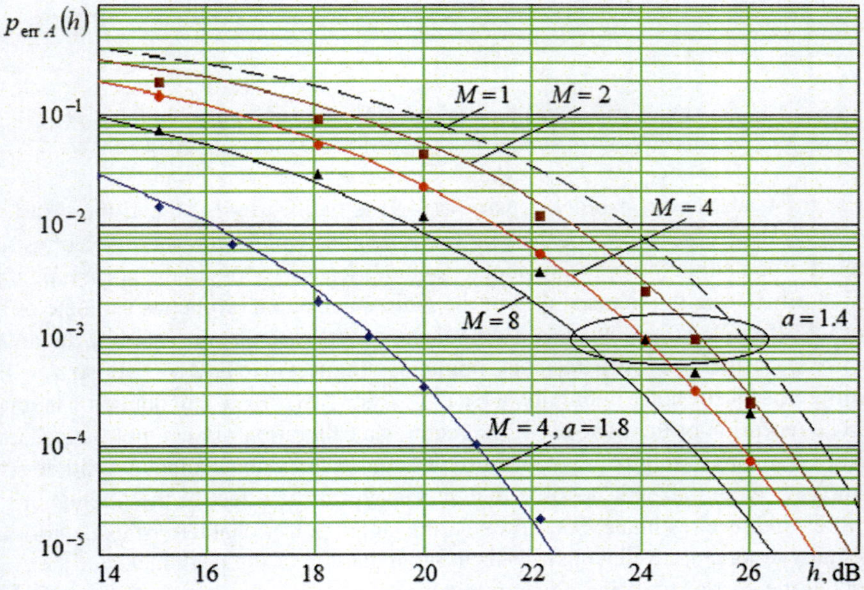

Fig. 7. The probability density of the ratio of the signals at the AD output and its approximation based on the Gaussian distribution

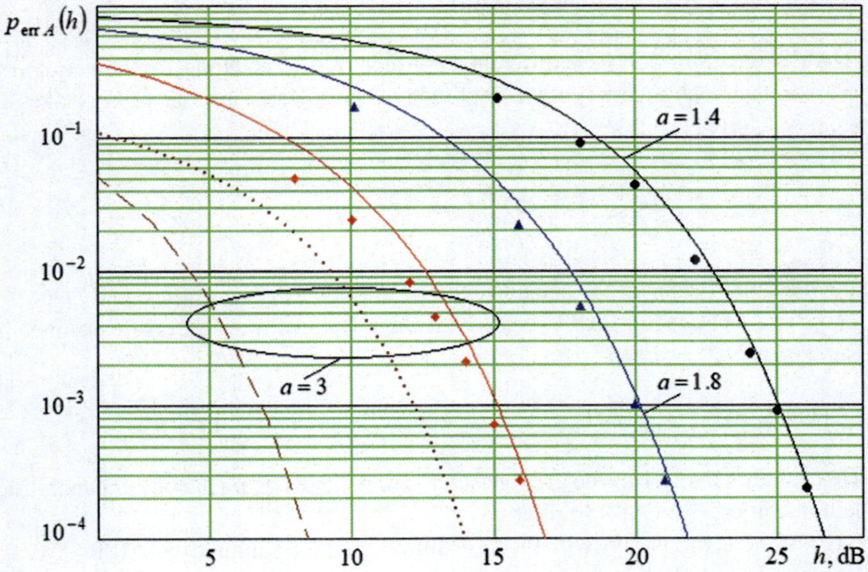

Fig. 8. The probability of erroneous determination of the amplitude ratio at $M = 2$

$$p_{err2} = \frac{1}{2}\left[\int\limits_{U_T}^{\infty} w(x, U_0)\,\mathrm{d}x + \int\limits_{0}^{U_T} w(x, aU_0)\,\mathrm{d}x\right],$$

where the Gaussian probability density $w(x, S)$ is defined by (15), and the threshold level is $U_T = 2U_0$. One can see that, with an ideal choice of the threshold, the transition to relative demodulation of the symbol amplitudes leads to a loss in the noise immunity by 7–8 dB. If, due to the instability of the radio channel, an erroneous estimate of the threshold $U_T = 2.5U_0$ occurs (the p_{err2} values for this case are presented by the dotted line in Fig. 8), then the loss in the noise immunity is expected to be 2.5–3 dB. At $a = 1.4$, the difference in the noise immunity of the DASK and ASK signal demodulation is small.

The reception of the MASK signals requires a calibration of the amplitude characteristics of the receiving path, and the use of a high ratio of the symbol amplitudes (for example, $a = 3$) makes it possible to compensate for the effect of the fluctuations in the threshold level estimate, but, however, the signal power characteristics deteriorate, especially at $M > 2$. Reducing a leads to the reduction of the peak factor of the ASK signal at the cost of reducing noise immunity. At $a = 1.4$, the loss in noise immunity compared to the case of ideal demodulation appears to be 20 dB.

7 The DASK Signal Power Characteristics

The DASK signal symbols have different amplitude U_m (3), and thus, for their equiprobable values, the mean square symbol amplitude (its normalized average power) takes the form.

$$U_{mean}^2 = \frac{1}{M}\sum_{m=0}^{M-1} U_0^2 a^{2m} = \frac{U_0^2}{M}\frac{a^{2M} - 1}{a^2 - 1},$$

and the maximum value of the squared symbol amplitude is $U_{\max}^2 = U_0^2 a^{2(M=1)}$.

The signal peak factor PAPR (Peak to Average Power Ratio) [22] is determined by the expression.

$$PARP = 10\lg\left(U_{\max}^2 \big/ U_{mean}^2\right) = 10\lg\left[Ma^{2(M-1)}\left(a^2 - 1\right)\big/\left(a^{2M} - 1\right)\right].$$

This value characterizes the average signal efficiency of using the power capabilities of the transmitter. The PAPR limit at $a \to \infty$ is $10\lg M$.

It is advisable, by analogy with PAPR, to introduce the magnitude of AMPR (Average to Minimum Power Ratio), that is equal to the ratio of the average symbol power to its minimum value necessary for the demodulation. It takes the form

$$AMPR = 10\lg\left(U_{mean}^2 \big/ U_0^2\right) = 10\lg\left[\left(a^{2M} - 1\right)\big/M\left(a^2 - 1\right)\right].$$

This magnitude characterizes the necessary redundancy of the average power of the transmitter (symbols with a minimum power must provide the required noise immunity).

The dynamic range of the DASK signal is $D = 20(M - 1)\lg a$. The PAPR, AMPR and D magnitudes are positive and are measured in dB, certainly: $D = PAPR + AMPR$. Their dependencies on a at $M = 2$ and $M = 4$ are shown in Fig. 9, where the solid line represents PARP, while the dashed one—AMPR and the dotted one—D. With a decrease in a, the considered power characteristics improve, but the minimum operating signal-to-noise ratio h increases too (Figs. 7 and 8).

When $M = 2$ and $M = 2$, a high noise immunity and acceptable power characteristics are provided, but there is a low data rate. With an increase in the number of signal positions ($M = 4$ in Fig. 9), the power characteristics are unacceptable, and it is necessary to reduce a.

Accounting for (16), the mean signal-to-noise ratio is equal to $h^2_{mean} = h^2 U^2_{mean} / U^2_0$, then, in dB, one gets $h^2_{mean}\,[dB] = AMPR + h^2\,[dB]$.

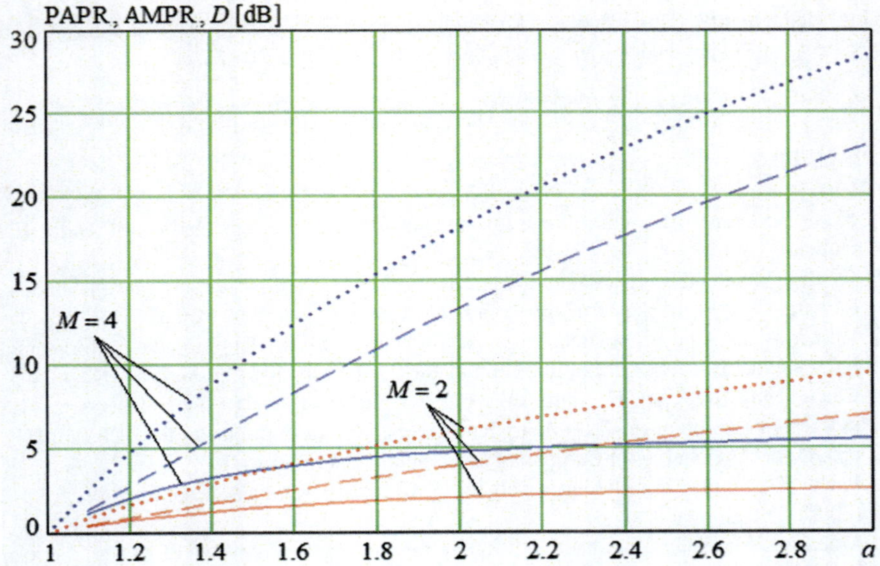

Fig. 9. The probability density of the ratio of the signals at the AD output and its approximation based on the Gaussian distribution

The signal-to-noise ratio per 1 bit of the communicated information is $h^2_B = h^2_{mean} / \log_2 M$, then $h^2_B\,[dB] = AMPR - 10\log_2 M + h^2\,[dB]$. For example, at $a = 1.4$ and $M = 4$, one gets $AMPR = 5.542$ dB and $h^2_B\,[dB] = 2.531 + h^2\,[dB]$. Thus, for the DASK signals, the noise immunity analysis can be carried out for the minimum signal-to-noise ratio h^2, followed by the recalculation to h^2_{mean} or h^2_B.

8 Conclusion

The algorithm for the noncoherent demodulation of a signal with a differential multi-level amplitude shift keying is considered, which does not require calibration of the amplitudes of the received symbols, thus simplifying the equipment required for its implementation and facilitating the information reception in an unstable communication channel. It is shown that a Gaussian approximation can be used for calculating the probabilistic characteristics of the amplitudes of the received signals. With high accuracy, they coincide with the corresponding characteristics obtained based of the exact Rice model in a wide range of signal parameters. An expression is obtained for the probability density of the parameters of the amplitudes that the adjacent received symbols have. A technique for calculating the noise immunity of a demodulator has been developed, and the results of it are consistent with what the statistical simulation of the demodulator demonstrates. It is established that, at $a < 2$, the DASK signals are characterized by higher operating signal-to-noise ratios compared to the ASK signals, and, at $a = 3$, the differences in operating signal-to-noise ratios are low (1–2 dB), if $M = 2$, while with an increasing number of positions, the power characteristics become unacceptable.

Acknowledgements. This work was financially supported by the Ministry of Education and Science of the Russian Federation (research project No. FSWF-2020–0022).

References

1. Sklar, B.: Digital communications: fundamentals and applications, 2nd edn. Prentice Hall, Hoboken (2017)
2. Proakis, J., Salehi, M.: Digital communications. McGraw-Hill, New York (2007)
3. Stuberg, G.L.: Principles of mobile communication. Springer, New York (2017)
4. Kim, Y., et al.: Analysis of noncoherent ASK modulation-based RF-interconnect for memory interface. IEEE J. Emerg. Sel. Top. Circuits Syst. **2**(2), 200–209 (2012)
5. Chang, M.C.F., Socher, E., Tam, S.W., Cong, J., Reinman, G.: RF interconnects for communications on-chip. In: 2008 ACM International Symposium on Physical Design, ISPD, pp. 78–83. Association for Computing Machinery, USA (2008)
6. Pavlov, D.: Single-chip RF transceivers for ISM bands (in Russian). Mod. Electron. **1**, 18–23 (2005)
7. Mohaisen, M., Yoon, H., Chang, K.: Radio transmission performance of EPC global Gen-2 RFID system. In: 2008 10th International Conference on Advanced Communication Technology, ICACT, pp. 1423–1428. IEEE, South Korea (2008)
8. Gureeva, O.: New radio frequency identification protocol Gen2 for radio frequency identification systems (in Russian). Compon. Technol. **1**, 112–116 (2006)
9. Chow, C.W., Yeh, C.H.: Using downstream DPSK and upstream wavelength-shifted ASK for Rayleigh backscattering mitigation in TDM-PON to WDM-PON migration scheme. IEEE Photon. J. **5**(2), 1–8 (2013)
10. Palmer, R., et al.: Silicon-organic hybrid MZI modulator generating OOK, BPSK and 8-ASK signals for up to 84 Gbit/s. IEEE Photon. J. **5**(2), 1–8 (2013)
11. Haris, M.: Advanced modulation formats for high-bit-rate optical networks. Ph.D. Thesis, Georgia Institute of Technology, Atlanta (2008)
12. Velichko, M.A., Naniy, O.E., Susian, A.A.: New modulation formats in optical communication systems (in Russian). LIGHTWAVE Russ. Ed. **3**, 21–36 (2005)
13. Liang, D., Ng, S.X., Hanzo, L.: Near-capacity turbo coded soft-decision aided DAPSK/Star-QAM for amplify-and-forward based cooperative communications. IEEE Trans. Commun. **61**(3), 1080–1087 (2013)

14. Wang, L., Hanzo, L.: Low-complexity near-optimum multiple-symbol differential detection of DAPSK based on iterative amplitude/phase processing. IEEE Trans. Veh. Technol. **61**(2), 894–900 (2012)
15. Chernoyarov, O.V., Glushkov, A.N., Litvinenko, V.P., Mironov, V.A., Salnikova, A.V.: Coherent demodulation of the two-level APSK signals with the symbol amplitude estimation. J. Phys: Conf. Ser. **2094**(2), 1–6 (2021)
16. Bukharin, S.V., Glushkov, A.N., Kostrov, N.A., Litvinenko, V.P.: Fast digital algorithms for detecting narrow-band signals. Telecommun. Radio Eng. **64**(2), 211–218 (2005)
17. Glushkov, A.N., Litvinenko, V.P., Matveev, B.V., Chernoyarov, O.V.: Basic algorithm for the noncoherent digital processing of the narrowband radio signals. Appl. Math. Sci. **9**(95), 4727–4735 (2015)
18. Bukharin, S.V., Litvinenko, V.P., Glushkov, A.N.: Detection of narrow-band signals with noise level assessment. Telecommun. Radio Eng. **67**(1), 33–39 (2008)
19. Levin, B.R.: Theoretical principles of radio engineering statistics. Defense Technical Information Center, Fort Belvoir (1968)
20. Recommendation ITU-R P.1057–2. (08/07) Probability distributions relevant to radio wave propagation modeling. https://www.itu.int/rec/R-REC-P.1057-2-200708-S/en, last Accessed 31 May 2022
21. Beylkin, G., Monzon, L., Satkauskas, I.: On computing distributions of products of non-negative independent random variables. Appl. Comput. Harmon. Anal. **46**(2), 400–416 (2019)
22. Palicot, J., Louët, Y.: Power ratio definitions and analysis in single carrier modulations. In: 13th European Signal Processing Conference, EUSIPCO, pp. 53–56. IEEE, Turkey (2005)

A Computational Error Method to Datapath Single Event Transient Analysis

Leila Golpaiegany[1(✉)] and Elena Chernoiarova[2]

[1] University of Ayatollah Ozma Boroujerdi, 69199-69411 Boroujerd, Iran
Golpaiegany@abru.ac.ir

[2] National Research University "MPEI", Krasnokazarmennaya Str. 14, 111250 Moscow, Russia

Abstract. The reduction of the size of the devices and the operating voltage are critical challenges in the design of digital circuits. This paper presents a computational error analysis approach for reliability evaluation in the circuit datapath. In this paper, Soft Error Rate (SER) is investigated in terms of dependency to the cell circuit structure and the input combination. Using this method, a sample cell library is evaluated to build up reliability profile for different implementation cases. Accordingly, error rate for different bits are extracted using Monte Carlo analysis, which have been reflected as a computational error. The proposed Method can be used besides other methods for analysing the reliability based computational error; instead of increasing the reliability via redundancy or other techniques, designers select a reliable design by using different implementations of a component. Considering reliability as a computational cost, during high-level synthesis (HLS) reliability can be traded with other implementation costs resulting in higher performance and lower implementation costs. In this paper, computational error is simulated with a Gaussian distribution. Again, in order to find the relationship between the coefficients of the Gaussian error model and error rate, the coefficients are fitted to a cubic polynomial. Results show this that fitting is suitable. #COMESYSO1120.

Keywords: Computational error · Soft error rate · Reliability · Cell library design

1 Introduction

Uncertainties are always a possibility with real-life digital computing tasks. These uncertainties are either inherited from data acquisition and/or any preceding computations, which are imported together with the input data in the form of input error or noise; on the other hand, they may arise from the current processing system because of computational limits and constraints. Due to the nature of the digital computing, in many applications one or both of these uncertainties are inevitable. Therefore, digital computation is considered to have a limited precision. System design should utilize the available information and systems with the best possible accuracy, relying on trade-offs between computational accuracy and hardware costs.

R. Silhavy et al. (Eds.): CoMeSySo 2022, LNNS 596, pp. 68–83, 2023.
https://doi.org/10.1007/978-3-031-21435-6_7

As VLSI systems scale down to the nano range, sensitivity to Single Event Effects (SEE) increases as a natural result of dimensions scaling, lower supply voltages, higher frequencies, and higher density circuits [1–3]. Although these effects have been examined for many years, reliability has now become a major design metric [4, 5]. In order to consider reliability as a design parameter, it is necessary to have a clear measure of what can be traded off with other implementation costs such as area, speed, and power consumption [6]. For reliable design, transient errors should be considered at different abstraction levels starting from the target technology cell library, which should be characterized in terms of latency, power and also reliability and transient error effects [7, 8].

This paper proposes a method for the analysis of the transient effects in VLSI circuits by using a computational error model in which the stochastic behavior of the effect is measured by a probabilistic computational error.

The paper is organized as follows. Section 2 presents a discussion of the basic method used in this paper and explains the approach used in this paper to evaluate the reliability. Section 3 presents the simulation result. Finally, the conclusion of the paper is presented in Sect. 4.

2 Computational Error Representation of the Transient Error Effect

The reliability evaluation of logic circuits is based on probabilistic models because most errors in logic circuits are inherently probabilistic [9]. Accordingly, a signal probability at the input or output of a gate is defined with the probability that the signal is a logical "1". In many previous works, the failure probability of the circuit, the error rate (ε), is considered a constant independent of the input [10–12]; therefore, any circuit has a constant SER.

In the proposed approach, transient error effects are modelled in the form of computational errors, regardless of their cause. From a mathematical point of view, a computational error can be modelled as an additive error, which means that uncertainty appears at the output as noise added to the original signal; therefore, a unit with a transient error can be modelled as an ideal unit followed by an additive noise source, as shown in Fig. 1.

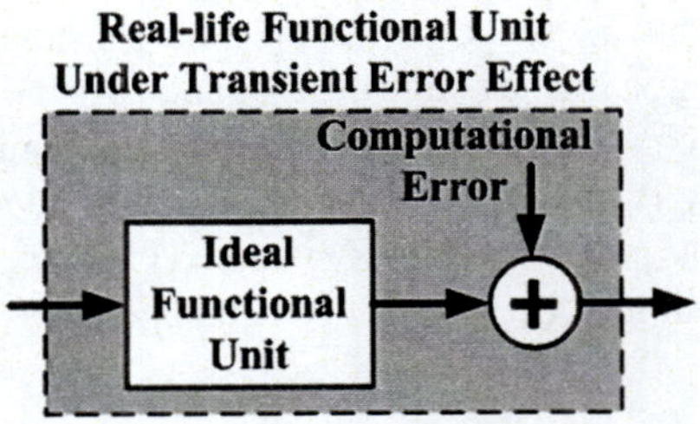

Fig. 1. Additive computational noise model for an arithmetic unit under transient error effect

Additive noise in the circuit can be analyzed by error propagation methods [13]. In order to perform a comprehensive analysis of the transient effects, it is required at the first step to provide a circuit level error model of the effect, then to investigate the effect at higher levels of abstraction such as gate level and functional level. In this section, two important issues need to be discussed: the transient error model and the input dependency of the error.

2.1 Transient Error Model

At the circuit level, a transient error corresponds to an unpredicted current spike. Therefore, it is possible to analyze the potential impact of the error by injecting a transient pulse into the high-level description of circuit blocks, early in the design flow. In general, charge storage on the transistor gate capacitor ($Q = C \cdot V$) determines the logical state [13]. A transient error can occur at a node in a circuit, when the collected charge Q at that node is greater than the critical charge ($Q_{critical}$) [14]. Hence, $Q_{critical}$ can be defined as the minimum collected charge that can cause a soft error. If the charge generated at a node is more than $Q_{critical}$, a latch may be flipped and store the wrong value [13, 15]. Therefore, an SEU occurs in a node when the collected charge in that node reaches a critical charge level and is propagated to a latch. $Q_{critical}$ can be used to estimate the SER [8, 9]:

$$SER \infty N_{flux} \times CS \times \{-Q_{critical}/Q_S\}, \tag{1}$$

where N_{flux} is the intensity of the neutron flux, CS is the area of cross section of the node, and Q_S is the collected charge, which strongly depends on doping. The neutron flux (N_{flux}) and the area of the node (CS) will be similar for different circuit implementations in given process technologies.

In order to calculate $Q_{critical}$, a current source is used to perturb the normal operation of the circuit. The value of $Q_{critical}$ can be obtained by measuring the current required to flip an inverter connected to the output of the circuit. When a high-energy particle strikes a device, it produces a current pulse with a rapid rise time, but a more gradual fall time. The shape of the pulse can be approximated by the current source model in [16, 17] with a double exponential current pulse, as

$$I(t) = I_0\left[\exp(-t/\tau_\alpha) - \exp(-t/\tau_\beta)\right], \tag{2}$$

where I_0 depends on the amount of injected charge. In fact, I_0 is the magnitude of the peak of the injected current, τ_α is the collection time constant of the junction, which depends on the contamination intensity and process and τ_β is the ion-track establishment time constant, which is technology independent. In circuit simulations, τ_β is considered smaller than τ_α, while τ_α is used as a variable parameter according to the empirical findings. For simulation purposes, the values given in [17] as $\tau_\alpha = 0.05$ ns, $\tau_\beta = 0.2$ ns are considered and, accordingly, $Q_{critical}$ can be calculated as

$$Q = \int_0^\infty I(t)\,\mathrm{d}t = I_0(\tau_\alpha - \tau_\beta). \tag{3}$$

In this study, a pulse current source is applied to different nodes of the circuits in the cell library in order to find $Q_{critical}$. To do so and to determine the sensitivity level by the current source, the amplitude of this short-term current is changed, until the output of inverter that is connected to the output of the circuit is flipped.

2.2 Input Dependency of the Error

In previous works, the failure probability of a circuit, ε, is considered independent from the input. In this work, since a soft error is a function of, the SER has a strong dependency on the circuit input pattern. Accordingly, different input combinations have been considered to calculate ε and the SER. In this investigation, a set of cells is considered for low power full adders, namely: CCMOS, CPL, TGA, TFA, Hybrid and 14T [18] and multipliers (Ripple carry save and Braun multipliers). In these simulations, all the combinations of the inputs are considered for each function. Accordingly, every gate (or functional unit) is considered to have an error rate (or failure probability) of ε, which depends on the inputs. $Q_{critical}$ is calculated for each transistor level implementation of the cells for all the combinations of the inputs, with a strike current source as explained in Sect. 2.1. This source is inserted to all the circuit points to find the weakest node. After finding the peak of the injected current in which an output flips, can be calculated using (3). Simulations are performed using HSPICE. ε is proportional to the inverse of: a larger result in smaller ε and it is more reliable against the soft error. At the third step, Monte Carlo analyses are performed at the RTL level by generating random bit inputs (as different input combinatorial) and random strike probabilities. This level of simulation is implemented in VHDL. For each bit of the input, an independent bit serial input is applied. Simulation results show different error rates for functional units as depicted in Fig. 2.

Figure 2 compares low power full adder implementations [18]. Overall according to the results, the hybrid circuit (with input "1 0 1") and the C-CMOS circuit (with input "1 1 1") are the weakest and the strongest circuits against soft errors strikes respectively. In this way, different cells based can be categorized on their reliability.

3 Simulation Results

The general process for estimating errors is shown in Fig. 3.

In order to evaluate this method, full adders, analyzed using Monte Carlo simulation as explained in the previous section, are used to build up larger circuits, such as an 8-bit ALU and 4 × 4 multipliers. Analytical results reflect computational error form of the output errors in the case of transient error occurrences. In this way, the reliability of the functional units can be measured in the form of stochastic computational errors.

In the proposed method, for each input, an independent bit pattern is produced along with the error occurrence probability. If the generated number is smaller than the error rate, ε, an error occurs and the output of the logic function is flipped representing the error. In a digital circuit with several sub units, this error propagates through the circuit creating errors in a specific bit of the output. These errors have different probabilities and importance (in terms of error value) depending on the position of the error bit in the output [18].

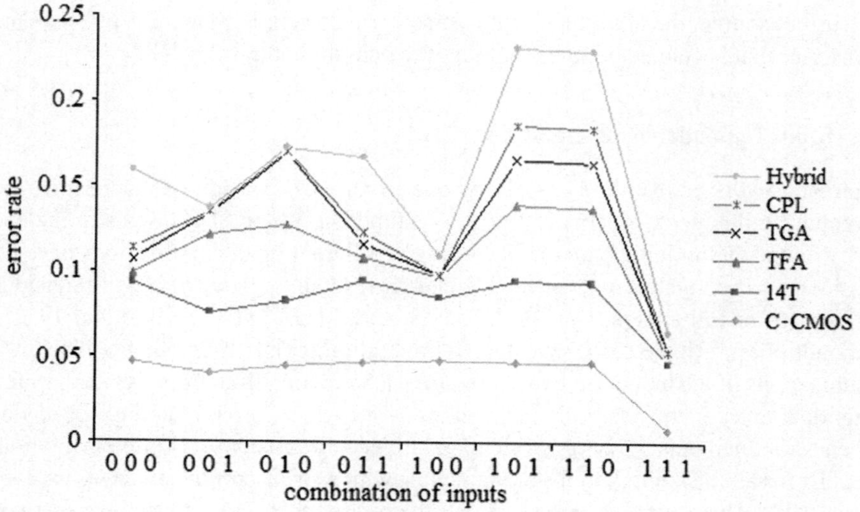

Fig. 2. Comparing full adder cells against transient error

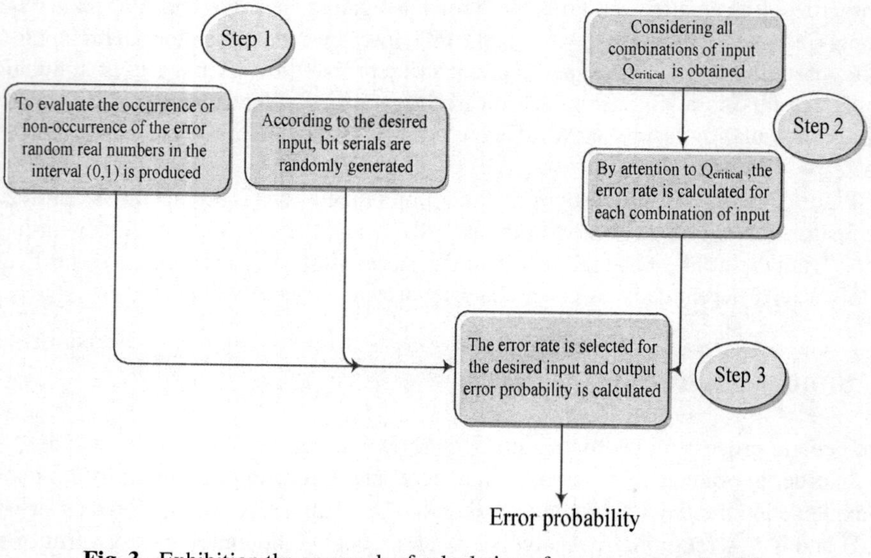

Fig. 3. Exhibition the approach of calculating of output error probability

In order to see this effect, simulation results for an 8-bit full adder are shown in Fig. 4. In case (a), ε is constant, which is independent from input combinations and also error does not propagate through the circuit. In case (b), ε is constant but the error propagates in the logic path. In case (c), ε depends on the input combinations but error does not propagate. In case (d), which is the most realistic case, ε is input dependent, and error propagates through the circuit. Error probabilities for different output bit positions

(S0–S7) are obtained in each case for six different implementations of the full-adder cells.

Results show that cases (a) and (b) considerably underestimate the error effect at the output, compared with case (d), which represents the most accurate error estimation. Moreover, as it can be seen in Fig. 4, generally, in the most significant bits, the error probability increases, which means that the error occurrence depends on the site of output and in valuable bits, the error probability increases.

All the simulations for adders and multipliers are performed with 65 nm technology and using a sequence length of 1000. The maximum relative error is around 0.1% for using a sequence length of 1000. The accuracy can be further improved by either incrementing the sequence length, or the number of experiments for each input combination.

Numerical errors, corresponding to the transient effect, are calculated for different 8-bit adders and the results are curve fitted as depicted in Fig. 5. This Figure shows (in accordance with Fig. 4) with utilizing full adder cell with smaller error rate, numerical error is distributed more around zero.

In order to generalize this approach, it is tried to model the computational error with a Gaussian distribution with formula as

$$F(x) = A\exp\left[-((x - B)/C)^2\right]. \tag{4}$$

Table 1 shows distribution fitting results for the output error of the 8-bit adders with different full adder cells and error rate of $\varepsilon = 0.05$.

In Table 1, the notations are as follows: CB is the confidence bound; SEE is the sum of squares due to error; R-square is the coefficient of determination; Adjusted R-square is the degree-of-freedom adjusted coefficient of determination; RMSE is the root mean squared error (standard error). According to these fitting outcomes, this fitting is appropriate.

For different error rate values ($\varepsilon = 0.01, 0.02, \ldots, 0.1$), the coefficients of the Gaussian error model (*A*, *B* and *C*) for adders with different FA cells are calculated. Again, in order to detect the relevance between coefficients of the Gaussian error model (*A*, *B* and *C*) and error rate (ε), coefficients are fitted to a cubic polynomial as

$$A, B, C(\varepsilon) = P_1\varepsilon^3 + P_2\varepsilon^2 + P_3\varepsilon + P_4. \tag{5}$$

Fitting results for the coefficients *A*, *B* and *C* are presented in Tables 2, 3 and 4, respectively. According to the fitting parameters, this fitting is suitable too.

Again, comparing error probability in several situation and numerical errors, relevant to the transient effect, are obtained for these multipliers and the results are shown in Figs. 6, 7 and 8.

Standard ITU RGB to CrYCb converter and a low pass FIR equiripple filter, Fig. 9, are also considered as case studies. Output error distributions are calculated using the proposed method and are depicted in Figs. 10 and 11. These results are similar to full adders too.

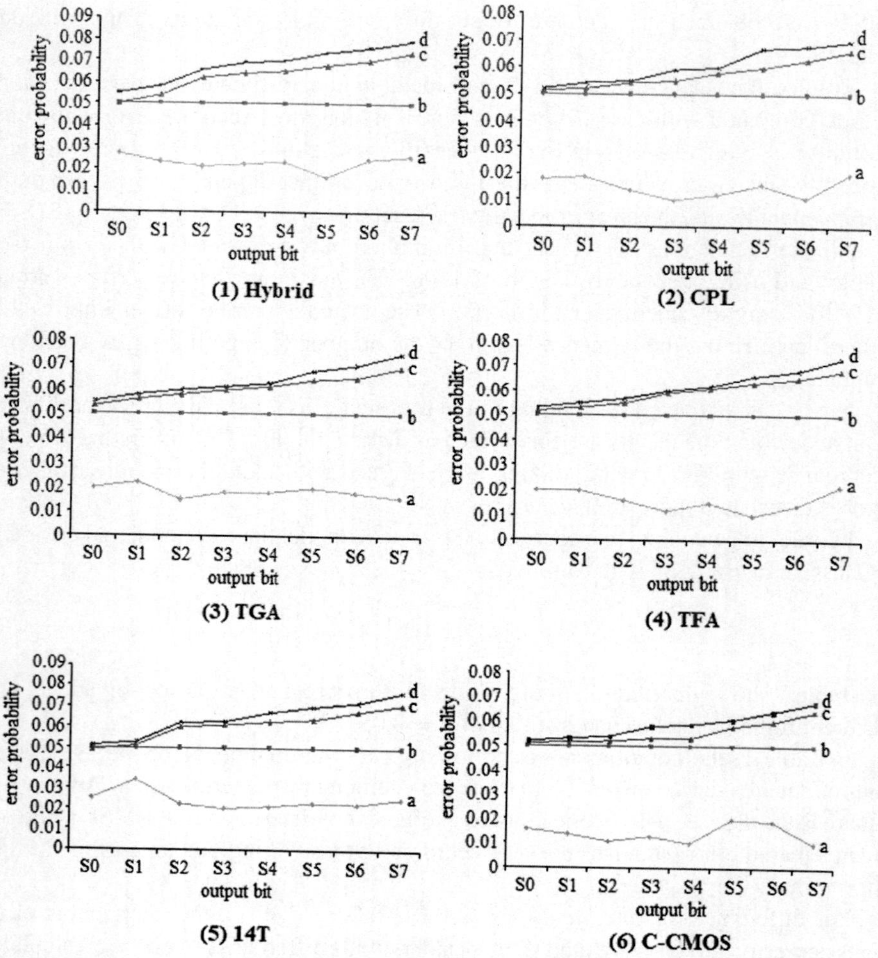

Fig. 4. Comparing error probability in several situation full adders: a) ε is not constant and error does not propagate b) ε is constant and error propagates c) ε is not constant and error does not propagate d) ε is not constant and error propagates

4 Conclusion

The technological advances emerge deep sub-micron and nanometer eras in digital circuits lead to higher sensitivity to the transient error effects. This paper presents a computational error model for transient error effects in datapath by modeling transient effects as computational error in the circuit output. This method takes into account the input dependency of the digital circuit reliability. The presented computational error model for the transient effect can be used in high-level synthesis algorithms to trade off computational accuracy or transient error reliability with other implementation costs. In the proposed method, the computational error modeled with a Gaussian distribution and the relationship between coefficients of the Gaussian error model and error rate is approximated

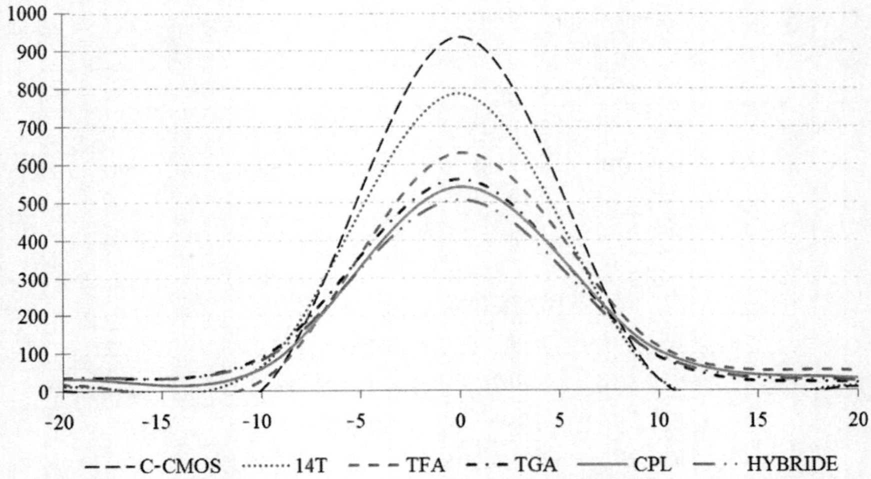

Fig. 5. Computational error representation for different 8-bit full adders

Table 1. Gaussian error model coefficients for adders with different building blocks (FAs), error rate = 0.05

		CCMOS	14T	TFA	TGA	CPL	Hybrid
Coeffi-cients	A	1004	796.7	615.5	539.9	538.3	506.1
	95% CB	(819.8, 1188)	(764.7, 828.7)	(527, 703.9)	(497.4, 582.3)	(480.9, 595.8)	(454.4, 557.7)
	B	1.231	−0.686	1.532	−0.017	0.7322	0.1602
	95% CB	(−0.1637, 2.625)	(−1.35, 0.015)	(0.1083, 2.956)	(−0.948, 0.9135)	(−0.529, 1.991)	(−1.022, 1.342)
	C	4.658	5.902	6.577	7.471	7.259	7.574
	95% C	(3.804, 5.512)	(5.468, 6.335)	(5.308, 7.846)	(6.757, 8.186)	(6.261, 8.257)	(6.65, 8.499)
Goodness of fits	SSE	257	1337	6089	4556	7306	6722
	R-square	0.9997	0.9976	0.9798	0.9822	0.9702	0.9695
	Adjusted R-square	0.9996	0.9972	0.9765	0.9792	0.9652	0.9644
	RMSE	4.628	10.56	22.53	19.48	24.67	23.67

with a cubic polynomial. The obtained result shows that both of them are convenient. According to the obtained results, offered way, relative has a desirable function, which can be used beside other approach for evaluation reliability.

Table 2. Cubic polynomial fitting result for *A* coefficient in the model of (4) for 8-bit adder

		A	B	C
Coefficients	P1	$2.161 \cdot 10^5$	2568	−691.9
	95% CB	$(-4.784 \cdot 10^5, 9.106 \cdot 10^5)$	(−2433, 7569)	(−2975, 1591)
	P2	$-6.905 \cdot 10^4$	−613.5	241.3
	95% CB	$(-1.849 \cdot 10^5, 4.678 \cdot 10^4)$	(−0.529, 1.991)	(−1.022, 1.342)
	P3	2990	40.48	−8.221
	95% CB	(−2626, 8605)	(0.05052, 80.91)	(−26.68, 10.23)
	P4	978.4	0.2445	4.636
	95% C	(903.4, 1053)	(−0.2949,0.7839)	(4.39, 4.882)
Goodness of fits	SSE	1493	0.07741	0.01613
	R-square	0.9714	0.731	0.9854
	Adjusted R-square	0.9571	0.5966	0.9781
	RMSE	15.78	0.1136	0.05185

Table 3 Cubic polynomial fitting result for *B* coefficient in the model of (4) for 8-bit adder

		A	B	C
Coeffi-cients	P1	$1.825 \cdot 10^{-19}$	$1.617 \cdot 10^{-6}$	$6.782 \cdot 10^{-7}$
	95% CB	$(-1.035 \cdot 10^{-18}, 1.4 \cdot 10^{-18})$	$(-4.29 \cdot 10^{-6}, 7.53 \cdot 10^{-6})$	$(-3.435 \cdot 10^{-6}, 4.792 \cdot 10^{-6})$
	P2	$-2.432 \cdot 10^{-16}$	−0.002154	−0.0007908
	95% CB	$(-1.847 \cdot 10^{-15}, 1.36\text{e} \cdot 10^{-15})$	(−0.009944, 0.00563)	(−0.006208, 0.00462)
	P3	1	0.9455	0.2879
	95% CB	(1, 1)	(−2.457, 4.348)	(−2.079, 2.654)
	P4	$-1.606 \cdot 10^{-11}$	−134	−21.39
	95% C	$(-1.175 \cdot 10^{-10}, 8.54 \cdot 10^{-11})$	(−627.1, 359)	(−364.3, 321.5)

(continued)

Table 3 (*continued*)

		A	B	C
Goodness of fits	SSE	$3.231 \cdot 10^{-26}$	0.7627	0.3689
	R-square	1	0.3467	0.8102
	Adjusted R-square	1	0.02	0.7153
	RMSE	$7.338 \cdot 10^{-14}$	0.3565	0.248

Table 4 Cubic polynomial fitting result for *C* coefficient in the model of (4) for 8-bit adder

		A	B	C
Coeffi-cients	P1	$-8.344 \cdot 10^{-5}$	$-2.13 \cdot 10^{-7}$	$-4.494 \cdot 10^{-7}$
	95% CB	(−0.000315, 0.000148)	$(-3.428 \cdot 10^{-6}$, $3.002 \cdot 10^{-6})$	$(-2.37 \cdot 10^{-6}$, $1.471 \cdot 10^{-6})$
	P2	0.1082	0.0002168	0.0006196
	95% CB	(−0.1975, 0.414)	(−0.004018, 0.004452)	(−0.00191, 0.003149)
	P3	−46.15	−0.06656	−0.2856
	95% CB	(−179.7, 87.41)	(−1.916, 1.783)	(−1.391, 0.8192)
	P4	7367	6.493	49.42
	95% C	$(-1.199 \cdot 10^{4}$, $2.672 \cdot 10^{4})$	(−261.6, 274.5)	(−110.7, 209.5)
Goodness of fits	SSE	1175	0.2254	0.08041
	R-square	0.5752	0.2974	0.5566
	Adjusted R-square	0.3628	−0.05393	0.3348
	RMSE	13.99	0.1938	0.1158

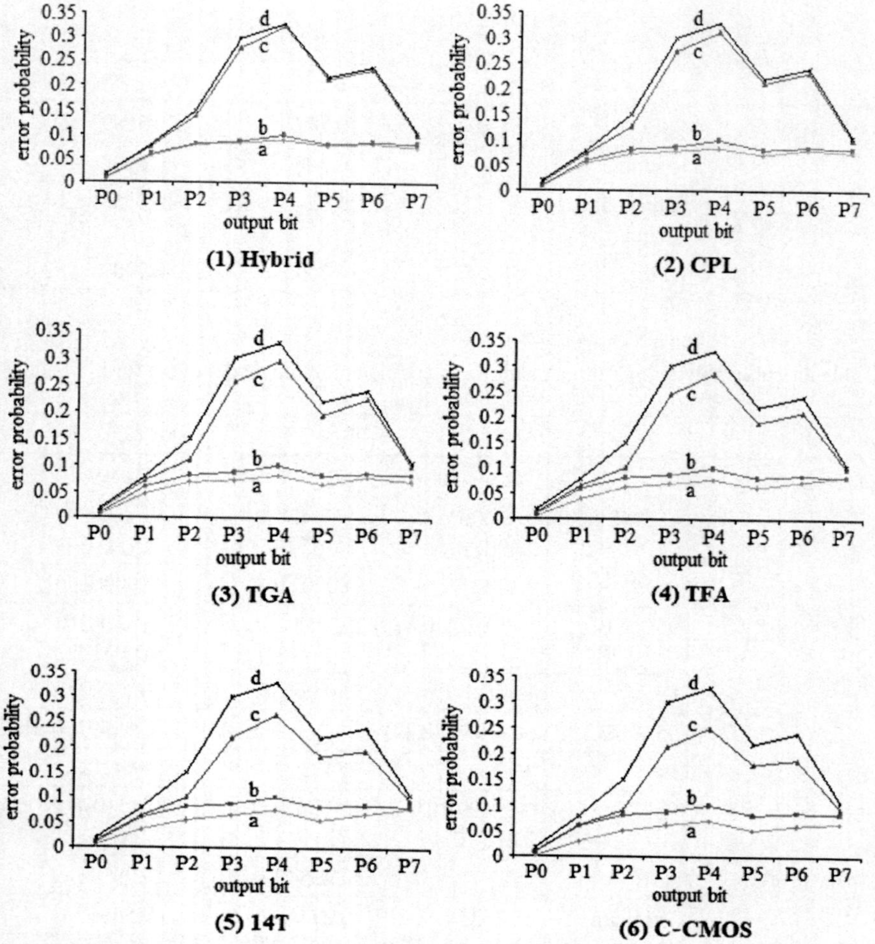

Fig. 6 Comparing error probability in several situation Braun multiplier: a) ε is not constant and error does not propagate b) ε is constant and error propagates c) ε is not constant and error does not propagate d) ε is not constant and error propagates

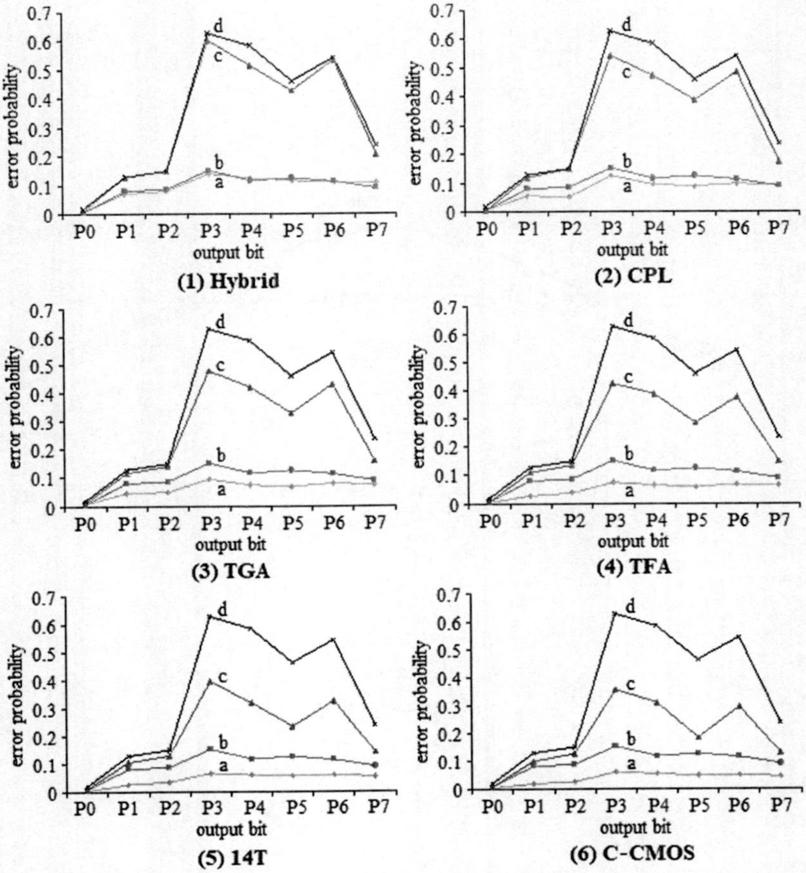

Fig. 7. Comparing error probability in several situation Ripple carry save multiplier: a) ε is not constant and error does not propagate b) ε is constant and error propagates c) ε is not constant and error does not propagate d) ε is not constant and error propagates

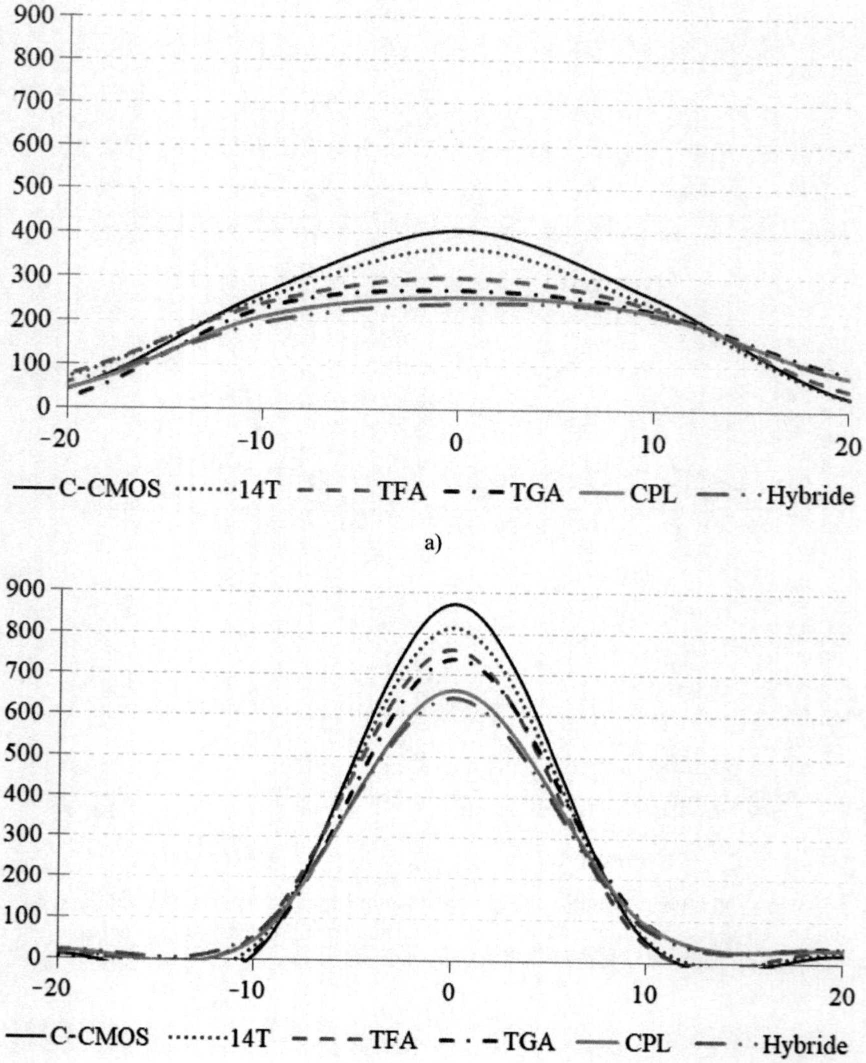

b)

Fig. 8. Computational error representation for a) Ripple carry save multiplier b) Braun multiplier with different full adders

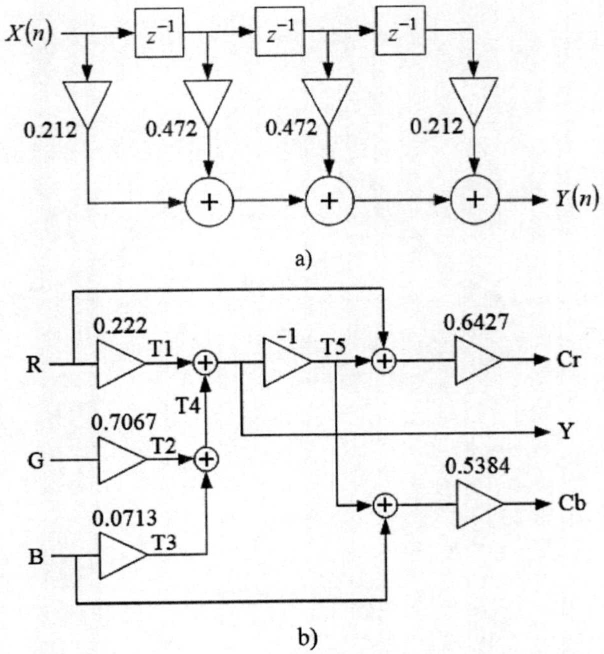

Fig. 9. a Standard ITU RGB to CrYCb converter; **b** The low pass FIR equiripple filter.

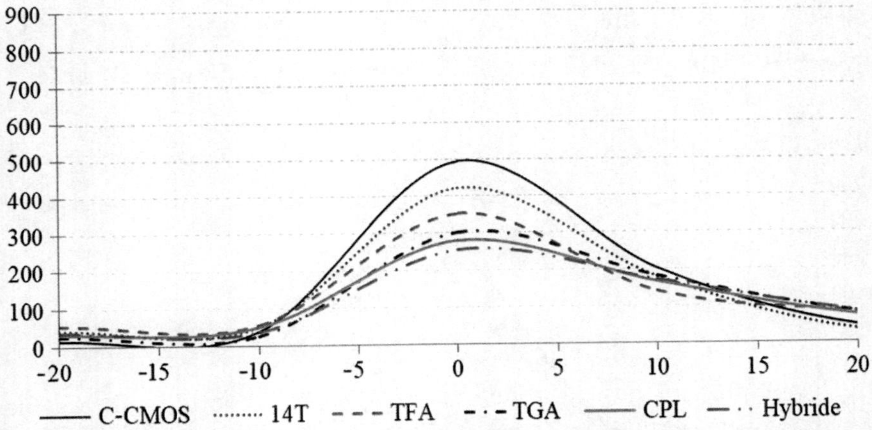

Fig. 10. Computational error representation for the low pass FIR equiripple filter with different full adders

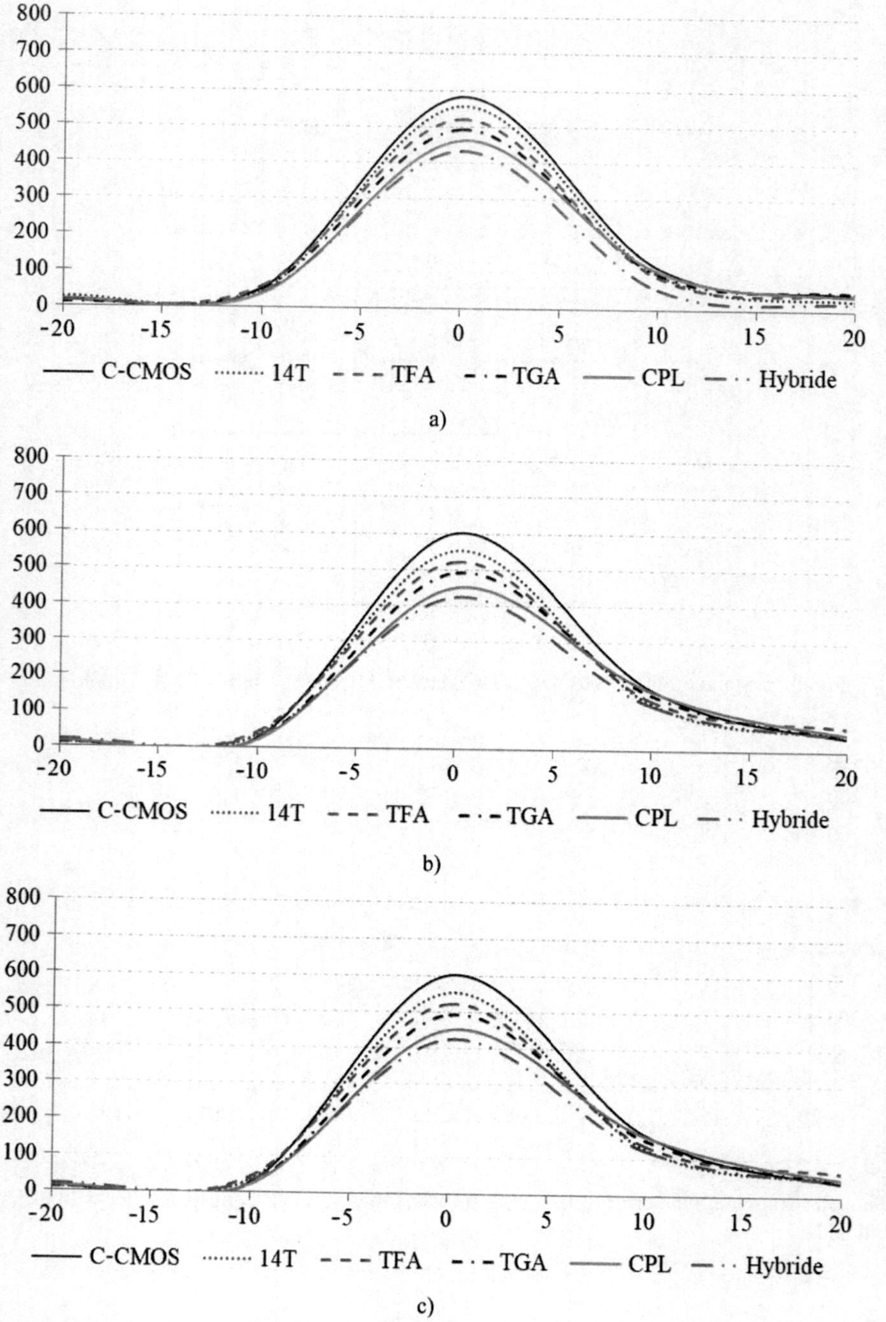

Fig. 11. Computational error representation for Standard ITU RGB to CrYCb converter filter with different full adders: a) cb output; b) y output; c) cr output

Acknowledgements. This work was financially supported by the Ministry of Education and Science of the Russian Federation (research project No. FSWF-2020–0022).

References

1. ITRS Homepage. http://www.public.itrs.net. last Accessed 31 May 2022
2. Beiu, V., Aunet, S., Nyathi, J., Rydberg, R., Ibrahim, W.: Serial addition: locally connected architectures. IEEE Trans. Circuits Syst. I Regul. Pap. **54**(11), 2564–2579 (2007)
3. Karnik, T., Hazucha, P., Patel, J.: Characterization of soft errors caused by single event upsets in CMOS processes. IEEE Trans. Dependable Secure Comput. **1**(2), 128–143 (2004)
4. Glaß, M., Lukasiewycz, M., Haubelt, C., Teich, J.: Towards scalable system-level reliability analysis. In: 7th Design Automation Conference, DAC, pp. 234–239. Association for Computing Machinery, USA (2010)
5. Glaß, M., Lukasiewycz, M., Reimann, F., Haubelt, C., Teich, J.: Symbolic reliability analysis and optimization of ECU networks. In: Design, Automation and Test in Europe, DATE, pp. 158–163. Association for Computing Machinery, USA (2008)
6. Schlichter, T., Lukasiewycz, M., Haubelt, C., Teich, J.: Improving system level design space exploration by incorporating SAT-solvers into multi-objective evolutionary algorithms. In: IEEE Computer Society Annual Symposium on Emerging VLSI Technologies and Architectures, ISVLSI, pp. 309–314. IEEE, Germany (2006)
7. Ackalloor, B., Gaitonde, D.: An overview of library characterization in semi-custom design. In: IEEE Custom Integrated Circuits Conference, pp. 305–312. IEEE, USA (1998)
8. Tosun, S., Mansouri, N., Arvas, E., Kandemir, M., Xie, Y.: Reliability-centric high-level synthesis. In: Design, Automation and Test in Europe, DATE, pp. 1258–1263. Association for Computing Machinery, USA (2005)
9. Rejimon, T., Lingasubramanian, K., Bhanja, S.: Probabilistic error modeling for nano-domain logic circuits. IEEE Trans. Very Large Scale Integration (VLSI) Systems 17 (1), 55–65 (2009)
10. Tosun, S., Ozturk, O., Mansouri, N., Arvas, E., Kandemir, M., Xie, Y., Hung, W.L.: An ILP formulation for reliability-oriented high-level synthesis. In: 6th International Symposium on Quality Electronic Design, ISQED, pp. 364–369. IEEE, USA (2005)
11. Han, J., Chen, H., Boykin, E., Fortes, J.: Reliability evaluation of logic circuits using probabilistic gate models. Microelectron. Reliab. **51**(2), 468–476 (2010)
12. Chen, H., Han, J.: Stochastic computational models for accurate reliability evaluation of logic circuits. In: 20th Great Lakes Symposium on VLSI, GLSVLSI, pp. 61–66. Association for Computing Machinery, USA (2010)
13. Ahmadi, A., Zwolinski, M.: Symbolic noise analysis approach to computational hardware optimization. In: 45th Design Automation Conference, DAC, pp. 391–396. Association for Computing Machinery, USA (2008)
14. Dodd, P.E., Shaneyfelt, M.R., Sexton, F.W.: Charge collection and SEU from angled ion strikes. IEEE Trans. Nucl. Sci. **44**(6), 2256–2264 (1997)
15. Dodd, P.E., Massengill, L.W.: Basic mechanisms and modeling of single-event upset in digital microelectronics. IEEE Trans. Nucl. Sci. **50**(3), 583–602 (2003)
16. Baumann, R.C.: Radiation-induced soft errors in advanced semiconductor technologies. IEEE Trans. Device Mater. Reliab. **5**(3), 305–316 (2005)
17. Binder, D., Smith, E.C., Holman, A.B.: Satellite anomalies from galactic cosmic rays. IEEE Trans. Nucl. Sci. **22**(6), 2675–2680 (1975)
18. Olsen, J., Fynbo, P.B.: Neutron-induced single event upsets in static rams observed at 10 km flight altitude. IEEE Trans. Nucl. Sci. **40**(2), 74–77 (1993)
19. Navi, K., Maeen, M., Foroutan, V., Timarchi, S., Kavehei, O.: A novel low-power full-adder cell for low voltage. Integr. VLSI J. **42**(4), 457–467 (2009)

Research and Development Corrective Automorphism of Repeated Code for Quantum Error Correction

Sergey Gushanskiy[1,2], Vladimir Bozhich[1,2], and Viktor Potapov[1,2](✉)

[1] Department of Computer Engineering, Southern Federal University, Taganrog, Russia
{smgushanskiy,vpotapov}@sfedu.ru

[2] Department of Natural Science and Life Safety, Rostov State Economic University, Taganrog, Russia

Abstract. One of the main problems to be overcome in the near future for the practical creation of quantum computer to make quantum computing resistant to interaction with the external environment and to eliminate inaccuracy of operations. In this paper we study quantum codes correcting errors, proposed as a way to protect against the occurrence errors due to the influence of the external environment. Results of work are considered software implemented model of the simplest quantum code. This work is a study of the influence of the environment on a quantum system of qubits. The article is supposed to describe the foundations of quantum information theory, as well as the place in it of the concept of quantum entanglement. A technique for correcting two main types of quantum errors has been developed, based on the implementation and execution of certain quantum circuits. The main difficulties of ensuring the protection of a quantum channel from various types of errors are analyzed and considered. Demonstrated are the dependences of data distortion on the noise level and the measure of decoherence on the noise level in one qubit.

Keywords: Qubit · Entanglement · Quantum scheme · Wave function · Quantum error

1 Introduction

When information is transmitted over communication channels, it can be distorted. The standard method for correcting errors is to build a table of syndromes, according to which its place is determined. Another method is an algorithm for functionally determining the location of an error by calculating combinations of check bits of the received message. In any case, protocols or correction algorithms require the introduction of additional computational registers. In this paper, we consider an automorphic version of the algorithm that does not require additional memory resources and allows using the available bits to translate an arbitrary sequence (having no more than one errors) to an undistorted state.

R. Silhavy et al. (Eds.): CoMeSySo 2022, LNNS 596, pp. 84–91, 2023.
https://doi.org/10.1007/978-3-031-21435-6_8

The problem of error correction is well known based on the theory of classical computers, where it is solved by coding and using for additional information. Since the classic information is easily in the simplest case, information two copies are added to the number of bits, i.e. values bits 0 and 1 are replaced by 000 and 111. In this case, the transmitted lines 000 and 111 plays the role of zero and unity respectively and called logical 0 and logical 1. Transmission of each of the three bits that make up Logic bit may be inverted the action of noise, and the change in bits is proceed independently from each other. In this case, restore lower source bit when decoding from three the received bits are carried out by most, i.e. strings like 000, 100, 010, 001 correspond to the initial bit is 0, and lines 011, 101, 110, 111 source is bit 1. How much error information can be easily identified? Encode and fix when decoding. Which with an increase in the intensity of the noise required complex methods of error correction, based on among methods, there is always the idea of adding extraneous information to messages during code leveling. It should be noted that the immediate using the idea of copying qubits when designing codes that correct commercial errors, which are not possible, since copying an arbitrary state qubit impossibility theorem cloning. Besides In addition, the measurement of the state of the qubit required for comparison with other qubits, completely destroys this state and restoring the input information is not possible. In order to deal with the methods of solving the known problems and ideas of quantum correction errors, first consider a simple case tea, when the only error that occurs when transmitting a quantum state over a channel connection, which is bit inversion. This means that, under the action of noise, an arbitrary quantum state

$$|\psi\rangle = \alpha|0\rangle + \beta|1\rangle \tag{1}$$

where the complex numbers α and β satisfy the conditions $|\alpha|^2 + |\beta|^2 = 1$ passes into $|\psi^-\rangle = \alpha|1\rangle + \beta|0\rangle$. Errors of this type are called classic by mistake. Required to encode state $|\psi\rangle$ so that it is possible to detect the possibility of new bit inversion, not measuring state qubit, and then restore its original status when an error is detected. Let us consider two additional qubits, first initially in the state $|0>$, and with using the quantum circuit shown in Fig. 1, we encode $|\psi\rangle$ with the following 3-qubit new state: $|\Psi>= \alpha|000> +\beta|111>$.

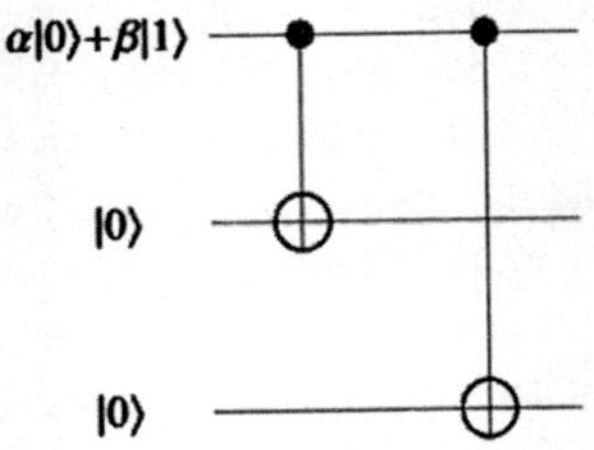

Fig. 1. Quantum circuit for encoding the state of a qubit $|\psi\rangle = \alpha|0\rangle + \beta|1\rangle$ 3-qubit state $\alpha|000> +\beta|111>$

Until recently, the possibility of creating an active QC (operating about 1000 qubits) was ruled out due to the strong exposure quantum particles to outside influence. However, as a solution to this problem were Quantum Error Correcting Codes (QECC) have been proposed. The main difference between QECC and classic codes is to correct the error "without delving into the meaning" of a quantum state is explained by the fact that measurement leads to the destruction of the quantum superposition of states. It is believed that the interaction of a qubit with the environment can lead to an error of one of three possible types:

1. Bit (or, as they are also called, amplitude) errors—X, i.e. leading to a qubit flip: $|0> \to |1>, |1> \to |0>$;
2. Phase errors—$Z : |0> \to |0>, |1> \to -|1>$;
3. Bit-phase (amplitude-phase) errors (i.e., simultaneous action of errors of the first and second types)—$Y : |0> \to |1>, |1> \to -|0>$.

Assuming that the probability of inversion of one qubit is small enough, it is possible to exclude from consideration of the case of a simultaneous inverter these two or three qubits and count what is most The least probable error that has arisen in the process state transmission $|\psi\rangle$, which is an inverted the absence of only one of the three qubits. For mod to correct such an error, it is sufficient to apply to the corresponding qubit logical element Pauli X. As a result, we get one from three states:

$$|\psi_0\rangle = \alpha|001\rangle + \beta|110\rangle,\ |\psi_1\rangle = \alpha|010\rangle + \beta|101\rangle,\ |\psi_2\rangle = \alpha|100\rangle + \beta|011\rangle$$

We recall that the numbering of qubits on a quantum the scheme is usually performed from the bottom to the top, starting from zero. Therefore, the state $|\psi\rangle$, (i = (0, 1, 2) corresponds to the inversion of the i-th qubit. It is easy to see that the vectors (1)–(4) belong to close to the mutually orthogonal two-dimensional subspaces, which together constitute the state space of three qubits of dimension $2^3 = 8$. Therefore, for detection and identification error that occurs when transmitting 3-qubit state, it suffices to find out, k which of the subspaces belongs to the semi-other state. This problem is solved by the choice of two switching operators U1 and U2, satisfying the conditions $U_1^2 = U_2^2 = 1$, in which states (1)–(4) are blow eigenvectors corresponding to with different sets of eigenvalues. Obviously, each of the operators U1 and U2 has These are eigenvalues ± 1, from which you can but add up equal to four different combination. Since the state of the computational battery qubit |0 > and |1 > are eigenvectors with the Pauli Z operator corresponding to eigenvalue ±1, operators U1 and U2 can be constructed from the operators Z0, Z1 and Z2, where index i of operator Zi corresponds to the cube number the bit it affects. It is easy to verify that the desired operator U1 and U2 can be selected as U1 = Z2Z1 and U2 = Z1Z0, moreover, their eigenvectors (1)–(4) corresponding four pairs of eigenvalues value (+1, +1), (+1, − 1), (−1, − 1), (−1, +1). Now the problem of identifying an error, during the transmission of a three-bit state (1), which reduces to the projection of the resulting standing on the substate space corresponding to the specified set of eigenvalues operators U1 and U2. If the operator U has two eigenvalues ± 1, then the projection operators on the spaces corresponding to these proper value, defined as 1/2(1 ± U). Quantum scheme for the implementation of the projection of the wave state |x > on the intrinsic space of the operator U is shown in Fig. 2. In addition to the control operator U, it contains There are two Adamar gates and an additional cooler bit that is initially in state $|0>$. Direct calculation shows that at the output of the quantum circuit state of qubits.

$$|x\rangle|0\rangle \rightarrow (\frac{1+U}{2}|x\rangle)|0\rangle + (\frac{1-U}{2}|x\rangle)|1\rangle \tag{2}$$

If the input state $|x>$ is an eigenstate new operator vector U corresponding to eigenvalue +1, the tone at the output is the state of $|0>$ And, by measuring the state of additional qubit, we get 0, the state $|x>$ corresponds to the eigenvalue lower −1, then the measurement of the additional qubit gives 1.

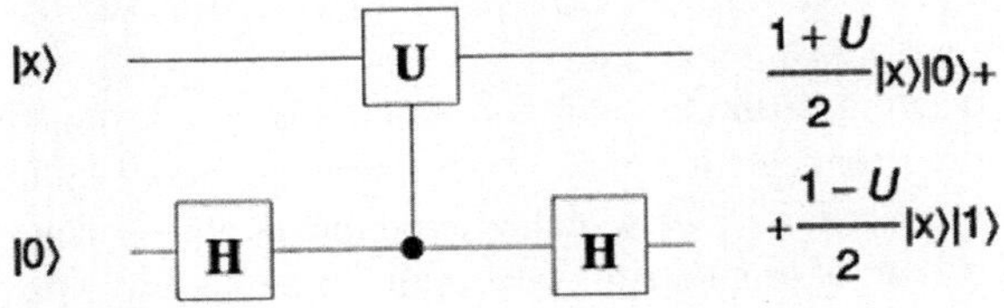

Fig. 2. Quantum circuit for projecting a qubit $|x>$ onto eigenspaces of the operator U

In both cases, the information qubit will remain in the initial state $|x>$. The measurement of an additional qubit is free to answer the question, which the value corresponds to the state of the information qubit $|x>$, without measuring this qubit. Measurements of the general type described by a set of operators $\{M_m\}$: The probability of obtaining the result m when measurement is equal to $p(m) = \langle\psi|M_m^* M_m|\psi\rangle$. After the measurement, the system goes into

$$\frac{M_m|\psi\rangle}{\sqrt{p(m)}}$$

It is quite easy to memorize these formulas: we consider the unnormalized vector $|m\rangle = M_m|\psi\rangle$. The square of the norm of this vector $\langle m \mid m\rangle$ is the probability, and it normalized version of the state after measurement. The only condition that must satisfies the operator $\{M_m\}$ is the condition width:

$$\sum_m M_m^* M_m = I. \tag{3}$$

2 Corrective Automorphism of Repeated Code

The simplest way to insure against errors is to repeat the transmitted character. For example, when character 1 is transmitted over a noisy information channel, it is distorted and the output can be 0. Therefore, instead of 1, the sequence {11} or {111} or {1111} is transmitted. For the repeated code P3 : (e) → (eee), there is a fairly simple scheme for detecting and correcting a single error. For each transmitted information symbol (e), an extended sequence is constructed S = (e, 0, 0) those three additional zero digits are added. After that, the value of the information symbol is entered by the operator U into the check bits according to the algorithm: S1 = US = (e, e ⊕ 0, 0, e ⊕ 0) = (e, e, e) and a message (eee) is sent over the information channel. At the output of the information channel, the sequence has the form O = (z1, z2, z3). Acting on it with the same operator U, we obtain O = UO = (z1, z1⊕z2, z1⊕z3) = (y1, y2, y3). The information symbol is allocated using the V operator:

$$V : z1 \rightarrow (y1\&y3) \oplus y1 = ((z1 \oplus z2)\&(z1 \oplus z3)) \oplus z1 \quad (4)$$

3 Development of Surface Code for Error Correction

The problem of creating quantum error correction codes is to find switching sets of stabilizers that allow error detection without disturbing the encoded information. Surface codes belong to a wider family of so-called topological codes. The general principle behind the development of topological codes is that the code is created by "correcting" co-repeating elements. This modular approach ensures that the surface code can scale in size in a straightforward manner, while ensuring that the stabilizer is commutative. In terms of actual implementations, a particular advantage of surface code for modern hardware platforms is that it only requires nearest neighbor interaction. For surface codes, it is useful to use a graphical representation of code qubits instead of schema notation.

On Fig. 3 shows a four-cycle surface code, the fundamental building block around which surface codes are built. Circles in Fig. 3 are code qubits, and squares are auxiliary qubits. Red edges represent controlled gate X, each controlled by an auxiliary qubit A and acting on a data qubit D. Similarly, blue edges represent controlled Z operations, each controlled by an auxiliary qubit and acting on a data qubit. These controlled operations are the gates with which four-stroke regulators are measured. The ancilla qubit A_1 connects to the data qubits D_1 and D_2 red edges, and therefore measures the stabilizer $X_{D_1}X_{D_2}$.

The four-cycle stabilizers $X_{D_1}X_{D_2}$ and $Z_{D_1}Z_{D_2}$ commute with each other since they non-trivially intersect on an even number of code qubits. A four-cycle codeword $|0\rangle_L$ can be prepared by setting the initial state of the code qubits to $|D_1D_2\rangle = |00\rangle$, and following the general encoding procedure. However, since the four-cycle has two code qubits n = 2 and two stabilizers m = 2, the number of logic qubits it encodes is k = n—m = 0. As a result, the four-cycle code is not a useful code by itself. However, we see that working detection and correction codes can be formed by combining several four-cycle codes together to form square grids. On Fig. 3 shows a five-qubit surface code formed by combining four four cycles into a square lattice. By checking which data qubits each auxiliary qubit is connected to, the code stabilizers in Fig. 1 can get to get

$$S_{[[5,1,2]]} = \langle X_{D_1}X_{D_2}X_{D_3}, Z_{D_1}Z_{D_3}Z_{D_4}, Z_{D_2}Z_{D_3}Z_{D_5}, X_{D_3}X_{D_4}X_{D_5}\rangle. \quad (5)$$

The first term above is the stabilizer as measured by the auxiliary qubit A_1, the second term is measured by the auxiliary qubit A_2, and so on. The stabilizers in $S_{[[5,1,2]]}$ commute with each other, as all X- and Z-type stabilizers intersect at an even number of code qubits. On Fig. 3 we see that there are five code qubits and four stabilizers, meaning that the code encodes one logical qubit. On Fig. 4 shows two examples of errors in surface code and how to detect them. In an error Z_{D_1} on the qubit D_1, it anti-commutates with the stabilizer $X_{D_1}X_{D_2}X_{D_3}$ and therefore triggers the "1" Syndrome. This is depicted by the red shading of the auxiliary qubit A_1. Similarly, the error X_{D_5} is anti-commutated with the stabilizer $Z_{D_2}Z_{D_3}Z_{D_5}$ and triggers the "1" syndrome measurement into the auxiliary qubit A_4.

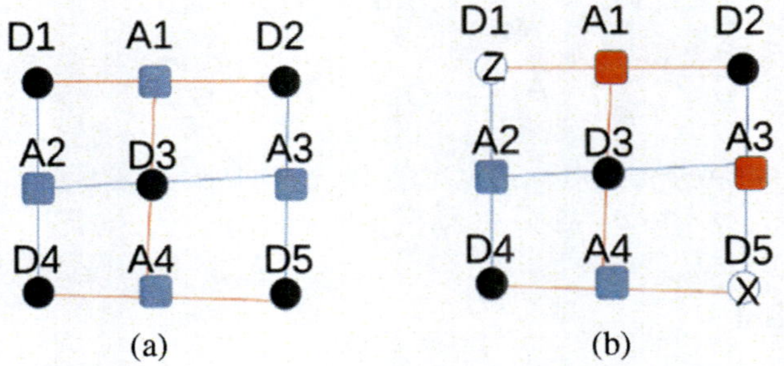

Fig. 3. **a** Surface code [[1, 2, 5]] formed by combining four cycles into a square lattice. **b** Examples of detection errors in surface code [[1, 2, 5]].

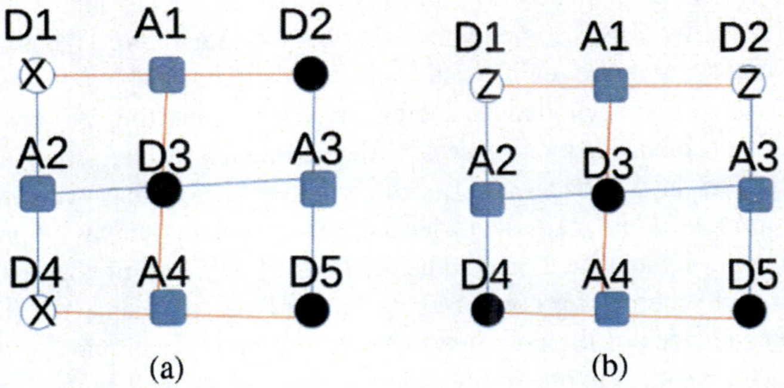

Fig. 4. The logical operators of the surface code can be defined as chains of Pauli operations that act along the boundaries of the lattice **a** Pauli-X logical operator **b** Pauli-Z logical operator

From Fig. 3shows that the surface code is a square lattice with two types of borders. The vertical borders are formed by the blue edges representing the measurements of the Z-type stabilizer. The logical operators of a surface code can be defined as chains of Pauli operators along the edges of these boundaries. On Fig. 4a shows a two-qubit Pauli chain $X_{D_1}X_{D_4}$ along the left boundary of a five-qubit. The operator $X_{D_1}X_{D_4}$ commutes with all stabilizers in $S_{[[5,1,2]]}$, in particular, the stabilizer $Z_{D_1}Z_{D_3}Z_{D_4}$ with which it shares two qubits. Similarly, Fig. 5b shows the operator $Z_{D_1}Z_{D_2}$, which acts across the vertex of the lattice. From the above, we can see that the minimum weight of logical operators is 2, which means [[1, 2, 5]] code is a detection code with d = 2.

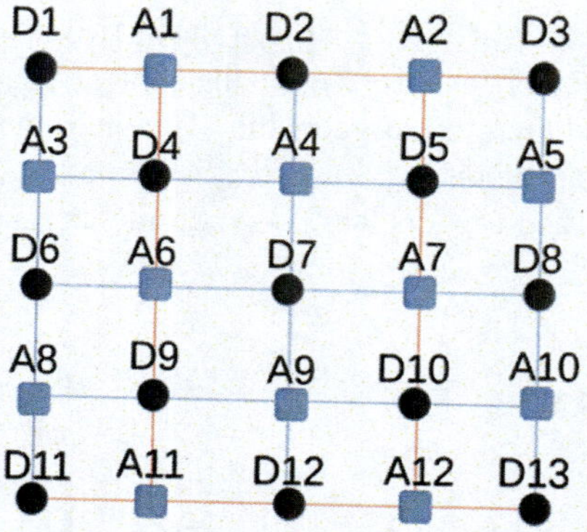

Fig. 5. Surface code with parameters [[13, 1, 3]]. The possible choice of logical operators of this code will be $X^{'} = X_{D_1}X_{D_6}X_{D_{11}}$ and $Z^{'} = Z_{D_1}Z_{D_2}Z_{D_3}$

4 Conclusion

Error correction is one of the main tasks facing quantum computing devices. And without solving this problem, further successful developments in this promising area will become ineffective. In this paper, codes for correcting various types of errors are numerically simulated. Analyzed the main obstacles and difficulties on the way and protection of the channel from noise, as well as some methods for overcoming them. The implementation of schemes for correcting the main types of quantum errors has been made. Correcting errors is one of the major challenges facing quantum computing devices. And without solving this problem, further successful developments in this promising area will become ineffective. In this paper, the codes for correcting various types of errors are numerically simulated. The main obstacles and difficulties in the way of protecting the channel from noise are analyzed, and some methods of overcoming them are proposed. Implementation of schemes for correcting two main types of quantum errors has been made. The dependences of the data distortion on the noise level and the decoherence measure on the noise level in one qubit are demonstrated, as well as the dependence of the error on the measure and purity of entanglement.

Acknowledgments. The research was funded by RFBR according to the project № 20-07-00368.

References

1. Guzik V., Gushanskiy S., Polenov M., Potapov V.: Complexity estimation of quantum algorithms using entanglement properties. In: 16th International Multidisciplinary Scientific GeoConference, pp. 20–26. Bulgaria, (2016)
2. Guzik V., Gushanskiy S., Polenov M., Potapov V.: Models of a quantum computer, their characteristics and analysis. In: 9th International Conference on Application of Information and Communication Technologies (AICT).—Institute of Electrical and Electronics Engineers, pp. 583–587 (2015)
3. Collier, D.: The Comparative Method. In: Finifter A.W. (ed.) Political Sciences: The State of the Discipline II. pp. 105–119: American Science Association. Washington, DC (1993)

4. Williams, C.P.: Explorations in quantum computing. Texts in Computer Science, Chapter 2. Quantum Gates, pp. 51–122. Springer, Heidelberg (2011)
5. Olukotun, K.: Chip Multiprocessor Architecture—Techniques to Improve Throughput and Latency. Morgan and Claypool Publishers, San Rafael (2007)
6. Milner, R.G.: A short history of spin. In: Contribution to the XV International Workshop on Polarized Sources, Targets, and Polarimetry—Charlottesville, Virginia, USA, September 9–13 (2013). arXiv:1311.5016
7. Hallgren H.S. An improved quantum Fourier transform algorithm and applications. In: Proceedings of the 41st Annual Symposium on Foundations of Computer Science, Redondo Beach, CA, p. 515
8. Potapov, V., Gushanskiy S., Polenov M.: The methodology of implementation and simulation of quantum algorithms and processes. In: Proceedings of 11th International Conference on Application of Information and Communication Technologies (AICT-2017). IEEE. pp. 437–441(2017)
9. Boneh D., Zhandry M.: Quantum-secure message authentication codes. In: Proceedings of Eurocrypt, pp. 592–608 (2013)
10. Potapov, V., Gushansky, S., Guzik, V., Polenov, M.: Architecture and Software Implementation of a Quantum Computer Model. In: Advances in Intelligent Systems and Computing, vol. 465. pp. 59–68. Springer (2016)

Minimax Filtering of the Complex Code Shift Keyed Signals

Oleg Chernoyarov[1,2,3](✉), Vladimir Ivanov[1], Leila Golpaiegany[4], and Alexey Golikov[5]

[1] National Research University "MPEI", Krasnokazarmennaya str. 14, 111250 Moscow, Russia
chernoyarovov@mpei.ru
[2] National Research Tomsk State University, Lenin Avenue 36, 634050 Tomsk, Russia
[3] Maikop State Technological University, Pervomaiskaya str. 191, 385000 Maikop, Russia
[4] University of Ayatollah Ozma Boroujerdi, 69199-69411 Boroujerd, Iran
[5] Voronezh State Technical University, Moscow Avenue 14, 394026 Voronezh, Russia

Abstract. A technique for separate estimation of the realizations of discrete normal random signals belonging to different ensembles with the specified normalized correlation matrices and unknown intensities is considered. This technique uses the method of minimax estimates in terms of the theory of statistical decisions and its principles are applied to the problem of time resolution of the pulse signals with a random initial phase and intensity and under an arbitrary complex shift keying. An iterative procedure for finding the processing vector is proposed, which provides for minimizing the most unfavorable of the interfering effects. The results of the efficiency analysis of the pulsed phase-shift keyed (PSK) and amplitude phase-shift keyed (APSK) signals filtering using the proposed processing procedure are presented. #COMESYSO1120.

Keywords: Synthetic aperture radar · Discrete signals · Minimax estimate · PSK and APSK signals · Side lobes · Unmatched filter · μ-criterion · Minimax filter

1 Introduction

Almost all airborne and space-based synthetic aperture radars (SAR) in operation and under design use a probing signal in the form of a sequence of identical radio frequency chirp pulses. The positive experience of operating SAR on the spacecraft "Venera-15" and "Venera-16" [1, 2] as well as a number of aircraft SAR using phase-code shift keyed probing signals (such as M-sequences, Barker codes, etc.) has showed the possibility of using discrete coherent signals in SAR. However, the phase-code shift keyed signals with a relatively small number of components and a fairly simple form (for example, a rectangular envelope and phase shift keying for two levels 0 and π) demonstrate autocorrelation functions (ACF) with rather large side lobes [3].

The use of discrete coding of SAR probing signals opens up the prospect of a significant improvement in the quality of a radar image in terms of the parameters related to the properties of the correlation functions. To reduce the side lobes of the ACF of the

R. Silhavy et al. (Eds.): CoMeSySo 2022, LNNS 596, pp. 92–103, 2023.
https://doi.org/10.1007/978-3-031-21435-6_9

probing SAR signal, there are widely applied ensembles of the phase-code shift keyed signals [4, 5], the orthogonal frequency division multiplexing (OFDM) signals [6], the linear frequency modulated and phase coded (LFM-PC) signals [7], the pulse signals with a zero autocorrelation zone [8], the unimodular sequences [9], etc.

A composition of random signals **s** reflected from the surface in the target area arrives at the input of the SAR reception path. As a consequence, the side lobes of the responses from the objects with a large scattering cross-section (SCS) hide the responses from the objects having a small SCS. Also, the side lobes can generate false responses that distort the final image. The high integral level of the side lobes causes background illumination, reduces the dynamic range, and makes it difficult to detect surface-distributed objects with a low backscatter coefficient and to recognize the objects with hardly distinguishable specific SCS [10].

When receiving a reflected signal, to suppress the side lobes, special (unmatched) filters [11] are used, selected in such a way that the filter output signal has the required side lobes. Unlike chirp pulses, the suppression of the side lobes when receiving the discrete signals does not lead to the expansion of the main maximum. A useful effect from the use of unmatched filters is obtained, if the filter delay line has K taps separated from each other by the time interval between the signal samples and K is not less than the number N of the signal samples, i.e. if $K \geq N$. In this case, the components $\lambda_1, \ldots, \lambda_M$, $M = N + K - 1$ of the output signal vector $\mathbf{\Lambda}$ can be presented in a matrix form as follows: $\mathbf{\Lambda}^T = \mathbf{H}\mathbf{s}^T$, where $\mathbf{H}$ is the pulse response of the filter, $\mathbf{s}$ is the discrete signal at the filter input with the components $s_1, \ldots, s_N$, and "T" is the transposition operation. Synthesis of non-optimal filters of discrete signals is performed in accordance with one of the criteria [12]:

$$\mu = min\left\{ \max(|\lambda_1|, \ldots, |\lambda_M|) / |\lambda_{\max}| \right\}; \quad (1)$$

$$\nu = \min\left\{ \sum_{i=1}^{M} \lambda_i \lambda_i^T \Big/ |\lambda_{\max}| \right\}, \quad (2)$$

where the choice of the maximum along with the summation apply only to the side components.

At the specified **S**, K, and the constraints on the side components of the output signal, the parameters μ and ν maximize the principal component $\lambda_{\max}$ of the filter output vector $\mathbf{\Lambda}$. The conditions (1) and (2) are used in the designing of μ and ν-filters, respectively, meanwhile with the same losses in terms of signal-to-noise ratio, the signals at the output of the μ-filter have half the level of the maximum side lobe compared to the signals at the output of the ν-filter. In the case of the real signal and filter components, the use of the μ-criterion leads to a linear programming problem.

The paper [13] proposes a method for separate estimation of the realizations of the discrete normal random signals belonging to different ensembles with the specified normalized correlation matrices and unknown variances, using the method of minmax estimates in terms of the theory of statistical decisions [14]. In the present study, the principles of this technique are applied to decide the problem of time resolution of the pulse signals the pulse signals with a random initial phase and intensity and under an arbitrary complex shift keying.

2 Time Resolution of the Arbitrary Complex Code Shift Keyed Signals

The result of observing oscillation $y(t)$ at a finite interval can be represented by a sequence of complex samples y_i, $i = \overline{1, M}$ that are equal to the sum of the samples of the normal noise $\mathbf{n}$ and the signals $\mathbf{s}_l = \varepsilon_l \mathbf{d}_l$, $l = \overline{1, L}$, where ε_l are independent normal complex random variables, $\mathbf{d}_l$ are shifted code sequences $\mathbf{d}$ with the duration N, so that $d_{li} = d_{i-l}$. The observation interval "cuts out" the sample sequence $\mathbf{y}$ with the duration $M \geq N$, $L \leq M + N - 1$. In this case, the estimation of the k-th signal realization is reduced to obtaining a Bayesian estimate $\hat{\varepsilon}_k = \mathbf{a}_k^+ \mathbf{y}$, where the symbol "+" means the transposition with a complex conjugation, and the processing vector $\mathbf{a}_k$ satisfies the equation [13]:

$$\left(\alpha_0 \mathbf{I} + \sum_{l=1}^{L} \alpha_l \mathbf{R}_l \right) \mathbf{a}_k = \alpha_k \mathbf{R}_k. \tag{3}$$

In (3), α_0 is the dispersion of the noise samples; $\mathbf{I}$ is the $M \times M$ unity matrix; $\mathbf{R}_l = \mathbf{d}_l \mathbf{d}_l^+$ is the normalized correlation matrix of dimension $M \times M$ constituted by the samples of the l-th signal in the observation interval; α_l is the signal intensity.

When normalizing the sum of the squared moduli d_i to unity, the partial signal-to-noise ratio is

$$q_i^2 = \alpha_l / \alpha_0 = 2E_l / N_0,$$

where E_l is the power energy; N_0 is the one-sided spectral density of white noise.

In terms of a quadratic loss function, the average risk for the estimate ε_k is equal to [14]

$$B_k = \overline{\left| \hat{\varepsilon}_k - \varepsilon_k \right|^2} = \alpha_k \left(1 - \mathbf{a}_k^+ \mathbf{y} \right). \tag{4}$$

Then one can state that:

$$\beta_l = \mathbf{a}_k^+ \mathbf{d}_l, \ (l \neq k), \ \beta_k = \mathbf{a}_k^+ \mathbf{d}_l - 1, \tag{5}$$

where $\mathbf{a}_k^+ \mathbf{d}_l$ is the value of the cross-correlation function of the source code $\mathbf{d}$ and the reference sequence $\mathbf{a}_k$ at L delay positions, or the "peaks" of the linear filter response; β_k is the difference between the main "peak" of the linear filter response from unity.

The use of the method of minimax estimates in the case of the unknown signal $\mathbf{s}_l$ intensities involves finding such values of α_l accounting for their fixed sum that, after substituting them into (1), one could get the maximum value of the average risk (4). In this case, the following conditions are met: the moduli of the side peaks of the filter response are equal for all delay positions corresponding to $\alpha_l \neq 0$, and they also coincide with the modulus of the difference between the main peak and unity:

$$|\beta_l| = |\beta_k|, \ \alpha_l \neq 0. \tag{6}$$

At other positions, the moduli of the side peaks may differ from this value, but not exceed it. A filter of this type will be called a minimax filter or, for brevity, an m-filter.

From (4), for the first partial derivatives of the average risk one can get $\partial B_k \big/ \partial \alpha_l = |\, \beta_l|^2$. However, taking into account the conditions

$$\alpha_l \geq 0, \ \sum_{l=1}^{L} \alpha_l = \alpha_s = const \tag{7}$$

the extreme values of α_l can be found by projecting the gradient of the objective function onto the admissible space of variables [15, 16]. In this case, this method leads to an iterative procedure for finding the next approximation α'_m at each step:

$$\alpha'_m = \alpha_m + \delta \left(|\, \beta_m|^2 - \sum_l |\, \beta_l|^2 \Big/ L_1 \right), \ \alpha'_m \geq 0. \tag{8}$$

Here δ is the chosen step length; under the sum sign, L_1 values of β_l are taken into account, to them non-negative projections of α'_l correspond. For the rest of α'_l, zero values are assigned.

When changing $\alpha_l = \sigma_l^2$, during searching for the maximum (4), the components σ_l can be considered as the variables. In this case, only the second of the conditions (7) is sufficed, and the iterations take the form:

$$\alpha'_m = \alpha_s \alpha_m \left(1 + \delta \,|\, \beta_m|^2 \right)^2 \Bigg/ \sum_{l=1}^{L} \alpha_l \left(1 + \delta \,|\, \beta_l|^2 \right)^2. \tag{9}$$

It should be noted that the procedure (7) is more convenient to implement and more stable, but it can lead to local maxima (4).

If $\alpha_0 = 0$, then another method can be applied, which in many cases allows one to obtain exact solutions for α_l. . Considering (5), the notation $\gamma_l = \alpha_l \beta_l$ should be introduced. Then, taking into account that $\gamma_k < 0$, one can conditionally assume that $\gamma_k = -1$ and write (3) in the form:

$$\sum_{l \neq k} \gamma_l \mathbf{d}_l = \mathbf{d}_k. \tag{10}$$

The vector equality (10) gives an indefinite system of the M linear equations with respect to γ_l (in a general case, the number of variables is $L - 1 > M$). If the solution is chosen correctly, then the substitution of $\alpha_l = |\, \gamma_l|$ into (3) ensures the fulfillment of the conditions (6). Taking into account (5), the average risk (4) is equal to

$$B_k = \alpha_k |\, \beta_k|. \tag{11}$$

Otherwise, while writing $\mathbf{y} = \mathbf{n} + \sum_{l=1}^{L} \varepsilon_l \mathbf{d}_l$ one gets

$$B_k = \overline{\left| \mathbf{a}_k^+ \mathbf{y} - \varepsilon_k \right|^2} = \alpha_0 \boldsymbol{\alpha}_k^+ \boldsymbol{\alpha}_k + \sum_{l=1}^{L} \alpha_l |\, \beta_l|^2. \tag{12}$$

At $\alpha_0 = 0$, , at the points of extremum, in accordance with (6), (11), (12), there are

$$|\beta_k| = \alpha_k \Big/ \sum_{l=1}^{L} \alpha_l, \; B_k = \alpha_k^2 \Big/ \sum_{l=1}^{L} \alpha_l = 1 \Big/ \sum_{l=1}^{L} |\gamma_l|.$$

There is a correspondence between the condition $B_k = \max$ and $\sum_{l=1}^{L} |\gamma_l| = \min$. This solution is the one that must be chosen from the set of solutions for the system (10).

The values of α_l obtained in this way help to get from (3) the vector $\mathbf{a}_k$ implementing processing of the μ-filtering type of the binary codes [11], but in contrast to this, they allow calculating the μ-filter coefficients for an arbitrary complex code and can be used as the initial approximations for (8) and (9) at $\alpha_0 \neq 0$.

In the case when $L \leq M$ (here one deals with a periodic code or a limited range of signal $\mathbf{s}_l$ delays) and $\alpha_0 = 0$, one solves the system (3) leads to the *m*-filter with a full suppression of the side peaks ($\beta_l = 0$) for any α_l. For a degenerate system, one should choose a normal solution.

3 The Efficiency of the Minimax Filtration of the PSK and APSK Pulse Signals

Now it is time to estimate the efficiency of the proposed processing procedure for resolving the code shift keyed pulse signals.

In Fig. 1, for example, the 61-element binary minimax code **d** [17] is demonstrated. In Fig. 2, at $\alpha_0 = 0$, $\alpha_k = 10$, one can see that for this code there are provided the following: (a) the filter response modulus $|\beta_l|$ (5) and (b) the weight coefficients a_i of the μ-filter ($N = 61$, $M = 127$) to which they belong; (c) the graph of the values of $|\beta_l|^2$ at all delay positions $l > k$ (when $l < k$, this graph is symmetrical). It is obtained that $|\beta_k|^2 \approx 1.7 \cdot 10^{-4}$ and the ratio of the level of the maximum side peak to the main peak is $\mu \approx 3.14 \cdot 10^{-2}$, which is 2.6 times less than when using a matched filter.

Figure 3 shows (a) the filter response modulus $|\beta_l|$ (at $\alpha_0 = 0$, $\alpha_k = 10$) and (b) the real (red points) and imaginary (blue points) parts of the weight coefficients a_i of the μ-filter ($N = 9$, $M = 9$) of the 9-element multiphase *p*-class pulse code [17], the symbol phases of which are determined as follows:

$$\theta_{kl} = -(\pi/p)\,(p - 2k - 1)\,(kp + l), \;\; k, l = 0, \ldots, p - 1.$$

For example, if $p = 3$, then one gets $\theta_{00} = 0$, $\theta_{01} = -2\pi/3$, $\theta_{02} = -4\pi/3$, $\theta_{10} = 0$, $\theta_{11} = 0$, $\theta_{12} = 0$, $\theta_{20} = 2\pi/3$, $\theta_{21} = 2\pi/3$, $\theta_{22} = 4\pi/3$.

The code sequence $d_n = \exp(j\theta_n)$, $n = \overline{1, N}$ of the μ-filter are as follows: 1.0; −0.5; −0.5 − 0.866*i*; −0.5 + 0.866*i*; 1.0; 1.0; 1.0; 1.0; −0.5 + 0.866*i*; −0.5 − 0.866*i*.

The ratio of the level of the maximum side peak to the main peak is $\mu \approx 9.6 \cdot 10^{-2}$, which is 13% less than when using a matched filter.

In Fig. 4, there are presented (a) the graph of the response modulus $|\beta_l|$ (at $\alpha_0 = 0$, $\alpha_k = 1$) and (b) the weight coefficients a_i of the μ-filter ($N = 21$, $M = 21$) when

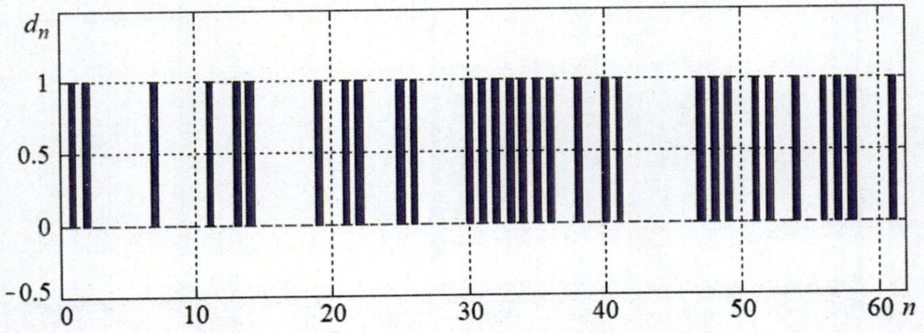

Fig. 1. The case of the binary code at $N = 61$

receiving the 21-element APSK signal [17] with the square phase spectrum. The symbols of such APSK signal are determined in the following way:

$$d_n = \sqrt{\frac{N}{D_0}}\left[(C(z_2) - C(z_1))\cos\left(\frac{\pi(n+0.5D_0)^2}{2D_0}\right) + (S(z_2) - S(z_1))\sin\left(\frac{\pi(n+0.5D_0)^2}{2D_0}\right)\right],$$

where $z_1 = (n - D_0/2)/\sqrt{D_0}$, $z_2 = (n + D_0/2)/\sqrt{D_0}$, and $C(z)$, $S(z)$ are Fresnel integrals.

The maximum side peak of the autocorrelation function of this signal at $D_0 = 16$ is $3 \cdot 10^{-2}$ [17]. For the μ-filter, the ratio of the level of the maximum side peak of the output signal to the main peak is $\mu \approx 2.15 \cdot 10^{-2}$, which is 1.4 times less than when using a matched filter.

In Fig. 5, for the 13-element Barker code [18], one can see (a) the response modulus $|\beta_l|$ (at $\alpha_0 = 0$, $\alpha_k = 10$) and (b) the weight coefficients a_i of the μ-filter ($N = 13$, $M = 13$). The ratio of the level of the maximum side peak to the main peak is $\mu \approx 5.7 \cdot 10^{-2}$, which is 1.35 times less than in the case of the matched filtration.

Figure 6 shows the transformation of the response modulus $|\beta_l|$ and the weight coefficients a_i of (a) the μ-filter [12] for the 5-element Barker code ($N = 5$, $M = 11$) when $\alpha_0 = 0$ the noise level is changing, (b) the *m*-filter when $\alpha_0 > 0$, and (c) the *m*-filter when $\alpha_0 > 1$. It is presupposed that $\alpha_k = 10$.

It can be seen that in the presence of the uncorrelated noise ($\alpha_0 > 0$), the *m*-filter retains the minimax filter property up to the certain values (Fig. 6b). A further increase in the noise level changes the *m*-filter response, keeping the conditions (6), and, at some α_0 (Fig. 6c), it becomes consistent ($\alpha_l = 0$ at $l \neq k$).

The noise immunity of various filters in this case can be compared by the value of the average risk B_k (the estimation variance ε_k) and under the most unfavorable distribution α_l. From (10), it follows:

$$B_{km} = \alpha_0 \mathbf{a}_k^+ \mathbf{a}_k + \alpha_s |\beta_{\max}|^2, \tag{13}$$

where α_s is determined according to (7) and $|\beta_{\max}|$ is the modulus of the maximum side peak of the filter response.

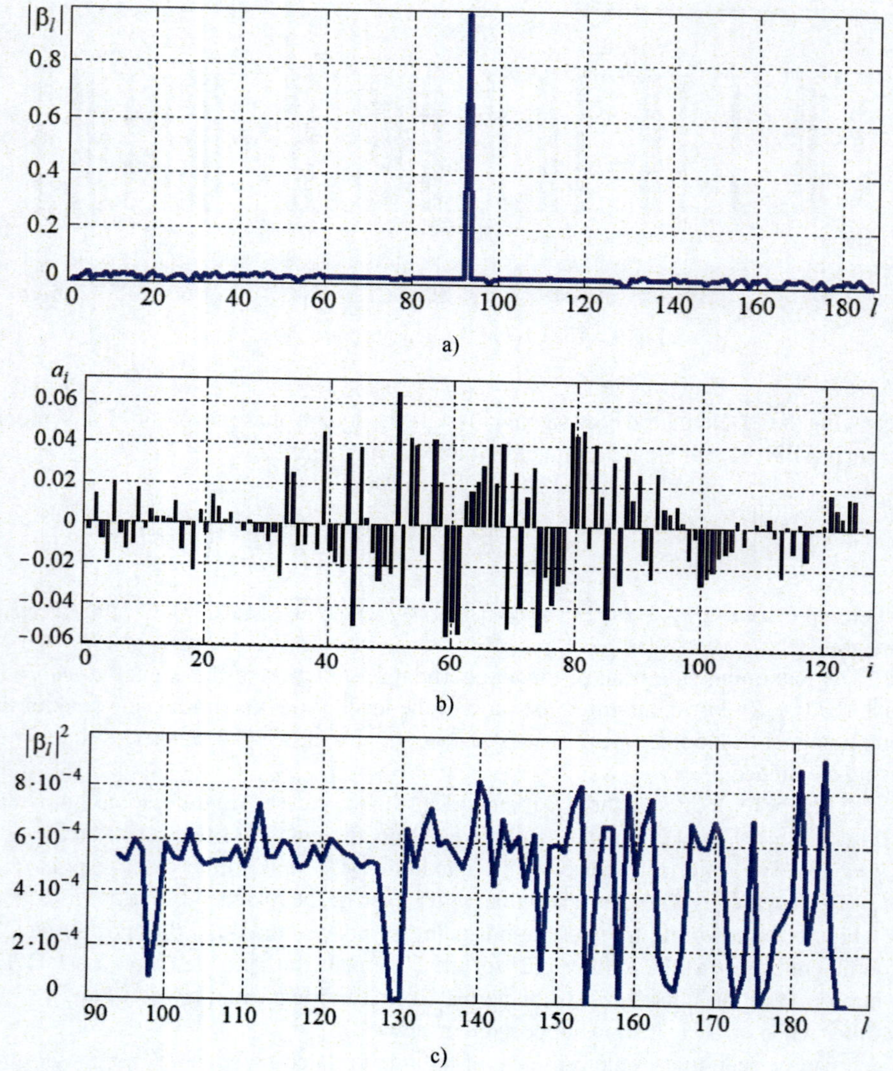

Fig. 2. The response modulus **a** and the weight coefficients **b** of the μ-filter, and the graph $|\beta_l|^2$ for $l > k = 93$ (c)

While carrying out the comparison, it is feasible to normalize (11) to $\alpha_s\left(\mathbf{a}_k^+\mathbf{d}_k\right)^2$ and, instead of B_{km}, consider

$$\eta = 1\Big/\rho q^2 + \mu^2,\ \rho = \left(\mathbf{a}_k^+\mathbf{d}_k\right)^2\Big/\left(\mathbf{a}_k^+\mathbf{a}_k\right)^2,\ q^2 = \alpha_s\big/\alpha_0. \tag{14}$$

In particular, it is important to estimate the efficiency of the μ-filtration.

In Fig. 7a, there is shown the ratio $\tilde{\eta}_\mu = \eta_\mu\big/\eta_{mf}$ of the normalized value of the average risk η_μ of the μ-filter and the normalized value of the average risk η_{mf} of the

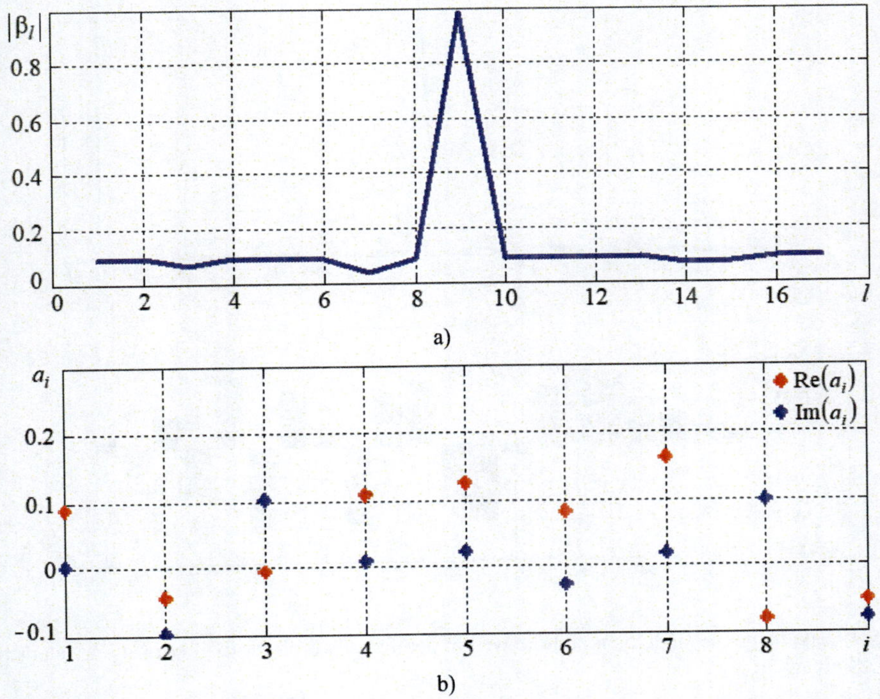

Fig. 3. The response modulus **a** and the weight coefficients **b** of the μ-filter of the 9-element multiphase *p*-class pulse code

matched filter for the case of the single Barker code message at $N = 13$, $M = 13$ and depending on q^2 (14) expressed in decibels.

The corresponding graph of $\tilde{\eta}_m = \eta_m / \eta_{mf}$ for the normalized average risks η_m and η_{mf} of the *m*-filter and the matched filter (Fig. 7b) practically differ only at $q^2 < -2$ dB. But in this case the results of the *m*-filter and the matched filter coincide. Thus, as it follows from Fig. 7, in a wide range of the signal-to-noise ratio values, the μ-filter is close in efficiency to the minimax one.

4 Conclusion

The proposed approach to the resolution problem expands the possibilities of the methods for developing the processing procedures that minimize the most unfavorable of the interfering effects. The introduced iterative procedure makes it possible to calculate the coefficients of the μ-filter and the minimax filter for a complex shift keyed pulse signal. Numerical experiments demonstrate the efficiency and convergence property of the introduced algorithm. The results of the study can be used in the design and development of the advanced remote sensing equipment.

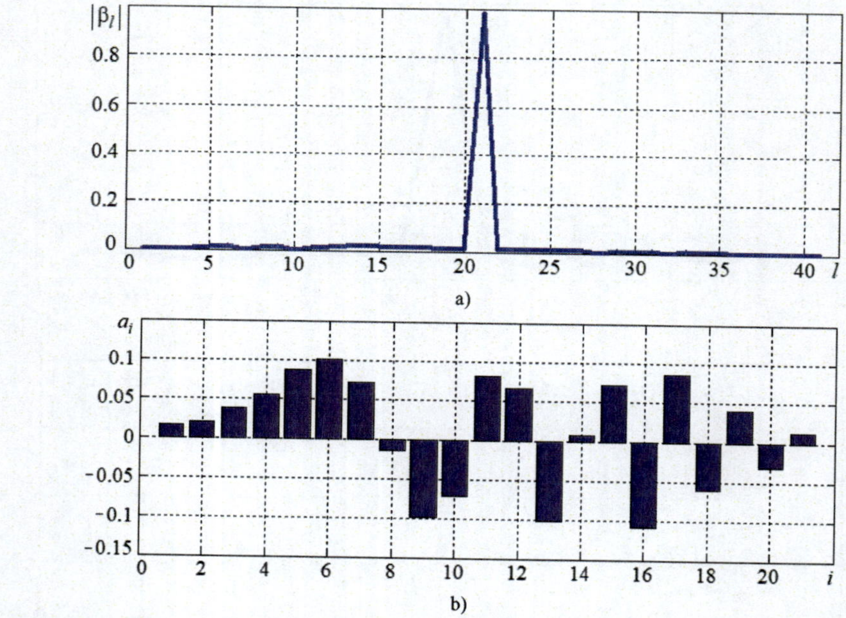

Fig. 4. The response modulus **a** and the weight coefficients **b** of the μ-filter of the 21-element APSK signal

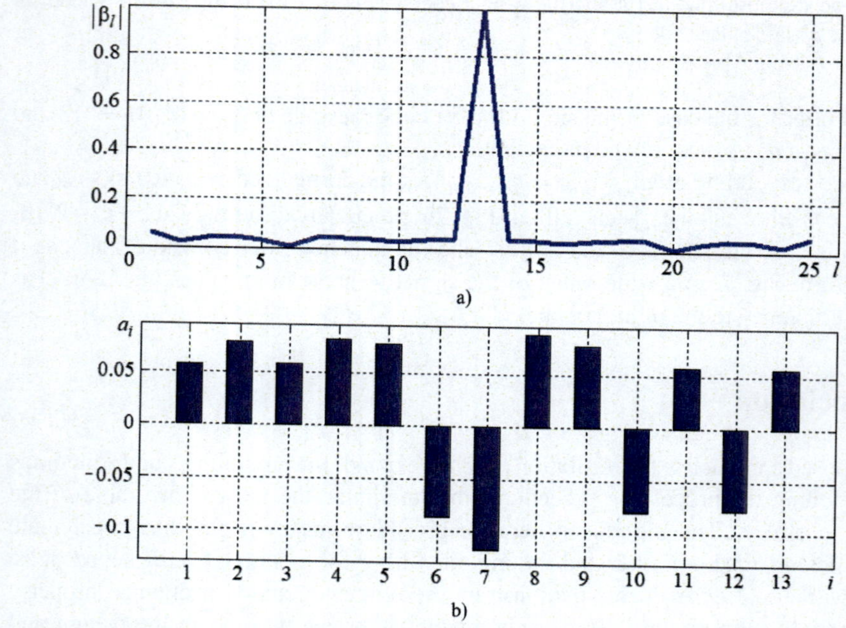

Fig. 5. The response modulus **a** and the weight coefficients **b** of the μ-filter of the 13-element Barker code

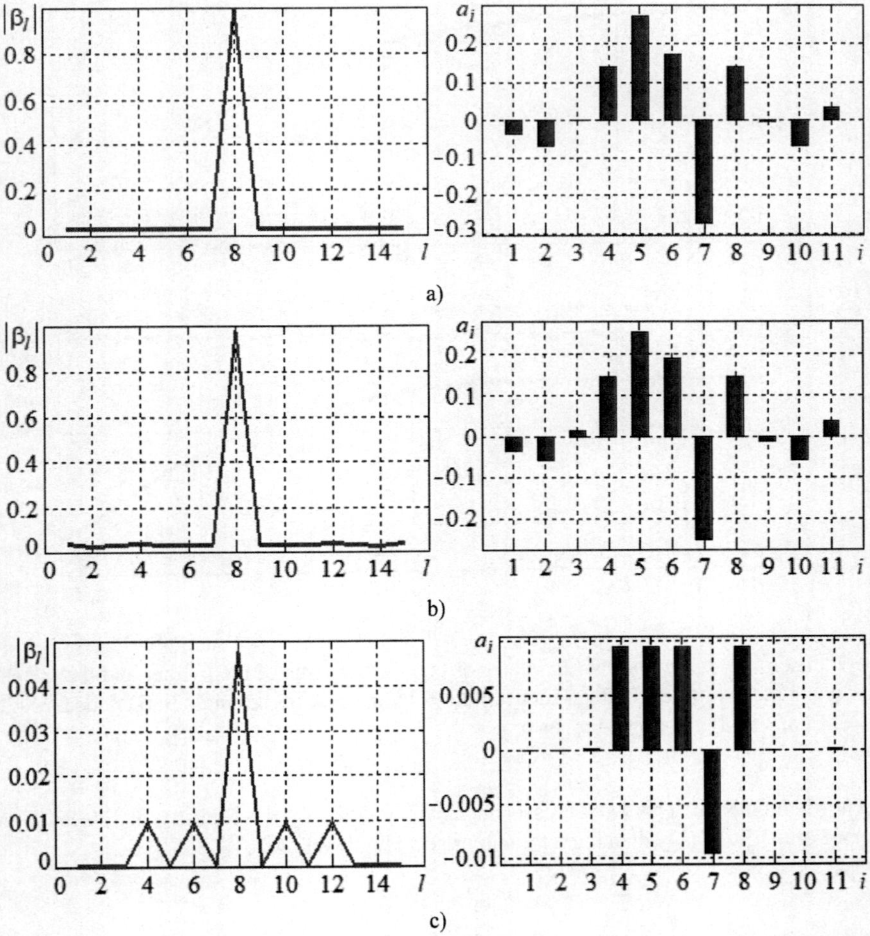

Fig. 6. The transformation of the response modulus and the weight coefficients of **a** the μ-filter at $\alpha_0 = 0$ and the *m*-filter at $\alpha_0 = 1$ **b** and $\alpha_0 = 1000$ **c** for the 5-element Barker code in the case when the noise level is changing

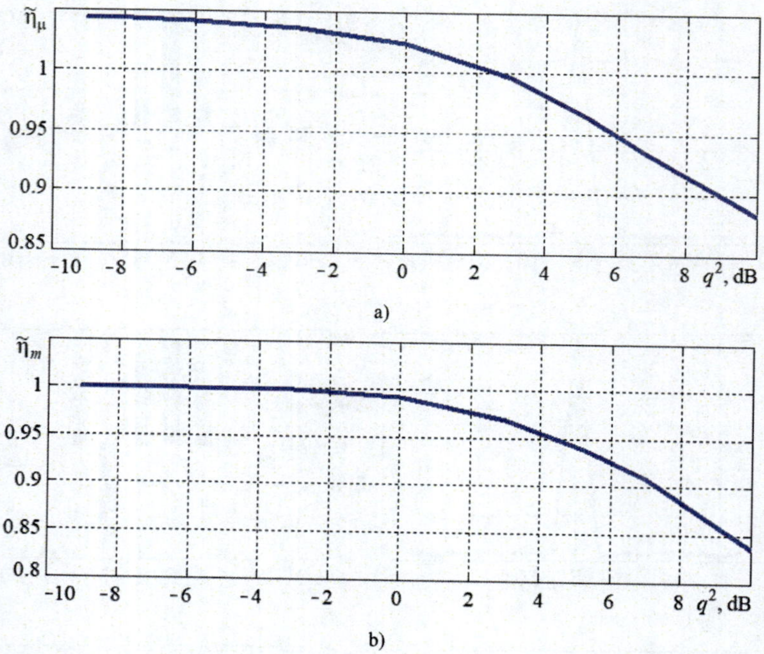

Fig. 7. The ratio of the normalized values of the average risks of the μ-filter and the matched filter **a** and of the *m*-filter and the matched filter **b** for the case of the single Barker code message ($N = 13$, $M = 13$) and depending on q^2.

Acknowledgements. This work was financially supported by the Ministry of Education and Science of the Russian Federation (research project No. FSWF-2020–0022).

References

1. Bakit'ko, R.B., Vasil'ev, M.B., Vinitskiy, A.S.: Radio systems of interplanetary spacecrafts. Radio i Svyaz', Moscow (1993)
2. Kotel'nikov, V.A., Barsukov, V.L., Akim, E.L. et al.: Surface map of venus. The Head Office of Geodesy and Cartography under the Council of Ministers of the USSR, Moscow (1989)
3. Cook, C.: Radar signals: an introduction to theory and application. Academic Press, Cambridge (2016)
4. Gruzdov, V.V., Kolkovskiy, Y.N., Krishtopov, A.V., Kudrya, A.I.: New technologies for remote sensing of the earth from space. Tekhnosfera, Moscow (2018)
5. Tarasenko, A.M.: Technique of digital formation of complex signals to improve the performance of radio equipment. Ph.D. Thesis, Vladimir State University Named After Alexander and Nikolay Stoletovs, Vladimir (2018)
6. Liu, K.: Monostatic airborne synthetic aperture radar using commercial WiMAX transceivers in the license-exempt spectrum. Electronic Thesis and Dissertation Repository, https://ir.lib.uwo.ca/etd/2611/. Last Accessed 31 May 2022
7. Li, S., Wang, Y., Zhang, J., Zhu, D.: Sequence set design for a new LFM-PC hybrid modulated radar signal. Sensors **21**(15), 1–18 (2021)

8. Ipanov, R.N.: Pulsed signals with zero autocorrelation zone for synthetic-aperture radars. J. Commun. Technol. Electron. **65**(9), 1022–1028 (2020)
9. Leukhin, A., Bezrodnyi, V., Parsaev, N.: Signals for synthetic aperture radar. Procedia Eng. **201**, 398–407 (2017)
10. Skolnik, M.I.: Radar Handbook, 3rd edn. McGraw-Hill Education, New York (2008)
11. Zhou, K., Li, D., Quan, S., Liu, T., Su, Y., He, F.: SAR waveform and mismatched filter design for countering interrupted-sampling repeater jamming. IEEE Trans. Geosci. Remote Sens. **60**, 1–14 (2022)
12. Amiantov, I.N.: Selected Issues of Statistical Communication Theory. Sovetskoe Radio, Moscow (1971)
13. Sokolov, G.A.: Resolution of discrete random signals in the case of limited a priori statistics. Telecommun. Radio Eng. **44**(5), 94–97 (1989)
14. Van Trees, H.L., Bell, K.L., Tian, Z.: Detection, Estimation, and Modulation Theory, Part I: Detection, Estimation, and Filtering Theory, 2nd edn. Wiley, New York (2013)
15. Vasilyev, F.P.: Numerical methods for solving extremal problems. Nauka, Moscow (1988)
16. Nocedal, J., Write, S.J.: Numerical optimization, 2nd edn. Springer, New York (2006)
17. Varakin, L.E.: Communication systems with noise-like signals. Radio i Svyaz', Moscow (1985)
18. Barker, R.H.: Group synchronizing of binary digital systems. In: Jackson, W. (ed.) Communication Theory, pp. 273–287. Academic Press, New York (1953)

Feature Extraction and Selection of Kidney Ultrasound Images Using a Deep CNN and PCA

Noor Hamzah Alkordy[1,2](✉), Hadeel K. Aljobouri[1], and Zainab Kassim Wadi[3]

[1] Biomedical Engineering Department, College of Engineering, Al-Nahrain University, Baghdad, Iraq
noor92_h@yahoo.com, hadeel_bme77@yahoo.com

[2] Diwaniyah Health Department, Diwaniyah General Teaching Hospital, Diwaniyah, Iraq

[3] Department of Radiology, Al-Imamain Al-Kadhimain Medical City Teaching Hospital, Baghdad, Iraq

Abstract. Chronic kidney disease (CKD) may be treated if diagnosed early, but recovery becomes difficult as the condition advances. Renal replacement treatment, such as transplantation or dialysis, will be required at some point. This paper proposes using VGG16 with principal component analysis (PCA) and only VGG16 for automatic kidney disease classification. A real dataset from Al-Diwaniyah General Teaching Hospital (Lithotripsy Center), Iraq is used in the presented work. The kidney classes are classified into a total of 1260 ultrasound images performed on normal, stone, hydronephrosis, and cyst images. Kidney abnormality detection involves three stages: feature extraction, feature selection, and then classification. The use of feature selection to decrease the retrieved features helps in quicker data training and machine learning. The accuracy, precision, recall, and F1 score of the suggested technique were all evaluated. The VGG16-PCA model has a classification accuracy of 95%.

Keywords: Feature selection · Multi-classification · PCA · VGG16

1 Introduction

Ultrasonography is a useful diagnostic technique that is non-invasive. It is particularly useful for detecting irregular fluid accumulation and abdominal organ pathology [1]. Ultrasonography is widely used since it is considered the safest medical imaging procedure [2]. Ultrasound is the mechanical vibration of the particles and molecules of a substance. Sound waves are classified as longitudinal or transverse based on how each particle moves in the same direction or the opposite direction. In diagnostic ultrasonographic applications, frequencies ranging from 1 to 30 MHz are used. For firm lateral resolution, needed narrow beam widths [3].

The kidney is the most essential organ in the urinary system. The renal cortex, renal column, renal medulla, and renal pelvis are the four distinct components of the kidney. The kidney regulates the acid-base balance, electrolyte concentrations, extracellular fluid volume, and blood pressure of the whole body [4, 5]. The occlusion of the urinary system, the aftereffects of urological treatments, and the mechanisms that lead to stone

R. Silhavy et al. (Eds.): CoMeSySo 2022, LNNS 596, pp. 104–114, 2023.
https://doi.org/10.1007/978-3-031-21435-6_10

formation may all cause damage to renal tissue. Nonobstructive kidney stones cause no other symptoms or indicators. Passage of a kidney stone causes renal colic, which often starts as minor pain and increases in intensity over 30–60 min [6]. The term obstructive hydronephrosis refers to the structural and functional abnormalities of the kidneys caused by obstructions in urine flow (difficulties in urinating) [7]. The appearance of hydronephrosis on concentrated emergency ultrasonography and hematuria on urinalysis is sufficient to identify and diagnose renal colic with accuracy [8]. Polycystic kidney disease (PKD) results from the creation of many fluid-filled cysts in the kidney area. Enlargement of cysts causes' nephron compression and impairment of kidney function. PKD is the most prevalent cause of chronic kidney disease and renal failure [9, 10].

The artificial neural network (ANN) attempts to mimic the human brain's learning processes [11]. Simonyan and Zisserman proposed the first VGG network design. VGG-16 is a dense CNN network that was released in 2015 [12]. The ImageNet database was used to train the VGG-16 network [13].

In recent, Srivastava et al., 2020 exhibited a typical VGG-16 model that has been fine-tuned utilizing its ultrasound image collection. The model can determine whether or not the ultrasound images reveal an ovarian cyst. The percentage of accuracy is 92.11% [14]. Hijab et al., 2019 presented three deep learning techniques for computer-assisted detection of malignant breast cancer in ultrasound images. Experiments have revealed that the fine-tuned variation of the pre-trained VGG16 model produces the best performance metrics. The accuracy ranges from 79 to 97% [15]. Menon et al., in 2015 presented mass in breast ultrasound images is categorized using textural, morphological, and histogram-oriented feature metrics. The acquired features are reduced in dimension using principal component analysis (PCA) and then categorized using the support vector machine (SVM) technique. With a 95.7% accuracy rate, a very high specificity rate, and a positive predictive value (PPV), the suggested strategy was found to be very good [16]. Mariam Wagih Attia et al., in 2015 proposed is a computer-aided categorization method for ultrasound kidney disorders. Consideration was given to images from the Normal, Cyst, Stone, Tumor, and Failure categories. Combining statistical and wavelet-based characteristics, neural network classifiers were developed. Principal component analysis was used to minimize the number of characteristics. An accurate classification rate of 97% has been achieved [17]. Priyanka et al., in 2020 introduced that the combination of GLCM and PCA for feature reduction results in good classification accuracy when utilizing an Artificial Neural Network to identify photos (ANN). There are substantial differences in categorization accuracy. Images are used with PCA an accuracy of 77.8% and 53.3% without PCA [18]. Balamurugan et al., in 2020 proposed system into four modules, including preprocessing, feature extraction, feature selection with an oppositional grasshopper optimization algorithm (OGOA), and illness prediction utilizing ANN. First, use the optimum wavelet and bilateral filters to get rid of noise in the image. Then, a collection of GLCM features is retrieved from each input image, and the (OGOA) is used to choose the most significant features. Using an artificial neural network, the suggested approach classifies an image as normal or abnormal (ANN). The experimental findings indicate that the suggested method achieves an accuracy of 95.83% [19].

In this research, real data collected from Al-Diwaniyah General Teaching Hospital (lithotripsy center) in Iraq was used to classify the ultrasound image of the kidneys using two models. In the first model, the features are extracted using VGG16 from the images and then they are categorized. In the second model, after extracting the features using VGG16, the important features are identified using PCA. After identifying the feature, classify the image as "normal, stone, hydrops, and cysts".

2 Methodology

In this part, the proposed model is presented in Fig. 1, beginning with ultrasound image acquisition, image preprocessing, feature extraction, feature selection, and disorder classification.

3 Image Acquisition

Real data was obtained from the Lithotripsy Center at Al-Diwaniyah General Teaching Hospital from a Philips HD11 XE device in DICOM format. The number of images reached 1260 for each classify images as show in Table 1. One radiologist and three urologists looked at images of each healthy and sick kidney from both men and women and made a diagnosis. According to the doctors' experience, a longitudinal cross-section of the kidney is taken to include the renal sinuses, medulla, and cortex for better ultrasonography. In addition, this improves the visual perception of healthy and diseased kidneys.

Table 1. Types of diseases and the number of images used for each type

Image class	Number of images
Normal	315
Stone	315
Hydronephrosis	315
Cysts	315
Total	1260

4 Preprocessing

Data preparation is crucial and, in many instances, required for conducting useful data analyses [20]. Initially, the image format is converted from DICOM to BITMAP. The image is cropped to define the region of interest to 512 × 512 and then the images are resized to 256 × 256. After that, the data is separated into 70% for training and 30% for testing and finally normalized. The primary goal of data normalization technology

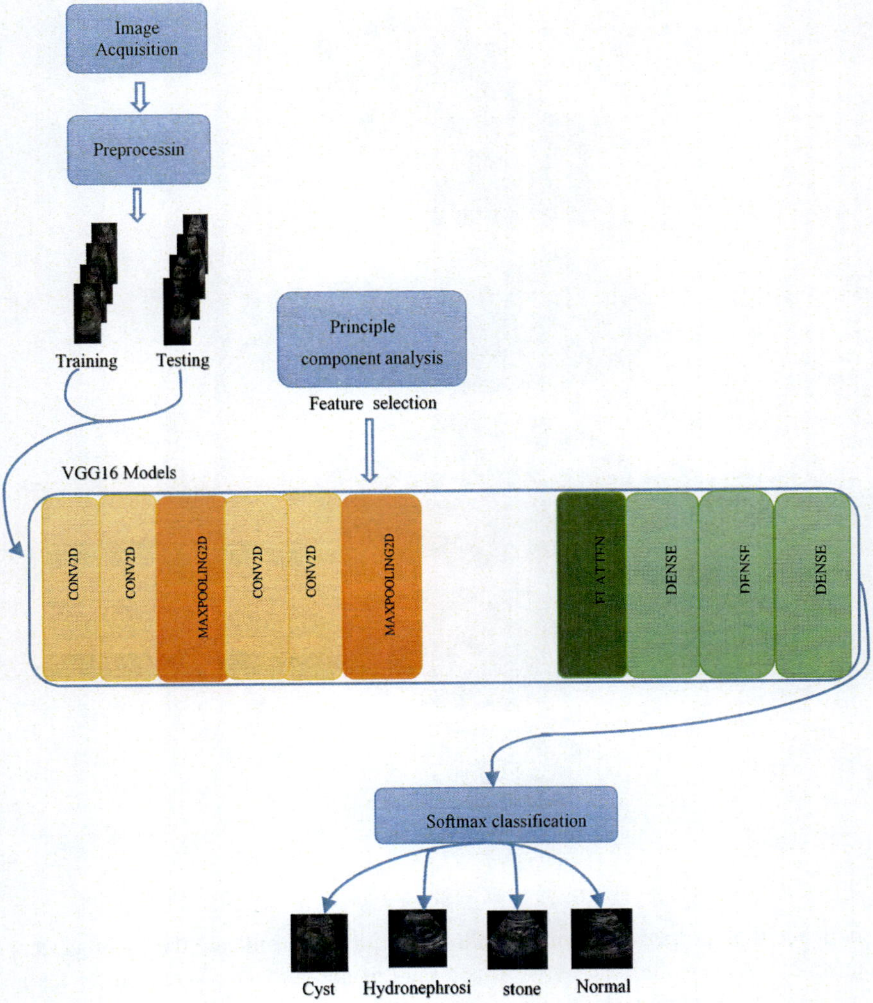

Fig. 1. Proposed models architecture

is to provide high-quality data that can be fed into any educational system. In order to speed up the learning process, it is necessary to extend the range of time series data to the same range of values [21]. Figure 2 illustrates the pre-processing procedures used on ultrasound images.

5 VGG16

Deep learning is a multilayer learning representation in an artificial neural network [22]. VGG16 is an extremely deep convnet model developed by Oxford's Visual Geometry Group (VGG) [12]. Feature extraction is the technique of collecting the visual information of images to reduce the needed amount of resources [23]. Features are sources

Fig. 2. Process of preprocessing

of evidence that are pertinent to a certain application and indicate the main aspects of images [24]. A convolutional network is a kind of neural network geared for learning an image's representation. It introduces two new hidden layer types: convolutional and subsampling/pooling [22]. The convolution and fully connected layers assess the likelihood of multiple classes/labels using the ReLu activation function Eq. 1, except for the output layer, which uses the SoftMax activation function Eq. 2 instead. Dropout is also used as regularization after each tanh fully-connected layer to prevent overfitting by randomly dropping/switching off (setting the value to zero) hidden layers [25].

$$y_i = log(1 + exp\, xi) \tag{1}$$

$$y_i = \frac{e^{xi}}{\sum_{k=1}^{k} e^{xk}} \tag{2}$$

for i = 1…k.

In this paper propose a deep convolutional neural network consisting of a pre-trained VGG16 for feature extractor and classify. VGG16 consists of sixteen hidden convolutional layers, maximum pooling layers, and fully connected layers.

6 PCA

Every method of feature extraction correlates to a number of surface feature descriptors, some of which are significant and others that are superfluous. Because the duplicate data could be misclassified, it needs to be thrown away [26]. A large number of features increases computation time and complexity, which wastes a considerable amount of storage memory, and occasionally makes classification more difficult, despite the importance of features for reaching high accuracy [27].

Image normalization, identifying the covariance matrix of the image data, computing the eigenvectors and eigenvalues of the covariance matrix, and finally translating image data into a new basis are the four primary processes of the PCA image dimension reduction approach. This process reduces the dimension of the original image collection by transforming it into a new axis by this equation [28]:

$$F_{transformed}(x, y) = U^T F_{transformed}(x, y) \tag{3}$$

where:

U^T is the transpose of the eigenvectors matrix

$F_{transformed}(x, y)$ is the adjusted original image dataset.

In this paper post VGG16 feature extraction, the next stage in image analysis is feature selection. Feature selection is used to minimize the dimensionality of imaging data. In the suggested method, PCA was used to reduce the amount of data. The PCA application takes high-dimensional image data and turns it into low-dimensional data based on the most important features. When classifying renal images with similar elements, PCA had a significant effect in decreasing feature vectors and transferring comparable items to more relevant feature vectors.

7 Result

7.1 Experimental Setup

Every experiment is performed on Windows10 Home ×64. The operating system on a machine with CPU Intel(R) Core(TM) i7-6600U CPU. The performance parameter of the suggested technique is calculated using Python tool version 3.9.7.

7.2 Performance Metrics

To evaluate the applicability of the proposed method is measured using accuracy, precision, specificity, sensitivity, and the F1-score. The performance of VGG16 with PCA and without PCA as shown in Tables 2 and 3.

The confusion matrix for each fold validation is computed for several parameters including the true positive rate (TPR), the true negative rate (TNR), and the accuracy. Accuracy is the proportion of all predictions correctly detected by the classifier. The accuracy of a classifier is determined by Eq. (4):

$$\text{Accuracy} = \frac{TP + FN}{TP + TN + FP + FN} \tag{4}$$

Table 2. The performance of VGG16 with PCA

No.	Precision%	Recall%	F1-score%	Support%
Normal	89	97	93	95
Stone	99	97	98	95
Hydronephrosis	97	97	97	95
Cysts	96	97	98	95

Table 3. The performance of VGG16 without PCA

No.	Precision%	Recall%	F1-score%	Support%
Normal	89	94	91	95
Stone	95	95	95	95
Hydronephrosis	95	93	94	95
Cysts	93	91	92	95

Precision refers to the proportion of relevant vs irrelevant results returned by a search. It is determined by Eq. (5):

$$\text{Precision} = \frac{TP}{TP + FP} \tag{5}$$

TPR is called as sensitivity or recall, the proportion of positive instances that are accurately recognized. It is determined by Eq. (6):

$$\text{sensitivity} = \frac{TP}{TP + FN} \tag{6}$$

TNR is called as specificity, this is the ratio of properly detected negative situations. It is determined by Eq. (7):

$$\text{specificity} = \frac{TN}{TN + FP} \tag{7}$$

F1-score measures the accuracy of a model on a dataset. It is determined by Eq. (8):

$$\text{F1-Score} = \frac{2 \times (\text{Recall} \times \text{Precision})}{(\text{Recall} + \text{Precision})} \tag{8}$$

7.3 Performance Analysis

The kidney dataset is used to train the neural network on how to classify kidney images into different groups, such as normal, stones, hydronephrosis, and renal cysts. The number of features chosen shows that the proposed method can significantly reduce the

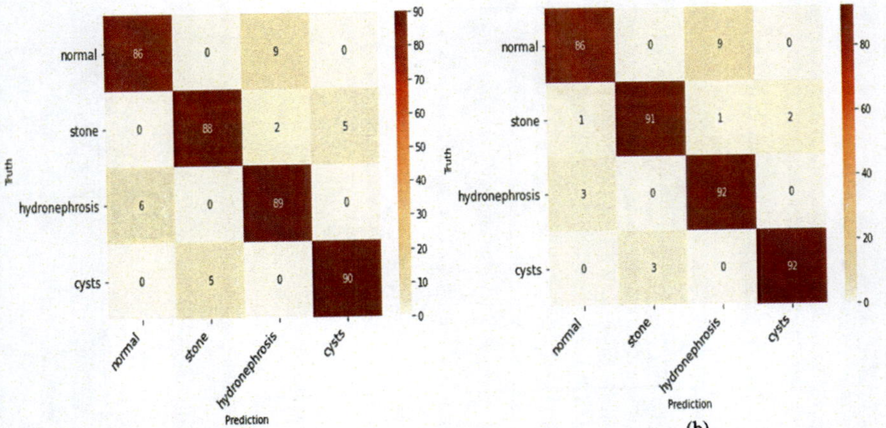

(a) (b)

Fig. 3. Confusion matrix of the true negative (TN), true positive (TP), false negative (FN), false positive (FP) values for (a) VGG16 without PCA, (b) VGG16-PCA model

number of dimensions in the original data while retaining only the most important features. The confusion matrix reflects the number of classification errors made by the classifier while trying to predict each class of kidney images, as shown in Fig. 3.

Principal component analysis PCA has been widely used in various fields. The main goal of PCA is to keep the variety of each data set, which is made up of as many related variables as possible while reducing the number of features to make the data set smaller. In this model, VGG16 was used for feature extraction, image classification, and comparison using PCA. When using PCA, the following results were observed: an increase in the performance speed of the training data and an increase in accuracy, where the total execution time with PCA is: 0:00:04.816946 and accuracy = 0.95%. While, by using only VGG16 without PCA, the training time for the data increased, with a total execution time without PCA: 0:00:54.415427 and accuracy = 0.92%. To find out how many features we need to get the most information according to the curve in this Fig. 4, where the overall characteristics of training and testing are 32768 features, and according to what is noticed in the curve, the highest information that can be obtained when the number of components is 880.

8 Discussion

This study provided a method for solving the classification issue of kidney ultrasound images, which included a data set of 315 images for each classify kidney image generated from an ultrasound equipment in Iraqi centers. The VGG-16 model has been fine-tuned using ultrasound image collecting and has an accuracy of 92.11% [14]. The combination of GLCM and PCA for feature reduction images are used with PCA an accuracy of 77.8% and 53.3% without PCA [18].

In this paper, kidney ultrasound images were classified into four types. Features from the renal ultrasound images are extracted with VGG16, followed by feature selection with PCA, then additional hidden dense layers, and a final output/classifier layer. The

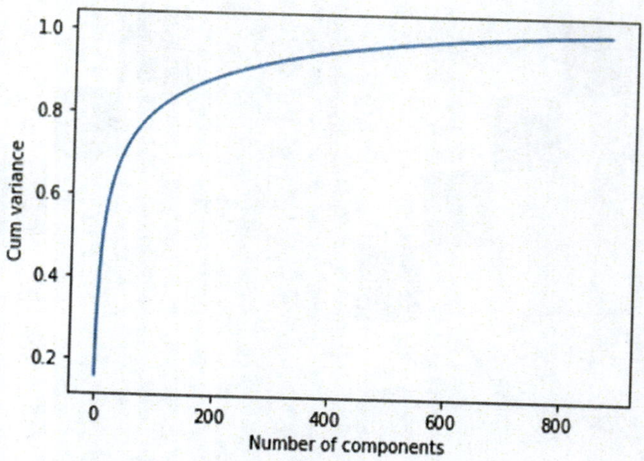

Fig. 4. The curve for ideal number of PCA components

proposed VGG16-PCA models have better classification accuracy than only VGG16. The accuracy of VGG16-PCA and VGG16 was 95% and 92%, respectively. PCA was used to reduce the features extracted from VGG16 in our proposed technique.

9 Conclusion

Two classification models are presented in this work. First one, VGG16-PCA model used VGG16 for feature extraction and PCA for feature selection, followed by additional hidden dense layers and an output/final classifier layer, and the second model (only VGG16) was used for feature extraction and classifying. It has been observed that when feature selection is used to reduce the extracted features, it results in faster training of the data and speeds up machine learning. The parameters used are appropriate to determine the images of normal, stone, hydronephrosis, and cyst images kidneys. Data was collected from the Lithotripsy Center at Al-Diwaniyah General Teaching Hospital, and a total of 1260 images were collected. These renal classification models may be useful for radiologists, especially in Iraq. These models provide a feasible answer to the stressful clinical situation in Iraq where there is a known dearth of clinical examination and diagnostic capabilities. Also, the proposed models can be used to find other problems with the kidneys, like kidney failure and tumors.

Acknowledgments. We thank the doctors of Al-Diwaniyah General Teaching Hospital (lithotripsy center) for their cooperation.

References

1. Woo, J.: A short history of the development of ultrasound in obstetrics and gynecology. Hist. Ultrasound Obstet. Gynecol. **3**, 1–25 (2008)

2. Akkasaligar, P.T., Biradar, S.: Classification of medical ultrasound images of kidney. In: International Conference on Information and Communication Technologies, pp. 24–28 (2014)
3. Ziskin, M.C.: Fundamental physics of ultrasound and its propagation in tissue. Radiographics **13**(3), 705–709 (1993)
4. Zhou, X.J., Laszik, Z.G., Nadasdy, T., D'Agati, V.D.: Silva's Diagnostic Renal Pathology. Cambridge University Press (2017)
5. Al-Timemy, A.H.A., Kamel, I., Abdul Ameer, H.K.: ANN-based prediction of kidney dysfunction using clinical laboratory data. Al-Nahrain J. Eng. Sci. Al-Nahrain Univ. Coll. Eng. J. **11**(1), 131–136 (2008)
6. Coe, F.L., Evan, A., Worcester, E.: Science in medicine kidney stone disease. J. Clin. Investig. **115**(10), 2598–2608 (2005)
7. Nuraj, P., Hyseni, N.: The diagnosis of obstructive hydronephrosis with Color Doppler ultrasound. Acta Inform. Med. **25**(3), 178–181 (2017)
8. Shrestha, R., Shakya, R.M., Khan, A.A.: Bedside ultrasound in the emergency department to detect hydronephrosis for the evaluation of suspected ureteric colic. Kathmandu Univ. Med. J. **14**(54), 172–176 (2016)
9. Bergmann, C., Guay-Woodford, L.M., Harris, P.C., Horie, S., Peters, D.J.M., Torres, V.E.: Polycystic kidney disease. Nat. Rev. Dis. Primers **4**(1) (2018)
10. Schrier, R.W., et al.: Predictors of autosomal dominant polycystic kidney disease progression. J. Am. Soc. Nephrol. **25**(11), 2399–2418 (2014)
11. Hasan, M.M., Hussein, A., Alhilli, A.M.: Modeling of neuropathic bladder lesions diagnosis using neural network algorithm. Al-Nahrain J. Eng. Sci. **11**(1), 38–42 (2008)
12. Simonyan, K., Zisserman, A.: Very deep convolutional networks for large-scale image recognition. In: 3rd International Conference on Learning Representations, ICLR 2015 - Conference Track Proceedings, pp. 1–14 (2015)
13. Pal, S.: Transfer Learning and Fine Tuning for Cross Domain Image Classification with Keras. GitHub (2016)
14. Srivastava, S., Kumar, P., Chaudhry, V., Singh, A.: Detection of ovarian cyst in ultrasound images using fine-tuned VGG-16 deep learning network. SN Comput. Sci. **1**(2), 1–8 (2020). https://doi.org/10.1007/s42979-020-0109-6
15. Hijab, A., Rushdi, M.A., Gomaa, M.M., Eldeib, A.: Breast cancer classification in ultrasound images using transfer learning. In: International Conference on Advances in Biomedical Engineering, ICABME, Oct 2019, pp. 1–4
16. Menon, R.V., Raha, P., Kothari, S., Chakraborty, S., Chakrabarti, I., Karim, R.: Automated detection and classification of mass from breast ultrasound images. In: 2015 5th National Conference on Computer Vision, Pattern Recognition, Image Processing and Graphics, NCVPRIPG 2015, pp. 3–6 (2016)
17. Wagih, M., Abou-Chadi, F.E.Z., El-Din, H., Mekky, N.: Classification of ultrasound kidney images using PCA and neural networks. Int. J. Adv. Comput. Sci. Appl. **6**(4), 53–57 (2015)
18. Priyanka, K.D.: Feature extraction and selection of kidney ultrasound images using GLCM and PCA. Procedia Comput. Sci. **167**(2019), 1722–1731 (2020)
19. Balamurugan, S.P., Arumugam, G.: A novel method for predicting kidney diseases using optimal artificial neural network in ultrasound images. Int. J. Intell. Enterp. **7**(1–3), 37–55 (2020)
20. Famili, A., Shen, W.M., Weber, R., Simoudis, E.: Data preprocessing and intelligent data analysis. Intell. Data Anal. **1**(1), 3–23 (1997)
21. Bhanja, S., Das, A.: Impact of Data Normalization on Deep Neural Network for Time Series Forecasting, pp. 5–10 (2018)
22. LeCun, Y., Bengio, Y., Hinton, G.: Deep learning. Nature **521**(7553), 436–444 (2015)
23. Mohanaiah, P., Sathyanarayana, P., GuruKumar, L.: Image texture feature extraction using GLCM approach. Int. J. Sci. Res. Publ. **3**(5), 1–5 (2013)

24. Guyon, I., Elisseefl, A.: An introduction to feature extraction. Stud. Fuzziness Soft Comput. **207**, 1–25 (2006)
25. Srivastava, N., Hinton, G., Krizhevsky, A., Sutskever, I., Salakhutdinov, R.: Dropout: a simple way to prevent neural networks from overfitting. J. Mach. Learn. Res. **15**(1), 1929–1958 (2014)
26. Balasubramanian, D., Srinivasan, P., Gurupatham, R.: Automatic classification of focal lesions in ultrasound liver images using principal component analysis and neural networks. In: Annual International Conference of the IEEE Engineering in Medicine and Biology - Proceedings, pp. 2134–2137 (2007)
27. Collins, R.T., Liu, Y., Leordeanu, M.: Online selection of discriminative tracking features. IEEE Trans. Pattern Anal. Mach. Intell. **27**(10), 1631–1643 (2005)
28. Ng, S.C.: Principal component analysis to reduce dimension on digital image. Procedia Comput. Sci. **111**, 113–119 (2017)

Process Control Laboratory

Daniel Honc(✉), Aleš Novotný, and Libor Havlíček

Faculty of Electrical Engineering and Informatics, University of Pardubice, Pardubice, Czech Republic
daniel.honc@upce.cz

Abstract. The authors are presenting Process Control Laboratory at the Faculty of Electrical Engineering and Informatics, University of Pardubice. The laboratory is equipped with six GUNT training systems covering common technological variables such as level, flow, pressure, temperature, speed, and position. The authors created SW support for MATLAB and Simulink environment. An internal part of the GUNT training systems is the LabJack U12 data acquisition card. LabJack Dynamic Link Library (DLL) is used to operate training systems from MATLAB. Authors created the M-S function to allow experimenting from Simulink. The possibilities of the proposed solution are demonstrated in several control applications.

Keywords: Automatic control · Laboratory · Experiments · GUNT training systems · MATLAB · Simulink

1 Introduction

Laboratory experiments help students to understand and practice theoretical knowledge. It is indisputable part of technical education together with derivations, calculations, and simulations. Simulation can be very complex with disturbances, noises, parameters, and structure uncertainties. Real experiments are usually more expensive, but they give students feeling and connection with the nature through visual impression, sounds, vibrations etc. They can also train their manual skills and patience. We should allow students to enjoy successful applications even though it does not bring anything new to the theoretical side of things. On the contrary reality is much more complex and unpredictable so we must balance complexity, time consumption of the experiments and their illustration and support of theoretical knowledge. The authors focus on laboratory experiments in control theory, process control and automation.

There are several possibilities how to build the control laboratory. One solution is to buy commercial products for process control education purposes. In this case, we get ready to use products. Some disadvantages could be the high price and the system is usually a closed solution from a hardware and software point of view. Examples of such products are e.g. Automation and process control engineering processes from German company GUNT [1], Process Control Training Systems from English company Tecquipment [2] or Teaching Controls and Mechatronics systems from National Instruments [3]. Another way is to construct your own laboratory equipment. This is

R. Silhavy et al. (Eds.): CoMeSySo 2022, LNNS 596, pp. 115–126, 2023.
https://doi.org/10.1007/978-3-031-21435-6_11

very time-consuming, and it is difficult to approach the quality of commercial products. On the other hand, we can get perfectly what we want. The authors decided to go the route of buying commercial products and creating their own software support for use in MATLAB and Simulink environments.

2 Materials and Methods

Simple process engineering control systems from company GUNT - level, flow, pressure, temperature, speed, and position control training systems create the main part of the laboratory – see Fig. 1. Description of the functionality together with wiring of integral LabJack U12 data acquisition card is given. It can help other users to connect training systems with own measuring and control systems. LabJack data acquisition card and authors software support are demonstrated at the end of the chapter.

Fig. 1. Process Control Laboratory

2.1 RT 010 - Level Control

A controlled pump (signal Y) pumps the liquid into the service tank and flows back to the storage tank via a controlled proportional solenoid valve (signal Z). The liquid level in the service tank is measured by a hydrostatic pressure sensor (signal X) (Fig. 2 and Table 1).

2.2 RT 020 - Flow Control

A controlled pump (signal Z) pumps the liquid in a loop through a controlled proportional solenoid valve (signal Y) back to the storage tank. The liquid flow is measured by a turbine flow meter (signal X) (Fig. 3 and Table 2).

2.3 RT 030 - Pressure Control

A diaphragm pump (signal Y) is operated to draw air into the pressure tank. The pressure in the tank is measured by a pressure sensor (signal X). The air leaves the tank via manually operated valves. One of them is connected to a solenoid-controlled valve (signal Z) (Fig. 4 and Table 3).

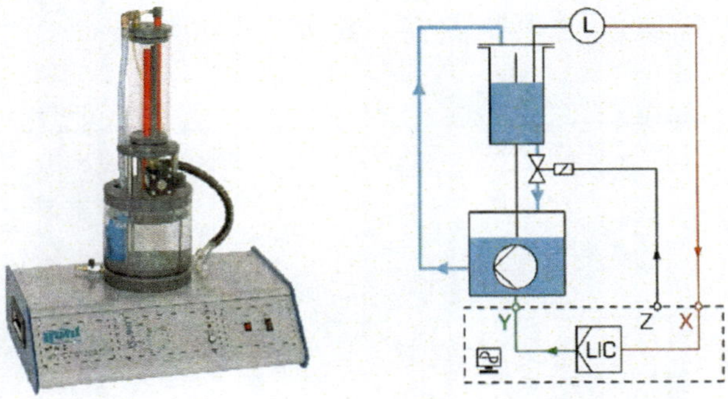

Fig. 2. RT 010 - level control [4]

Table 1. Acquisition card wiring

Terminal	Meaning	Value
AI0	Level X	0.6 ÷ 2.6 V
AO0	Pump Y	0 ÷ 5 V
AO1	Valve Z	0 ÷ 5 V
IO2	On/off switch - pump	0/1
IO3	On/off switch - valve	0/1

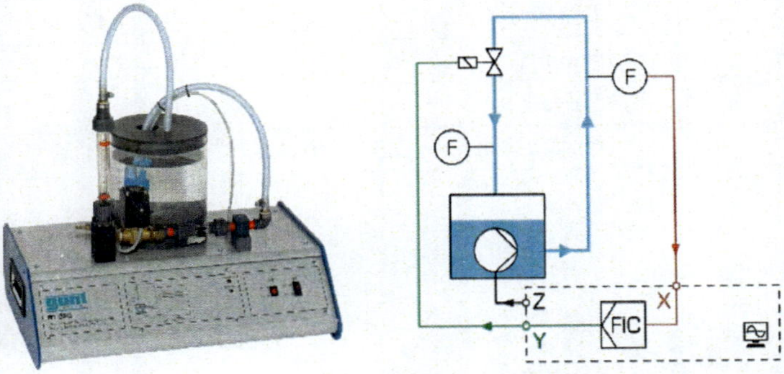

Fig. 3. RT 020 - flow control [5]

2.4 RT 040 - Temperature Control

A metal rod is heated or cooled by a Peltier cell (Y signal). The temperature of the rod is sensed in three positions by the temperature sensors (signals X1, X2 and X3). The fan (signal Z) on the Peltier cell can be controlled (Fig. 5 and Table 4).

Table 2. Acquisition card wiring

Terminal	Meaning	Value
AI0	Flow X	0 ÷ 8.1 V
AO0	Valve Y	0 ÷ 5 V
AO1	Pump Z	0 ÷ 5 V
IO2	On/off switch - pump	0/1

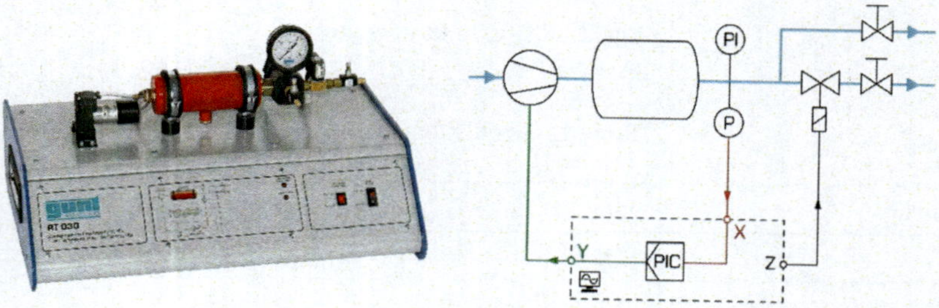

Fig. 4. RT 030 - pressure control [6]

Table 3. Acquisition card wiring

Terminal	Meaning	Value
AI0	Pressure X	0.1 ÷ 4.9 V
AO0	Pump Y	0 ÷ 5 V
IO1	Valve Z	0/1
IO2	On/off switch - pump	0/1

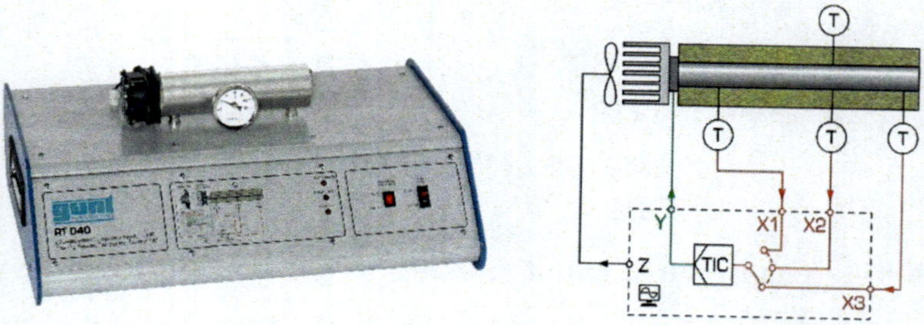

Fig. 5. RT 040 - temperature control [7]

2.5 RT 050 - Speed Control

The voltage of the DC motor (signal Y) controls its speed. The motor is connected to the generator, and a flywheel is placed on the axis. The speed (signal X) is measured by

Table 4. Acquisition card wiring

Terminal	Meaning	Value
AI0	Temperature X1	$0 \div 5$ V
AI1	Temperature X2	$0 \div 5$ V
AI2	Temperature X3	$0 \div 5$ V
AO0	Peltier cell Y	$0 \div 5$ V
IO1	On/off switch – Peltier	0/1
IO2	Heating/cooling	0/1
IO3	Fan Z	0/1

an inductive sensor. The generator (signal Z) can be used to change the load (Fig. 6 and Table 5).

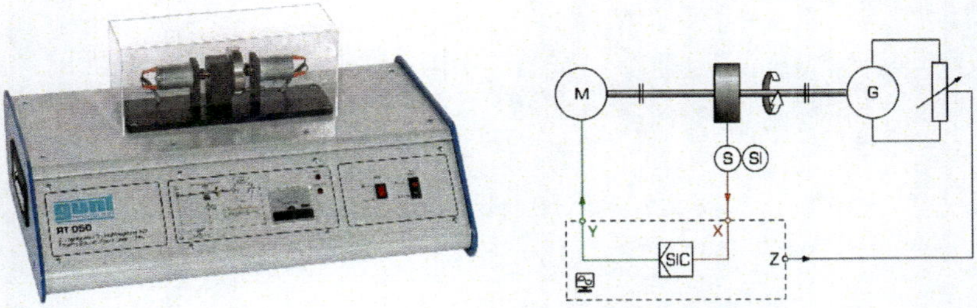

Fig. 6. RT 050 - speed control [8]

Table 5. Acquisition card wiring

Terminal	Meaning	Value
AI0	Speed X	$0 \div 8$ V
AO0	Motor Y	$0 \div 5$ V
AO1	RPM indicator	$0 \div 5$ V
IO1	On/off switch - motor	0/1
IO2	Generator Z1	0/1
IO3	Generator Z2	0/1

Two digital signals control load of the generator at four levels – see Table 6.

Table 6. Generator load

Generator	0%	33%	66%	100%
Z2	0	0	1	1
Z1	0	1	0	1

2.6 RT 060 - Position Control

The motor (signal Y) and the toothed belt control the position of the trolley, which is sensed by the rotary encoder (signal X). The system is equipped with limit switches that switch off the motor when the trolley reaches the end of the track (Fig. 7 and Table 7).

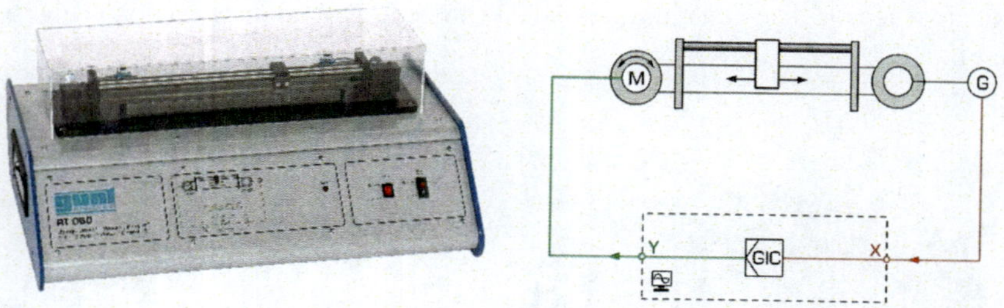

Fig. 7. RT 060 - position control [9]

Table 7. Acquisition card wiring

Terminal	Meaning	Value
AI0	Position X	$0 \div 4$ V
AO0	Motor Y	$0 \div 5$ V
IO1	Direction of motor rotation	0/1
IO2	On/off switch - motor	0/1

2.7 LabJack U12 Acquisition Card

The GUNT RT 010-060 training systems use the LabJack U12 data acquisition card [10]. It is a USB multifunction data acquisition card with elementary software support for many environments: MATLAB, LabVIEW, Python, Java, C++, Visual Basic, VBA Excel, Perl, .NET, Delphi and others. The card can measure 8 analog inputs in the ± 10 V range with 12-bit resolution, control 2 analog outputs in the 0–5 V range with 10-bit resolution, and measure and control 20 digital inputs and outputs (programmatically define whether they are input or output). There is a 32-bit counter on the card. Multiple

Fig. 8. LabJack U12 measurement card [10]

LabJack cards can be connected to the computer (they are addressed). The card does not require an external power supply (Fig. 8).

For experiments with GUNT training systems from the MATLAB environment authors created M-S function which can be used in Simulink scheme (Fig. 9).

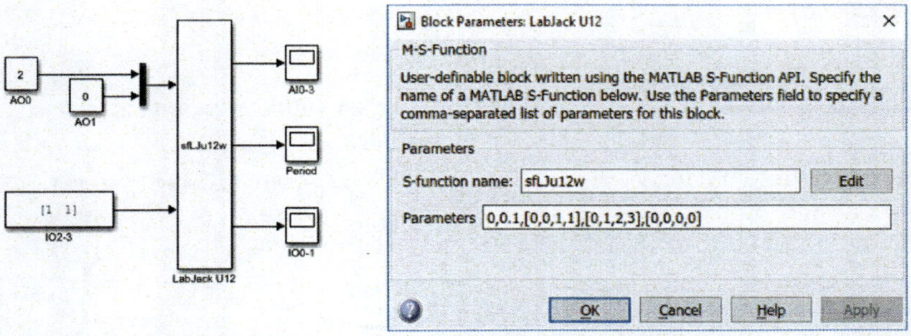

Fig. 9. Simulink scheme and LabJack U12 M-S function dialog

In the LabJack U12 block dialog we can set the card parameters LocalID, Period, DIO, Channel and Gain. LocalID is the device identifier 0–255. A period is the sampling period >0.1 s. DIO determines whether the first four digital channels 0–3 are being set or read (0 means read, 1 means set). A channel is representing by numbers of the four analog inputs 0–4 that we are measuring (0–7 in single-ended mode and 8–11 in differential mode). Gain is the gain when using differential mode (0 for no gain for ±20 V measurement range to 7 for 20× gain for ±1 V measurement range). For example, the parameters 0, 0.1, [0, 0, 1, 1], [0, 1, 2, 3], [0, 0, 0, 0] mean that we use the first LabJack card, sample with a period of 0.1 s, read the first two digital channels and set the second two, measure the first 4 analog inputs, and since this is a single-ended circuit, the gain parameters are meaningless and are set to 0. In addition to the simulation model, the following files must be in the MATLAB working directory for proper operation: ljackproto.m, ljackuw.dll, ljackuw_thunk_pcwin64.dll, sfLJu12w.m.

In MATLAB, we can use the functions of the ljackuw.dll DLL library for the .NET environment. The following DLL libraries must be present in the MATLAB working directory: ljackuw.dll and LJDotNet.dll. First, we must create an instance of the .NET class using the command

```
ljAsm = NET.addAssembly(strcat(cd,'\LJDotNet.dll'));
```

Then the DLL library function can be used to measure the analog input with the command

```
%[errorCode,
idnum, overVoltage, voltage] = lj.LabJack.EAnalogIn(idnum, demo,
channel, gain, overVoltage, voltage);
```

where idnum −1 means the first LabJack card, demo 0 means the card is used and it is not demo mode, channel is the analog input channel and gain is the gain used. For example, to measure the first analog input, we use the command (AI is the value of the measured voltage)

```
[errorCode, idnum, overVoltage, AI] = lj.LabJack.EAnalogIn(-1, 0,
0, 0, 0, 0);
```

The analog outputs AO0 and AO1 can be similarly set with the command

```
%[errorCode, idnum] = lj.LabJack.EAnalogOut(idnum, demo, AO0,
AO1);
[errorCode, idnum] = lj.LabJack.EAnalogOut(-1, 0, 5, 0);
```

We can also work with digital inputs and outputs in a similar way. A digital input can be read with the command

```
%[errorCode, idnum, state] = lj.LabJack.EDigitalIn(idnum, demo,
channel, readD, state);
```

where channel is the digital input channel 0–3, readD 0 means that we read the inputs on the card terminal and state is the state (>0 = high, 0 = low)

```
[errorCode, idnum, state] = lj.LabJack.EDigitalIn(-1, 0, 0, 0, 0);
```

The digital output can be set with the command

```
%[errorCode, idnum] = lj.LabJack.EDigitalOut(idnum, demo, chan-
nel, writeD, state);
[errorCode, idnum] = lj.LabJack.EDigitalOut(-1, 0, 0, 0, 1);
```

3 Results and Discussion

In the Process Control Laboratory students measure e.g. basic characteristics of the GUNT training systems - static characteristic, step, impulse, and frequency responses, design and apply control strategies like logic control, on-off, PID, or state-space controllers both in continuous and discrete-time domains. Following experiments are carried out to demonstrate functionality of the proposed solution identification experiment is prepared and realized to excite RT 050 - Training system: speed control to approximate the dynamical response with the proportional second order linear model. On-off controller, PID and state-space model predictive controller are applied – firstly as real-time simulations for initial tuning of the parameters and then as real experiments as well.

3.1 Experimental Identification

Input u, response y of the system RT 050 with zero generator load is plotted in Fig. 10. Last graph is the load calculated as a percentage of the time spend in the program from whole sampling period (T = 0.1 s). The working point on input is 2.5 V and on output 4 V.

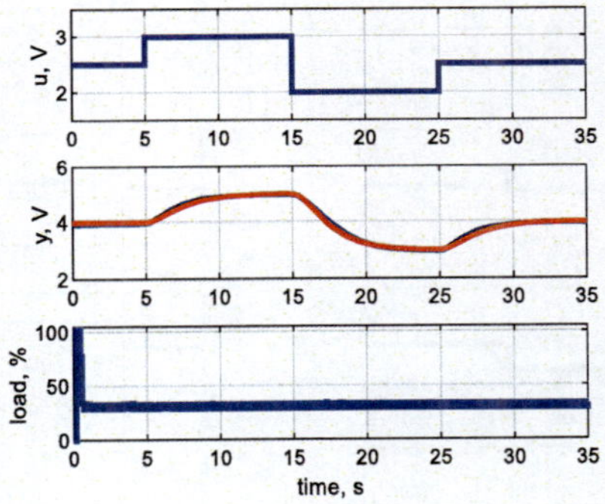

Fig. 10. Experimental identification of RT 050

Model was identified as transfer function

$$F(s) = \frac{2}{(1.3s + 1)^2} \tag{1}$$

and the response is plotted as red line in Fig. 10.

3.2 On-Off Controller

The on off controller is applied to the RT 050 system. The simulated control response is on the left side of Fig. 11 while the real experiment is on the right side.

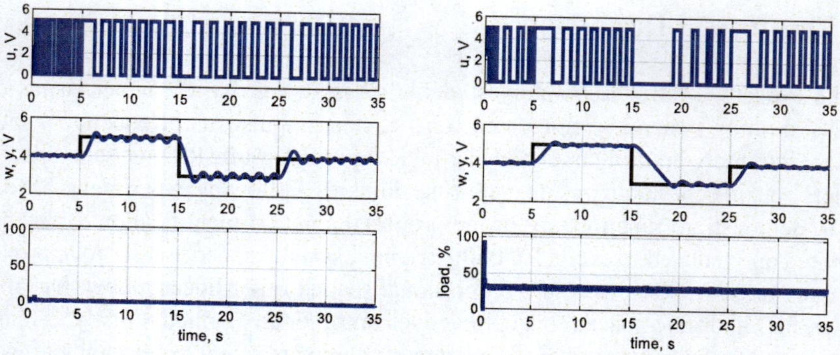

Fig. 11. On-off controller simulation and real experiment response

Table 8. PID parameters

Parameters	Value
Gain	1.5
Proportional time constant, s	3
Derivative time constant, s	0

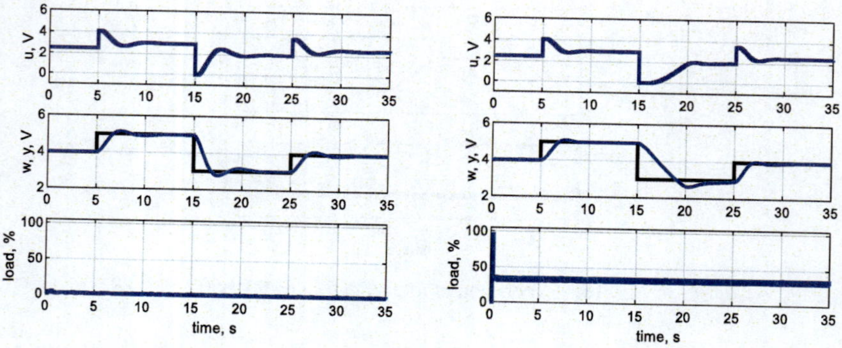

Fig. 12. PID controller simulation and real experiment response

3.3 PID Controller

The digital PID controller was applied to the RT 050 system. The simulated control response is on the left side of Fig. 12 while the real-world experiment is on the right side. Parameters of the controller were tuned by the trial-error method and are in Table 8.

3.4 Model Predictive Controller

The state-space Model Predictive Controller (MPC) was applied to the RT 050 system. The simulated control response is on the left side of Fig. 13 while the real-world experiment is on the right side. Parameters of the controller are in Table 9.

Table 9. MPC parameters

Parameters	Value
Prediction horizon, s	5
Control horizon, s	5
Control error penalization	1
Control change penalization	0.1
State observer poles	[0.6 0.6]

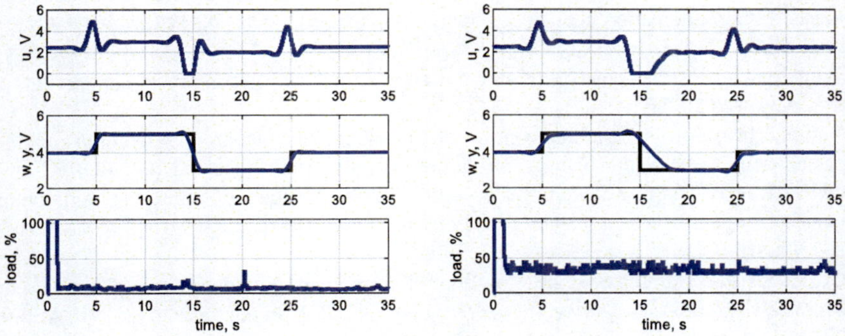

Fig. 13. MPC controller simulation and real experiment response

4 Conclusion

Authors presented Automatic Control Laboratory at the Faculty of Electrical Engineering and Informatics, University of Pardubice. The laboratory is equipped with six GUNT training systems mimicking common technological processes - level, flow, pressure, temperature, speed, and position control. The authors created their own software solutions to be able to operate GUNT training systems from MATLAB and Simulink through internal LabJack U12 data acquisition cards. Identification experiment, on-off, PID and MPC control are presented to demonstrate the functionality of the proposed solution.

Not all training systems are perfectly suitable for every laboratory experiments. GUNT RT 030 pressure control is quite nonlinear. GUNT RT 020 flow control is very fast compared to the sample time of 0.1 s. On the other hand, GUNT RT 040 temperature control is a slow system with a settling time around 1.5 h. Model and simulation come in handy for this system because it is impossible to use trial-error control design strategies, and everything must be planned and verified in advance. GUNT RT 010 level control time behavior is also quite slow and because of the dominant time constant easy to control. Our favorite systems are GUNT RT 050 speed control with second-order proportional behavior and GUNT RT 060 position control as a representative of an unstable system with an integrative character.

Thanks to the authors software support, the field of application is very broad and various laboratory tasks can be created not only in the field of process control, but also modelling and experimental identification.

Acknowledgment. This research was supported by the grant SGS_2022_014 at the Faculty of Electrical Engineering and Informatics, University of Pardubice.

References

1. GUNT: Automation and process control engineering. https://www.gunt.de/en/products/mechatronics/automation-and-process-control-engineering/glct-1:pa-148:ca-78. Accessed 2022/06/01
2. Tecquipment: Process control training systems. https://www.tecquipment.com/process-control. Accessed 2022/06/01
3. National Instruments: Teaching controls and mechatronics. https://www.ni.com/cs-cz/solutions/academic-research/teaching-controls-mechatronics.html. Accessed 2022/06/01
4. GUNT RT 010 Training System: Level control. http://www.gunt.de/index.php?option=com_gunt&task=gunt.list.category&lang=en&category_id=83&product_id=1042. Accessed 2022/06/01
5. GUNT RT 020 Training System: Flow control. http://www.gunt.de/index.php?option=com_gunt&task=gunt.list.category&lang=en&category_id=83&product_id=1043. Accessed 2022/06/01
6. GUNT RT 030 Training System: Pressure control. http://www.gunt.de/index.php?option=com_gunt&task=gunt.list.category&lang=en&category_id=83&product_id=1044. Accessed 2022/06/01
7. GUNT RT 040 Training System: Temperature control. http://www.gunt.de/index.php?option=com_gunt&task=gunt.list.category&lang=en&category_id=83&product_id=1045. Accessed 2022/06/01
8. GUNT RT 050 Training System: Speed control. http://www.gunt.de/index.php?option=com_gunt&task=gunt.list.category&lang=en&category_id=83&product_id=1046. Accessed 2022/06/01
9. GUNT RT 060 Training System: Position control. http://www.gunt.de/index.php?option=com_gunt&task=gunt.list.category&lang=en&category_id=83&product_id=1047. Accessed 2022/06/01
10. LabJack - U12: https://labjack.com/products/u12. Accessed 2022/06/01

Kinect-Based Evaluation of Severity of Facial Paresis: Pilot Study

Jan Kovarik(✉), Martin Schätz, Jakub Ciler, Jan Kohout, and Jan Mares

Department of Computing and Control Engineering, UCT Prague, Technicka 5, 16628 Prague 6 Dejvice, Czech Republic
jan.kovarik@vscht.cz
http://uprt.vscht.cz/en

Abstract. Facial paresis, temporary or permanent loss of facial muscle mobility, seriously affects the patient's physical and mental health. Patients suffer from difficulties with eating, salivating or speaking. Facial rehabilitation is usually the solution to facial paresis. However, during the rehabilitation facial muscle mobility must be monitored. Unfortunately, the assessment of facial mobility usually depends on the subjective opinion of the doctor. The aim of this pilot study was to design a method of objective identification of a healthy patient and a patient with facial paralysis. Using a depth sensor Kinect a few records of facial movements of health patient were taken for both symmetric and asymmetric movement. Using statistical methods the most suitable of 1347 facial landmarks were chosen for the classification model.

Keywords: Facial paresis · Kinect · Statistical methods · Biomedical data

1 Introduction

Facial paresis is a common disease caused by a number of reasons (stroke, brain tumor, Vestibular tumor, etc.). The patient has difficulty performing some routine activities. Moreover, it strongly affects patient's mental and social health.

Therefore, it is necessary to evaluate the facial muscular and nerve function as accurate as possible. There are some research groups dealing with this topic. [1] used the extracted facial temperature distribution features in the particular facial regions to evaluate facial nerve function. Some references deal with this topic, such as in the analysis of muscle fatigue detection the monitoring of the motor rehabilitation [2] or deep learning in muscle activity segmentation [3]. In addition, it has also been used to correlate muscle activity to EEG or EMG data [4].

The aim of this study is to find a set of facial landmarks which can be suitable for classification patients into groups according to the severity of paresis.

R. Silhavy et al. (Eds.): CoMeSySo 2022, LNNS 596, pp. 127–138, 2023.
https://doi.org/10.1007/978-3-031-21435-6_12

1.1 Biomedical Background

Facial muscles provide facial movements which play important role in ordinary life. For example, they are used for facial expressions, speaking or swallowing [5]. Facial muscles are innervated by the facial nerve (*nervus facialis* - cranial nerve VII) [6]. Facial paresis is caused by facial nerve palsy. It is usually one-sided. Causes of facial nerve palsy include stroke, injuries, brain tumors, but also infectious diseases (different types of *herpes* viruses). Another cause is a psychological condition of patient after trauma, depression or anxiety. We distinguished the central paresis which only affects the lower part of the face and the peripheral that affects the whole face [7,8]. The most frequent variant of the peripheral facial paresis is Bell's palsy. Fortunately, facial muscle mobility is partially maintained in most cases. Complete facial paresis is called facial plegia. Evaluation of severity of facial paresis is based on clinical monitoring using House-Breckmann scale [5]. Facial paresis is often treated by rehabilitation or medically. There is a high probability of cure or significant improvement within 6 weeks to three months [9].

2 Materials and Methods

2.1 Data Acquisition

Kinect v2 (including color camera, IR sensor and emitters) was used to obtain facial data. 1347 facial landmarks were tracked during the recording. A volunteer was asked to perform several exercises, once naturally and then to repeat the exercise simulating paresis of one part of the face in order to get two sets of recordings for further comparative analysis. A complete overview of the performed exercises is shown in Table 1, where N means health exercise and R refers to a simulated right paresis.

Table 1. Overview of performed exercises

Exercise	Variants
Baring teeth	N, R
Closing eyes	N, R
Static	N

All calculations and analyzes were performed using Matlab software and its add-on toolboxes. A new graphical user interface has been created to obtain recordings. Helping this GUI it is possible to completely operate the recording device. Moreover, some personal and health details about a volunteer and about a performed exercise may be written (see Fig. 1) and folders named after the special key containing this information will be created.

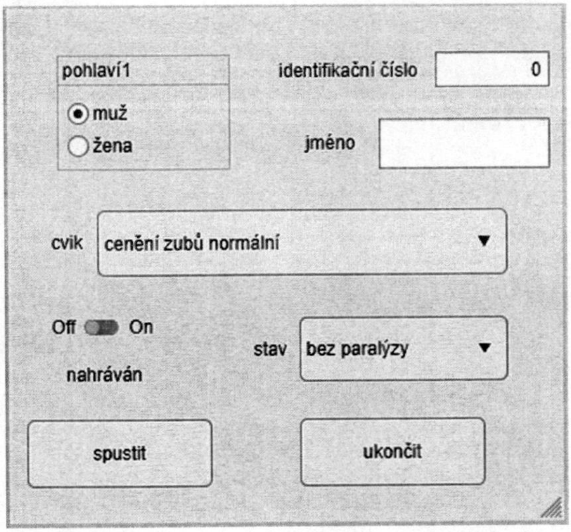

Fig. 1. Example of a dialogue in created GUI (Czech version). We can choose sex, wirte id number, type of exercise and choose whether a volunteer is heath, has facial paresis or simulates paresis

2.2 Data Pre-processing

In this step, it was important to find some key landmark pairs whose properties differ significantly in the case of healthy patients and patients suffering from facial paresis in order to determine which pairs are appropriate for use in the classification model. First, Euclidean 3D distances of all 1347 facial landmarks against all were calculated for each time frame. Thus, a matrix of size 1347*1347 symmetric according to the main diagonal was obtained for further analysis. I assumed that during exercises there are statistically more significant differences in variability of Euclidean 3D distances between landmark pairs than in central tendency. Due to the large differences in distances between pairs a variability needed to be standardized. Therefore, the coefficient of variation was chosen as an appropriate characteristic for assessing of the suitability of including the pair. The coefficient of variation of the Euclidean distance over time was calculated for each landmark pairs (see Eq. 1). The described calculation procedure was performed for both normal facial movement and movement with simulated paresis. The ratio of the coefficients of variation of movement simulating facial paresis and normal movement (CVR) was then calculated for each landmark pair (see Eq. 2).

This procedure is visualised in Fig. 2.

$$V(ed_3(j,k)) = \frac{sd(ed_3(i,j,k))}{mean(ed_3(i,j,k))}. \tag{1}$$

$$CVR(j,k) = \frac{V(ed_3(j,k))[asym]}{V(ed_3(j,k))[sym]}. \tag{2}$$

where $ed_3(i,j,k)$ are Euclidean 3D distances between selected landmarks (indexed by j and k) for ith time frame.

The characteristic proposed in this way is therefore supposed to reach a value of 1 for healthy movement.

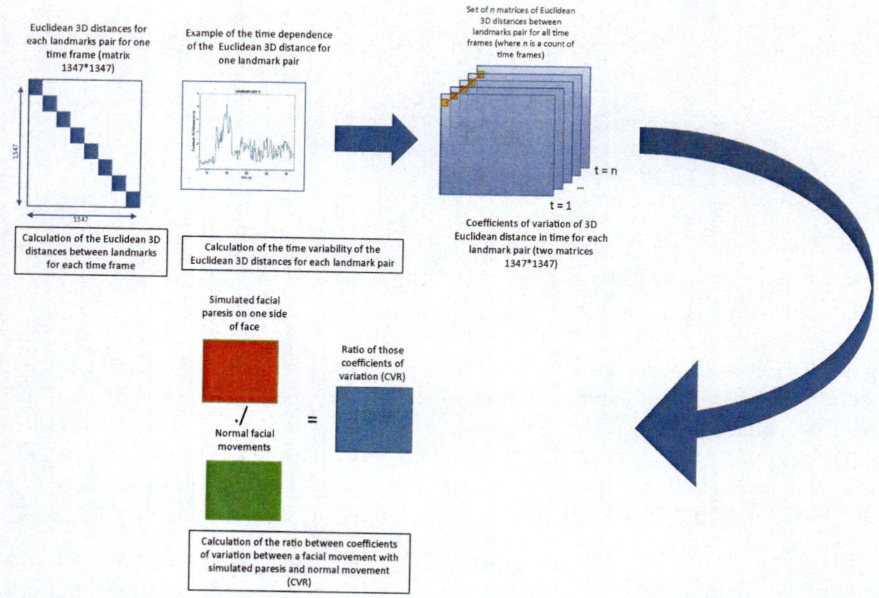

Fig. 2. Process of finding suitable landmark pairs for use as variables in the classification model

2.3 Data Processing

Sets of 30 pairs with the highest and 30 pairs with the lowest CVR for each type of exercise were selected as suitable for use in the classification model.

2.4 Statistical Modeling

The last row of the matrix contained a logical variable 0 resp. 1, depending on whether the respective column contained values from the time series symmetric, resp. asymmetric motion (see Fig. 3). Some of the 30 variables were experimentally included (using step-wise method) in several types of classification models.

One of methods is discriminant analysis which aims to classify an observation, or several observations, into the clusters that are known *a priori*. For classification a discriminant rule is used. Each cluster has its probability density depending on predictors. In general, a discriminant rule is given by allocating

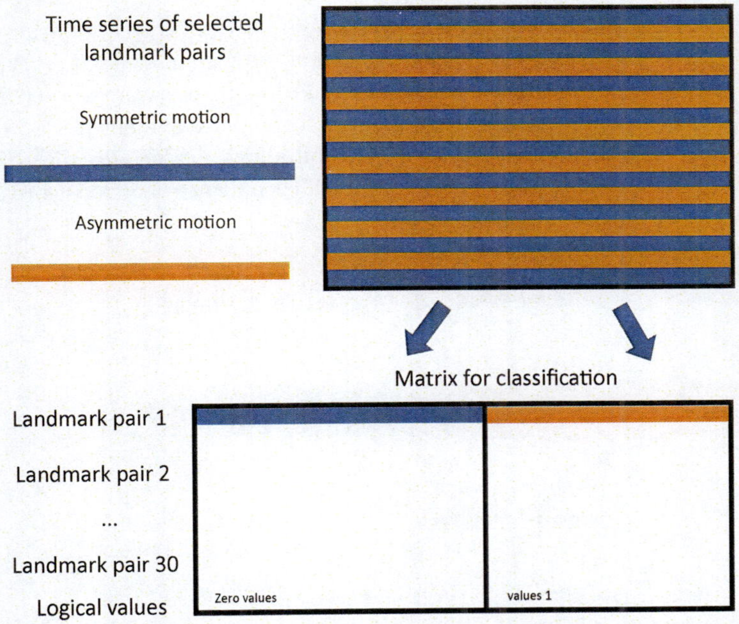

Fig. 3. Diagram that describes the formation of a matrix for classification models

an observation to the cluster by maximizing the density probability. There are many definitions of a discriminant rule. For instance, in case of maximum likelihood rule (ML rule) we aim to find a cluster with a maximum likelihood [10] (see Eq. 3).

$$L_j(x) > L_i(x). \tag{3}$$

where $L_j(x)$ and $L_i(x)$ are values of likelihood for clusters i and j. As it is evident, $L_j(x)$ is higher, so x will be allocated to the cluster j.

Next method is a method of k-nearest neighbours. In this method, there is a set of objects arranged in Euclidean space and some metrics (like Euclidean distances) are calculated. A fixed number k is voted and for new unknown objects k nearest objects are found. Unknown object is assigned to the category with maximum of objects represented between these neighbours. Finally, the model with the highest accuracy was selected for further analysis [11]. Nevertheless, after classification it was obvious that some landmark pairs were formed by landmarks in opposite dense clusters and their characteristics during movements were very similar. For selected landmarks, the correlation matrix of coefficients of variation was calculated.

3 Results and Discussion

At first, we received matrices of CVR for all exercises. As it was said, the landmark pairs those showed the highest or lowest CVR during the motion were supposed to be the best indicators of facial paresis. However, teeth baring seemed to be the best indicator of asymmetry of facial movements. As it is evident in boxplots (see Fig. 4), both extremely high and low values were observed in the matrix. Obtained matrices visualised using colormap are evident in Fig. 9. Warm colours mean higher values of ratio, cold colours lower values. In the colormaps there are a few dark red resp. Blue crosses which indicate that most of values in this column or row are higher resp. Lower.

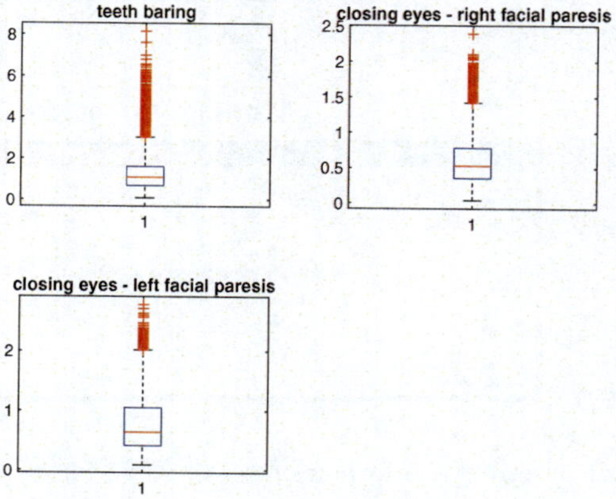

Fig. 4. Boxplots showing distribution of the ratio of standardized variability for all exercises

Teeth baring

In case of teeth baring, in particular, some pairs on the right side of the chin, the pairs connecting the right corner of the lips and the lower part of the face and the pairs between right cheek and forehead or right cheek and the nose were considered as the most suitable landmark pairs (see Fig. 5). Moreover, in the colormap (see Fig. 9) the lines corresponding to the landmarks located near the right corner of the lips and under the right eye were mostly colored blue, which means that almost all pairs with these landmarks had low CVR (see Fig. 6). This is consistent with the anatomy of facial muscles. Teeth baring is performed mainly with oral muscles (*musculus depressor anguli oris* or *musculus depressor labii inferior*) [8]. Thus, in the case of asymmetric motion, the lower mobility of these points must be evident.

On the other hand, the highest ratio of variability was observed at the pairs located on the lips around the left corner and further on the left cheek close to the junction of the left corner of the lips and the left nostril (see Fig. 5). This may correspond to the patients effort to compensate the asymmetry with increased effort on the other side of the face. Nevertheless, a few short pairs in the right nostril and on the lips around the right corner of the lips were also included

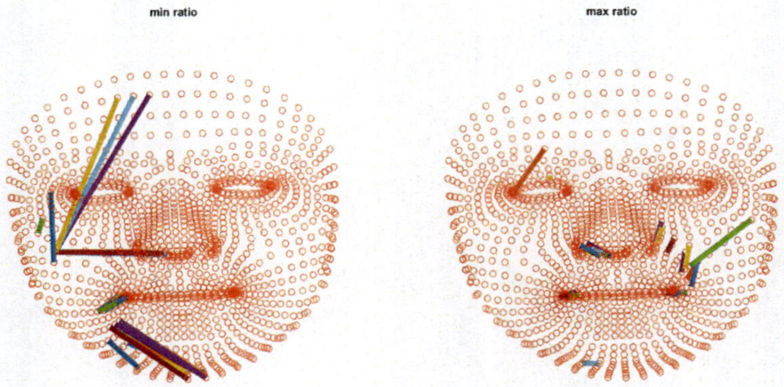

Fig. 5. Landmark pairs those reported the highest and the lowest ratio of coefficients of variation for both asymmetric and symmetric motion during baring teeth

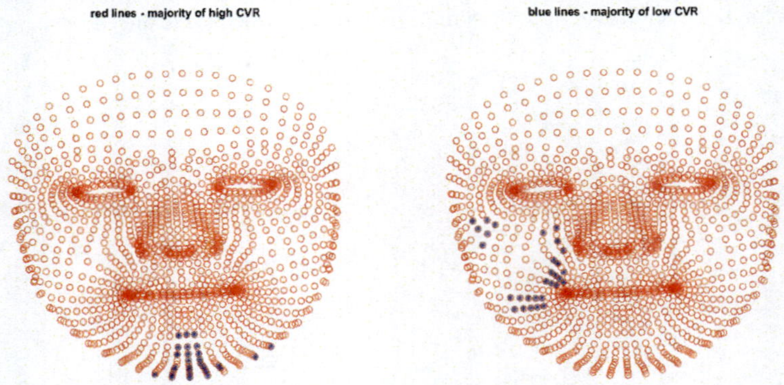

Fig. 6. Landmarks corresponding to those rows of the colormap that contained most of minimum resp. maximum values of CVR in case of baring teeth with rightside paresis

among the most suitable pairs (see Fig. 5) while the right side of the face was supposed to be motionless. This surprising fact can be explained by pulling the nostril and the corner of the lips through the oral muscle (*orbitularis oris*) from the left facial side. In the colormap red lines were observed at the landmarks in the area of the chin (see Fig. 6). These landmarks did not participate in the symmetric movement. But, they were slightly deviated from their position during asymmetric baring teeth. As a result, most of landmark pairs form with them showed higher variablity in case of the asymmetric movement.

Closing eyes with rightside facial paresis

During closing eyes with rightside facial paresis the landmark pairs above the right eyelid showed the lowest CVR (see Fig. 7). Hence, they seemed to be the best indicator of the rightside facial paresis. Furthermore, in the colormap the lines corresponding to the landmarks located above the right eyelid were mostly

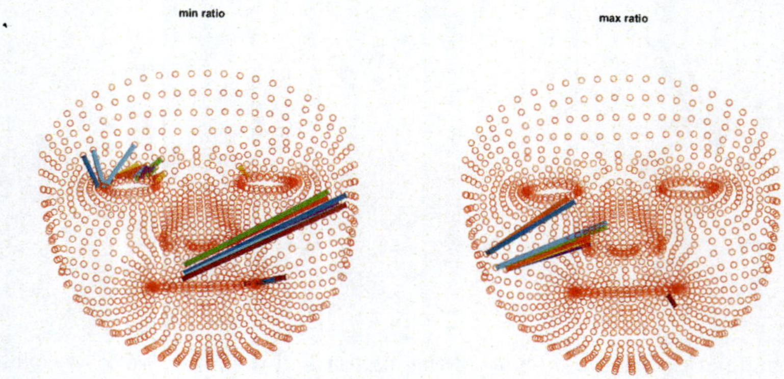

Fig. 7. Landmark pairs those reported the highest and the lowest ratio of coefficients of variation for both asymmetric and symmetric motion during closing eyes - right facial paresis

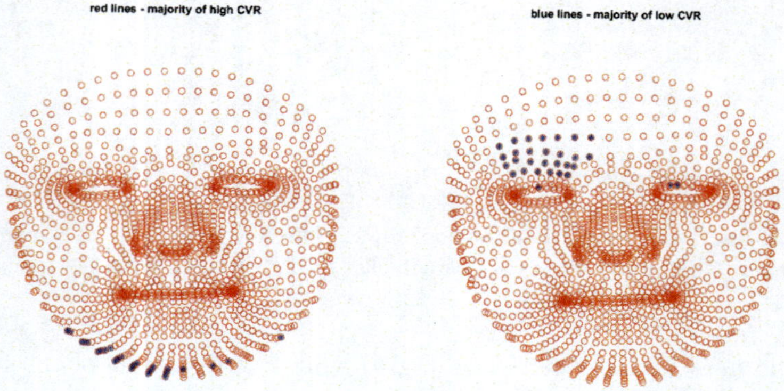

Fig. 8. Landmarks corresponding to those rows of the colormap that contained most of minimum resp. maximum values of CVR in case of closing eyes with rightside paresis

Table 2. Classification models

Exercise	Variables	Method	Accuracy (%)
Baring teeth	2	k-NN	97.2
	2	Quadratic discriminant	97.2
Closing eyes	2	k-NN	100

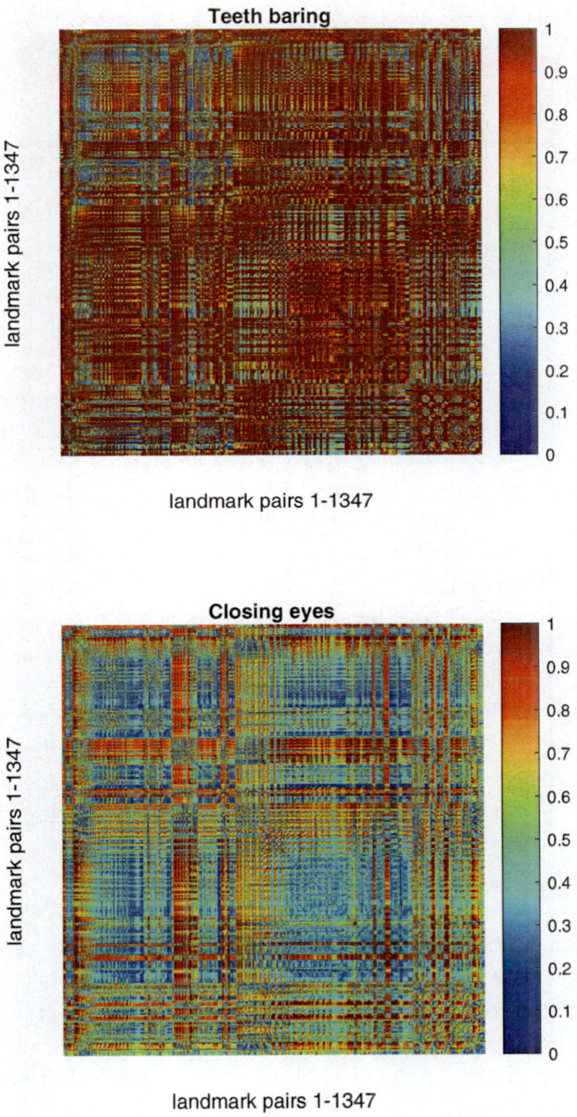

Fig. 9. Colormaps in which the intensity of CVR for all the landmark pairs can be seen for baring teeth and closing eyes

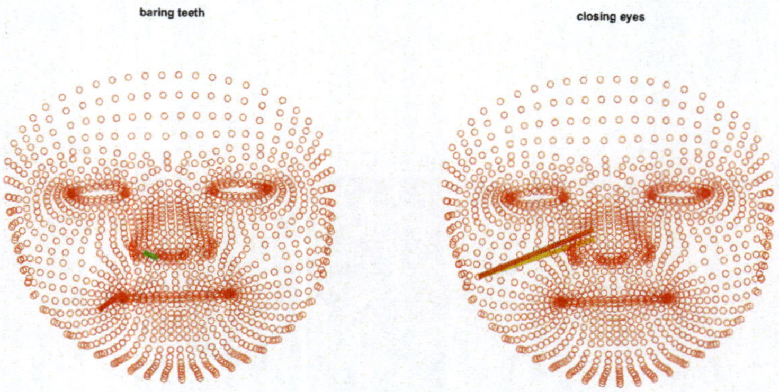

Fig. 10. Landmark pairs which were used to build a classification model with the highest accuracy

colored blue (see Fig. 8). This corresponds to the facial anatomy. However, the pairs connecting the area slightly left of the corner of the left eye and the midline groove in the upper lip (*philtrum*) also seemed as suitable landmark pairs.

Moreover, the maximum CVR occured particularly at the pairs between the nose and the right cheek (see Fig. 7). These facts are not consistent with facial anatomy because during the second exercise the right side of the face was considered motionless. Nevertheless, these results can be affected by wrong simulation of facial paresis performing by a health person. As in the previous exercise, in the colormap, red lines were observed at the landmarks located in the area of the chin (see Fig. 8). However, in contrast to the baring teeth during closing eyes a chin was not involved in either symmetrical or asymmetrical movement and these landmarks were indifferent to other landmarks. Contrarily, many landmarks were involved in the asymmetrical movement while during the symmetrical movement they were static. This may explain red lines in matrix for landmarks on the chin.

Classification models

Some of landmark pairs which were considered suitable in the previous analysis were included in the statistical models. Correlation matrix between individual variables was calculated in order to remove variables that are strongly correlated with each other. As a result, we reached the model with accuracy over 95 % for both baring teeth and closing eyes (Table 2). Chosen landmark pairs are evident in Fig. 10.

4 Conclusion

Helping a depth sensor Kinect one volunteer was recorded while performing several facial exercises, both healthy movement and movement with simulated paresis. Euclidean 3D distances between all landmarks during the exercises were calculated and they were analyzed in terms of time variability. Based on this analysis, a set of landmark pairs with the largest differences in temporal variability for symmetric and asymmetric motion was selected. Results were mostly explained by facial anatomy. Irregularities could be caused by wrong simulation of facial paresis. Moreover, it should be noted that the extreme value of CVR does not necessarily mean lower or higher mobility of selected landmark pair. Finally, a classification model with a high accuracy was compiled. In conclusion, suggested method using comparison between standardized variability between symmetric and asymmetric motion proved to be suitable for statistical modelling. For further analysis it would be interesting to suggest other indicators for finding suitable variables for classification model and capture recordings of more health volunteers.

Acknowledgement. The work was funded by the Ministry of Education, Youth and Sports by grant 'Development of Advanced Computational Algorithms for evaluating post-surgery rehabilitation' number LTAIN19007 and by grant A2-FCHI-2022-013. This support is gratefully acknowledged.

References

1. Liu, X., Dong, S., An, M., Bai, L., Luan, J.: Quantitative assessment of facial paralysis using infrared thermal imaging. In: 2015 8th International Conference on Biomedical Engineering and Informatics (BMEI). BMEI. pp. 106–110 (2015)
2. Elshafei, M., Costa, D.E., Shihab, E.: Toward the personalization of biceps fatigue detection model for gym activity: an approach to utilize wearables' data from the crowd. Sensors. **22**, 1454 (2022)
3. Nishiyama, D., Iwasaki, H., Taniguchi, T., Fukui, D., Yamanaka, M., Harada, T., Yamada, H.: Deep generative models for automated muscle segmentation in computed tomography scanning. PLoS One. **16**, e0257271 (2021)
4. Asghar, A., Jawaid Khan, S., Azim, F., Shakeel, C. S., Hussain, A. and Niazi, I. K.: Review on electromyography based intention for upper limb control using pattern recognition for human-machine interaction. Proc. Inst. Mech. Eng. **236**, 628–645 (2022)
5. Kohout, J., Verespejova, L., Kriz, P., Cervena, L., Sticha, K., Crha, J., Trnkova, K., Chovanec, M., Mares, J.: Advanced statistical analysis of 3D Kinect data: mimetic muscle rehabilitation following head and neck surgeries causing facial paresis. Sensors **21**, 103 (2020)
6. Rosahl, S., Bohr, C., Lell, M., Hamm, K., Iro, H.: Diagnostics and therapy of vestibular schwannomas-an interdisciplinary challenge. GMS current topics in otorhinolaryngology, head and neck surgery. **16** (2017)
7. Cervena, L., Kriz, P., Kohout, J., Vejvar, M., Verespejova, L., Sticha, K., Crha, J., Trnkova, K., Chovanec, M., Mares, J.: Advanced statistical analysis of 3D Kinect data: a comparison of the classification methods. Appl. Sci. **11**, 10 (2021)
8. Nguyen, J. D., Duong H.: Anatomy, head and neck, eye nerves. In: StatPearls. StatPearls Publishing, Treasure Island (FL) (2021)
9. Finsterer, J.: Management of peripheral facial nerve palsy. Eur Arch Otorhinolaryngol **265**, 743–752 (2008)

10. Härdle, W.K., Simar, L.: Applied multivariate statistical analysis. Springer Nature, Berlin (2019)
11. Clements, J.: K-Nearest Neighbors (K-NN) Explained. https://towardsdatascience.com. Last Accessed 19 June 2022

Explanation of the Predictive Controller and the Effect of Its Tuning on the Control Quality

Daniel Honc(✉), Aleš Novotný, and Libor Kupka

Faculty of Electrical Engineering and Informatics, University of Pardubice, Pardubice, Czech Republic
daniel.honc@upce.cz

Abstract. The Model Predictive Control (MPC) concept and its realization is explained together with some real-world laboratory application examples. Usually, future control errors and control action changes are penalized in the cost function (objective) by predictive controllers. The cost function can be seen as a function of future control actions utilizing the process model in a form of the predictor. A derivation of the predictors for the transfer function (external) and state-space (internal) model is indicated in the paper. An analytical solution to the given optimization problem is possible in an unconstrained case. A quadratic programming strategy must be used in case of the occurrence of the process constraints that should be respected by the controller. The authors apply both algorithms to two types of dynamical systems—to proportional (stable) system and to integrating (unstable) system and they demonstrate the influence of the penalization parameters (weights) in the cost function on the control quality. The authors aim to highlight the MPC strategy and its potential, and on the other hand, mention some bottlenecks and risks associated with model-optimisation-based methods.

Keywords: Model predictive control · Laboratory · Experiments · GUNT training systems

1 Introduction

Model Predictive Control is an optimisation-based control strategy in discrete time on a final horizon [1–3]. The aim of the control is to minimize a given cost function with respect to constraints. From this point of view, it is a feedforward control law respecting initial conditions (initial state). The receding horizon concept is applied to get a feedback feature. This means that state estimation (state measurement) and optimization procedure are repeated every sampling instant again and again, and only first from the calculated optimal control actions is used.

The MPC is able to handle complex, multivariable, nonlinear and time-varying systems under respecting process constraints [4–6]. It can guarantee stability and robustness or optimize economically oriented control objectives. Applications are no longer limited to slow processes like refineries, chemical industry, pulp and paper as it was in the 80s.

R. Silhavy et al. (Eds.): CoMeSySo 2022, LNNS 596, pp. 139–151, 2023.
https://doi.org/10.1007/978-3-031-21435-6_13

For example, autonomous vehicles, energy systems or robotics are new areas where MPC is successfully applied.

The process model is a key part of successful applications. It has not to be perfect, but its quality is linked to the requirements of the control response (for the tight control the model quality has to be higher, for lower requirements the approximation model could be fine). First-principles models and data-driven models are two extreme cases—the reality will be somewhere in between. Controller tuning is another decision which must be done. Trial-error method on a real system could be impossible for slow or yet not realized systems. Here, simulation and maybe another optimization come in handy to get reasonable initial parameters. The complexity of the processes can be very high, so some reduction is needed to be able to control fast systems online. Not only continuous decisions have to be made, but also logical control actions are considered frequently. Artificial intelligence and machine learning methods have the potential to solve some of the bottlenecks connected with predictive control.

The aim of the paper is to present simple MPC formulations, derivations and solutions for Single-Input Single-Output (SISO) processes. Then we show applications for two typical types of dynamical systems—stable and unstable and discuss tuning of the penalization coefficients (weightings) in the cost function (objective). The paper can serve as an introduction to MPC methods based on authors' experience with MPC teaching [7].

2 Materials and Methods

The cost function will be defined first. Predictors for two types of dynamical models will be derived. Optimisation problem solutions will be described for unconstrained and constrained cases. Two GUNT training systems will be introduced briefly at the end of the chapter.

2.1 Cost Function Definition

For a SISO system, the cost function (control objective) can be formulated as

$$J = \sum_{j=1}^{N_2} r_j \big(\hat{y}(k+j) - w(k+j)\big)^2 + \sum_{j=1}^{N_u} q_j \Delta u(k+j-1)^2 \tag{1}$$

Future control error (difference between predicted controlled process output and setpoint) is penalized in the first term, and the control action changes in the second term. We have to penalize control changes. Otherwise, the controller will leave a steady-state control error for proportional systems (stable systems). The cost function in matrix form is

$$J = (\mathbf{Y} - \mathbf{W})^T \mathbf{R} (\mathbf{Y} - \mathbf{W}) + \mathbf{U}^T \mathbf{Q} \mathbf{U} \tag{2}$$

where

$$\mathbf{Y}=\begin{bmatrix}\hat{y}(k+1)\\ \hat{y}(k+2)\\ \hat{y}(k+3)\\ \vdots\\ \hat{y}(k+N_2)\end{bmatrix},\quad \mathbf{W}=\begin{bmatrix}w(k+1)\\ w(k+2)\\ w(k+3)\\ \vdots\\ w(k+N_2)\end{bmatrix},\quad \mathbf{U}=\begin{bmatrix}\Delta u(k)\\ \Delta u(k+1)\\ \Delta u(k+2)\\ \vdots\\ \Delta u(k+N_u-1)\end{bmatrix}$$
$$\mathbf{R}=\begin{bmatrix}r_1 & 0 & 0 & \cdots & 0\\ 0 & r_2 & 0 & \cdots & 0\\ 0 & 0 & r_3 & \cdots & 0\\ & & \vdots & & \\ 0 & 0 & 0 & \cdots & r_{N_2}\end{bmatrix},\quad \mathbf{Q}=\begin{bmatrix}q_1 & 0 & 0 & \cdots & 0\\ 0 & q_2 & 0 & \cdots & 0\\ 0 & 0 & q_3 & \cdots & 0\\ & & \vdots & & \\ 0 & 0 & 0 & \cdots & q_{N_u}\end{bmatrix} \tag{3}$$

2.2 Predictions and Predictor for State-Space Model

It is easy to deal with Multi-input Multi-output (MIMO) systems in state-space, so the derivations are general for a system with nu inputs, nx states and ny outputs. For a SISO system $nu = ny = 1$.

The process model is

$$\begin{aligned}\mathbf{x}(k+1)&=\mathbf{A}\,\mathbf{x}(k)+\mathbf{B}\,\mathbf{u}(k)\\ \mathbf{y}(k)&=\mathbf{C}\,\mathbf{x}(k)\end{aligned} \tag{4}$$

We have to transform this model into a form with control action changes as input

$$\begin{aligned}\underbrace{\begin{bmatrix}\mathbf{x}(k+1)\\ \mathbf{u}(k)\end{bmatrix}}_{\mathbf{x}_p(k+1)}&=\underbrace{\begin{bmatrix}\mathbf{A} & \mathbf{B}\\ \mathbf{0}_{nu\times nx} & \mathbf{I}_{nu\times nu}\end{bmatrix}}_{\mathbf{M}}\underbrace{\begin{bmatrix}\mathbf{x}(k)\\ \mathbf{u}(k-1)\end{bmatrix}}_{\mathbf{x}_p(k)}+\underbrace{\begin{bmatrix}\mathbf{B}\\ \mathbf{I}_{nu\times nu}\end{bmatrix}}_{\mathbf{N}}\Delta\mathbf{u}(k)\\ \mathbf{y}(k)&=\underbrace{\begin{bmatrix}\mathbf{C} & \mathbf{0}_{ny\times nu}\end{bmatrix}}_{\mathbf{O}}\underbrace{\begin{bmatrix}\mathbf{x}(k)\\ \mathbf{u}(k-1)\end{bmatrix}}_{\mathbf{x}_p(k)}\end{aligned} \tag{5}$$

This means that we forward in front of the system discrete-time integrator, and the state of the system will be extended with the last control action.

One-step ahead state and output predictions can be expressed as

$$\begin{aligned}\mathbf{x}_p(k+1)&=\mathbf{M}\,\mathbf{x}_p(k)+\mathbf{N}\,\Delta\mathbf{u}(k)\\ \hat{\mathbf{y}}(k+1)&=\mathbf{O}\,\mathbf{x}_p(k+1)=\mathbf{OM}\,\mathbf{x}_p(k)+\mathbf{ON}\,\Delta\mathbf{u}(k)\end{aligned} \tag{6}$$

Two-step ahead state and output predictions can be expressed as

$$\begin{aligned}\mathbf{x}_p(k+2)&=\mathbf{M}\,\mathbf{x}_p(k+1)+\mathbf{N}\,\Delta\mathbf{u}(k+1)=\mathbf{M}^2\,\mathbf{x}_p(k)+\mathbf{MN}\,\Delta\mathbf{u}(k)+\mathbf{N}\,\Delta\mathbf{u}(k+1)\\ \hat{\mathbf{y}}(k+2)&=\mathbf{O}\,\mathbf{x}_p(k+2)=\mathbf{OM}^2\,\mathbf{x}_p(k)+\mathbf{OMN}\,\Delta\mathbf{u}(k)+\mathbf{ON}\,\Delta\mathbf{u}(k+1)\end{aligned} \tag{7}$$

k-step ahead state and output predictions can be expressed as

$$\mathbf{x}_p(k+j) = \mathbf{M}^j\,\mathbf{x}_p(k) + \sum_{i=0}^{j-1} \mathbf{M}^{j-i-1}\mathbf{N}\,\Delta\mathbf{u}(k+i)$$

$$\hat{\mathbf{y}}(k+j) = \mathbf{OM}^j\,\mathbf{x}_p(k) + \sum_{i=0}^{j-1} \mathbf{OM}^{j-i-1}\mathbf{N}\,\Delta\mathbf{u}(k+i) \tag{8}$$

Predictor in matrix form is

$$\underbrace{\begin{bmatrix} \hat{\mathbf{y}}(k+1) \\ \hat{\mathbf{y}}(k+2) \\ \hat{\mathbf{y}}(k+3) \\ \vdots \\ \hat{\mathbf{y}}(k+N_2) \end{bmatrix}}_{\mathbf{Y}} = \underbrace{\begin{bmatrix} \mathbf{ON} & \mathbf{0}_{ny\times nu} & \mathbf{0}_{ny\times nu} & \cdots & \mathbf{0}_{ny\times nu} \\ \mathbf{OMN} & \mathbf{ON} & \mathbf{0}_{ny\times nu} & \cdots & \mathbf{0}_{ny\times nu} \\ \mathbf{OM}^2\mathbf{N} & \mathbf{OMN} & \mathbf{ON} & \cdots & \mathbf{0}_{ny\times nu} \\ & & \vdots & & \\ \mathbf{OM}^{N_2-1}\mathbf{N} & \mathbf{OM}^{N_2-2}\mathbf{N} & \mathbf{OM}^{N_2-3}\mathbf{N} & \cdots & \mathbf{OM}^{N_2-N_u}\mathbf{N} \end{bmatrix}}_{\mathbf{G}}$$

$$\underbrace{\begin{bmatrix} \Delta\mathbf{u}(k) \\ \Delta\mathbf{u}(k+1) \\ \Delta\mathbf{u}(k+2) \\ \vdots \\ \Delta\mathbf{u}(k+\mathbf{N}_u-1) \end{bmatrix}}_{\mathbf{U}} + \underbrace{\begin{bmatrix} \mathbf{OM} \\ \mathbf{OM}^2 \\ \mathbf{OM}^3 \\ \vdots \\ \mathbf{OM}^{N_2} \end{bmatrix}}_{\mathbf{F}_p} \mathbf{x}_p(k) \tag{9}$$

$$\boldsymbol{Y} = \underbrace{\boldsymbol{GU}}_{\text{forced response}} + \underbrace{\boldsymbol{F}_p\boldsymbol{x}_p(k)}_{\text{free response } \mathbf{f}} \tag{10}$$

The first term is called forced response because it describes the response to actual and future control actions. The second term is called free-response because this response is a response to the actual state of the system—response of the system if we keep the last control action constant on a whole prediction horizon.

2.3 Predictions and Predictor for Transfer Function Model

With a transfer function, we also model the effect of the prediction error of the output predictions—we expect correlated prediction error with the polynomial C. The result is that we can use polynomial C as a controller parameter or filter (Fig. 1).

The model (process model and prediction error model) is

$$Y(z) = \frac{B}{A}U(z) + \frac{C}{\Delta A}E(z) \tag{11}$$

where delta means increment operator $\Delta = 1 - z^{-1}$, prediction error is $e(k) = y(k) - \hat{y}(k)$ and polynomials are $A = 1 + a_1z^{-1} + a_2z^{-2} + \cdots + a_nz^{-n}$, $B = b_1z^{-1} + b_2z^{-2} + \cdots + b_nz^{-n}$ and $C = 1 + c_1z^{-1} + c_2z^{-2} + \cdots + c_{nc}z^{-nc}$.

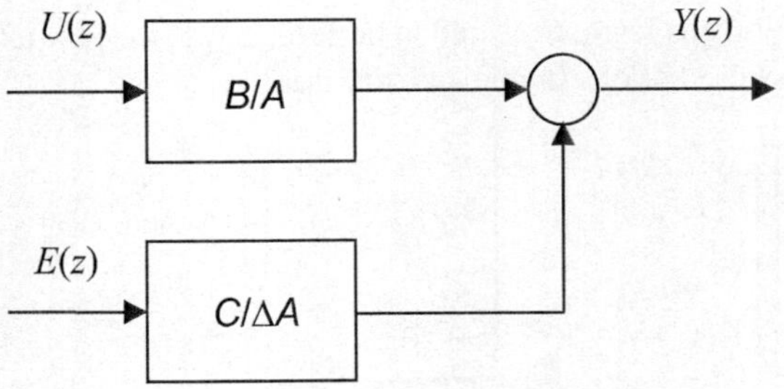

Fig. 1. Prediction model

The prediction model can be written as

$$\underbrace{\Delta A}_{\tilde{A}} Y(z) = B\Delta U(z) + CE(z) \tag{12}$$

where $\tilde{A} = \Delta A = 1 + \tilde{a}_1 z^{-1} + \tilde{a}_2 z^{-2} + \cdots + \tilde{a}_n z^{-n} + \tilde{a}_{n+1} z^{-(n+1)}$.

One-step ahead the output prediction can be expressed as

$$\begin{aligned} &\hat{y}(k+1) + \tilde{a}_1 y(k) + \tilde{a}_2 y(k-1) + \cdots + \tilde{a}_{n+1} y(k-n) \\ &= b_1 \Delta u(k) + b_2 \Delta u(k-1) + \cdots + b_n \Delta u(k-n+1) \\ &\quad + \underbrace{e(k+1)}_{0} + c_1 e(k) + c_2 e(k-1) + \cdots + c_{n_c} e(k-n_c+1) \end{aligned} \tag{13}$$

Two-step ahead the output prediction can be expressed as

$$\begin{aligned} &\hat{y}(k+2) + \tilde{a}_1 \hat{y}(k+1) + \tilde{a}_2 y(k) + \cdots + \tilde{a}_{n+1} y(k-n+1) \\ &= b_1 \Delta u(k+1) + b_2 \Delta u(k) + \cdots + b_n \Delta u(k-n+2) \\ &\quad + \underbrace{e(k+2)}_{0} + c_1 \underbrace{e(k+1)}_{0} + c_2 e(k) + \cdots + c_{n_c} e(k-n_c+2) \end{aligned} \tag{14}$$

N-step ahead the output prediction can be expressed as

$$\begin{aligned} &\hat{y}(k+N) + \tilde{a}_1 \hat{y}(k+N-1) + \tilde{a}_2 \hat{y}(k+N-2) + \cdots \\ &\quad + \tilde{a}_{n+1} \hat{y}(k+N-n-1) \\ &= b_1 \Delta u(k+N-1) + b_2 \Delta u(k+N-2) + \cdots \\ &\quad + b_n \Delta u(k+N-n) + \underbrace{e(k+N)}_{0} + c_1 \underbrace{e(k+N-1)}_{0} \\ &\quad + c_2 \underbrace{e(k+N-2)}_{0} + \cdots + c_{n_c} \underbrace{e(k+N-n_c)}_{0} \end{aligned} \tag{15}$$

Future prediction errors are expected to be zero. To get the predictor in a matrix form we have to express predictions in matrix form first

$$\underbrace{\begin{bmatrix} 1 & 0 & \cdots & 0 \\ \tilde{a}_1 & 1 & \cdots & 0 \\ & & \vdots & \\ 0 & 0 & \cdots & 1 \end{bmatrix}}_{A_p} \begin{bmatrix} \hat{y}(k+1) \\ \hat{y}(k+2) \\ \vdots \\ \hat{y}(k+N) \end{bmatrix} = \underbrace{\begin{bmatrix} b_1 & 0 & \cdots & 0 \\ b_2 & b_1 & \cdots & 0 \\ & & \vdots & \\ 0 & 0 & \cdots & b_1 \end{bmatrix}}_{B_p} \begin{bmatrix} \Delta u(k) \\ \Delta u(k+1) \\ \vdots \\ \Delta u(k+N-1) \end{bmatrix}$$
$$+ \underbrace{\begin{bmatrix} -\tilde{a}_1 & -\tilde{a}_2 & \cdots & -\tilde{a}_{n+1} \\ -\tilde{a}_2 & -\tilde{a}_3 & \cdots & 0 \\ & & & \\ 0 & 0 & \cdots & 0 \end{bmatrix}}_{A_m} \begin{bmatrix} y(k) \\ y(k-1) \\ \vdots \\ y(k-n) \end{bmatrix} + \underbrace{\begin{bmatrix} b_2 & b_3 & \cdots & b_n \\ b_3 & b_4 & \cdots & 0 \\ & & \vdots & \\ 0 & 0 & \cdots & 0 \end{bmatrix}}_{B_m} \begin{bmatrix} \Delta u(k-1) \\ \Delta u(k-2) \\ \vdots \\ \Delta u(k-n+1) \end{bmatrix}$$
$$+ \underbrace{\begin{bmatrix} c_1 & c_2 & \cdots & c_{n_c} \\ c_2 & c_3 & \cdots & 0 \\ & & \vdots & \\ 0 & 0 & \cdots & 0 \end{bmatrix}}_{C_m} \begin{bmatrix} e(k) \\ e(k-1) \\ \vdots \\ e(k-n_c+1) \end{bmatrix} \tag{16}$$

Then we have to multiply the whole equation with $\mathbf{A}_p{}^{-1}$

$$\underbrace{\begin{bmatrix} \hat{y}(k+1) \\ \hat{y}(k+2) \\ \vdots \\ \hat{y}(k+N) \end{bmatrix}}_{\mathbf{Y}} = \underbrace{\mathbf{A}_p^{-1}\mathbf{B}_p}_{\mathbf{G}} \underbrace{\begin{bmatrix} \Delta u(k) \\ \Delta u(k+1) \\ \vdots \\ \Delta u(k+N-1) \end{bmatrix}}_{\mathbf{U}} + \underbrace{\mathbf{A}_p^{-1}\mathbf{A}_m}_{\mathbf{F}_y} \underbrace{\begin{bmatrix} y(k) \\ y(k-1) \\ \vdots \\ y(k-n) \end{bmatrix}}_{\mathbf{Y}_m}$$
$$+ \underbrace{\mathbf{A}_p^{-1}\mathbf{B}_m}_{\mathbf{F}_u} \underbrace{\begin{bmatrix} \Delta u(k-1) \\ \Delta u(k-2) \\ \vdots \\ \Delta u(k-n+1) \end{bmatrix}}_{\mathbf{U}_m} + \underbrace{\mathbf{A}_p^{-1}\mathbf{C}_m}_{\mathbf{F}_e} \underbrace{\begin{bmatrix} e(k) \\ e(k-1) \\ \vdots \\ e(k-n_c+1) \end{bmatrix}}_{\mathbf{E}_m} \tag{17}$$

The first term is the forced response. The second, third and fourth terms create the free response.

$$\mathbf{Y} = \mathbf{G}\mathbf{U} + \mathbf{F}_y\mathbf{Y}_m + \mathbf{F}_u\mathbf{U}_m + \mathbf{F}_e\mathbf{E}_m$$

$$= \underbrace{\mathbf{GU}}_{\text{forced response}} + \underbrace{\underbrace{\left[\mathbf{F}_y\ \mathbf{F}_u\ \mathbf{F}_e\right]}_{\mathbf{F}_p} \cdot \underbrace{\begin{bmatrix} \mathbf{Y}_m \\ \mathbf{U}_m \\ \mathbf{E}_m \end{bmatrix}}_{\mathbf{x}_p}}_{\text{free response } \mathbf{f}} \tag{18}$$

2.4 Problem Solution

If we substitute the predictor (10) or (18) into the cost function (2), we get

$$\begin{aligned} J &= (\mathbf{GU} + \mathbf{f} - \mathbf{W})^T \mathbf{R}(\mathbf{GU} + \mathbf{f} - \mathbf{W}) + \mathbf{U}^T \mathbf{QU} \\ &= \mathbf{U}^T \underbrace{\left(\mathbf{G}^T \mathbf{RG} + \mathbf{Q}\right)}_{\mathbf{H}} \mathbf{U} + \mathbf{U}^T \mathbf{G}^T \mathbf{R} \underbrace{(\mathbf{f} - \mathbf{W})}_{\mathbf{g}} + \underbrace{(\mathbf{f} - \mathbf{W})^T \mathbf{RG}}_{\mathbf{g}^T} \mathbf{U} \\ &\quad + \underbrace{(\mathbf{f} - \mathbf{W})^T \mathbf{R}(\mathbf{f} - \mathbf{W})}_{k} \end{aligned} \tag{19}$$

which is quadratic form

$$J = \mathbf{U}^T \mathbf{HU} + 2\mathbf{g}^T \mathbf{U} + k \tag{20}$$

It can be optimised by standard quadratic programming methods in the case of the inequality process constraints $\mathbf{AU} \leq \mathbf{b}$ or analytically for unconstrained cases—we have to calculate a derivation according to the vector $\mathbf{U}$, put it to $\mathbf{0}$ and solve the equation

$$\frac{dJ}{d\mathbf{U}} = \mathbf{HU} + \mathbf{g} = 0 \tag{21}$$

The result is

$$\mathbf{U} = -\mathbf{H}^{-1}\mathbf{g} = \left(\mathbf{G}^T \mathbf{RG} + \mathbf{Q}\right)^{-1} \mathbf{G}^T \mathbf{R}\big(\mathbf{W} - \mathbf{F}_p \mathbf{x}_p(k)\big) = \mathbf{L}\big(\mathbf{W} - \mathbf{F}_p \mathbf{x}_p(k)\big) \tag{22}$$

Actual control move is

$$\Delta u(k) = \mathbf{K}\big(\mathbf{W} - \mathbf{F}_p \mathbf{x}_p(k)\big) \tag{23}$$

where $\mathbf{K}$ is the first row of matrix $\mathbf{L}$ for the SISO case or the first nu rows for the MIMO case.

The controller algorithm can be seen as negative feedback, which is proportional to the difference between the future setpoint and the free response. In simple words, if the process will behave without our intervention as we want the control change can be zero—we can keep the same control action as it was in the previous sample time instant (Fig. 2).

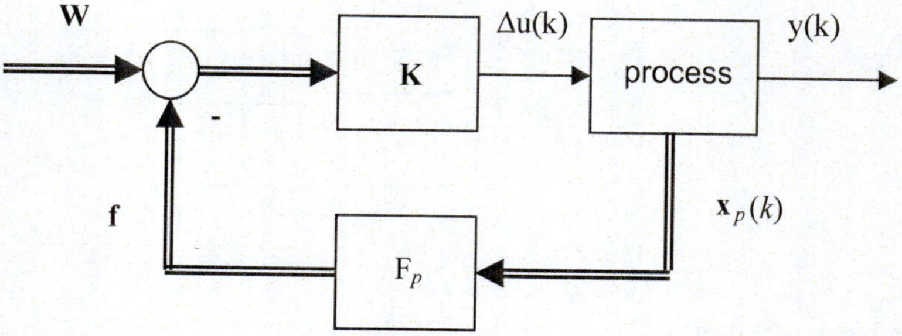

Fig. 2. MPC control law

2.5 GUNT Training Systems

The predictive controllers will be applied to two GUNT training systems. RT 050—speed control [8] is representative of the proportional system (stable system, with steady-state) and, on the contrary RT 060—position control [9] is an integrating system (unstable system).

In RT 050 the voltage of the DC motor (signal Y) controls its speed. The motor is connected to the generator, and a flywheel is placed on the axis. The speed (signal X) is measured by an inductive sensor. The generator (signal Z) can be used to change the load (Fig. 3).

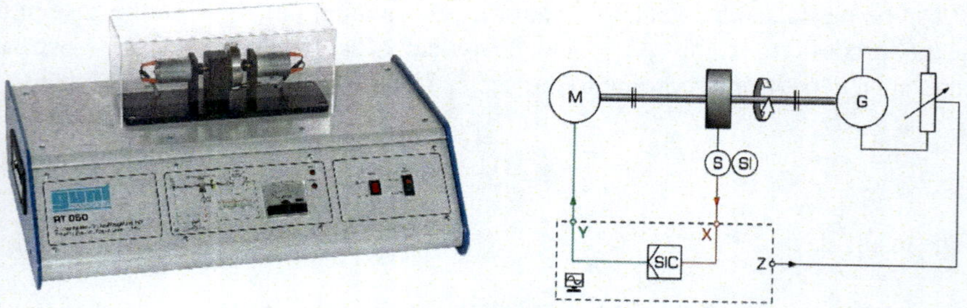

Fig. 3. RT 050—speed control [8]

In RT 060 the motor (signal Y) and the toothed belt control the position of the trolley, which is sensed by the rotary encoder (signal X). The system is equipped with limit switches that switch off the motor when the trolley reaches the end of the track (Fig. 4).

3 Results and Discussion

Both the MPC algorithms were applied to the GUNT training systems RT 050—speed control and RT 060—position control. Parameters of the controllers are in Table 1. The influence of the control change penalization (weighting in the control objective) to the control quality was tested in the following experiments. Two values of the penalization

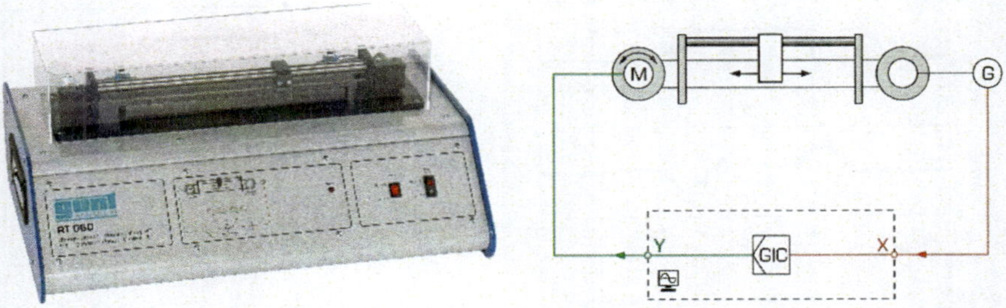

Fig. 4. RT 060—position control [9]

were used 1 and 0.01. They are quite far from each other to demonstrate clearly their influence.

Table 1. MPC parameters

Parameters	Value
Sample time, s	0.1
Prediction horizon, s	2
Control horizon, s	2
Control error penalization	1
State observer poles	[0.6 0.6]
Filtering polynomial	$(1-0.8z-1)^2$

The simulated control responses are on the left side of the figures, while the real-world experiments are on the right side. The bottom part graph is the load calculated as a percentage of the time spent in the program from the whole sampling period.

3.1 RT 050—Speed Control—State-Space MPC

See Figs. 5 and 6.

3.2 RT 050—Speed Control—Transfer Function MPC

See Figs. 7 and 8.

3.3 RT 060—Position Control—State-Space MPC

See Figs. 9 and 10.

3.4 RT 060—Position Control—Transfer Function MPC

See Figs. 11 and 12.

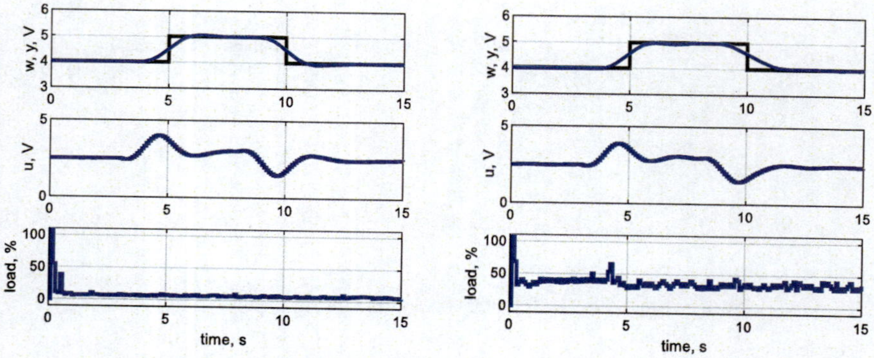

Fig. 5. RT 050, State-space MPC, control change penalization 1

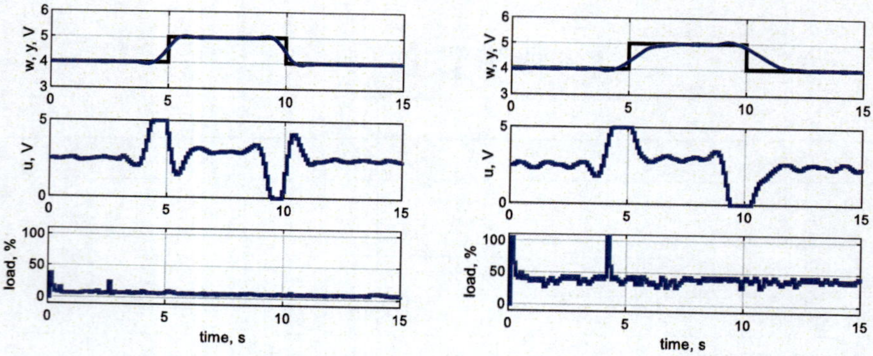

Fig. 6. RT 050, State-space MPC, control change penalization 0.01

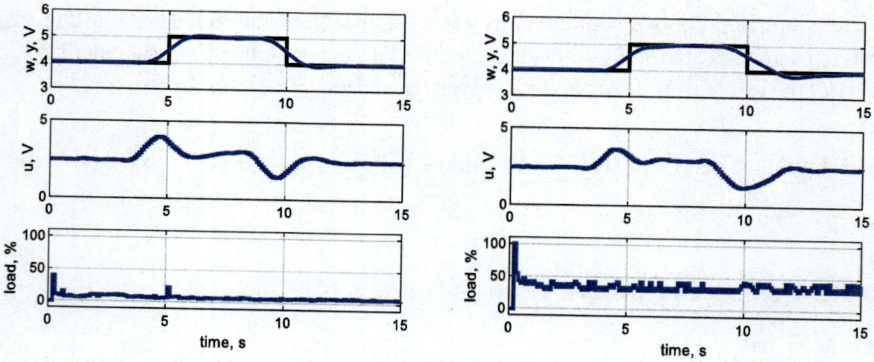

Fig. 7. RT 050, Transfer Function MPC, control change penalization 1

4 Conclusion

Two types of predictive controllers are explained, derived and applied to two training systems. The cost function and the solution is the same for both controllers. The predictors based on state-space and transfer function process models are different. In the

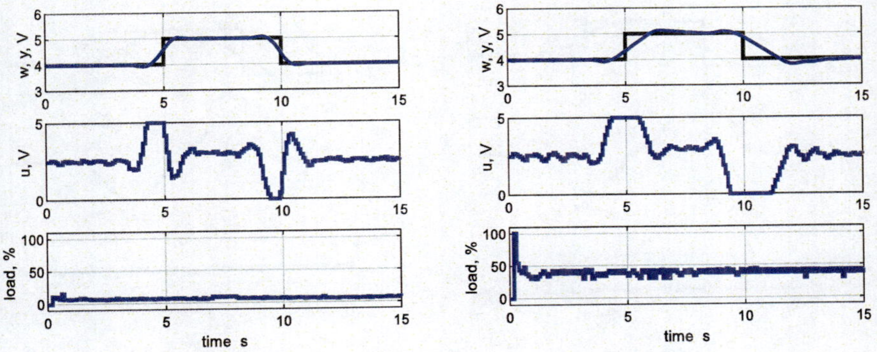

Fig. 8. RT 050, Transfer Function MPC, control change penalization 0.01

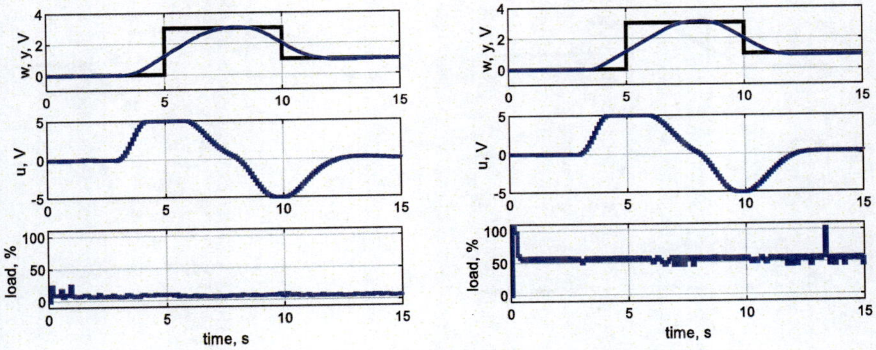

Fig. 9. RT 060, State-space MPC, control change penalization 1

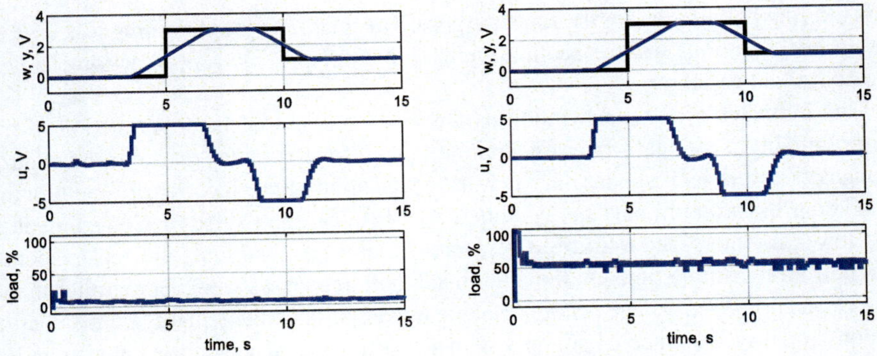

Fig. 10. RT 060, State-space MPC, control change penalization 0.01

state-space version, the state used for free response calculation consist of state variables and the last control action. A transfer function predictor uses a history of outputs, control changes and prediction errors instead of the state vector. Noise filtering is possible because of the state estimation in the state-space version. The transfer function controller

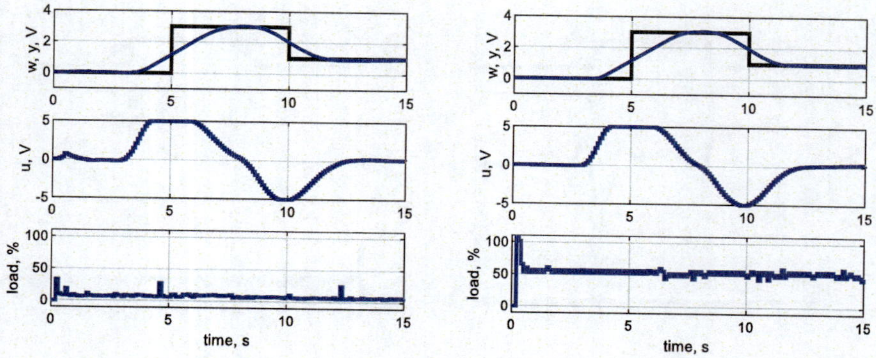

Fig. 11. RT 060, Transfer Function MPC, control change penalization 1

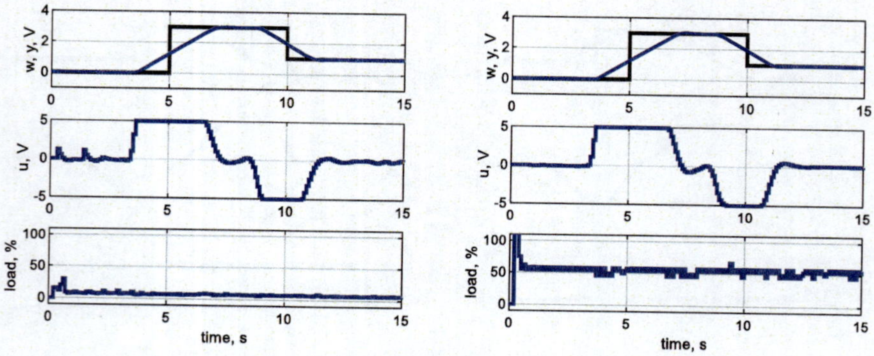

Fig. 12. RT 060, Transfer Function MPC, control change penalization 0.01

uses filtering polynomial for the same purpose. The quadratic programming was used to get optimal solutions with respecting limits on control actions, control action changes and controlled variables.

The influence of the penalization parameters (weights) in the cost function of the control quality is straightforward in simulations. A smaller penalization of the control changes gives a faster response. In real-world applications, the effect was different for the speed control (proportional) and position control (integrating system). Control quality was lower for the speed control system and smaller penalization—the real response differs from the simulation. On the other hand, both responses were very similar in the case of position control. The result is that the impact of the model quality on the control response quality is connected not only with a cost function penalization but also with a type of the system. The speed control is a second-order system while the position control behaves like a first-order system with an integrator and generally it is easier to control a system with lower order.

Acknowledgment. This research was supported by grant SGS_2022_014 at the Faculty of Electrical Engineering and Informatics, University of Pardubice.

References

1. Camacho, E.F., Bordons, C.: Model Predictive Control. Springer, Berlin (2007)
2. Maciejowski, J.: Predictive Control with Constraints. Prentice Hall, Hoboken (2002)
3. Rossiter, J.A.: Model-based Predictive Control—A Practical Approach. CRC Press, Boca Raton (2003)
4. Qin, S.J., Badgwell, T.A.: An overview of nonlinear model predictive control applications. In.: Nonlinear Predictive Control, pp. 369–392. Springer, Berlin (2000)
5. Allgöwer, F., Findeisen, R., Nagy, Z.K.: Nonlinear model predictive control: from theory to application. J. Chin. Inst. Chem. Eng. **35**(3), 299–315 (2004)
6. Magni, L., Raimondo, D.M., Allgöwer, F.: Nonlinear Model Predictive Control—Towards New Challenging Applications, Lecture Notes in Control and Information Sciences, vol. 384. Springer, Berlin (2009)
7. Honc, D., Sharma, K.R., Abraham, A., Dušek, F., Pappa, N.: Teaching and practicing model predictive control. IFAC-Pap. **49**(6), 34–39 (2016)
8. GUNT RT 050: Training system: speed control. http://www.gunt.de/index.php?option=com_gunt&task=gunt.list.category&lang=en&category_id=83&product_id=1046. Last accessed 1 June 2022
9. GUNT RT 060: Training system: position control. http://www.gunt.de/index.php?option=com_gunt&task=gunt.list.category&lang=en&category_id=83&product_id=1047. Last accessed 1 June 2022

Learningchain. A Novel Blockchain-Based Meritocratic Marketplace for Training Distributed Machine Learning Models

Enrico Zanardo(✉)

Department of Mathematics and Computer Science, Via Pietro Bucci, Università Della Calabria, Rende, (CS), Italy
enrico.zanardo@mcast.edu.mt

Abstract. Beez blockchain is a decentralized AI model network that encourages peer collaboration to improve machine learning accuracy. It's a peer-to-peer network where peers can submit models with improved precision. The system maintains a decentralized, append-only ledger to record critical information, such as who trained and improved the model, when and by how much, and where to find the most recent model. Beez gives cryptographic tokens to contributors. To demonstrate it, in this paper, we designed a simple Wallet API that isolates low-level node communication across the temporal cluster of compute nodes and enables fine-grained parameter-sharing behavior customization. We created a low-overhead communication layer using a custom blockchain to help developers build computational clusters. This API connects model training and parameter pooling. We want that Beez operations to be stateless and independent of ML. Medicine, reconnaissance, and resource-limited situations benefit from decentralized learning. Decentralization lets nodes learn independently and cooperatively without sharing data. To capitalize on this developing technology, decentralized learning algorithms may differ in clustering connection and individual node sharing. A high-level learning framework can improve codebase uniformity and readability. This open-source library accelerates decentralized learning research. Decentralized learning is still young, so many issues must be addressed before cooperative learning technologies become mainstream.

Keywords: Blockchain · Artificial intelligence · Distributed systems · Decentralized consensus

1 Introduction

AI and blockchain are hot topics. Its popularity is well-deserved. AI can automate and model complex situations better than humans. Blockchain technology protects data and reduces centralized authority. Combining blockchain and AI may seem like a scam ICO (Initial Coin Offering). However, these two technologies can coexist. This paper suggests prospective AI may use blockchain-based

R. Silhavy et al. (Eds.): CoMeSySo 2022, LNNS 596, pp. 152–169, 2023.
https://doi.org/10.1007/978-3-031-21435-6_14

datasets and distributed computing. Blockchains could use AI to monetize user-controlled data and create autonomous organizations. Today's paper compares these two technologies. In the coming decades, we'll see and discuss technology differently. Blockchain's data storage and retrieval limitations affect AI. Smart contracts and decentralized organizations are easier to code manually than AI.

Divergent Philosophies. Before analyzing blockchain and AI together, let's first examine why it's so difficult. Blockchain and AI are very different. They pursue similar goals in diametrically opposed ways. Contrasting philosophies makes possible combinations more appealing. It's the biggest barrier to creating new apps.

Centralized Versus Decentralized. First, massive data stores and computing power are used to train AI algorithms an AI's power structure differs from blockchain's. AI benefits companies with more data and resources. They benefit from AI advancements and can invest more in better AI. AI is centralized and it consolidates data and computing. AI will enrich the wealthy and increase income inequality for the poor, who can't use it. Blockchain may solve AI's philosophical dilemma. Blockchain ledgers decentralize data and computational resources while maintaining network access. Decentralization increases network latency, so blockchain ledgers must be sped up to work with AI. The idea that customers could rent their data or processing power to large businesses is encouraging.

Distributed Computing for Artificial Intelligence. Combining blockchain and AI by using mining networks to train algorithms is promising. In AI and machine learning, training algorithms requires a lot of CPU power to calculate dozens or millions of training sessions. Machine learning is slowed by algorithm training. AI researchers may pay more than crypto miners. Future algorithm training may benefit from GPU networks that power proof-of-work. If these training networks are successful and cost-effective, everyone could learn algorithms. AI algorithms need centralized server farms. Through decentralized GPU mining networks, everyone could train algorithms on a supercomputer.

An Online Market for Algorithms. AI offers community-driven growth when computation and data are democratized. Blockchain-based data and computing may accelerate this trend, creating a market for algorithms. Businesses may pay data scientists for algorithm insights or buy the algorithm for ongoing use. Artificial intelligence development could be an open marketplace where anyone can create value. Open source and as-a-service software and infrastructure promote innovation more than private development.

Integration Trends for Blockchain and Artificial Intelligence. AI and blockchain could enable future innovations. Understanding blockchain and AI trends may help justify the combination. AI trends would improve blockchain functionality by adding the following:

- Protection of data;
- Smart computing power;
- Monetization of data;
- Creation of diverse data sets;

Blockchain technology may help understand AI. It debugs machine learning decision-making. Blockchain technology may help AI algorithms document every data point and variable used in decision-making. Blockchain AI convergence boosts blockchain efficiency. AI can automate blockchain management tasks like audit trail monitoring.

2 Research Objective

This paper has two objectives. Our goal is to create a scalable and efficient custom blockchain by improving machine learning model submission to network nodes. This goal addresses the introduction's costly and ongoing transaction. Second, we want to give users a well-trained model trained by a group of peers who complete the task. The paper addresses how datasets can be used securely in the blockchain and how peers can train machine learning models using them.

Research Question. As discussed, numerous studies have examined AI integration into blockchain technology. These two technologies haven't been combined thoroughly yet. So, here's the research question:
[RQ]: Is it possible to train machine learning models without having to hold the complete dataset or owning dedicated computational power?

Machine learning develops application prediction models. These models' accuracy increases when trained on a large amount of data from multiple sources. Large-scale data collection raises privacy concerns. This paper provides a blockchain-based learning platform for machine learning models so peers can create accurate models interactively.

Significance of the Study. Machine learning requires optimization. In many cases, it must be distributed to finish on time. In machine learning, minimizing a loss function helps find optimal model parameters given data. First order optimization methods like stochastic gradient descent (SGD) are often used to improve the loss function of a deep neural network. By dividing gradient computation into minibatches, we can speed up these algorithms. Each minibatch update transfers the full model parameters to and from each node in the network.

3 Literature Review

This research aims to show how blockchain technology and AI models can be combined to benefit both. In this section, we'll summarize blockchain research, its relevance, and its interaction with AI.

Li et al. [8] proposed a decentralized blockchain-based federated learning framework (BFLC). The system uses blockchain for model storage and exchange. To allow BFLC, they developed a novel committee consensus method that reduces consensus computing and harmful attacks. They tested the BFLC architecture using real-world datasets. Otoum et al. [10] combined federated learning and blockchain to protect data privacy and network security. They decentralize machine learning on end devices. A blockchain-based consensus mechanism adds a second layer of privacy to fog training. Blockchain consensus enables on-device machine learning without centralized data training or coordination. Korkmaz et al. [5] suggests ChainFL, a decentralized federated learning strategy that uses blockchain technology to delegate model storage to network nodes. Haque et al. [3] presents a decentralized transfer learning mechanism using smart contracts and private blockchain. Data scientists, deep learning experts, dataset holders, and deep learning infrastructure providers will be integrated into a trustless deep learning analytics solution. The platform will allow the decentralized exchange of data, deep learning expertise, and processing capacity. Short et al. [12] developed a blockchain network, smart contract, and model verification algorithm to coordinate training, record user performance, and deliver transparent rewards. Li et al. [7] investigates privacy and security in blockchain-based FL methods on scientific datasets to provide a road map of the issue's current state. Blockchain-based FL has grown significantly over the past five years, and blockchain technology is increasingly used to solve issues involving patient healthcare records, image retrieval, cancer datasets, industrial equipment, and economic data in IoT applications and smart environments. Kusi et al. [6] created a decentralized Proof-of-Work (aPoW) method for blockchain-based learning. In aPoW, differential privacy-based techniques are used to protect each party's data, and a secure domain against Byzantine attacks is proposed. Through testing, they introduced deepLearning Chain on the blockchain and proved its performance. Habib ur Rehman et al. [11] proposed fine-grained FL to decentralize ML models on edge servers. Customization, decentralization, fine-grained FL, incentive mechanisms, trust, activity monitoring, heterogeneity and context-awareness, and model synchronization were also outlined. To provide reliable collaborative training, they proposed blockchain-based reputation-aware fine-grained FL for mobile edge computing systems. Tibúrcio and Spohn [13] describes an adaptive P2P computer grid. Ad hoc Grid uses open-source P2P grid middleware OurGrid (OG). OG's centralized administration prevents the deployment of ad hoc computer grids. New strategies for Peer discovery, failure management, and failure recovery made the P2P grid adaptable and self-organizing. Grid nodes communicate via multicast: a set of grid members are logically connected to a single Peer, and they coordinate through a local multicast group. Continuous data production and transmission to the cloud for model training and data processing raises the issue of data security. Zhang et al. [17] presented GPU-based ParSecureML to improve two-party secure machine learning algorithms. Complex computation patterns, frequent intra-node data transmission between CPU and GPU, and intricate inter-node data reliance are ParSecureML's main challenges. To address these issues, they proposed profiling-guided adaptive GPU use, fine-grained double pipeline for intra-node CPU-GPU cooperation and compressed inter-node communication. Xiao et al. [16] presented TuX^2, a distributed graph engine that combines graph computing and machine learning. TuX^2 inherits an elegant graph computation model, an efficient graph layout, and balanced parallelism to scale to billion edge graphs, while being optimized for distributed machine learning to support heterogeneity in data model, Stale Synchronous Parallel in scheduling, and a new Mini-batch, Exchange, GlobalSync, and Apply (MEGA) programming model. Nadeem et al. [9] proposed two machine learning-based ensemble workflow prediction techniques. Two Grid scenarios illustrate their method. Their method accurately estimates scientific process application execution time for many problem sizes, Grid sites, and runtime settings. Using process structure and execution environment attributes, they measured Grid workflow performance. Three strong learners balanced each other's weaknesses by modeling process execution durations. Wang et al. [15] presented Turbo for geo-distributed analytics with time-varying resources and network bandwidth across data centers. Turbo uses machine learning to predict query execution time. Turbo requires no changes to existing data analytics software. They built a real world Turbo prototype and tested it in eight Google Cloud Platform regions. Turbo estimates time costs with 95% accuracy and cuts query completion time by 41%. Tuncer et al. [14] suggested using machine learning to diagnose runtime performance issues. Their technology retrieves abnormalities from past resource data. They convert time series data into statistical features and extract anomaly signatures and determine required features. These characteristics detect real-time abnormalities. Using a real-world HPC supercomputer, they show their framework detects 98% of injected anomalies and outperforms existing methods. Cheng et al. [1] showed that dataset sample size affects concurrent machine learning scalability. They suggested comparing sample sizes using statistical properties like variance, sparsity, diversity, and sampling sequence similarity. They studied asynchronous parallel SGD (Hogwild), parallel model average SGD (minibatch SGD), decentralization optimization, and dual coordinate optimization (DADM algorithm).

4 Related Work

One of the reasons we need effective distributed computing is that we are accumulating more and more diverse data at a faster rate. The storage systems, data processing, and analytics engines that have arisen during the past decade are critical to the success of a great number of businesses. Surprisingly, the most majority of "big data" systems are designed for and controlled by (data) engineers, who are responsible for data collecting and processing activities. The goal is to enable data scientists to focus on their areas of expertise. To accomplish this, your chosen distributed system must support all these fine-grained "large compute" operations, possibly on specialized hardware. Ideally, it will also be compatible with the big data tool chain you employ and meet your latency needs. In other words, distributed computing must be powerful and flexible enough to handle complex data science workloads, and the purpose of this paper is to combine blockchain technologies and machine learning models to allow users to obtain better models and guarantee the integrity of the data and shared computational power using a dedicated blockchain and a fair rewarding system.

Data Integrity. Blockchain's digital record gives visibility into AI's basic structure and the source of the data it utilizes, so overcoming the difficulty of explaining Artificial Intelligence. This increases trust in the integrity of the data and, by extension, the Artificial Intelligence's suggestions. Using blockchain to store and distribute AI models generates an audit trail, and the combination of blockchain with AI can enhance data security.

Scaling. AI can rapidly and comprehensively read, comprehend, and correlate data, bringing a new level of intelligence to business networks based on blockchain technology. Blockchain helps AI to expand in order to give more meaningful insights, govern data consumption and model sharing, and develop a transparent and trustworthy data economy by enabling access to vast amounts of data from within and outside the company.

Efficiency. Multiple-party business processes can derive new value from artificial intelligence, automation, and blockchain by reducing friction, adding speed, and improving efficiency. For instance, AI models embedded in blockchain transactions executed in temporary distributed clusters can recommend recalling expired products, executing transactions such as reorders, payments, or stock purchases based on predetermined thresholds and events, resolving disputes, and selecting the most environmentally friendly shipping method

5 System Model

When most people hear the phrase "Data Science", they instantly think of Machine Learning and its increasingly sophisticated algorithms; nevertheless, Data Science would not be feasible without data, and the first criterion for broad usage is safe data storage and access. Prior to around ten years ago, Blockchain appeared as a new choice among the several pre-existing options. This notion requires serious consideration; in fact, because to its rising popularity, an increasing number of individuals are using the word "blockchain" without understanding its technical intricacies. Blockchain is a way for storing data securely via a peer-to-peer network of nodes that function without a central authority.

5.1 Formal Notions of the Beez Blockchain

Consider that that H represents the `SHA256` hash function we can than define:

Definition 1 (Transaction). *A transaction is the primary piece of data stored in a blockchain by definition. To paraphrase, the purpose of a blockchain is to securely store entries. Currently, a generic transaction is indicated by* `Tx`*. We picked this notation because, on our blockchain, a transaction involves an exchange between two parties.*

Since the Beez Blockchain is being considered, a transaction would consist of a simple record, such as when Enrico gives Rossella 22 tokens. By writing this, however, we highlight the initial challenges a blockchain such as blockchain faces: Does Enrico truly possess 22 tokens? Are we certain that Enrico began this exchange? To answer these problems, the next section will present the notion of validation procedure. In addition, we will see that it is possible to generate diverse types of transactions, such as Challenges, that allow users to request the network to train a particular machine learning model. Clearly, a blockchain is a sequence of blocks; thus, we must define what a block is.

Definition 2 (Block). *The definition of a block B is a vector of transactions. We denote its size B_{size}:*

$$\mathcal{B} = \{Tx_1, Tx_2, ..., Tx_{B_{size}}\}$$

To form a block, we therefore stack transactions. Using a parameter, the size $Bsize$ of transactions that can be contained within a Block in our blockchain can be specified. The default value for this parameter p, is 1, therefore:

$$\mathcal{B} = \{Tx\}$$

In order to "chain" the blocks, another concept is crucial: the consensus algorithm, in our case a customized version of `proof-of-stake`, sometimes denoted `PoS`. The first `Proof-of-Stakes` (PoS) network [4], `Peercoin`, was created as a `Proof-of-Exchange (PoX)` consensus mechanism to reduce the computational requirements of Proof-of-Work (PoW). Participants with a greater coin age, that is, the product of network tokens and their holding time, have a greater chance of being chosen. Each node in the `Peercoin` network solves a Proof-of-Work puzzle with its own difficulty, which can be reduced by consuming coin age. In modern PoS networks, solution searching has been eliminated entirely, and forgers are no longer chosen based on computational ability. Instead, they are chosen according to the stakes they hold. In our version of the PoS, the forger is also selected using a lottery function.

As seen in Table 1, with the stake-based forger selection process, a node's likelihood of being picked as a leader is no longer reliant on its computing capability; accordingly, the energy consumption of PoS mechanisms is significantly lower than that of PoW mechanisms. Moreover, PoW networks maintain relatively modest and consistent block creation and transaction confirmation rates

Table 1. PoW Versus PoS

	PoW	Beez PoS
Forger selection	Based on hash rate	Based on stake + lottery
Energy consumption	Significant	Negligible
Hardware requirement	High	None
Block generation speed	Slow	Fast
Transaction confirmation speed	Slow	Fast
Applications	Bitcoin, ethereum, etc	Algorand, EOS, etc

to ensure security, as miners offer a range of blocks. In contrast, because only one block is formed in each PoS round, block creation and transaction confirmation times are often substantially faster, and PoS techniques are therefore gaining popularity.

Definition 3 (Proof-of-Stake). *In spite of the lack of computational power resources, forgers are chosen based on their contributions to the blockchain network, or stakes. Particularly in the PoS consensus mechanism, a node's stake is the number of digital tokens it holds or deposits, such as coins in a cryptocurrency.*

In PoS networks, the $Block_{size}$ is typically larger, and the block time is significantly shorter, resulting in a transaction throughput of up to 875 Tx/s [2]. Moreover, in our blockchain it may attain immediate finality because the default parameter p is set to 1, resulting in a substantially reduced transaction confirmation time. In addition to network synchronization, the incentive mechanism is essential for the Proof-of-Stake consensus method's security.

Definition 4 (Hash Lottery). *Our customized consensus depends on the number of stakes that a node has staked and also on the chance-based lottery election model to pick the next forger in the network. Here, the next forger will broadcast his block to rest of the network.*

Identifying the "players" is the first stage in building the Beez blockchain. At the beginning of the chapter, we described a blockchain as a way for storing data without a central authority over a network of peer-to-peer nodes. Therefore, the main players are the nodes, and they are structured distributively (Fig. 1).

The objective of blockchain is to facilitate the implementation of secure distributed networks. It is difficult for such a network to be socially or financially acceptable without security. Blockchain is frequently referred to as a "trust revolution" owing to the absence of a central or semi-central authority in this context. In both centralized and decentralized systems, efforts are made to avoid the behavior of the determining node(s): in an ideal world, they would be impervious to corruption, cyber-attacks, and any other threat. The challenge in Blockchain is not that a node may be attacked, but rather how to ensure that the network

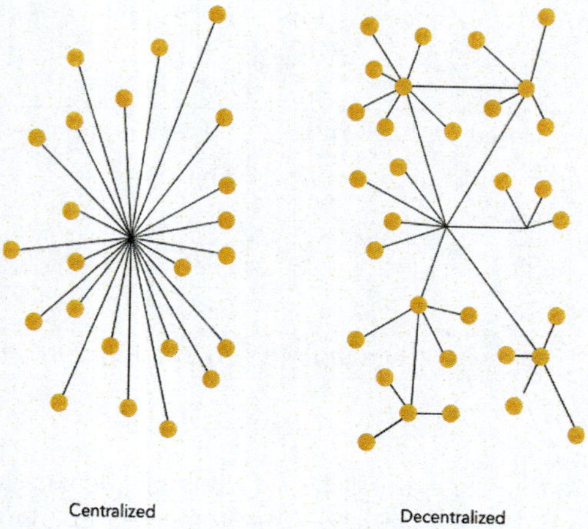

Fig. 1. Centralized, decentralized, distributed

continues to operate even if certain nodes are hacked. Simply said, Blockchain cannot inspire confidence; computational approaches must replace it. After this lengthy introduction, we may go into the building details of the Beez blockchain.

Definition 5 (Node). *Each orange dot in Fig. 1 symbolizes a node, which is the core element of a distributed network. Physically, a node is only an individual with many computers. There are various types of nodes: simple nodes, also known as wallets, whose function is to transmit information, generate transactions or challenges without holding a local copy of the blockchain; complete nodes, whose function is to mint new blocks and join to challenges coming from wallets to distributely train the given machine learning models; they maintain a local copy of the entire blockchain and share their computational power.*

Since there is no central authority, there is no node that would "wisely" preserve the real version of the blockchain; otherwise, the entire system would collapse: hacking the "wise" node would be sufficient to compromise the entire system. Each node retains its own copy of the blockchain; yet, owing to the architecture of the network, these nodes establish consensus.

STEP 1: A New Transaction Entry in the System. Let Tx a new transaction entry, such as when Enrico sends 22 tokes to Rossella; it enters the network through a node, denoted n, which could be a wallet or a full node. Each node has its own validation protocol (part of the Beez Blockchain), which is a program that verifies whether or not a transaction entry is valid.

Definition 6 (Validation Protocol). *A node does not trust the information it receives (there is no room for trust in Blockchain), so it employs its own validation protocol to perform a few checks. This can be viewed as a function:*

$$\begin{aligned} Validation_{protocol}\colon \mathbf{N} \times \mathbf{E} &\rightarrow \{True, False\} \\ (n, Tx) &\mapsto Validation_{protocol}(n, Tx) \end{aligned}$$

where $\mathbf{N}$ *is the set containing all the nodes, and* $\mathbf{E}$ *the set of all transactions.*

In our situation, n thus computes $Validation_{protocol}(n, Tx)$. If the response is True, the transaction is deemed genuine by n, and Tx is consequently sent to the neighboring nodes. If not, n will reject the item, and it will not be communicated. The present version of our blockchain validates that the transaction is covered (has enough tokens) and comes from the proper sender. If n is a node according to our notations, it will do certain checks: for our model, we suppose a node validates merely if the transaction is signed by its initiator and covered. Indeed, a node can guarantee that there is no duplicate spending since it maintains its own local copy of the `Account State Model` of the blockchain. By examining the history of all transactions, a node can respond to the following query. Does Enrico can transmit 22 tokens? If the transaction Tx is legitimate, n adds it to its local pool of valid transactions (a list). $L^n_{pool_{tx}}$:

$$L^n_{pool_{tx}}.append(Tx)$$

and then Tx is broadcast using the p2p protocol to the known nodes.

STEP 2: Check the Kind of Transaction. Currently, the Beez blockchain supports four types of transactions:

1. `Exchange`: The genesis node may send tokens to a node's wallet;
2. `Transaction`: Nodes may transmit tokens among themselves;
3. `Stake`: Nodes that want the opportunity to become the next forger node must stake tokens;
4. `Challenge`: Wallets that need to train machine learning models can generate a challenge transaction that requests the network to join the challenge and compute the training;

The first three types of transactions on a PoS Blockchain are the traditional transactions. In Beez Blockchain, the challenge transaction is used to ask nodes to share their computational power not to solve quizzes in order to mine a block (as in PoW), but to collaborate on training a particular machine learning model. A wallet w will send a challenge transaction Tx_c to the first connected node n. The connected node n, after the validation, will add this Tx_c to its pool of transactions and start to broadcast it. Moreover, each node n has its local challenge keeper that keep and track the changes that will be received on a given challenge c.

STEP 3: A New Block Has Been Emitted. If the node n is elected to be the next forger $n = n_{forger}$, it generates a new block by including some of the valid transactions in its local $L^n_{pool_{tx}}$ list:

$$B_n = (Tx_1, ..., Tx_p)$$

where $Tx_i \in L^n_{pool_{tx}}$ for $1 \leq i \leq p$ where by default $p = 1$. Then n_{forger} will add this block to its copy of the blockchain and broadcast it to other peers: since there is no space for trust in Blockchain, nodes do several checks. Let's designate n_r as the node that has just received B_{new}: It begins by validating that each transaction in B_{new} is legitimate. In the event that a transaction is invalid, B_{new} is refused. It is also easy to verify that the Merkle root and hash value of the new block are valid. In this instance, a node n must determine if B_{new} will be added to its local copy of the blockchain. The block is added if the hash of the previous block in the transmitted header $Head(B_{new})$ matches the hash of the most recent block on the local blockchain. In this instance, n deletes the transactions included within B_{new} from its local pool transaction list, and the forger who generated the new block gets rewarded.

STEP 4: Work on a Challenge. If the b_{new} contains a challenge transaction Tx_c than the node n will check if it can join the challenge.

$$J_{c_i} = \begin{cases} 1 & if \sum_n cpu + gpu \geq c_r \\ 0 & \text{otherwise} \end{cases} \tag{1}$$

where J_c represent the Boolean result to join or not the challenge, i represent the id of the challenge, *cpu* represent the node available CPU, *gpu* the available GPU and c_r the minimum challenge requirements that are set when the challenge is created. If a node (aka worker) accepts the challenge, it will initiate an iteration (epoch) of the specified machine learning model. The dataset is provided to the nodes via a link allowing them to download it. After the iteration is complete, the node updates its local version of the challenge that is stored by the challenge keeper and broadcasts the updated version to the temporal cluster of nodes that joined the challenge (within the model).

When all iterations/epochs have been completed, a final transaction will be added to the blockchain, and all nodes that participated in the challenge will be rewarded. Additionally, rewarding transactions are recorded on the blockchain. The wallet that generated the challenge will be able to decrypt and begin using the model. The pseudocode of the rewarding algorithm is reported in the Algorithm 1:

6 Experiments and Obtained Results

This section addresses the primary challenges that this paper wanted to demonstrate. A substantial amount of work must be completed in order to have a fully

Algorithm 1 Reward algorithm pseudocode

Input: Closed Challenge
Output: Reward tokens for each node that join the challenge

get the Challenge reward;
check that total iterations $= \sum_{1..n}^{Worker} iterations$
while $Worker$ in $Challenge_{workers}$ **do**
  calculate worker reward;
  create a Tx_{reward};
  broadcast Tx_{reward};

functional system. However, the primary functionality and objective have been attained and thoroughly tested. Here, we provide a practical response to the research question that this paper addresses in its totality.

Experiment 1: Forger Election. The verification of our customized version of the next forger election based on the lottery extraction is one of the first experiments that we have conducted. We conducted our tests by extracting 100 times 100 tokens from two wallets that had staked tokens. The wallets were balanced in the first trial, but not in the second.

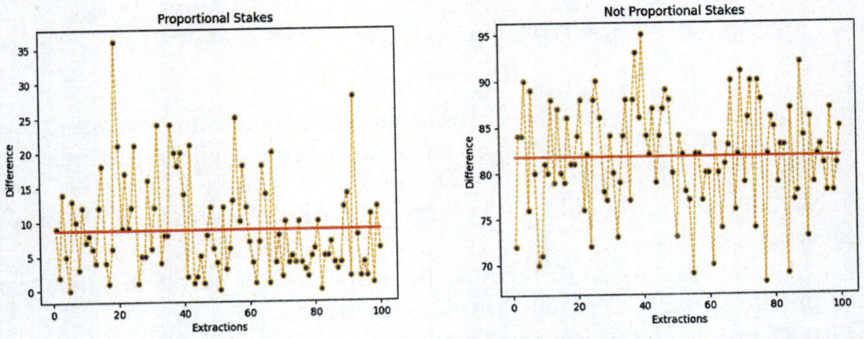

(a) Proportional Forger Election (b) Not Proportional Forger Election

Fig. 2. Forger election

As shown in Figs. 3a and 2b, the red line representing the mean is significantly lower in Fig. 3a than it is in Fig. 2b. This is due to the fact that in the figure on the left, the proportion of stakes stored by the two analyzed wallets is identical (100 tokens). In the right figure, the ratio between the two wallets is 1/10, which indicates that wallet 1 has staked 10 tokens while wallet 2 has staked 100 tokens. In any case, it is evident that, due to the lottery, the selection of the next forger is based not only on the amount of tokes a wallet stake, but also on the randomness of the lottery.

Experiment 2: Single Node Training Versus Distributed Training. This is the most relevant test performed within the beez blockchain. It represents the project's primary objective. We have trained the same simple machine learning model in two different ways, first on a single machine and then on blockchain nodes that have executed epochs until the defined number of epochs has been reached. The final model is then sent back to the wallet that requested it, and the transaction is recorded in the blockchain. In addition, based on the computational contribution of each node participating in the challenge (training the machine learning model), a reward transaction is sent to their wallets and recorded in the blockchain.

Train/Test Split. Initially, as demonstrated in Listing 1.1, we must divide the Iris dataset into target and features, or X and y. The column Name will serve as the target variable, while everything else will function as a feature (or predictor). I will also use a random seed so that you can replicate my results.

```
X = iris.drop("Name", axis=1).values
y = iris["Name"].values
X_train, X_test, y_train, y_test = train_test_split(X, y,
    test_size=0.2, random_state=42)
X_train = torch.FloatTensor(X_train)
X_test = torch.FloatTensor(X_test)
y_train = torch.LongTensor(y_train)
y_test = torch.LongTensor(y_test)
```

Listing 1.1. Train/Test Split

Defining a Neural Network Model. Regarding the model's architecture, it will be very straightforward. Consider the network's organizational structure:

1. Fully Connected Layer (4 input features (number of features in X), 16 output features (arbitrary));
2. Fully Connected Layer (16 input features (number of output features from the previous layer), 12 output features (arbitrary));
3. Output Layer (12 input features (number of output features from the previous layer), 3 output features (number of distinct classes));

We use `ReLU` (rectified linear unit) as activation function. In PyTorch there's a module `torch.nn.ReLU` gives us that we need. We use a learning rate of 0.01 as a setting for hyper-parameters, then observe its effect on the training procedure. We also utilize a Cross Entropy criterion (basically how we measure loss) and an Adam optimizer.

```
class ANN(nn.Module):
    def __init__(self):
        super().__init__()
        self.fc1 = nn.Linear(in_features=4, out_features=16)
        self.fc2 = nn.Linear(in_features=16, out_features=12)
```

```
        self.output = nn.Linear(in_features=12, out_features=3)

    def forward(self, x):
        x = F.relu(self.fc1(x))
        x = F.relu(self.fc2(x))
        x = self.output(x)
        return x

model = ANN()
criterion = nn.CrossEntropyLoss()
optimizer = torch.optim.Adam(model.parameters(), lr=0.01)
```

Listing 1.2. Machine Learning Model

Listing 1.2 show the basic machine learning model that is used to perform the test using the IRIS[1]

Model Training. As demonstrated in Listing 1.3, this part is incredibly simple. We train the model for one hundred epochs while monitoring time and loss. Every 10 epochs, we output the current status to the console, indicating which epoch we are on and the current loss.

```
epochs = 100
loss_arr = []
for i in range(epochs):
    y_hat = model.forward(X_train)
    loss = criterion(y_hat, y_train)
    loss_arr.append(loss)

    if i % 10 == 0:
        print(f'Epoch: {i} Loss: {loss}')

    optimizer.zero_grad()
    loss.backward()
    optimizer.step()
```

Listing 1.3. Training

The last three lines refer to the `backpropagation` consequently updating of weights and biases so that the model can "learn".

In Fig. 3b is a visual representation of our loss (epoch number on the x-axis and loss on the y-axis). Left represents single node training, while right represents distributed model learning. As we can see, there are no big differences whatsoever.

Model Evaluation. During the evaluation process, we wish to keep track of the model's predictions. It will be necessary to iterate over the `X_test` and

[1] source: https://datahub.io/machine-learning/iris/r/iris.csv dataset.

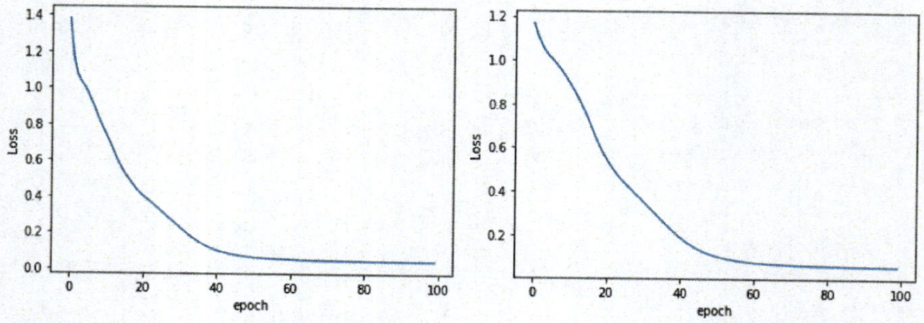

(a) Single node training

(b) Distributed training

Fig. 3. Loss propagation

make a prediction, which will then be compared to the actual value. We use `torch.no_grad()` because we are only evaluating and do not need to update weights and biases.

Conclusions. Wallets are able to generate requests in order to obtain a well-trained machine learning model for the IRIS dataset because the aforementioned machine learning model is accessible through the `BeezKeeper` on each beez Node. This experiment demonstrates that the computational power of a blockchain can be used to fairly train and reward the nodes that collaborate to solve a given problem.

As shown in Fig. 4, the training performance of the machine learning model on both systems is comparable. The distributed system requires more time to train the model because, for testing purposes, a sleep period is added to each epoch executed by nodes participating in the challenge. Moreover, based on the scores, we can see that the models are quite similar; there are no important distinctions between them. With these results, we can assume that training a distributed learning model is feasible and that the obtained results are very close to those obtained when the training is performed in a single machine (node), in the so-called classical way.

7 Conclusion and Future Work

There are a variety of fascinating potential applications for the coupling of artificial intelligence and blockchain technology. However, it should be noted that these technologies are still in their infancy. The development of most of these hybrid applications will take years, if at all. While these are undeniably potent ideas, researchers and developers working on these kinds of projects must exercise considerable prudence. Both blockchain and Artificial Intelligence have enormous ethical implications, and the stakes for the business, government, and society are exceedingly high.

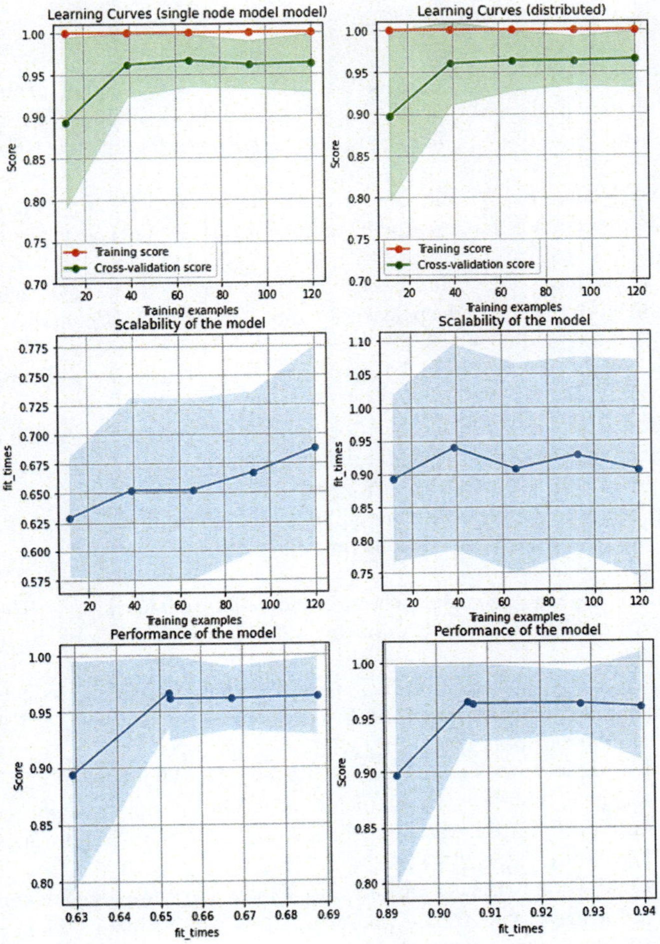

Fig. 4. Single node training versus distributed training

These technologies can be combined to improve anything from the logistics of the food supply chain and the sharing of medical records to media royalties and financial security.

Artificial Intelligence is capable of efficiently sifting through massive datasets to generate novel scenarios and identify data-driven patterns. Blockchain facilitates the elimination of defects and fake data sets. New classifiers and patterns generated by AI can be authenticated via a decentralized blockchain architecture. This is applicable to all consumer-facing businesses, including retail transactions. Using client data obtained through blockchain infrastructure, marketing automation may be constructed using artificial intelligence, and the Beez Blockchain might be one of these future uses.

References

1. Cheng, D., Li, S., Zhang, H., Xia, F., Zhang, Y.: Why dataset properties bound the scalability of parallel machine learning training algorithms. IEEE Trans. Parallel Distrib. Syst. **32**(7), 1702–1712 (2021). https://doi.org/10.1109/TPDS.2020.3048836
2. Gilad, Y., Hemo, R., Micali, S., Vlachos, G., Zeldovich, N.: Algorand: Scaling byzantine agreements for cryptocurrencies. pp. 51–68 (2017). https://doi.org/10.1145/3132747.3132757
3. Haque, A., Ghani, M., Mahmood, T.: Decentralized transfer learning using blockchain and IPFS for deep learning. pp. 170–177 (2020). https://doi.org/10.1109/ICOIN48656.2020.9016456
4. King, S., Nadal, S.: Ppcoin: Peer-to-peer crypto-currency with proof-of-stake (2022)
5. Korkmaz, C., Kocas, H., Uysal, A., Masry, A., Ozkasap, O., Akgun, B.: Chain fl: Decentralized federated machine learning via blockchain. pp. 140–146 (2020). https://doi.org/10.1109/BCCA50787.2020.9274451
6. Kusi, G., Xia, Q., Cobblah, C., Gao, J., Xia, H.: Training machine learning models through preserved decentralization. pp. 465–472 (2020). https://doi.org/10.1109/MSN50589.2020.00080
7. Li, D., Luo, Z., Cao, B.: Blockchain-based federated learning methodologies in smart environments. Cluster Computing (2021). https://doi.org/10.1007/s10586-021-03424-y
8. Li, Y., Chen, C., Liu, N., Huang, H., Zheng, Z., Yan, Q.: A blockchain-based decentralized federated learning framework with committee consensus. IEEE Network PP (2020). https://doi.org/10.1109/MNET.011.2000263
9. Nadeem, F., Alghazzawi, D., Mashat, A., Faqeeh, K., Almalaise, A.: Using machine learning ensemble methods to predict execution time of e-science workflows in heterogeneous distributed systems. IEEE Access **7**, 25138–25149 (2019). https://doi.org/10.1109/ACCESS.2019.2899985
10. Otoum, S., Al Ridhawi, I., Mouftah, H.: Blockchain-supported federated learning for trustworthy vehicular networks. pp. 1–6 (2020). https://doi.org/10.1109/GLOBECOM42002.2020.9322159
11. Habib ur Rehman, M., Salah, K., Damiani, E., Svetinovic, D.: Towards blockchain-based reputation-aware federated learning (2020). https://doi.org/10.1109/INFOCOMWKSHPS50562.2020.9163027
12. Short, A., Helen, L., Theocharis, E.: Execution of a federated learning process within a smart contract. pp. 1–4 (2021). https://doi.org/10.1109/ICCE50685.2021.9427734
13. Tibúrcio, P., Spohn, M.: Ad hoc grid: An adaptive and self-organizing peer-to-peer computing grid. pp. 225–232 (2010). https://doi.org/10.1109/CIT.2010.504
14. Tuncer, O., Ates, E., Zhang, Y., Turk, A., Brandt, J., Leung, V.J., Egele, M., Coskun, A.K.: Online diagnosis of performance variation in HPC systems using machine learning. IEEE Trans. Parallel Distrib. Syst. **30**(4), 883–896 (2019). https://doi.org/10.1109/TPDS.2018.2870403
15. Wang, H., Niu, D., Li, B.: Turbo: dynamic and decentralized global analytics via machine learning. IEEE Trans. Parallel Distrib. Syst. **31**(6), 1372–1386 (2020). https://doi.org/10.1109/TPDS.2020.2964667
16. Xiao, W., Xue, J., Miao, Y., Li, Z., Chen, C., Wu, M., Li, W., Zhou, L.: Distributed graph computation meets machine learning. IEEE Trans. Parallel Distrib. Syst. **31**(7), 1588–1604 (2020). https://doi.org/10.1109/TPDS.2020.2970047

17. Zhang, F., Chen, Z., Zhang, C., Zhou, A.C., Zhai, J., Du, X.: An efficient parallel secure machine learning framework on GPUs. IEEE Trans. Parallel Distrib. Syst. **32**(9), 2262–2276 (2021). https://doi.org/10.1109/TPDS.2021.3059108

Hiding Simple Text Files in Multimedia Files: A New Method and Its Implementation

Marcin Lawnik(✉), Artur Pełka, and Adrian Kapczyński

Faculty of Applied Mathematics, Silesian University of Technology, Gliwice, Poland
marcin.lawnik@polsl.pl, artur260@op.pl, adriank@polsl.pl

Abstract. Specific data, due to their nature, must be adequately secured. It can be done, among others, by the use of steganographic methods, whose purpose is to hide information by embedding hidden information in a selected object (carrier). The most common way of hiding text in a carrier is the Least Significant Bit (LSB) method. Nevertheless, this method requires a sufficiently large medium to host hidden information. This article proposes a new method to hide simple text files. Firstly, secret data is converted with a simple affine cipher (which can be replaced with another form of encryption). Then the content is transformed directly into the output file without using a carrier. The output file can be a graphic file with an RGB structure or a sound file. The method is discussed in the example of selected test files. As the analysis shows, the resulting data obtained occupy less disk space than those obtained using the LSB method. Additionally, the implementation of the method in the form of a Python script is presented. Finally, the modifications and limitations of the proposed method of hiding the text are discussed.

1 Introduction

In the 21st century, information plays a key role in many areas of life. Confidential or even secret data is particularly important. Such data cannot be publicly available because then they lose their importance or may result in serious problems for entities related to them. For this purpose, the data is properly secured against access performed by unauthorized persons. This applies to various levels—from governmental and military, where the data is extremely important, to the ordinary citizen who uses (sometimes even unconsciously) appropriate techniques to secure the data.

These security countermeasures come in two forms—cryptography and steganography. Both forms have been known since antiquity [1]. With the advancement of computers in the 20th century, both cryptography and steganography were significantly developed. Roughly, the first relies on modifying the

R. Silhavy et al. (Eds.): CoMeSySo 2022, LNNS 596, pp. 170–188, 2023.
https://doi.org/10.1007/978-3-031-21435-6_15

information in such a way that a person unauthorized to read it could not do it without proper knowledge. Examples of such methods are encryption algorithms [2]. On the other hand, steganography is about hiding the very fact that you are dealing with important information. This concealment consists in immersing the information in some other storage medium, e.g., hiding text in a digital image [3].

It is worth emphasizing that data security is often beyond the control of the data owners, e.g., data in the cloud. This leads to situations where an extract from such a server causes the disclosure of information. There are a lot known cases of such issue, for example [4].

The data in question is often stored in the form of text files. There are many ways how such files can be protected, for example by using appropriate encryption algorithms [2] or hiding in another medium [5].

One of the solutions for this problem was proposed in [6]. The solution provided not only the minimum level of confidentiality by use of steganography, but also resulted in the smallest size of the output file.

In this article a new way to hide text data in various multimedia media is shown. The root concept uses the results presented in [6]. The proposed approach is a significant extension of [6]. It assumes, inter alia, adding the possibility of hiding text files in sound carriers, extending the allowed image file formats (not only *png*), and changing the shape from square to rectangular in the case of the image file itself. Additionally, the text is secured with simple affine encryption before being changed to another multimedia medium. This approach, combining simple encryption with conversion to another multimedia medium, makes it very difficult to read the resulting file contents quickly. To the best knowledge of the authors, no one has proposed such an approach to securing information in this context before.

This article is organized as follows. The first section (Introduction) covers the topic of the article as well as the outline of solutions. The result of the literature review on information classification using steganography is presented in the second section (Related research). The next sections cover the description of the method, an example of its application, and a discussion on the strengths and weaknesses of the proposed method. The article ends with conclusions.

2 Related Research

This article is thematically related to hiding the existence of information (steganography) and cloud computing, which is where the steganographically transformed files are stored. In this section of the article, the primers related to steganography and cloud computing will be provided. For completeness, in addition to the basic issues, selected results of research in the area of steganography, cloud computing, and work combining steganography and cloud computing are also presented.

The main task of steganography is to hide information. For this task, steganographic programs are used, which are technical implementations of steganography algorithms. Implementations are made with the medium in which a given type of information will be hidden. To perform tasks using the classical steganographic system, two files should be delivered as input data: the steganographic medium and the file with the message to be hidden. In commonly used steganographic systems, a steganographic key may be entered, although it is not mandatory. As a result of the work of the steganographic system, steganographic files are created, i.e. multimedia files containing hidden message files. Those files might be stored locally or remotely, for example in cloud storage.

Cloud computing is is a model for enabling network access on-demand to a shared pool of configurable digital resources [7]. Essential Characteristics of cloud computing consist of on-demand self-service, broadband network access, resource pooling, rapid elasticity, and metering capability. There are three main service models: Software as a Service (abbreviated SaaS), Platform as a Service (abbreviated PaaS), and Infrastructure as a Service (abbreviated IaaS). It is worth mentioning that the are defined four deployment models of cloud computing: private cloud, community cloud, public cloud, and hybrid cloud.

Basic steganography concepts and different kinds of steganography (text, image, audio, video) can be found in [8]. Monahar et al. proposed secure steganography methods [9], while S. Katzenbeisser and F.A.P. Petitcolas [10], provided in their work valuable information about the model of steganographic communication with a detailed focus on various steganographic protocols. Authors of [11] described the digital techniques related to steganography. It's worth mentioning that inside of this chapter one can find a substantial number of examples as well as information about the security of the steganographic system. There are research papers devoted to information hiding in different carriers: text [12], image [13], audio [14] and even QR codes [15]. More information about the current state of the art related to steganography can be found in: [16–19].

If it's going about the basic information about cloud computing, it is worth referencing the following bibliography items: [20–22]. Due to the fact, that cloud computing is a highly explorable area of scientific interest, there are documented research results worth recalling here. Kaufman [22] provided a general overview of data security in cloud computing, while Kajal et al. [23] described in detail security threats to resources processed and stored using cloud computing. There are scientific achievements related to secure storage, which is well-documented in [24,25]. It's worth emphasizing, that there exists a cloud computing data storage security framework, presented in [26]. The importance of cloud computing security requirements has been shown in [27–29]. More information about the current state of the art related to cloud computing can be found in [30,31].

According to the performed literature review, there exist scientific challenges related to both steganography and cloud computing, among others [32–34]. There are papers documenting the exploration of the area of cloud computing and steganography combined - for more information please reference [35,36].

The conducted literature search indicated the timeliness of the subject matter discussed in this article.

3 Method

This part of the article describes the method of hiding data in multimedia files. It is divided into two parts. The first part shows how to hide data in images. In the second part the target medium is sound. The general scheme of the method for both variants, where the output file is an image file and an audio file, is shown in Fig. 1.

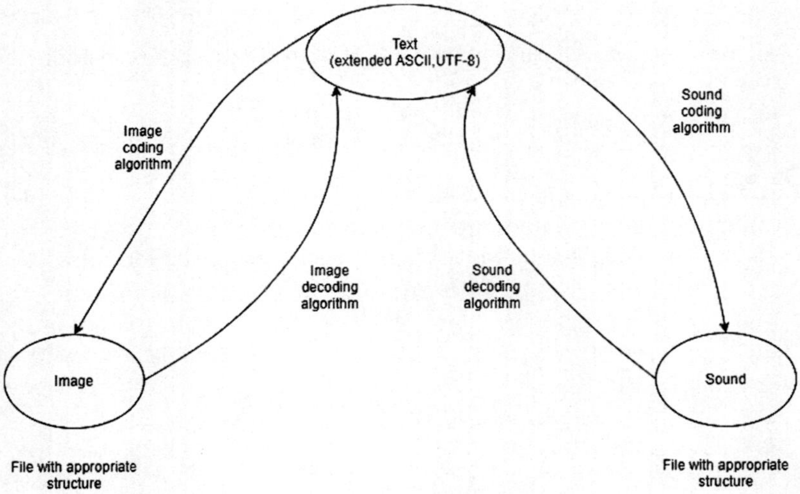

Fig. 1. General scheme of the proposed method.

3.1 Text to Image

The given algorithm has originated from [6]. Its modifications mainly concern the size of the image that can be created and additionally introduced simple text encryption by specifying parameter values a and b belonging to $\mathbb{Z}_{256}$ on condition $\gcd(a, 256) = 1$. In addition, the number of graphic formats that can be obtained as a consequence of the method's operation has been significantly increased. With these modifications in mind, the text-to-image method can be defined in one of two text file encoding variants: extended ASCII and UTF-8.

1. The text file is read into the buffer.
2. (optional) Additionally, the following are added to its content: ETX, filename with the extension, ETX,
   where ETX is *end of text* and is encoded with ASCII code equal to 3.
3. The text characters are converted to numeric values and added to the one-dimensional array Tn according to:

- for ASCII text encoding: numeric value corresponding to the ASCII character v_i is transformed by the equation

$$c_i = av_i + b \mod 256, \tag{1}$$

where a and b are hidden key values. c_i value is stored in the array Tn;
- for UTF-8 text encoding: each of the character values of the text t_i is stored in the positional system on base 256 according to the equation (2):

$$t_i = v_i \cdot 256 + w_i, \tag{2}$$

where v_i and w_i are the coefficients of t_i in the base 256 system. Values

$$\begin{aligned} c_{i1} &= av_i + b \mod 256 \\ c_{i2} &= aw_i + b \mod 256 \end{aligned} \tag{3}$$

are added to the Tn array. In turn, the values a and b are hidden key values. c_{i1} and c_{i2} values are stored in the array Tn.

4. The image size (height and width) is calculated using the formulas (4)–(6):

$$size_s = \left\lceil \sqrt{\frac{len}{3}} \right\rceil, \tag{4}$$

$$width = random\left(\frac{3size_s}{4}, \frac{5size_s}{4}\right), \tag{5}$$

$$height = \left\lceil \frac{size_s^2}{width} \right\rceil, \tag{6}$$

where $\lceil \cdot \rceil$ stands for ceiling function, len stands for the length of the array Tn, while $random()$ stands for a random integer from a given range.
5. If the length of the array Tn is less than

$$3 \cdot width \cdot height, \tag{7}$$

then Tn is padded with random values so that its length is exactly (7).
6. From array Tn a two-dimensional array is created $T_{width \times height}$, whose elements will be 3-element tuples.
7. The two-dimensional array of tuples is treated as an array of pixels and saved to a graphics format selecting one of the allowed image file extensions.
8. (optional) The name of the graphic file is the abbreviation obtained from the file name using the selected hashing algorithm, e.g. SHA3.

A detailed text to image transition procedure is shown in Fig. 2.

The above presented method is reversible. This is due to saving to a graphic file, where individual pixels contain fragments of encrypted text. The encryption algorithm used allows for a quick and easy decryption operation, and thus it is possible to recover the original message. It means that the hidden text can be 100% recovered from a previously obtained image file. This can be done by implementing the following steps:

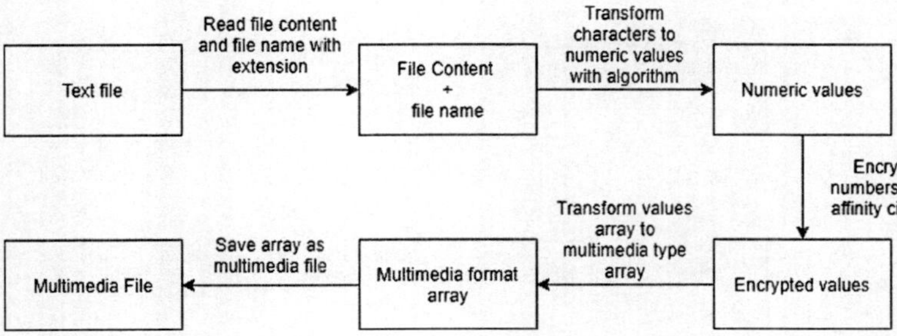

Fig. 2. Coding scheme of the proposed method.

1. Successive pixels of the graphic file are loaded until a pixel with an RGB color component equal to 3 is found.
2. The RGB components of the loaded pixels c_i are
   - ASCII encoding: using secret keys a and b, c_i value is transformed by mapping
   $$a^{-1}(c_i - b) \mod 256, \tag{8}$$
   where a^{-1} is the inverse of a in $\mathbb{Z}_{256}$ and written as ASCII characters to text
   - UTF-8 encoding: grouped in pairs (c_{i1}, c_{i2}) and transformed by expression
   $$\begin{aligned} v_i &= a^{-1}(c_{i1} - b) \mod 256, \\ w_i &= a^{-1}(c_{i2} - b) \mod 256 \end{aligned} \tag{9}$$
   where a^{-1} is the inverse of a in $\mathbb{Z}_{256}$. Then the values v_i and w_i are written in the number system with a base of 256, i.e.
   $$t_i = v_i \cdot 256 + w_i. \tag{10}$$
   Value t_i is written into the text.
3. (optional) Successive pixels are loaded to obtain values representing the name and extension of the file until the next color component value is 3.
4. (optional) The RGB components of the read pixels are converted to ASCII or UTF-8 characters according to the procedure from step 2. The resulting string of characters is the name and extension of the text file.

A detailed image to text transition procedure is shown in Fig. 3.

3.2 Text to Sound

Converting a text file to an audio file is slightly different than that of the resulting image file. For example the parameter values a and b in this case they belong to the set $\mathbb{Z}_{65,536}$ on condition $\gcd(a, 65{,}536) = 1$. The procedure has the following steps:

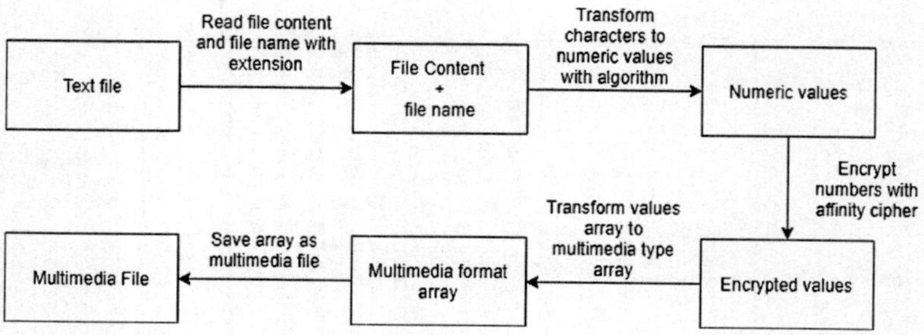

Fig. 3. Decoding scheme of the proposed method.

1. The text file is read into the buffer.
2. (optional) The following are added to its content: *ETX*, filename with the extension, *ETX*, where *ETX* stands for *end of text* and is encoded with ASCII code equal to 3.
3. Text characters are converted to their corresponding numeric values in UTF-8 encoding v_i and transformed by mapping.

$$c_i = av_i + b \mod 65{,}536, \tag{11}$$

   where a and b are treated as secret keys.
4. The obtained numerical values c_i are saved to the sound file by selecting one of the allowed extensions.
5. (optional) The name of the graphic file is the abbreviation obtained from the file name using the selected hashing algorithm, e.g. SHA3.

A detailed text to sound transition procedure is shown in Fig. 2.

The above-presented method is reversible. This is due to the recording of the audio file, where individual samples contain fragments of ciphertext. The encryption algorithm used allows for a quick and easy decryption operation, and thus it is possible to recover the original message. It means that you can recover 100% of the hidden text from a previously received sound file. It is possible by implementing the following steps:

1. Consecutive samples are read from the audio file until encountering a sample with a numeric value of 3.
2. The samples are saved as consecutive UTF-8 characters c_i and transformed by expression

$$v_i = a^{-1}(c_i - b) \mod 65{,}536, \tag{12}$$

   where a^{-1} is the inverse of a in $\mathbb{Z}_{65{,}536}$. Then items v_i are saved in the text file.
3. (optional) Samples are read in until it reaches the next sample value of 3.
4. (optional) The resulting samples are converted to UTF-8 characters, and the resulting string consists of the name and extension of the text file.

A detailed sound to text transition procedure is shown in Fig. 3.

3.3 Modifications

The described method assumes that both the name and the extension of the original file are directly incorporated into the new medium. This operation is performed at the stage of preparing the text buffer in point 2 of both algorithms and is marked as optional. This process can be skipped.

Naming the created medium with the hash function may also be omitted in favor of the original name of the file. This step (8 for image and 5 for audio, respectively) has been marked optional in the presented algorithms.

Additionally, if the output medium is an image, its size determined by the formulas (5)–(6) does not have to be a random size. The user can define one of the dimensions himself and then calculate the other.

The processed text file can also be divided into several parts, and each of them can then be handled with the proposed method. This approach can be useful for very large text files, where splitting them into chunks can speed up the algorithm (less use of the computer's RAM).

It is also possible to modify the data encryption process itself. In the proposed version, a simple affine cipher was used for this purpose. This process aims to make it difficult to read the original data by using appropriate text editors. It is, therefore, a data hiding action and not the main element of data security. However, this process can be replaced with another algorithm (e.g., AES [37]) while keeping the whole method relatively fast.

The algorithm also requires specifying the exact values of the a and b keys. These values could be put in the output file to have the script read the key values itself. Keys can be placed in any appropriately specified place in the output file, e.g., as the first or last bit values.

3.4 Implementation

The presented variants of the transition from text to image or sound have been implemented using the Python language with use of the following libraries:

- *Pillow* library [38] for writing and reading a graphic file;
- *Pydub* library [39] for writing and reading an audio file;
- *hashlib* library [40] to obtain the hash value of the name of text file by use of SHA3 algorithm.

The script has been developed to work with popular multimedia carriers such as graphics and sound. The user gets a considerable amount of file extensions that can become output files. However, due to the script's principle of operation, the pool of file formats has been limited for a particular group for each type of medium. It makes it obligatory to remember the available file formats for every kind of medium.

The graphic files that can be obtained may have the extensions from the following list (13):

$$\{png,\ bmp,\ sgi,\ ppm,\ tga\}. \tag{13}$$

This list results from the solutions adopted for the processing of graphic files. In turn, the sound files that can be obtained may have the extensions from the following list (14):

$$\{wav,\ ogg,\ m4a,\ wma,\ flac\}. \tag{14}$$

As in the case of graphic files, the (14) list results from the adopted solutions for sound processing.

When converting a text file into an audio file, it is essential to specify how many channels of such a file should have (whether it is mono or stereo) and the sampling frequency. Typically this frequency for a large part of audio formats is set 44,100 Hz.

Moreover, a valuable modification of the script for users includes adding the possibility of having a configuration file. At the moment, the user can define the script startup parameters at the time of script execution or use the default values for some parameters. In some cases, this is a satisfactory solution. However, for users who would use the script more often and the defaults would not be acceptable, having the configuration file would be appropriate. Paths to directories for output and input files can be an example of parameters included in a configuration file. This amount of information already allows keeping order on the machine on which the script is executed. The script would look for a configuration file first and then fill it with values from the command line or a set of default values.

The Python code that implements the presented method can be found on the Internet at: https://github.com/MMI-research-team/mmi/tree/master/PSSIT.

3.5 Applications

The discussed method allows you to convert text files to other multimedia carriers. Thus, it can be used to hide data in areas such as Big Data or IoT, where these areas' continuous development requires certain forms of security. Due to the massive amount of data, it is worth mentioning that the areas above require advanced algorithms that ensure a certain security level. However, due to the viability and nature of these data, this level is not necessarily very high. This assumption is fulfilled by the discussed method. Moreover, the scripted implementation is easy to use in cloud hearings where this type of data can end. As shown in the following section, the running time of scripts on a personal computer is so small that larger text data can also be converted into graphic or sound files with appropriate computing power. Besides, converting a text file into a graphic file in the *png* format allows you to save disk space, which is a significant factor in the applications mentioned above.

Apart from Big Data or IoT, the discussed method can also be used to secure data transmission. An unauthorized listener can easily intercept a text message sent over an unsecured channel. If, instead of a text message, it receives a graphic file obtained due to the scripts of the proposed method, the content of the file will appear random.

At the same time, it should be emphasized that in the case of a required high level of security, the use of the proposed solution may not give appropriate results.

3.6 Example

The following frequently used text formats were selected as test files:

1. TXT: plain text in UTF-8 encoding; file size is 252,640 bytes;
2. JSON: file containing e-mail data (name and surname, e-mail address, IP address) in ASCII encoding; file size is 117,773 bytes;
3. XML: a file containing company employees' data (name and surname, job, telephone number) in ASCII encoding; the file size is 146,097 bytes;
4. CSV: file containing credit cards details (name and surname, credit card number, credit card type) in ASCII encoding, the file size is 46,071 bytes.

The script implementing the method was run on a computer Microsoft Windows 10 equipped with an Intel (R) Core (TM) i7-8565U CPU and 16 GB of RAM. In all cases, values were taken as the values of the secret keys in the affine cipher $a = 7$ and $b = 3$. The results for the individual test files are shown in the Tables 1, 2, 3 and 4.

Table 1. Selected result files obtained by running the method for the TXT test file. The following columns are: the output file, text file encoding, size of the output file, compression ratio coefficient, time needed to get the result file, time needed to recover the original file.

Output	Coding	Output size (B)	CR	Time (s)	Reverse time (s)
png	UTF-8	127,402	0.5043	0.14	0.96
bmp	UTF-8	450,270	1.7823	0.10	0.89
wav	UTF-8	447,336	1.7706	0.09	0.13

3.7 Case Study

The case study for the above test files was carried out for the disk space occupied by the files before and after processing, and for the time needed to obtain the result file and the original file using the reverse method.

Disk space analysis

When analyzing the disk space used by the source file and processed by a particular algorithm, it is useful to determine the compression ratio (CR) [41]:

$$\text{Compression Ratio (CR)} = \frac{\text{size of the output stream}}{\text{size of the input stream}}. \tag{15}$$

Table 2. Selected result files obtained by running the method for the JSON test file. The following columns are: the output file, text file encoding, size of the output file, compression ratio coefficient, time needed to get the result file, time needed to recover the original file.

Output	Coding	Output size (B)	CR	Time (s)	Reverse time (s)
png	ASCII	46,054	0.391	0.03	0.21
	UTF-8	43,342	0.368	0.07	0.50
bmp	ASCII	119,526	1.0148	0.03	0.21
	UTF-8	237,606	2.0175	0.06	0.46
wav	UTF-8	235,612	2.0006	0.04	0.07

Table 3. Selected result files obtained by running the method for the XML test file. The following columns are: the output file, text file encoding, size of the output file, compression ratio coefficient, time needed to get the result file, time needed to recover the original file.

Output	Coding	Output size (B)	CR	Time (s)	Reverse time (s)
png	ASCII	29,723	0.2034	0.04	0.26
	UTF-8	63,464	0.4344	0.08	0.62
bmp	ASCII	14,7654	1.0107	0.04	0.30
	UTF-8	294,838	2.0181	0.06	0.58
wav	UTF-8	292,260	2.0005	0.06	0.08

Table 4. Selected result files obtained by running the method for the CSV test file. The following columns are: the output file, text file encoding, size of the output file, compression ratio coefficient, time needed to get the result file, time needed to recover the original file.

Output	Coding	Output size (B)	CR	Time (s)	Reverse time (s)
png	ASCII	34,613	0.7513	0.03	0.10
	UTF-8	27,352	0.5937	0.06	0.20
bmp	ASCII	45,846	0.9931	0.03	0.12
	UTF-8	91,446	1.9849	0.03	0.20
wav	UTF-8	90,208	1.958	0.03	0.03

The interpretation of the CR ratio is as follows: if CR < 1, the result file takes up less disk space than the source file; CR > 1 means that the output file is larger than the source file.

Analysis of test files shows that the CR coefficient is lower than 1 in the case of the output files with extension *png*. It is because the *png* format is built using lossless compression algorithms [42]. It means that the text files processed in this way take up less space, and therefore the transfer of such a file requires the use of much fewer data. The other acceptable formats do not have built-in compression algorithms, so their CR is close to 1 (extended ASCII encoding) or close to 2 (UTF-8 encoding).

Time

The time analysis for the test files shows that both the method itself and the reverse method are high-speed. This means that both can be successfully applied in home and professional environments.

4 Discussion

The discussion is divided into several parts. They concern the presented method's advantages and disadvantages, the encryption process, and comparison with the LSB method.

4.1 Hiding in Another Multimedia Files

The main advantage of applying the presented method is hiding information in a multimedia medium. This is done differently from the LSB method, in which the carrier is known to both parties involved in the communication. In this method, there is no need to use a separate carrier, since the text is directly converted to the output format.

The result, as a consequence of text transformation, looks random.

For example, it can be seen in the Fig. 4. In case of audio format output file, the result is noise. Thus, the content of the text file is invisible and inaudible to the eavesdropper. This approach can be called *coverless steganography*. Besides, the output files obtained in this way can be subjected to additional processing, e.g., by hiding the file using the classic steganographic method (e.g., LSB). This extra security feature will make reading the original value much more difficult.

An additional advantage of the presented approach is the occupied disk space issue, which in some cases will be much smaller than the original content. The resulting output files, depending on the format, will be at most twice the size. If the output file is the *png* format, it is possible that thanks to the lossless compression algorithms built into this format, its size will be smaller than the text input file.

Fig. 4. Image *png* obtained from the test file TXT.

4.2 Comparison with LSB Method

The proposed method of hiding the text is not a classic steganographic method that requires an additional medium to convey the message. However, it is reasonable to compare it with the LSB method's simplest variant to assess the difference in the disk space occupied by a specific file. Compared to the classic LSB method, where 1 bit of text is encoded on 8 bits of the data carrier, the resulting output file's size is significantly smaller. Analyzing the LSB method case carefully: if the text file is X bits long, then to encode the whole text, the data carrier size must be at least $8X$ bits. If this medium is a graphic file and there is no lossless compression (e.g. *bmp* format), the number of pixels with RGB components of such a file must be equal to $\frac{8}{3}X$ pixels, which, assuming the proportion $\frac{4}{3}$ of the image leads to image width and height of $\frac{4}{3}\sqrt{2X}$ and $\sqrt{2X}$ pixels, respectively. For example: for a text that consists of 1000 characters encoded on 8-bits each, the text size is equal to 8000 bits and the data carrier for the classic LSB method must be at least 21,334 pixels in size. This means that assuming the proportions of the size of the medium as $\frac{4}{3}$ it gives 169 and 127 pixels, respectively.

The above analysis concerns the theoretical comparison of how much disk space is needed to hide a file in a medium in the form of an RGB image. It shows that the resulting file obtained from the described method takes up much less disk space than the file obtained using the LSB method. In turn, the Table 5 shows the time needed to obtain the result file using the classic LSB method and how much space the result files take up. For this purpose, a ready-made script [43], also written in Python, was used. The files *png*, *bmp* and *wav* were chosen as the media, which are large enough to fit the contents of the test file. These files are 614,179, 117,9702, and 2,146,166 bits, respectively. Additionally, for the medium not to be too large, the content of the test file was saved on the two least significant bits. The times shown apply to the processing and reading of the medium and saving the resulting file. The obtained results show that our method is comparably fast in the process of hiding the input file. In recovering the original content from the resulting file due to additional decryption operations, the obtained results are worse than the classic LSB method but still satisfactory for various applications. However, in the disk space occupied by the files, the results obtained are much better. This also applies to the observation of the CR coefficient for the medium and the result file. This factor is often greater than 1, which means that the resulting file takes up much more space.

Table 5. Results for test files embedded in multimedia formats by the operation of the LSB method. The following columns are: the text file, output file, output file size, compression ratio coefficient, time needed to get the result file, time needed to recover the original file.

Text file	Output	Output size (B)	CR	Time (s)	Reverse time (s)
TXT	png	739,627	1.2043	0.24	0.15
	bmp	117,9702	1	0.23	0.14
	wav	2,145,940	0.9999	0.04	0.01
JSON	png	714,977	1.1641	0.19	0.11
	bmp	1,179,702	1	0.18	0.10
	wav	2,145,940	0.9999	0.04	0.01
XML	png	718,113	1.1692	0.20	0.14
	bmp	1,179,702	1	0.19	0.11
	wav	2,145,940	0.9999	0.03	0.01
CSV	png	701,934	1.1429	0.17	0.09
	bmp	1,179,702	1	0.16	0.08
	wav	2,145,940	0.9999	0.02	0.01

4.3 Encryption

In the proposed algorithm, a simple affine encryption [2] is proposed. Such encryption in itself is, of course, not a security solution. A possible frequency analysis or a brute force attack to check all possible keys will effectively break this [2]. However, the encryption process in the proposed method should be viewed as an additional layer of security. In addition, the affine encryption used may be replaced by another algorithm designed for this purpose. Such an approach will undoubtedly be more effective in securing the data content, but it may slow down the entire method's performance. However, using an already simple encryption algorithm will change the message medium from text to multimedia, making it impossible to read the text content easily. Using simple encryption will make it impossible to read the message by well-known editors such as Notepad++ [44]. An exemplary screen of the audio file obtained from the proposed method is shown in Fig. 5. Thus, the security measures used may turn out to be much more difficult to break than their property suggests.

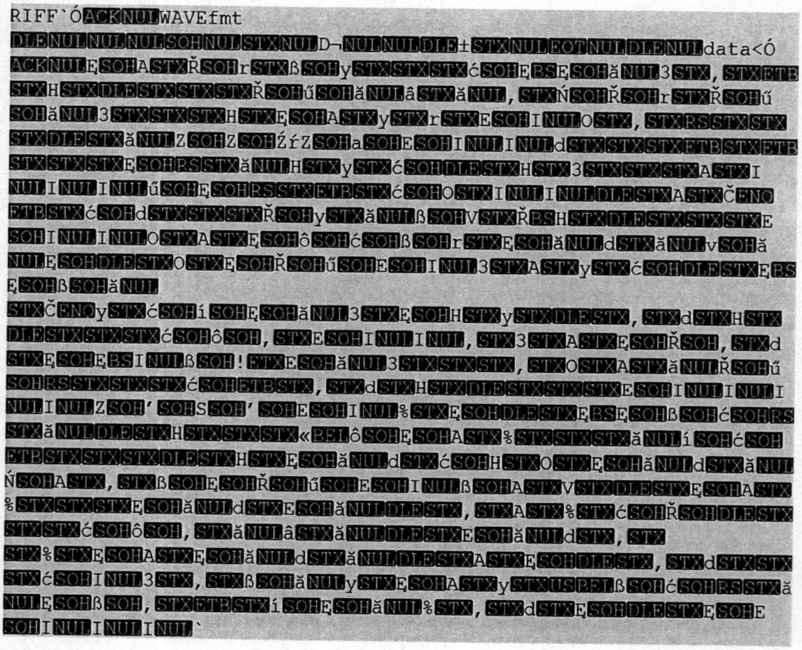

Fig. 5. Sample screenshot from the audio file received after opening it in the Notepad++ editor.

4.4 Limitations

The proposed steganography method requires appropriate file encoding. Depending on the encoding of the text, i.e., extented ASCII or UTF-8, we have different variants in the conversion. In this case, it is possible to process the file in extended ASCII encoding with an algorithm that assumes UTF-8 encoding. The reverse situation, i.e., a file encoded in UTF-8 and processed with the extended ASCII algorithm, will cause the file to be processed incorrectly. The problem that arises here is related to writing the appropriate values in pixels of the graphic file. These values must be in the range of 0–255. If any value is outside this range, it will be incorrectly encoded (or there will be errors in the script executing the method).

In this case, it will not be possible to recover the original text value after performing the reverse method. A similar situation will be in the case when the text encoding is outside the UTF-8 range — such files cannot be encoded using the described method.

Also, the output files cannot have any extensions, but must fit into a predetermined set. For image files, this set is {*png, bmp, sgi, ppm, tga*}. This is because files with such extensions have a similar structure (they are built on the RGB color principle) and can be processed using an accepted programming tool (in the case of this article, Python). Also, extensions related to lossy compression like *jpg* cannot be processed, as described in the article. This is because the original text message cannot be recovered from the image or sound obtained in this way.

Moreover, not every text file can be used as an input file for a script. The algorithm is based on files, which content can be read directly, and there is no complex formatting. The files allowed to work with the program are "Plain text" files, where it is possible to read the content of the file, and there is no need for any special programs to display the content with additional formatting. An example of a file extension with complex formatting is *doc*. In this type of file, there is additional formatting, such as properties of elements that affect the text's appearance. This does not mean, however, that such files cannot be converted into image or sound – it will be possible if the text is separated from the formatting, and each of them is treated separately with the proposed method.

5 Conclusions

The article presents a new method of text steganography using multimedia files. Concerning the techniques of hiding text known from the literature, which are in no small extent, variants of the LSB method, the proposed method does not use an additional carrier (graphic or sound file). The resulting file contains the entire content of the hidden text and has a randomized structure. The article also deals with the implementation aspects of the method described earlier.

The proposed method was implemented in Python programming language. Time and disk space of the output files were examined. As the presented analysis shows, the obtained output files take up much less space than the variants of the LSB method used so far. In some cases, the output files take up even less disk space than the processed text file. Additionally, possible modifications and limitations were discussed. This method can be successfully used to hide text in well-known multimedia carriers.

The method may be applied in the field of IoT devices. The described way is simple and does not require complicated calculations that can be easily implemented in various electronic devices.

Further research work on this subject will concern the use of a similar approach of hiding text files to the video format.

References

1. Kahn D.: The Codebreakers: The Comprehensive History of Secret Communication from Ancient Times to the Internet. Rev sub ed. Scribner (1996)
2. Stallings, W.: Cryptography and Network Security: Principles and Practice, 5th edn. Prentice Hall Press, USA (2010)
3. Yahya A.: Steganography Techniques for Digital Images. 1st ed. Springer International Publishing (2019)
4. Bloomberg (2019) ECB Says One of Its Websites Was Hacked, Data Possibly Captured. https://www.bloomberg.com/news/articles/2019-08-15/ecb-says-one-of-its-websites-was-hacked-data-possibly-captured
5. Yahya, A.: Steganography Techniques for Digital Images. Springer International Publishing AG, Switzerland (2018)
6. Lawnik, M., Pełka, A., Kapczyński, A.: A new way to store simple text files. Algorithms **13**(4) (2000). https://doi.org/10.3390/a13040101
7. Mell, P.M., Grance, T.: SP 800–145. The NIST Definition of Cloud Computing. Gaithersburg, MD, USA (2011)
8. Rana, M.S., Sangwan, B.S., Jangir, J.S.: Art of hiding: an introduction to steganography. Int. J. Eng. Comput. Sci. **1**(1), 11–22 (2012)
9. Manohar N, Kumar PV. (2020) Data encryption amp; decryption using steganography. In: 2020 4th International Conference on Intelligent Computing and Control Systems (ICICCS); pp. 697–702
10. Katzenbeisser, S., Petitcolas, F.: Information Hiding Techniques for Steganography and Digital Watermaking. vol. 28 (1999)
11. Marvel, L.M.: In: Javidi, B.,(ed.) Information Hiding: Steganography and Watermarking, pp. 113–133. Springer, New York (2005). https://doi.org/10.1007/0-387-25096-4_6
12. Gupta, S., Jain, R.: An innovative method of Text Steganography. In: 2015 3rd International Conference on Image Information Processing (ICIIP) pp. 60–64 (2015)
13. Altaay, A.A.J., Sahib, S.B., Zamani, M.: An Introduction to Image Steganography Techniques. In: 2012 International Conference on Advanced Computer Science Applications and Technologies (ACSAT), pp. 122–126 (2012)
14. Zamani, M., Manaf, A.A., Ahmad, R.: Current problems of substitution technique of audio steganography. In: The 2009 International Conference on Artificial Intelligence and Pattern Recognition, pp. 154–160 (2009)
15. Hajduk, V., Broda, M., Kováč, O., Levický, D.: Image steganography with using QR code and cryptography. In: 26th International Conference Radioelektronika (RADIOELEKTRONIKA), pp. 350–353 (2016)
16. Jain, R., Boaddh, J.: Advances in digital image steganography. In: 2016 International Conference on Innovation and Challenges in Cyber Security (ICICCS-INBUSH), pp. 163–171 (2016)
17. Mukhedkar, M., Powar, P., Gaikwad, P.: Secure non real time image encryption algorithm development using cryptography amp; steganography. In: 2015 Annual IEEE India Conference (INDICON), pp. 1–6 (2015)
18. Kumar, R., Singh, A.: Understanding steganography over cryptography and various steganography techniques. Int. J. Comput. Sci. Mob. Comput. **4**(3), 253–258 (2015)
19. Phadte, R.S., Dhanaraj, R.: Enhanced blend of image steganography and cryptography. In: 2017 International Conference on Computing Methodologies and Communication (ICCMC), pp. 230–235 (2017)

20. Nazir, M.: Cloud computing: overview and current research challenges. IOSR J. Comput. Eng. **8**(1), 14–22 (2012)
21. Alhenaki, L., Alwatban, A., Alamri, B., Alarifi, N.: A survey on the security of cloud computing. In: 2019 2nd International Conference on Computer Applications Information Security (ICCAIS), pp. 1–7 (2019)
22. Kaufman, L.M.: Data security in the world of cloud computing. IEEE Secur. Priv. **7**(4), 61–64 (2009). https://doi.org/10.1109/MSP.2009.87
23. Spillner, J., Müller, J.: Tutorial on distributed data storage: from dispersed files to stealth databases. In: 2014 IEEE/ACM 7th International Conference on Utility and Cloud Computing, pp. 535–536 (2014)
24. Nepal, S., Friedrich, C., Henry, L., Chen, S.: A secure storage service in the hybrid cloud. In: 2011 4th IEEE International Conference on Utility and Cloud Computing, pp. 334–335 (2011)
25. Anitha, R., Pradeepan, P., Yogesh, P., Mukherjee, S.: Data storage security in cloud using metadata. In: 2nd International Conference on Machine Learning and Computer Science (IMLCS'2013), Kuala Lumpur (Malaysia), pp. 26–30 (2013)
26. Sirohi, P., Agarwal, A.: Cloud computing data storage security framework relating to data integrity, privacy and trust. In: 2015 1st International Conference on Next Generation Computing Technologies (NGCT), pp. 115–118 (2015)
27. Albugmi, A, Alassafi, M.O., Walters, R., Wills, G.: Data security in cloud computing. In: 2016 5th international conference on future generation communication technologies (FGCT). IEEE, pp. 55–59 (2016)
28. Sun, Y., Zhang, J., Xiong, Y., Zhu, G.: Data security and privacy in cloud computing. Int. J. Distrib. Sens. Netw. **10**(7), 190903 (2014)
29. Kamboj, L., Luthra, P.: Multi-layer data security in cloud computing. Int. J. Computat. Eng. Res. (IJCER) **7**, 1–7 (2017)
30. Chen, D., Zhao, H.: Data security and privacy protection issues in cloud computing. In: 2012 International Conference on Computer Science and Electronics Engineering. vol. 1, , pp. 647–651. IEEE (2012)
31. Sharath, M.N., Rajesh, T.M., Patil, M.: Analysis of secure multimedia communication in cloud computing. In: 2019 2nd International Conference on Intelligent Computing, Instrumentation and Control Technologies (ICICICT), vol. 1, pp. 136–144 (2019)
32. Ajala, J.A., Singh, S., Mukherjee, S., Chakraborty, S.: Application of steganography technique in cloud computing. In: 2019 International Conference on Computational Intelligence and Knowledge Economy (ICCIKE), pp. 532–537 (2019)
33. Sarkar, M.K., Chatterjee, T.: Enhancing data storage security in cloud computing through steganography. Int. J. Netw. Secur. **5**(1), 13 (2014)
34. Pant, V.K., Prakash, J., Asthana, A.: Three step data security model for cloud computing based on RSA and steganography. In: 2015 International Conference on Green Computing and Internet of Things (ICGCIoT), pp. 490–494 (2015)
35. Murakami, K., Hanyu, R., Zhao, Q., Kaneda, Y.: Improvement of security in cloud systems based on steganography. In: 2013 International Joint Conference on Awareness Science and Technology Ubi-Media Computing (iCAST 2013 UMEDIA 2013), pp. 503–508 (2013)
36. Ranjan, A., Bhonsle, M.: Advanced technics to shared amp; protect cloud data using multilayer steganography and cryptography. In: 2016 International Conference on Automatic Control and Dynamic Optimization Techniques (ICACDOT), pp. 35–41 (2016)

37. Daemen, J., Rijmen, V.: The Block Cipher Rijndael. In: Quisquater JJ, Schneier B, editors. Smart Card Research and Applications, pp. 277–284. Springer, Berlin, Heidelberg (2000)
38. Pillow (2022). https://python-pillow.org/
39. pydub (2022). http://pydub.com/
40. hashlib (2022). https://docs.python.org/3/library/hashlib.html
41. Salomon, D.: In: Introduction, pp. 1–12. Springer, Berlin Heidelberg (2000)
42. Portable Network Graphics (PNG): Specification, 2nd edn (2003) https://www.w3.org/TR/2003/REC-PNG-20031110/#F-Relationship
43. LSB coding (2022). https://github.com/ragibson/Steganography
44. Notepad++ (2022). https://notepad-plus-plus.org/

How to Optimally Reconfigure Average Consensus with Maximum-Degree Weights in Bipartite Regular Graphs

Martin Kenyeres[1(✉)] and Jozef Kenyeres[2]

[1] Institute of Informatics, Slovak Academy of Sciences, Dubravska cesta 9, 84507 Bratislava, Slovakia
martin.kenyeres@savba.sk

[2] Frequentis AG, Innovationsstraße 1, 1100 Vienna, Austria
jozef.kenyeres@frequentis.com

http://www.ui.sav.sk/w/en/dep/mcdp/

Abstract. The average consensus algorithm is a frequently applied iterative scheme to efficiently estimate various aggregate functions in many modern multi-agent systems. This paper addresses the average consensus with the Maximum-degree weights in bipartite regular graphs, where this algorithm does not converge as the third convergence condition is broken in these graphs. We apply a distributed mechanism (determined by five conditions) for detecting whether or not the algorithm is executed in this critical graph topology and analyze how to reconfigure it the most efficiently. The latest research suggests that its high performance is likely not to be achieved with the same values of the mixing parameter as in non-bipartite non-regular topologies. Hence, our goal is to identify the reconfiguration of the weight matrix ensuring the optimal performance of the algorithm under various conditions. Its performance is examined in randomly generated bipartite regular graphs with various connectivity and under three scenarios differing from each other in the rounding accuracy. Also, the asymptotic convergence factor of the weight matrices after reconfiguration is examined in the experimental section. Besides, the performance of a reconfigured average consensus with the Maximum-degree weights in bipartite regular graphs is compared to the Metropolis-Hastings algorithm, which also diverges in these topologies with its optimal initial configuration.

Keywords: Average consensus algorithm · Bipartite regular graphs · Consensus · Data aggregation · Distributed algorithms · Maximum-degree weights · Multi-agent systems

1 Introduction

Multi-agent systems (referred to as MASs) are one of the most straightforward and generally-acknowledged technology for addressing various topical issues such

R. Silhavy et al. (Eds.): CoMeSySo 2022, LNNS 596, pp. 189–204, 2023.
https://doi.org/10.1007/978-3-031-21435-6_16

as artificial intelligence, bioinformatics, grid computing, military, controlling, etc. [1–3]. Typically, MASs are formed by numerous autonomous decision-making entities (referred to as agents) that can interact with each other over a shared geographical area or with their adjacent environment in order to effectively solve various common problems [4]. As shown in the literature [5–8], the term "agent" can represent many different physical forms in real-world scenarios, e.g., computers, robots, sensor devices, animals, people, etc. As stated in [5], the most popular definition of this term is considered the one from [9] defining an agent of MAS as an independent flexible entity that can perceive the adjacent environment via its sensors. Moreover, the agents of MAS are capable of interacting with their adjacent environment through their actuators [5] (see Fig. 1 for how agents of MASs interact with their adjacent environment).

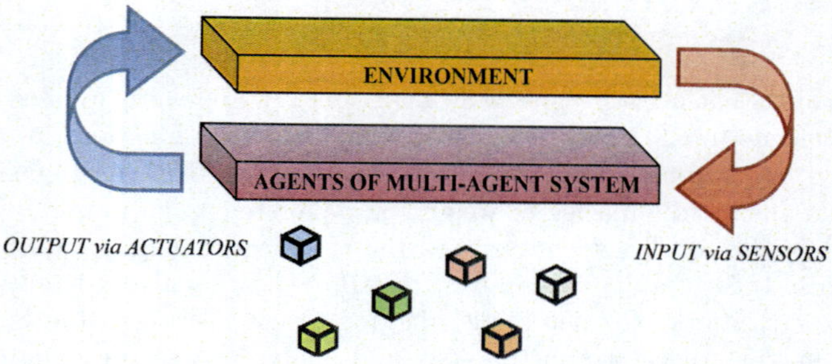

Fig. 1. Graphical demonstration of how agents of multi-agent systems interact with their adjacent environment

Generally speaking, the application of MASs is recommended wherever solving an issue is difficult and inefficient for individuals [10,11]. The application of MASs ensures splitting an issue of one's interest into a set of subgoals that can be effectively solved by different agents with their tasks [5,12]. As a consequence, the execution of problem-solving can be optimized in numerous aspects such as the execution rate, collapse risk, parallelism, etc. [12]. As further stated in [5], an agent of MAS can be characterized by several features that are depicted in Fig. 2.

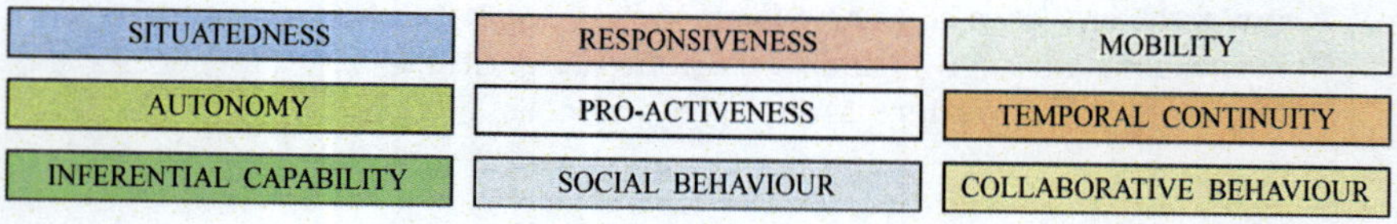

Fig. 2. Characterical (resp. potential) features of agents forming multi-agent systems

As seen from [13–16], many modern MAS-based applications employ mechanisms for aggregating data from multiple independent sources of information.

The term "data aggregation" is generally understood as a process of integrating large information amounts from numerous datasets into a single output credibly representing this intransparent information mass [17]. It can be carried out either manually or by applying specific software tools designed for this purpose [18]. The application of data aggregation in MASs is particularly intended to increase the value of information by summarizing data from disparate sources as well as enabling simple access to large information amounts [19]. One of its most important goals is to provide more information about a set of elements based on specific variables [20].

As seen in [21–23], consensus-based mechanisms for aggregating data have been gaining great attention in many industries over the last decades. A consensus problem is a scientific discipline addressing how a set of independent entities with any initial states can achieve an agreement on some value, provided that neither central coordination nor global communication is allowed [17]. Agents in MASs communicate with each other, update their local states accordingly, and iteratively approach the value of the desired aggregate function [17]. See a general operation of a consensus algorithm depicted in Fig. 3.

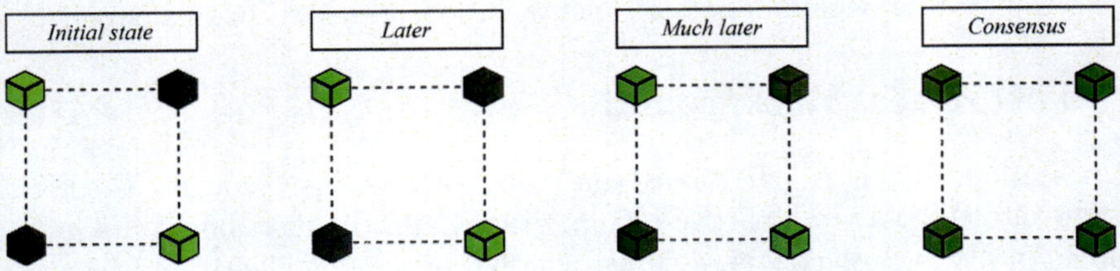

Fig. 3. Example of execution of consensus algorithm—four stages of this algorithm are depicted

In this paper, we focus our attention on the average consensus algorithm with the Maximum-degree weights (referred to as AC MD), which is a distributed consensus algorithm for estimating various aggregate functions (arithmetic mean, graph order, sum) [24]. More specifically, we analyze how to optimally reconfigure the algorithm, provided that it is detected to be executed over bipartite regular graphs[1] by applying the mechanism from [24]. As identified in [25], values of the mixing parameter resulting in a higher convergence rate in non-bipartite non-regular graphs do not always guarantee the optimal performance in bipartite regular graphs after reconfiguration if the Metropolis-Hastings algorithm is executed in this critical topology. This motivates us to investigate how various values of the mixing parameter affect the convergence rate of AC MD after its reconfiguration in bipartite regular graphs – based on the findings presented in [25], the reconfiguration provided in [24] is assumed not to always result in the optimal performance. Our primary contribution presented in this paper is to

[1] AC MD diverges in this critical topology, as identified in [26].

identify the mixing parameter that ensures the fastest algorithm execution in bipartite regular graphs after its reconfiguration. The algorithm is tested over randomly generated bipartite regular graphs with various connectivity and under three scenarios differing from one another in the rounding accuracy of the inner states processed by the applied mechanism. Besides, in the experimental part, the asymptotic convergence factor after reconfiguration is examined, and AC MD is compared to the Metropolis-Hastings algorithm.

This paper is organized as follows:

- Section 2: is concerned with theoretical insight into the topic of interest (i.e., the definition of bipartite regular graphs, the definition of AC MD, and its convergence conditions), introduces the applied distributed mechanism for detecting AC MD in bipartite regular graphs (namely, five conditions for classifying graphs as bipartite regular/non-bipartite non-regular), and provides how to reconfigure the algorithm after completed detection.
- Section 3: deals with the used research methodology and metrics for evaluating the performance of AC MD.
- Section 4: consists of the experimentally obtained results and a discussion about observable phenomena.
- Section 5: briefly summarizes our contribution presented in this paper.

2 Theoretical Background

This section, which is partitioned into two subsections, provides theoretical insight into AC MD in bipartite regular graphs and the distributed mechanism defined in [24], whose goal is to detect whether or not the algorithm is executed in this critical topology.

2.1 Average Consensus with Maximum-Degree Weights in Bipartite Regular Graphs

As stated above, this paper addresses AC MD in MASs of bipartite regular topology, where this algorithm diverges [24,26]. The algorithm is applicable to estimating various aggregate functions, as mentioned earlier. In this paper, we focus our attention on estimating the arithmetic mean from all the initial inner states, which is graphically demonstrated in Fig. 4.

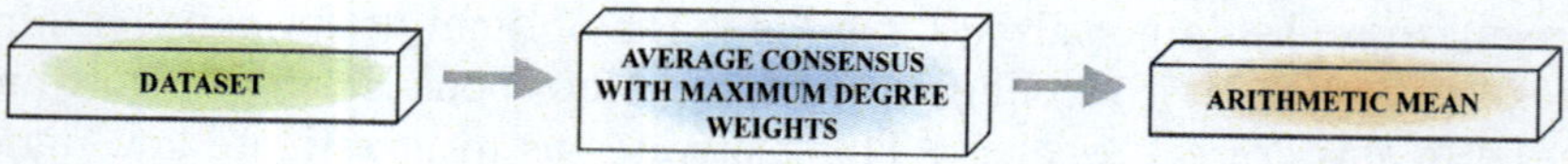

Fig. 4. Application of average consensus algorithm with maximum-degree weights for distributed averaging

In this paper, MASs are represented by undirected simple finite connected graphs G determined by the vertex set $\mathbf{V}$ and the edge set $\mathbf{E}$ ($G = \mathbf{V}, \mathbf{E}$) [17].

As its name evokes, the vertex set $\mathbf{V}$ contains all the graph vertices (i.e., $\mathbf{V} = \{v_1, v_2, ..., v_n\}$),[2] representatives of the agents of MAS with unique index i. The edge set $\mathbf{E} \subset \mathbf{V}\times\mathbf{V}$ gathers all the graph edges in G, which represent a direct link between two vertices from $\mathbf{V}$. The edge linking v_i and v_j[3] is labeled as e_{ij}. As mentioned earlier in this paper, we analyze the algorithm in bipartite regular graphs (see Fig. 5 for an example), which have to meet these two conditions to be classified as these graphs [25]:

1. *their vertex set* $\boldsymbol{V}$ *can be partitioned into two disjoint subsets* $(\boldsymbol{V}_1, \boldsymbol{V}_2)$, *i.e., the vertices from* $\boldsymbol{V}_1$ *can only be adjacent to the vertices from* $\boldsymbol{V}_2$ *and vice versa.*
2. *all the vertices from* $\boldsymbol{V}$ *have the same degree equal to d.*

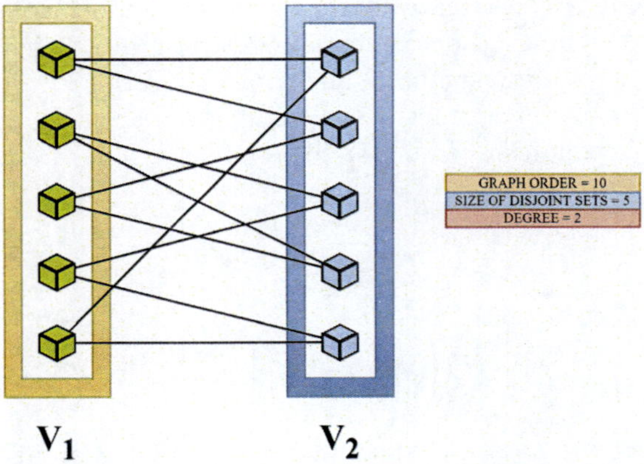

Fig. 5. Example of bipartite regular graph formed by ten vertices

In this paragraph, we turn the readers' attention to the definition of AC MD and related relevant information. The algorithm is based on iterative exchanges of the inner states among the adjacent agents in MAS and applying a local update rule [24]. Let $\mathbf{x}(k)$ be a variant column vector formed by n entries gathering the inner states of all the agents in MAS at a kth iteration [27]. The value $k = 0$ poses the initial inner states, which can represent locally measured values for instance. As stated in [24], the weight matrix $\mathbf{W}$ of AC MD is defined as the following matrix:

$$[W]_{ij} = \begin{cases} \frac{1}{\Delta}, & \text{if } e_{ij} \in \mathbf{E} \\ 1 - \frac{d_i}{\Delta}, & \text{if } i = j \\ 0, & \text{otherwise} \end{cases} \quad (1)$$

[2] Here, n represents the graph order.

[3] Note that we say that such two vertices are adjacent or neighbors.

Here, the parameter d_i represents the degree of the corresponding vertex v_i, and Δ is the maximum degree of the corresponding graph. As we analyze AC MD in bipartite regular graphs, its weight matrix can be simplified as follows:

$$[W]_{ij} = \begin{cases} \frac{1}{d}, & \text{if } e_{ij} \in \mathbf{E} \\ 0, & \text{otherwise} \end{cases} \tag{2}$$

As shown in [28], the update rule of the analyzed algorithm expressed as the following difference equation is likely to be its most common definition:

$$\mathbf{x}(k+1) = \mathbf{W}\mathbf{x}(k) \tag{3}$$

As further stated in [28], the algorithm operates in such a way that all the inner states are iteratively updated according to (3), thereby converging to the value of the estimated aggregate function. In this paper, we assume that AC MD is applied to estimate the arithmetic mean from all the initial inner states; therefore, the inner states asymptotically converge to this arithmetic mean [24]. As shown in [28], the convergence of the algorithm is ensured, provided that the three convergence conditions from (4)–(6) are met.

$$\mathbf{1}^{\mathrm{T}}\mathbf{W} = \mathbf{1}^{\mathrm{T}} \tag{4}$$

$$\mathbf{W}\mathbf{1} = \mathbf{1} \tag{5}$$

$$\rho\left(\mathbf{W} - \frac{1}{n}\mathbf{1}\mathbf{1}^{\mathrm{T}}\right) < 1 \tag{6}$$

Here, the parameter ρ poses the spectral radius, and $\mathbf{1}$ is an all-ones vector. As identified in [26], the weight matrix $\mathbf{W}$ from (1) guarantees the convergence of AC MD in each arbitrary topology except for bipartite regular graphs.

2.2 Five Conditions for Detecting Bipartite Regular Topology in Distributed Fashion

This part is focused on the distributed mechanism (from [24]) for detecting that AC MD is/is not executed in bipartite regular topologies. As mentioned earlier in this paper, it is crucial to detect AC MD in bipartite regular graphs and properly reconfigure it in order to ensure its convergence in this problematic graph topology. As lately identified in [24], the inner states do not approach the desired aggregate function in bipartite regular graphs but oscillate between two values instead. Hence, the algorithm does not operate properly, and its application is therefore counterproductive as the arithmetic mean from the initial inner states of all the agents in MAS cannot be estimated. In our previous research published in [24], we propose a mechanism determined by five conditions for distributively detecting with impressive precision that AC MD is/is not executed in bipartite regular graphs. Besides, it is identified in the mentioned paper that proper reconfiguration of AC MD after completed detection ensures the convergence of the algorithm whereby the algorithm can operate properly also in bipartite regular graphs. Namely, these five conditions are applied to detect this critical graph topology in a distributed fashion [24]:

$$\nabla_2^1\, x_i(k+1) = 0 \tag{7}$$

$$\nabla_2^1\, x_i(k) = 0 \tag{8}$$

$$\triangle^1\, x_i(k) = \nabla^1 x_i(k-1) \tag{9}$$

$$\triangle^1\, x_i(k) \neq 0 \tag{10}$$

$$\nabla^1\, x_i(k-1) \neq 0 \tag{11}$$

Here, the symbols $\triangle^1$ and ∇^1 are the first-order forward finite difference and the first-order backward finite difference, respectively [24]. The lower symbol 2 means the difference between the inner states two iterations away from each other, and no provided symbol represents the difference between two subsequent inner states [24]. The detection is carried out on each agent in the network separately, and its result does not affect other agents in detection. Besides, in our previous paper [24], the mixing parameter ϵ is reconfigured to the value theoretically outperforming other reconfigurations of this parameter. However, as later identified in [25], better performing mixing parameters ϵ in non-bipartite non-regular topologies do not always ensure higher performance in bipartite regular graphs after reconfiguration, provided that the generalized Metropolis-Hastings algorithm is applied. More specifically, the mixing parameter ϵ is equal to 0.9 $\frac{1}{\Delta}$ after reconfiguration in [24], but we do not assume according to the findings from [25] that this configuration always results in the best performance of AC MD. Therefore, the goal of our research presented in this paper is to vary the mixing parameter ϵ in order to identify whether its other values can outperform the theoretically optimal mixing parameter, as seen in the case of the Metropolis-Hastings algorithm [25]. Thus, the weight matrix $\mathbf{W}$ is reconfigured in our experiments after completed detection as follows:

$$[W]_{ij} = \begin{cases} \frac{\psi}{d}, & \text{if } e_{ij} \in \mathbf{E} \\ 1-\psi, & \text{if } i = j \\ 0, & \text{otherwise} \end{cases} \tag{12}$$

Here, ψ is a weighting coefficient of the mixing parameter ϵ and affects its value after reconfiguration. In order to ensure the convergence of the algorithm after reconfiguration, it has to take a value from the following interval (see Fig. 6 for examples of three different reconfigurations):

$$\psi \in (0,1) \tag{13}$$

The algorithm is reconfigured once all the agents in MAS classify the graph as bipartite regular [25].

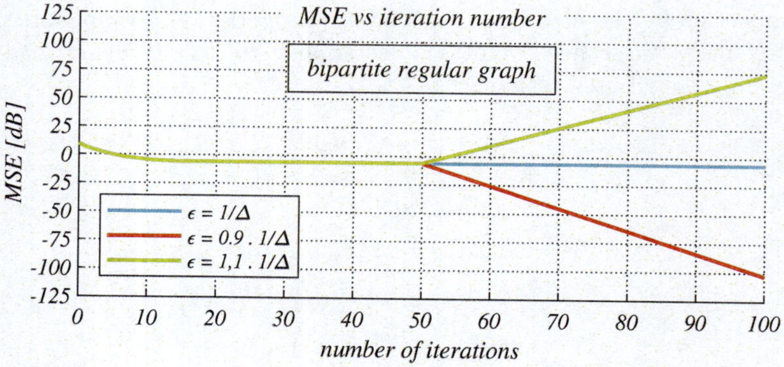

Fig. 6. Evolution of inner states in bipartite regular graph after algorithm is reconfigured—weighting coefficient ψ is equal to either 1.1 (mean square error (MSE) [29] grows after reconfiguration), 1 (MSE neither grows nor declines), or 0.9 (MSE declines)

3 Research Methodology and Metrics for Performance Evaluation

This section deals with the research methodology exploited in our research and the metrics applied for the performance evaluation of AC MD. To evaluate the performance of the algorithm, we apply the convergence rate expressed as the number of iterations necessary for achieving the consensus among the agents. Note that we bound the algorithm by applying the stopping criterion defined in (14) since the inner states asymptotically converge to the arithmetic mean after reconfiguration [24]. In our experiments, the consensus is considered to have been achieved at the iteration when the following condition is met for the first time [25]:

$$\left| x_i(k) - \frac{\mathbf{1}^{\mathrm{T}}\mathbf{x}(0)}{n} \right| < P, \ for\ \forall v_i \in \mathbf{V} \tag{14}$$

Here, the parameter P determines the performance of the algorithm. In this paper, we set its value to 0.0001 just like in [25] in order to directly compare our experimental results with those from the cited paper. Its value conditions both the estimation precision and the convergence rate in such a way that the precision can be increased at a cost of decelerating the algorithm. As mentioned earlier in this paper, the weight matrix $\mathbf{W}$ is reconfigured according to (12) once AC MD is detected to be executed in bipartite regular graphs. In this paper, the weighting coefficient ψ is varied as follows in order to identify which setup of the weight matrix $\mathbf{W}$ results in the best performance after reconfiguration:

- $\psi \in [0.1, 0.9]$ (with step = 0.1)

Like in our previous paper [25], the performance of AC MD is tested over randomly generated bipartite d-regular graphs with the graph order $n = 30$

(note that 100 graphs with unique topology are generated for each d). In our experiments, we analyze AC MD in bipartite regular graphs of various connectivity that is ensured by changing the value of the parameter d as follows:

- $d = \{2, 3, 4, 5\}$

In the figures provided in Sect. 4, we only depict and analyze the average convergence rate (averaged over 100 unique graphs for each d), which is chosen as a representative of the performance of AC MD in the given graphs.

In this paper, AC MD is applied to estimate the arithmetic mean from all the initial inner states, which are independent and identically distributed random values of the standard Gaussian distribution in the experimental part, i.e.,

$$x_i(0) \sim N(0,1),\ for\ \forall v_i \in \mathbf{V} \tag{15}$$

We use the same inner states as in [25], allowing a direct comparison with the results from the cited paper.

In [24], it is stated that the mechanism determined by (7)–(11) has to work with rounded values of the inner states. Thus, we examine these three scenarios (like in [25]) differing from each other in the rounding accuracy as follows:

- Scenario 1 (abbreviated as S1): the applied mechanism rounds the inner states to decimals.
- Scenario 2 (abbreviated as S2): the applied mechanism rounds the inner states to hundreds.
- Scenario 3 (abbreviated as S3): the applied mechanism rounds the inner states to thousands.

In the experimental part, we also focus our attention on the asymptotic convergence factor r_{asym}, a measure to evaluate the performance of the algorithm [28] (see (16) for its definition). We analyze the weight matrices $\mathbf{W}$ for each examined weighting coefficient ψ in order to identify whether lower values of r_{asym} [4] result in better performance in bipartite regular graphs.

$$r_{asym} = \sup_{\mathbf{x}(0) \neq \overline{x}} \lim_{k \to \infty} \left(\frac{||\mathbf{x}(k) - \overline{x}||_2}{||\mathbf{x}(0) - \overline{x}||_2} \right)^{\frac{1}{k}} \tag{16}$$

Here, the symbol $||\cdot||$ is used to refer to the L2 norm (in some sources, it is findable as its equivalent "the Euclidean norm"), a vector norm defined for complex vectors to determine the distance of the vector from the origin. The symbol $\overline{x}$ represents the arithmetic mean calculated from the inner states of all the agents in MAS. In the experimental part, r_{asym} averaged over 100 unique graphs for each d and each ψ is examined.

[4] Note that lower values of the asymptotic convergence factor r_{asym} indicate higher performance of the algorithm [28].

4 Experiments and Discussion about Observed Phenomena

In this part, we present the results obtained in Matlab2018a and provide a subsequent discussion about observable phenomena. As already stated in this paper, we examine the convergence rate of AC MD in bipartite regular graphs, provided that the algorithm in this topology is detected by (7)–(11) and reconfigured according to (12). As further mentioned, our goal is to identify the optimal reconfiguration of the weight matrix **W** in various bipartite regular graphs and under three different scenarios since reconfigurations ensuring higher performance are assumed to vary from high performing configurations in non-bipartite non-regular topologies. Also, the asymptotic convergence factor r_{asym} after reconfiguration is analyzed, and a comparison of AC MD with the Metropolis-Hastings algorithm in terms of the maximum convergence rate is provided.

The first experiment is concerned with the asymptotic convergence factor r_{asym} of the reconfigured weight matrices **W**, i.e., we analyze this metric for each examined weighting coefficient ψ and each selected d. From the results depicted in Fig. 7, one can see that higher ψ results in lower values of the asymptotic convergence factor r_{asym}; therefore, the performance after reconfiguration is assumed to be greater for higher values of ψ. Also, the difference among the configurations increases for higher values of d, as seen in the figure. Moreover, it is observed that an increase in d ensures lower values of the asymptotic convergence factor r_{asym} for each ψ. Hence, greater performance after reconfiguration is expected in graphs of higher connectivity.

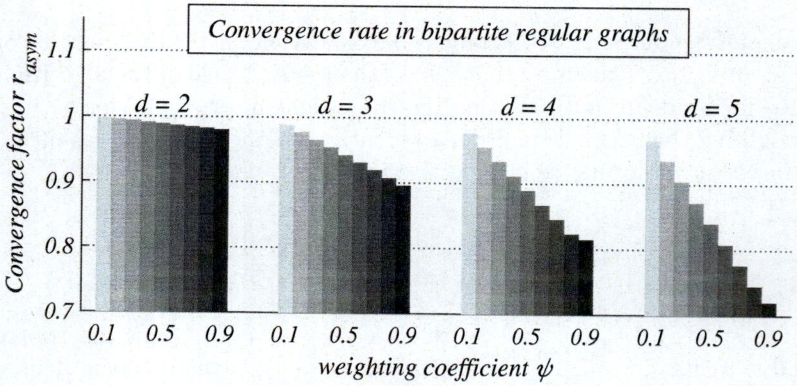

Fig. 7. Asymptotic convergence factor r_{asym} of reconfigured weight matrices **W** in various bipartite regular graphs and for various values of weighting coefficient ψ

In the following paragraphs, we turn the readers' attention to how to optimally reconfigure the weight matrix **W** of AC MD after the applied mechanism detects a bipartite regular topology. First, let us see Fig. 8, where the results for $d = 2$ are shown. We can see from this figure that an increase in the weighting coefficient ψ ensures higher performance of the algorithm after reconfiguration. Besides, it can be observed that the best performance is achieved in Scenario 3, while, the worst one is in Scenario 1. However, the difference among the scenarios is decreased as ψ grows. The maximum convergence rate is equal to 422.9 it. in Scenario 1, 419.4 it. in Scenario 2, and 408.3 it. in Scenario 3. Compared to the Metropolis-Hastings algorithm analyzed in [25], AC MD performs worse in terms of the maximum convergence rate in all the three examined scenarios – the maximum convergence rate of the concurrent algorithm takes 367. it. in Scenario 1, 366.1 it. in Scenario 2, and 361.1 it. in Scenario 3.

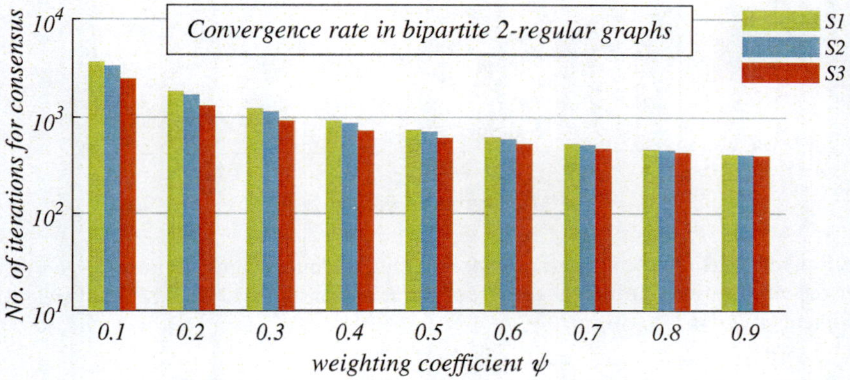

Fig. 8. Convergence rate of average consensus algorithm with maximum-degree weights after reconfiguration in all three examined scenarios and for various values of weighting coefficient ψ – $d = 2$

In the next figure (Fig. 9), we analyze the convergence rate if $d = 3$ (i.e., in better-connected graphs). From the results, we can see that the algorithm performs better in graphs of higher connectivity. Also, we can see that an increase in the weighting coefficient ψ results in a decrease in the number of iterations for consensus in Scenario 1/2 just like in the previous analysis. However, in Scenario 3, we can see that the maximum convergence rate is achieved for $\psi = 0.8$, and the convergence rate then begins to decline with further growth of ψ. The best performance is achieved in Scenario 3 and the worst in Scenario 1, however only for $\psi \in [0.1 - 0.8]$ (on this interval, we can further see that the difference among the scenarios is decreased as ψ is increased like in the previous analysis). In the case of $\psi = 0.9$, Scenario 1 outperforms the two other scenarios, meanwhile, Scenario 3 achieves the worst performance for this configuration—a very similar phenomenon is observed for the Metropolis-Hastings algorithm [25]. The maximum convergence rate is equal to 78.75 it. in Scenario 1 (for $\psi = 0.9$), 79.67 it. in Scenario 2 (for $\psi = 0.9$), and 78.99 it. in Scenario 3 (for $\psi = 0.8$); therefore, we can see that the difference in this aspect is just marginal for this d. At the end of this paragraph, again, we focus on a comparison of AC MD with the Metropolis-Hastings algorithm [25] achieving the following maximum convergence rates in bipartite 3-regular graphs: 73.77 it. in Scenario 1, 73.75 it. in Scenario 2, and 73.68 it. in Scenario 3. Thus, as seen, AC MD is worse in terms of the maximum convergence rate than the mentioned concurrent algorithm like in the previous analysis.

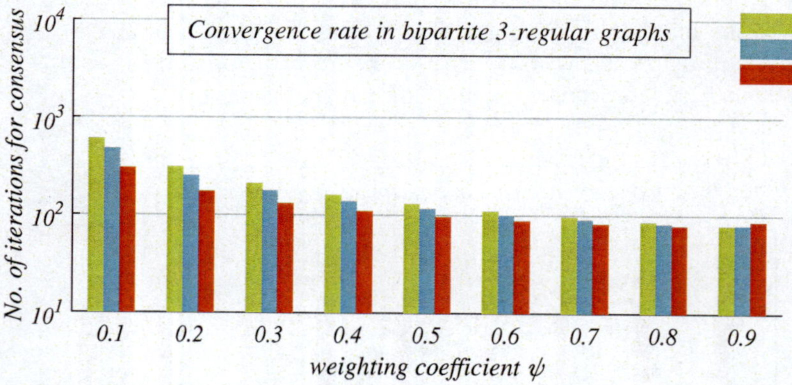

Fig. 9. Convergence rate of average consensus algorithm with maximum-degree weights after reconfiguration in all three examined scenarios and for various values of weighting coefficient ψ – $d = 3$

In this paragraph, we turn our attention to the results when $d = 4$ (see Fig. 10). From the figure, we see that an increase in the weighting coefficient ψ ensures an increase in the convergence rate until the maximum convergence rate is reached (on this interval, we can see again that the difference among the scenarios is decreased as ψ increases). However, both Scenario 1 and 2 reach their maximum convergence rate for $\psi = 0.8$ in contrast to the two previous analyses. In Scenario 3, the maximum is achieved for $\psi = 0.6$ and therefore for a lower value than in the two other scenarios. In all the three scenarios, the convergence rate starts declining after the maximum is reached, as seen in the figure. For lower values of the weighting coefficient ψ, the best performance is achieved by Scenario 3 and the worst one by Scenario 1. In contrast, Scenario 1 outperforms the two other scenarios, meanwhile, Scenario 3 is the worst one for higher values of the weighting coefficient ψ. The maximum convergence rates take the following values: 40.16 it. in Scenario 1, 39.87 it. in Scenario 2, and 39.44 it. in Scenario 3. Again, the difference among the scenarios in this aspect is just marginal like in the previous analysis. In bipartite 4-regular graphs, the Metropolis-Hastings algorithm minimally requires 36.99 it. in Scenario 1, 36.78 it. in Scenario 2, and 36.12 it. in Scenario 3, as seen from [25]; therefore, AC MD is outperformed by it in this aspect once again.

In the last figure (Fig. 11), we see that the functions for $d = 5$ have the same character as in the previous analysis. The maximum convergence rates are, however, lower and equal to 27.66 it. in Scenario 1 (for $\psi = 0.8$), 26.62 it. in Scenario 2 (for $\psi = 0.7$), and 24.54 it. in Scenario 3 (for $\psi = 0.5$). Thus, it is

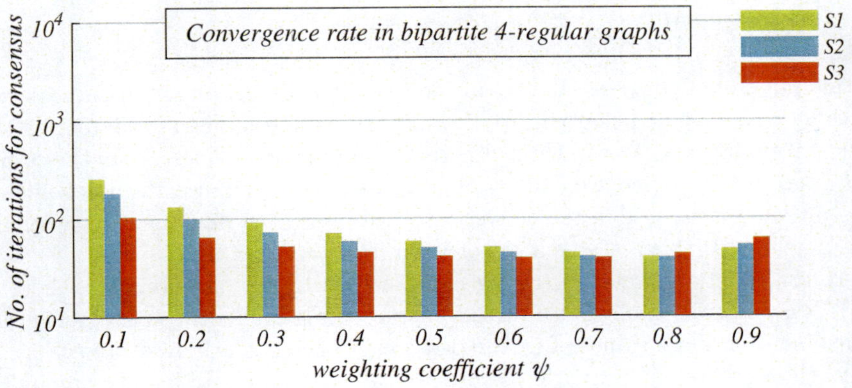

Fig. 10. Convergence rate of average consensus algorithm with maximum-degree weights after reconfiguration in all three examined scenarios and for various values of weighting coefficient $\psi - d = 4$

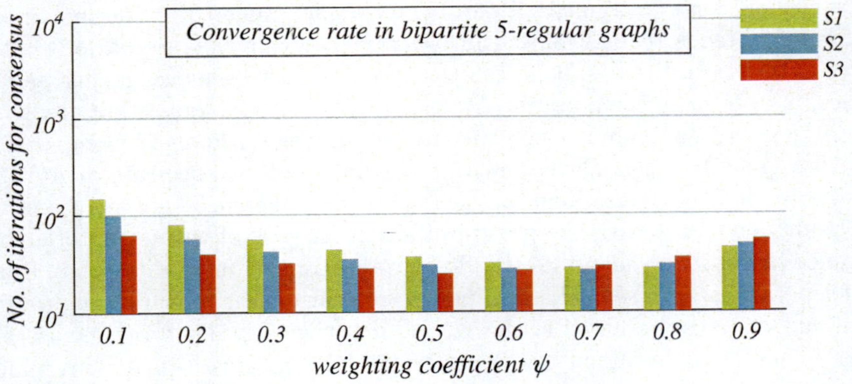

Fig. 11. Convergence rate of average consensus algorithm with maximum-degree weights after reconfiguration in all three examined scenarios and for various values of weighting coefficient $\psi - d = 5$

also seen that an increase in d causes that the maximum convergence rate tends to be achieved for lower ψ. Regarding comparison with the Metropolis-Hastings algorithm, the maximum convergence rate of AC MD is worse than its again (24.88 it. in Scenario 1, 24.49 it. in Scenario 2, and 22.46 it. in Scenario 3). Thus, we can see from the comparisons for each examined graph connectivity that the difference between AC MD and the Metropolis-Hastings algorithm is decreased as d increases.

From Figs. 8, 9, 10 and 11, we can further see that the algorithm performs better for higher values of d in each scenario. As also seen from Figs. 7, 8, 9, 10 and 11, a higher asymptotic convergence factor r_{asym} ensures a higher convergence rate in bipartite regular graphs after reconfiguration only until the maximum convergence rate is achieved. Besides, both r_{asym} and the convergence rates after reconfiguration are greater if d takes higher values in each scenario.

5 Conclusion

In this paper, we address AC MD in bipartite regular graphs with various connectivity and provided that the applied distributed mechanism for detecting this critical topology operates in three different rounding accuracies. From the results presented in this paper, we can see that an increase in the weighting coefficient ψ results in an increase in the convergence rate until the maximum convergence rate is achieved (this statement is not valid for $d = 2$ and Scenario 1/2 if $d = 3$, where the functions have a growing character on the whole interval $\psi \in [0.1 - 0.9]$). It is then observed that a further increase in ψ causes the convergence rate to decline. For lower values of ψ, the best performance is achieved in Scenario 3, while, the worst one in Scenario 1. Here, we can see that the difference among the scenarios decreases as ψ grows. The maximum convergence rate is also achieved in Scenario 3, and the worst one is in Scenario 1 for each examined d except for $d = 3$, when the maximum convergence rate is achieved in Scenario 1. Also, we can see that the difference among all the three scenarios in this aspect is just marginal for $d = 3$ and 4. However, for higher values of ψ, the maximum performance tends to be achieved in Scenario 1, meanwhile, Scenario 3 is the worst in this case. Besides, it is seen that the algorithm achieves higher performance in better-connected graphs. Moreover, better graph connectivity tends to ensure that the maximum convergence rate is reached for lower ψ. Compared to the Metropolis-Hastings algorithm, the maximum convergence rate of AC MD is lower in all the scenarios and for each d, but the difference between these two algorithms decreases with the growth of d. An analysis of the asymptotic convergence factor r_{asym} identifies that lower values of this metric do not indicate higher performance of AC MD in bipartite regular graphs after reconfiguration in all our analyses presented in this paper. It is shown in our experiments that lower values r_{asym} results in higher performance of the algorithm after reconfiguration only until the maximum convergence rate is reached. Also, it is identified that r_{asym} is higher in better-connected graphs just like the convergence rate after reconfiguration.

Acknowledgment. This work was supported by the VEGA agency under the contract No. 2/0155/19 and by the project Cyber-Physical System for smart telemonitoring and tele-medicine for patients with COVID-19 (BAS-SAS-21-03). Since 2019, Martin Kenyeres has been a holder of the Stefan Schwarz Supporting Fund.

References

1. Calegari, R., Ciatto, G., Mascardi, V., Omicini, A.: Logic-based technologies for multi-agent systems: a systematic literature review. Autonomous Agents and Multi-Agent Systems **35**(1), 1–67 (2020). https://doi.org/10.1007/s10458-020-09478-3
2. Cardoso, R.C., Ferrando, A.: A review of agent-based programming for multi-agent systems. Computers **10**, 1–15 (2021). https://doi.org/10.3390/computers10020016

3. Melgarejo, D.C., Da Costa Filho, L.Q.R., De Medeiros, Á.A.M., Neto, C.L., Figueiredo, F.L., Rodríguez, D.Z.: Dynamic algorithm for interference mitigation between cells in networks operating in the 250 MHz band. IEEE Access **10**, 33803–33815 (2022). https://doi.org/10.1109/ACCESS.2022.3162618
4. Chen, F., Ren, W.: On the control of multi-agent systems: a survey. Found. Trends Syst. Control **6**, 1–164 (2019). https://doi.org/10.1561/260-000-0019
5. Balaji, P.G., Srinivasan, D.: An introduction to multi-agent systems. Stud. Comput. Intell. **310**, 1–27 (2010). https://doi.org/10.1007/978-3-642-14435-6_1
6. Iskandarani, M.Z.: Effect of error probability, data rates, output power and noise floor on link quality of WSN channels. Int. Rev. Electr. Eng. **16**, 596–607 (2021). https://doi.org/10.15866/iree.v16i6.20742
7. Jaichandran, R., ShyamalaBharathi, P., Rani, D.P., Kalaiselvi, V.K.G., Lal, S.P.: Consensus based data transfer in wireless sensor network for reliability interpretation. In: First International Conference On Computing, Communication and Control System (I3CAC 2021), pp. 1–7. Core Publishers, London (2021). https://doi.org/10.4108/eai.7-6-2021.2308782
8. Melgarejo, D.C., Pokorny, J., Seda, P., Narayanan, A., Nardelli, P.H.J., Rasti, M., Hosek, J., Seda, M., Rodríguez, D.Z., Koucheryavy, Y., Fraidenraich, G.: Optimizing flying base station connectivity by RAN slicing and reinforcement learning. IEEE Access **10**, 53746–53760 (2022). https://doi.org/10.1109/ACCESS.2022.3175487
9. Russell, S., Norvig, P.: Artificial Intelligence: A Modern Approach. Prentice Hall, New Jersey (2002)
10. Sabo, R., Krammer, P., Mojzis, J., Kvassay, M.: Identification of spontaneous spoken texts in Slovak. Jazykoved. čas. **70**, 481–490 (2019). https://doi.org/10.2478/jazcas-2019-0076
11. Yousif, N.A., Naji, D.J., Al-Khazraji, A., Raafat, S.M.: Consensus multi-health care systems with optimized quality of measurement. Int. J. Comput. Digit. Syst. **11**, 1147–1155 (2022). https://doi.org/10.12785/ijcds/110193
12. Ge, X., Han, Q.L., Ding, D., Zhang, X.M., Ning, B.: A survey on recent advances in distributed sampled-data cooperative control of multi-agent systems. Neurocomputing **275**, 1684–1701 (2018). https://doi.org/10.1016/j.neucom.2017.10.008
13. Ullah, A., Azeem, M., Ashraf, H., Alaboudi, A.A., Humayun, M., Jhanjhi, N.Z.: Secure healthcare data aggregation and transmission in IoT–a survey. IEEE Access **9**, 16849–16865 (2021). https://doi.org/10.1109/ACCESS.2021.3052850
14. Dorri, A., Kanhere, S.S., Jurdak, R.: Multi-agent systems: a survey. IEEE. Access **6**, 28573–28593 (2018). https://doi.org/10.1109/ACCESS.2018.2831228
15. Mojžiš, J., Kvassay, M.: Review Rating Prediction Based on Text Polarity in the Czech and Slovak Languages. In: Silhavy, R. (ed.) CSOC 2021. LNNS, vol. 229, pp. 251–261. Springer, Cham (2021). https://doi.org/10.1007/978-3-030-77445-5_22
16. Pourghebleh, B., Navimipour, N.J.: Data aggregation mechanisms in the Internet of things: A systematic review of the literature and recommendations for future research. J. Netw. Comput. Appl. **97**, 23–34 (2017). https://doi.org/10.1016/j.jnca.2017.08.006
17. Kenyeres, M., Kenyeres, J.: Comparative study of distributed consensus gossip algorithms for network size estimation in multi-agent systems. Fut. Internet **13**, 134 (2021). https://doi.org/10.3390/fi13050134
18. Deshpande, P.: Predictive and perspective analytics in big data era. In: Iyer, B., Nalbalwar, S.L., Pathak, N.P. (eds.) Computing, Communication and Signal Processing. AISC, vol. 810, pp. 123–132. Springer, Singapore (2019). https://doi.org/10.1007/978-981-13-1513-8_14

19. Abbasian Dehkordi, S., Farajzadeh, K., Rezazadeh, J., Farahbakhsh, R., Sandrasegaran, K., Abbasian Dehkordi, M.: A survey on data aggregation techniques in IoT sensor networks. Wireless Networks **26**(2), 1243–1263 (2019). https://doi.org/10.1007/s11276-019-02142-z
20. Ghai, S., Kumar, V., Kumar, R., Vaid, R.: Optimized Multi-level Data Aggregation Scheme (OMDA) for Wireless Sensor Networks. In: Marriwala, N., Tripathi, C.C., Kumar, D., Jain, S. (eds.) Mobile Radio Communications and 5G Networks. LNNS, vol. 140, pp. 443–457. Springer, Singapore (2021). https://doi.org/10.1007/978-981-15-7130-5_35
21. Zhou, Y., Liu, Y., Zhao, Y., Huang, P.: Appointed-time average consensus over directed networks. IEEE Trans. Circuits Syst. II Exp. Briefs IEEE T CIRCUITS-II **69**, 2922–2926 (2022). https://doi.org/10.1109/TCSII.2022.3152521
22. Olfati-Saber, R., Fax, J.A., Murray, R.M.: Consensus and cooperation in networked multi-agent systems. Proc. IEEE **95**, 215–233 (2007). https://doi.org/10.1109/JPROC.2006.887293
23. Abdulghafor, R., Abdullah, S.S., Turaev, S., Othman, M.: An overview of the consensus problem in the control of multi-agent systems. Automatika **59**, 143–157 (2018). https://doi.org/10.1080/00051144.2018.1492688
24. Kenyeres, M., Kenyeres, J.: Distributed mechanism for detecting average consensus with maximum-degree weights in bipartite regular graphs. Mathematics **9**, 3020 (2021). https://doi.org/10.3390/math9233020
25. Kenyeres, M., Kenyeres, J.: Convergence of metropolis-hastings algorithm with optimal mixing parameter in bipartite regular graphs. Lect. Notes Netw. Syst. **501**, 484–498 (2022). https://doi.org/10.1007/978-3-031-09070-7_40
26. Schwarz, V., Hannak, G., Matz, G.: On the convergence of average consensus with generalized metropolis-hasting weights. In: 2014 IEEE International Conference on Acoustics, Speech, and Signal Processing, (ICASSP 2014), pp. 5442–5446. IEEE Press, New York (2019). https://doi.org/10.1109/ICASSP.2014.6854643
27. Kishida, M., Ogura, M., Yoshida, Y., Wadayama, T.: Deep learning-based average consensus. IEEE. Access **8**, 142404–142412 (2020). https://doi.org/10.1109/ACCESS.2020.3014148
28. Xiao, L., Boyd, S.: Fast linear iterations for distributed averaging. Syst. Control. Lett. **53**, 65–78 (2004). https://doi.org/10.1016/j.sysconle.2004.02.022
29. Pereira, S.S., Pages-Zamora, A.: Mean square convergence of consensus algorithms in random WSNs. IEEE Trans. Signal Process **58**, 2866–2874 (2010). https://doi.org/10.1109/TSP.2010.2043140

Blockchain as an Object of Value Assessment

I. Z. Yarygina, A. A. Pomulev(✉), and A. A. Bakhramov

Financial University Under the Government of the Russian Federation, Moscow, Russia
sasha-pomulev@yandex.ru

Abstract. The article reveals the features of blockchain technology in contemporary situation of the Russian financial market. The problems of economic and value assessment of blockchain technology, which is the object of non-financial digital assets are discovered. The authors consider the typology of blockchain and their value for business. The types of effects are formulated and the structure of the financial model to determine the economic efficiency is specified. The results confirm the correct choice of methods that justify the cost estimation. The calculation of the cost of NFT-marketplace is carried out. Based on the analysis, proposals aimed at expanding the use of new technologies in business processes have been made.

Keywords: Evaluation · Blockchain technology · Cost estimation · Value for business

1 Introduction

The new breakthrough technologies of the fourth industrial revolution are accelerating the development of the digital economy. The pandemic, caused by COVID-19, contributed to the adoption of digital technology to the economy. One of the most sought-after areas of digital technology is blockcha as a continuous sequence of blocks containing information, built according to certain rules. The Russian scientists consider the blockchain technology as a combination of several time-tested technologies, such as cryptography, using a hash function; data storage; public key encryption. The following properties of blockchain are highlighted: unified, continuous, immutable, decentralized, protected by cryptography and mathematics, open code registry.

According to the Russian practice, the following varieties of blcochain networks worthwhile mentioning: public (Bitcoin, Etherium, Iota), where any external participant can connect to edit the base; consortium (Hyperledger), where special permission to join the base is required; private (Rospatent)—internal use, external participants are allowed to observe database changes.

Private institutions like banks have realized that they can take the basic idea of blockchain and create a chain (private or consortium/federal) where the validator is a member of a legal entity of the same institution. The term “blockchain” in this context is assessed as highly controversial. Thus, the term “distributed ledger technology” (DLT) has become more general in the Russian Federation.Open DLTs are the open-source

R. Silhavy et al. (Eds.): CoMeSySo 2022, LNNS 596, pp. 205–215, 2023.
https://doi.org/10.1007/978-3-031-21435-6_17

protocols based on consensus algorithms. It is extremely useful for financial institutions as anyone can join the system, for example, run a full node, make transactions across the network or expect them to be included into blockchain. Anyone can read the transaction in the block explorer. Transactions are transparent but anonymous or pseudo-seed, backed by BTC, ETH, LTC, XMR, DOGE, etc.

It is worthwhile mentioning, that new technologies influence financial relations of any market participants. They disrupt existing business models by rejecting mediation, reduce infrastructure costs to zero, delete the necessity of maintaining servers or sysadmins, drastically reducing the cost of building and running decentralized applications (dApps).

Federated distributed ledgers, as a type of a blockchain consortia, are systems that operate under the general principles of a linked group. Unlike public ones, they do not allow anyone with Internet access to participate in the verification process of transactions. Such registries are faster and more confidential. They are mainly used in the banking sector. Pre-selected nodes control commitment to consensus. For example, a consortium of 15 financial institutions, each of which holds a node, and 10 of them must sign each block (R3, EWF, B3i, Corda). As a result, financial institution as a participant of a consortium, reduces transaction costs and data redundancy, replaces outdated systems, simplifies workflow and reduces manual labor. Generally, such kind of adherence complements the ecosystem. Private blockchains are not DLT because permissions are stored centrally for one organization. Read permissions can be issued as public or limited (MONAX, Multichain). Actually, they are not Russian banks' priorities. Assessment demonstrates, that clients consider, that private or federated registries can collapse, repeating the Internet fortune of the 90s. Back then, private companies built their own local area networks or WANs instead of using the public Internet, and then became obsolete, especially with the advent of SAAS in Web2. Anyway, the most attractive is the implementations of distributed ledgers with its store transactions sequentially, one after the other. The private network is supported by observer nodes, which are usually miners. The activity is to a certain extent secure, but inherently slow way to process transactions. This inherent limitation is often referred to as the «scaling problem». Actually, several projects are moving away from blockchain-style building, creating entirely new types of DLT. The most common alternative is the Directed Acyclic Graph (DAG) system. It looks like a network of blocks instead of a chain. Having gone beyond the restriction in the recording of blocks (sequentially, along the chain), it became possible to conduct a larger number of transactions and solve the scaling problem. These systems are mainly aimed at zero-reward instant transactions.

While evaluating blockchain as object of value assessment, it is important to mention cloud technologies, that allow us to access resources (from storage systems to business services) that can be quickly provided, scaled and optimized with minimal operational costs. The use of cloud technologies allows us to quickly locate the number of resources depending on changing requirements, and allows small companies to increase resources as needed at a certain price. The cloud infrastructure creates the conditions for the implementation of joint initiatives between financial organizations, fintech companies and institutions from other sectors of the economy, that allows creating new business models and accelerates the introduction of new products to the market. Robotization of blockchain activity is also very attractive at the moment. Optimization and automation of repetitive processes through programmable algorithms is frequently used in Russia at the moment. As a result, the workflow gets quantitative and qualitative optimization, the algorithm can work almost around the clock and consider incoming tasks of a market participants in a short time. In this case, it is important to control the metrics of task execution by algorithms, manage data, and adapt the algorithm to changing conditions.

Blockchain is actually presented in different types of platforms and business. Gartner [1] estimates that blockchain productivity is expected to reach a plateau in 2–5 years for cryptocurrencies, blockchain wallets, and 5–10 years for decentralized applications and make different types of processes faster, easier, and cheaper.

Russian experts conclude, that blockchain technology grows in popularity and the number of intellectual property and, therefore, objects of value will increase [2]. We consider blockchain as a digital intellectual asset, then for its valuation it is advisable to follow the International Valuation Standards [3], the Federal Standard for the Valuation of Intangible Assets and Intellectual Property (FSO № 11) [4].

However, methodological recommendations are required to consider the specificity of the blockchain technology with its objects and types of determinable value. From our perspective, the publications on the value assessment of blockchain is extremely important for consumers of new technologies for its better understanding and usage.

2 Methodology

The purposes of the valuation determine the type of value to be determined by the valuation object. According to the methodology of valuation, which is based on economic theory, it is customary to divide the value into two categories:

1. Value in use (reflects the point of view of a particular individual on the usefulness of the object);
2. Value in exchange (reflects the point of view on the utility of the object by the set of individuals who form the demand on the market).

Consider the goals of blockchain technology valuation and the type of value that corresponds to:

(a) for making management decisions in projecting the effects of the intended use of the technology (under Federal Valuation Standard № 2—investment [5]);
(b) for the purposes of conclusion of an exclusive or non-exclusive license agreement (in the frameworks of the Federal Valuation Standard № 2—market value [5]);
(c) for the purpose of accounting (for the purposes of the Russian accounting regulation—book value, for the purposes of international financial reporting standards—fair value);
(d) for pledge purposes when obtaining financing (under the Federal Valuation Standard № 9 [6]—market, investment or liquidation value).

When calculating blockchain value, the appraiser must take into account the following internal and external factors, such as:

- legal aspect (existence and protection);
- level of costs for technology creation (purchase of technical means, labor remuneration);
- economic effect (creation of a new service/product, optimization of time for performing, risk management, etc.);
- restrictions;
- quality of operations;
- market peculiarity (availability of demand, volume, supply and demand ratio).

For the whole list of assets created on the basis of blockchain technology, it is possible to distinguish different legal forms of ownership of these assets [7].

The main ones are exclusive rights to blockchain-based IPRs, such as: ownership rights to various complete objects (assets), such as various platforms, networks, private blockchains, and complete technical business solutions; rights of access to various know-how and proprietary algorithms and code structures, various software licenses.

Then, it is reasonable to mention contractual obligations of participants: a contract to develop certain software and various contracts and agreements for the employ of various off-the-shelf objects.

Since blockchain technology is used to create computer networks, applications and other software, any development based on blockchain falls under 3 types of intangible assets: software itself; blockchain networks, specifics of algorithms and code structures applicable to the development [7].

Due to the active development and proliferation of blockchain technology, by 2022 all major types of blockchain assets have formed their own markets.

When conducting an appraisal, the appraiser must examine all market segments in which the appraised object is positioned, as well as all segments in which the project products that can be produced using the appraised object are positioned. At the same time, the appraiser should analyze only the markets of similarly structured objects.

When choosing approaches and methods for blockchain valuation, the Discounted Cash Flow method (DCF) will be preferable, as it reflects the expected utility of the object. The second in importance will be the cost approach (Table 1).

3 Results

The consideration of the most common purposes of technology evaluation: investment decisions and sale of technology/business/development has led to the following results.

The effectiveness of implementation depends on the level of digitalization of processes in an organization. Technology evaluation is a step in the overall process of managing intellectual property in an organization.

Table 1. Approaches and methods of blockchain technology cost estimation (based on)

Method	The possibility of method application to evaluation
The cost approach (the method of direct calculation, the method of indexation)	In some cases is the only applicable, if there are documented costs
Comparative approach (transaction method)	Is not applicable as there are no comparable analogues
Income approach (method of real license payments calculation)	Applicable if the effect can be estimated quantitatively and expressed in a monetary equivalent
Profit-benefit method, cost advantage method	
The method of the advantage in sales volume, the method of economy in operating costs	
Direct capitalization method	
Discounted cash flow method	
Venture method	
Surplus income method	

Source Calculated and worked out by the authors

The intangible asset management process, in the context of the overall concept of business intellectual property management, consists of the following steps:

(a) decision of the appropriateness of development;
(b) development;
(c) evaluation of quality and effects;
(d) patent protection (legal identification);
(e) evaluation of the value;
(f) registration as an intangible asset (IA);
(g) implementation into operational process or sale.

When assessing the effects, resulting as the introduction of technology, one should not forget about other types of both positive and negative social effects that may arise from its introduction. Table 2 presents main areas of a large-scale effect, expected by implementation of the Blockchain technology in the coming years (Table 2).

Well-known performance evaluation metrics, such as NPV (Net Present Value), IRR (Internal Rate of Return), PBP (Payback Period), DBP (Discounted Payback Period), PI (Payback Index) are suitable for technology development investment planning purposes.

The structure of the financial model, which should be used to justify the economic effect, is standard, but a number of specific aspects should be taken into account in its development. Let's consider these peculiarities of structural parts of financial model in details.

Table 2. Blockchain-technology effects

Sphere of application	Economic effect	Social effect
Platforms for blockchain systems	Creating additional value by providing capacity for developers	Promoting the technology, expanding abilities for its use
Public administration and online voting	Decreasing expenses of face-to-face voting	Increasing public trust in election processes
Typical transactions with acceptance	Reduction of time and labor hours to entry transactions	Acceleration of accounting and settlement processes, raise in the number of these processes
Digital national currencies	Reduction of costs for issuance and cash control Getting rid of the shadow sector	Transparency of cash flows
Cross-border payment processing	Reduce time to process payments and transfers [8]	Simplify and make transfers more accessible and transparent
Optimization of metering and generation of energy	Reduction of costs for metering energy production, consumption and payments	An ability to redistribute surplus energy from renewable sources
Transport	Accounting for vehicle condition and status	Ability to track full vehicle history
Production control	Provides tracking and adjusting production volumes to prevent overproduction	Optimizes consumption of limited resources
Quality control	Offers real-time tracking of both defective product and defective machine	Reduces the number of bad or defective products
Food industry	Enables product tracking from the producer to the end consumer [9]	Enables tracking of contaminated product sources
Labor market	Enables a single, balanced labor market	Reduces an unemployment
Financial sector	Unification, speeding up and simplifying transactions, reducing costs [10]	Simplifying and making financial services more accessible and transparent

Source Calculated and worked out by the authors

Income

Table 2 shows the main values that technology can provide to business, which should be transformed into specific economic indicators.

Expenses

Expenses that are required to develop the technology:

- Programmer training;
- Software development, purchase of equipment for software development, expense of registering the project company and registration of security documents;
- Operating costs: rent of space, domain, server, maintenance of the product.

Forecasting Period

The forecasting period is 3 years from the introduction of the technology, due to the high dynamics of new technological solutions and legal constraints. In some cases, the evaluation period can be increased or decreased on a case-by-case basis if there is sufficient reason to do so.

Risks and Uncertainty

There are two main types of uncertainty when evaluating such technologies: technological and demand uncertainties. In addition to these types of uncertainty, there is regulatory risk. It is suggested to take these risks into account when calculating the discount rate in the form of an additional risk premium of 5% for each component:

- Degree of technology sophistication: 0–5%;
- Formation of demand: 0–5%;
- Regulatory risk in the country: 0–5%.

Specifics of Calculating the Discount Rate

The discount rate reflects a basic certain level of asset profitability at a corresponding risk level. The most commonly used methods of calculating the discount rate:

- CAPM (Capital Assets Pricing Model, proposed by W. Sharpe in 1964 [11]);
- cumulative building up method (Build up);
- ROE (Return on Equity);
- WACC (Weighted average cost of capital).

The choice of discount rate depends on the cash flow model. For cash flow to equity (Free Cash Flow Equity) is used the rate of return required by the owners, determined by the method of CAPM, cumulative construction, by the current rate of return ROE.

When attracting external financing to the project, it is advisable to calculate debt-free cash flow (Free Cash Flow to Firm) and apply the weighted average cost of capital (WACC).

The most preferable method to calculate the discount rate for valuation of technology implementation effect for nonpublic companies is cumulative construction method, for public companies—W. Sharpe's method.

4 Discussion

As an example of evaluation, let us consider a development based on blockchain and TON technologies, the NFT-marketplace: https://beta.disintar.io.

The object is the latest development and one of the first marketplaces using the TON technology. As an assumption, it is worth noting that this asset is in commercial use, but not yet legally registered.

Today there are many different NFT-marketplace in the world, using different technologies. This market is international without any local boundaries. Only the legal status and conditions of use depend on the place of use.

What makes the chosen platform unique is that it is based on TON technology. Among the competitors in the market there is only one, working on this technology—getgems.io.

Let's consider the most successful NFT-marketplace:

- Binance NFT—marketplace from the world's largest cryptocurrency exchange. Uses its own network Binance Smart Chain. In the measure of its fame and ubiquity, has the ability to regularly enter into new partnerships with major brands and creators.
- CryptoPunks—the largest NFT-marketplace at the moment. It is realized on Ethereum ERC-721 network. It is a collection of 10,000 unique image-tokens, which can belong to only one participant in the Ethereum blockchain. All-time trade turnover: $1,930,000,000. Last month's trading turnover: $36,020,000.
- SuperRare is one of the first platforms for artists to create and post their NFT work. Allows creators to receive a token certificate of authorship—a digital signature of the work. All-time trade turnover: $240,000,000. Last month's trade turnover: $12,000,000.
- Sorare is a football-themed game in which the player is the manager of his soccer team. NFT tokens are digital cards of real players, varying in rarity and characteristics. The platform has among its investors a group of global soccer stars such as Antoine Griezmann (Atletico Madrid) and Gerard Pique (Barcelona). All time trade turnover: $30,000,000. Last month's trading turnover: $8,000,000.
- Getgems.io is an NFT marketplace based on TON technology. It is based on the next-generation web3 blockchain ecosystem The Open Network, which is the best tool for NFT artists. The platform, in addition to the possibilities of placing, buying and selling and storing tokens, also offers the creation of its own exposition of user-owned tokens. All-time trade turnover: $4,000,000. Last month's trade turnover: $400,000.

The most applicable Cost Approach, the Reproduction Cost Method, was chosen to evaluate the object using the Cost Approach.

Necessary expenses for development and launching of the project are shown in Table 3.

Profit Approach

For the object valuation by the income approach, the most applicable method was chosen—discounted cash flows. In connection with the specifics of the project, was used a discount rate of 50%, which corresponds to the maximum risk rate of venture projects in the fourth stage of development, according to the research of James Plummer [12]

Table 3. Expenses for development of NFT-marketplace

Costs	Term, months	Cost, rubles	Total, rubles
Staff training (4 people)	6	130,000/month + 30% social security deductions	4,056,000
Salaries and wages of employees (4 people)	6	150,000/months + 30% social security deductions	4,680,000
Computer purchase (4 pcs.)	1	100,000	400,000
Rent of a premise and equipment	12	150,000/мес	1,800,000
Business registration and patent	1	1,000,000	1,000,000
Total			11,936,000

Source Calculated and worked out by the authors

and Walter Sahlman [13]. In Table 4 the calculations on the income from the assessed asset are demonstrated.

Table 4. Calculation of NFT-marketplace revenues

Year	No. of users, thousands of people	No. of transactions, thousands	Turnover, hundreds of thousands $/month	Average cost of transaction, $	Revenue, hundreds of thousands of $/month	Average quarterly revenue, hundreds of thousands $	Total revenue, hundreds of thousands of $
2022	107.912	71.95	143.89	200	7.19	20.33	56.48
2023	138.02	92.01	184.03	200	9.20	27.02	156.15
2024	155.63	103.75	207.50	200	10.38	30.75	274.05

Source Calculated and worked out by the authors

In Table. 5 the cost calculations are given.

Table 6 shows the final calculations on the value of the asset.

5 Conclusion

It is concluded, that blockchain technology poses a certain value, that should be assessed by market participants, especially when they are supposed to use smart contracts. The possible recommendation for the regulators of financial markets in adoption of the

Table 5. Calculation of costs of NFT-marketplace

Year	Operating expenses, rubles	Capital expenses, rubles	Depreciation, rub
2022	9,394,000.00	2,800,000.00	237,500.00
2023	15,094,200	700,000.00	331,250.00
2024	18,750,160	100,000.00	371,875.00

Source Calculated and worked out by the authors

Table 6. Calculation of the NFT-marketplace value

Name of index	Forecast period		
	2022	2023	2024
Revenues, $	5,648,211.28	9,966,558.61	11,790,511.66
Revenue, rub	414,533,521.90	731,465,669.18	865,329,231.75
Operating expenses, rub	9,394,000.00	15,094,200.00	18,750,160.00
EBITDA, rub	405,139,521.90	716,371,469.18	846,579,071.75
EBIT, rub	404,902,021.90	716,040,219.18	846,207,196.75
Capital expenses, rub	2,800,000.00	700,000.00	100,000.00
Depreciation, rub	237,500.00	331,250.00	371,875.00
Free cash flow to firm (FCFF), rub	321,359,117.52	572,463,425.34	677,237,632.40
Discounting rate (%)	50	50	50
Discounting index	0.67	0.44	0.30
Present value of FCFF	214,239,412	254,428,189	200,663,002
Accumulated discounted cash flow, rub	669,330,603		

Source Calculated and worked out by the authors

blockchain assessment will facilitate risk management in a creation of a prototype payment system using distributed ledger technology by consortia of international banks when bank users can exchange currency for the blockchain by placing a crypto currency in the DLT platform. This is also a solutions of secure usage of unregulated cyber coins to facilitate settlement of transactions by creating permissible networks for settlements.

Special attention should also be paid to the creation of new financial products, such as tokens and others, that pay out a percent of revenue. The digital blockchain-based securities are able to create a liquid secondary market, there could be a significant increase in institutional interest. Thus the relevant blockchain value assessment is extremely important for the principals and endusers of financial innovations.

It is also possible to draw the conclusion, that approaches and methods suggested in the article are applicable in evaluating business in any field of blockchain technology. It is worth noting that the values calculated by the approaches are considerably different. It happens because of the specifics of the object estimated, which is able to bring very high revenues at a very low cost. Unique knowledge and skills are used in the development of blockchain theory and practice of value assessment, the mastering of which is extremely difficult.

Finally, it is worthwhile mentioning, that evaluation of any investment or sale decision based on new technologies is linked with the results of the income approach and blockchain value assessment.

References

1. Hype Cycle for Blockchain: https://www.gartner.com/en/documents/4003463 (2021). Date of application 10 June 2022
2. Andrianova, Yu.V., Belyaeva, I.Yu., Bogatyrev, S.Yu., et al.: In: Fedotova, M.A., Tazikhina, T.V., Kosorukova, I.V. (eds.) Property Value in the Digital Economy: Assessment and Management : Monograph. Limited Liability Company Publishers "Knorus", Moscow (2021), 442c. ISBN 978-5-406-07966-9
3. International Valuation Standards: Date of application 10 June 2022
4. Order of the Ministry of Economic Development of Russia: Evaluation of Intangible Assets and Intellectual Property (FSO № 11). Dated 22 June 2015, p. 385
5. Order of the Ministry of Economic Development of the Russian Federation: On Approval of the Federal Valuation Standard, "Purpose of Valuation and Types of Value (FSO No. 2)". Dated 20 May 2015, p. 298
6. Order of the Ministry of Economic Development of the Russian Federation: On Approval of the Federal Valuation Standard, Valuation for the Purposes of Collateral (FSO No. 9). Dated 1 June 2015, p. 327
7. Bakulina, A.A., Grigoryev, V.V.: Blockchain as an Object of Evaluation: A Monograph. Rusains, Moscow (2021), 198c. ISBN 978-5-4365-5238-5. https://book.ru/book/939733. Date of accession: 23 May 2022
8. Toorajipour, R., Oghazi, P., Sohrabpour, V., Patel, P.C., Mostaghel, R.: Block by block: a blockchain-based peer-to-peer business transaction for international trade. Technol. Forecast. Soc. Change **180**, 121714 (2022). ISSN 0040-1625. https://doi.org/10.1016/j.techfore.2022.121714. https://www.sciencedirect.com/science/article/pii/S0040162522002402
9. David, A., et al.: Blockchain technology in the food supply chain: empirical analysis. IJISSCM **15**(3):1–12 (2022). http://ezpro.fa.ru:2725/10.4018/IJISSCM.290014
10. Oh, J., Shong, I.: A case study on business model innovations using blockchain: focusing on financial institutions. Asia Pacific J. Innov. Entrepr. **11**(3), 335–344 (2017). https://ezpro.fa.ru:2696/10.1108/APJIE-12-2017-038
11. Sharpe, W.F.: CaPital asset prices: a theory of market equilibrium under conditions of risk. J. Finance **19**(3), 425–442 (1964). https://onlinelibrary.wiley.com; https://doi.org/10.1111/j.1540-6261.1964.tb02865.x
12. Plummer, J.: QED Report on Venture Capital Financial Analysis. QED Research Inc, Palo Alto (1987)
13. Sahlman, W.A., Scherlis, D.R.: A Method for Valuing High-Risk, Long-Term Investments: The "Venture Capital Method". Harvard Business School. Background Note 288-006 (2009), p. 52

Analysis of Mimetic Muscle Rehabilitation: Multi-platform Facemarks Detection

Jakub Tomeš(✉), Ondřej Vácha, Karel Šticha, Jan Kohout, and Jan Mareš

Department of Computing and Control Engineering, University of Chemistry and Technology in Prague, Technická 1905/5, 16628 Praha 6, Prague, Czech Republic
jakub.tomes@vscht.cz

Abstract. The paper deals with frameworks for the description of mimetic muscle rehabilitation of patients with facial palsy. Detection models on different platforms (Android, iOS, Windows) are compared: ARCore Augmented Faces, ARKit Face Tracking, and MediaPipe Face Mesh. They can provide data for further quantitative analysis of mimetic muscle dysfunction. The real-time nature of these facemark detectors allows for direct feedback in the form of automatic mimetic muscle detection to the patient during the rehabilitation session.

Keywords: Facemarks · ARCore · ARKit · MediaPipe · Mimetic exercises · Rehabilitation · Facial palsy

1 Introduction

The inability to complete mimetic muscle movement (e.g. closing the eye, smiling, or raising the eyebrow) in half the face is typical for patients with facial palsy. It has physical and emotional toll on affected individuals. Quantitative analysis of facial motion is crucial to determine the degree of facial palsy, make a decision on the appropriate treatment, and also monitor the rehabilitation process [10].

The quality of the analysis is predetermined by the quality of the measured data. The three-dimensional digital analysis of facial motion is still of research interest. Tzou et al. [13] presents video analysis system, where physical facemarks and mirror arrangement were used for 3D analysis. After the Microsoft Kinect depth camera was released, the need for physical facemarks was eliminated [11]. Advanced analysis of Kinect v2 facemark data, collected from patients after head and neck surgeries, can be found in [8,12].

Currently, it is possible to gain real-time 3D facemark data from common webcams using appropriate face mesh models and provide immediate mimetic muscle exercise detection. In this paper, we focus on state-of-the-art facemark detection models, including ARCore Augmented Faces (Android), ARKit Face Tracking (iOS), and MediaPipe Face Mesh (multi-platform).

R. Silhavy et al. (Eds.): CoMeSySo 2022, LNNS 596, pp. 216–224, 2023.
https://doi.org/10.1007/978-3-031-21435-6_18

2 Materials and Methods

The rehabilitation session consists of exercises for the mimetic muscles presented in Table 1. The exercises were standardized by Prof. Janda, and each exercise is usually repeated four times [14].

Table 1. Set of mimetic muscles exercises by Prof. Janda methodology

Order	Exercise
1	Raising eyebrows
2	Frowning
3	Closing eyes tightly
4	Smiling
5	Baring the teeth
6	Pursing the lips
7	Blowing out the cheeks
8	Closing eyes tightly and baring the teeth
9	Raising eyebrows and pursing the lips

We deal with two rehabilitation conditions: hospital control session and home rehabilitation. For both of them, automatic detection of exercises in the mimetic muscles is important as *(i)* decreases the complexity of work with the data collection application *(ii)* provides data of better quality since real-time feedback can be provided to the patient and the measurement can be repeated immediately and *(iii)* increases the probability of long-term use of the application, as it is easy to use.

Table 2. Platform overview

Platform	Framework	Precision	Notes
Android	ARCore	468	3D facemarks
iOS	ARKit	1220	Vertexes
Windows	MediaPipe Face Mesh	478	3D facemarks

Our interest is to provide multi-platform support for facemarks data collection application. We focus on three platforms, Android, iOS, and Windows. Comparison of used frameworks is shown in Table 2.

2.1 Automatic Exercises Detection

The basic activity model for the rehabilitation unit is shown in Fig. 1. The patient has the ability to start or pause exercise. Only after a predefined set of completion times for the exercise can he move to the next exercise.

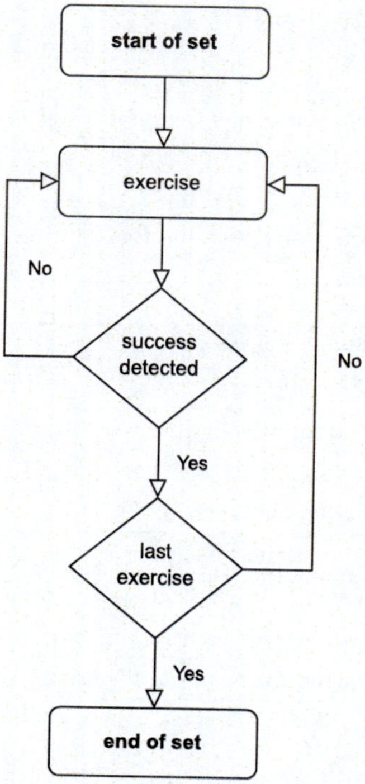

Fig. 1. Activity diagram for exercise automatic detection

2.2 Android Platform

On Android platform for facemark detection we have decided to use Google's augmented reality solution called *ARCore*. With a normal (no depth camera is needed) front-facing camera we can track 468 3D facemarks (see Fig. 2). Each facemark is labeled with numbers 1 through 468. Each 3D facemark contains the distance for the axis X, Y and Z to the middle of the face. Middle of the face is inside the skull, between the eyes.

2.3 iOS Platform

The iOS and iPadOS platforms provide a native library called *ARKit*, which facilitates the processing of front camera and face ID data. The data are then accessible in the form of an ARFaceAnchor structure [1]. Facial geometry can be accessed as a mesh containing 1220 vertices. Triangle indices and texture coordinates are also provided to simplify the visualization of the face mesh (Fig. 3) [3]. Furthermore, the structure provides 52 BlendShapes that represent current facial expression as a dictionary of weights between 0 and 1 each corresponding to its BlendShapeLocation [2,4].

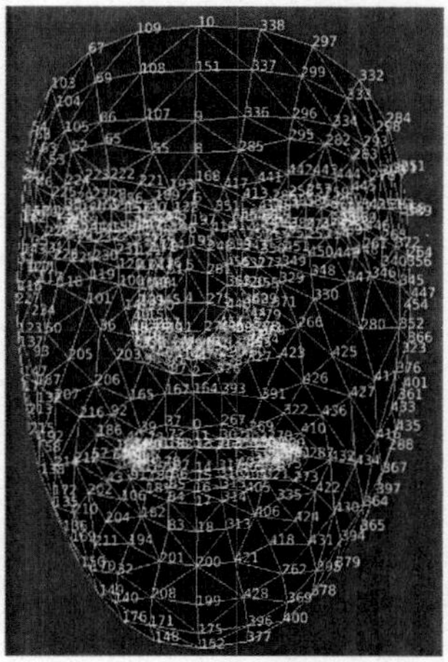

Fig. 2. Android: 468 facemarks from ARCore [9]

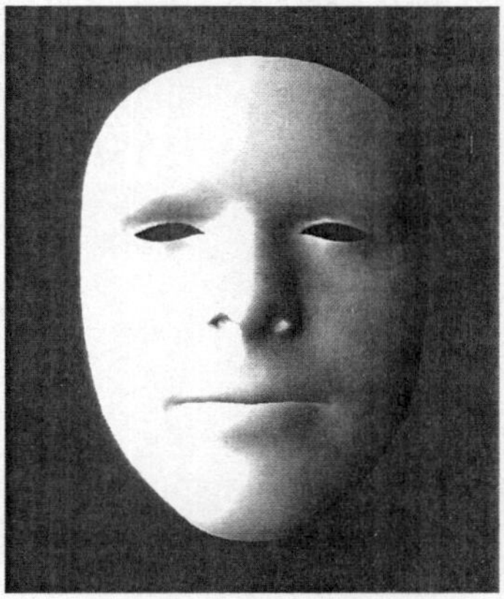

Fig. 3. iOS: face mesh extracted from the application

2.4 Windows Platform

For real-time facemark detection on the Windows platform, we have chosen an open source (and also multiplatform) solution MediaPipe, specifically MediaPipe Face Mesh. It estimates 478 3D facemarks from a single RGB camera input (see Fig. 4).

MediaPipe Face Mesh is based on a two-stage architecture, a face detector followed by a facemark regression network. The facemark regression network processes 256 $\times$ 256 face image which was obtained from the face detector or by tracking the previous frame. Unlike the previous model presented in [7], this model provides better facemarks predictions in the eyes and mouth regions, where attention submodels are used [6]. The training data set contained about 30k photos from the mobile camera, which were annotated with 2D facemarks.

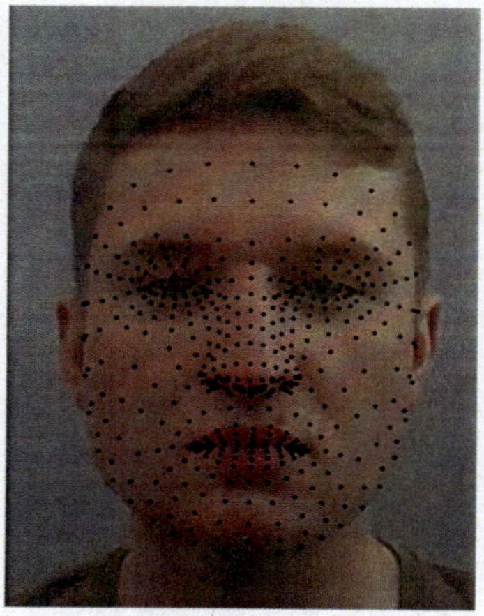

Fig. 4. Windows: 478 facemarks from MediaPipe Face Mesh

Z coordinate was approximated using a morphable model for the synthesis of 3D faces [5–7]. The whole model accepts RGB or RGBA images of any size and provides a list of facemarks. The facemarks are normalized to the screen coordinates X and Y and the coordinate Z is scaled under the weak perspective projection.

3 Results

3.1 Android

Using Android Studio, we have created an app that can be used on any Android phone with Android version 9 (Fig. 5) (the app was tested on Samsung s10e). Using the front-facing camera, we can exercise with the patient and go through a predefined set of exercises. With the addition of custom information about the patient, for example: the patient can use only the left half of the face, and we can modify the needs for each exercise. With this, we were able to achieve consistent and satisfying results.

Because ARCore does not provide enough trackable areas, we have to use facemarks and calculate the distances between them. For each exercise, it is crucial to select the most representative ones. For example, for smiling, we calculate if corners of mouth and upper and downer lips reached a set threshold. Then we can decide whether the exercise was performed correctly. Using Android Studio, we have developed an app for Android phones with a UI like this.

3.2 iOS

The application running on an iPad Pro 11 uses facial expression data provided by ARKit to determine the current state of the exercise. The rate of change is detected from the BlendShape values for BlendShapeLocations, which are

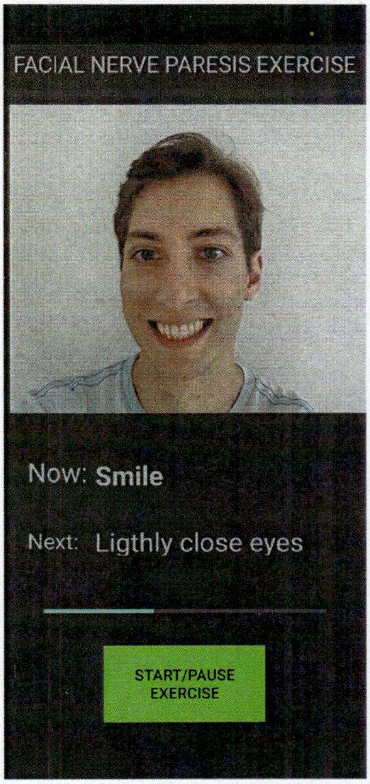

Fig. 5. Android: app UI

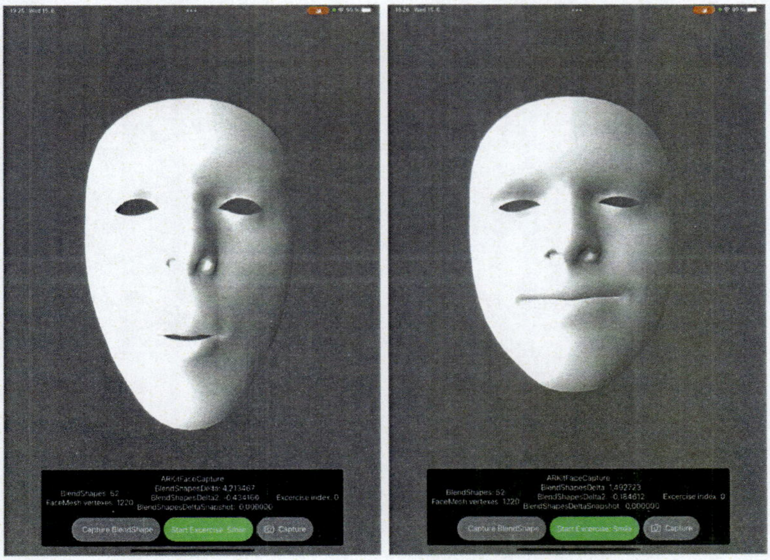

Fig. 6. iOS: app UI

relevant to the current exercise. The exercise is automatically started after a change in users' facial expression occurs. The exercise then remains active until the facial expression remains unchanged for a specified amount of time or until the user inputs the exercise. Key points of the face mesh are stored for patient evaluation when the exercise is completed.

The iOS application is capable of running on any face ID-equipped device. It is able to automatically go through all exercise states and save the face mesh when the exercise is completed. Each exercise defines key areas that are monitored to ensure that exercise states are detected correctly. Furthermore, the application can compare the current facial expression with a snapshot captured previously. The user interface shown in Fig. 6 shows all current metrics and contains manual controls for exercise, facial mesh capture, and facial expression snapshot. The figure additionally shows two different expressions captured via the application.

3.3 Windows

The screenshot of the Windows app user interface is shown in the Fig. 7. Technically it is WPF (Windows Presentation Foundation) application developed with .NET Framework. The application is deployed in *University Hospital Královské Vinohrady* and currently is using Kinect v2 depth camera as a facemarks source. Kinect camera is now obsolete, therefore new solution which is using MediaPipe Face Mesh is being developed with *.NET MAUI* (Multi Platform User Interface) framework.

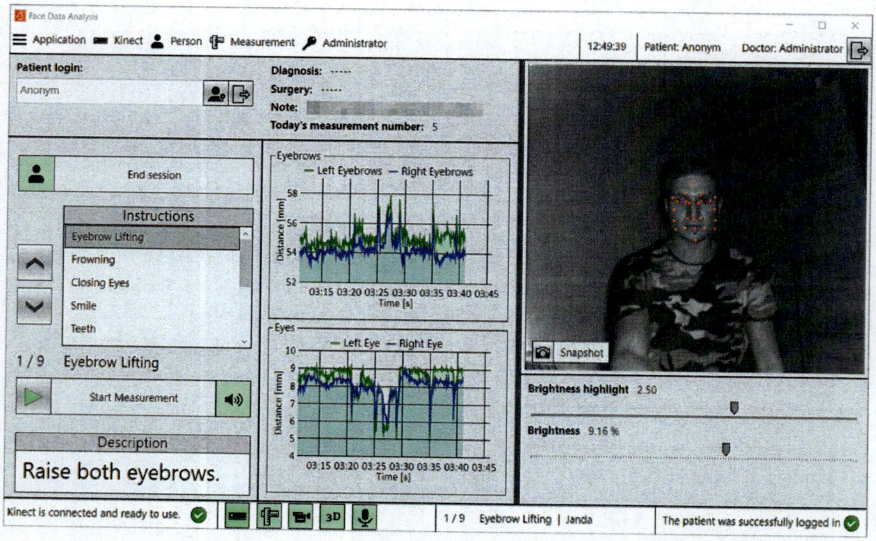

Fig. 7. Windows: Kinect v2 data collection app UI

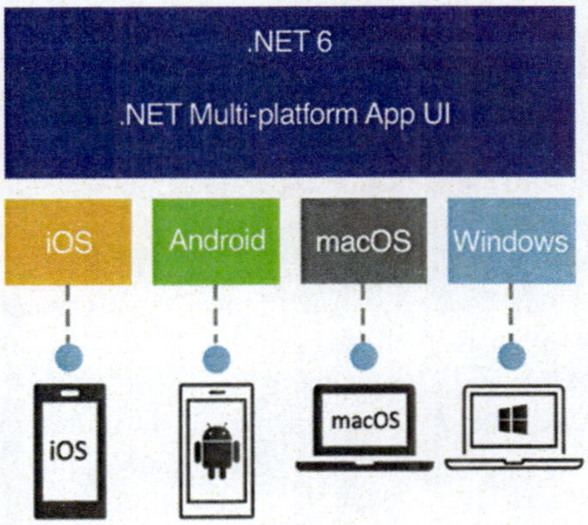

Fig. 8. .NET MAUI schema

4 Future Work

Next work will be focused on development migration to .NET MAUI which is currently (summer 2022) in preview. .NET MAUI enables simultaneous development for multiple platforms, such as iOS, Android, macOS and Windows (see Fig. 8).

5 Conclusion

Three robust facemark detection models were presented: ARCore Augmented Faces, ARKit Face Tracking, and MediaPipe Face Mesh. This triplet can provide a reliable source of data for a further detailed analysis of mimetic muscle dysfunction on the most common platforms Android, iOS, and Windows. Furthermore, online exercise detection simplifies application usage and can provide higher-quality data thanks to immediate exercise feedback.

Acknowledgement. The work was funded by the Ministry of Education, Youth and Sports by grant 'Development of Advanced Computational Algorithms for the Evaluation of Post-surgery Rehabilitation' number LTAIN19007 and by grant A1-FCHI-2022-001. This support is gratefully acknowledged.

References

1. Apple: Documentation—ARFaceAnchor. https://www.developer.apple.com/documentation/arkit/arfaceanchor. Accessed on 20 Apr 2022
2. Apple: Documentation—ARFaceAnchor.BlendShapeLocation. https://www.developer.apple.com/documentation/arkit/arfaceanchor/blendshapelocation. Accessed on 20 Apr 2022
3. Apple: Documentation—ARFaceGeometry. https://www.developer.apple.com/documentation/arkit/arfacegeometry. Accessed on 20 Apr 2022

4. Apple: Documentation—BlendShapes. https://www.developer.apple.com/documentation/arkit/arfaceanchor/2928251-blendshapes. Accessed on 20 Apr 2022
5. Blanz, V., Vetter, T.: A morphable model for the synthesis of 3d faces. In: Proceedings of the 26th Annual Conference on Computer Graphics and Interactive Techniques, pp. 187–194. SIGGRAPH '99, ACM Press/Addison-Wesley Publishing Co., USA (1999). https://doi.org/10.1145/311535.311556
6. Grishchenko, I., Ablavatski, A., Kartynnik, Y., Raveendran, K., Grundmann, M.: Attention mesh: high-fidelity face mesh prediction in real-time. CoRR abs/2006.10962 (2020). https://arxiv.org/abs/2006.10962
7. Kartynnik, Y., Ablavatski, A., Grishchenko, I., Grundmann, M.: Real-time facial surface geometry from monocular video on mobile gpus. CoRR abs/1907.06724 (2019). http://arxiv.org/1907.06724
8. Kohout, J., Verešpejová, L., Kříž, P., Červená, L., Štícha, K., Crha, J., Trnková, K., Chovanec, M., Mareš, J.: Advanced statistical analysis of 3d kinect data: mimetic muscle rehabilitation following head and neck surgeries causing facial paresis. Sensors 21(1) (2021). https://www.mdpi.com/1424-8220/21/1/103
9. ManuelTS: Manuelts/augmentedfacemeshindices: Google arcore sceneform 1.10.0 augmented face mesh indicies from their.fbx file as pictures. http://www.github.com/ManuelTS/augmentedFaceMeshIndices
10. Owusu, J.A., Stewart, C.M., Boahene, K.: Facial nerve paralysis. Med Clin North Am **102**(6), 1135–1143 (2018)
11. Schätz, M., Procházka, A., Ťupa, O., Vyšata, O., Sedlák, V.: Face movement analysis with ms kinect. In: 2016 International Workshop on Computational Intelligence for Multimedia Understanding (IWCIM), pp. 1–5 (2016)
12. Shayestegan, M., Kohout, J., Štícha, K., Mareš, J.: Advanced analysis of 3d kinect data: supervised classification of facial nerve function via parallel convolutional neural networks. Appl. Sci. 12(12) (2022). https://www.mdpi.com/2076-3417/12/12/5902
13. Tzou, C.H.J., Pona, I., Placheta, E., Hold, A., Michaelidou, M., Artner, N., Kropatsch, W., Gerber, H., Frey, M.
14. Zemanová, M., Janda, V., Ondráčková, Z.: Rehabilitace po obrně lícního nervu. Technical report. Státní zdravotní ústav, Praha (1998)

Suitability of MQTT and REST Communication Protocols for AIoT or IIoT Devices Based on ESP32 S3

Eduard Nemlaha, Peter Střelec, Tibor Horák(✉), Szabolcs Kováč, and Pavol Tanuška

Institute of Applied Informatics, Automation and Mechatronics, Faculty of Materials Science and Technology in Trnava, Slovak University of Technology, Bratislava, Trnava, Slovakia
{eduard.nemlaha,peter.strelec,tibor.horak,szabolcs.kovac,pavol.tanuska}@stuba.sk

Abstract. Internet of Things (IoT) devices are not just simple devices anymore. The introduction of industry IoT (IIoT) and Artificial Intelligence of Things (AIoT) creates new challenges in terms of communication. The presented comparation shows the advantages and disadvantages of two communication protocols, which are suitable for new coming generation of IoT devices. Message Queuing Telemetry Transport (MQTT) and Representational State Transfer (REST) can provide extended functions for data acquisition and they offer different options in terms of data transfer. A demonstration of the integration of IoT devices is shown, based on both protocols and there are classified from the functional perspective in terms of Industry 4.0. The experiment is built on the ESP32 S3 development board and data are processed by Node-Red flow. Data are acquired and saved into relation database. Verification of data transfer is performed and is visible on the dashboard. The transfer of data and connections were also recorded. The IoT device sends three data types, which are produced by sensors. Data types were selected to demonstrate a possibility how to work with standard data types like char variables or numerical values but main topic is to show how to transfer binary or complex data structures like JavaScript Object Notation (JSON). The experiment also demonstrates how communication protocol can impact the power consumption of a device.

Keywords: IoT · AIoT · ESP32 S3 · MQTT · REST

1 Introduction

IoT devices are being deployed in industrial applications and cover a large number of tasks. IoT devices were single-purpose devices that typically had limited computing power and also limited communication capabilities [1].

IIoT or AIoT classifications are being pushed in the industry, which have much greater computing power and have integrated computing units to support the acceleration of matrix computations that are used to accelerate AI [2]. This makes it possible to deploy them for a larger number of tasks and to implement even more challenging tasks. There

R. Silhavy et al. (Eds.): CoMeSySo 2022, LNNS 596, pp. 225–233, 2023.
https://doi.org/10.1007/978-3-031-21435-6_19

may be issues related to the design of the communication that the IoT device has to handle. The available communication protocols that standard IoT devices support are Message Queuing Telemetry Transport (MQTT), Advanced Message Queuing Protocol (AMQP), REST etc. An overview is given in Table 1 [3].

Table 1. Overview of Communication Protocols

	Transport	Paradigm	Scope	Security	Data prioritisation	Fault tolerance
MQTT	TCP/IP	Publish/subscribe	Device-to-device	TLS	Yes	Broker is the single point of failure
REST	HTTP over TCP/IP	Request/reply	Device-to-device	SSL/TLS	None	Replication of server
AMQP	TCP/IP	Point-to-point Message Exchange	Device-to-device Device-to-cloud Cloud-to-cloud	TLS	None	Depend on implementation

2 Background

AIoT and IIoT devices generate various data. Variable types are not only text or numeric datatypes anymore. Due to more complex tasks, the data provided by IoT devices can contain a lot of data types including binary and structured data types. Acquired data from IoT needs to be analyzed and evaluated in case of debugging [4]. Given these requirements, reliable data transmission needs to be addressed. MQTT and REST protocol has been selected for review in the industrial application. Both of them work in the application network layer and have been tested in terms of capabilities and reliability of sending data [5].

MQTT is a Client Server publish/subscribe messaging transport which is designed also for real-time communication that is suitable for efficient data distribution. Using microservices, it is possible to convert data into a real-time streaming service. It is developed by IBM and used since 1999. Current specification is known as version 5. MQTT topics have a hierarchical structure and support wildcards + and #. MQTT messaging principle is for publishing messages and subscribing to topics [6].

Topis identify messages, and clients determine which topics are addressed to them and client subscribe them. MQTT supports first in first out (FIFO) communication. There is guaranteed the sequence of data. This feature is supported since version 3 and a message queue needs to be defined to guarantee FIFO processing. It is also possible to specify other buffer types and it is possible to change the priority (Quality of Service QoS) according to the topic(s) and thus achieve higher priority delivery of preferred messages as required. Many IoT devices connect over unreliable networks [7]. MQTT support for persistent sessions and is able to reconnect the client with the broker. MQTT Publish Subscribe principle and communication with broker is shown in Fig. 1.

Broker is server and has assigned TCP service port of 1883 or a secure encrypted port of 8883. The Broker is an intermediary and performs the function of a post office

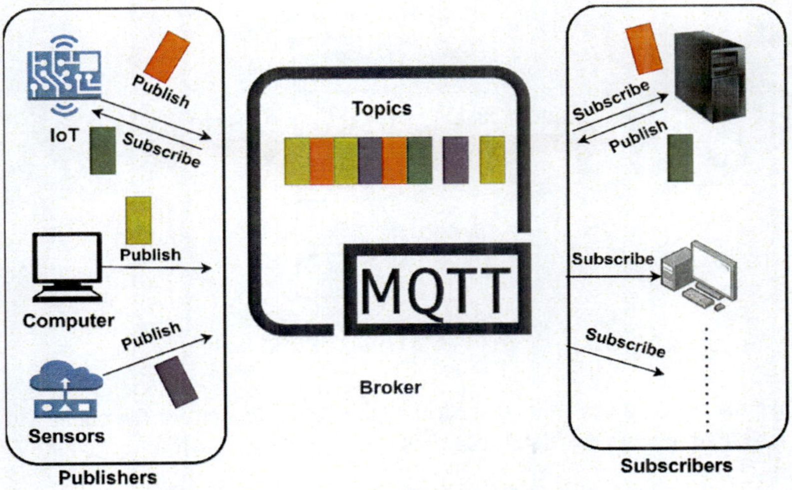

Fig. 1. MQTT broker and publish subscriber principle

with only one difference. It does not use the address of the intended recipient, but uses a subject called "Topic". Anyone who wants a copy of the message subscribes to the appropriate one. Multiple clients can receive the same message from a single broker. Similarly, multiple publishers can publish Topics for a single Client—multiple for one [8].

Client is the software that interacts with Broker. It can act as a Subscriber (Subscribe) or a Sender (Publish). Clients are typically devices that can publish sensor data and are also capable of receiving configuration information or processing control commands. MQTT is a bidirectional communication protocol. This provides both data sharing and device management and control. The design of the MQTT has been emphasized for data that is sent frequently and is small in size. The main disadvantage of MQTT is the inability to deliver larger binary data. These can only be delivered after the content has been split into smaller blocks. The main advantage is small cost for the data delivery and full duplex communication [9, 10].

REST APIs communicate via HTTP requests. REST API supports all basic HTTP methods—POST, GET, PUT, PATCH, DELETE. Requests are sent using the HTTP protocol and the information is processed in JSON or XML format [11].

The mapping according to the functionality for securing create, update, read, delete operations is determined as follows. Creation of a new record is provided by the POST method. The GET method is used to read data. If there is a record that needs to be altered or a new record needs to be created, the PUT method is used. The DELETE method is used to delete a record [12]. The principle of how REST works is shown in Fig. 2.

REST is a one-way communication. The client sends the data and receives an acknowledgement or response, as appropriate. It is not possible to send data from the server to the client [13]. Major disadvantage of REST is also latency in request processing time and bandwidth usage. This is due the fact of using HTTP protocol, which is delivering REST messages to the server. The client connects to the server. Then the client sends the required data gets the response back. The server waits for client connections

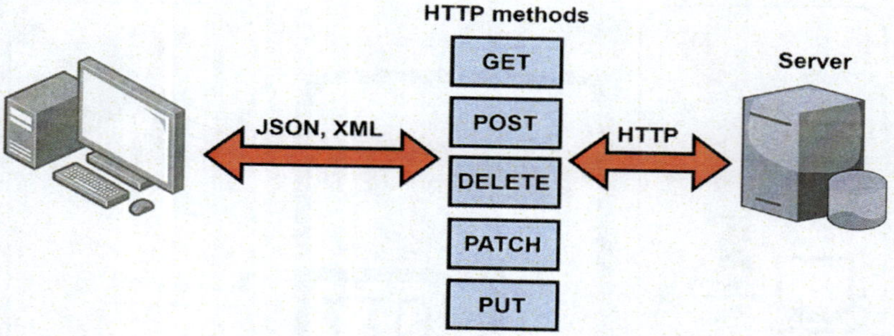

Fig. 2. REST API methods

to receive the data, process it, and send back the data intended for the client. However REST is able to process any data structures and binary data [14].

3 Materials and Methods

A comparison of the MQTT and REST protocols for AIoT and IIoT devices was performed on a development kit equipped with the ESP32 S3 processor. The ESP32-S3 is a low-power MCU-based system-on-chip (SoC) that supports 2.4 GHz Wi-Fi and Bluetooth® Low Energy (Bluetooth LE). It consists of a high-performance dual-core microprocessor (Xtensa® 32-bit LX7), a low-power coprocessor, a Wi-Fi baseband, a Bluetooth LE baseband, an RF module, and peripherals. It has sufficient power and capacity to store and process audio visual information. Xtensa LX7 Processors support DSP for imaging and convolutional neural network (CNN) processing and DSP for multi-purpose, fixed, and floating-point DSP applications [15].

The IoT device has a micropython installed and connected to the temperature and humidity sensors DHT22, ultrasonic sensor HC-SR04 and a camera that captures the image information. The test was conducted on sensing the growth of the plant, where the ambient temperature and humidity were monitored, the ultrasonic sensor measured the growth rate of the plant and the camera recorded the growth process and the AI identified the plant if possible [16].

Data acquisition is in the Node-Red. Node-Red was chosen because it provides simple implementation of both protocols—REST and MQTT. The fact that performance and scalability are not an object of comparison, this solution suits the needs of the assignment.

4 Experiment Design

The ESP32 S3 development kits were fitted with sensors. Sensors are connected to General Purpose Input/Output (GPIO) and acquired data are stored as JSON string. Binary data are encoded by base64. Due the fact that JSON is supported only ascii there is one more step needed to convert it via binascii to correct format. A JSON structure is sent via MQTT and REST to the server. Node-Red runs on the independent server and parse ad store received data. A schematic representation is shown in Fig. 3.

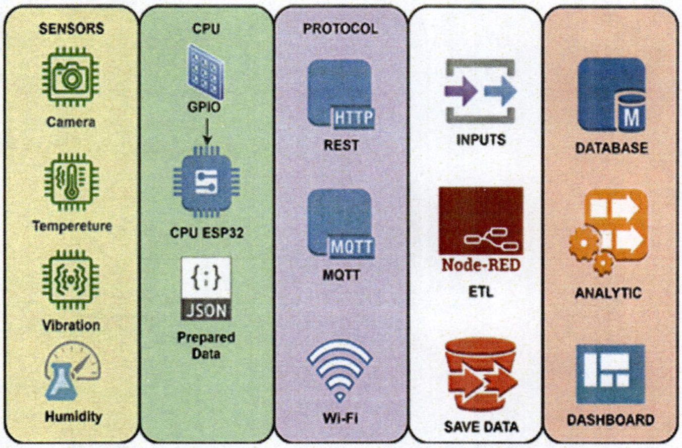

Fig. 3. The block scheme of the experiment design

The comparation of MQTT and REST focuses on the delivery of numerical and character data types and as an additional objective is transfer of images. As a test scenario is analysed the growth development of the plant. The aim is to compare the MQTT and REST protocols with respect to reliability and the effect of the chosen protocol on battery life. All produced data fit into the main objective to test various data types from AIoT or IIoT devices.

The tests were conducted without and with image information to get better overview. Every failed connection or failed sent request has been stored on ESP32 S3 and in the next successful delivery, this information has been delivered as a part of the dataset. Additional information is a battery lifetime and consumption of ESP32 S3. The technical specification promises a long battery life time but we attempted to measure the impact of the protocol on power consumption in long-term usage.

The final real applied measurement was done based on the battery voltage. Fully charged 3.2 V LiFePO4 batter has 3.65V. Based on these specifications, 3.28 V is equal to 40% of the battery capacity. This voltage has been selected as a referential point while the battery is discharged and has sent a message that ESP32 S3 is out of power. The batteries have been used from the same range and they have the same age. Altogether there were connected ten AIoT devices. Each IoT device was created photos and doing classification image and were using both protocols.

5 Experiment Implementation and Result

The communication of the connected devices is shown in Fig. 4. The devices were connected to a Wi-Fi router, which was also connected to a server running the MQTT broker and Node-Red. Data were sent from each device at set intervals and were stored. Each AIoT device had its time synchronized according to the ntp server every 5 min to avoid any possible deviations.

The data flow was implemented in the Node-Red. Data acquisition flow is shown in Fig. 5.

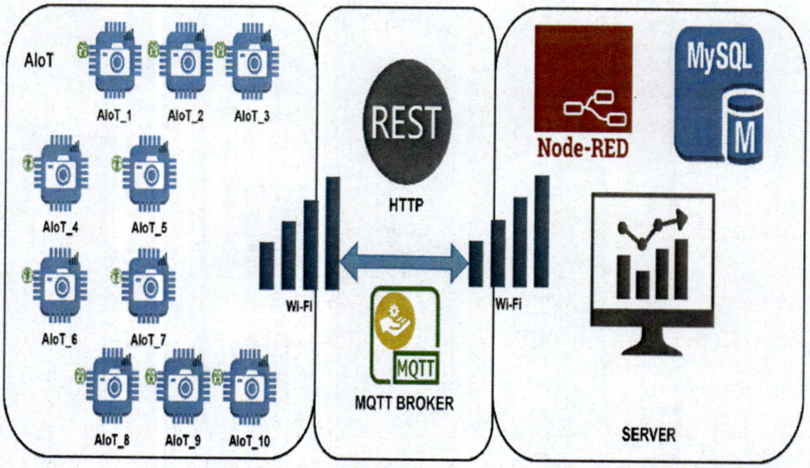

Fig. 4. Block diagram of IoT device communication

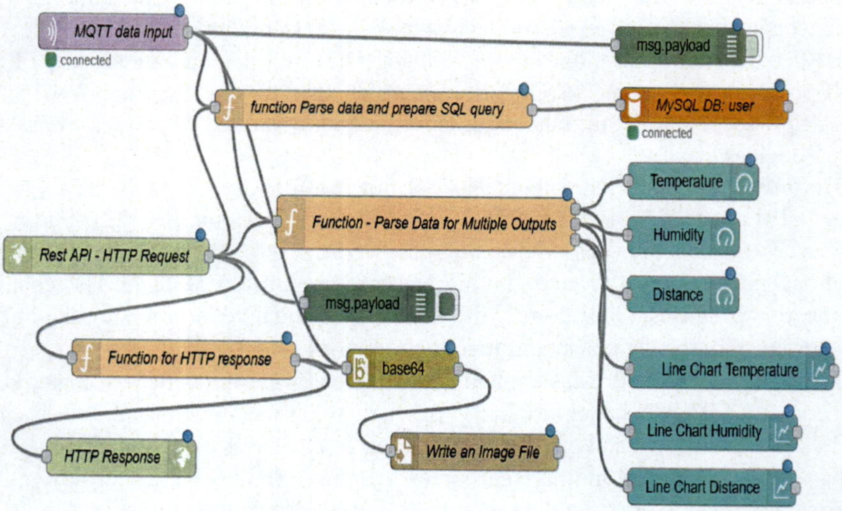

Fig. 5. Node-red flow diagram for MQTT and REST data acquisition

The data are coming to Node-Red object Inputs. MQTT objects can acquire MQTT requests and they are parsed in function. Data are sent to inputs, to the decode function and to the base64 decoder. Finally, binary data are written into a file, where the name of the file is id information, which is in parallel stored in the database. All the data collected were important for data analysis. All timestamps and content have been reported in dashboard and line graphs.

The data stored in the database have the following attributes, shown in Table 2.

Humidity is an integer data type, temperature and distance are float data type. Images are binary. Power monitoring is done by CPU it self. There is a reference voltage of 1.1 V,

Table 2. Attributes for data acquisition from AIoT

Value	Description
ID	Unique record number
DEVICE_ID	Device identification
TEMPERATURE	Value of temperature
HUMIDITY	Value of humidity
DISTANCE	Value of distance
IMAGE_ID	Identification of image
BATTERY_VOLTAGE	Battery voltage
TIMESTAMP	Timestamps provided by device
REPEAT	The number of sending attempts
RECEIVED	Timestamp when data have been received
STORED	Timestamp when data have been stored in the DB
DEBUG	Debug report on the reading process

which can be used for comparing the voltage value. This is good enough and an analog pin of ESP 32 S3 can provide the wanted measured value. An ADC conversion is to convert the input analog voltage to a digital value.

The ADC conversion results provided by the ADC driver APIs are raw data. The resolution of ESP32-S3 ADC raw results under Single Read mode is 12-bit-4095 under Single Read mode. There is a need to calibrate each device and set the correct offset [17]. Voltage data was read before at the end of studying all sensors. When the voltage enters a level equal to or less than 3.28 V device send final message that the battery is out of power. Programing has been done in the micropython. Five AIoT devices have been using MQTT as a communication protocol and the next five have REST. They have been collecting the same data and the programs have been identical in all parts of code except communication protocol.

6 Conclusions

Both these protocols are capable of covering the requirements of modern IoT devices. During testing, it was possible to demonstrate good reliability of both solutions, where MQTT is a better choice for data that do not contain encoded binary data blocks. Here, the speed of sending data was done in less time and it was on average 45% better than REST. For sending binary data, the REST protocol performed better, sending the image information 30% faster than when the data was sent over MQTT. For MQTT, a setup without REST support was used. But solutions provided the REST API and MQTT, both of which are integrated into MQTT Broker.

This option has not been investigated. However, the consumption of ESP32 S3 was significantly different in the two cases investigated, for sending a standard data set without or even with image information. MQTT had on average 20% longer battery life than when using the REST protocol. However, the biggest differences occurred in the reliability of sending data. MQTT demonstrated much better reliability in this regard compared to REST. The difference was 6%. While investigating the reasons, it was found that REST, if the keep alive parameter is not enabled on the server side, can make these parameters even worse.

The issue occurred while connecting to the server and then disconnecting. The main disadvantage for REST is the latency and, according to the measured values, the processing time on the server side. However, this parameter can be influenced by the Node-Red implementation of the web server and possibly the available network bandwidth. MQTT is always-connected against the intermittent REST Calls. Protocol MQTT supports bi-directional full duplex communication with persistent connections. REST is suitable for sending larger amounts of data and should be used where there is no need to change the parameters of the IoT device.

For standard IoT devices, the MQTT protocol is a better choice. For AIoT or IIoT devices, both protocols can be considered and the appropriate protocol can be selected according to the requirements and data to be processed.

Acknowledgments. This research was funded by the Scientific Grant Agency of the Ministry of Education, Science, Research, and Sport of the Slovak Republic and the Slovak Academy of Sciences, grant number VEGA 1/0176/22 Proactive control of hybrid production systems using simulation-based digital twin and VEGA 1/0193/22 Proposal of identification and monitoring of production equipment parameters for the needs of predictive maintenance in accordance with the concept of Industry 4.0 using Industrial IoT technologies. This article was also written thanks to the generous support under the Operational Program Integrated Infrastructure for the project: "Strategic research in the field of SMART monitoring, treatment and preventive protection against coronavirus (SARS-CoV-2)", Project no. 313011ASS8, co-financed by the European Regional Development Fund.

References

1. Šimon, M., Huraj, L.: A study of DDoS reflection attack on Internet of Things in IPv4/IPv6 networks. In: Computer Science On-line Conference. Springer, Cham (2019)
2. Haqiq, N., et al.: AIoT with I4. 0: the effect of Internet of Things and Artificial Intelligence technologies on the industry 4.0. In: ITM Web of Conferences, vol. 46. EDP Sciences (2022)
3. How Does DDS Compare to other IoT Technologies? DDS outperforms traditional messaging solutions in Data Centricity, https://www.whoishostingthis.com/resources/ravendb/. Accessed 11 June 2022
4. Malik, P.K., et al.: Industrial internet of things and its applications in industry 4.0: state of the art. Comput. Commun. **166**, 125–139 (2021)
5. Froiz-Míguez, I., et al.: Design, implementation and practical evaluation of an IoT home automation system for fog computing applications based on MQTT and ZigBee-WiFi sensor nodes. Sensors **18**(8), 2660 (2018)
6. Yassein, M.B., et al.: Internet of things: survey and open issues of MQTT protocol. In: 2017 International Conference on Engineering & MIS (ICEMIS). IEEE (2017)
7. Jaloudi, S.: Communication protocols of an industrial internet of things environment: a comparative study. Fut. Internet **11**(3), 66 (2019)

8. Pipatsakulroj, W., Visoottiviseth, V., Takano, R.: muMQ: a lightweight and scalable MQTT broker. In: 2017 IEEE International Symposium on Local and Metropolitan Area Networks (LANMAN). IEEE (2017)
9. Houimli, M., Kahloul, L., Benaoun, S.: Formal specification, verification and evaluation of the MQTT protocol in the Internet of Things. In: 2017 International Conference on Mathematics and Information Technology (ICMIT). IEEE (2017)
10. Manandhar, S.: MQTT Based Communication in IoT. MS thesis (2017)
11. Masse, M.: REST API Design Rulebook: Designing Consistent RESTful Web Service Interfaces. O'Reilly Media, Inc. (2011)
12. Subramanian, H., Raj, P.: Hands-on RESTful API Design Patterns and Best Practices: Design, Develop, and Deploy Highly Adaptable, Scalable, and Secure RESTful Web APIs. Packt Publishing Ltd (2019)
13. Colitti, W., et al.: REST enabled wireless sensor networks for seamless integration with web applications. In: 2011 IEEE Eighth International Conference on Mobile Ad-Hoc and Sensor Systems. IEEE (2011)
14. Jabbar, S., Khan, M., Silva, B.N., Han, K.: A REST-based industrial web of things' framework for smart warehousing. J. Supercomput. **74**(9), 4419–4433 (2016). https://doi.org/10.1007/s11227-016-1937-y
15. ESP32-S3 Designed for AIoT Applications 2.4 GHz Wi-Fi and Bluetooth 5 Features. https://www.espressif.com/en/products/socs/esp32-s3/. Accessed 1 June 2022
16. Raikar, M.M., Desai, P., Naragund, J.G.: Active learning explored in open elective course: internet of things (IoT). In: 2016 IEEE Eighth International Conference on Technology for Education (T4E). IEEE (2016)
17. Analog to Digital Converter (ADC). https://docs.espressif.com/projects/esp-idf/en/latest/esp32s3/api-reference/peripherals/adc.html. Accessed 12 June 2022

Goal Modeling for Linked Data Exploitation of Municipalities Data Access in South Africa

Gerald Ovono(✉) and Sihle Moyo

Vaal University of Technology, Private Bag x 021, Vanderbijlpark, South Africa
{geraldo,sihlem}@vut.ac.za

Abstract. Ontology validation has been an important part of measuring the quality and accuracy of knowledge encoded in that ontology. Since ontologies are machine-readable representations of knowledge, the use of formal languages, such as SPARQL can help query ontology. In this article, we apply Tropos as the goal modelling methodology to enable the definition of Competency Questions (CQs). We used a combination of qualitative and quantitative methods to collect data using a literature review, and online related resources from the local, district, and metropolitan municipalities. The data were analyzed with the Topos methodology and then translated into SPARQL queries using the Competency Query Translation Approach (CQT).

Keywords: South Africa municipalities · Information system · Linked data · Topos methodology · Competency question · Competency questions translation · SPARQL

1 Introduction

1.1 Governmental Structure

The issues of sustainable development continue to be of global importance even into the 21st century. The active participation and involvement of people with their government in decision-making are central to sustainable development in South Africa, the constitution sets the rules for how government works [1]. In further reviewing of the constitution, a set of three spheres of government namely national government, the provincial government, and local government have been established.

The whole of South Africa is divided into local government or local municipalities, and there are three different types of municipalities in South Africa which are; The Metropolitan municipalities (the metropolitan municipality coordinates the delivery of services to the whole area, and there is a total of 8 metropolitan municipalities), the Local municipalities (those are the areas that fall outside of the six metropolitan municipal areas. There is a total of 231 of these local municipalities and each municipality is broken into wards), and the district municipalities (district municipalities are made up of several local municipalities that fall in one district. There is a total of 44 districts and are usually between 4 and 6 local municipalities that come together in a district council).

R. Silhavy et al. (Eds.): CoMeSySo 2022, LNNS 596, pp. 234–248, 2023.
https://doi.org/10.1007/978-3-031-21435-6_20

The constitution in Section 156 (1), mentioned in Part B of Schedules 4 and 5 indicates the type of services or competencies that municipalities should provide [2, 3]. According to [3] municipalities have for main objective to provide an effective and sustainable service to various stakeholders. In this regard, different services or competencies rendered by South African municipalities have shown two broad categories of services [1]. Some services or competencies are regarded as support services while others are classified as operational services.

1.2 Information Technology Adoption in South African Municipalities

Several initiatives have emerged to improve the interface between government and stakeholders through the effective use of Information and Communication Technology, and specifically through the use of web standards [4–10].

With the Open Government Partnership (OGP) which South Africa had joined in 2011, the government had been publishing data on data.gov.za for use in both commercial and non-commercial ways [11]. According to [12] South African municipalities, data access has been made available by the use of advanced information technology to improve their services to stakeholders. However, many of the municipalities within South Africa have not automated their information systems for better public service delivery to their shareholders, particularly those that serve rural areas. It is important to mention that the South Africa portal for public services does not provide many of those municipalities' services or competencies [13].

1.3 Background of Linked Data

The W3C [14] defines Linked Data as the exposition, sharing, and connecting of pieces of data, information, and knowledge on the Semantic Web using (Uniform Resource Identifier) URIs and (Resource Description Framework) RDF. Linked Data is about using the Web to connect related data that wasn't previously linked, or using the Web to lower the barriers to linking data currently linked using other methods [15].

Linked Data represents an opportunity for the South African municipalities to bring a rich network of knowledge by standardizing municipalities' datasets in a common format. Linked Data can enable an intelligent chain of transactions on the web without human intervention, allowing publication in the open data to link with other datasets for reuse. In fact, Linked Data technologies are applied in various domains (e-health, e-commerce, e-government, etc.) to build a new generation of intelligent web applications [16–18].

1.4 Objectives

This study attempts to provide a solution to fill in the gap between the transactions within and between municipalities and the Linked Data technology. For that, we rely on goal modeling. Goal modeling has been effectively applied in requirements engineering in the past years to deliver Information Systems for organizations in general [19, 20]. The main objective of this study is to elaborate a conceptual goal modeling for municipalities transactions in South Africa, as a basis for municipalities to interlink their Information Systems and provide a sustainable service to their stakeholders.

This goal modeling is designed to establish a set of competency questions with the Tropos methodology using Goal-Oriented Modeling Tools, which will then be used for querying the Linked Data model of the municipalities' datasets using SPARQL Query.

2 Goal Modeling to Capture Competency Questions

It is important to mention that the goal modeling represented in this study is from transactions within and between municipalities in the Gauteng province of South Africa. According to [21], Gauteng has three metropolitan municipalities namely Ekurhuleni, Johannesburg, and Tshwane, two district municipalities namely Sedibeng and Rand West, and six local municipalities namely Emfuleni, Lesedi, Midvaal, Merafong City, Mogale City, and Rand West City. A depiction of the municipalities' regional is given in Table 1.

Table 1. Gauteng municipalities

Metropolitans	Districts	Locals
City of Ekurhuleni City of Johannesburg City of Tshwane	Sedibeng West Rand	Emfuleni Lesedi Midvaal Merafong City Mogale City Rand West City

Upon a close evaluation of the different municipal services that should be rendered, two broad categories of such services are recognizable [3]. Support Services refer to the operation of the overall municipal, and Operational Services refer to services rendered by a municipality to its local community [1, 3].

Studies assessing the determinants of productivity efficiency in South African municipalities have pointed out that the performance of the municipalities is mostly related to the effectiveness of Support Services [22, 23]. Municipalities Support Services include services such as Information and technology services. According to [3], no municipalities can operate effectively and sufficiently without modern technological development. Therefore, data access from the Information and technology services is used for the goal modeling within and between the 11 selected municipalities.

The Tropos methodology was carried out using a combination of web-based and systematic literature reviews to investigate the use of Information and Technology for transactions within and between municipalities. Tropos is an agent-oriented software engineering methodology that relies on i* notation to offer the notions of actor, goal, plan, and resource, and uses these as a foundation to model early, late requirements and ontology modeling [24–26]. The ontology modeling phase of the Tropos Methodology is not applied in this study. Instead, competency questions obtained in the Late Requirements phase were translated into SPARQL Query using the Competency Questions Translation Approach.

Modeling municipalities' data access provided a suitable concept for the exchange of asset information. Hence, directly applying a Tropos concept to visually model municipalities' information communication [27]. The next subsection presents the early requirement for the goal-oriented model of the municipalities.

2.1 Early Requirements Analysis: The Goal-Oriented Model of the Municipality

The Early Requirement analysis provides an overview of the municipality's data access for its stakeholders and is concerned with understanding the municipalities' Information System domain settings as it is used before a solution is conceived [27].

In this Early Requirement analysis, we firstly conducted a domain knowledge acquisition on the municipalities' information system, thereafter we further elaborated on the needs, preferences, and responsibilities of the stakeholders and transform them into models of Early Requirements. The output of this activity is an organizational model, which includes relevant actors' diagrams and goals diagrams to define their interdependencies. Figure 1 gives an illustration of the different processes involved.

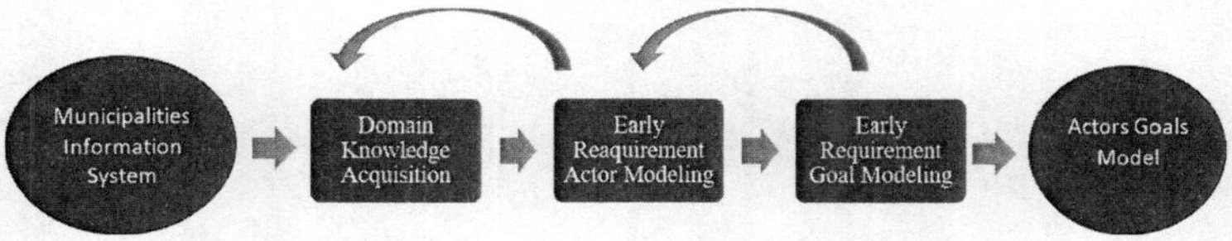

Fig. 1. Flow of activity in the early requirement analysis

2.1.1 Domain Knowledge Acquisition

Domain knowledge acquisition consisted of gathering information from domain stakeholders and municipalities Information System domain. We conducted it by reviewing the literature on the 11 South African municipalities and the application of Information and Communication Technology within the municipalities' Support Services and Operational Services to capture their fundamental organizational setting, such as domain actor and their responsibilities.

2.1.2 Early Requirement Actor and Goal Modeling

In this phase, the identified stakeholders are modeled in Tropos in terms of actors and actor dependencies. As mentioned early, the Support Services, Operational Services, and the 11 municipalities represent the actors and, we analyzed the basis of the domain documentation if a goal is achievable by the actor itself or if the actor has to delegate it to another actor, see Figs. 2 and 3 depictions.

In this Early Requirement phase, we elaborated two-goal diagrams for the municipalities, the diagram in Fig. 2 illustrates the actors as Support Services and Operational

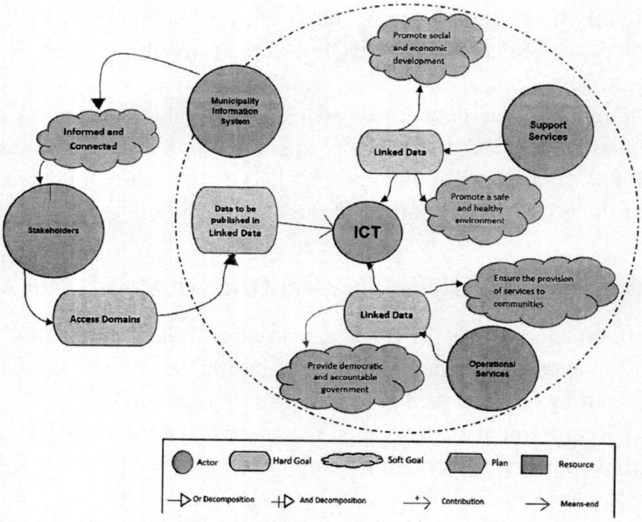

Fig. 2. Early requirement of support services and operational services within municipalities (constructed with piStar 2.0)

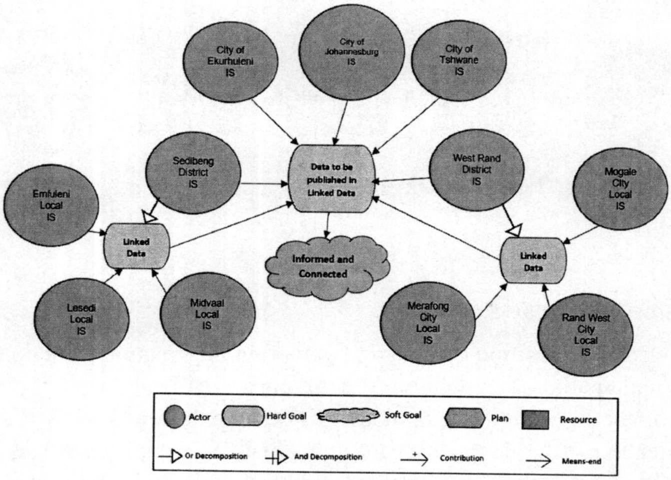

Fig. 3. Early requirement of municipalities in Gauteng (constructed with piStar 2.0)

Services with the dependency relationship in publishing data as Linked Data and, with Information and communication technology services as an actor to provide data access to the stakeholders' actor via a Linked Data relationship. The municipalities' Information systems would have to provide information and connectivity through Linked Data Technology, and the stakeholders access the municipalities' domains through a dependency goal relationship. The diagram in Fig. 3 illustrates the actors as municipalities from the Gauteng province, with all municipalities' goal to link their data as Linked Data, and have a pendency relationship among themselves.

2.2 Late Requirements Analysis: Modelling the Competency Questions

To model municipalities and shareholders' information requirements into Tropos resources which are represented by Competency Questions, the necessary information for the goals to be achieved is provided by the resources that represent the competency questions [28]. Thus, this late requirement refines the municipalities' data access by modeling their competencies into Competency Questions and their most important actors, goals, plans, and resources. We accomplished this Late Requirement with Requirement Analysis. During this phase, the Requirement Analysis was in charge of:

- The system to be is represented as an actor
- Actor goals to be delegated through dependency links
- Top goals to be refined as leaf-level goals
- Model, all capabilities of the system to be as actions to be queried.

2.3 Late Requirement Actor Goals Modeling

In this phase, we introduced the system to be an actor. We then assigned the stakeholder needs and preferences to the system to be by establishing goals and dependency links between stakeholder actors and system actors. Thereafter, we refined into more fine-grained goals the top-level goals of the system to be. We analyzed those goals from the system actor perspective, and are AND or OR linked into more fine-grained goals.

The systematic analysis of goals provided by Tropos gradually led to understand the objectives involved in linking the municipalities' domains environment. And from this point, understanding that to satisfy these goals, the right kind of information is needed to define the competency questions.

According to the Johannesburg municipality walk-in customer service centre [29], municipalities in South Africa attend General Inquiries related to:

1. Accounts payments
2. Accounts inquiries
3. Apply for special cases
4. Credit control arrangements
5. Customer education on understanding statements
6. Full and final settlement of accounts
7. Lodge a customer query
8. Meter investigation requests
9. Obtain information on accounts
10. Reconnection request of services after the cut-off
11. Refund request
12. Remissions on accounts for senior citizens
13. Termination of services by owners
14. New connections to electricity, water, and sewer.

Those General Inquiries formed the soft goals of our Late Requirement, which then was expended with the systematic analysis of goals provided by Tropos to formulate the relevant queries. Figure 4 illustrates the Late Requirement actor goals modeling phase of the municipalities.

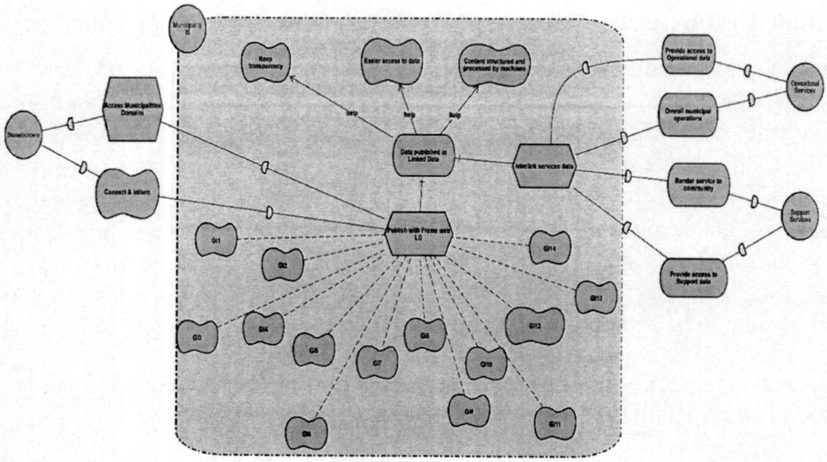

Fig. 4. Late Requirement municipality's goal diagram (constructed with piStar 2.0)

2.4 Empirical Evaluation of Competency Questions

Following the 14 General Inquiries collected from the municipality walk-in customer service center which represents our soft goals, we formulated and captured the resources by reviewing the literature from articles and web publications on South African municipalities. Those resources were identified as Competency Questions to reflect on stakeholders' presuppositions of questions. We aimed at understanding the different kinds of queries that are asked by stakeholders in real-world scenarios, to ensure that the municipality's Linked Data model can respond optimally. Hence, the 26 Competency Questions are formulated based on the 11 General Inquiries and then linked to their respective goals on the actor goal model. Table 2 lists the General Inquiries (GI) and Competency Questions (CQs) that are encoded each with identifiers. Depending on the complexity of the goals, including the information needed, several competency questions have been linked to the same goal [27]. Thus, a depiction of municipalities' Information System actor goal model of data access to capture Competency Questions is illustrated in Fig. 5.

Tropos was used as a goal-oriented modeling technique to capture and analyze South African municipalities' data access into Competency Questions (CQs), then translated into executable form with the Competency Question Translation (CQT) approach [27, 30]. In the next section, we present the Competency Questions Translation approach used to translate those CQ into SPARQL-OWL queries.

3 Translating Natural Language Competency Questions into SPARQL Queries

Competency Questions in ontology engineering studies are natural language questions that describe the scope of knowledge represented by an ontology-based information system to be answered [31, 32]. Ontology in the domain of knowledge sharing is an explicit specification of a conceptualization, and attempt to express an exhaustive conceptual scheme within a given domain [33, 34]. In the context of this study, the CQs

Table 2. Competency questions from general inquiries

Identifier	General inquiry	Competency questions	Identifier
GI1	Accounts payments	Where can I pay my municipal account?	CQ1
		How do I pay my municipality account?	CQ2
		Is it better to pay bills with credit or debit?	CQ3
		When should I pay a bill?	CQ4
GI2	Accounts inquiries	How do I get my municipal account via email?	CQ5
		Which municipality does my account fall under?	CQ6
		Does my account have any outstanding bills?	CQ7
		How to register for a municipality account?	CQ8
GI3	Apply for special cases	Where to apply for municipality special cases?	CQ9
GI4	Credit control arrangements	How to make a payment arrangement plan?	CQ10
GI5	Customer education on understanding statements	What is my water and/or electricity consumption from my account statement?	CQ11
		What is my tax and property rate from my account statement?	CQ12
GI6	Full and final settlement of accounts	How is the full and final settlement calculated?	CQ13
		How long does it take to get a full and final settlement?	CQ14
GI7	Lodge a customer query	What are the types of customer complaints?	CQ15
GI8	Meter investigation requests	How does a prepaid meter work?	CQ16
		How to report a meter malfunctioning?	CQ17
GI9	Obtain information on accounts	What type of account for services do I get from the municipality?	CQ18
		Where do I inquire for any of my municipality accounts?	CQ19

(continued)

Table 2. (*continued*)

Identifier	General inquiry	Competency questions	Identifier
GI10	Reconnection request of services after the cut-off	How to apply for service after cut-off?	CQ20
GI11	Refund request	How to request any refund on an account?	CQ21
		Where to request a refund on an account?	CQ22
GI12	Remissions on accounts for senior citizens	What criteria pensioners can apply for rates and tax rebates?	CQ23
GI13	Termination of services by owners	What happens when a contract is terminated?	CQ24
		How do you end a service contract?	CQ25
GI14	New connections to electricity, water, and sewer	How to apply for a new connection in electricity, water, and sewer?	CQ26

are translated into SPARQL-OWL queries using the Competency Questions Translation approach for querying the ontology of the Information System domain of municipalities.

3.1 Competency Questions Translation Approach

Several Competency Questions Translation Approaches have been proposed for natural language ontology editors such as CNL editor and OWLPATH, or natural language query systems such as PANTO and DEANNA [35–38]. Other approaches address this issue for a specific knowledge domain, such as the medical domain [39]. All mentioned approaches presented shared a scalability limitation as the ontology used for test purposes is relatively small [30]. Additionally, approaches such as PANTO and DEANNA suppose that for every queried knowledge base, there exists a dictionary that maps questions' concepts to potential knowledge bases' semantic items, which makes it challenging to correctly map users' words to vocabularies of the knowledge bases, due to the control from the vocabularies of large knowledge bases such as DPEDIA, Freebase and Yago [37, 40].

In this study, we adopted the Competency Questions Translation Approach proposed by [30]. We decided to slightly adapt it to respond to the objectives of this study, and the resulting approach can be summarized as illustrated in Fig. 6:

- Categorizing the competency questions to the type of the expected answer such as;
- Extracting entity or entities from questions and their corresponding expected answers;
- Constructing the appropriate SPARQL query that gives the closest answer to the ideal answer.

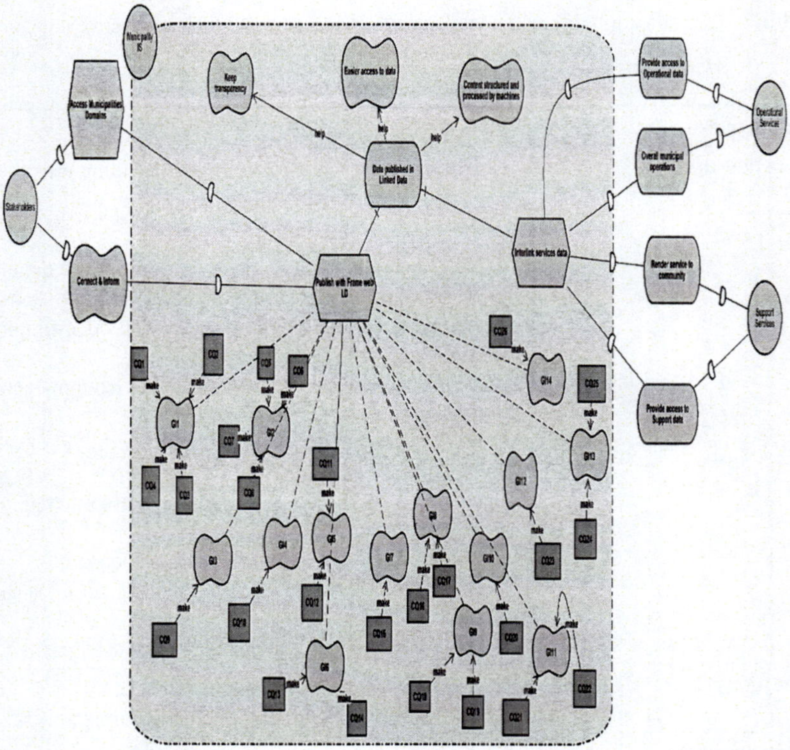

Fig. 5. Late Requirements municipality capturing and delegating the competency questions (constructed with piStar 2.0)

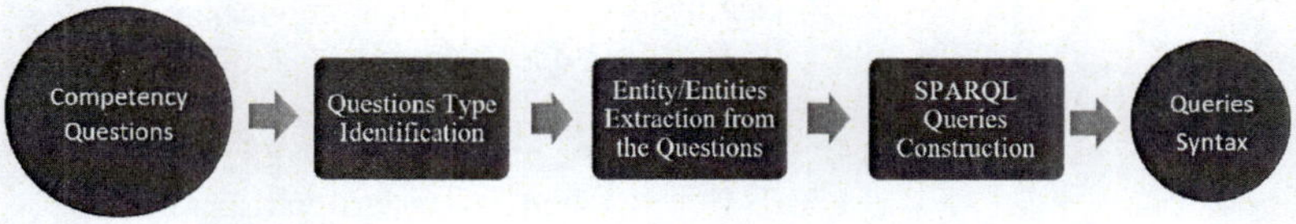

Fig. 6. Competency questions translation approach

3.2 Categorizing the Competency Questions

As the first step in translating the Competency Questions into SPARQL query syntax, we classified these CQs according to the answer categories, as mentioned in the previous section, see Table 3.

With these sorted Competency Questions, we have the facility for the construction of the corresponding SPARQL OWL queries. As a result, from this sorting, we found 0 definition questions, 2 Boolean questions, 10 factual questions, 2 list questions, and 12 complex questions. According to [40], a query of a factual question has to target a specific entity that might be a class, an instance, or whatever. A query of a list question targets several entities as a single answer. A query of complex questions targets a detailed answer

Table 3. Classification of competency questions according to their respective categories

Competency questions categories	Competency questions
Definition questions	NONE
Boolean questions	CQ7. Does my account have any outstanding bills? CQ3. Is it better to pay bills with credit or debit?
Factual questions	CQ4. When should I pay a bill? CQ6. Which municipality does my account fall under? CQ9. Where to apply for municipality special cases? CQ11. What is my water and/or electricity consumption from my account statement? CQ12. What is my tax and property rate from my account statement? CQ18. What type of account for services do I get from the municipality? CQ19. Where do I inquire about any of my municipality accounts? CQ22. Where to request a refund on an account? CQ23. What criteria pensioners can apply for rates and tax rebates? CQ24. What happens when a contract is terminated?
List questions	CQ1. Where can I pay my municipal account? CQ15. What are the types of customer complaints?
Complex questions	CQ2. How do I pay my municipality account? CQ5. How do I get my municipal account via email? CQ8. How to register for a municipality account? CQ10. How to make a payment arrangement plan? CQ13. How is the full and final settlement calculated? CQ14. How long does it take to get a full and final settlement? CQ16. How does a prepaid meter work? CQ17. How to report a meter malfunctioning? CQ20. How to apply for service after cut-off? CQ21. How to request any refund on an account? CQ25. How do you end a service contract? CQ26. How to apply for a new connection in electricity, water, and sewer?

or corresponding ontology annotations. In the next section, the extraction of entity or entities from competency questions is conducted.

3.3 Extracting Entity or Entities from Questions

According to [41], entity extraction is an information extraction task that identifies from a question the different key element of predefined categories. Thus, it helps with machine-readable and standard processing for retrieving information, extracting facts, and answering questions [42].

This process was applied to the competency questions in Table 3 to build SPARQL OWL queries for each question. An excerpt of this extraction is shown in Table 4 for the competency questions CQ1, CQ2, CQ4, and CQ7.

Table 4. Entities' extraction from competency questions

Competency questions categories	Competency questions	CQs' relevant terms	Corresponding ontology terms
Definition questions	None	None	None
Boolean questions	CQ7. Does my account have any outstanding bills?	Account Bill Outstanding Bill Customer	Class: account Datatype property: outstanding bill, customer
Factual questions	CQ4. When should I pay a bill?	Pay Bill Date Customer	Class: account Datatype property: bill, date, customer
List questions	CQ1. Where can I pay my municipal account?	Pay Municipality Account Location	Class: account Datatype property: bill, location
Complex questions	CQ2. How do I pay my municipality account?	Pay Municipality Account Transaction method	Class: account Datatype property: bill, transaction method

In this section, we extracted the entity or entities by mapping between relevant terms from the question and their corresponding terms in the ontology. In the next section, we constructed the SPARQL-OWL query of the CQs.

3.4 Constructing the Appropriate SPARQL Queries

The SPARQL query language is the standard pattern-matching for querying the RDF dataset, and a powerful Semantic Web tool to extract information from large datasets [43]. To construct the appropriate SPARQL queries, the corresponding ontology terms from the Competency Questions' relevant terms needed to be categorized such as the classes, data properties, object properties, and instances, see Table 4.

From the corresponding ontology terms, we constructed the appropriate SPARQL queries for each of the 26 CQs. This process was applied to the set of CQs in Table 5 to build the SPARQL queries for each category.

In Table 5, the SPARQL query CQ7 select (SELECT?) all outstanding bills from the Account Class (Ont:Account) in the ontology (Onto). This result in the list of bills from

Table 5. Competency questions to SPARQL queries

Competency questions	SPARQL queries
CQ7. Does my account have any outstanding bills?	SELECT ?bill WHERE {?account a SAM:Id ?customer SAM customer.Id ?Id.}
CQ4. When should I pay a bill?	SELECT ?date WHERE {?account a SAM:Id ?customer SAM customer.Id ?Id.}
CQ1. Where can I pay my municipal account?	SELECT ?location WHERE {?account a SAM:Id ?customer SAM customer.Id ?Id.}
CQ2. How do I pay my municipality account?	SELECT ?transactionType WHERE {?account a SAM:Id ?customer SAM customer.Id ?Id.}

the ontology. Similarly, the SPAQL query CQ4 returns the date from (SELECT*), i.e., details on the query class (Ont:Pay) in the ontology (Onto). The CQ1 and CQ2 address from the Property class (Ont:Municipality) in the ontology (Onto). The query also has a WHERE clause. The query will result in a location for CQ1 and a transaction type for CQ2.

4 Conclusion and Future Work

Translating Competency Questions into SPARQL queries is a sine qua non-condition for automatic evaluation of ontology requirements satisfaction. We reviewed the literature and municipalities' general inquiries to define and elaborate a set of CQs. The trop Methodology was used to elaborate on the CQs. A well-defined CQs translation process was critical for the evolution of ontologies for machine-readable questions answered with a general perspective. Thus, we translated CQs into SPARQL queries. The future direction of this research would be to extend the study with the application of municipalities' linked data datasets to provide a more comprehensive performance evaluation of the evaluation of CQs for querying ontologies of public services in knowledge graph representation.

References

1. Republic of South Africa: Constitution of the Republic of South Africa, 1996 15th Amendment Act of 2008, vol. 243 (1996)
2. Lehohla, P.: The State of Basic Service Delivery in South Africa (2016)
3. Bekink, B.: Municipal Services and Service Delivery and the Basic Functional Activities of Municipal Governments (2006)
4. Bizer, C., Heath, T., Berners-Lee, T.: Linked data: the story so far. Int. J. Semant. Web Inform. Syst. **5**, 1–2 (2009)

5. Janev, V., Mijovic, V., Vranes, S.: Using the linked data approach in European e-government systems: example from Serbia. Int. J. Semant. Web Inf. Syst. **14**, 27–46 (2018)
6. Sarker, M.N.I., Hossin, M.A., Frimpong, A.N.K., Xiaohua, Y.: Promoting information resource management for e-government through big data approach. In: Proceedings of the Proceedings of the 2018 International Conference on Information Management & Management Science; Association for Computing Machinery, New York, NY, USA, pp. 99–104 (2018)
7. Agbozo, E., Spassov, K.: Establishing efficient governance through data-driven e-government. In: Proceedings of the Proceedings of the 11th International Conference on Theory and Practice of Electronic Governance, Association for Computing Machinery, New York, NY, USA, pp. 662–664 (2018)
8. Schmitz, P., Francesconi, E., Landercy, S., Batouche, B., Touly, V.: A Knowledge Organization System for E-Participation in Law-Making (2017)
9. Nečaský, M., Klímek, J., Škoda, P.: Practical Use Cases for Linked Open Data in E-Government Demonstrated on the Czech Republic, pp. 80–96 (2017)
10. Amalanathan, A., Acouncia, S.M., Vairamuthu, S., Vasudevan, M.: A framework for e-governance system using linked data and belief-desire-intention agent. Indian J. Sci. Technol. **8** (2015)
11. OGP Homepage: www.opengovpartnership.org. Last accessed 21 Nov 2018
12. Mogotlane, K.D., Fonou Dombeu, J.V.: Development and semantic exploitation of a relational data model for service delivery in South African municipalities. Afr. J. Inform. Syst. (2016)
13. Gov.za Homepage: www.gov.za. Last accessed 15 Apr 2020
14. W3.org: What Is Linked Data? What Is Linked Data Used for?, pp. 2–3 (2016)
15. Guerrini, M.: Introduction to the seminar global interoperability and linked data in libraries. JLIS.it **4**, 21–24 (2013)
16. Fonou-Dombeu, J.V.: Ranking e-government ontologies on the semantic web. In: Proceedings of the Electronic Government and the Information Systems Perspective. Springer International Publishing, Cham, pp. 18–30 (2020)
17. Gagnon, M.P., Ndiaye, M.A., Boucher, M., Dequanter, S., Buyl, R., Gorus, E., Bourbonnais, A., Giguère, A., Rahimi, S.A.: Suitable e-health solutions for older adults with dementia or mild cognitive impairment: perceptions of health and social care providers in Quebec city. In: Proceedings of the ICT4AWE, pp. 172–175 (2020)
18. Hua, N., Hight, S., Wei, W., Ozturk, A.B., Zhao, X.R., Nusair, K., DeFranco, A.: The power of e-commerce. Int. J. Contemp. Hospital. Manage. (2019)
19. Guizzardi, R.S.S., Ludermir, P.G., Sona, D.A: Recommender agent to support knowledge sharing in virtual enterprises. In: Agent and Web Service Technologies in Virtual Enterprises; IGI Global, pp. 115–134 (2008)
20. Liaskos, S., McIlraith, S.A., Sohrabi, S., Mylopoulos, J.: Integrating preferences into goal models for requirements engineering. In: Proceedings of the 2010 18th IEEE International Requirements Engineering Conference, pp. 135–144 (2010)
21. South Africa Municipalities Homepage: https://municipalities.co.za/. Last accessed 21 Oct 2020
22. Kanyane, M.: Exploring challenges of municipal service delivery in South Africa (1994–2013). In: Africa's Public Service Delivery and Performance Review, pp. 90–110 (2014)
23. Monkam, N.F.: Local municipality productive efficiency and its determinants in South Africa. Dev. South. Afr. **31**, 275–298 (2014)
24. Mylopoulos, J., Castro, J.: Tropos: a framework for requirements-driven software development. Inform. Syst. Eng. State Art Res. Themes **8**, 261–273 (2000)
25. Castro, J., Kolp, M., Mylopoulos, J.: A Requirements-driven development methodology. In: Proceedings of the LNCS, vol. 2068, pp. 108–123 (2001)
26. Silva, C.T.L.L., de Castro, J.B., Mylopoulos, J., Castro, J.F.B., Mylopoulos, J.: Detailing Architectural Design in the Tropos Methodology. Straw, pp. 85–93 (2003)

27. Fernandes, P.C.B., Guizzardi, R.S.S., Guizzardi, G.: Using goal modeling to capture competency questions in ontology-based systems. J. Inform. Data Manage. **2**, 527 (2011)
28. Bresciani, P., Perini, A., Giorgini, P., Giunchiglia, F., Mylopoulos, J.: Tropos: an agent-oriented software development methodology. Auton. Agent. Multi-Agent Syst. **8**, 203–236 (2004)
29. City of Johannesburg Homepage. http://www.joburg.org.za. Last accessed 20 Oct 2018
30. Zemmouchi-Ghomari, L., Ghomari, A.: Translating Natural Language Competency Questions into SPARQL Queries: A Case Study, pp. 81–86 (2013)
31. Gruninger, M., Uschold, M.: Ontologies: principles, methods and applications. Knowl. Eng. Rev. **11**, 93–136 (1996)
32. Wiśniewski, D., Potoniec, J., Ławrynowicz, A., Keet, C.M.: Analysis of ontology competency questions and their formalizations in SPARQL-OWL. J. Web Semant. (2019)
33. Karyda, M., Balopoulos, T., Dritsas, S., Gymnopoulos, L., Kokolakis, S., Lambrinoudakis, C., Gritzalis, S.: An ontology for secure e-government applications. In: Proceedings of the First International Conference on Availability, Reliability and Security, ARES, vol. 2006, pp. 1033–1037 (2006)
34. Gruber, T.R.: Towards principles for the design of ontologies used for knowledge sharing. In: Proceedings of the formal ontology in conceptual analysis and knowledge representation. Kluwer Academic Publishers, Deventer (1993)
35. Namgoong, H., Kim, H.: Ontology-based controlled natural language editor using CFG with lexical dependency. In: Proceedings of the ISWC/ASWC (2007)
36. Valencia-, R., García, F., Castellanos-Nieves, D., Fernández-Breis, J.T.: OWL path: an OWL ontology-guided query editor. IEEE Trans. Syst. Man Cybern. Part A Syst. Hum. **41**, 121–136 (2011)
37. Wang, C., Xiong, M., Zhou, Q., Yu, Y.: Panto: a portable natural language interface to ontologies. In: Proceedings of the European Semantic Web Conference, pp. 473–487 (2007)
38. Yahya, M., Berberich, K., Elbassuoni, S., Ramanath, M., Tresp, V., Weikum, G.: Deep answers for naturally asked questions on the web of data. In: Proceedings of the Proceedings of the 21st International Conference on World Wide Web, pp. 445–449 (2012)
39. Ben Abacha, A., Zweigenbaum, P.: Medical question answering: translating medical questions into SPARQL queries. In: Proceedings of the Proceedings of the 2nd ACM SIGHIT International Health Informatics Symposium, pp. 41–50 (2012)
40. Zemmouchi-Ghomari, L., Ghomari, A.R.: Translating natural language competency questions into SPARQL queries: a case study. In: Proceedings of the First International Conference on Building and Exploring Web Based Environments, pp. 81–86 (2013)
41. Das, P., Pandey, M., Rautaray, S.S.: A CV parser model using entity extraction process and big data tools. Int. J. Inform. Technol. Comput. Sci. **10**, 21–31 (2018)
42. expert.ai Homepage: https://www.expert.ai/blog/entity-extraction-work. Last accessed 15 Nov 2020
43. Cyganiak, R.: A Relational Algebra for SPARQL. Digital Media Systems Laboratory HP Laboratories Bristol (2005)

Application of the K-medians Clustering Algorithm for Test Analysis in E-learning

Ayman Aljarbouh[1], Roman Tsarev[2,3,4](✉), Adrián Segura Robles[5], Sergei Elkin[6,7], Irina Gogoleva[8], Irina Nikolaeva[9], and Ivetta Varyan[10,11]

[1] University of Central Asia, Naryn, Kyrgyz Republic
[2] MIREA—Russian Technological University (RTU MIREA), Moscow, Russia
rtsarev@acadst.org
[3] V.A. Trapeznikov Institute of Control Sciences of Russian Academy of Sciences, Moscow, Russia
[4] International Academy of Science and Technologies, Moscow, Russia
[5] University of Granada, Ceuta, Spain
[6] PT Taplink Digital Global, Kuta, Indonesia
[7] Institute of Economics of the Ural Branch, Russian Academy of Sciences, Yekaterinburg, Russia
[8] Yakut State Agricultural Academy, Yakutsk, Russia
[9] M.K. Ammosov North-Eastern Federal University, 58, Belinsky Street, Yakutsk 677027, Russia
[10] N.M. Emanuel Institute of Biochemical Physics, Russian Academy of Sciences, Moscow, Russia
[11] Plekhanov Russian University of Economics, Moscow, Russia

Abstract. E-education allows significantly automating the learning process. One of the key elements of e-learning is testing. The use of tests allows an objective and unbiased assessment of student knowledge. However, this automated approach requires high-quality tests that are relevant to the theoretical material and practical skills acquired by the student while taking an e-learning course. This paper proposes the use of the K-medians clustering algorithm, which makes it possible to divide the test questions into separate clusters in order to analyze them and further improve the tests, and possibly the entire e-course as a whole. The paper presents fragments of program code that can be used by other researchers in solving clustering problems.

Keywords: K-medians clustering · Clustering analysis · Clustering algorithm · E-learning · Manhattan distance

1 Introduction

The development of educational and information technologies has led to the emergence of new methods, tools, and approaches to learning [1–6]. The use of information technology and, in particular, learning management systems has transformed the usual educational process [1–9]. The use of computers, the Internet, and specialized software led to

R. Silhavy et al. (Eds.): CoMeSySo 2022, LNNS 596, pp. 249–256, 2023.
https://doi.org/10.1007/978-3-031-21435-6_21

the development of e-learning. Chernyshov in his work [10] presents a detailed analysis of transformations in education both in terms of digitalization and informatization, as well as in pedagogical and didactic terms. The author provides a detailed analysis of recent changes in methodological recommendations in the field of information support of the educational process [10].

E-learning implies automation of the learning process [11–13]. Besides, e-learning can take place both in the classroom and remotely. An important aspect of e-learning is the possibility to study most of the material independently. It is of great importance in the situations when the traditional form of learning is not available or has some limitations. This was particularly demonstrated by the situation with learning during the COVID-19 pandemic [14–16]. However, even under normal conditions e-learning has a number of advantages that extend the traditional model of learning.

Electronic tools used in learning allow to automate the process of studying theoretical material, as well as performing tests on the results of studying a module or the entire course.

Tests allow you to evaluate the quality of learning theoretical material and assimilation of practical skills [17]. As a rule, tests involve the teacher's exclusion from the assessment of test results, which has the undoubted advantages of an objective and unbiased assessment of student results. On the other hand, the use of tests forces to pay special attention to the quality of test questions, their comprehensibility, consistency, and relevance to the material studied. Evaluation of test results can be used to assess the quality of the e-course itself, in particular the tests, theoretical material, practical tasks, and their relevance.

This paper proposes an approach that allows to analyze the results of the answers to the test questions, to assess how well the questions are composed and to improve the bank of test questions if necessary. For this purpose, cluster analysis is used [18, 19], in which questions are divided into clusters based on such parameters as the complexity of the question, the number of correct and incorrect answers to the question.

Clustering allows to analyze questions, to identify ambiguous and contradictory questions, questions that have a weak correlation with the material studied, too complex or, on the contrary, trivial questions.

On the basis of the cluster analysis of test questions it is possible to improve the e-course, refining test questions, and possibly the theoretical material. After that, the tests will more adequately reflect the results of the course.

2 Method

We will use the K-medians clustering algorithm [20, 21] to cluster test questions. Each question represents a point in space (Fig. 1).

All questions (points in space) are divided into several homogeneous groups (clusters). To analyze test questions, the questions are divided into four clusters corresponding to the four grades of achievement: excellent, good, satisfactory, unsatisfactory. To do this, four points ($K = 4$) are randomly selected in the whole point space (Fig. 2).

After that we group around these points the points that have the minimum distance to the first points. The K-medians clustering algorithm uses the Manhattan distance, which

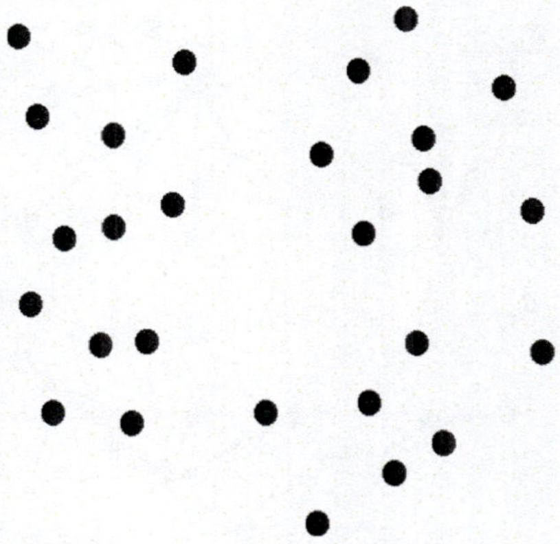

Fig. 1. Each point corresponds to a test question.

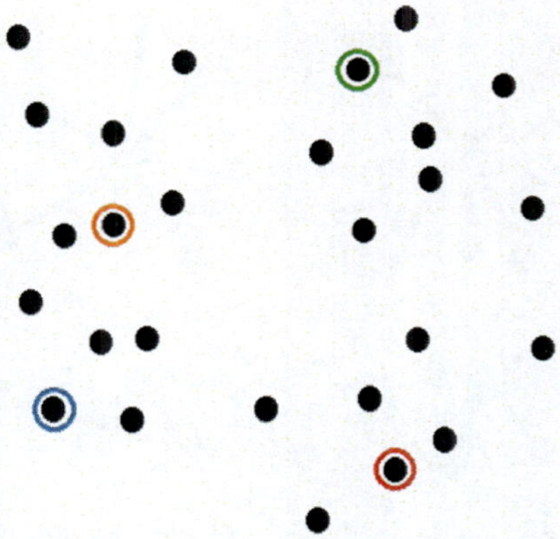

Fig. 2. Randomly selected four points (K = 4).

represents the distances in each dimension, to calculate the distance between points in n-dimensional space [22]:

$$d(x, y) = \sum^{n} |x_i - y_i|.$$

Thus, all points are divided into initial clusters (Fig. 3).

After that, we iteratively select the center of clusters until the center is detected or the number of iterations is maximal. Figure 4 shows clusters obtained on a given point space.

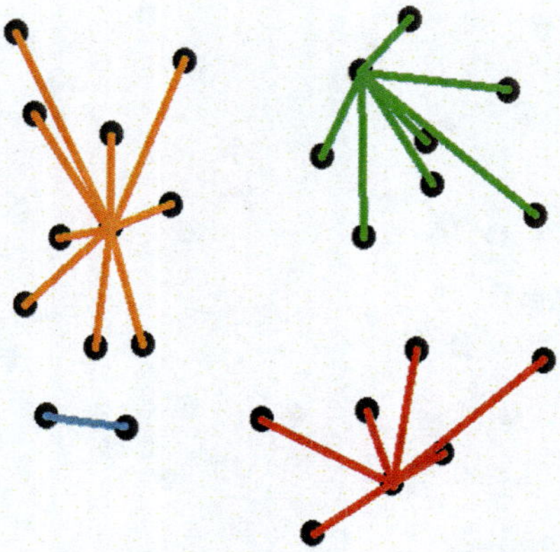

Fig. 3. Distribution of points in clusters (K = 4).

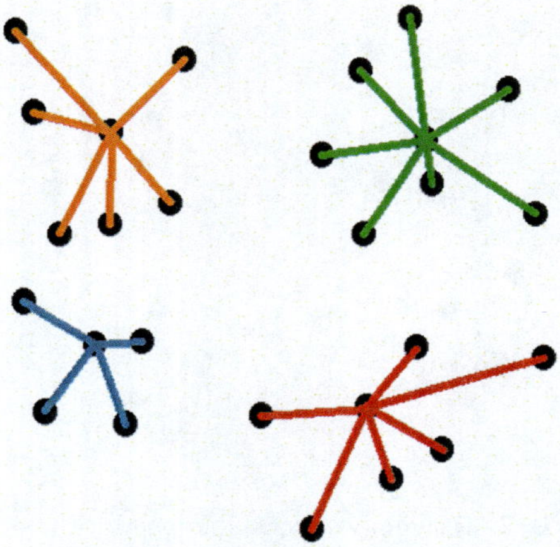

Fig. 4. Four clusters.

3 Program Implementation

A program was developed to divide the test questions of the e-course into separate clusters. Figure 5 shows a fragment of program code, which divides the original point space into clusters.

After the initial clusters are formed, the center of the clusters is searched for. The clusters are reconfigured and some points may move to another cluster. The program code for finding the cluster centers is shown in Fig. 6.

```
cc = []
for p in P:
        pp = -1
        minimal_distance = np.inf
        for i,j in enumerate(center):
                distance = self.__distance(p, j)
                if distance < minimal_distance:
                        pp = i
                        minimal_distance = distance
        cc.append(pp)
```

Fig. 5. Dividing the original point space into clusters.

```
cls = {}
for x in set(cc):
        cls[x] = []

for p,c_cur in zip(P, cc):
        cls[c_cur].append(p)

        centers = []
        for p in cls.values():
                tmp =  p[0]
                for x in p[1:]:
                        tmp = np.cc_[tmp,i]
                if tmp.n_dimention == 1:
                        tmp = tmp.reshape(-1,len(tmp))
                tmp = np.centers(tmp, axis=1)
                centers.append(tmp)
```

Fig. 6. Finding cluster centers.

4 The Effect of the Obtained Results on the Test Questions of the E-course

After running the program and clustering test questions, it is possible to analyze test questions in essence, to evaluate what answers were obtained to certain test questions, to analyze what these results may be related to.

Application of the developed program made it possible to identify questions that were not fully consistent with the material presented. Thanks to this the questions were rephrased or changed meaningfully. In some cases, theoretical material was expanded and improved, including with the help of graphs, tables, and illustrations. Some questions were simple, but they could only be answered by studying the course material. Such questions remained unchanged. No trivial questions were identified. A few questions that were too difficult were broken down into two or three simpler questions. Several questions required small calculations before they were answered. Although it is difficult to call them simple, they remained in their original form, as they clearly reflect the practical skills acquired by students.

5 Conclusion

Clustering of test questions in e-learning allows to evaluate how adequately students' knowledge is assessed, how relevant the test questions are to the theoretical material of the e-learning course and the practical skills acquired by the student while taking the course.

Practical application of cluster analysis in assessing the test results of a particular e-course allowed to identify deficiencies in the tests, to improve the addition and, in some cases, removal of test questions, to improve the quality of testing and learning of the e-course as a whole.

Thus, the application of the computer program for clustering questions in the evaluation of tests allowed to analyze not only how well the tests are composed, but also how they correspond to the theoretical material and practical assignments of the e-course. The analysis made it possible to improve the quality of the e-course as a whole that will have a positive impact on the study of e-course in the future.

References

1. Baftiu, N.: Application of intelligent applications for controlling software and hardware devices. Technium Rom. J. Appl. Sci. Technol. **3**(7), 1–13 (2021)
2. Hassouni, B.E., et al.: Realization of an educational tool dedicated to teaching the fundamental principles of photovoltaic systems. J. Phys: Conf. Ser. **1399**(2), 022044 (2019). https://doi.org/10.1088/1742-6596/1399/2/022044
3. Kaniadakis, A., Padumadasa, E.U.: Can e-learning enable the transition to university for computing and electronic engineering students from low socio-economic status? A socio-cultural approach. J. Inf. Syst. Educ. **33**(1), 87–97 (2022)

4. Khazaal, Y.M., Hammo, A.Y.: Survey on software code clone detection. Technium Rom. J. Appl. Sci. Technol. **4**(3), 28–36 (2022). https://doi.org/10.47577/technium.v4i3.6361
5. Rozhkov, A., Bulatova, M., Noda, L.: Mobile blogging as a mean to improve journalism education. Int. J. Interact. Mob. Technol. (iJIM) **14**(21), 38–50 (2020). https://doi.org/10.3991/ijim.v14i21.18465
6. Tyrkba, K.V., Asmyatullin, R.R., Ruzina, E.I.: Drivers of e-learning market: case of China. Res. Econ. Anthropol. **42**, 175–185 (2022). https://doi.org/10.1108/S0190-128120220000042017
7. Gamage, S.H.P.W., Ayres, J.R., Behrend, M.B.: A systematic review on trends in using Moodle for teaching and learning. Int. J. STEM Educ. **9**(1), 1–24 (2022). https://doi.org/10.1186/s40594-021-00323-x
8. Tsarev, R.Y., Yamskikh, T.N., Evdokimov, I.V., Prokopenko, A.V., Rutskaya, K.A., Everstova, V.N., Zhigalov, K.Y.: An approach to developing adaptive electronic educational course. Adv. Intel. Syst. Comput. **986**, 332–341 (2019). https://doi.org/10.1007/978-3-030-19813-8_34
9. Zaini, A., Santoso, H., Sulistyanto, M.P.T.: Fault tolerance strategy to increase Moodle service reliability. J. Phys: Conf. Ser. **1869**(1), 012095 (2021). https://doi.org/10.1088/1742-6596/1869/1/012095
10. Chernyshov, P.S.: Analysis of the latest changes in methodological recommendations in the field of information support of the educational process. Int. J. Human. Nat. Sci. **9–2**(60), 61–66. https://doi.org/10.24412/2500-1000-2021-9-2-61-66
11. Alnasraween, M.S., Shahadab, F.H.: Obstacles of teaching distance universities courses in light of E-learning quality standards. Cypriot J. Educ. Sci. **17**(4), 1244–1257 (2022). https://doi.org/10.18844/cjes.v17i4.7145
12. Long, R.W., III., Watanabe, H.: Adapting e-learning for Japanese EFL students. Int. J. Inform. Educ. Technol. **12**(4), 299–304 (2022). https://doi.org/10.18178/ijiet.2022.12.4.1618
13. Putro, H.P.N., Hadi, S., Rajiani, I., Abbas, E.W.: Mutiani: adoption of e-learning in Indonesian higher education: innovation or irritation? Educ. Sci. Theor. Pract. **22**(1), 36–45 (2022). https://doi.org/10.12738/jestp.2022.1.0004
14. Kansal, A.K., Gautam, J., Chintalapudi, N., Jain, S., Battineni, G.: Google trend analysis and paradigm shift of online education platforms during the COVID-19 pandemic. Infect. Dis. Rep. **13**(2), 418–428 (2021). https://doi.org/10.3390/idr13020040
15. Maatuk, A.M., Elberkawi, E.K., Aljawarneh, S., Rashaideh, H., Alharbi, H.: The COVID-19 pandemic and e-learning: challenges and opportunities from the perspective of students and instructors. J. Comput. High. Educ. **34**(1), 21–38 (2021). https://doi.org/10.1007/s12528-021-09274-2
16. Odeh, A., Keshta, I.: Impact of COVID-19 pandemic on education: moving towards e-learning paradigm. Int. J. Eval. Res. Educ. **11**(2), 588–595 (2022). https://doi.org/10.11591/ijere.v11i2.21945
17. Nikolaeva, I., Sleptsov, Yu., Gogoleva, I., Mirzagitova, A., Bystrova, N., Tsarev, R.: Statistical hypothesis testing as an instrument of pedagogical experiment. In: AIP Conference Proceedings (2022)
18. Chen, B.: Application of cluster analysis algorithm in the construction of education platform. Lect. Notes Data Eng. Commun. Technol. **98**, 424–430 (2022). https://doi.org/10.1007/978-3-030-89511-2_54
19. Hu, L.: Research on the construction of college English online education resources based on cluster analysis algorithm. In: ACM International Conference Proceeding Series, pp. 2763–2767 (2021). https://doi.org/10.1145/3482632.3487510
20. Ushakov, A.V., Vasilyev, I.: A computational comparison of parallel and distributed K-median clustering algorithms on large-scale image data. Commun. Comput. Inform. Sci. **1090**, 119–130 (2019). https://doi.org/10.1007/978-3-030-33394-2_10

21. Zhu, H., Shi, Y.: Brain storm optimization algorithms with k-medians clustering algorithms. In: Proceedings of the 7th International Conference on Advanced Computational Intelligence, ICACI 2015, 7184758, pp. 107–110. IEEE, Wuyi (2015). https://doi.org/10.1109/ICACI.2015.7184758
22. Craw, S.: Manhattan distance. In: Sammut, C., Webb, G.I. (eds.) Encyclopedia of Machine Learning. Springer, Boston (2011). https://doi.org/10.1007/978-0-387-30164-8_506

Pursuit on the Plane by Objects Moving Away from the Trajectory of the Defending Object, with Simulation of the Detection Area

Dubanov Alexander(✉)

Buryat State University, Ulan-Ude, Russian Federation
alandubanov@mail.ru

Abstract. This article describes the model of pursuit of a single target by the method of chasing a group of objects. All objects participating in the pursuit model move with a constant modulo speed. The pursuing object moves along a certain trajectory and releases objects at specified intervals, the task of which is to overtake the target by the chase method. Released objects have restrictions on the curvature of the motion path. A single target, in turn, is tasked with overtaking the pursuer by the method of parallel approach. For each pursuing object, a detection area is formed. The detection area is formed by two beams. The object's velocity vector is the bisector of the angle formed by such rays. If the target enters the detection area, then the object begins the pursuit by the chase method. If the target leaves the detection area, then the object makes a uniform and rectilinear movement. The task is to implement a dynamic model of multiple group pursuit, where each object has its own tasks, which implements the chase method. The model was developed using computer mathematics systems. Based on the results of the research, animated images were created. As an example, where the model developed in the article could be in demand, the following can be given. Consider the movement of a low-maneuverable object that is overtaking a faster target. As a means of protection, instead of releasing passive heat traps, it is proposed to drop a multitude of autonomously controlled weapons of destruction. From the analysis of the media, descriptions of such means of protection for aircraft have not been found. Modeling was carried out on a plane, there are no obstacles for modeling in space. The results of the research can be in demand in the design of unmanned aerial vehicles with elements of autonomous control and artificial intelligence.

Keywords: Parallel pursuit · Chase method · Target · Pursuer · Trajectory · Correction

1 Introduction

The methods of pursuit and parallel approach, as well as the method of proportional approach are widely used methods of aircraft guidance.

In this article, a model of group pursuit is proposed for consideration, when the pursuer moves along a certain trajectory. The target approaches the pursuer by the method

R. Silhavy et al. (Eds.): CoMeSySo 2022, LNNS 596, pp. 257–265, 2023.
https://doi.org/10.1007/978-3-031-21435-6_22

of parallel approach. From the trajectory of the pursuer, objects are released at certain intervals of time that will pursue the target using the chase method.

This model is given in the article as an example. The target can approach the pursuer by the chase method, the parallel approach method, or the proportional method. The pursuer has little maneuverability, but releases objects that have the ability to homing. In the model of the article, homing objects are released perpendicular to the pursuer's trajectory.

The model presented in this article aims not to show the advantages of the methods of pursuit, parallel and proportional approach, but to show the possibility of coordinated group capture of targets. Moreover, the feature is the autonomous management of protective equipment. That is, the choice of the decision about who to pursue, and how to pursue, and what to do when capturing a target is made by the pursuing device.

In this case, the question of the target recognition system becomes acute. In military aviation, a "friend-foe" recognition system is common, therefore, the same feature would be desirable for identifying targets in our model.

The article model can be transferred to virtual reality systems with different scenarios, with different goals and strategies of pursuit. The peculiarities the consistency of group behavior.

In the model of the article, the pursuing objects descend perpendicular to the pursuer's trajectory. The descent angle in the model is selected as an example.. The descent angle can be any. Descending from the trajectory in the model, sequential and at regular intervals are selected.

Each pursuing object has a detection area formed. In the model, for example, the area is formed as an angle with a vertex at the point where the object is located. The bisector of this angle coincides with the direction of the object's velocity.

2 Problem Statement

Consider the movement of the pursuer along a certain trajectory on the plane:

$$P(t) = \begin{bmatrix} X_P(t) \\ Y_P(t) \end{bmatrix}.$$

At time t_n, an object breaks off from the path of the pursuer in the direction of:

$$N(t_n) = \begin{bmatrix} -\frac{dY_p}{dt}(t = t_n) \\ \\ \frac{dX_p}{dt}(t = t_n) \end{bmatrix}. \quad (1)$$

Figure 1 shows that in the model considered in the article, five objects come off perpendicular from the pursuer's trajectory, which after separation will move uniformly and rectilinearly with a speed equal to modulo V_G.

Figure 1 is supplemented with an animated image [14], where it will be possible to see how objects $G_n(t)$ are released from the pursuer's trajectory at certain intervals.

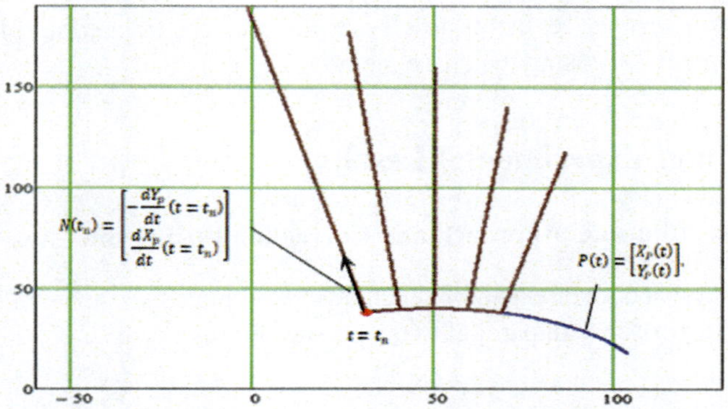

Fig. 1. Simulation of the movement of objects descending perpendicular to the pursuer's trajectory.

Modeling is performed on a section of the plane [−60:130] × [0:190]. The measurement is made in meters. The pursuer moves at a constant speed $V_p = 20$ m/s. Objects that break off perpendicular to its trajectory have a velocity $V_G = 40$ m/s.

Figure 2 shows a net of parallel lines. As you can see, the initial positions of the pursuer and the target, the initial speed of the pursuer determine the entire course of the iterative process. The pursuer's trajectory completely and unambiguously determines the target's trajectory.

The target $T(t)$ pursues $P(t)$ by the method of parallel approach. Figure 2 is supplemented with an animated image [15], where it will be possible to see the movement on the plane of the target and the pursuer.

Figure 2 shows the trajectory along which the target $T(t)$ moves uniformly with a velocity modulus $V_T = 25$ m/s.

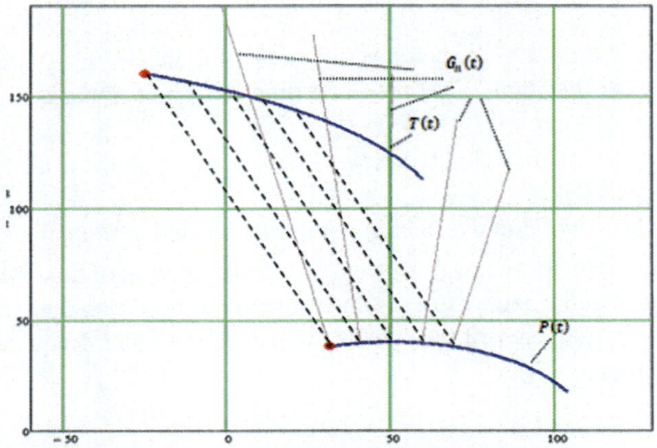

Fig. 2. Modeling of the trajectory of the target moving towards the pursuer by the method of parallel approach.

The task that we have set in this article is to simulate the trajectories of objects $G_i(t)$ pursuing the target $T(t)$ using the chase method.

3 Theory and Experimental Results

3.1 Methods of Parallel Approach and Correction During the Chase

The parallel approach method can be schematically depicted as shown in Fig. 3a, when the target T(t) approaches in parallel to the pursuer P(t).

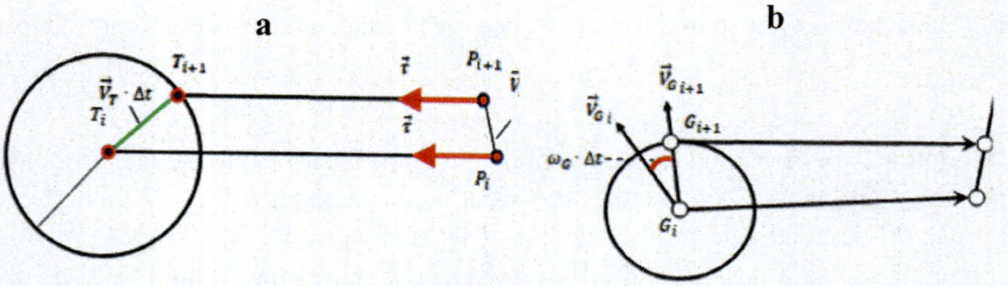

Fig. 3. Methods of parallel approach and correction during the chase.

Following the iterative scheme shown in Fig. 3a, the step of the target trajectory T_{i+1} satisfies the solution of the system of Eq. (2), with respect to the parameter h:

$$\begin{cases} (T_{i+1} - T_i)^2 = \left(\left|\overrightarrow{V}_T\right| \cdot \Delta t\right)^2 \\ T_{i+1} = P_{i+1} + h \cdot \frac{T_i - P_i}{|T_i - P_i|} \end{cases}. \tag{2}$$

The next step of the goal T_{i+1} belongs to a circle of radius $|\overrightarrow{V}_T| \cdot \Delta t$, centered at the point of the previous location T_i. At the same time, the point of the next position T_{i+1} belongs to the straight line $P_{i+1} + h * \overrightarrow{\tau}$ (Fig. 3a).

In the chase method, the velocity vector of that catching object is always directed at the object being overtaken.

In our case, this is not. Let the catching object be located at some time t_i at the point G_i, while having a velocity vector $\overrightarrow{V}_{Gi}$ (Fig. 3b). After a period of time Δt, the catching object makes a turn at the angle $\omega_G * \Delta t$ and moves to a distance $V_G * \Delta t$, where ω_G is the angular rotation frequency of the catching object. The angular frequency of rotation can be interpreted as:

$$\omega_G = \frac{V_G}{R_G},$$

where R_G is the minimum curvature radius of the trajectory of the catching object, that is, the curvature limit.

3.2 The Descent Function of Catching Objects from the Pursuer's Trajectory

Consider the motion function of catching up objects G_n (Fig. 1), when they move along the trajectory of the pursuer $P(t)$ before the moment of time t_n.

If at the moment of time t_n the direction changes to the direction $N(t_n)$ specified in (1), then the coordinates of the object G_n are determined as follows:

$$G_n(t) = \begin{cases} \text{if } t < t_n, & \text{then } G_n(t) = P(t) \\ \text{if } t \geq t_n, & \text{then } G_n(t) = P(t_n) + V_G \cdot (t - t_n) \cdot \frac{N(t_n)}{|N(t_n)|} \end{cases}$$

Based on the results of modeling the process of group pursuit of an object by the chase method, a program was written in a computer mathematics system, the results of which are shown in Fig. 4.

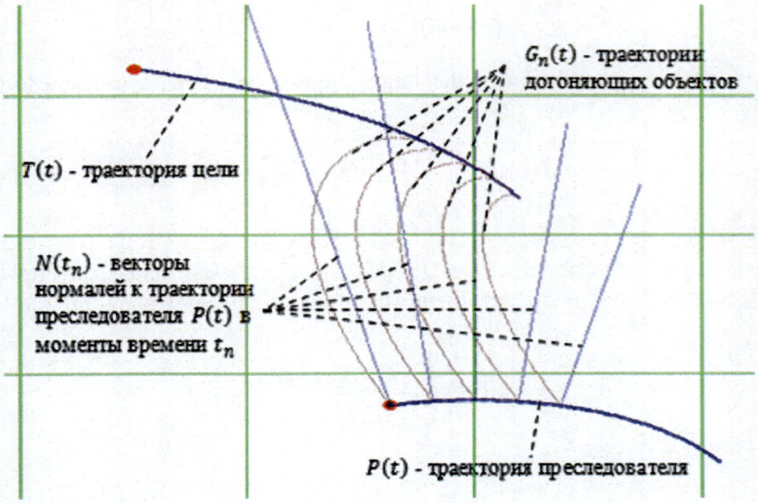

Fig. 4. The process of chasing a target by a group of objects using the chase method.

An animated image of the group pursuit of a single target by objects that descend perpendicular at certain intervals from the pursuer's trajectory was also produced [16].

3.3 Formation of a Tracking Area for Pursuing Objects

The formation of the tracking area of objects G_n for the target T is performed as follows.

A local coordinate system is being created $(E_1\ G_i\ E_2)$ (Fig. 5), where G_i is the location of the pursuing object at the moment t_i. The abscissa vector E_1 of the object is co-directed to the velocity vector $\overrightarrow{V}_{Gi}$. Accordingly, the ordinate vector E_2 is orthogonal to the velocity vector $\overrightarrow{V}_{Gi}$.

The tracking area is given by an angle of magnitude 2α, the direction of the velocity vector $\overrightarrow{V}_{Gi}$ is the bisector of this angle. In the coordinate system $(E_1\ G_i\ E_2)$ the vectors

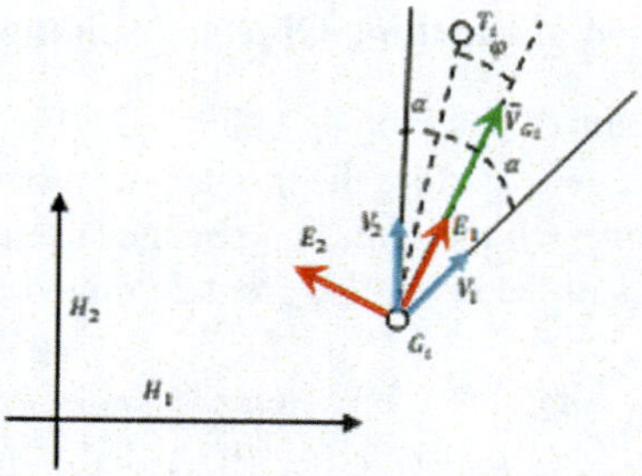

Fig. 5. Formation of the tracking area.

V_1 and V_2 defining the tracking area are determined:

$$V_1 = \begin{bmatrix} \cos(\alpha) \\ -\sin(\alpha) \end{bmatrix}, V_2 = \begin{bmatrix} \cos(\alpha) \\ \sin(\alpha) \end{bmatrix}.$$

The coordinate conversion of the target point T_i is carried out according to the formulas:

$$T_i^* = \begin{bmatrix} (T_i - G_i) \cdot E_1 \\ (T_i - G_i) \cdot E_2 \end{bmatrix}.$$

If the angle φ between the vectors $T_i - G_i$ and $\vec{V}_{Gi}$ is less than α, then the target T at time t_i is in the tracking area of the pursuing object. The angle φ is equal to:

$$\varphi = \left| \arccos\left(\frac{(T_i - G_i) \cdot \vec{V}_{Gi}}{|T_i - G_i| \cdot \left|\vec{V}_{Gi}\right|} \right) \right|.$$

3.4 The Behavior Model of the Pursuing Object

Modeling of tracking angles of pursuing objects G_n in the world coordinate system $(H_1\ H_2)$ is reduced to the transformation of vectors V_1 and V_2 from the coordinate system $(E_1\ G_i\ E_2)$ to the world.

Transformation of vectors V_1 and V_2 from the coordinate system $(E_1\ G_i\ E_2)$ to the world $(H_1\ H_2)$ it looks like this:

$$v_1 = \begin{bmatrix} V_1 \cdot h_1 \\ V_1 \cdot h_2 \end{bmatrix} + G_i, v_2 = \begin{bmatrix} V_2 \cdot h_1 \\ V_2 \cdot h_2 \end{bmatrix} + G_i.h_1 = \begin{bmatrix} H_1 \cdot E_1 \\ H_1 \cdot E_2 \end{bmatrix},$$

$$h_2 = \begin{bmatrix} H_2 \cdot E_1 \\ H_2 \cdot E_2 \end{bmatrix}, H_1 = \begin{bmatrix} 1 \\ 0 \end{bmatrix}, H_2 = \begin{bmatrix} 0 \\ 1 \end{bmatrix}$$

Figure 6 shows how the tracking areas of the pursuing objects are formed, Fig. 6 is supplemented with an animated image [17]. In Fig. 5, the pursuing objects catch up with

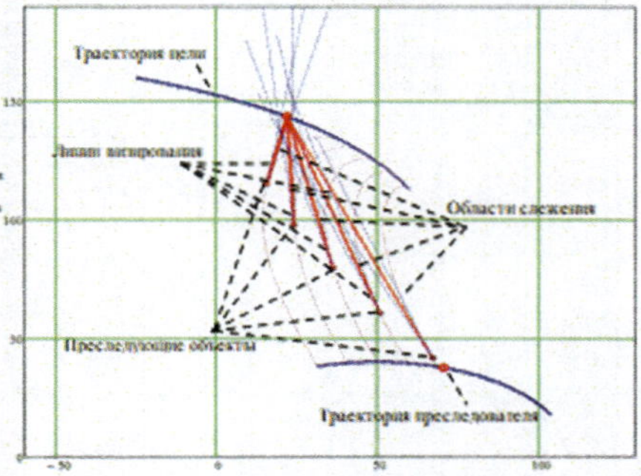

Fig. 6. Dynamic tracking areas of pursuing objects.

Fig. 7. The target is out of the detection zone.

the target by the chase method without changing behavior, depending on whether the target enters the tracking area. Tracking areas are displayed for each object. The lines of sight connecting the pursuing object with the target are also displayed.

Consider the behavior of the pursuing object.

Figure 7 shows that if at some point in time the target does not enter the detection area, the pursuing object moves in a straight line. If the target has entered the detection area, then the behavior of the pursuing object corresponds to the chase method. Figure 6 is supplemented with an animated image [18].

4 Discussion of Experiments

In modeling the pursuit process in this article, the method of chasing objects descending perpendicular to the pursuer's trajectory was used. In the model described in the article, nothing prevents us from replacing the chase method with the parallel approach method for catching up objects. And the descent perpendicular to the trajectory of the pursuer is replaced by a tangent descent.

Based on the results of the research presented in this article, modeling was performed in a rectangular area [−60:130] × [0:190], measurement in meters.

Figure 8 shows the simulation results. The pursuer's speed is 20 m/s, the target's speed is 20 m/s. The speed of the pursuing objects is 60 m/s. The curvature radius of

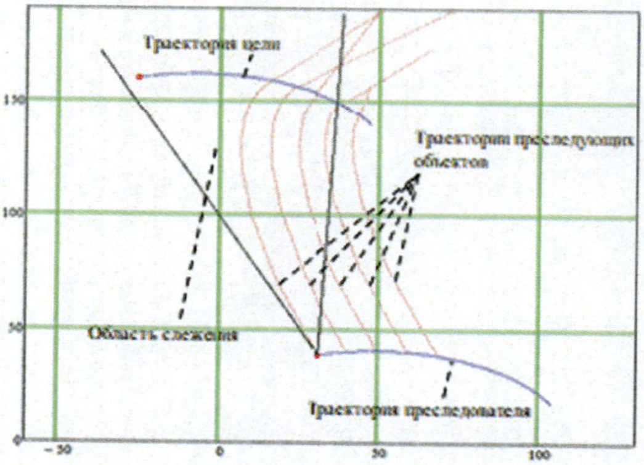

Fig. 8. Group pursuing model.

the pursuer's trajectory should not be less than 50 m. The target pursues the method of parallel approach, the curvature radius of the trajectories of the pursuing objects should not be less than 10 m. The pursuing objects descend perpendicular from the pursuer's trajectory at regular intervals of 0.02 s. Figure 7 is supplemented with an animated image [19], where it will be possible to get acquainted with the results of such a group pursuit.

In the simulation presented in this article, all objects released from the trajectory reach the target. This result depends on several factors. From the angle of the detection zone, from the speed of the pursuing objects, from the value of the minimum radius of curvature of the object trajectories.

During experimentation, it was found out that in order for the target to leave the tracking area faster, it is necessary to correct the movement to the direction perpendicular to the speed of the pursuing object.

For the pursuing object, the guaranteed result will be the capture of the target if it switches to the movement direction, the vector of which would be co-directed to the vector of the target's speed. Based on the results of the program, a certificate of state registration of the computer program No. 2020614336 "Modeling of trajectories from the pursuer to the target with curvature restrictions and with specified boundary conditions" was issued [20].

References

1. Isaacs, R.: Differential Games. Mir, Moscow (1967)
2. Krasovsky, N.N., Subbotin, A.I.: Positional Differential Games. Nauka, Moscow (1974)
3. Petrosyan, L.A.: Differential Pursuit Games. LSU Publishing House (1977)
4. Bannikov, A.C.: Some non-stationary problems of group pursuit. Proc. Inst. Math. Comput. Sci. UdSU **1**(41), 3–46 (2013)
5. Khachumov, M.V.: Solving the problem of following the target with an autonomous aircraft. Artif. Intel. Decis.-Mak. **2**, 45–52 (2015)
6. Khachumov, M.V.: Tasks of group pursuit of a target in conditions of disturbances. Artif. Intel. Decis.-Mak. **2**, 46–54 (2016)
7. Abramyants, T.G., Maslov, E.P., Yakhno, V.P.: Evasion of a group target in three-dimensional space. Autom. Remote Control **69**(5), 737–747 (2008)

8. Gusyatnikov, P.B.: The escape of one nonlinear object from several more inert pursuers. Differ. Eqn. **12**(2), 1316–1324 (1976)
9. Gusyatnikov, P.B.: Differential escape game of m persons. Izv. AN USSR. Tech. Cybern. **6**, 22–32 (1978)
10. Gusyatnikov, P.B.: Differential escape game. Cybernetics **4**, 72–77 (1978)
11. Bogdanov, A.V., Filonov, A.A., Kovalev, A.A., Kuchin, A.A., Lyutikov, I.V.: Methods of Homing Fighters and Air-to-Air Missiles to a Group Air Target. Siberian Federal University Publishing House, Krasnoyarsk (2014)
12. Nikitchenko, S.N., Bassauer, A.A.: Simulation model of the mutual pursuit problem. In: Proceedings of the International Conference "Regional Informatics and Information Security", vol. 5, pp. 479–483. SPOISU, SPb (2018)
13. Kuzmina, L.I., Osipov, Y.: Calculation of the trajectory length for the pursuit task. Vestnik MGSU **12**, 20–26 (2013)
14. Video, Simulation Results of the Pursuit Task. https://www.youtube.com/watch?app=desktop&v=t9cxOgk6bdk&feature=youtu.be. Last accessed 7 July 2022
15. Video, Simulation Results of the Pursuit Task. https://www.youtube.com/watch?v=keZ5fzd2o3Q. Last accessed 7 July 2022
16. Video, Simulation Results of the Pursuit Task. https://www.youtube.com/watch?v=ODS75MCwjGg. Last accessed 7 July 2022
17. Video, Simulation Results of the Pursuit Task. https://www.youtube.com/watch?v=oHowdT2n5-U. Last accessed 7 July 2022
18. Video, Simulation Results of the Pursuit Task. https://www.youtube.com/watch?v=5_-0TurWvwQ. Last accessed 7 July 2022
19. Video, Simulation Results of the Pursuit Task. https://www.youtube.com/watch?v=JcBrYjfaXTg. Last accessed 7 July 2022
20. Certificate of State Registration of the Computer Program No. 2020614336: Modeling of trajectories from the pursuer to the target with curvature restrictions and with specified boundary conditions

Prediction of the Failure of a Business Process Event Using Event Log and Deep Learning

Mamadou Diahame(✉), Mouhamed Diop, and Mamadou Samba Camara

Superior Polytechnical School of Dakar, University Cheikh Anta Diop of Dakar (UCAD), Dakar, Senegal
{diahame.mamadou,mouhamed.diop,mamadou.camara}@esp.sn

Abstract. Event logs are a collection of events that contain the execution of a business process. In them, each event is considered an observation and describe the feature of its execution. These events are regrouped into instance or case. Ina given instance, we can find that some events have not been executed properly and therefore, there is usually a feature that describes the status of the event. This type of execution is considered as failure and many publications usually propose prediction model to improve of the business model. In this work, we propose a prediction model for the failure of a business process event while considering the loops as failure. The loop can be considered as a repeated execution on an event during an instance. For the proposed model, we use two deep learning algorithms and then compared the best one. And finally, since, while using deep learning, it is hard to determine which feature are important for the prediction model, we have given an explainability of the model. For this work, we use the event log's dataset of a loan application done in a financial institution.

Keywords: Process mining · Failure prediction · Event log · Event loop · Deep learning · Event failure

1 Introduction

Business process prediction failure is an important topic that many researchers are studying [3]. This prediction can help a company improve his business or create better business process model. The event log is a record of events that concerns a business process [1]. Within the event log, there are some situations where an event is repeated multiple times during the execution of a business process. These repetitions, that we considered as loop, can greatly affect the performance of the process. So, predicting the loop can have a benefit in the execution of a business process.

Therefore, the aim of this work is to propose a prediction model for the failure of a business process event while considering these loops as failure. We will use some deep learning algorithm in order to do the prediction. The dataset we used is an event log of loan applications made in a financial institution [2].

R. Silhavy et al. (Eds.): CoMeSySo 2022, LNNS 596, pp. 266–276, 2023.
https://doi.org/10.1007/978-3-031-21435-6_23

We defined the research around these questions:

RQ1: Which dataset to use and how to implement the loop concept into it?

RQ2: Which deep learning method should we use in order to make the prediction model?

RQ3: How explainable is that model?

2 Related Works

2.1 Business Process Failure

There are multiples researches that are around the prediction of process failure. We have [3] who propose a prediction model using Long Short-Term Memory Recurrent Neural Network (LSTM RNN) in the context of the construction industry. They use the proposed model in order to develop a model by using a set of accounting data, the proxies for the construction market condition and data from the macroeconomic environment. In [4], they work on the exploitation of events in order to find errors and predict potential failures during a distributed process execution. They used a combination of convolutional neural network and recurrent neural network for the prediction model. In [5], they propose a novel approach for prediction of abnormal termination during real-time business process monitoring. For that, they use a Local Outlier Factor (LFO) which is according to them a widely used fault detection algorithm with a KNNI (K Nearest Neighbor Imputation) prediction algorithm for the prediction model. For [6], they propose a novel method for predicting the next process event. For that novel method, they use a deep learning model which consists of using the Long Short-Term Memory.

2.2 Loop Based Process

According to [7, 8], in the context of process mining, a loop causes a task to be executed multiple times during a given case. In order to define a representation of loops inside the event log, one can add the element "task_instance". In [9], they propose a methodology to improve the representation of event log to enhance the definition of the variables and models. They address the loop issue by proposing a methodology based on the CRISP-DM. In [10], they used a similar approach to represent the loop in existent event log and named that element "iteration". The purpose of their work is to propose a prediction model for business process failure while considering the loops as failure. They use machine learning algorithm for the proposed model.

2.3 Critics

In [3–6], they use some predictive model for different objectives and did not treat the loop issue. Their propositions show different ways to identify the failure of a business process.

In [7, 8], they made a good explanation of the loop but there wasn't an implementation of the said loop as a failure. In [9], through their proposed methodology, there wasn't a representation or an implementation of the said methodology while mentioning the loop issue. In [10], they identify the loop issue and represented the loop within an existent event log. The proposed method didn't use the deep learning which may provide better performance.

3 Description and Preprocessing of the Dataset

The dataset, that we used, is an event log from a loan application made in a financial institution [2]. The dataset contains all loans application filed between 2016 and 2nd February 2017. It has the following characteristic:

- Number of features: 19 among them, we have 10 categorical, 6 numerical, 2 Booleans and 1 of the date type.
- Number of events: 1,202,267.
- Number of cases: 31,509 it is represented by the feature "case:concept:name".

In addition to that, we added a variable iteration just like in [10] which is of the numerical type. So, the features and the characteristics of the data are show in Table 1.

Among the variables of the dataset, we can find that there are categorical variables that need some preprocessing to use them in the deep learning model. Therefore, among the preprocessing technique, we decide to encode the categorical one. For the encoding, we use the label encoder which is explained in Fig. 1.

There are others variables that we didn't use so we drop them. These variables are selected because of their high % of cardinalities. Those variables are:

- EventID
- time:timestamp.

This preprocessed data will be used as input for our prediction model. As for the target variable, we defined it as in [10]. Therefore, we consider that for the target variable, it is equal to 1 when:

- The "lifecycle:transition" variable has the value "ate_abort"
- The "Iteration" variable is superior or equal to 3.

And the target variable is equal to 0 otherwise.

The "lifecyce:transition" represents the status of the execution of an event. The "Iteration" represents how many times an activity has been executed during an instance.

Table 1. Characteristics of the dataset

Feature name	Type	# of cardinalities	% of cardinalities
Action	Categorical	5	< 0.1
org:resource	Categorical	149	< 0.1
concept:name	Categorical	26	< 0.1
EventOrigin	Categorical	3	< 0.1
EventID	Categorical	1,202,267	100
lifecycle:transition	Categorical	7	< 0.1
Time:timestamp	DateTime	1,202,209	> 99.9
case:LoanGoal	Categorical	14	< 0.1
case:ApplicationType	Categorical	2	< 0.1
case:concept:name	Categorical	31,509	2.6
case:RequestedAmount	Numerical	701	< 0.1
FirstWithdrawalAmount	Numerical	5930	0.5
NumberOfTerms	Numerical	147	< 0.1
Accepted	Boolean	2	< 0.1
MonthlyCost	Numerical	5816	0.5
Selected	Boolean	2	< 0.1
CreditScore	Numerical	520	< 0.1
OfferedAmount	Numerical	663	< 0.1
OfferID	Categorical	42,995	3.6
Iteration	Numerical	123	< 0.1

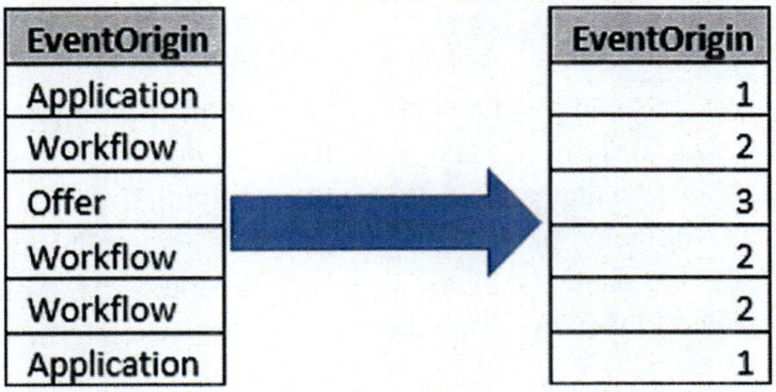

Fig. 1. How label encoder works

4 Prediction Model

Deep learning is a subdomain of machine learning based on neural network to design, build and evaluate a prediction model. To do deep learning, several libraries have been created such as TensorFlow, Torch, Theano etc. These libraries work with the python programming language. Each library has its own way of way of representing neural networks. Among these libraries, we have used the TensorFlow [10]. Each layer of neural networks is composed of nodes that TensorFlow calls internal units. Among the deep learning technique that TensorFlow has, we used two different one which are the Recurrent Neural Network (RNN) and the Convolutional Neural Network (CNN). In addition to these techniques, TensorFlow, also have some optimization functions that are the dropout and the batch normalization.

According to [11], the dropout is effective preventing overfitting as it continuously alters the nodes and weights in the training process. The batch normalization help improve the training time of the model.

4.1 Recurrent Neural Network (RNN)

The RNN is one of the basic architectures of the deep neural network. The concept of it is that the result of prediction of one layer would be enter as input in the next layers. Therefore, some techniques have been created based on this concept. TensorFlow implemented two among the techniques that existed which are the Long Short-Term Memory (LSTM) first propose in [12] and the gated recurrent unit GRU first proposed in [13]. In our RNN prediction model, we use a mix of these techniques to build the hidden layer of our model. Therefore, the structure of the model is proposed in Fig. 2.

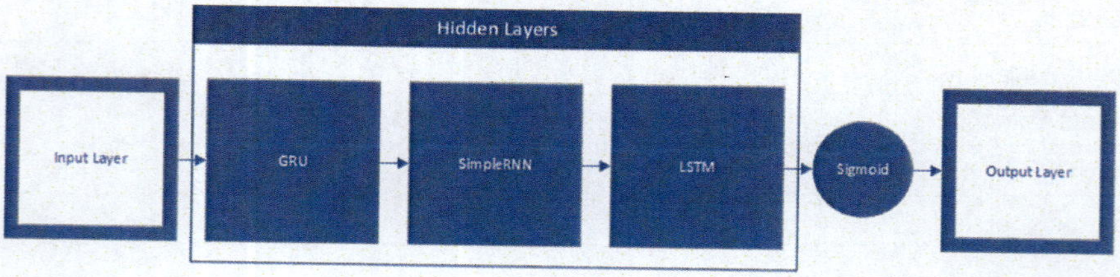

Fig. 2. The structure of our RNN model

In our hidden layers, we can see that there is a mix of the techniques that TensorFlow proposed. Table 2 shows us the summary of the RNN model.

In Table 2, We can see that after each layer, we applied the dropout and batch normalization layer. That helps us to greatly improve our training with a time around 220 s. We can also see that there is a Dense layer which represent our output layer. This layer is a fully connected layer in which each node is connected to every node of the previous layer.

4.2 Convolutional Neural Network (CNN)

The CNN is a type of deep learning that focuses on imitating functionality and behavior of our visual cortex [13]. It is useful for issues where the original input features are not individually informative. That's why it is mostly use for image prediction because each pixel of an image is not very informative. For our problem, TensorFlow propose a

Table 2. Summary of the RNN model

Layer (type)	Output shape	Params #
Embedding	(None, none, 128)	123,112,064
Dropout	(None, none, 128)	0
Batch_normalization	(None, none, 128)	512
GRU	(None, none, 64)	37,248
Dropout	(None, none, 64)	0
Batch_normalization	(None, none, 64)	256
Simple RNN	(None, none, 32)	3104
Dropout	(None, none, 32)	0
Batch_normalization	(None, none, 32)	128
LSTM	(None, none, 32)	8320
Dropout	(None, none, 32)	0
Batch_normalization	(None, none, 32)	128
Dense	(None, none, 1)	33
Total params: 123, 161, 793		
Trainable params: 123, 161, 281		
Non-trainable params: 512		

layer named "Conv1D" that is useful for predicting time-series data. Table 3 shows the summary of the CNN model.

For our CNN model, we can see that after the Conv1D layer, we have the pooling layer. In the convolutional model, these two are essential for the structure of the model. We can also see that after the layer Conv1D, we have the dropout and batch normalization layers. So, we have an average training time for each step of 56 s.

5 Results

As for the results, among the usual metric that deep learning method used, we focus on the accuracy to made the comparison of the two proposed model.

Figure 3 shows the accuracy graph of the RNN model. We can see that during the training process, we have an evolution of the accuracy all along the process for each epoch. The reason why we have a low number of epochs is because for both models, we added an early stopping. This early stopping occurs when the accuracy value doesn't surpass a certain value for 12 epochs. Therefore, for the RNN model, the training process terminated in 13 epochs and has an accuracy of 95.59%.

Figure 4 shows the accuracy graph of the CNN model. We can see that during the training process, we have an evolution of the accuracy as well for each epoch. As earlier, we also added an early stopping with the same configuration. There-fore, for the CNN model, the training process terminated in 16 epochs and has an accuracy of 95.87%.

Table 3. Summary of the CNN model

Layer (type)	Output shape	Params #
Conv1D	(None, 15, 128)	512
Dropout	(None, 15, 128)	0
Batch_normalization	(None, 15, 128)	512
Conv1D	(None, 13, 64)	24,640
Dropout	(None, 15, 64)	0
Batch_normalization	(None, 15, 64)	256
Conv1D	(None, 11, 32)	6176
Dropout	(None, 11, 32)	0
Batch_normalization	(None, 11, 32)	128
Flatten	(None, 352)	0
Dropout	(None, 352)	0
Batch_normalization	(None, 352)	1408
Dense	(None, 64)	22,592
Dropout	(None, 64)	0
Batch_normalization	(None, 64)	256
Dense	(None, 1)	65
Total params: 56, 545		
Trainable params: 55, 265		
Non-Trainable params: 1280		

Comparing these results to the result in [10], we can say that using deep learning method give better performance and result.

6 Explainability

According to [15], explainability provides us with algorithms that give insights into trained model predictions. Therefore, many libraries have been implemented with the help of the Explainable AI Toolkit [16]. Among the existent libraries, we use the Alibi Explain tool [15]. This tool is an open-source Python library aimed at machine learning model inspection and interpretation. Figure 5 shows how the tool works.

We can see that the tool needs the model as well as the data in order to get insights about the model prediction. We did the explanation for both models. As for the explanation, the tool uses the data in order to explain one instance of the data and then show how each feature contributes to the prediction for the instance of interest. The method we use is the integrated gradient originally proposed in [17]. So, we made a mean of a sample of the data for each model to show the features contributions for the prediction. This selection of the sample is random and the sample size is 2000.

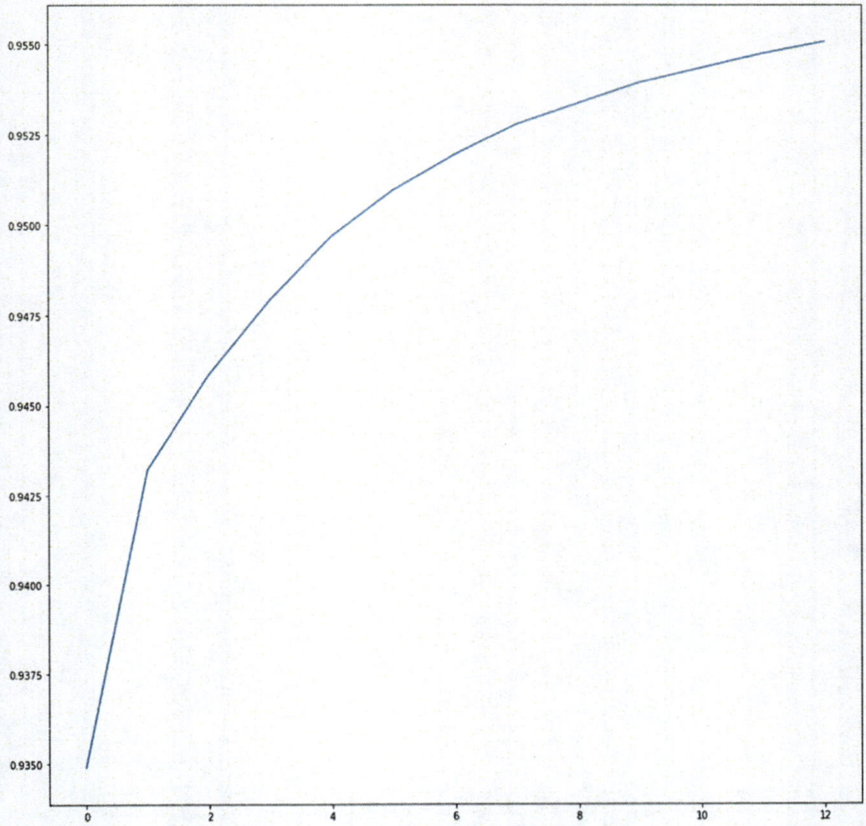

Fig. 3. Accuracy graph of the RNN model

Thus, Fig. 6 show the features that have affected the RNN model during the training.

Based on the sample that we selected, we can say that for the RNN model, the features that are contributing in the prediction are "Action" and "org:resource". That mean that the modification of their value can have a positive or negative impact on the accuracy of the model. Therefore, the feature "Action" has a strong positive impact on the prediction while the other feature "org:resource" has a negative one.

Figure 7 shows the features that have affected the CNN model during training. We use the same sample data as earlier.

Figure 7 shows us that the features "Action", "concept:name", "lifecycle:transition", "case:concept:name", "case:ApplicationType", "case:RequestedAmount", "FirstWithdrawalAmount", "NumberOfTerms", "Accepted", "Selected", "CreditScore" have an positive impact on the prediction of the CNN model. The other features have a negative impact on it.

Compare to the earlier model, we can say that there is more feature that contribute to the prediction of the model and that contribution is more diverse. These results show that the feature that are needed for the prediction are different from one model to another. Therefore, compared to the features that are selected in [10], we can say that the there are

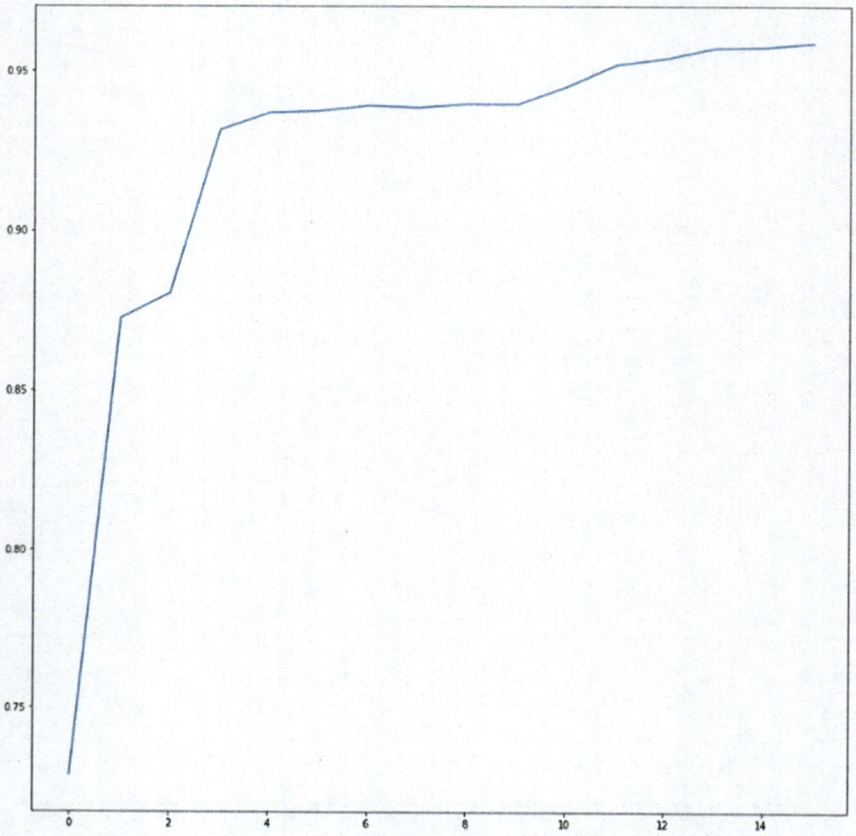

Fig. 4. Accuracy graph of the CNN model

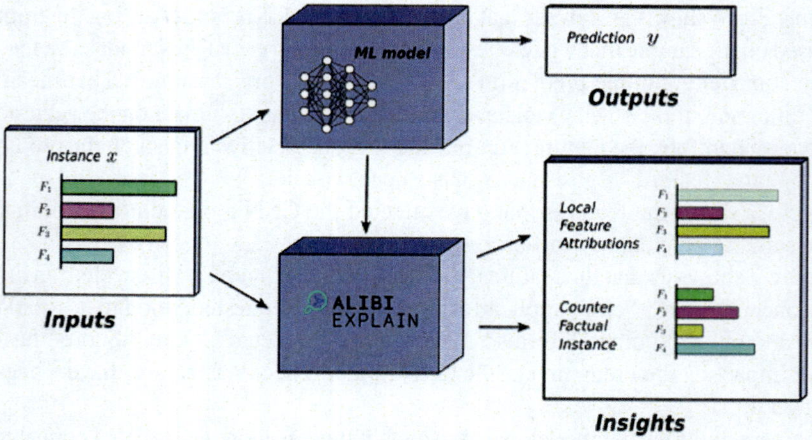

Fig. 5. How Alibi explain works?

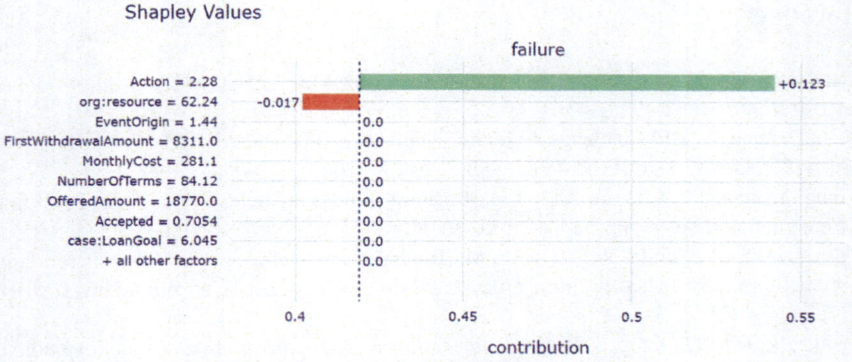

Fig. 6. Explanation of our RNN model

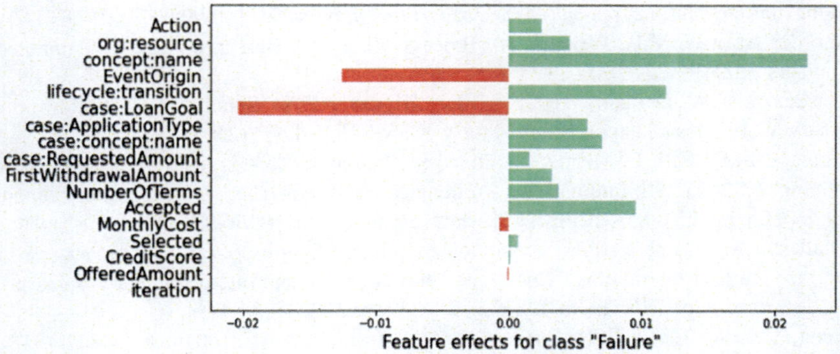

Fig. 7. Explanation of our CNN model

some similarities in the selection of the features. On the other hand, we can say that while using much less features than [10], our RNN model got better accuracy with 95.59% compared to 91.53%. The CNN model has used the most feature out of the dataset and has got the best accuracy out of the 3 models which is 95.87%.

7 Conclusion

The prediction the failure of a business process event is always useful for improving the efficiency of a business process. And, considering unnecessary repetition as failure can greatly enhance the performance of the business process. Based on the comparison of these models and the dataset we used, we can say that the CNN model has better performance in predicting the failure of a process with an accuracy of 95.87%. We have also an explanation of which feature has an importance for the model. Therefore, we can also try other method to further explain our model. Considering the summary of the models, we can also learn about the architecture of a deep learning model in order to know how the change of that architecture has an effect on the prediction.

References

1. Van der Aalst, W.: Process mining software. Process Min. 325–352 (2016). https://doi.org/10.1007/978-3-662-49851-4_11
2. Van Dongen, B.: BPI Challenge. Eindhoven University of Technology (2017). https://doi.org/10.4121/uuid:5f3067df-f10b-45da-b98b-86ae4c7a310b
3. Jang, Y., Jeong, I., Cho, Y., Ahn, Y.H.: Business Failure Prediction with LSTM RNN in the Construction Industry, pp. 114–121 (2019). https://doi.org/10.1061/9780784482438.015
4. Borkowski, M., Fdhila, W., Nardelli, M., Rinderle-Ma, S., Schulte, S.: Event-based failure prediction in distributed business processes. Inform. Syst. (2017). https://doi.org/10.1016/j.is.2017.12.005
5. Kang, B., Kim, D., Kang, S.-H.: Real-time business process monitoring method for prediction of abnormal termination using KNNI-based LOF pre-diction. Exp. Syst. Appl. **39**(5), 6061–6068 (2012). ISSN 0957-4174. https://doi.org/10.1016/j.eswa.2011.12.007
6. Evermann, J., Rehse, J.-R., Fettke, P.: A Deep Learning Approach for Predicting Process Behaviour at Runtime, pp. 327–338 (2017). https://doi.org/10.1007/978-3-319-58457-7_24
7. van der Aalst, W.M.P., Weijters, A.: Process Mining: A Research Agenda. Computers in Industry (2004)
8. van der Aalst, W., Dongen, J., Herbst, B.F.V., Maruster, G.W., Schimm, L.: Workflow mining: a survey of issues and approaches. Data & Knowl. Eng. **47**(2) (2003)
9. Camara, M.S., Fall, I., Mendy, G., Diaw, S.: Activity failure prediction based on process mining. In: 2015 19th International Conference on System Theory, Control and Computing (ICSTCC), pp. 854–859. IEEE (2015). https://doi.org/10.1109/ICSTCC.2015.7321401
10. Diahame, M., Camara, M.S.: Prediction of process failure approach using process mining. In: 5th EAI International Conference on Innovation and Interdisciplinary Solutions for Underserved Areas (INTERSOL 2022). 22–24 Mar 2022 (In press)
11. Zeng, X., Long, L.: Basic TensorFlow (2022). https://doi.org/10.1007/978-1-4842-7915-1_4
12. Subasi, A.: Machine learning techniques. In: Practical Machine Learning for Data Analysis Using Python, pp. 91–202. Academic Press (2020). ISBN 9780128213797. https://doi.org/10.1016/B978-0-12-821379-7.00003-5
13. Hochreiter, S., Schmidhuder, J.: Long short-term memory. Neural Comput. **9**(8), 1735–1780 (1997)
14. Cho, K., Van Merriënboer, B., Gulcehre, C., Bahdanau, D., Bougares, F., Schwenk, H., Bengio, Y.: Learning Phrase Representations Using RNN Encoder-de-Coder for Statistical Machine Translation (2014). arXiv preprint arXiv:1406.1078. https://doi.org/10.48550/arXiv.1406.1078
15. Klaise, J., Van Looveren, A., Vacanti, G., Coca, A.: Alibi explain: algorithms for explaining machine learning models. J. Mach. Learn. Res. **22**(181), 1–7 (2021). http://jmlr.org/papers/v22/21-0017.html
16. Hu, B., Tunison, P., Vasu, B., Menon, N., Collins, R., Hoogs, A.: XAITK: the explainable AI toolkit. Appl. AI Lett. **2**(4), e40 (2021)
17. Sundararajan, M., Taly, A., Yan, Q.: Axiomatic attribution for deep networks. In: International Conference on Machine Learning, pp. 3319–3328. PMLR (2017). https://doi.org/10.48550/arXiv.1703.01365

The Demand Dynamics Forecasting for Perishable Products

Ulyana Firyago, Ivan Kocherov, Yaroslavna Pankratova, and Inna Trofimova(✉)

St. Petersburg State University, St. Petersburg, Russia
isolovyeva@mail.ru

Abstract. In this paper, models for forecasting the dynamics of demand for products with a short expiration date are constructed. Here we propose to construct models for time series forecasting using the decomposition method and taking into account the assumptions of experts about the influence of certain factors on the behavior of product consumers. These models were implemented and tested on real data, and the obtained forecast were evaluated. The decomposition method and Holt-Winters method were used, and comparative analysis of these models was carried out, and a forecast are made.

Keywords: Time series forecasting · Demand forecasting · Holt-Winters model · Decomposition method

1 Introduction

Demand forecasting is an important tool in the operation and management of retail networks, because company leaders need to plan a large number of values: from sales volumes of one specific unit of goods to profits of the company in the following periods. The flexible mathematical model that most accurately would describe this or that economic process in the company and demand forecast could help leaders in inventory management and decisions making. For a retail networks, the main prerequisites for the model constructing are expert assumptions and a set of data.

On the other hand, data analysis methods in demand forecasting problems enable the company to increase the efficiency of the following processes: strategic planning, assortment management, inventory optimization at all nodes of the supply chain and selection the optimal number of stores [1]. These processes are especially important for grocery chains, since a large number of products have a short shelf life. A large number of demand forecasting techniques and different classes of mathematical models: linear and nonlinear are discussed in contemporary literature [2–4]. Including in particular forecasting demand for products with a short shelf life using the Holt-Winters method could be seen in works [5–8], different regression methods in [6,9,10] the decomposition method in [8].

R. Silhavy et al. (Eds.): CoMeSySo 2022, LNNS 596, pp. 277–287, 2023.
https://doi.org/10.1007/978-3-031-21435-6_24

The use of the classical Holt-Winters method allows to easy and efficiently obtain a forecast. However, the decomposition method allows to construct the mathematical model based on both data and the subjective opinion of experts. In this paper, we propose to construct a time series forecasting model using the decomposition method and taking into account the assumptions of experts about the influence of certain factors on the behavior of product consumers. Then on the real data we will test the additive and multiplicative models, and compare results with the Holt-Winters model.

2 Problem Statement

Let us consider one retail network that has a grocery stores, each store is allocated several departments, and everyone of them contains perishable products. In constructing the demand dynamics model of products with a short shelf life, it is recommended to take into account expert assumptions about the influence of important factors on demand, and a set of sales data for the previous time for various product names. The initial data is stored in the table that was created using a database management system, it contains columns: date, store number, department code, product code, volume of sales on this date, unit price of the product, the total returns from sales on that day, the inventory level of the product in the warehouse at the end of the day.

The purpose of this work is construction of the mathematical model and the time series forecast for the sales volume of goods, using the decomposition method. At the same time, in the original nonstationary time series model, taking into account expert assumptions about significant factors, it is necessary to include a linear (piecewise linear) trend, three types of seasonality and the influence of holidays. Then to conduct a comparative analysis of the constructed model with Holt-Winters model and Prophet model.

3 Mathematical Model

Through time series studies, it is customary to single out several constituents

$$y(t) = u(t) + s(t) + v(t) + \varepsilon(t). \tag{1}$$

Here $u(t)$ is a variable representing long-term trend (trend) of changes in a number of features, $s(t)$ is a seasonal component, reflecting the periodic repetition of processes, within one period (year), $v(t)$ is a cyclic component reflecting the periodic recurrence of processes, but with a large period of fluctuations, and $\varepsilon(t)$ is a random component, containing the influence of other factors, often hidden and not obvious.

The selection of various types of seasonal components is carried out using analysis of the graph of the autocorrelation function. However, when analyzing non-stationary time series, it is important to take into account that only one part of the peaks on the graph may be due to cyclical fluctuations, other part due to internal dependencies of the terms of the series [11]. In this case, the key part is the recommendations of experts who have been observing processes over time, because they could highlight significant factors that are not visible at the data.

Taking into account the specifics of the products and according to the assumptions of experts: significant factors are the trend, holidays, as well as annual, monthly and weekly seasonality. In addition, weather conditions (if presented in an indicator form) and staff work schedules can also be significant factors.

Thus, in this case the model (1) is represented in the form:

$$y(t) = trend(t) + h(t) + ys(t) + ms(t) + ws(t) + \varepsilon(t). \tag{2}$$

In (2) the following components are selected: $trend(t)$ is a trend, $h(t)$ are holidays, $ys(t)$, $ms(t)$, $ws(t)$ are annual, monthly and weekly seasonality, respectively.

Consider the log-likelihood function (LLF) of the following form [12]:

$$\log L = -\frac{N}{2}\log(2\pi\sigma^2) - \frac{1}{2}\sum_{i=1}^{n}\frac{(y_i - \beta_1 - \beta_2 x_i)^2}{\sigma^2} \tag{3}$$

Then

$$\begin{cases} \frac{\partial logL}{\partial \beta_1} = \sum\limits_{i=1}^{N} \frac{(y_i-\beta_1-\beta_2 x_i)}{\sigma^2} = 0, \\ \frac{\partial logL}{\partial \beta_2} = \sum\limits_{i=1}^{N} \frac{(y_i-\beta_1-\beta_2 x_i)}{\sigma^2} x_i = 0. \end{cases} \tag{4}$$

Then let us transform (4) to the form:

$$\begin{cases} \sum\limits_{i=1}^{N} x_i y_i - \beta_1 \sum\limits_{i=1}^{N} x_i^2 - \beta_2 \sum\limits_{i=1}^{N} x_i = 0, \\ \sum\limits_{i=1}^{N} y_i - \beta_1 \sum\limits_{i=1}^{N} x_i - N\beta_2 = 0. \end{cases}$$

Thus, we obtain equations for the parameters β_1 and β_2:

$$\begin{cases} \beta_1 = \frac{1}{N}\left(\sum\limits_{i=1}^{N} y_i - \frac{\sum\limits_{i=1}^{N} x_i y_i \sum\limits_{i=1}^{N} x_i - 1/N(\sum\limits_{i=1}^{N} x_i)^2 \sum\limits_{i=1}^{N} y_i}{\sum\limits_{i=1}^{N} x_i^2 - 1/N(\sum\limits_{i=1}^{N} x_i)^2}\right), \\ \beta_2 = \frac{\sum\limits_{i=1}^{N} x_i y_i - 1/N \sum\limits_{i=1}^{N} x_i \sum\limits_{i=1}^{N} y_i}{\sum\limits_{i=1}^{N} x_i^2 - 1/N(\sum\limits_{i=1}^{N} x_i)^2}. \end{cases}$$

The determined parameters values of the trend component we remove from the data and obtain a new series of values $\overline{y}_i = y_i - (\beta_1 + \beta_2 x_i)$. Then we will use this time series to obtain other components.

The selection of several seasonal components in the model is based on the expert hypotheses that, in addition to the annual seasonality, there are monthly and weekly cyclic components. The assumption of monthly seasonality in the data is based on the fact that there are fixed days for receiving wages in a month, and they could be the same for many people, and therefore, local monthly peaks can be observed on these days sales. The expert assumption of a weekly cyclical component is based on the fact that people shop more on weekends and less on weekdays.

Then taking into account that the differences between the seasonal and cyclical components are in different periods of fluctuations, as well as in a different number of terms of the Fourier series, the sum of which describes the fluctuations, the general form of the seasonal component is:

$$s(t) = \sum_{n=1}^{N} (a_n \cos(2\pi nt/T) + b_n \sin(2\pi nt/T)),$$

where T is the fluctuation period. Then the LLF will take the following form:

$$\log L(\beta, \sigma^2) = -\tfrac{1}{2} \textstyle\sum_{i=1}^{N} \dfrac{(y_i - \sum\limits_{j=1}^{k} (\beta_{2j-1} \sin(2\pi ij/T) + \beta_{2j} \cos(2\pi ij/T)))^2}{\sigma^2} - \tfrac{N}{2} \log(2\pi\sigma^2). \tag{5}$$

Here k is the number of pairs of Fourier series terms required to determine a specific type of seasonality. In (5) we have $2k$ parameters.

Next, to find the maximum of LLF (5) and obtain the desired parameters, we compose the system of equations:

$$A\beta = c. \tag{6}$$

Here, the parameter vector $\beta = (\beta_1, \ldots, \beta_{2k})^T$, and the elements a_{jl} of the matrix A, and the elements c_j of the vector of right-hand sides $c = (c_1, \ldots, c_{2k})^T$ have the following form:

$$a_{jl} = \sum_{i=1}^{N} \theta(j)\theta(l), \; c_j = \sum_{i=1}^{N} y_i \theta(j),$$

$$\theta(x) = \begin{cases} \sin(2\pi i(x+1)/2T), \; if \; x \; is \; the \; odd \; number, \\ \cos(2\pi ix/2T), \; if \; x \; is \; the \; even \; number. \end{cases}$$

Then, $2k$ parameters could be defined from the system (6). At the same time, to model the annual seasonality the number of pairs of terms of the Fourier series $k = 10$ and $T = 365.25$, for the monthly cyclical component $k = 5$ and $T = 30.4375$, for the weekly cyclical component $k = 3$ and $T = 7$.

To select the holidays component, we will tease out those days in which sales are strong different from the main trend. If a holiday (event) begins to influence the time series in advance, for example, New Year's holidays, then greater demand is due to preparations for the "holiday" and the fact that after it shops do not work for several days. Therefore, it is proposed to take the holidays component as a vector of length N, in which on the holidays places are the average sales at that day for several years, and zeros on other days.

We describe in detail the case of the additive model. In construction process of multiplicative models, we use the same algorithm after model linearization. In addition, we consider a modification of the multiplicative model, one of the components is a piece-wise linear trend with a single break point.

4 Realization

In this section, we construct three models: one classical model (Holt Winters) and two decomposition models based on ideas describing above.

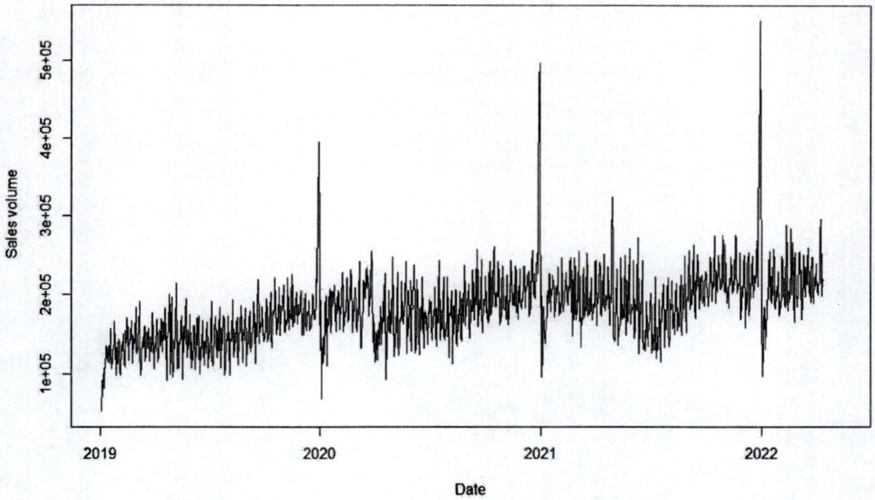

Fig. 1. Data of sales 01.01.2019–13.12.2021

Figure 1 shows real data of the sales volume of the product department for the period from January 2019 to December 2021. We can see that there is a linear trend on the chart, and the above seasonal components (annual, monthly and weekly).

4.1 Additive Decomposition Model

Here, we construct the additive decomposition model using (2). As we mentioned earlier, the model is based on daily data for 3 years, starting from January 1, 2019. We demonstrate the simulation result in Fig. 2.

In Fig. 3, we could see the prediction value (Fig. 4).

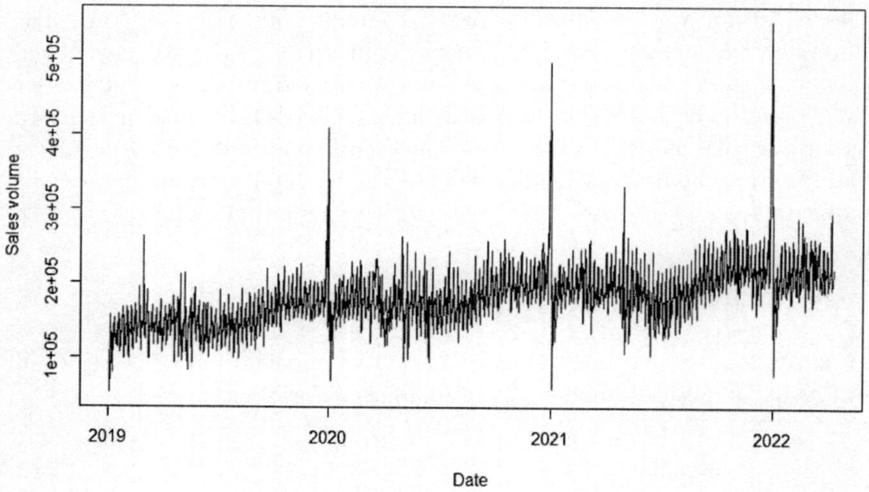

Fig. 2. Additive decomposition model, red line—fitted value, black line—real data

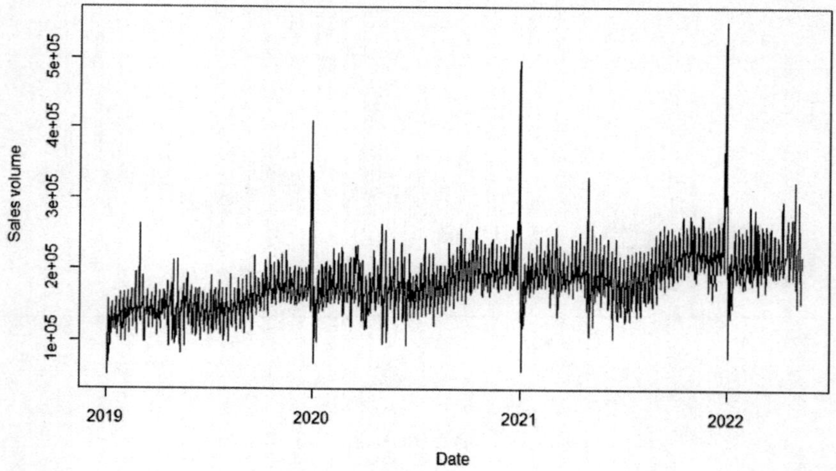

Fig. 3. Prediction value for the additive decomposition model, red line—fitted value, black line—real data, blue line—prediction value

4.2 Multiplicative Decomposition Model

We note that almost all components depend on the linear trend: the amplitude of annual fluctuations increases with the growth of the trend, the same happens with weekly ones, the New Year's peak also increases every year. Moreover, we remark that the weekly seasonality depends on the annual one: in addition to the division into summer and non-summer seasonality (their combination is contained in the $ws(t)$ component), the difference of which is in the shift of the weekly peak, we also add multiplication by annual seasonality, because judging

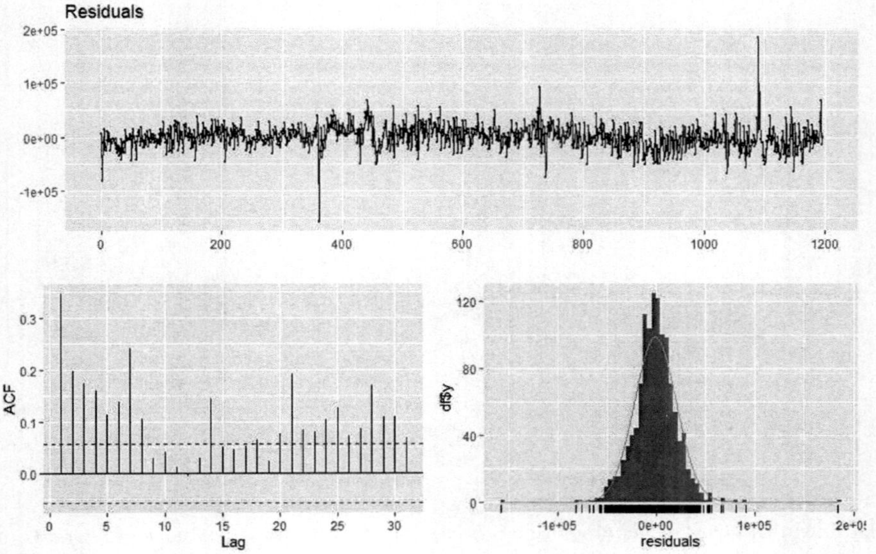

Fig. 4. Residuals of the additive decomposition model

by the data, the amplitude of fluctuations also changes during the year. After all the manipulations, we decide to add another regressor with the weather, which contains ABS data ("air temperature" – "feels like ... (in degrees)").

Finally, we have the following model

$$y(t) = trend(t) + h(t) \cdot trend(t) + ys(t) \cdot trend(t)$$

$$+ys(t) \cdot trend(t) \cdot ws(t) + trend(t) \cdot ws(t) + weather(t).$$

After that we decide change linear trend and use piece-wise linear trend with a single break point, that additionally improve the model.

We present the comparison of models predictions on the following figures (Figs. 5, 6, 7, 8 and 9; Table 1).

5 Conclusion

The aim of this paper was to propose the model based on demand dynamics forecasting approaches and taking into account the assumptions of experts about the influence of certain factors on the behaviour of product consumers. The additive decomposition model, multiplicative decomposition model with linear trend, multiplicative decomposition model with piece-wise linear trend, Holt-Winters model were realized and tested on the the given data set and used to obtain prediction values. After the prediction process, the comparative analysis of considered models was done. On the basis of which it could be concluded that

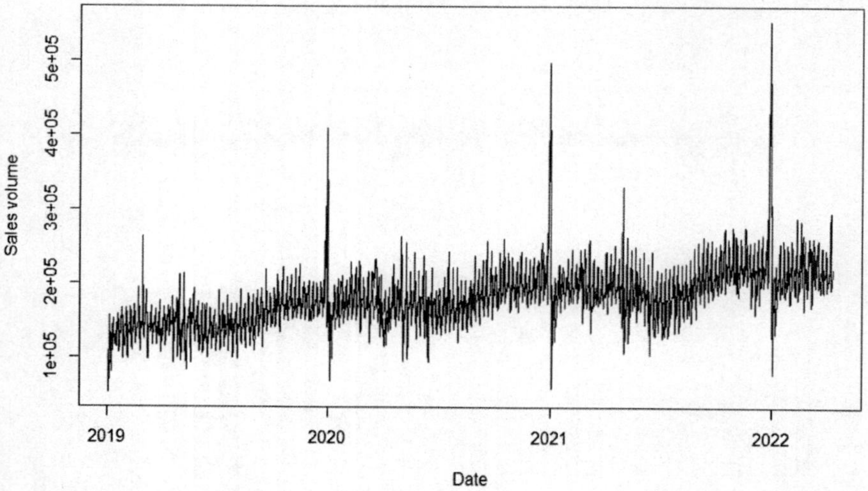

Fig. 5. Multiplicative decomposition model, red line—fitted value, black line—real data

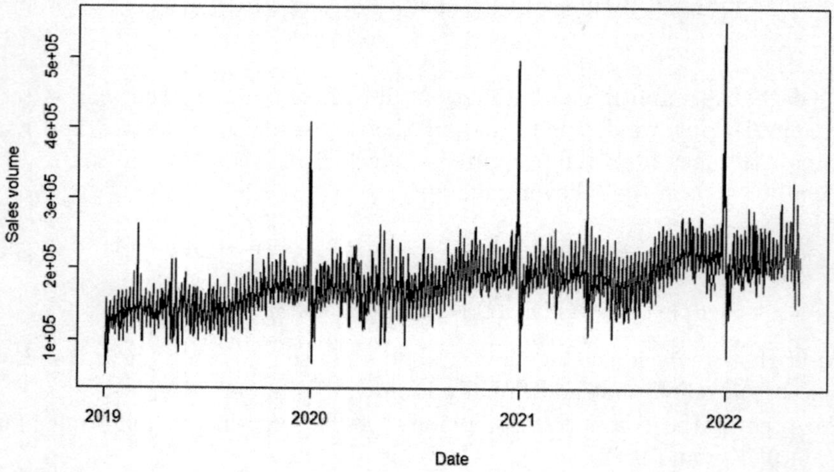

Fig. 6. Prediction of multiplicative decomposition model, red line—fitted value, black line—real data, blue line—prediction value

the multiplicative decomposition model with piece-wise linear trend is better than others for the given data set. A reasonable choice of a model for demand dynamics forecasting for perishable products can be used in order to develop an auto-order systems or other decision support systems.

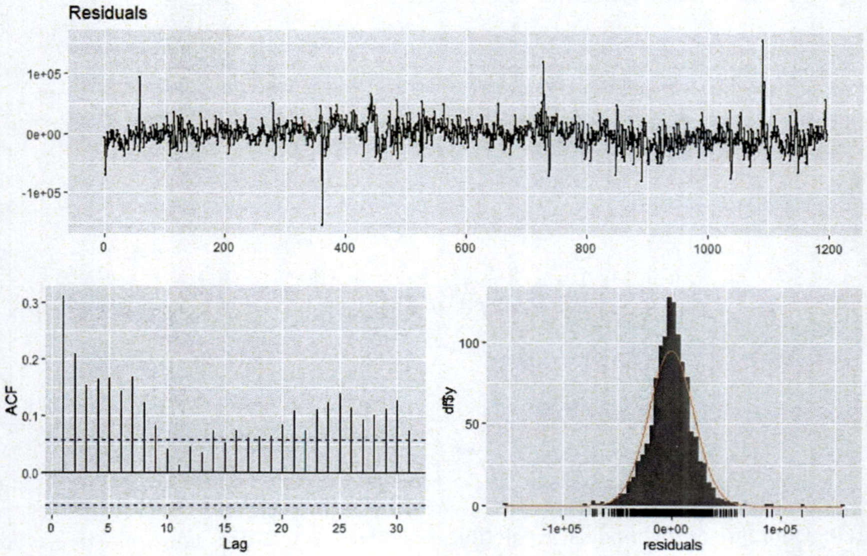

Fig. 7. Residuals of multiplicative decomposition model

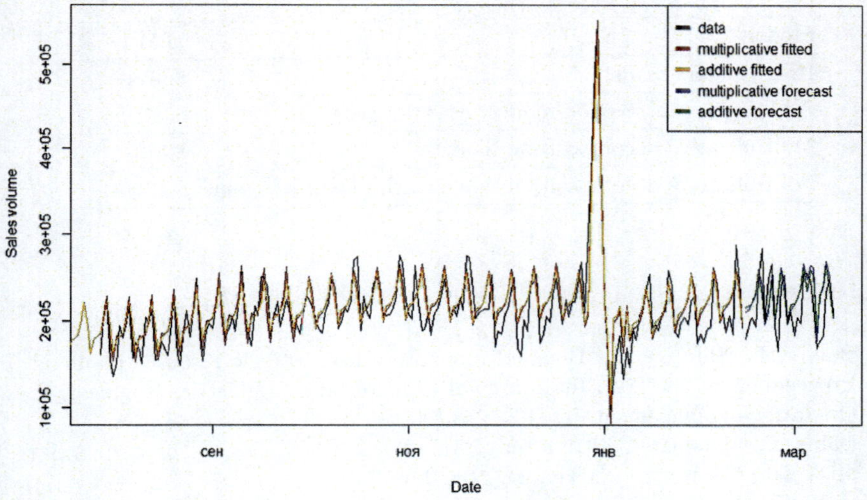

Fig. 8. Comparison of models predictions (additive model and multiplicative model)

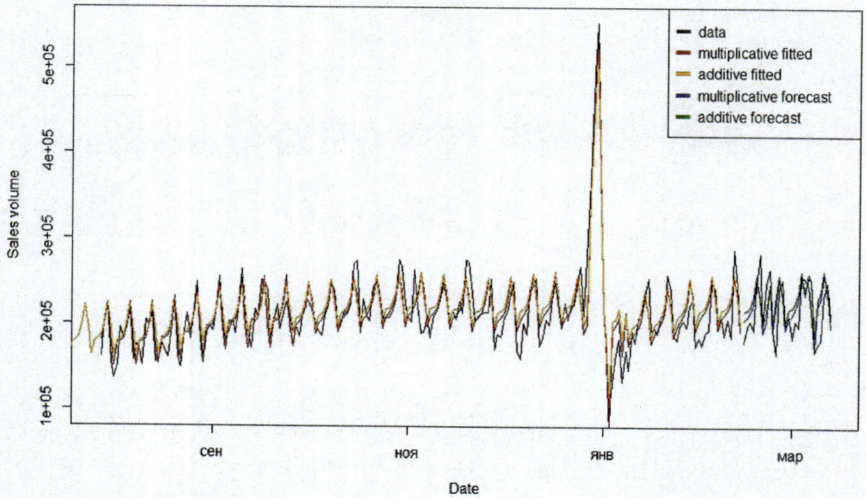

Fig. 9. Comparison of models predictions (additive model and multiplicative model with piece-wise linear trend with one break point)

Table 1. Comparison

Model	MAPE (%)
Holt-Winters model	8.97
Additive decomposition model	8.93
Multiplicative decomposition model	8.51
Multiplicative decomposition model with one break point	8.43

References

1. Bon, A.T., Shi Jun, O.: Demand forecasting for strategic resource planning. In: Proceedings of the 2016 International Conference on Industrial Engineering and Operations Management, pp. 196–208 (2016)
2. Dama, F., Sinoquet, C.: Time Series Analysis and Modeling to Forecast: A Survey, pp. 1–76 (2021). http://arxiv.org/2104.00164pdf
3. Hosseini, R., Yang, K., Chen, A., Patra, S.: A Flexible Forecasting Model for Production Systems, pp. 1–27 (2021). http://arxiv.org/2105.01098pdf
4. Trull, O., García-Díaz, J.C., Troncoso, A.: Initialization methods for multiple seasonal Holt-Winters forecasting models. Mathematics **8**(268), 1–16 (2020). https://doi.org/10.3390/math8020268
5. Barbosa, N., da Silva Christo, E., Costa, K.A.: Demand forecasting for production planning in a food company. ARPN J. Eng. Appl. Sci. **10**(16), 7137–7141 (2015)
6. Veiga, C., Catapan, A., Tortato, U., Silva, W.: Demand forecasting in food retail: a comparison between the Holt-Winters and ARIMA models. WSEAS Trans. Bus. Econ. **11**, 608–614 (2014)

7. Wassana, S., Thomas, L., Chamnein, C., Nittaya, M.: Long-term sales forecasting using Lee-Carter and Holt-Winters methods. J. Appl. Bus. Res. **27**(1), 87–102 (2011)
8. Tirkeş, G., Güray, C., Çelebi, N.: Demnd forecasting: a comparison between the Holt-Winters, trend analysis and decomposition models. Tehnički vjesnik. 24(2), 503–509 (2017). https://doi.org/10.17559/TV-20160615204011
9. Lon-Mu, L., Siddhartha, B., Stanley, L.S., Rong, C., William, J.L.: Data mining on time series: an illustration using fast-food restaurant franchise data. Comput. Stat. Data Anal. **37**(4), 455–476 (2001). https://doi.org/10.1016/S0167-9473(01)00014-7
10. Caruana, A.: Step in forecasting with seasonal regression: a case study from the carbonated soft drink market. J. Prod. Brand Manag. **10**(2), 94–102 (2001). https://doi.org/10.1108/10610420110388654
11. Neusser, K.: Time Series Analysis in Economics. Springer Nature, Switzerland (2016)
12. Verbeek, M.: A Guide to Modern Econometrics. Wiley, Chichester (2004)

On the Software Projects' Duration Estimation Using Support Vector Regression

Vo Van Hai[1](✉), Mohsin Javed[2], Zuhair Abbas[2], Meryem Ari[2], and Michaela Bílá[3]

[1] FAI, Tomas Bata University, Nad Stráněmi 4511, 76005 Zlín, Czech Republic
vo_van@utb.cz

[2] FAME, Tomas Bata University, Mostní 5139, 76001 Zlín, Czech Republic
{javed,abbas,a_meryem}@utb.cz

[3] FHS, Tomas Bata University, Štefánikova 5670, 76001 Zlín, Czech Republic

Abstract. Estimating the project's duration is one of the critical steps in helping to ensure project success. It helps to allocate resources and personnel appropriately during project development. This study aims to look for a more suitable algorithm between two selected algorithms for estimating project duration. Two machine learning algorithms, Multiple Linear Regression and Support Vector Regression, were used to estimate the project's duration. The data used here is an ISBSG dataset with intelligent preprocessing to give an ideal fit to the algorithm used. The dependent variables used in the test are project size, maximum team size, and resource level. With the two algorithms selected, the estimated value of the project's duration is relatively close to the actual duration of the project. Through the six evaluation criteria, R-square, MAE, MAPE, RMSE, MBRE, MIBRE and the pair-wise t-test statistical method, the Support Vector Regression algorithm gives a much better estimate of the project's duration than the Multiple Linear Regression algorithm.

Keywords: Software duration estimation · Machine learning · Multiple linear regression · Support vector regression

1 Introduction

Software engineering is a category that involves many stages in developing a software system. It is related to requirements gathering, design, selection of tools, and languages for implementation, testing, deployment, and maintenance. In parallel with these stages, tasks related to management should also be focused on, such as software configuration management, software engineering management, and other management tasks [1].

Ensuring a successful project is not easy work. Partial or complete failure of projects is related to many causes. One of them is an inaccurate prediction of resource allocation, personnel, and duration time [2].

There have been many studies and approaches to improve the accuracy of the software estimation process. These accuracy improvement approaches can be classified in relation to analogy-based and algorithmic-based [3, 4]. In algorithmic-based estimation

R. Silhavy et al. (Eds.): CoMeSySo 2022, LNNS 596, pp. 288–298, 2023.
https://doi.org/10.1007/978-3-031-21435-6_25

techniques, the FPA method [5] is considered a widely used method in the software industry. The goal when using the FPA method is to estimate the size of the software. This size is represented by the counting process's Unadjusted Function Points (UFP) value.

Machine learning has gradually become a widespread technique in recent decades and is applied in many human service fields. The field of software estimation is no exception. Many studies have used machine learning techniques to improve the estimation process's accuracy. It can even be seen as an alternative to the two approaches mentioned [6].

In the process of software estimation using machine learning techniques, the independent and dependent variables involved in the estimation process are of primary concern. Usually, software size is considered the primary independent variable for effort estimation and then uses this estimated effort variable as the independent variable to estimate the project's duration [7, 8]. However, the project's duration depends not entirely on the variable effort but also on many other factors [8, 9]. One of the critical factors is the number of people involved in the development. In addition, another equally important factor is the Resource Level of the project. The Resource Level concept can be understood as the degree of involvement of external factors in the development team [10]. For example, involve the customer or end-user in the development process. Clearly, with this participation, the development process will be vastly different than without it. For that reason, in this study, we use three independent variables to estimate the project's duration: Project size (used through the Unadjusted Function Size variable—UFP), Resource Level (RL) and the maximum number of developers involved in project development (maximum team size—MTS).

Finding non-linear relationships between independent and dependent variables is a prevalent task in software estimation [11]. In this study, two regression algorithms, Multiple Linear Regression (MLR) [12] and Support Vector Regression (SVR) [25], are used to estimate the project's duration.

Besides, the dataset used for training and testing the algorithms here was obtained from the International Software Benchmarking Standards Group (ISBSG) Repository R1 2020 [13]. In addition, these two selected algorithms will be trained and tested using k-folds cross-validation [14] (k = 5).

The objective of this study is to test a hypothesis during the project's duration estimation:

H1: The project's duration estimation accuracy using the SVR algorithm is significantly better than that of the MLR algorithm when using three dependent variables MTS, RL, and UFP.

2 Related Works

Many studies have been conducted with the desire to improve the accuracy of software duration estimation. In a study, Pospieszny et al. [15] used three machine learning techniques, Support Vector Machines, Neural Networks and Generalized Linear Models, on the ISBSG dataset to create an ensemble model for predicting software effort and duration. This ensemble model is achieved by taking the mean from 3 machine learning models to reduce bias and variance error. The study results show that this ensemble model has better results than an individually single model.

López-Martín [16] used Feedforward Neural Networks to estimate the duration of maintained software projects. The author uses two independent variables, software size (UFP) and the number of developers involved in the project's development, to estimate the project's duration.

According to Morgenshtern et al. [17], effort estimation and duration estimation are driven by relatively different processes, and their inaccuracies are influenced by many factors that could reflect the stakeholders' priorities in the project. Accordingly, effort estimation and duration estimation are driven by relatively different processes, and their inaccuracies are influenced by many factors that could reflect the stakeholders' priorities in the project.

In a study, Tan et al. [18] proposed a model based on the Fuzzy Inference Rules for semi-automatic prediction to reduce the potential negative aspects of expert judgment-based estimation. Fourteen fuzzy inference rules are presented to elicit and revise expert tacit knowledge. Fuzzy inference rules are a complement to expert judgment-based estimation findings. Through surveys and one-on-one interviews with project managers from various disciplines, the outcomes of expert judgment and fuzzy inference rules are compared with those of the expert judgment-based approach. The comparisons are made using data from previous software projects. When fuzzy inference rules are included, the expert judgment-based approach's estimation accuracy is increased by 39.35%.

Connor [19] presented a framework for representing uncertainty in the estimates for software design projects for use throughout the entire project lifecycle. The framework is adaptable to account for project uncertainty and uses Monte Carlo simulation to compute uncertainty propagation in effort estimates towards the overall project uncertainty. As a result, it gives a project manager the tools necessary to make well-informed decisions throughout the project.

Besides, in the relation between effort and duration, many studies have used machine learning to improve estimation accuracy. In a previous study [20], we proposed an approach that combines machine learning and IFPUG FPA methods. In this method, machine learning infers the complexity weight for computing the UFP value, and then the FPA model conducts effort estimation. In another study [21], we evaluate the effect of clustering on effort estimation. As a result, we found the Industry Sector categorical variable to be the best fit for the clustering criteria. And the effect of clustering on accuracy is significant.

3 Background

3.1 Multiple Linear Regression

Studying the relationship between a responding variable $Y \in R^1$ and a large number of independent variables $X = [X_1, X_2, ..., X_n] \in R^n$ is a common task in data analysis. The regression model is the representation of each relationship between Y and X as a mathematical equation, $Y = f(X)$. It is known as Simple Linear Regression when there is only one independent variable. Multiple Linear Regression, or MLR, is used when there are two or more independent variables [22]. This is the shape of the fundamental model [12]:

$$Y = \beta_0 + \beta_1 X_1 + \cdots + \beta_n X_n + \varepsilon \quad (1)$$

where Y is the estimated dependent variable, n is the number of variables, β_0 is the intercept, X_i are the independent variables and $\beta_i i \in 1, n$ are called Partial Regression Coefficients, and ε is the error residual.

In this study, the class sklearn.linear_model.LinearRegression was used with the default parameters.

3.2 Support Vector Regression

The support vector machine (SVM) concept was first introduced by Vapnik [24, 25]. The approach was initially designed for binary object classification and then adapted to predict numerical values (supported vector regression, SVR).

SVR is a statistical learning technique that maps the vectors into a higher dimensional feature space using a kernel function, such as the linear kernel function (LKF), polynomial kernel function (PKF), or radial basis function (RBF) kernel function. The target variable can be linearly regressed using SVR by introducing an alternate loss function and kernel function [26].

In this study, the linear kernel function is used for the experiment. The implementation of this algorithm and the parameters are given in Table 1.

Table 1. SVM algorithm parameters

Algorithm	Implemented	Parameters
Support vector regression	sklearn.linear_svm.SVR	Kernel = 'linear', gamma = "auto", tol = 1e-6, epsilon = 0.01

4 Experiment

4.1 Data Preprocessing

The data used in this experiment is the ISBSG dataset R1, August 2020 [13]. The ISBSG is a dataset contributed by software companies worldwide with information about their projects. It has a tremendous role to play in studying metrics about projects as well as its role in training/testing machine learning models.

In this dataset, the quality of projects is rated from A to D for best (A) and worst (D). In the dataset manual, ISBSG recommend users should use records with the Data Quality Rate of A and B. Therefore, in this study, we only used projects of this quality.

In the Counting Approach feature of this dataset, there are different counting methods like COSMIC, NESMA, LOC, IFPUG 4 +, IFPUG Old, and Components. We are only interested in counting the IFPUG 4 + method in this study. The reason is that we want to evaluate the duration estimation that affects this method.

Two other essential properties in this study are Project Elapsed Time and Project Inactive Time. Project Elapsed Time is the property indicating the total elapsed time for the project in calendar months; Project Inactive Time is the number of months in which no activity occurred. Thus, the subtraction of Project Elapsed Time with Project Inactive Time derives the actual time spent working on the project. Thus, projects with empty values in this attribute will be removed.

The UFP, Max Team Size (MTS), and Resource Level (RL) attributes will also be investigated for relevance in this duration estimation process. Therefore, projects that lack these values (empty) will also be omitted.

After an analysis, we recognized that some sets of values of the selected attributes are too far from the mean group values. That implies that the data has a certain amount of noise. This noisy data adversely affects the experimental results. Therefore, we proceed to eliminate these tuples. The interquartile range (IQR) technique [26, 28] with lower bound 0.15 and upper bound 0.85 is applied in detecting and removing these outliers. Figure 1 shows the UFP values before and after removing these outliers.

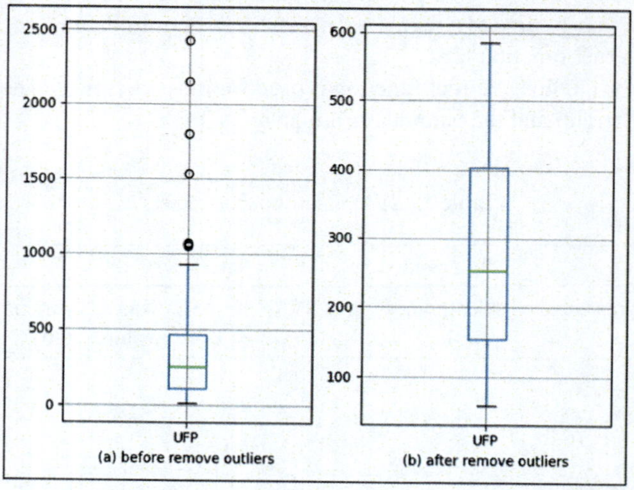

Fig. 1. Dataset before and after removing outliers

4.2 Experiment Setup

In this experiment, the ISBSG dataset will be preprocessed and filtered out outliers as described in the previous section. This data then uses a 5-fold cross-validation technique to ensure the model's reliability. Then, two machine learning algorithms, MLR and SVR, will estimate the project's duration. The results from this estimation will be compared and evaluated through six criteria mentioned in Sect. 4.3 (Fig. 1).

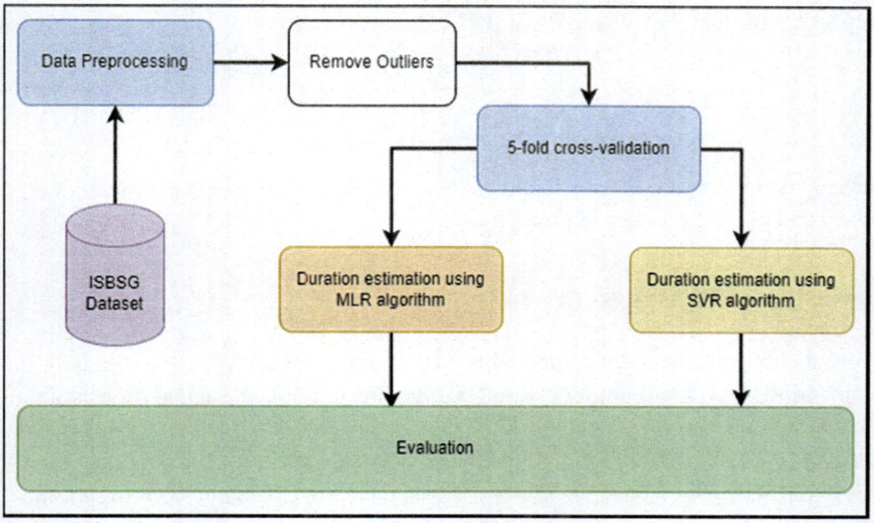

Fig. 2. Experiment setup

4.3 Evaluation Criteria

In this study, the following evaluation criteria were used. They are unbiased evaluation criteria and are recommended [29–31].

R^2—Determination coefficient

$$R^2 = \frac{\sum_{i=1}^{n}(\hat{y}_i - \overline{y})^2}{\sum_{i=1}^{n}(y_i - \overline{y})^2} \tag{2}$$

Mean Absolute Error (MAE):

$$MAE = \frac{1}{n}\sum_{i=1}^{n}|y_i - \hat{y}_i| \tag{3}$$

Root Mean Square Error (RMSE):

$$RMSE = \sqrt{\frac{\sum_{i=1}^{n}(y_i - \hat{y}_i)^2}{n}} \tag{4}$$

Mean Balance Relative Error (MBRE):

$$MBRE = \frac{1}{n}\sum_{i=1}^{n}\frac{|(y_i - \hat{y}_i)|}{min(y_i - \hat{y}_i)} \tag{5}$$

Mean Inverted Balance Relative Error (MIBRE):

$$MIBRE = \frac{1}{n}\sum_{i=1}^{n}\frac{|(y_i - \hat{y}_{i)}|}{max(y_i - \hat{y}_i)} \tag{6}$$

Mean Absolute Percentage Error (MAPE):

$$MAPE = \frac{1}{N}\sum_{i=0}^{N}\left|\frac{y_i - \hat{y}_i}{y_i}\right| * 100 \quad (7)$$

where y_i is the actual value, $\hat{y}_i$ is the estimated value, $\overline{y}$ represents the mean of all the values, and n is the number of projects.

5 Result and Discussion

Figure 3 shows the average estimated project duration of the two selected algorithms compared to the actual duration value. The blue line presents the actual project duration (DU_ACT), and the orange line introduces the project duration predicted using the MLR algorithm (DU_MLR). The green line shows the estimated project duration using the SVR algorithm (DU_SVR). We can readily see that when using the SVR algorithm, the estimated duration value is much closer to the actual value than the MLR algorithm.

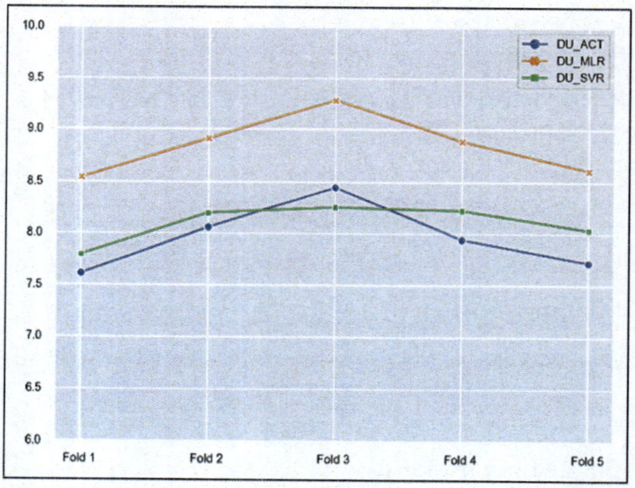

Fig. 3. Duration estimation using MLR, and SVR algorithms in compared with the actual duration

In terms of evaluation criteria, Table 2 presents the result values obtained after five times of model execution on 5-fold cross-validation. The first five rows of this table represent five-folds, and the last line represents the mean value of 5 folds. Each evaluation criterion is performed on two algorithms to serve the comparison. Figure 4 shows a graphical representation of the evaluation criteria for easy observation.

As we can easily observe, in the evaluation criteria MAE, MAPE, RMSE, MBRE, and MIBRE, the estimated error value of the model using the SVR algorithm is always smaller than the estimated error of the model using the MLR algorithm. These results prove the model using the SVR algorithm in this test has a higher estimation accuracy than the other model.

For the R-square criterion, the mean R-square value reaches 0.831 for the MLR algorithm and 0.885 for the SVR algorithm, which means that these regression models fit the dataset above 80%. The R-square values of the SVR method are consistently better than those of the MLR algorithm in the five runs that correspond to the experiment.

Table 2. The evaluation results

Fold	R^2		MAE		MAPE		RMSE		MBRE		MIBRE	
	MLR	SVR	MLR	SVR	MLR	SVR	MLR	SVR	MLR	SVR	MLR	SVR
Fold 1	0.831	0.889	1.51	1.28	1.316	1.124	1.73	1.40	0.244	0.223	0.185	0.172
Fold 2	0.803	0.847	1.70	1.44	1.954	1.766	2.12	1.87	0.361	0.336	0.203	0.189
Fold 3	0.846	0.903	1.38	1.21	1.231	1.100	1.82	1.45	0.230	0.214	0.159	0.157
Fold 4	0.849	0.901	1.47	1.23	1.307	1.148	1.70	1.38	0.239	0.217	0.178	0.165
Fold 5	0.828	0.884	1.56	1.24	1.451	1.189	1.83	1.51	0.269	0.227	0.188	0.161
Mean	**0.831**	**0.885**	**1.52**	**1.28**	**1.452**	**1.265**	**1.84**	**1.52**	**0.268**	**0.243**	**0.183**	**0.169**

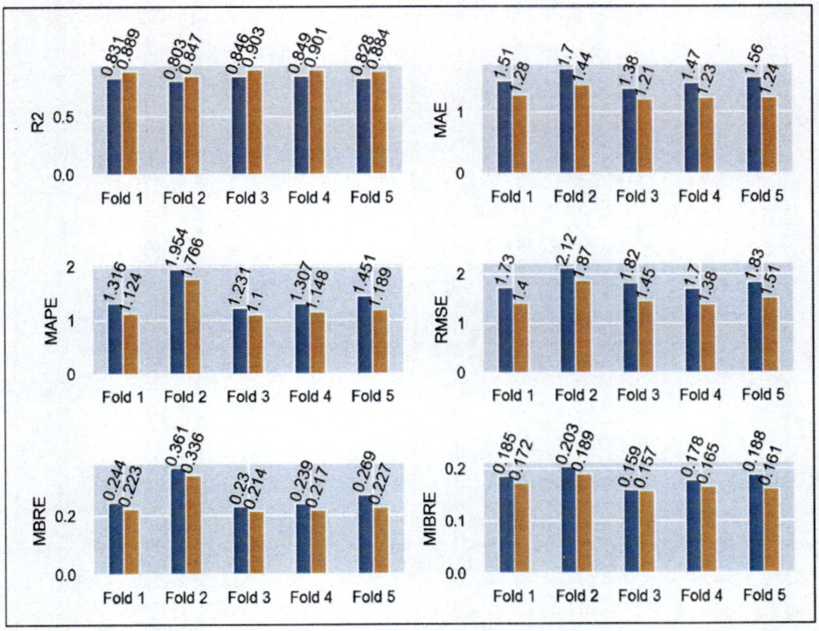

Fig. 4. Evaluation results

Table 3 shows the intercept values and corresponding coefficients for the variables of the two models.

We also conducted a pair-wise t-test method on the experiment's results using MLR and SVR algorithms. Table 4 shows the t-test results. The notation > > presents the

Table 3. Coefficients for algorithms

	intercept	β_1	β_2	β_3
MLR	6.727	0.0084	−0.7859	0.0295
SVR	6.392	0.0076	−0.6990	0.0106

SVM algorithm's statistical superiority in duration estimation compared to the MLR algorithm. The p-values of this test are always smaller than 0.05 means that the duration estimation using the SVR method is statistically significant at the 95% confidence level compared to the duration estimation using the MLR algorithm.

Thus, the H1 hypothesis has been accepted. It means that the project's duration estimation accuracy using the SVR algorithm is significantly better than that of the MLR algorithm when using three dependent variables MTS, RL, and UFP.

Table 4. The statistical t-test based on evaluation results

Pairs of methods		DU SVR versus DU MLR
MAE results	Avg. MAE	1.276 versus 1.524
	Avg. p-value	0.00022
	Statistical conclusion	> >
MAPE results	Avg. MAPE	1.265 versus 1.452
	Avg. p-value	0.00052
	Statistical conclusion	> >
RMSE results	Avg. RMSE	1.521 versus 1.842
	Avg. p-value	0.00004
	Statistical conclusion	> >
MBRE results	Avg. MBRE	0.243 versus 0.268
	Avg. p-value	0.00258
	Statistical conclusion	> >
MIBRE results	Avg. MIBRE	0.169 versus 0.183
	Avg. p-value	0.01317
	Statistical conclusion	> >

6 Conclusion

In this study, we have estimated the duration of projects based on two machine learning algorithms, MLR and SVR. The data for this test is taken from the ISBSG dataset. The dataset has been processed using an intelligent mechanism to bring the data suitable to selected machine learning algorithms. The dependent variables used in this test are the project size (via the FPA method's UFP variable), resource level, and the maximum team size.

Experimental results show that the project's duration estimate is relatively accurate compared to the project's duration actual value. Of the two machine learning algorithms employed, the SVR algorithm provides outstanding accuracy compared to the MLR algorithm. This result is confirmed based on six evaluation criteria and through a statistical method.

Besides, this method has many limitations, such as the inability to systematically analyze the variables that may affect the project's duration or the choice of only two experimental algorithms.

We will investigate other factors influencing the project's duration estimation in future work. We will also search for different machine learning algorithms suitable for this estimation.

Acknowledgment. This work was supported by Tomas Bata University in Zlín under the project no. IGA/CebiaTech/2022/001, and by Trinity Bank funding under the project no. IGA-K-TRINITY/005.

References

1. Bourque, P., Fairley, R.: Guide to the software engineering body of knowledge. SWEBOK V3.0 IEEE Computer Society (2014)
2. The Standish Group: CHAOS chronicles. MA, USA, Boston (2018)
3. Vera, T., Ochoa, S.F., Perovich, D.: Survey of software development effort estimation taxonomies. Technical report, computer science department, University of Chile, Chile (2017)
4. Khan, B., Khan, W., Arshad, M., Jan, N.: Software cost estimation: algorithmic and non-algorithmic approaches. Int. J. Data Sci. Adv. Anal. **2** (2020)
5. ISO/IEC 20926:2009, (IFPUG) Software and systems engineering—software measurement—IFPUG functional size measurement method (2009)
6. Sharma, P., Singh, J.: Systematic literature review on software effort estimation using machine learning approaches. Int. Conf. Next Gen. Comput. Inf. Syst. (ICNGCIS) **2017**, 43–47 (2017)
7. Berlin, S., Raz, T., Glezer, C., Zviran, M.: Comparison of estimation methods of cost and duration in IT projects. Inf. Softw. Technol. Elsevier **51**, 738–748 (2009)
8. Bourque, P., Oligny, S., Abran, A., Fournier, B.: Developing project duration models in software engineering. J. Comput. Sci. Technol. Springer-Verlag **22**(3), 348–357 (2007)
9. Kitchenham, B.A., Pfleeger, S.L., McColl, B., Eagan, S.: An empirical study of maintenance and development estimation accuracy. J. Syst. Softw. Elsevier **64**(1), 57–77 (2002)
10. IFPUG. Function point counting practices manual (2010)
11. Chao-Jung, H., Chin-Yu, H.H.: Comparison of weighted grey relational analysis for software effort estimation. Softw. Qual. J. Springer-Verlagpp **19**(1), 165–200 (2011)
12. Hair, J.F.: Multivariate data analysis. Upper Saddle River, NJ, Pearson Prentice Hall (2006)
13. International software benchmarking standards group. ISBSG Repository August 2020 R1 (2020) https://www.isbsg.org/
14. Krogh, A., Vedelsby, J.: Neural network ensembles, cross validation, and active learning. Adv. Neural Inf. Process. Syst. **7** (1994)

15. Pospieszny, P., Czarnacka-Chrobot, B., Kobyliński, A.: An effective approach for software project effort and duration estimation with machine learning algorithms. J. Syst. Softw. **137**, 184–196 (2018)
16. López-Martín, C.: Feedforward neural networks for predicting the duration of maintained software projects. In: 2016 15th IEEE international conference on machine learning and applications (ICMLA), pp. 528–533 (2016)
17. Morgenshtern, O., Raz, T., Dvir, D.: Factors affecting duration and effort estimation errors in software development projects. Inf. Softw. Technol. **49**(8), 827–837 (2007)
18. Tan, C.H., Yap, K.S., Ishibuchi, H., Nojima, Y., Yap, H.J.: Application of fuzzy inference rules to early semi-automatic estimation of activity duration in software project management. IEEE Trans. Human-Mach. Syst. **44**(5), 678–688 (2014)
19. Connor, A.M.: Probabilistic estimation of software project duration. ArXivabs/1606.05926 (2016)
20. Hai, V.V., Nhung, H.L.T.K., Prokopova, Z., Silhavy, R., Silhavy, P.: A new approach to calibrating functional complexity weight in software development effort estimation. MDPI Comput. **11**(15) (2022)
21. Hai, V.V., Nhung, H.L.T.K., Prokopova, Z., Silhavy, R., Silhavy, P.: Towards improving the efficiency of software development effort estimation via clustering analysis. IEEE Access (2022). https://doi.org/10.1109/ACCESS.2022.3185393
22. Khatibi Bardsiri, V., Jawawi, D.N.A., Hashim, S.Z.M., Khatibi, E.: A flexible method to estimate the software development effort based on the classification of projects and localization of comparisons. Empir. Softw. Eng. **19**(4), 857–884 (2013). https://doi.org/10.1007/s10664-013-9241-4
23. Mendenhall, W.: A second course in statistics: regression analysis. Pearson Education Inc, Boston, MA, USA (2012)
24. Vapnik, V.: Estimation of dependencies based on empirical data. Springer, New York (1982)
25. Cortes, C., Vapnik, V.: Support-vector networks. Mach Learn **20**, 273–297 (1995)
26. Gunn, S.R.: Support vector machines for classification and regression. Department of Electronics and Computer Science. University of Southampton. Report No.: ISIS-1-98 (1998)
27. Upton, G., Cook, I.: Understanding statistics. Oxford University Press. p. 55. ISBN: 0-19-914391-9
28. Zwillinger, D., Kokoska, S.: CRC standard probability and statistics tables and formulae. CRC Press. ISBN 1-58488-059-7 p. 18 (2000)
29. Azzeh, M., Nassif, A.B., Attili, I.B.: Predicting software effort from use case points: a systematic review. Sci. Comput. Program. (2021)
30. Nhung, H.L.T.K., Van Hai, V., Silhavy, R., Prokopova, Z., Silhavy, P.: Parametric software effort estimation based on optimizing correction factors and multiple linear regression. IEEE Access **10**, 2963–2986 (2022)
31. de Myttenaere, A., Golden, B., Le Grand, B., Rossi, F.: Mean absolute percentage error for regression models. Neurocomputing **192**, 38–48 (2016)

Modeling Tasks of Patient Assistance and Emergency Management in Medical Cyber-Physical Systems

Dmitriy Levonevskiy(✉) and Anna Motienko

St. Petersburg Institute for Informatics and Automation of the Russian Academy of Sciences, St. Petersburg Federal Research Center of the Russian Academy of Sciences (SPC RAS), 39, 14Th Line, St. Petersburg 199178, Russia

DLewonewski.8781@gmail.com

Abstract. Medical cyber-physical systems represent an important integration of a network of medical devices in healthcare and open a wide range of possibilities in the field of personalized medicine. There is a variety of separate solutions in this area, but no comprehensive approach to organizing the cyber-physical environment of patients in hospitals and other medical institutions that meet actual requirements. The aim of this paper is to expand the capabilities of MCPS in terms of integrating the behavior of heterogeneous components in various situations that arise in the process of caring for a patient. In the paper a set of regular and emergency scenarios for the functioning of medical cyber-physical systems have been studied and modeled at the logical level. To specify processes, activity diagrams are constructed, data structures are described using class diagrams. The general architecture of a medical cyber-physical system is provided. The developed approach can be used for design and development of medical cyber-physical systems and their components that monitor the health status of patients, interact with them, collect and visualize health data, and solve auxiliary problems. The application of the proposed solutions will allow medical staff to obtain and evaluate data more quickly, which, in turn, will lead to an increase in the quality of medical service.

Keywords: Medical cyber-physical systems · Smart space · Smart ward · Human-computer interaction · Activity diagram · Sequence diagram · Class diagram · Software design · Software architecture

1 Introduction

The modern world is increasingly relying on the successful combination of digital and physical systems to perform complex automation and control tasks. The development of cyber-physical systems and ensuring their efficiency, reliability and security is a very important area of research. Such systems include smart grids, smart transport systems, smart manufacturing and agriculture systems, smart homes, buildings and communities, as well as smart medical equipment. There are even such large-scale and complex solutions as "smart" cities.

R. Silhavy et al. (Eds.): CoMeSySo 2022, LNNS 596, pp. 299–308, 2023.
https://doi.org/10.1007/978-3-031-21435-6_26

A big leap in the development of cyber-physical systems has ensured the future of personalized medicine, which is also one of the priority areas in the Strategy for the Development of Medical Science in the Russian Federation for the period up to 2025. Medical cyber-physical systems (MCPS) represent an important integration of a network of medical devices for healthcare. Medical Cyber-Physical Systems are context-sensitive, life-saving systems with patient safety as their primary concern, requiring rigorous validation processes to ensure user compliance and specification-driven accuracy. They ensure the development of personalized medicine within the framework of the 4P medicine, which is built on the principles of a personalized approach to human health.

MCPS in developed countries are increasingly being used in hospitals to provide various aspects of continuous high-quality medical care. Such technologies enable monitoring of human health and alert health professionals in case of emergency in order to offer optimal care at nominal management.

An integrated approach to organizing the cyber-physical environment of wards in hospitals, clinics, boarding schools, nursing homes and other inpatient institutions, taking into account modern requirements for organizing the stay of patients in a hospital, as well as at the patient's home, is based on the use of modern technologies in the field of human-machine interaction, intelligent video surveillance, heterogeneous data processing, intelligent spaces, etc.

The aim of this study is to expand the capabilities of MCPS in terms of integrating the behavior of heterogeneous components in various situations that arise in the process of caring for a patient. To achieve this goal, the development of formal scenario models for the functioning of MCPS, based on possible options for changing the physical state of patients and the response of specialists, is being carried out. The stated goal involves the design of technological solutions for the implementation of the proposed approach in accordance with the new generation production paradigms adopted by the European Consortium for Peripheral Computing (EECC), "Industry 4.0" [1–3].

2 Related Work

During the study, certain aspects of MCPS were considered including physiological sensors, solutions for smart spaces, and, in particular, smart wards, as well as various integrated solutions for building MCPS. Existing patient monitoring systems are designed, as a rule, to analyze the physical (physiological) state of various population groups, primarily the elderly ones [4–10]. Thus, in [4], a system for determining the fact of falling of a person is proposed based on multimodal wireless sensor networks, including temperature sensors, motion and biosensors, video cameras, audio sensors and RFID tags. Sensors are used together to improve the accuracy of fall detection and minimize false alarms. RFID tags are used to track the location of people and activate video cameras in the premises where the latter are located. The system allows remote monitoring of its users through mobile and local area networks [5].

The paper [6] presents the concept of a smart ward with a fall detection system, which can alarm medical workers when a fall is detected. It consists of a Kinect depth camera and a neural network based fall detection algorithm.

The work [7] explores the benefits of using mobile three-dimensional (3D) imaging technology to enhance environmental assessment interventions to address external risk factors for falls in the home.

The wireless home monitoring system [8] includes a human condition monitor worn on the wrist and optional pulse oximeters or blood pressure monitors connected wirelessly, and allows collecting data on the activity and location of a person, recognizing falls and issuing panic alerts.

In [9], a system for remote monitoring of the health of older people based on a smart home is proposed, consisting of "smart clothes" that read the parameters of a person's heartbeat and movement, and a "smart home" gateway designed for multiplexing and transmitting data to a server healthcare. The proposed system has good scalability and ease of operation.

In Russia, an analogue of the considered systems is the automated system of remote monitoring of human health "Health Monitor" [10], which allows automatically reading data from various specialized devices for analyzing human health parameters. If necessary, the user can provide the doctor with access to the measurements stored in the system.

More complex and integrated solutions include the concept of a smart ward proposed in [11]. It will allow implementing network monitoring of the patient's condition using physiological sensors (real-time monitoring on a doctor's medical tablet), and two-way communication with medical staff using a mobile telemedicine diagnostic system. The development of a medical information system for managing a smart ward and monitoring patients in real time and software will enable the doctor to monitor their condition using medical tablets or desktop computers.

Another example of an integrated system is Bundang Hospital Electronic System for Total Care implemented at Seoul National University (South Korea) Bundang Hospital, which is a recognized leader among smart hospitals. The BESTCare digital system implemented in this hospital is an integrated system that includes an electronic medical record, computer-aided ordering systems for prescriptions and warning recommendations (when medical prescriptions carry certain risks) for doctors and nurses, a clinical decision support system, administration closed-loop drug movement, clinical data storage, health information exchange, and disaster recovery [12].

The software product Smart ward SMART ROOM from the company "Analitika M" allows you to manage a set of actions that can take place in the patient's ward, namely: to strengthen personal control over the patient, reduce the risk of force majeure situations and at the same time reduce the burden on medical staff. With the help of a smart control panel - a tablet, both the patient and the medical staff can control various functions in the ward. Thanks to a special patient fitness bracelet, a medical specialist can remotely monitor his condition and immediately respond to sudden changes in indicators. Special call buttons "water", "pain relief", "sharp pain", etc. allow medical personnel to quickly respond to a specific call and be ready for a specific patient request [13].

Artificial intelligence has revolutionized medical technologies. Its application allows solving complex problems with many applications in areas with a huge amount of data [14]. Intelligent medical technologies based on artificial intelligence make it possible to use the concept of 4P medicine, which requires the use of modern diagnostic technologies, the use of decision-making algorithms based on their results by doctors in their activities.

Thus, in part, the problems associated with the construction of medical cyber-physical systems are the subject of research by many leading scientific teams. For this, there are many technologies, ideas and approaches, conceptual models and architectures for MCPS for various purposes have been developed [15]. At the same time, some of the available solutions have closed implementations, and the issue of modeling scenarios for the functioning of integrated medical cyber-physical systems is also not sufficiently reflected. Thus, in the presence of separate developments in this area, there is no holistic, comprehensive, reflected in scientific publications approach to the organization of the cyber-physical environment of patients in hospitals, clinics, boarding schools, nursing homes, other inpatient institutions and other locations of patients, which would take into account modern requirements for the organization of patient care and supported the integration of methods and systems for monitoring the health of patients, tracking emergency situations, analytics of medical data and support for auxiliary service functions.

3 Methods and Results

A necessary condition for MCPS functioning is the specification of its behavior. The behavior of the system can be described using the functioning scenarios, which, in turn, are specified by formal models. In this work, models in the UML 2.0 notation are used. Possible scenarios for the functioning of an information system for collecting and visualizing data in medical cyber-physical systems can be divided into the following categories:

- scenarios for patients while staying at home;
- scenarios for patients in hospitals, clinics, boarding schools, nursing homes and other inpatient institutions.

Each of the categories above, in turn, involves functioning in a normal and emergency situation. Figure 1 shows a general sequence diagram for the preliminary interaction between the patient and the doctor, taking into account the use of sensors of physiological indicators of the body and the multimodal input system for the patient. Interaction involves regular preliminary collection of data on the patient's condition in automatic (using sensors) and semi-automatic (using multimodal input tools in the patient's client application) mode. These data are sent to the server, systematized and visualized in the doctor's client application, which allows him/her to more quickly assess the patient's condition and make a decision on how urgently a face-to-face consultation is needed, and on the need for emergency care.

Emergency scenarios include the occurrence of emergency conditions of a patient, which are understood as a set of symptoms (clinical signs) that require emergency first aid or hospitalization of the patient [16]. In case of an emergency scenario, when the patient is at home, it is necessary to call an ambulance team. In a hospital, a specialist is called directly to the patient's room. Reasons for calling an ambulance in an emergency are:

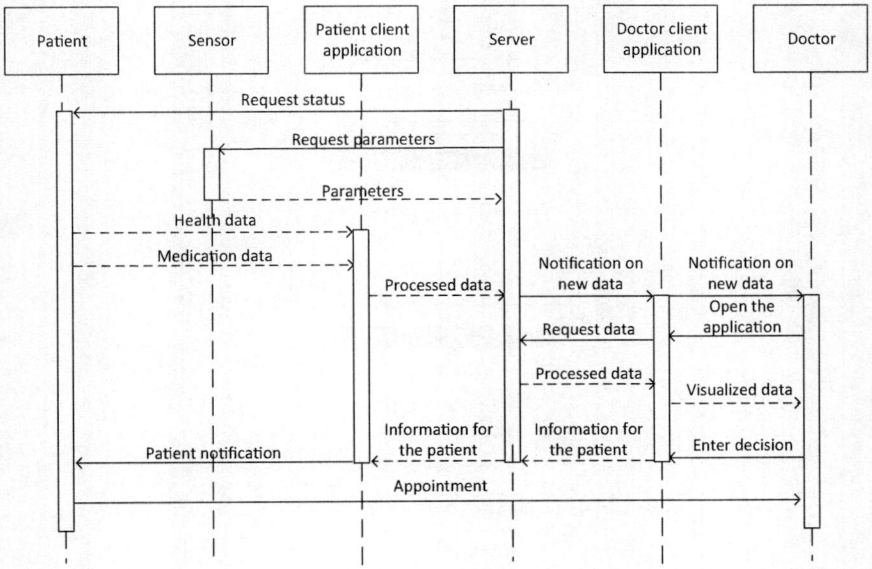

Fig. 1. Sequence diagram for the preliminary patient-doctor interaction

- sudden acute diseases (conditions) without obvious signs of a threat to life, requiring urgent medical intervention;
- sudden exacerbations of chronic diseases without obvious signs of a threat to life, requiring urgent medical intervention.

Obvious signs of a threat to life are understood as strongly marked signs of a disease (condition) that can lead to the death of the patient. The necessary actions on the server side can be represented by an activity diagram in Fig. 2. This diagram also reflects the process of monitoring indicators on the server, which allows identifying emergency scenarios for the development of the situation, for example, a strong manifestation of an allergic reaction, such as angioedema (Quincke's edema), which often requires emergency care.

The regular scenario of the system operation when the patient is at home involves keeping a patient diary with regular records of his condition. The system can be configured to remind the patient to enter information about the current condition, and it also provides for an appointment with a doctor at the earliest available time. If new symptoms appear, the patient can enter them in the diary at any time. At the same time, it is possible to connect the system with a voice assistant that can make entries about the user's condition in the patient's diary, periodically (the period can be configured) clarify information about changes in the condition, remind the patient to make an appointment with a doctor or take medicines, etc. In addition to symptoms, it is possible to record the actions that the patient performs in order to improve or stabilize his condition: taking drugs and non-drugs, various manipulations. With a significant deterioration in the condition, it is possible to call an ambulance team and transfer all status records.

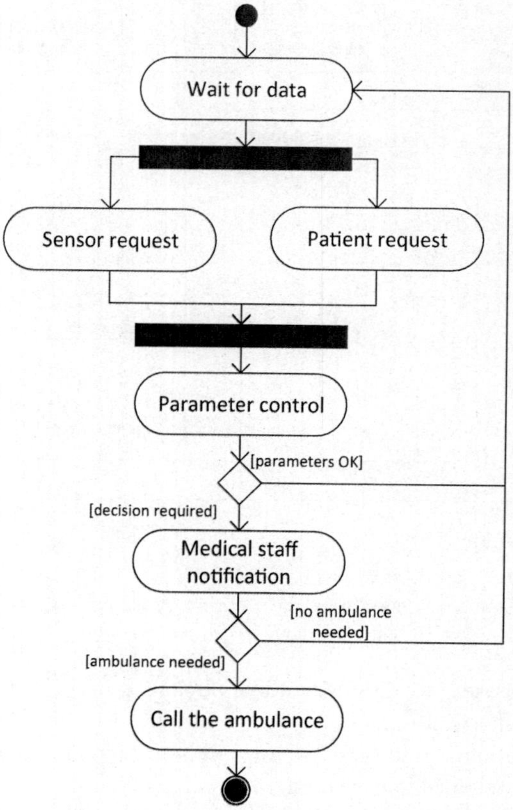

Fig. 2. Activity diagram for the server during the preliminary interaction between the patient and the doctor

Regular scenarios include the actions of the MCPS in the hospital, performed at the direction of the patient (for example, regulation of a comfortable stay in the ward—light, temperature, ventilation, calling staff, etc.). At the same time, if in some situations the scripts conflict, then the script is executed according to the instructions of the doctor. For example, ventilation of the room at the request of the patient can be performed only in the absence of a ban on ventilation by the doctor. The sequence diagram for such a case is shown in Fig. 3. Here, the doctor, using his client application, manages the permissions for the climate control system, and the patient can request certain actions from the MCPS, which are performed if allowed. The corresponding activity diagram is shown in Fig. 4.

Thus, for some scenarios of functioning of the MCPS, formal models were built in the form of diagrams of sequences and activities. Such models, as well as similar models of other scenarios and situations, can be used to design the components of integrated patient care solutions. In particular, class diagrams are built for software development, and component diagrams for system design, which may include known and proposed solutions. Thus, in [17], a number of solutions for the subsystem for collecting and visualizing medical data are proposed to optimize the process of patient management

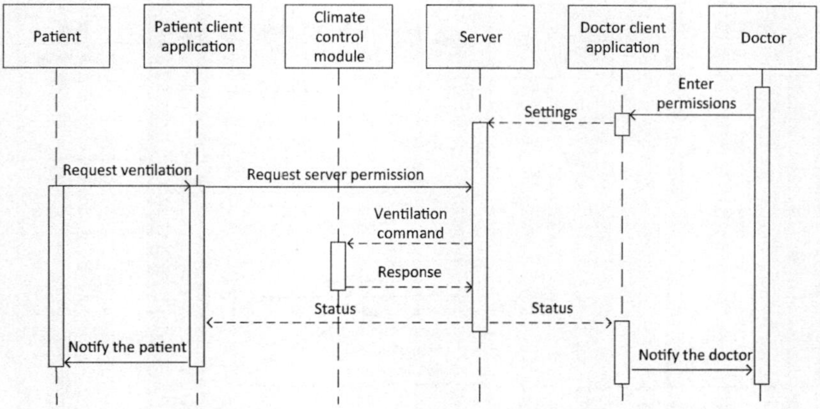

Fig. 3. Sequence diagram for room ventilation control

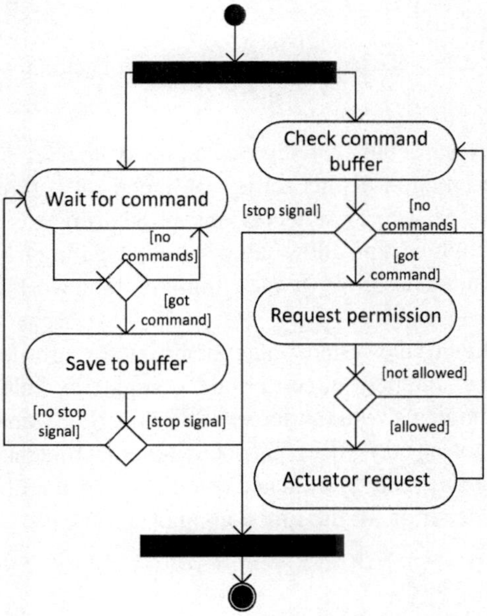

Fig. 4. Activity diagram for processing climate control commands

and a client-server architecture is presented within which these and similar subsystems can be implemented. However, a broader view of MCPS involves not only collection and visualization of medical information, but also presence of a control loop. Thus, Fig. 5 shows an object-oriented model in the form of a class diagram for the implementation of the process of managing climate control modules.

The diagram describes a set of MCPS components characterized by parameters that users can adjust. At the same time, users with necessary roles (medical staff) can set policies that limit the ability to adjust settings for a particular user group. This

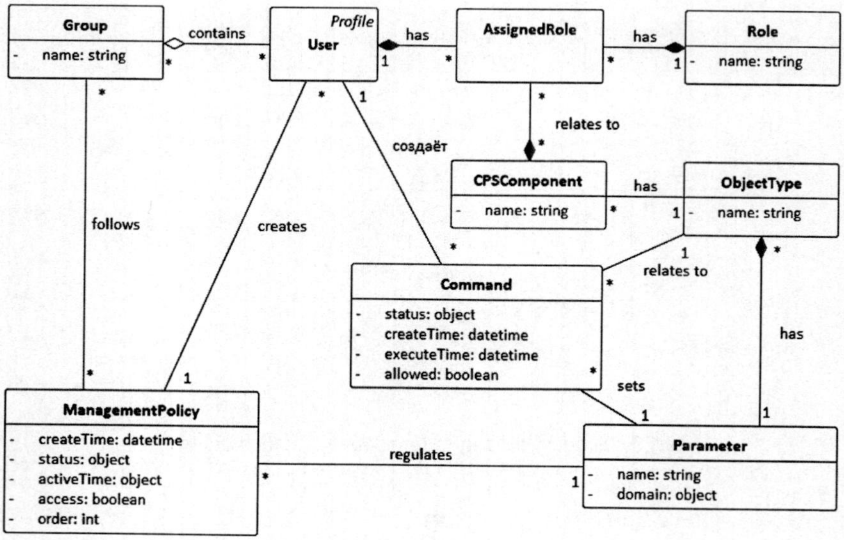

Fig. 5. Class diagram for managing climate control modules

diagram does not provide details for representation of the set of MCPS component states, since the representation of this set is of a technical nature and is provided by standard data types and structures (integers, intervals, arrays). Such diagrams can be used for software development and allow automatically creating frameworks of software modules in computer-aided design systems within the framework of the general MCPS architecture shown in Fig. 6.

This architecture divides the system components into two groups of client and server modules. The client part includes edge devices: computers, tablets, sensors, physical components that communicate with the server part via API or web interface. The server part is responsible for storing and processing data, implementing the described scenarios, analytics and visualization. The coordinated interaction of the client and server parts allows implementing scenarios for the functioning of the MCPS.

4 Conclusion

In this study, a set of regular and emergency scenarios for the MCPS functioning were considered, and their modeling was performed at the logical level using UML. Activity diagrams were used to specify processes, data structures were described using class diagrams. The generalized MCPS architecture was provided.

The obtained solutions can be used in the design and development of MCPS and their components that monitor the health status of patients, interact with them, collect and visualize health data for medical staff. The inclusion of these components in the circuit of medical decision-making will allow specialists to receive and evaluate information more quickly, which, in turn, will lead to an increase in the quality of medical services provided. Use of MCPS also makes it possible to simplify the performance of a wide

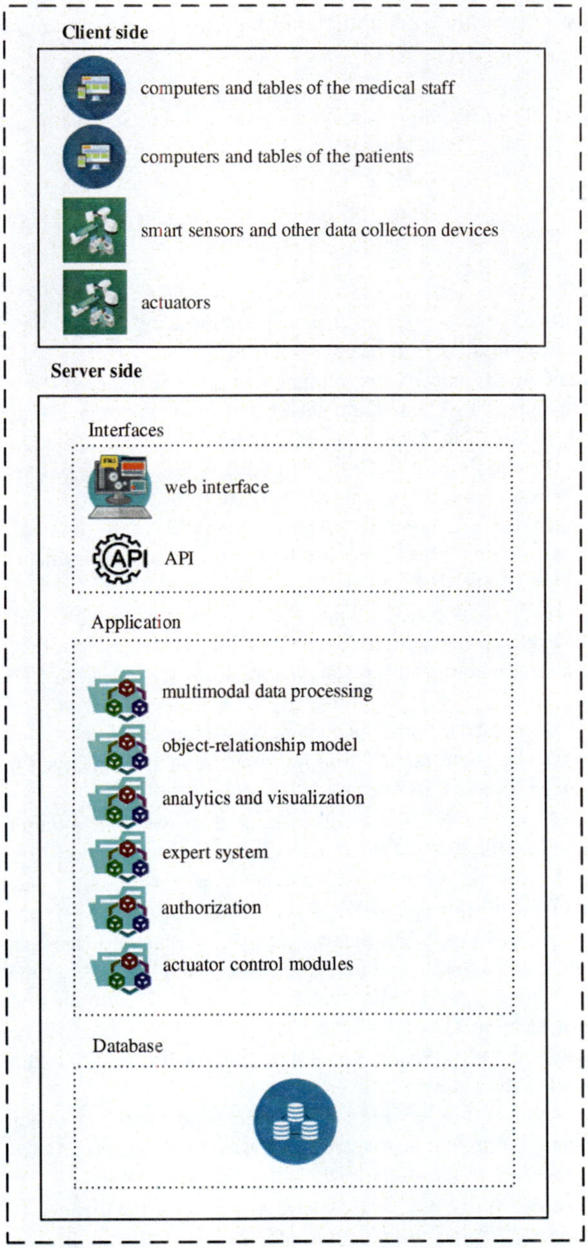

Fig. 6. Architecture of the medical cyber-physical system

range of auxiliary tasks by means of a "smart ward", including climate control, ward equipment, operational communication with staff, delivery, cleaning, etc.

Further directions of work may include the creation of a general classifier of scenarios for the functioning of MCPS, the development of software models and architectures for such systems not only at the conceptual and logical, but also at the physical level, and assessment of their effectiveness.

Acknowledgement. The research is supported by the Russian Science Foundation, project No. 22-71-10092.

References

1. Klitou D., Conrads J., Rasmussen M., et al.: Digital transformation monitor germany: industrie 4.0, European Commission. European Union (2017)
2. Schwab, K.: The fourth industrial revolution. Currency, P. 192 (2017)
3. European edge computing Consortium (EECC). Available at: https://ecconsortium.eu/
4. Alemdar H. Ö., Yavuz G. R., Özen M. O. et al.: Multi-modal fall detection within the WeCare framework. In: Proceedings of the 9th ACM/IEEE international conference on information processing in sensor networks pp. 436–437 (2010)
5. Saitov, I., Motienko, A., Astapov, S., Basov, O.: Synthesis of the topological structure of distributed terminal system for audio monitoring of users of local information spaces. SPIIRAS Proc. **18**(6), 1357–1380 (2019)
6. Su M. C., Liao J. W., Wang P. C., Wang C.H.: A smart ward with a fall detection system. In: 2017 IEEE international conference on environment and electrical engineering and 2017 IEEE industrial and commercial power systems Europe (EEEIC/I&CPS Europe), pp. 1–4 (2017)
7. Hamm, J.J.: Technology-assisted healthcare: exploring the use of mobile 3D visualisation technology to augment home-based fall prevention assessments, Ph.D. Thesis, Brunel University London, p. 316 (2018)
8. Papadopoulos, A., Crump, C., Wilson, B.: Comprehensive home monitoring system for the elderly. Wireless Health **2010**, 214–215 (2010)
9. Guan K., Shao M., Wu S.: A remote health monitoring system for the elderly based on smart home gateway. J. Healthcare Eng. **2017**, 10 (2017)
10. Shalkovsky, A.G., Kuptsov, S.M., Berseneva, E.A.: Topical issues of person health remote monitoring automated system creation. Vrach i inf. Tehnol. **1**, 67–69 (2016). (in Russian)
11. Orsaeva, A.T., Tamrieva, L.A., Mischvelov, A.E., et al.: Development of a prototype of a «smart ward» as an element of a digital polyclinic. Pharmacophore **11**(1), 142–146 (2020)
12. Kobrinskii, B.A.: Smart hospital as a tool of digital medicine. Inf. Tehnol. i vychislitel'nye sistemy **4**, 3–14 (2018). (in Russian)
13. The company "ANALITIKA M" offers management and integration of complex solutions for the efficient operation of your healthcare facility, available at: http://www.analitika-m.ru/services/ (in Russian)
14. Peng, Y., Zhang, Y., Wang, L.: Guest editorial: Artificial intelligence in biomedical engineering and informatics: an introduction and review. Artif. Intell. Med. **48**(2–3), 71–73 (2010)
15. Chen F., Tang, Y., Wang, C. et al.: Medical cyber-physical systems: a solution to smart health and the state of the art, In: IEEE Trans. Comput. Soc. Syst. 1–21 (2021)
16. Emergency Medical Guide. In: Bagnenko, S.F., Vertkina, A.L., Miroshnichenko A.G., M.Sh. Hubutiya, GJeOTAR-Media, p. 783 (2012) (in Russian)
17. Levonevskiy, D.K., Motienko, A.I.: Information system for collection and visualization of medical data for optimizing the patient management process. Inf. Commun. **7**, 21–29 (2021)

Analysis of Collaborative Filtration Parameters

R. S. Kulshin(✉), A. A. Sidorov, P. V. Senchenko, and G. A. Volokitin

Tomsk State University of Control Systems and Radioelectronics, 40, Lenina Ave, Tomsk 634050, Russia
kulshinroman@mail.ru

Abstract. This article examines the influence of the main parameters of collaborative filtering on the quality of the formation of personalized recommendations and the running time of the algorithm, and also determines the effective values of the parameters and their ratios that ensure high quality recommendations with a short time.

Keywords: Collaborative filtering · Recommendation systems · Personalization · Similarity of vectors · Quality of recommendations

1 Introduction

Modern consumers are migrating more and more to the digital space, making purchases there and engaging in the selection of interested goods through market research, since it is convenient and does not require time to visit physical spaces. Due to the fact that such services are very popular, the business is actively developing this niche (some in parallel with offline sales, and some completely going online). All this intensifies competition among entrepreneurs in the struggle for consumers, in which different marketing models, technologies, and tools are used.

In the current situation, information technologies come to the aid of the user. Huge corporations and small companies are starting to implement recommendation systems everywhere. Recommendation systems are used to improve the quality of products offered to the user, thereby increasing the company's profit. The most high-quality recommendation systems form personalized recommendations that are focused on each individual user.

The high demand of companies for recommendation systems has led to rapid development and great interest in algorithms for generating user recommendations.

The basics of recommendation generation technologies appeared quite recently, at the time of the advent of machine learning. It was then that the basic mathematical principles and algorithms were laid. Since then, a lot of changes have taken place, a large number of hybrid approaches to the formation of recommendations have appeared.

To ensure the effective operation of the algorithms for the formation of recommendations, a deep understanding of the influence of parameters on the final result and the speed of work is required. In the conditions of forming recommendations in real time, the operating time of the algorithm should not exceed 500 ms, so that the total request

R. Silhavy et al. (Eds.): CoMeSySo 2022, LNNS 596, pp. 309–318, 2023.
https://doi.org/10.1007/978-3-031-21435-6_27

time does not exceed 2000 ms [1], which is a comfortable waiting time for the user. Further in the work, the process of forming recommendations will be considered and the analysis of the parameters affecting it will be described.

2 Formation of Recommendations Using Collaborative Filtering

2.1 Collaborative Filtering Algorithm

The collaborative filtering method, as well as its hybrids and modifications, is most common in the field of personalized recommendations formation. The principle of operation of the method is based on the assumption of the conservativeness of user preferences (if users have evaluated a certain product in the same way, then it is quite likely that users will similarly evaluate other products with similar characteristics).

The collaborative filtering method generates personalized recommendations based on the behaviors of users similar to it. Two types of collaborative filtering methods are most common: collaborative filtering by analyzing the preferences of user groups with similar characteristics (User-User CF) and collaborative filtering by identifying the relationships between products (Item—Item CF) [2].

The main advantage of this method is the high theoretical accuracy of user recommendations, which leads to the achievement of high latency.

Within the framework of this work, research will be carried out on the basis of the method of collective filtering (User-Based).

2.2 Finding Similarities Between Objects Using Cosine Similarity

The most important task in forming a personalized recommendation is finding similar users for the User-Based approach or finding similar objects for the Item-Based approach, relative to the user to whom the recommendation will be formed. Many metrics are used to solve this problem: cosine similarity, correlation, Jacquard index and Hamming distance [3].

In the tasks of forming recommendations, an important parameter is the effectiveness of the evaluation measure for discharged vectors. No user will be able to evaluate all objects and, in turn, all objects cannot be evaluated by all users [4]. The method of cosine similarity copes with this task perfectly, which solves the problem described above.

Therefore, the cosine similarity method was chosen for this study. It is a measure of similarity between two nonzero vectors and is defined as equal to the cosine of the angle coinciding with the inner product of the vectors.

The cosine similarity method is represented by the formula (1), where A is the first vector, B is the second vector and n is the length of the vectors.

$$Similarity = \frac{\sum_{i=1}^{n}(A_i \times B_i)}{\sqrt{\sum_{i=1}^{n}(A_i)^2} \times \sqrt{\sum_{i=1}^{n}(B_i)^2}} \tag{1}$$

2.3 Determination of the Quality of the Generated Recommendations

To determine the quality of recommendations, two main approaches are used: the root mean square error (RMSE) method and the mean absolute error (MAE) method. Both approaches have proven themselves well in the tasks of evaluating the quality of recommendation systems [5, 6].

RMSE is the square root of the mean of quadratic errors. But RMSE is very sensitive to large errors that have a disproportionate impact on the final score. RMSE is also dependent on the scale of the numbers used [7].

RMSE is represented by formula (2), where *N* is the length of samples, *Predicted* is the selection of predicted and *Actual* is a sample of real data.

$$RMSE = \sqrt{\frac{\sum_{i=1}^{N}(Predicted_i - Actual_i)^2}{N}} \tag{2}$$

MAE is the arithmetic mean of absolute errors. This method is also dependent on the scale of the numbers used, like RMSE, but unlike it is not sensitive to large errors, due to which the final assessment becomes more objective [8].

MAE is represented by formula (3), where *N* is the length of samples, *Predicted* is the selection of predicted and *Actual* is a sample of real data.

$$MAE = \frac{\sum_{i=1}^{N}(Predicted_i - Actual_i)}{N} \tag{3}$$

The method of arithmetic mean of absolute errors was chosen for the study.

3 Investigation of Parameters

3.1 Dataset for the Study

To study the effect of collaborative filtering parameters on the quality of recommendations and the algorithm's running time, a data set presented by GroupLens Researcher in 2019 called MovieLens was used. This data set contains 25 million records of user ratings of films, 62 thousand films and 162 thousand users [9]. The data set also contains a large number of other records that will not be taken into account in this study.

Within the framework of this study, it is the ratings records that will be considered, in a sample from 5 thousand to 150 thousand, depending on the research being carried out. The size of the test sample from the total number is 20%.

Prior to the start of the study, permission was obtained to use the dataset from GroupLens Researcher.

3.2 Selection of Research Parameters

To conduct the study, the main parameters affecting the quality of the generated recommendation and the operating time of the collaborative filtering algorithm were selected.

The main parameters to study:

- The number of closest users who are similar to the user for whom recommendations are being generated;
- Number of users in the dataset.

Thanks to the study of the above parameters, you can get the following information:

- The number of closest users for a certain total number of users with the highest performance;
- Influence of Parameters on the Quality of Recommendations;
- Influence of Parameters on the Speed of the Algorithm;
- Effective ratio of parameters to achieve maximum quality recommendations with minimal time.

3.3 Effective Number of Nearby Users for a Specific Total Number of Users

The algorithm of joint filtering strongly depends on the number of nearby users, on the basis of which the recommendation will be calculated. Using a minimum number of nearby users, as well as a large number of nearby ones, leads to a rapid growth of MAE. According to the results of our experiment, the dependence of the quality of the recommendation on the change in the number of nearby users was revealed. This dependency is shown in Fig. 1.

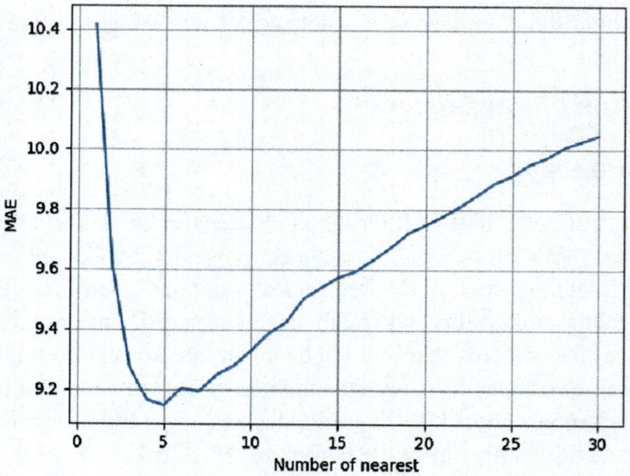

Fig. 1. Change of MAE depending on the number of nearest

The study of changes in MAE depending on the number of nearest ones took place on a sample of 20,000 thousand estimates, among 166 users and 4418 objects. The conducted research has shown that the 5 closest users are effective for this sample.

As the sample size increases, it can be seen that the trend described above persists. The results of the study with other sample sizes are described in Table 1.

Table 1. The effective number of the nearest in accordance with the total number of users

Sample size	Number of users	Number of objects	Effective number of nearest
10,000	75	3287	4
20,000	166	4418	5
30,000	227	5432	5
40,000	318	5976	6
50,000	406	6489	6
60,000	476	7018	6
70,000	548	7925	6
80,000	610	8969	6
90,000	689	9618	8
100,000	757	9786	8

Based on the data described in Table 1, it is possible to identify the dependence of the effective number of nearest ones on the number of users. Having calculated the average value between all pairs, it turns out that with an increase in the number of users by about 67, the effective number of the nearest increases by 1.

3.4 Analysis of the Influence of Parameters on the Quality of Recommendations

To determine the influence of the parameters on the size of the average absolute error, 6720 calculations of recommendations were carried out for various parameters of the number of users (from 10 to 1137 with a step of 4) and the number of nearest ones (from 1 to 30 with a step of 1).

The calculation results were combined and for clarity shown on the heat map [10], Fig. 2.

On the heat map, the darkest areas express the smallest average absolute error. The lightest areas indicate a high error. Based on the data obtained, we can identify the area of effective values of the parameters of the total number of users and the number of the closest users, which give us a low error when using them.

You can also notice small light stripes located parallel to the axis of the number of nearest ones. They appeared due to the limitation of the sample size of the ratings records. There was a situation that some users had few objects. As the sample of the dataset grows, these bands disappear, so we will not take them into account when determining the final area.

The area of effective values is shown in Fig. 3.

The resultant values are highlighted on the heat map by an area bounded by a blue contour. The average absolute error in this area does not exceed 9.

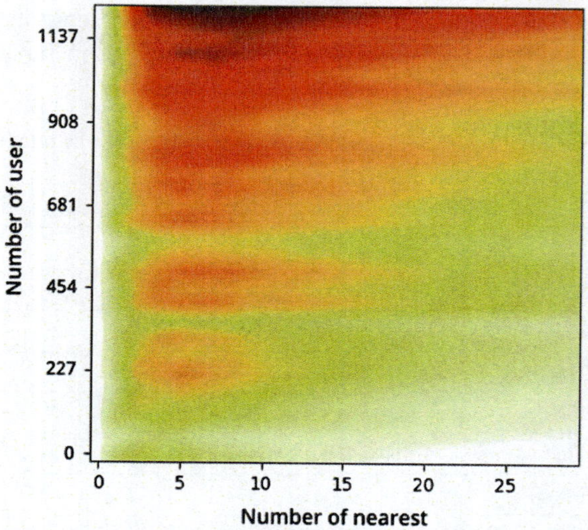

Fig. 2. Heat map of the dependence of MAE on the number of users and the number of nearby

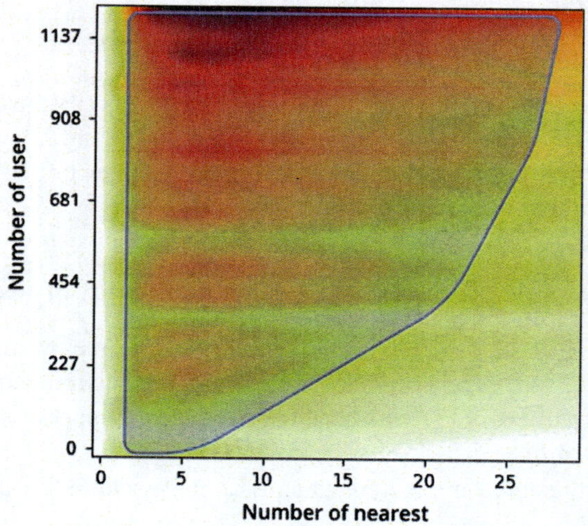

Fig. 3. Low value area of MAE

3.5 Analysis of the Influence of Parameters on the Speed of the Algorithm

To determine the influence of parameters on the running time of the algorithm collaborative filtering, as well as in the previous study was conducted 6720 calculations of recommendations for various parameters of the number of users (10 to 1137 from step 4) and the next number (1–30 in increments of 1).

The results of the calculations have been combined for clarity depicted in the heat map shown in Fig. 4.

On the heat map, the darkest areas express the shortest execution time. The lightest areas indicate the longest working time. Based on the data obtained, we can identify the

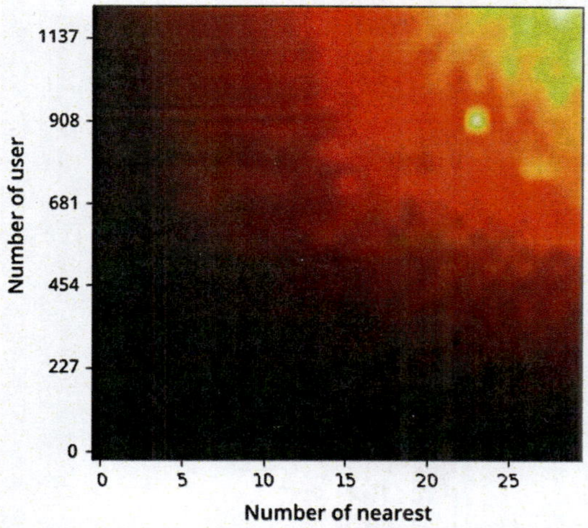

Fig. 4. Heat map of the dependence of the algorithm's operating time on the number of users and the number of nearby

area of the effective values of the parameters of the total number of users and the number of nearby users, which give us the shortest operating time of the collaborative filtering algorithm.

The area of effective values is shown in Fig. 5.

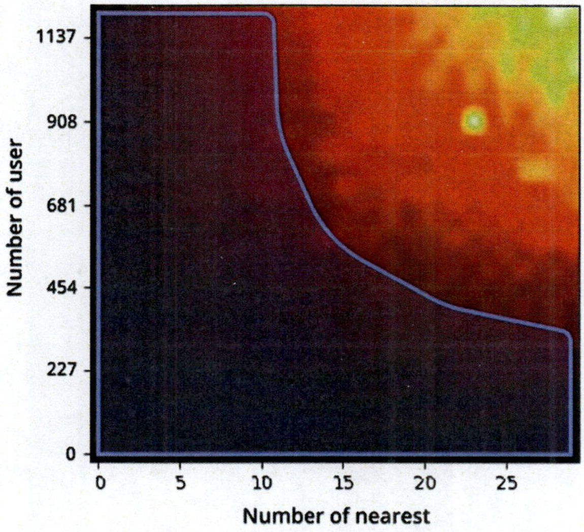

Fig. 5. Low time value area

Effective values are highlighted on the heat map with a blue area. In this area, the time spent calculating recommendations for one user does not exceed 0.08 s.

3.6 Determining the Effective Ratio of Parameters

To determine the effective ratio of the parameters of the total number of users and the number of nearby users to find the area at the same time effective for MAE and for the time spent on the calculation, data from the previous sections were used.

The calculation of the effective ratio of parameters was carried out by summing the normalized matrices described in the previous sections. The values of the matrix elements were normalized in the range from 0 to 1.

The results of the calculations were combined and for clarity shown on the heat map shown in Fig. 6.

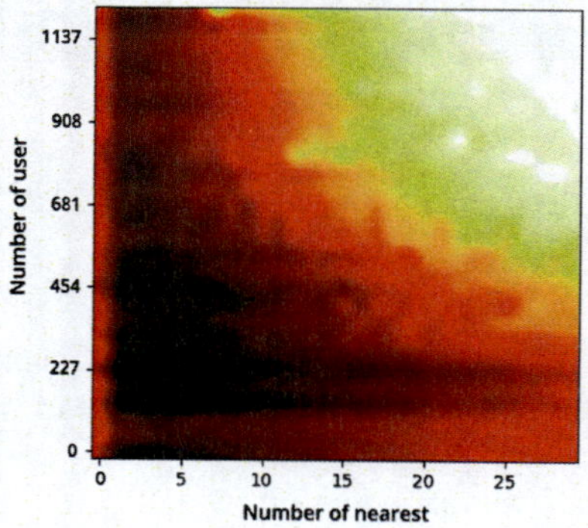

Fig. 6. Heat map of summation of normalized matrices

On the heat map, the darkest areas express the greatest effectiveness of the ratio of parameters. The brightest areas indicate the least effectiveness. Based on the data obtained, we can identify the area of the effective ratio of parameters by time and by error. The area of effective values is shown in Fig. 7.

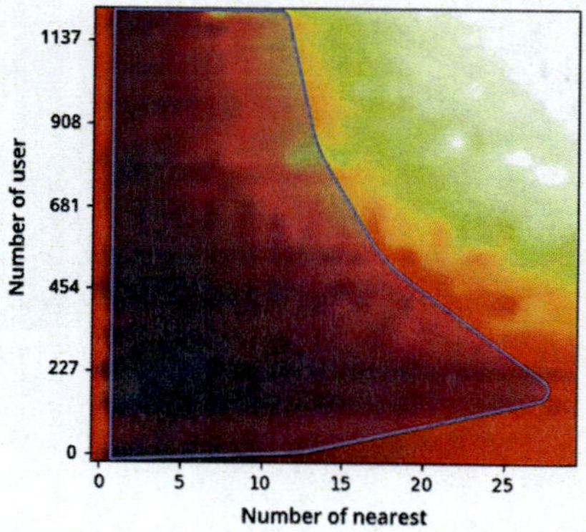

Fig. 7. The area of effective parameter values by time and by error

The resultant values are highlighted on the heat map with a blue area. In this area, the time spent calculating recommendations for one user does not exceed 0.11 s, and the average absolute error does not exceed 10.

4 Conclusions

Within the framework of this work, the work and the influence of the parameters of the algorithm for the formation of personalized recommendations based on the method of collective filtering (User-Based) were considered.

In the course of the study, the parameters were considered: the number of users in the data set and the number of nearby users. According to the results of the study, the results were obtained on the effective number of the closest users for a specific total number of users, the influence of parameters on the quality of recommendations, the influence of parameters on the speed of the algorithm, the effective ratio of parameters to achieve the maximum quality of recommendations with minimal time. Heat maps were also constructed and areas of effective values were identified.

The results of this work can be used in various information systems when implementing recommendation systems based on collaborative filtering for more precise parameter settings, to achieve a rapid increase in the effectiveness of the recommendation system.

Acknowledgements. This paper is designed as part of the state assignment of the Ministry of Science and Higher Education; project FEWM-2020-0036.

References

1. Vora, M.N., Shah, D.: Estimating effective web server response time. Sec. Int. Conf. Inf. Syst. Eng. (ICISE) **2017**, 37–44 (2017). https://doi.org/10.1109/ICISE.2017.9
2. He, X., Liao, L., Zhang, H., Nie, L., Hu, X. and Chua, T.S.: Neural collaborative filtering. In: Proceedings of the 26th international conference on world
3. Su, X. and Khoshgoftaar, T.M.: A survey of collaborative filtering techniques a survey of collaborative filtering techniques. Hindawi publishing corporation, advances in artificial intelligence archive, pp. 1–19. USA : журнал (2009)
4. Salton, G., Wong, A., Yang C.S.: A vector space model for automatic indexing, communications of the ACM, vol. 18, no. 11, pp. 613–620
5. Chai1, T., Draxler, R.: Root mean square error (RMSE) or mean absolute error (MAE)?—Arguments against avoiding RMSE in the literature. Copernicus publications on behalf of the European Geosciences Union
6. Jonathan, B., Rahim, Z., Barzani, A., Oktavega, W.: Evaluation of mean absolute error in collaborative filtering for sparsity users and items on female daily network. In: Proceedings—1st international conference on informatics, multimedia, cyber and information system. ICIMCIS 2019 (2019)
7. Hyndman, R.J., Koehler, A.B.: Another look at measures of forecast accuracy. Int. J. Forecasting **22** (4), 679–688. CiteSeerX 10.1.1.154.9771. https://doi.org/10.1016/j.ijforecast.2006.03.001
8. Willmott, Cort, J., Matsuura, K.: Advantages of the mean absolute error (MAE) over the root mean square error (RMSE) in assessing average model performance. Climate Res. **30**, 79–82. https://doi.org/10.3354/cr030079

9. MovieLens. https://grouplens.org/datasets/movielens/. Last Accessed 12 Feb 2021
10. Leland Wilkinson and Michael Friendly. The History of the Cluster Heat Map // The American Statistician

A Modelling of Novel Approach for Stock Portfolio Management Using Unsupervised Algorithm and Deep Reinforcement Technique

D. Ramya(✉) and Suresha

DoS in Computer Science, University of Mysore, Mysuru, India
ramyad@compsci.uni-mysore.ac.in

Abstract. Recent developments in Machine Learning have attracted the attention of the financial community for the development of an optimal strategy in stock markets. The majority of existing approaches introduced a model to maximize return. However, these methods often fail to account to adapt to the constantly changing market circumstances. This paper develops an efficient learning model based on the unsupervised learning and deep reinforcement learning technique to determine an optimal stock portfolio. An unsupervised algorithm is employed to select suitable stocks in the portfolio. An LSTM (Long Short-Term Memory) based agent mechanism is proposed to address the weight assignment problem to each stock in order to minimize the risk and increase portfolio value. The study also presents an intelligent trading algorithm that is flexible to trade on any Indian stock portfolio. The experimental analysis showcases the proposed scheme performs better than the baseline and secured portfolio.

Keywords: Stocks · Reinforcement learning · Unsupervised learning · Portfolio optimization · Risk control

1 Introduction

The stock market provides a glimpse of future growth, although many other factors affect its ups and downs, such as macroeconomic factors, market expectations, and company operations [1, 2]. In order to perform stock analysis, huge amounts of data needs to be processed in a very short period due to the advancement of technology. This has made the task of stock analysis increasingly challenging [3, 4]. One of the hottest issues in finance is portfolio management, which is all about choosing the right combination of securities for a specific profit goal [5]. In line with the development in the AI field, deep learning techniques have become widely used in many fields and practical scenarios [6]. The neural networks used in these applications have evolved as deep learning approaches which is more popular and have been put to use in the financial prediction field [7, 8]. Apart from this, the adoption reinforcement learning also contributes significantly to the stock market analysis due to its self-exploration abilities [9]. Therefore, this paper explores the effectiveness of the deep learning and reinforcement learning methods on

R. Silhavy et al. (Eds.): CoMeSySo 2022, LNNS 596, pp. 319–332, 2023.
https://doi.org/10.1007/978-3-031-21435-6_28

stock portfolio management to control the financial risk factor and achieve a better return. In this research work, a clustering approach is employed to select a stock set. The study further proposes a model based on deep reinforcement learning to diversify the capital fund's investment in selected stocks. This method employs a buy-and-hold strategy based on the weight assigned to each stock. An optimal strategy is derived for higher return but maintaining lower risks.

The manuscript's organization is as follows: Sect. 2 discusses existing portfolio management approaches. The system design (proposed methodology) is briefed in Sect. 3. Implementation Strategy/procedure is discussed in Sect. 4 while Sect. 5 discusses result analysis, and Sect. 6 concludes the overall work.

2 Related Work

When investing in the stock market, the first and most crucial issue investors must deal with is selecting the right stock. The work of Chou et al. [10] focused on the problem of financial risk due to adverse stock selection. Here, the authors have introduced a risk estimation technique to evaluate portfolio risk, and a genetic algorithm (GA) is incorporated to determine optimal portfolio. By defining critical roles for cooperatively executing stock pricing and portfolio selection decisions, Lee et al. [11] demonstrated an effective method for dividing up the stock trading problem based on multiple Q-learning agents. Lim et al. [12] applied GA to the Portfolio Design based on market trends and asset estimates. The study by Yue et al. [13] provided a comprehensive overview of the financial data at three levels: risk-factor level, multiple-portfolio level, and the single-portfolio level. A discrete-continuous differential evolution algorithm is developed specifically for optimizing a stock selection model using both discrete and continuous decision variables by Yu et al. [14]. The work of Maa et al. [15] used deep multilayer neural networks to develop prediction-based portfolio optimization models for each stock built by integrating prediction returns and semi-absolute deviations of the forecast. Tunc et al. [16] developed a portfolio selection system that minimizes the expected cumulative wealth in markets where stocks are bought and sold at proportional transaction costs.

Nonconvex Cardinality Constraint is used in the study of Khan et al. [17] to determine the Optimal Portfolio Management scheme. The study of Athihal et al. [18], identifies the influence of macroeconomic indicators on national stocks, which are then aggregated to determine the correlation matrix. He et al. [19] propose a multi-period mean-variance optimization scheme with high order coupled assets analysis. The work carried out by Gao et al. [20] investigated how to hedge impact the financial assets to risk exposer an imperfectly liquid stock. The authors have applied Bayesian statistical methods to determine the optimal strike price and the minimal value at risk. Hence, there are a variety of schemes presented in the literature. However, most of the existing systems are associated with a significant problem, such as the existing learning models are prone to sampling errors and over-fitting problems. The statistical based approaches are not much scalable to adopt dynamic scenarios. Actually, most of the existing systems are designed specific to the particular context, limiting their scope in the real-time.

3 Proposed Methodology

The proposed study focuses on addressing the issue of appropriate stock selection and portfolio management to achieve higher returns under controlled financial risk. The proposed study employed a deep reinforcement learning approach to derive a high-level decision policy for long-term profit and short-term return with minimal trading cost. The schematic architecture of the proposed system is depicted in Fig. 1. Initially, data listed on the national stock exchange is downloaded from the yahoo finance website. The data are further analyzed to extract significant features to understand the data's volatile characteristics. The k-means algorithm is used to cluster data according to the market capitalization in this process. A particular stock cluster is chosen as a stock set for portfolio construction based on investor risk apatite. The selected stocks are further subjected to the weight assignment using the deep Reinforcement Learning (RL) technique. In this phase, the proposed study utilizes LSTM a deep learning technique that is integrated with the RL agent to derive an optimal strategy for the weight allocation. Later the money available to trade is divided into a total number of stocks according to its weighted value. In this way, an effective stock portfolio is managed, and trades are executed at the optimal price for long-term return without posing a higher risk to the investor.

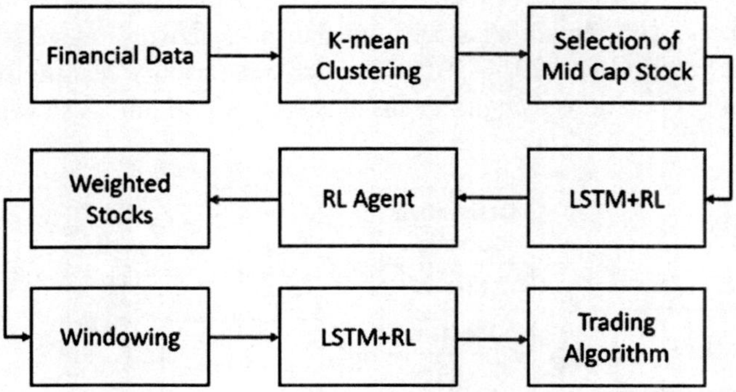

Fig. 1. Proposed scheme of stock portfolio management

On the other hand, this study also proposes an intelligent trading algorithm to trade for a short time. The design of the proposed trading algorithm is also based on the deep RL technique, which considers past closing price patterns to derive an optimal price from executing long step, short step, and holding of the stocks. To validate agent-based mechanisms for portfolio management and trading, a customized environment is designed using a function of the Open AI Gym library. Also, two different environments are constructed that imitate scenarios of secured portfolios like fixed deposit and baseline portfolios like NIFTY-50.

4 Implementation Strategy

This section discusses the procedure adopted for implementing and executing the proposed system. The dataset adopted for this study includes 1617 × 8 stocks listed on NSE as of March 2019. Figure 2 illustrates the description of the downloaded stock data.

```
RangeIndex: 1617 entries, 0 to 1616
Data columns (total 8 columns):
 #   Column            Non-Null Count  Dtype
---  ------            --------------  -----
 0   SYMBOL            1617 non-null   object
 1   NAME OF COMPANY   1617 non-null   object
 2    SERIES           1617 non-null   object
 3    DATE OF LISTING  1617 non-null   datetime64
 4    PAID UP VALUE    1617 non-null   int64
 5    MARKET LOT       1617 non-null   int64
 6    ISIN NUMBER      1617 non-null   object
 7    FACE VALUE       1617 non-null   int64
```

Fig. 2. NSE stock data description

Stocks with long-term suspensions are removed from the data set to ensure continuity and completeness. The remaining stocks are the most actively traded stocks during the entire sample period i.e. (2014–2019). Further, a descriptive feature is extracted to perform grouping operations using k-means clustering algorithm, as shown in Fig. 3.

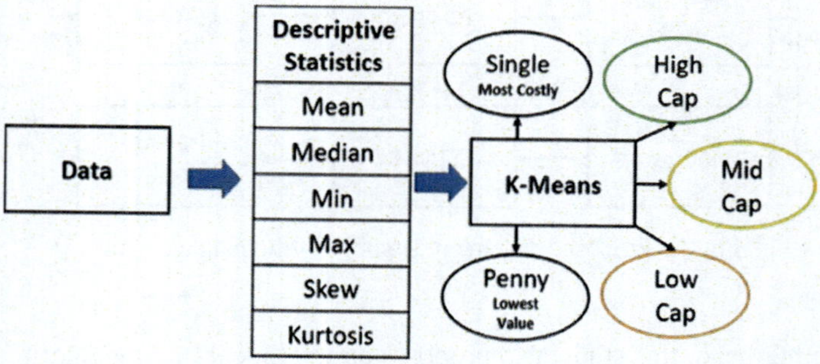

Fig. 3. Mechanism of clustering

The important features are extracted using mean, standard deviation, minimum and maximum values, skewness, and Kurtosis statistical measures. According to the analysis, there are two methods to measure volatility in a stock. The first one is by measuring the stock's standard deviation. The second method is to measure the stock's skewness. Furthermore, the proposed study assumes that, only the closing price is relevant for the analysis as it is the most significant part of the stock price. Volatile stocks can earn higher profits but, pose a huge risk to investors. Long-term investors who keep stocks for many days usually use the closing price data. Since this model is not meant for intraday trading, it only considered the closing price. Further, k-means clustering is applied to obtain features in a vector format to construct different groups of stocks according to their market value. Stocks are classified as: low-cap stocks, mid-cap stocks, and large-cap stocks. Low-cap stocks have a market cap of fewer than 5000 crores, mid-cap stocks have a market cap between 5000 crores and 20000 crores, and large-cap stocks have more than 20000 crores. Based on the cluster analysis, the proposed study considers a mid-cap stock as it has medium risk with good return scope. Large-cap stocks are already saturated, having low risk and consumes more time to make a considerable return, and low cap stocks are very volatile having higher risk.

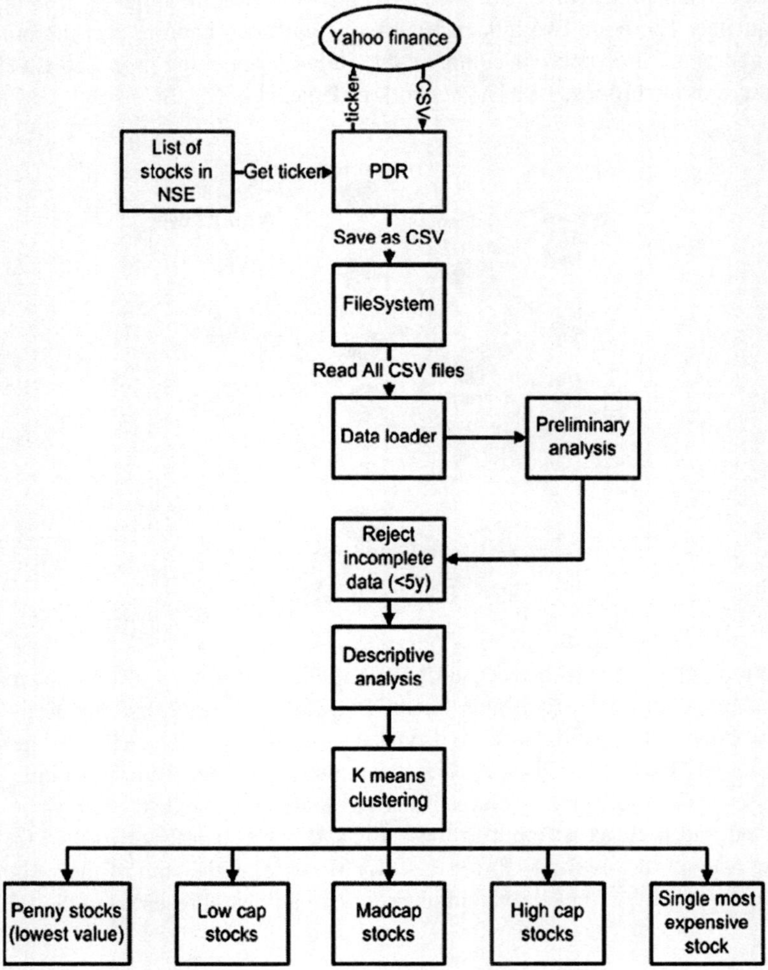

Fig. 4. Flow chart of the stock selection process using K-means clustering

Figure 4 presents the flow chart of the k-means clustering process to classify stock according to market capital. Based on the risk appetite, mid-cap stocks are extracted for the portfolio management based on the weight assignment.

4.1 Portfolio Management

This phase involves the assignment of a weight to each stock using machine learning to create an optimal portfolio for a long-term investment. The weight is based on the interaction between the RL agent and the environment. The actor and critic neural networks are trained using the LSTM network in the agent model by determining an initial weighted stock. Three neurons are used as a place holder at the input layer to hold data in a sequence. There are two hidden layers, one with two neurons and the other with twenty neurons. Three neurons comprise the output layer in the proposed LSTM. The architecture of the proposed LSTM network is depicted in Fig. 5.

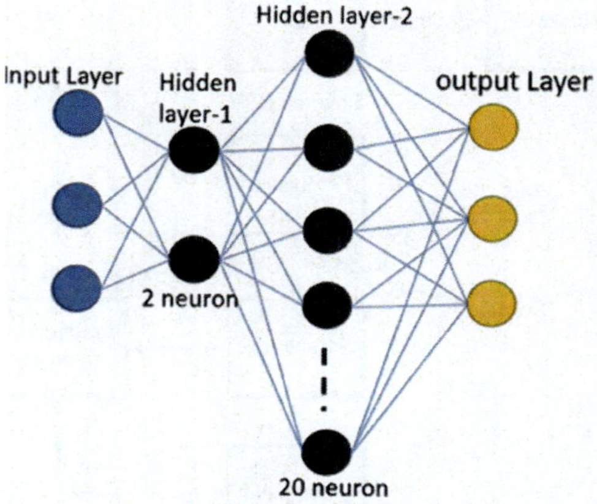

Fig. 5. Architecture of LSTM

Figure 5 depicts how the proposed LSTM model is designed to account for long-term investment and short-term investment. The LSTM, is used to compute the initial weighted stocks by learning patterns from an input dataset (selected midcap stocks) and predicting next values based on previous patterns, as each pattern depends on the previous one. Consequently, a data set that is highly correlated to each other can be constructed and used as an input to the actor and critic neural network in the agent modeling. The portfolio value (P) for this LSTM model is the sum of the weight over time (one day) and the stocks' total number numerically is represented in the following Eq. 1

$$P = \sum_{i=1}^{N} S \times w_{i,t} c'_{i,t} \tag{1}$$

Initial weights to each stock are assigned randomly, taking into account that the assignment of weights is static for the entire trading period. In this regard, we obtain different portfolio values such that {P1, P2, P3 ..... PN}, where N signifies portfolio value on day basis on the closing price. The LSTM is trained only with 80% of the data. After training process, the model is tested with the remaining 20% of data to assess its usefulness. For an example if we give the model an input value P78, then the model predicts the next sequence of input i.e., P79 by itself based on the learned time-varying sequence from the time varying dataset. To maintain a good portfolio for long-term investment, the trained LSTM model is applied to the agent architecture to compute the final weights assigned to each stock. Figure 10 illustrates the interaction between a portfolio agent and its environment for the weight assigned to each stock by the agent (Fig. 6).

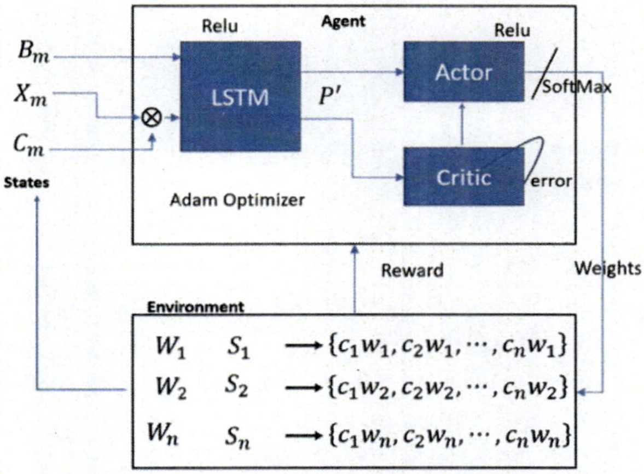

Fig. 6. Agent and environment interaction for assignment of weights

The proposed study uses the Open AI-gym library function to build a customized environment; the agent is built on the actor and critical neural networks. The actor provides a strategy for action, whereas the critic provides an optimum policy. The trained LSTM model is incorporated in the agent modeling to facilitate better generalization of the pattern based on data. The agent observes the state of the stock after interaction with the environment and acts accordingly. The action of the agent is the assignment of weights to each stock (w), state (S) perceived by agent total number of stocks, money in the hand and closing price of each stock expressed in the equation as follows

$$S = \{X_m, B_m, C_m\} \tag{2}$$

where m denotes time steps of each N stocks.

The reward (R) can be negative or positive. However, the agent always strives to obtain higher rewards, which is expressed as follows:

$$R = (X_m \times C_m) + B_m - B_0 \tag{3}$$

where, m is cash or money available before trading.

The environment consists of *Cm* and *Xm.* An agent adjusts the weights so that, the value of the portfolio is always higher than the closing price of the previous day at each step based on the closing price of the day divided by the closing price of the previous day. Since this is not feasible in practice, weights will be assigned to increase the value of the portfolio as much as possible. The weights given by agent to each stock can be formulated as follows:

$$S_1 == \{C_1 \times w_1 + C_2 \times w_1 + C_3 \times w_1 \cdots + C_m\} \tag{4}$$

By assigning weights to stocks, the available money is divided over time and based on these, the portfolio policy decides which stocks to purchase. This process can be numerically expressed as follows

$$Accu = \left(\frac{B_0}{M}\right) \tag{5}$$

where M denotes time stamp (or day) and if Accu $\geq P_i$, then purchase. The remaining balance can be computed as follows:

$$remaining = Accu - P_i \tag{6}$$

This is based on the fact that available money is greater or equal to present-day portfolio value (P_i), then stocks are purchased according to their weights. The procedural flow of the stock portfolio management using proposed agent mechanism is depicted in Fig. 7.

4.2 Trading Algorithm

A trading algorithm based on the LSTM-agent structure is developed to trade periodically over a proposed stock portfolio structure. Based on the interaction with the environment, the agent is modeled to engage in three actions: to buy, sell, and hold, based on the situation on the stock market. The agent algorithm observes a state C_i (closing price) and B_i (remaining balance) and get rewards (R) for its every action numerically formulated as follows:

$$R = [Q_i C_i + B_i] - [Q_{i-1} C_{i-1} + B_{i-1}] \tag{7}$$

where, Q_i refers to action policy. The procedural flow of the trading algorithm is depicted in Fig. 8.

The implementation procedure adopted in developing a trading algorithm is shown in Fig. 8. Initially, the LSTM is constructed with a neural input layer with seven neurons, a hidden layer with 256 neurons, and an output layer with three neurons. To train the LSTM, the stock's previous closing price is passed as input using sliding windowing with a window size of three. An LSTM network is trained using the balance max optimizer. The RL agent executes real-time stock trading with the help of the trained LSTM network.

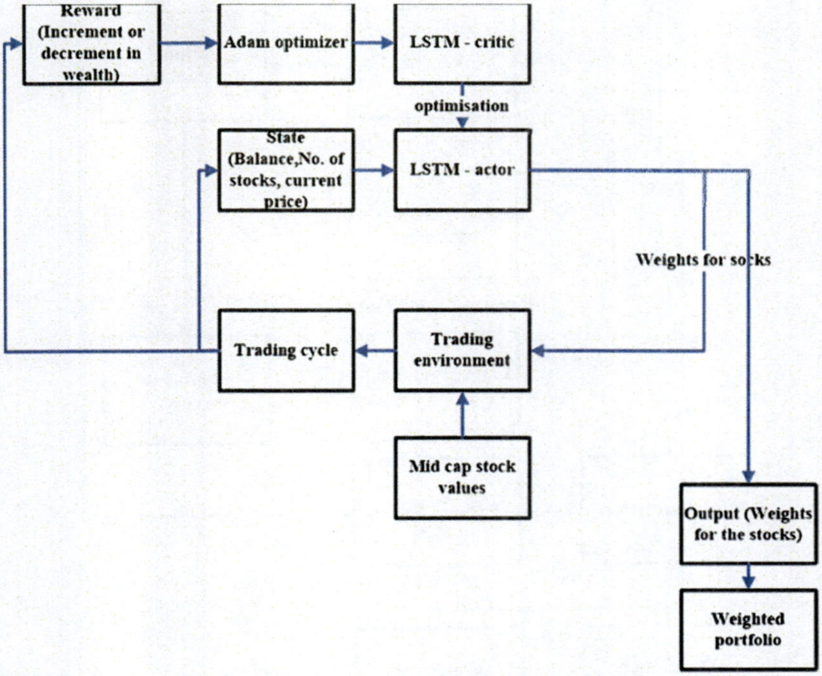

Fig. 7. Flow chart of the portfolio management using agent mechanism

5 Result and Discussion

The implementation procedure adopted in developing a trading algorithm is shown in Fig. 8. Initially, the LSTM is constructed with a neural input layer with seven neurons, a hidden layer with 256 neurons, and an output layer with three neurons. To train the LSTM, the stock's previous closing price is passed as input using sliding windowing with a window size of three. An LSTM network is trained using the balance max optimizer. The RL agent executes real-time stock trading with the help of the trained LSTM network.

As observed in the graph shown in Fig. 9, Gillete, Laxmi Mittal, Maruti are among the top priorities in the industry, and these stocks are the leaders. Also, some of the stocks that are given a low priority are well-known. For example, Bajaj FinServ is a well-known company, but it has consistently booked losses over the given time period.

Figure 10 illustrates the comparative analysis regarding different portfolio values. Agent portfolio value is the proposed one, comprised of several individual stocks, extracted from k-means clustering and weighted by deep learning technique. It is observed from the graph trend, the proposed Agent portfolio outperforms another portfolio.

Figure 11 shows the buy and sell performance of the proposed algorithm evaluated against a benchmark portfolio (NIFTY50). The trade signals shows that there are higher numbers of trades (buy and sell), which lead to a higher degree of slippage in which investors' intended prices are not met.

Figure 12 shows the trend of change in wealth at the end of a trading cycle (365 days).

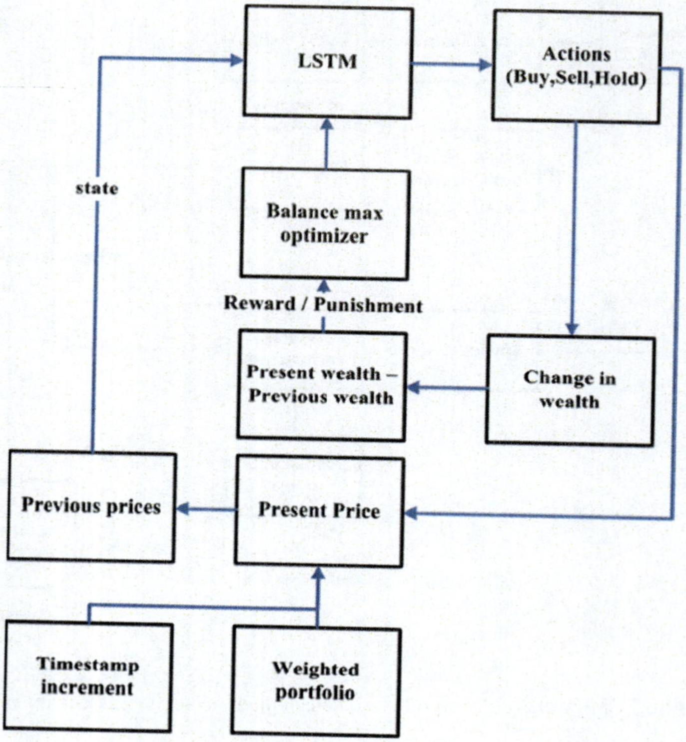

Fig. 8. Flow chart of trading algorithm

As shown in Fig. 13, the proposed trading algorithm is compared to a hypothetical portfolio (mid-cap stocks). The number of trades affected by the RL agent for a particular portfolio is smaller than the number given by the proposed algorithm. The system is more efficient by generating more profit with a lower number of trades.

According to the graph trend in Fig. 14, the wealth increased around 35% in the trading cycle of 365 days. The graph illustrates that the wealth is higher than the baseline portfolio (NIFTY50).

6 Conclusion

The focus of the study is on portfolio selection and managing trading costs through deep learning and reinforcement learning techniques. In order to build an optimal portfolio of stocks, a combination of LSTM and reinforcement learning is used to assign static weights to the selected stocks. Long-term investment decisions are made in this process according to a high-level policy. Additionally, a trading algorithm is proposed that manages purchase, sale, and hold decisions daily or weekly. The proposed system influence that the trades are carried out at the lowest cost, and orders are placed rapidly and precisely. A reduction in transaction expenses is achieved by processing trades on time and automating checking on multiple market conditions simultaneously. The outcome validates the scope of the proposed system.

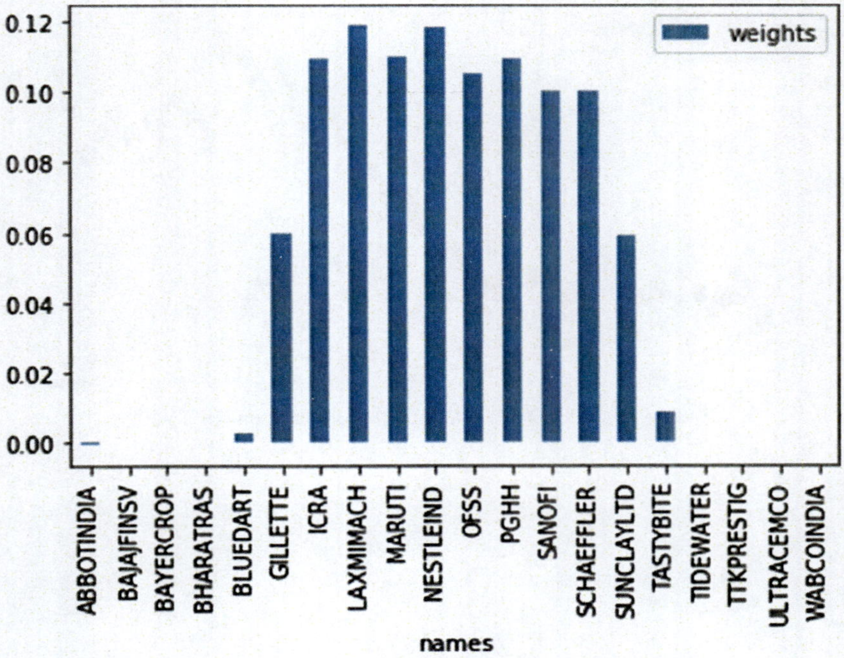

Fig. 9. Analysis of weights

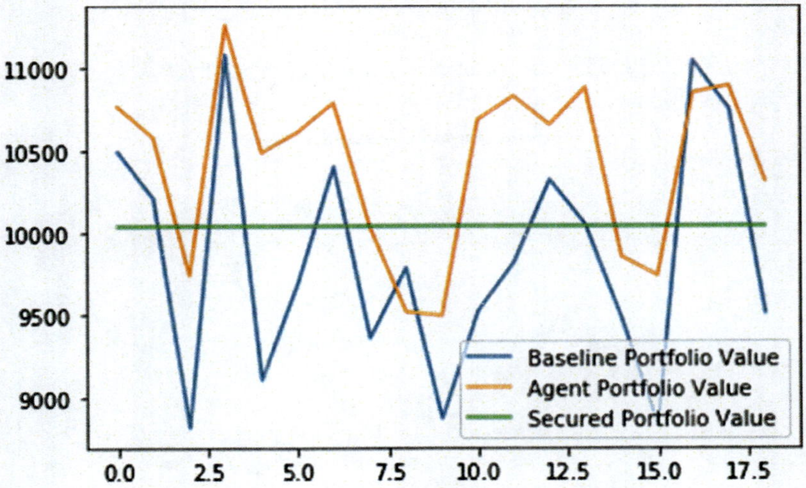

Fig. 10. Comparative analysis based on the portfolio

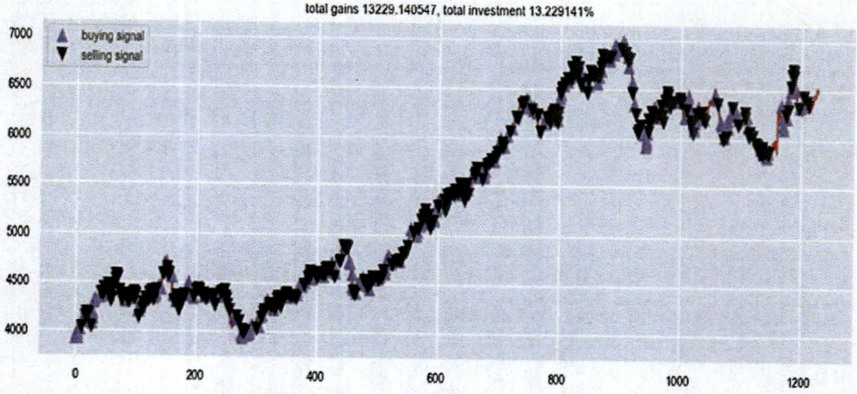

Fig. 11. Analysis of buy and sell performance on NIFTY50

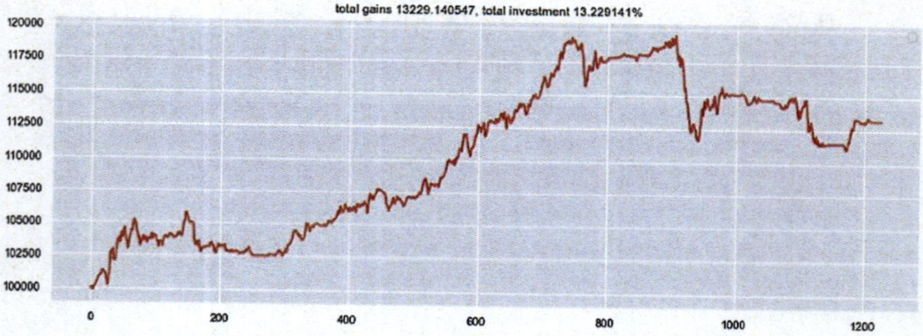

Fig. 12. Analysis of change in wealth on NIFTY50

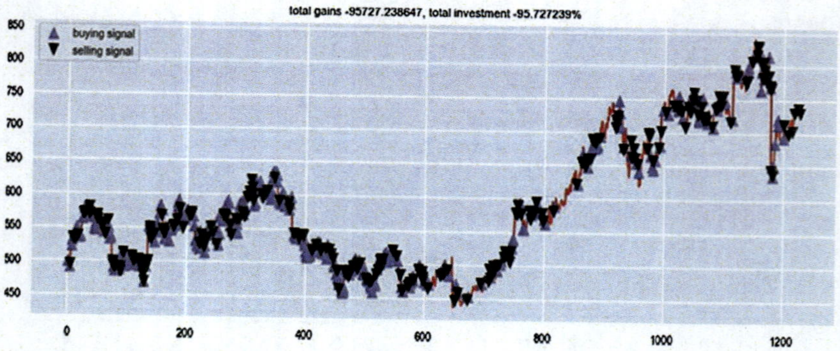

Fig. 13. Analysis of buy and sell performance on proposed portfolio

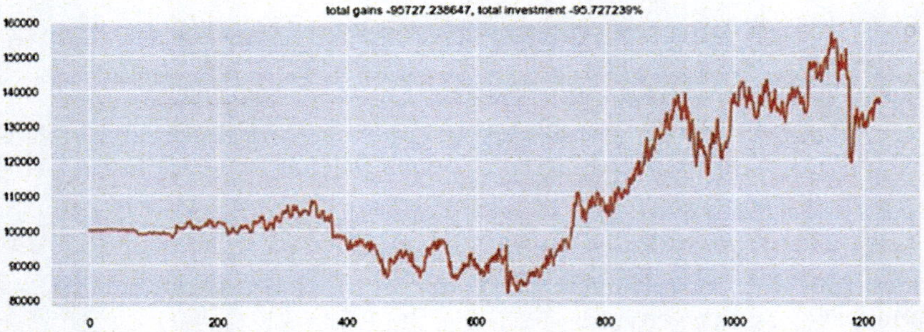

Fig. 14. Analysis of change in wealth on proposed portfolio

References

1. Das, A.K., Mishra, D., Das, K.: Currency exchange prediction for financial stock market: an extensive survey. In: Mallick, P.K., Bhoi, A.K., Marques, G., Hugo C. de Albuquerque, V. (eds.) Cognitive informatics and soft computing. AISC, vol. 1317, pp. 697–709. Springer, Singapore (2021). https://doi.org/10.1007/978-981-16-1056-1_54
2. Kumar, G., Jain, S. and Singh, U.P.: Stock market forecasting using computational intelligence: a survey. Archives of computational methods in engineering, pp.1–33 (2020)
3. Beyaz, E., Tekiner, F., Zeng, X.J. and Keane, J.: Comparing technical and fundamental indicators in stock price forecasting. In: 2018 IEEE 20th international conference on high performance computing and communications; IEEE 16th international conference on smart City; IEEE 4th international conference on data science and systems (HPCC/SmartCity/DSS) (pp. 1607–1613). IEEE (2018)
4. Hu, Z., Zhao, Y., Khushi, M.: A survey of forex and stock price prediction using deep learning. Appl. Syst. Innov. **4**(1), 9 (2021)
5. Sato, Y.: Model-free reinforcement learning for financial portfolios: a brief survey (2019). arXiv:1904.04973
6. Dargan, S., Kumar, M., Ayyagari, M.R., et al.: A survey of deep learning and its applications: a new paradigm to machine learning. Arch Comput. Methods Eng. **27**, 1071–1092 (2020)
7. Ertenlice, O. and Kalayci, C.B.: A survey of swarm intelligence for portfolio optimization: algorithms and applications. Swarm Evol. Computat. **39**, 36–52 (2018)
8. Thakkar, A., Chaudhari, K.: A comprehensive survey on portfolio optimization, stock price and trend prediction using particle swarm optimization. Arch. Comput. Methods Eng. **28**(4), 2133–2164 (2021)
9. Sato, Y.: Model-free reinforcement learning for financial portfolios: a brief survey. arXiv:1904.04973 (2019)
10. Chou, Y., Kuo, S., Lo, Y.: Portfolio optimization based on funds standardization and genetic algorithm. IEEE Access **5**, 21885–21900 (2017). https://doi.org/10.1109/ACCESS.2017.2756842
11. Lee, J.W., Park, J., Lee, J.O.J., Hong, E.: A multiagent approach to Q-learning for daily stock trading. IEEE Trans. Syst. Man Cybern. Part A Syst. Humans **37**(6), 864–877 (2007). https://doi.org/10.1109/TSMCA.2007.904825
12. Lim, S., Kim, M-J., Ahn, C.W.: A genetic algorithm (ga) approach to the portfolio design based on market movements and asset valuations. IEEE Access **8**, 140234–140249 (2020)
13. Yue, X., et al.: sPortfolio: stratified visual analysis of stock portfolios. IEEE Trans. Visual. Comput. Graphics **26**(1), 601–610 (2020)

14. Yu, L., Hu, L., Tang, L.: Stock selection with a novel sigmoid-based mixed discrete-continuous differential evolution algorithm. IEEE Trans. Knowl. Data Eng. **28**(7), 1891–1904 (2016). https://doi.org/10.1109/TKDE.2016.2545660
15. Ma, Y., Han, R., Wang, W.: Prediction-based portfolio optimization models using deep neural networks. IEEE Access **8**, 115393–115405 (2020). https://doi.org/10.1109/ACCESS.2020.3003819
16. Tunc, S., Donmez, M.A., Kozat, S.S.: Optimal investment under transaction costs: a threshold rebalanced portfolio approach. IEEE Trans. Signal Process **61**(12), 3129–3142 (2013)
17. Khan, A.H., et al.: Optimal portfolio management for engineering problems using non-convex cardinality constraint: a computing perspective. IEEE Access **8**, 57437–57450 (2020)
18. Aithal, P.K., Dinesh, A.U., Geetha, M.: Identifying significant macroeconomic indicators for indian stock markets. IEEE Access **7**, 143829–143840 (2019)
19. He, J., Wang, Q., Cheng, P., Chen, J., Su, Y.: Multi-period mean-variance portfolio Optimization with high-order coupled asset dynamics. IEEE Trans. Autom. Control **60**(5), 1320–1335. https://doi.org/10.1109/TAC.2014.2370236
20. Gao, R., Li, Y., Bai, Y., Hong, S.: Bayesian inference for optimal risk hedging strategy using put options with stock liquidity. IEEE Access **7**, 146046–1460

Development of Motion Analysis Tool for Orthopedic Diseases: Inertial Measurements

Pavel Jicinsky(✉) and Jan Mares

Department of Process Control, Faculty of Electrical Engineering and Informatics, University of Pardubice, Pardubice, Czech Republic
pavel.jicinsky@student.upce.cz, jan.mares@upce.cz

Abstract. Rehabilitation in orthopedics is one of the most important processes that influence the patient's quality of life. In today's practice, the orthopedic assessments still depend mainly on subjective view of physician. To further objectify the results, a wearable electronic device was developed based on inertial sensors.

Keywords: IMU · Orthopedics · Signal acquisition and processing · Motion analysis

1 Introduction

Orthopedic surgeries have a significant impact on the quality of the patient's life, especially the large joint surgeries (hip, knee, ankle). Rehabilitation is prescribed after most of these surgeries and its course can have a big effect on the ability to use the affected limb.

However, objective analysis of a patient's condition plays the same role as rehabilitation. But there is a lack of tools for such an analysis. The evaluation of patient's condition is usually based on the clinician's (expert) opinion.

Therefore, the aim of the work is to join information engineering with clinical praxis and develop a complex tool for precise measurement of the movements of individual joints and subsequent evaluation of the patient's condition. This paper describes the development of an inertial measurement unit (IMU) and application for data acquisition.

1.1 State of the Art

Today it is a common practice to enhance a subjective medical assessment using sensors and wearable electronics to further objectify the results. It allows the use of complex analysis tools later in the process like the gait features extraction using neural networks. The authors of the paper [1] used the data from electromyography to extract the gait phases using NN as well as to find out the muscles' contribution to motion.

Many of the medical projects use optical sensors to estimate a 3D model of patient's movement. These projects typically use any of the commercially available sensors like Kinect which is very useful out of the box, although with limited accuracy and it requires

R. Silhavy et al. (Eds.): CoMeSySo 2022, LNNS 596, pp. 333–338, 2023.
https://doi.org/10.1007/978-3-031-21435-6_29

quite some space for its operation. The following projects [2, 3] use Kinect in conjunction with the inertial measurements to further enhance the results of the tests. Both projects achieved similar or better accuracy using the inertial measurements compared to the optical approach.

Very few projects use different means of measurements than those mentioned. The authors of the paper [4] used the inertial sensor containing an accelerometer and a gyroscope to identify the gait phases. The following project [5] used a velostat which is in this case used as a pressure sensitive material to create a pressure matrix to identify the gait footprint pressure and to investigate the pressure distribution during movement.

2 Inertial Measurements Unit - IMU

A battery powered wearable solution was designed to estimate the spatial orientation as well as an absolute position of the major joints of the lower limbs during assessed movement. The device was designed according to the need to maintain safety and comfort in use. This is achieved using battery power and wireless communication using Wi-Fi.

Figure 1 depicts the block diagram of the IMU.

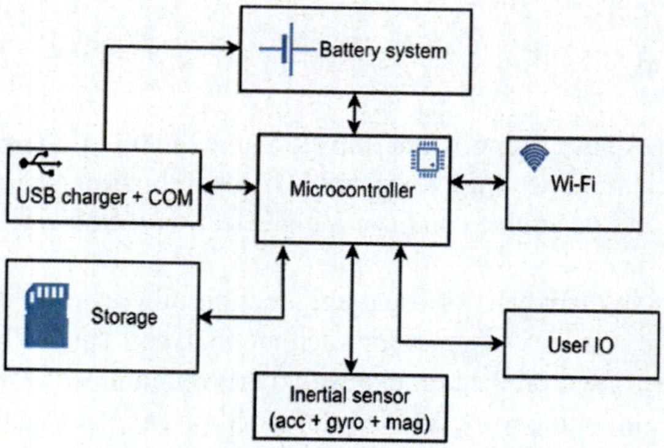

Fig. 1. IMU block diagram

The IMU is powered by a single Li-Pol battery cell, and it can be charged using a micro-USB port. The battery system provides reverse polarity protection as well as over and under voltage protection using dedicated integrated circuit. A dedicated analog input pin of the microcontroller is used to monitor the battery voltage in the application code. The power supply is separated into two power domains. The main power domain is used to power the MCU and the inertial sensor, and it cannot be turned off – turning it off sets the whole board into an OFF state. The second power domain is used to power the rest of the systems – storage and wireless com. This domain is turned off for most of the time to save power as the wireless module draw a significant amount of current during its operation. Powering the board using a battery is an obvious choice as it provides inherent safety and portability. There are no cables attached to the board that could limit the patient's movement ability.

The storage system consists of an SD card operated in SDIO 4-bit mode and the MCU is offloaded using DMA. The storage is used for datalogging of all the measured and computed values as well as system settings and the calibration constants (offsets and rotation matrices for each sensor). The measured data can be easily retrieved through USB serial com port. The IMU can also operate without an SD card inserted. In this mode, the default calibration constants are used (preprogrammed) and the measured values are transmitted over USB or Wi-Fi to the host application.

The wireless communication module Atwinc1500 is used and is connected using an SPI bus, enabling sufficient throughput and it is also offloaded using DMA (for transfers only). The Wi-Fi module implements the TCP/IP stack to further offload a host MCU and it is configured as an UDP client.

Fig. 2. Assembled PCB - battery is on the bottom side

The ST's LSM9DS1 inertial sensor was used in this project as it integrates an accelerometer, gyroscope and a magnetometer in a single chip, although it is obvious that it is merely a merge of two separate ICs on a single die (LSM6DS and LIS3MDL). It is also connected to the MCU using a separate SPI bus without DMA as the amount of data transferred isn't that big (3 x 16 bit values for each sensor – 18B at 119Hz for acc/gyro, 80Hz for magnetometer). There are two main difficulties using this IC, that is the need to merge data sampled at different sample rates (119 and 80Hz used) and the flipped X axis of the magnetometer compared to the other two sensors, as is clearly visible in Figure 2 (bottom left silkscreen). The LSM9DS1 can be interfaced using both the SPI and I2C – it uses separate data output lanes on the SPI for acc/gyr and magnetometer respectively and it has two separate chip select pins. The two separate output lanes were connected into a single lane in this project and the data flows are mutually excluded using chip selects.

To enable user control and to observe the application status during headless operation (without any external control connection – USB/Wi-Fi), the user input-output section is used. It consists of two user buttons to navigate through the state machine states. It also consists of three user LEDs with different colors to indicate various events. The last part of the IO system is the piezo buzzer that serves as a heartbeat indicator (indicates the running state of the application) as well as an error siren.

In order to synchronize multiple devices, the synchronization interface is used. It consists of a several GPIOs and a single UART bus. The GPIOs are used as an interrupt request pins (sort of an event trigger). A synchronization unit will be created in the following period. The IMUs will then connect to this unit to receive a timestamp as well as to reset their relative position to zero (sync unit will be the origin).

An STM32F405 MCU was used as a main controller for the IMUs. It was chosen for its high system clock (168 MHz), plenty of IOs and communication interfaces and most importantly a floating-point accelerator (FPU).

3 IMU Application

An MCU application was written in C++ to control the flow of operation of the IMU unit. The application runs a real-time operating system which allows user to separate the program into independent tasks between which the OS preemptively switches. This way, even a single core MCU can run multiple tasks at once, or more accurately switch between those tasks at a very high rate, at which it seems as if they were running concurrently. This approach also ensures that all the tasks are serviced in a deterministic manner and that they are not starved by other tasks (a slow interface task for instance may block other tasks in a cooperative multitasking).

3.1 Architecture

The application was divided into several tasks according to the need to service different communication interfaces used as well as GPIOs and other routines. The main task is IMU task which retrieves the data from the IMU sensor and runs the estimation engine. Its outputs in form of Euler angles, quaternion, gravity vector and linear acceleration are passed into the Storage, Wi-Fi or the USB task according to the application state using shared memory and mutexes. Figure 3 depicts a flow chart of the MCU application.

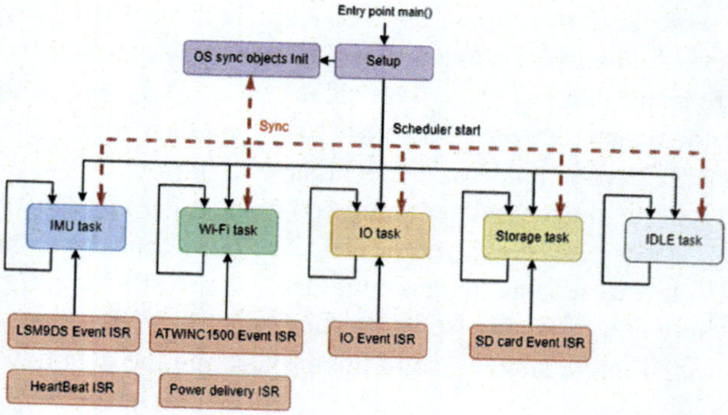

Fig. 3. IMU application

3.2 AHRS Algorithm

There are several approaches of spatial orientation estimation available. All those algorithms use data fusion to merge data from different kinds of sensors. These sensors are typically accelerometers, gyroscopes, magnetometers, barometers, humidity sensors, temperature sensors and GPS. This project uses ST's MotionFX fusion algorithm [6]. The MotionFX uses an extended Kalman filter to estimate orientation and linear acceleration without earth gravity vector. It also handles gyroscope and magnetometer calibrations.

4 PC Application

A PC application was written in C++ using wxWidgets to visualize the IMU output data in real time. The output visualization is done using OpenGL and GPU rendering of a mockup that resembles the physical dimensions of the IMU board.

A 3D orientation is represented using Euler angles in the output panel of the application, but the rotations are calculated using quaternions in the vertex shader. The input vector representing the object vertices positions is first multiplied by the scaling matrix. After this the quaternion rotation is done and finally the translation is applied.

The application also provides a convenient way to load basic settings into the IMU unit (e.g., calibration constants) (Fig. 4).

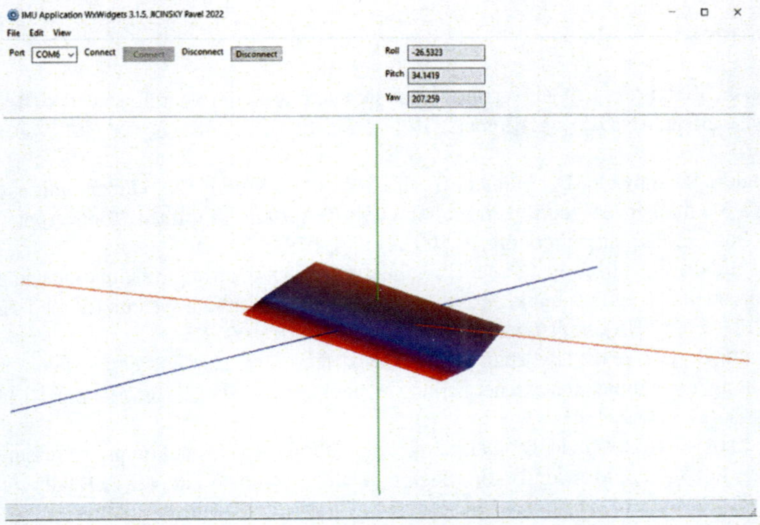

Fig. 4. PC application

5 Conclusion

The aim of this paper is to describe the development process of the proposed inertial measurements unit used to evaluate the feasibility of sensored diagnostics in orthopedic patients. This paper sums up the second development stage of out proposed system [7].

A wearable battery powered device was created using inertial measurement unit containing accelerometer, gyroscope and magnetometer to estimate the spatial orientation of the large joints of lower limbs during motion. It is also able to extract the linear acceleration of the joints without the Earth gravity content, enabling a relative 3D position estimation. This device was developed according to our previous work [7].

6 Future Work

The created IMUs and the PC application will be tested in the following period in clinical practice with real patients. The used AHRS algorithm will be further tuned to further improve its performance.

The outputs of this development stage will be used in conjunction with our electromyography (EMG) acquisition system [8] which is used to capture the skeletal muscles activity during movement.

The complete system will be tested in clinical trials in University Hospital Královské Vinohrady

Acknowledgment. This research was funded by an SGS grant from the Faculty of Electrical Engineering and Informatics, University of Pardubice, Czech Republic. This support is very gratefully acknowledged.

References

1. Wang, J., Zielinska, T.: Gait features analysis using artificial neural networks - testing the footwear effect. Acta Bioeng. Biomech. **19** 17–32 (2017). https://doi.org/10.5277/ABB-00490-2015-05
2. Rodrigues, T., Salgado, D., Catháin, C., O'Connor, N., Murray, N.: Human gait assessment using a 3D marker-less multimodal motion capture system. Multimed. Tools Appl. **79**(3–4), 2629–2651 (2019). https://doi.org/10.1007/s11042-019-08275-9
3. Sharif Bidabadi, S., Murray, I., Lee, G.: Validation of foot pitch angle estimation using inertial measurement unit against marker-based optical 3D motion capture system. Biomed. Eng. Lett. **8**(3), 283–290 (2018). https://doi.org/10.1007/s13534-018-0072-5
4. Gujarathi, T., Bhole, K.: Gait analysis using IMU sensor. In: 10th international conference on computing, communication and networking technologies. (2019). https://doi.org/10.1109/ICCCNT45670.2019.8944545
5. Muhedinovic, H., Boskovic, D.: Design of IoT Solution for Velostat Footprint Pressure Sensor System. In: Ahmed, Mobyen Uddin, Begum, Shahina, Raad, Wasim (eds.) HealthyIoT 2016. LNICSSITE, vol. 187, pp. 171–176. Springer, Cham (2016). https://doi.org/10.1007/978-3-319-51234-1_30
6. STMicroelectronics. Getting started with MotionFX sensor fusion library in X-CUBE-MEMS1 expansion for STM32Cube, p. 25 (2021)
7. Jičinský, P., Mareš, J.: Motion Analysis Tool for Diagnosis of Orthopedic Disorders. In: RANSACTIONS of the VŠB—technical university of Ostrava, mechanical series, pp. 9–16. Ostrava: VŠB–Technical University of Ostrava. https://doi.org/10.22223/tr.2020-1/2058
8. Jičinský, P., Mareš, J.: Development of motion analysis tool for orthopedic diseases. In: International conference on applied electronics (AE), pp. 1–5. Pilsen: University of West Bohemia. https://doi.org/10.23919/AE51540.2021.9542895

Simulation of Multiple Autonomous Mobile Robots Using a Gazebo Simulator

Stanislav Alexovič(✉), Milan Lacko, and Ján Bačík

Department of Electrical Engineering and Mechatronics, Technical University of Košice, Kosice, Slovakia
stanislav.alexovic@tuke.sk

Abstract. Simulation is an indivisible part of the development and testing of robotic systems. This is also true for the simulation of interactions among multiple mobile robots. This paper presents a design of a multi-mobile robot simulation that is built on ROS and a gazebo simulator. Focus is given to an interaction between the ROS node which encapsulates access to hardware and the gazebo. In this design communication with hardware is redirected to the gazebo server with the usage of its transport protocol. All robots are executed inside docker containers and communicate with the gazebo server which also runs inside a docker container. Further in the paper is a description of the gazebo model plugin which acts as a virtual PLC and provides a behavior of hardware. Since all robot models are spawned in one virtual environment they can interact with each other. This is demonstrated in the conclusion where is an example of a factory environment that holds three mobile robots which could see each other.

Keywords: ROS · Gazebo · AMR · Simulation · Fleet

1 Introduction

A simulation is an essential tool for building complex systems such as robots. It is almost unimaginable to build a robot without simulation. Maybe small hobby robots could be created without simulation but real-world robots which are used in industry or for any critical application are created with help of simulation [1]. Of course, the simulation could not substitute real-world testing but it can dramatically help with the implementation and testing of various robotics components which do not directly interact with the real-world environment.

This article focuses on the design of simulation for multiple mobile robots which is an important area for testing interactions between mobile robots. Mainly, synchronized movement, crossroad resolution, following another robot, and resolving the deadlocks [2–5]. However, solutions for these problems are not included in this article. The article's main goal is to provide design specifications on how to build a simulation for multiple robots using well know tools such as ROS [6] and Gazebo simulator [7].

There are many robotics projects which are provided with working a simulation [8–10]. However, it is always dedicated to only one robot. Immediately when some

R. Silhavy et al. (Eds.): CoMeSySo 2022, LNNS 596, pp. 339–351, 2023.
https://doi.org/10.1007/978-3-031-21435-6_30

multi-robot problems arose is a need for multi-robot simulation. However, this is not as common as one robot simulation. The creation of this article was also motivated by this need. When single robot development changed to multi-robot cooperation.

The base requirement is to have all robots inside one virtual environment which is in opposition to the principle where each robot has its dedicated virtual environment [11]. This will mimic the reality where one robot can interact with another one. In this case, one robot would be able to see another one with its sensor, specifically a lidar laser scan.

2 Methods

Gazebo is a powerful 3D simulator based on the OGRE 3D engine [12], features several physics engines which can be chosen according to the needs of a given project [13]. A physics engine is the most important part of the simulator because it can recreate the behaviour of the real world. It mimics the dynamics of an object in various environments, be it movement of a mobile robot on the ground and its surface friction, flight of an aircraft with the simulation of wind forces, or even boats and submarines which are mostly affected by the dynamics of fluids.

3D rendering is not only important for visualization of progress and results of simulation, but also for generating inputs for sensors related to vision, such as cameras and depth sensors. Many of these virtual sensors have already been created by various manufacturers, who sell real-world sensors and who, if need be, also provide sensor plugins that can be integrated into a simulation. For more realistic results, some plugins support noise and errors which are common for real-world sensors. Sensors are plugins that are used in simulation and can be created by any user of simulation to meet its needs.

In addition, Gazebo provides a visual of some additional information that could not be seen with a real-world object. Two examples of such include: a value of internal state which can be visualized by a custom GUI plugin, or visualization of rays invisible for a real-world lidar.

2.1 Components of the Robotic System

Robots are made up of many different parts related to software, electronics, and construction [14]. From these, only software components are related to simulation. Others, even if can be simulated, are not the case of the Gazebo simulator and are beyond the scope of this paper.

Software components are further divided into two groups. Components that interact with hardware and components that do not. Those responsible for processing data from sensors or components generating actions for actuators, mainly motors, fall within the first group. The second one includes the logic and behaviour of the whole system which could be developed, debugged, and tested in a simulator.

To make simulation efficient, it is recommended that the two categories remain separate. It should be done using a clear interface with well-defined data types which are passed from one group to another (see Fig. 1). Such an interface is the place where interaction with real-world components can be replaced by a simulation. Yet, in an

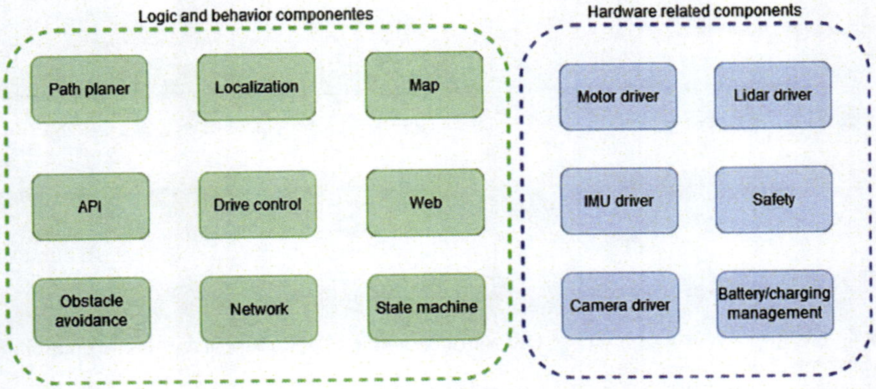

Fig. 1. Some of the components used in a robotics system

abstract hierarchy of a robotic program, it is much more unclear as to where exactly should such a border separating the two be placed.

The goal of this research is to test as much code in simulation as possible and the interface should be as close to the hardware as possible. However, this paper describes a little bit different approach which has been chosen. It is based on a real mobile robot with two control units. One is an Intel NUC computer holding the main logic of the robot and the second one is a PLC which is a real-time system responsible mainly for controlling motors. This means that the hardware interface is placed at the bottom of the abstract hierarchy of logic components and is called the Hardware Abstract Layer (HAL). One drawback of such an arrangement is that some code needs to be implemented twice. Once in a simulated environment and once in PLC. For example, the computation of angular velocities of the robot's wheels must be calculated in real PLC as well as in simulation (Sect. 2.6). This, of course, leads to differences between the real and simulated robots in their driving performance but it still can be used for the purposes of a multi-robot simulation.

Simply put, a simulation contains a virtual PLC among other sensors which are not directly connected to PLC, such as cameras. The simulation communicates with the hardware abstract layer in such a way that the rest of the robotic system is not aware what is running inside the simulation (see Fig. 2).

2.2 Hardware Abstract Layer

The hardware abstract layer is one of the ROS nodes which are used in this project. The main purpose of this node is to provide uniform communication with the hardware. Other nodes higher in hierarchy do not need to know any specific details about the hardware. This design is valuable also in case some hardware components are changed. This is the only one place where the hardware difference will be handled without requiring to change other nodes.

On top of that, it is also a place where communication with simulation is performed. It is achieved by creating a transport object implementation of which depends on the configuration. It can be implemented as an object communicating with a real PLC or as

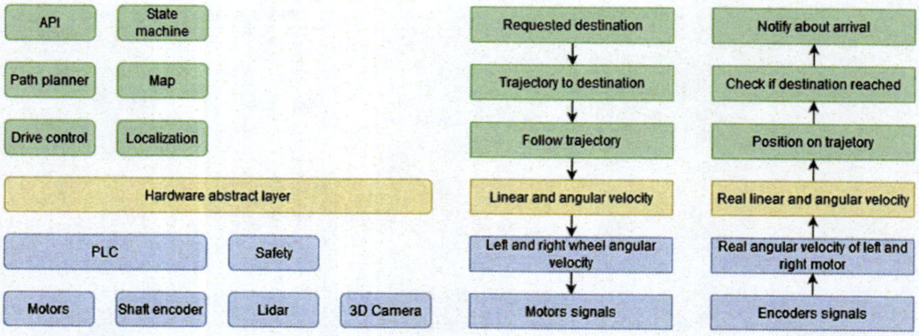

Fig. 2. Abstract layers of a robotic system with an example of layer responsibilities for controlling the movement of mobile robot

an object communicating with the Gazebo simulator (see Fig. 5). This way, the transport object provides a uniform way of sending and receiving PLC data (Sect. 2.4). These data are processed by a set of objects which refer to the specific hardware components. These objects implement an interface of base class *IComponent* which provides two purely virtual methods, namely *collect* and *update* (see Fig. 4). The responsibility of the *collect* method is to gather requests from the upper logic, say, the desired velocity, and set these setpoints into PLC data. Receiving is performed asynchronously by registering a callback method to a specific topic. This callback will be executed whenever a new message arrives on this topic. The *update* method serves as a publisher of the received PLC data. Implementation of *collect* and *update* methods is optional. While some components use only *collect* (LEDs on the robot which do not provide feedback), others use only the *update,* (a battery notifying the system about the information on its remaining battery life). Every component object has access to shared PLC data and operates only with parts that directly relate to specific hardware. These two methods are executed for each component in an infinite loop described in the flow chart diagram below (see Fig. 3).

Such an algorithm has the following advantages:

- The definition of components and a transport object is done at the start of a node according to the value defined in the configuration. This is the only place where the program decides whether it will use simulation or real hardware. The rest of the program flow is the same in both cases.
- Every component encapsulates specific hardware details and hides them from other parts of the system.
- The main node loop algorithm is not aware of any details related to hardware. It only operates with the uniform interface of the component object.
- Transferring data to and from the hardware is also encapsulated and a specific endpoint is hidden inside the implementation of the transport object.

2.3 Communication with the Gazebo

As mentioned above, communication with the Gazebo simulator is performed in the child class of *ITransport* called *TransportSimPLC*. This class overwrites the abstract

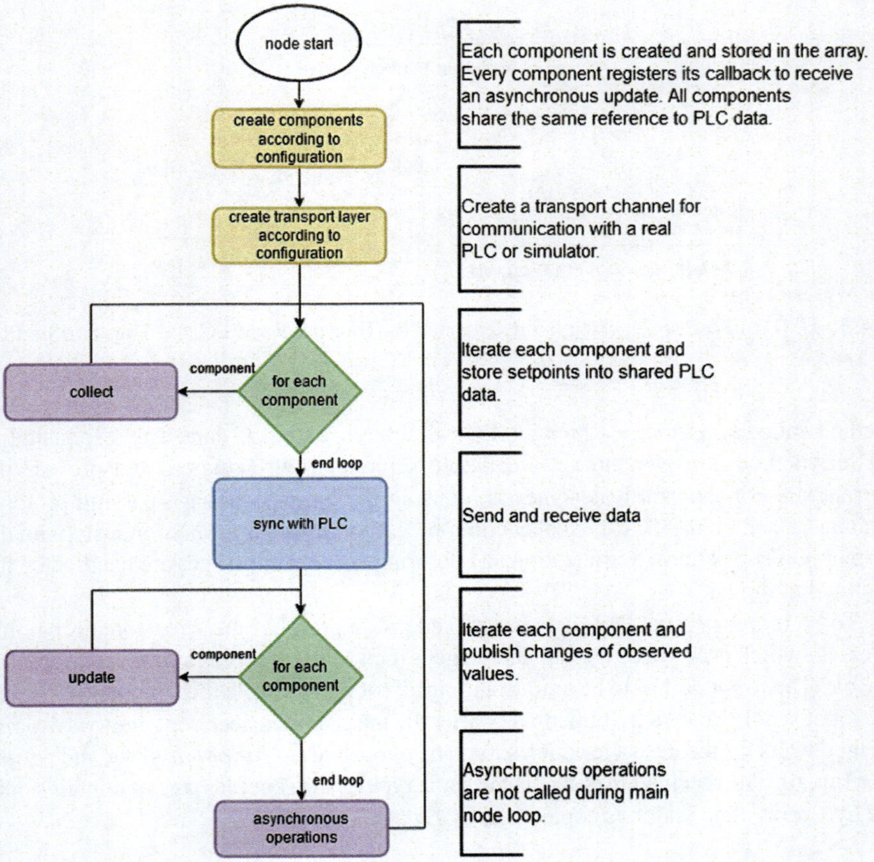

Fig. 3. Flow chart of the main loop in the Hardware abstract layer

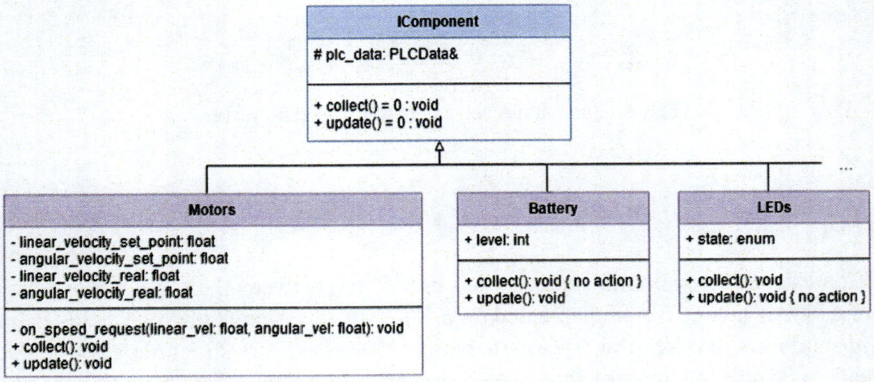

Fig. 4. UML of IComponent base class and inherited child classes

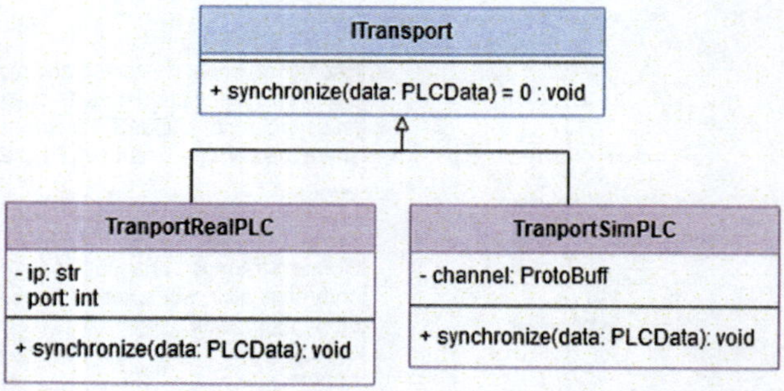

Fig. 5. UML of ITransport which is implemented by TransportRealPLC and TransportSimPLC classes

method called *synchronize* which works with reference to PLC data. The responsibility of this method is to synchronize requested setpoints with actuators and provide the current state of sensors. It is achieved by using the gazebo message-passing protocol which is based on the TCP/IP socket communication. It works asynchronously with the help of pub/sub pattern. Serialization and deserialization of messages are handled by the protobuf compiler.

To be more specific, PCL data are copied into a protobuf message containing only one field which is a byte array. This message is then sent to a topic called / < *robot_id* > */to_plc*. The *robot_id* has to be unique among all robots to enable the simulation identify the robot sending the data. Immediately after the message has been sent, the *synchronize* method waits for the message which should be published to */<robot_id/from_plc>* topic (see Fig. 6). The received message is the same type as the sent message, it contains only the byte array field which is copied into PCL data.

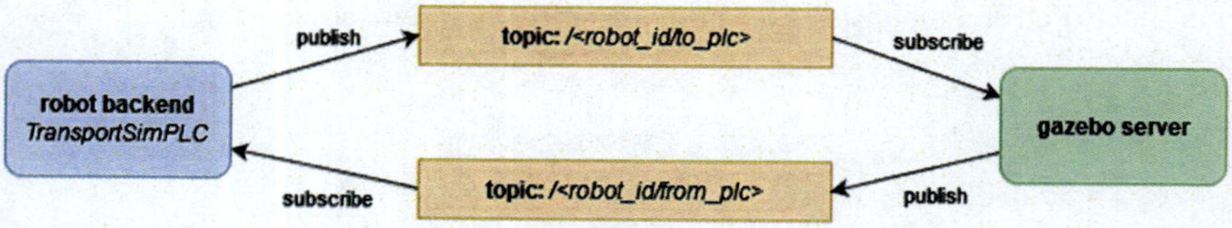

Fig. 6. Send and receive data to the gazebo server

2.4 PLC Data Structure

PLC data act as the main message which is transferred between the simulator and hardware abstract layer. It is implemented as a C++ struct and contains two fields called *to_plc* and *from_plc* (see Fig. 7). Again, both of these fields are structures composed of a header and a set of sub-structures called packets. The header contains only a count of packets. Every packet is identified by id which is located at the beginning of the struct. There are two kinds of packets. One is for data that are sent to PLC and contain requested setpoints, while the second is for data that come from the PLC and contain the actual state of sensors.

The whole structure of PLC data is created at compile time. It would be possible to create this structure dynamically at run time, although it would require creating a buffer where all packets would be copied. Despite flexibility as its main advantage, it is unnecessary because the goal is to always send all setpoints and receive all sensor data in each iteration. Moreover, this approach greatly simplifies serialization and deserialization which is just the casting of a pointer that points to the transferred buffer. It should be mentioned that this structure has to be packed with alignment to one byte and both sides have to share the same declaration of this structure. Otherwise, the received data do not have to fit into the struct type.

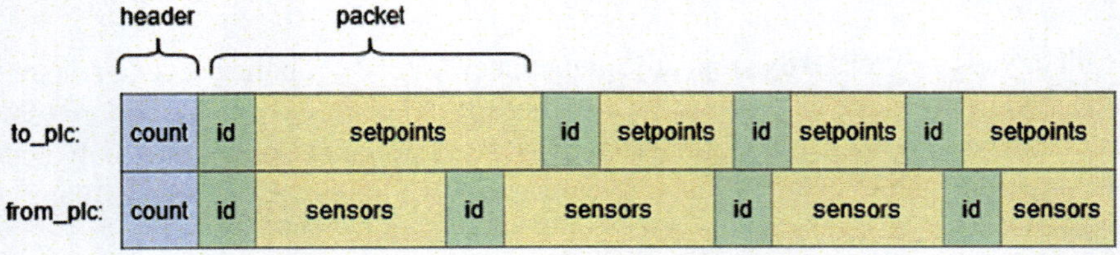

Fig. 7. PLC data structure

2.5 Multiple Robots Inside the Simulator

The Gazebo simulator is usually composed of two parts. A core of the simulation is called The Gazebo server together with an optional client which serves as a GUI interface for the server. The term server is accurate here which means that the Gazebo can run on a separate computer without any graphical interface and the client can be executed on a different computer. If necessary, it might even be a web-based client. This way, there would be no need to install the Gazebo client on a different computer and only use a web browser. Running Gazebo on a dedicated server proves to be ideal for tackling its big computation load.

However, this paper describes an approach without a dedicated server for easier setup in several ways. Firstly, all entities are executed inside a sandboxed environment with the help of docker containers [15] (see Fig. 8). This is a convenient way of running applications in one host system without the risk of them affecting each other. Secondly. The ROS, being is the main robotic framework used in this project, does not support multiple instances of ROS core executed in one host. Therefore, containers provide a convenient way of running multiple mobile robots on one host system. Docker can also create a separate network and assign IP addresses to each container. This results in a configuration where each container has its IP and acts as a separate application, running on dedicated hardware with everything running on one host computer.

Gazebo acts as a virtual PLC for a specific mobile robot. Now, the important question is how to create virtual PLC for each mobile robot in one instance of Gazebo simulation. This would enable one robot to see another robot in the same way as in reality. This has been achieved by creating a model of a robot in the Gazebo with an attached plugin that acts as a virtual PLC.

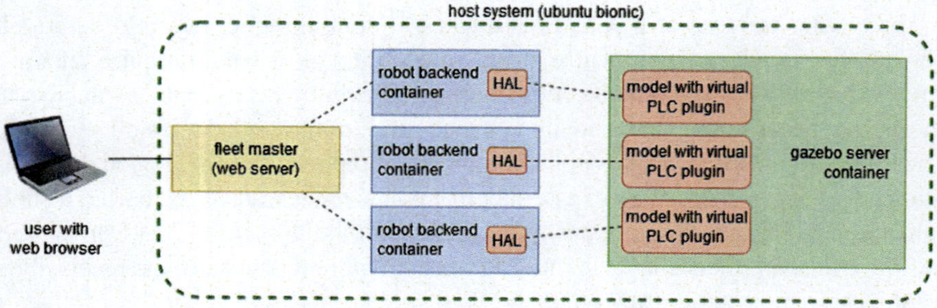

Fig. 8. Host system with docker containers

All models are created dynamically, immediately after the robotic backend is started. It is the first step inside the HAL node. For this, the simulator provides a factory topic, where a message with an SDF definition [16] of the model can be published. SDF model also contains a path to a library that implements a model plugin acting as a virtual PLC as in the example of such a factory message shown down below:

```
<sdf version="1.4">
 <model name="mobile_robot_3">
  <include>
   <uri>model://mobile_robot</uri>
  </include>
  <plugin name="vplc_plugin" filename="libvplc.so">
   <serial_number>3</serial_number>
  </plugin>
 </model></sdf>
```

This will effectively spawn the model inside the simulator. In the next step, HAL starts the synchronization loop, which has already been described (see Fig. 3). Spawning a model also relates to a position where this model will be spawned. This has to be pre-configured in advance to prevent collision with other models. Every model has to be spawned in its dedicated position (see Fig. 9).

2.6 Virtual PLC Inside a Gazebo Model

The responsibility of virtual PLC is the same as real PLC. It is supposed to read data from sensors and control speed of the wheels. However, virtual PLC is placed inside the model of the robot controlled by the gazebo simulator, as opposed to a real PLC placed physically inside the mobile robot. Some of the algorithms, for example spinning of the wheels, are the same in both the real and the virtual PLC. Ultimately, the movement of the robot should be controlled by spinning the left and right wheels and not just always redefining the position of the robot model.

Input for movement is linear velocity v_r and angular velocity ω, the two values have to be transformed according to differential drive equations Eqs. 1 and 2 into angular velocity for the left wheel v_l and angular velocity for the right wheel v_r. L is distance

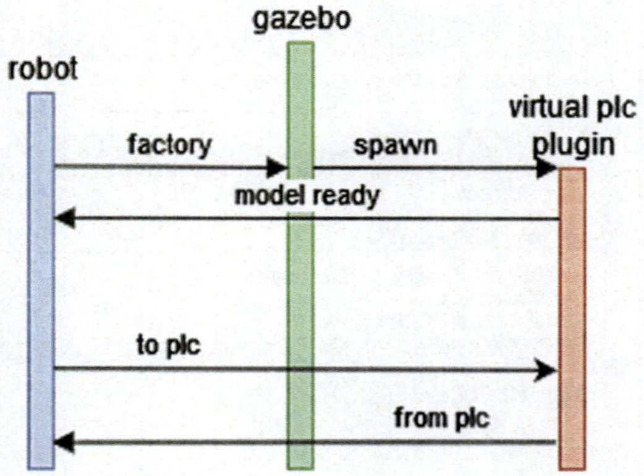

Fig. 9. Process of creating a new model in the Gazebo

between wheels and R is the radius of a wheel [17]. Only these two values are then applied to wheel joints to stir wheels which will move the whole robot. Its main advantage is that it is closer to reality which means that breaking or skidding of the robot may also be simulated.

$$v_r = \frac{2v + \omega L}{2R} \tag{1}$$

$$v_l = \frac{2v - \omega L}{2R} \tag{2}$$

Virtual PLC is implemented as a C++ dynamic library according to the rules for the Gazebo model plugin (see Fig. 10). The main rule is to create a class that will inherit from *ModelPlugin* class. This class has to overwrite the *Load* method which is automatically called when the plugin is loaded by the simulator. This method also provides pointers to an instance of model and world which enables the possibility to stir wheel joints which are accessible through a model pointer. Inside the *Load* method are also created subscriptions for *to_plc* topic and lidar data which are provided by 3rd party plugin for a virtual lidar. Moreover, the topic *from_plc* is advertised and the pre-last step in the process is to register the *OnUpdate* callback method for the Gazebo *update* event which is called with each simulation iteration. *OnUpdate* method changes speed for the left and right wheel joints according to data provided by the *to_plc* topic. The data from sensors to *from_plc* topic, mainly lidar is then published.

Publishing lidar data satisfies the main goal of this project. Since all robots are in one simulated environment, their lidar data also contain the information about other robots which act as obstacles.

3 Results and Discussion

To run the whole simulation, docker containers for the Gazebo simulator and mobile robots must be started. For this, a docker-compose tool can be used. It provides a convenient way of starting multiple containers and holding the whole configuration inside one file, usually called a *docker-compose.yaml*.

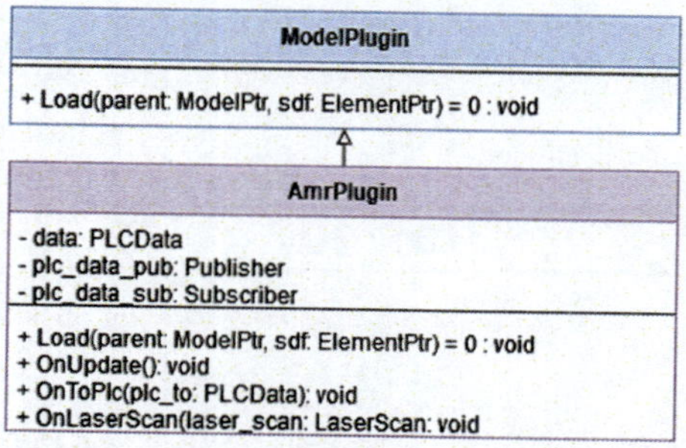

Fig. 10. Gazebo plugin UML class diagram

When all containers are up, only the Gazebo simulator shows GUI. Figure 11 shows a running Gazebo simulation with three mobile runners inside a virtual factory [18]. The fact that all mobile runners are inside one simulated environment fulfils the goal of this paper. Now, all interactions among robots are possible. Robots can see each other or, in case of wrong navigation, crash into each other. To inspect what is seen by robots, a RViz tool can be used. If docker container for a mobile robot is pre-configured to run GUI application, a user can attach into the container's console and start RViz application (see Fig. 12). RViz can visualize standard ROS messages such as lidar data, map, robot model, etc. In this case, lidar data show what is seen by a mobile robot where among static environment which fits into the map contours of other robots are detected. This is initial information required by the obstacle avoidance algorithm which could be tested by this simulation. In this particular case, it is an even moving obstacle but it would be beyond scope of this article to describe the strategy for solving this problem.

Fig. 11. Gazebo simulator with mobile robots inside a simulated factory

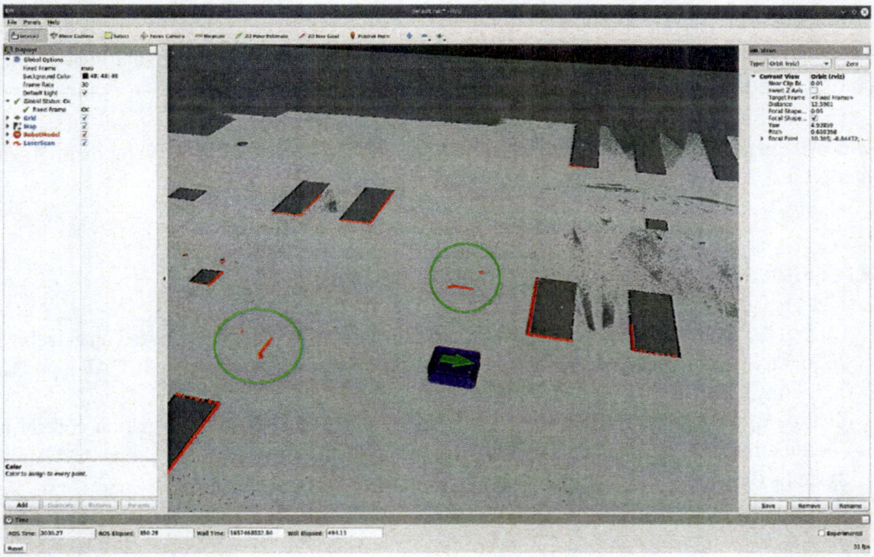

Fig. 12. RViz application which shows what is seen by the robot. Green circles show laser scans reflecting other robots.

Controlling a specific robot was achieved by attaching to its console running inside a docker container and executing ROS command-line tools for sending messages to the robot. This is not very convenient and therefore some fleet master system should be added to this project. Currently, this system is not part of the whole simulation but the design is already known. The fleet master system will be a web application running in a separate docker container. It will interact with all robots by using the ros_bridge protocol which tunnels selected topics outside the robot. This application will be accessible through the web browser and it will show the overall status of all robots. It will also enable sending some commands to robots which will speed up preparation for various test scenarios [19].

4 Conclusion

This paper provides a description of running a multi-robot simulation using ROS and Gazebo. It answers the need for an algorithm development tool related to navigating a fleet of mobile robots which are becoming increasingly common in modern warehouses and factories.

The provided solution does not require any special hardware, other than that for ROS, which is usually a computer with Linux operating system with installed docker. The usage of docker images enables the possibility to run everything on one computer which is very convenient for development and debugging. However, this configuration opens also the possibility to run the simulation on multiple computers to decrease the overall load. These cases require the network to be configured in a way that all containers could connect to a server running a Gazebo simulation. Even in one computer setup, all containers communicate over TCP/IP communication but only on localhost.

Further work will be dedicated to building a management system for the whole fleet. This is not a mandatory part but it will be convenient to be able to control all robots from one place and to see the overall status of the whole fleet.

Acknowledgement. This work was supported by Slovak Research and Development Agency under project APVV-18–0436.

References

1. Huang, Z., Li, F., Xu, L.: Modeling and simulation of 6 DOF robotic arm based on Gazebo. In: 2020 6th International Conference on Control, Automation and Robotics (ICCAR), pp. 319–323. https://doi.org/10.1109/ICCAR49639.2020.9107989
2. Zhang, Y., Mehrjerdi, H.: A survey on multiple unmanned vehicles formation control and coordination: normal and fault situations. In: International Conference on Unmanned Aircraft Systems (ICUAS)
3. Hoshino, S., Seki, H.: Multi-robot coordination for jams in congested systems. Robot. Auton. Syst.
4. LaValle, S., Hutchinson, S.: Optimal motion planning for multiple robots having independent goals. IEEE Trans. Robot. Autom. **14**(6), 912–925 (1998)
5. Digani, V., Sabattini, L., Secchi, C., Fantuzzi, C.: Hierarchical traffic control for partially decentralized coordination of multi AGV systems in industrial environments. In: 2014 IEEE International Conference on Robotics and Automation (ICRA), pp. 6144–6149 (2014, May)
6. ROS Homepage. https://www.ros.org/. Last accessed 12 July 2022
7. Gazebo Homepage. https://classic.gazebosim.org/. Last accessed 12 July 2022
8. Sagitov, A. Gerasimov, Y.: Towards DJI Phantom 4 realistic simulation with gimbal and RC controller in ROS/Gazebo environment. In: 2017 10th International Conference on Developments in eSystems Engineering (DeSE), pp. 262–266 (2017). https://doi.org/10.1109/DeSE.2017.40
9. Watanabe, T., et al.: The Rock-Gazebo integration and a real-time AUV simulation. In: 2015 12th Latin American Robotics Symposium and 2015 3rd Brazilian Symposium on Robotics (LARS-SBR), pp. 132–138 (2015). https://doi.org/10.1109/LARS-SBR.2015.15
10. Takaya, K., Asai, T., Kroumov, V., Smarandache, F.: Simulation environment for mobile robots testing using ROS and Gazebo. In: 2016 20th International Conference on System Theory, Control and Computing (ICSTCC), pp. 96–101 (2016). https://doi.org/10.1109/ICSTCC.2016.7790647
11. Robot fleet simulation. https://github.com/aws-samples/multi-robot-fleet-sample-application. Last accessed 12 July 2022
12. Object Oriented Graphics Rendering Engine Homepage. http://www.ogre3d.org/. Last accessed 12 July 2022
13. The Open Dynamics Engine (ODE) Homepage. https://bitbucket.org/odedevs/ode/src/master/. Last accessed 12 July 2022
14. Jia, J., Chen, W., Xi, Y.: Design and implementation of an open autonomous mobile robot system. In: IEEE International Conference on Robotics and Automation, 2004. Proceedings. ICRA'04. 2004, vol. 2, pp. 1726–1731 (2004). https://doi.org/10.1109/ROBOT.2004.1308073
15. Docker Homepage. https://www.docker.com/. Last accessed 12 July 2022
16. SDFormat Homepage. http://sdformat.org/spec. Last accessed 12 July 2022
17. Chitta, S., Marder-Eppstein, E., Meeussen, W., Pradeep, V., Rodríguez, A., Tsouroukdissian, A., et al.: ros_control: a generic and simple control framework for ROS. J. Open Source Softw. 2(20):456–456 (2017). https://doi.org/10.21105/joss.00456 (hal-01662418)

18. Gazebo world models. https://automaticaddison.com/useful-world-files-for-gazebo-and-ros-2-simulations/. Last accessed 12 July 2022
19. Ros web tools Homepage. http://robotwebtools.org/. Last accessed 12 July 2022

AVL Based on Fusion of IMU, GPS and Camera Sensors

Frederik Valocky[1], Andrej Grnac[1], Milos Orgon[1], and Antonin Bohacik[2](✉)

[1] Slovak University of Technology, FEI, Ilkovicova 3, Bratislava 812 19, Slovakia
orgon@ktl.elf.stuba.sk

[2] Department of Telecommunications, Brno University of Technology, FEEC, Technicka 12, 616 00 Brno, Czech Republic
Antonin.Bohacik@vut.cz

Abstract. The subject of this paper is the fusion of measurement data from IMU (Inertial Measurement Unit), GPS (Global Positioning System) sensors and cameras used in autonomous vehicles to locate an autonomous land vehicle. The experimental part presents the design of a fusion filter, which based on the fusion of data from the IMU, GPS, and camera can determine the position, orientation and speed of the land vehicle. The design of the fusion filter was tested using several simulations in the Matlab environment, the results of which are analyzed at the end of this paper.

Keywords: Autonomous vehicles · GPS · IMU · MVO

1 Introduction

Intelligent transportation faces many challenges that include not only autonomous driving but also tasks like traffic modelling [1,2], cyber-security [3–7], infrastructure resiliency and reliability [8–11], and planning [12–14]. Due to price affordability, high accuracy and global signal coverage, GPS is the most widely used sensor for navigating and locating vehicles, robots, people, and other various objects on Earth [15,16]. The great economic power of car manufacturers is increasingly pushing for the development and improvement of the capabilities of vehicles and the systems responsible for their autonomous driving, which is considered to be the standard of the future. This causes more and more driving tasks to be transferred from the driver or pilot to the vehicle, respectively a system which is able to independently, automatically and constantly observe the surrounding environment, determine the position of the vehicle and, on the basis of this data, evaluate and anticipate possible situations. Subsequently, according to the programmed logic, the system should make such decisions as to prevent any danger on the road, but at the same time manage to transport a vehicle with or without passengers from point A to point B as efficiently as possible.

R. Silhavy et al. (Eds.): CoMeSySo 2022, LNNS 596, pp. 352–375, 2023.
https://doi.org/10.1007/978-3-031-21435-6_31

Due to the increasing density of vehicles on the roads, road infrastructure is becoming increasingly complex, especially in dense urban environments, which abound in tall buildings along roads and contain tunnels, even tunnel junctions, various underpasses, overpasses, bridges, and other objects that obstruct satellite signals. These objects may interfere with the direct view of the GPS receiver on any of the at least 4 satellites needed to accurately locate the receiver in the vehicle, causing this GPS receiver to temporarily lose its position awareness.

The most common example is an underground junction in a tunnel. If the driver is using on-board navigation, in places such as a tunnel where no GPS signal is available, the GPS-based navigation system has no vehicle position awareness, at most it only assumes the speed of the vehicle when passing through the tunnel and assumes that it will follow the route planned by the navigation. However, if for any reason the driver turns at an underground junction differently than the navigation expected, the navigation will not capture this information until the vehicle leaves the tunnel and the receiver in the vehicle reestablishes connection to the 4 satellites. If the navigation only serves as an assistant, but the control is up to the person, there will be no problem, maybe just a little detour. However, if the vehicle were to drive autonomously and the said unpredictable situation occurs in the tunnel (if the vehicle was not precisely located due to the unavailability of the GPS signal), the vehicle would not be able to respond adequately, which would of course pose a high safety risk. Therefore, it is very necessary that vehicles, as well as any other autonomous systems, be able to accurately track their movement even in situations of short-term unavailability of GPS signals.

Now, it is already a standard in the automotive industry to include several cameras in the equipment of vehicles, which, depending on their design, provide many assistance elements for the driver. However, the camera can also be used to monitor and calculate the vehicle's own motion by capturing an image from one or more cameras located on the vehicle – this method is called visual odometry (VO). This is a method that can temporarily replace the GPS system when locating the vehicle in situations where the GPS signal is unavailable. In our paper, we proposed a method of fusing data from several sensors located on a land vehicle. The paper describes the various methods and tools cooperating in a robust way of locating the vehicle, which provides reliable and sufficiently accurate data needed for autonomous vehicle control even in an environment with insufficient GPS signal coverage through the implementation of visual odometry. In the following part of the paper there is a description of the simulation of the proposed design, performed in the Matlab environment.

2 Matlab Simulation

The Matlab programming environment from MathWorks provides very specific tools and functions from various technical areas in the form of various additional "toolboxes", enabling the implementation, simulation, analysis, and evaluation of a really wide range of system designs and algorithms. All the functionalities and tools provided by Matlab, whether basic or advanced are also sufficiently documented and described in the official manuals, which allows even less experienced programmers to work with them. The program used as part of the simulation, was also created in this environment, and uses several tools from the "Sensor Fusion and Tracking Toolbox" extension. It consists of a main function called VOINScar.m and several auxiliary functions, which are described in more detail below. In this article, only fragmentary parts of the code are presented for illustration purposes.

The program as a whole simulates the system of an autonomous vehicle, which has a GPS receiver, a monocular camera located on the vehicle, that captures images in the direction of the vehicle movement and an IMU (Inertial Measurement Unit) containing 3-D accelerometers and gyroscopes. The data from the individual sensors are sampled at different frequencies, while the Kalman error status filter processes this data, weighs it and then uses it to estimate the current position, speed and orientation of the vehicle. The GPS receiver provides the vehicle position and speed information, the monocular camera can provide a relatively reliable vehicle position and orientation information for a short time using the visual odometry method, and the IMU provides basic information from inertial sensors such as accelerations and angular velocities.

As part of the analysis and evaluation of the results of the presented proposal, attention is primarily paid to locations, i.e. the determination of the vehicle position, but the accuracy of the vehicle orientation estimation provided by the filter is also monitored.

2.1 Driving Scenario

In the Matlab environment, the driver scenario is a programming object that represents the 3-D arena, or the environment, which includes vehicles and their trajectories, roads, buildings, pedestrians and other aspects of road traffic. This object is suitable for modeling realistic traffic situations and for generating synthetic sensor data and measurements, which can then be used to test the proposed control systems and fusion algorithms. Within this work, driving scenarios 1 and 2 were created, the visual interpretation of which can be seen from a bird's eye view in Figs. 1 and 2, and from a third-person perspective behind the vehicle in Fig. 3.

For the purpose of our simulation, a simplified driving scenario was created, sufficient to test the proposed system. The scenario contains several road aspects, such as the road the vehicle travels, the buildings around the road, which in the actual deployment of visual odometry would be important objects for the in-image feature comparison algorithm, but in virtual simulation they only complete the environment and have no direct impact on simulation.

The most important elements of the driving scenario object are the subobjects of the two vehicles, the definition of which is shown in Listing 1.1. The first vehicle shown in Fig. 3 as a blue cube, is a vehicle representing a real vehicle in traffic (so-called Ground Truth Vehicle), i.e. the characteristics of this vehicle, such as its trajectory, position, orientation and speed, are the reference characteristics that would describe the actual condition of the real vehicle in traffic. The second vehicle shown in Fig. 3 in green, represents an Estimated Vehicle, whose properties are given by the output of the fusion filter, i.e. this green cube shows how the fusion filter we designed estimates the position, orientation, and speed of the blue (real) vehicle, based on data received from modeled GPS, IMU and camera sensors.

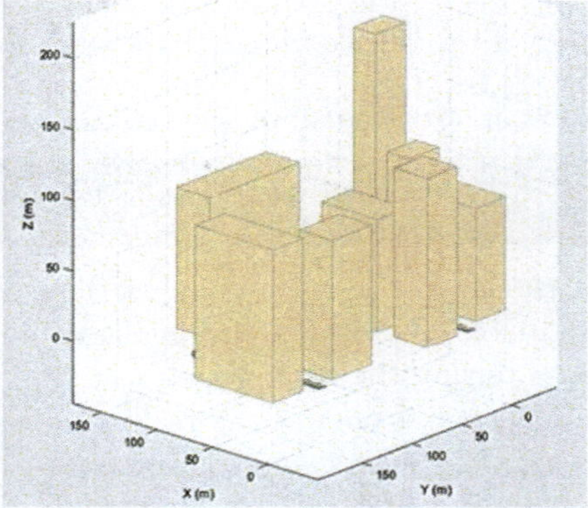

Fig. 1. Driving scenario 1

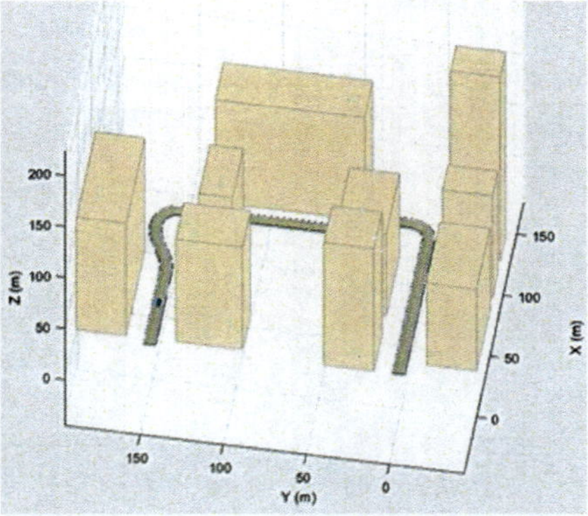

Fig. 2. Driving scenario 2

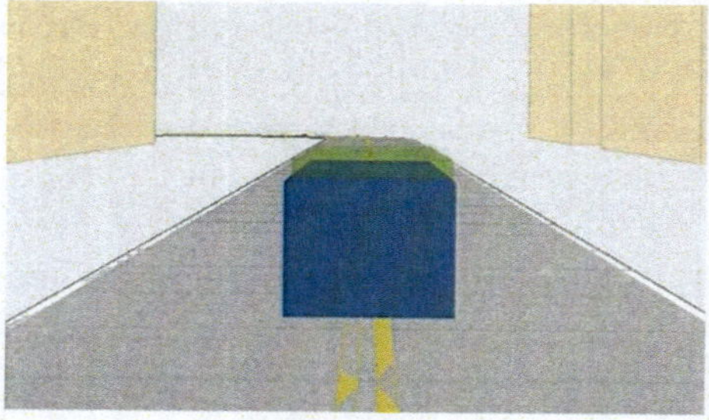

Fig. 3. A view of the vehicle from a third-person perspective

Listing 1.1. Creating a scenario with a real vehicle and an estimated vehicle in Matlab

```
scene = drivingScenario;
groundTruthVehicle = vehicle(scene, 'PlotColor', [0 0.44701 0.7410]);
estVehicle = vehicle(scene, 'PlotColor', [0.4660 0.6740 0.1880]);
```

It is worth noting that the more the green cube overlaps with the blue one, the more accurate the estimate of the proposed fusion algorithm is, with the accuracy varying at different times during the simulation.

The whole driving scenario was created manually, i.e. by code. However, Matlab also provides an interactive graphical interface *drivingScenarioDesigner* that allows you to create driving scenarios. As can be seen in Listing 1.2, the vehicle trajectory object waypoint. Trajectory is created using the so-called waypoints, which are defined as points in a three-dimensional matrix with x, y, z coordinate values. To create a trajectory using waypoints, it is necessary to determine the travel time to the given points, calculated from the beginning of the simulation, as well as the speed in [m/s] at which the vehicle moves in the sections between the individual waypoints.

Listing 1.2. Defining the actual vehicle trajectory (Ground Truth) in Matlab

```
sampRate = 100;
wayPoints =[0   0   0;
            50  0   0;
            100 0   0;
            100 0   0;
            100 50  0;
            100 100 0;
            50  150 0;
            0   150 0];
t = [0 10 20 30 40 50 60 70].';
speed = 15;
velocities =[speed  0     0;
             speed  0     0;
             speed  0     0;
             0      speed 0;
             0      speed 0;
             0      speed 0;
             -speed 0     0;
             -speed 0     0;
             -speed 0     0];
traj = waypointTrajectory(wayPoints, 'TomeOfArrival', t, 'Velocities', velocities,
'SampleRate', sampleRate);
```

The *supportPopulateScene* function is used to populate the arena with objects such as roads, buildings, and their rendering:

supportPopulateScene(scene, groundTruthVehicle);

which is called with two arguments; the first is the object of the driving scenario itself, and the second argument is the object of the real, blue vehicle (Ground Truth Vehicle). The method of creating building objects can be seen in Listing 1.3, where the parameters of a particular building are defined – position, dimensions, and color.

Listing 1.3. Creating buildings in the Matlab code

```
buildingColor = [250 235 215] / 255;
actor(scene, ...
    'Length', 50, ...
    'Width', 30, ...
    'Height', 80, ...
    'Position', [25 -30 0], ...
    'PlotColor', buildingColor);
```

3 Fusion Filter

The base of the system is a fusion filter, which processes the measurement data from the sensors, received at different frequencies, and on their basis produces an estimate of the position, orientation, and speed of the vehicle. In this simulation, an extended Kalman error filter was used, which implements the fusion of data from the IMU, GPS and camera with the so-called "loosely coupled" method.

3.1 Loosely-Coupled Method

The Loosely-coupled Kalman filter method represents the most widely used architecture for integrating data from multiple sensors; in navigation applications, this is the most common fusion of INS / GNSS sensors. With this method, position and velocity data obtained from GNSS or another low-frequency and more accurate sensor are taken as more reliable. But since they are only available at frequencies several times lower than those from the IMU, they are used to update the system states whose values are estimated by the filter based only on measurements from IMU. In order to achieve higher accuracy of this navigation method, error states are returned to the INS mechanization algorithm by means of feedback, which suppresses or mitigates the cumulative error rate affecting the IMU output. A block diagram of this method is shown in Fig. 4 [17].

The second architecture for fusion data from multiple sensors using a Kalman filter is the Tightly-Coupled Method. Although the results of the tight connection method achieve higher accuracy, the computational complexity rate is significantly lower when a loosely coupled method is used, which reduces the energy consumption of such a system, while achieving a satisfactory level of accuracy for most navigation systems, and we all know that energy consumption is a hot topic with autonomous vehicles. In works [17] and [18], the authors discuss the performance and suitability of both methods, but the results show that the loosely coupled method is more advantageous for applications in land vehicles.

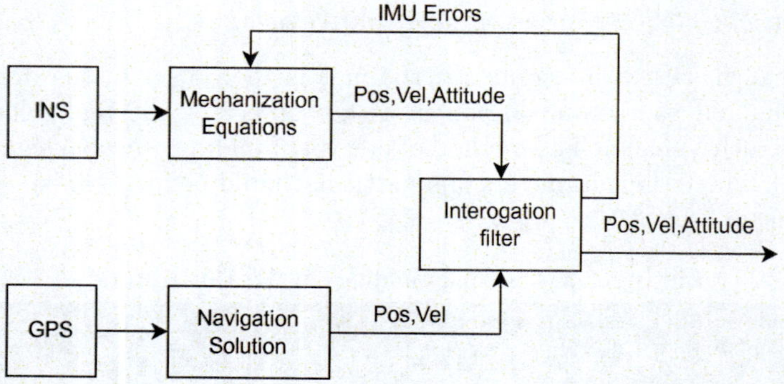

Fig. 4. Loosely-coupled method

3.2 Orientation as Quaternion

Before the next part, it is necessary to state that the orientation of the vehicle in this simulation is in the form of a quaternion, so is the object of the Kalman filter defined in Matlab. Quaternion is a hypercomplex number consisting of 4 members, used in three-dimensional rotations and orientations. The quaternary number is expressed in the form $\boldsymbol{a} + \boldsymbol{b}\mathbf{i} + \boldsymbol{c}\mathbf{j} + \boldsymbol{d}\mathbf{k}$, where **a**, **b**, **c** and **d** are real numbers, and **i**, **j** and **k** are unit quaternions, meeting the equation:

$$i^2 = j^2 = k^2 = ijk = -1 \tag{1}$$

All rotations in the three-dimensional view can be described by the axis of rotation and the angle around this axis. The advantage of quaternions over rotating matrices is a simpler interpretation of the axis and angle of rotation. For better orientation, let us look at the example shown in Fig. 5 and consider the point in R^3. If we want to rotate a point, we define the axis of rotation and the angle of rotation.

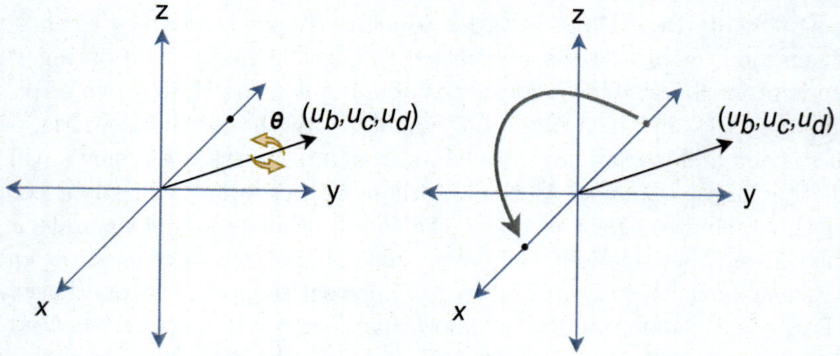

Fig. 5. Quaternion expression of rotation

The representation of a given rotation in the form of a quaternion can be expressed as:

$$q = cos(\Theta/2) + sin(\Theta/2)(u_b i + u_c j + u_d k) \tag{2}$$

where Θ is the angle of rotation and $[u_b, u_c, u_d]$ is the axis of rotation.

For the purposes of this simulation, it is sufficient to understand only the basics of the quaternary representation of the orientation, because Matlab performs calculations and transformations between Euler angles and quaternions automatically and working with them is as simple as possible.

4 Matlab Implementation

To implement the above fusion filter, the ***insfilterErrorState*** object was used in the Matlab environment, which combines data from IMU, GPS and monocular visual odometry (MVO), and estimates vehicle conditions with respect to the ENU reference framework. This object uses a 17-element status vector in which it monitors the orientation, speed, position of the vehicle, distortion of IMU sensors, i.e. gyroscopes and accelerometers, and the monocular visual odometry (MVO) scaling factor. This filter object provides many functions, of which the following three are used in this simulation.

The ***predict*** function is a prediction function whose inputs are accelerometer and gyroscope measurements from the IMU. The prediction function is called whenever data from the IMU is sampled. The prediction estimates the value of the vehicle states one epoch ahead and updates the covariance matrix of the filter error states. This function predicts values for vehicle orientation, speed, and position.

The ***fusemvo*** function is a function responsible for correcting the status vector based on camera data. It is called every time data from MVO are sampled, so its inputs are data from visual odometry. Kalman's gain is calculated for these inputs, which weights the individual inputs based on their uncertainty. Thus, this function, like the prediction function, updates the covariance matrix of error states, but this time based on estimates from visual odometry and with respect to Kalman's gain. The MVO updates the status values of the vehicle's orientation and position.

The ***fusegps*** function is a function responsible for correcting the status vector based on data from the GPS receiver. The function is called every time GPS data is sampled. This feature also updates the covariance matrix of error conditions, with Kalman gains at high values for GPS data, with at least 4 satellites visible. GPS updates vehicle status and speed status values.

When creating a filter (***filt*** variable), the sampling frequency of the IMU unit and the reference frame with which the filter interprets its output are defined. The ***supportInitialize*** function shown in Listing 1.4, is called only once, immediately after the filter is created. The initialization values are based on the ***Ground Truth*** trajectory of the vehicle.

Listing 1.4. Matlab filter state initialization function

```
function supportInitialize(filt, traj)

% Read initialization values of positions, orientations, speeds and reset internal states
[pos, orient, vel] = traj();
reset(traj);

% Setting the initialization values of filter states
filt.State(1:4) = compact(orient(1)).';
filt.State(5:7) = pos(1,:).';
filt.State(8:10) = vel(1,:).';

% Gyroscopes noise and MVO scaling factor adjustment
filt.StateCovariance(10:12,10:12) = 1e6;
filt.StateCovariance(end) = 2e2;
end
```

4.1 IMU Sensor Model

The ***imuSensor*** system object from the "***Sensor Fusion and Tracking Toolbox***" extension was used to simulate the IMU unit measurements. This object made it possible to model an IMU unit containing individual combinations of gyroscopes, accelerometers, and magnetometers. For these sensors, it was possible to individually define their properties through sub-objects such as ***gyroparams*** for gyroscopes, ***accelparams*** for accelerometers and ***magparams*** for magnetometers.

In our simulation, the IMU contains (as is typical for land vehicles) only three gyroscopes and three accelerometers, whose 6 axes capture the movement of the vehicle in the 3-D space. The IMU sensor model was defined directly in the main function of the program, as seen in Listing 1.5. Defined sensor parameter values shown in Listing 1.6, reflect the values of consumer-friendly, affordable IMU sensors based on the MEMS technology.

Listing 1.5. IMU sensor model

```
imu = imuSensor('SampleRate', sampleRate, 'ReferenceFrame', 'ENU');

% Accelerometers
imu.Accelerometer.MeasurementRange = 19.6; % m/s·s
imu.Accelerometer.Resolution = 0.0024; % m/s·s/LSB
imu.Accelerometer.NoiseDensity = 0.01; % (m/s·s/sqrt(Hz)

% Gyroscopes
imu.Gyroscope.MeasurementRange = deg2rad(250); % rad/s
imu.Gyroscope.Resolution = deg2rad(0.0625); % rad/s/LSB
imu.Gyroscope.NoiseDensity=deg2rad(0.0573);%(rad/s)/sqrt(Hz)
imu.Gyroscope.ConstantBias = deg2rad(2); % rad/s
```

Listing 1.6. IMU sensor model properties

```
imu = imuSensor with properties:
          IMUType: 'accel-gyro'
       SampleRate: 100
      Temperature: 25
    Accelerometer: [1x1 accelparams]
        Gyroscope: [1x1 gyroparams]
     RandomStream: 'Global stream'
```

4.2 GPS Receiver Model

The ***gpsSensor*** system object, which is part of the "***Sensor Fusion and Tracking Toolbox***" extension, is used in Matlab to model data from the GPS receiver. In our simulation, the GPS receiver model was defined in the main function, as shown in Listings 1.7 and 1.8.

Listing 1.7. GPS receiver model

```
gpsSampleRate = 10;
gps = gpsSensor('SampleRate',gpsSampleRate,'ReferenceFrame', 'ENU');
gps.DecayFactor = 0.5; % Random walk noise parametr
gps.HorizontalPositionAccuracy = 1.0;
gps.VerticalPositionAccuracy = 1.0;
gps.VelocityAccuracy = 0.1;
```

Listing 1.8. GPS receiver model properties

```
gps = gpsSensor with properties:
                  SampleRate: 10              Hz
           ReferenceLocation: [0 0 0]         [deg deg m]
  HorizontalPositionAccuracy: 1               m
    VerticalPositionAccuracy: 1               m
            VelocityAccuracy: 0.1             m/s
                RandomStream: 'Global stream'
                 DelayFactor: 0.5
```

The object thus defined simulated the vehicle position and speed data obtained from a GPS receiver, which were sampled every tenth of a second and related to the ENU reference frame. Other parameters of the object determined the properties of the noise in the output signal, and the initial position of the vehicle, which in real terrestrial applications was determined in terms of longitude, latitude, and altitude. In our case, however, the origin was at the point [0, 0, 0] of the 3-D space in which our driving scenario was located.

4.3 Monocular Visual Odometry Model

The system of monocular visual odometry is modeled in the simulation using the ***paramsVO*** structure, which simulates MVO parameters. A structure is a data type that groups related data based on data fields. Each field can contain any type of data. The data in the structure is accessed by a dot notation.

The parameters of the object, which can be seen in Listing 1.9, model a system of monocular visual odometry based on an image comparison and tracking algorithm. The ***scale*** parameter represents the unknown scale between successive image frames captured by the camera. As the cameras cannot reliably determine this scale, it is the main limiting factor for the camera sensors as such. Other parameters model the deviations in the visual odometry measurements, these deviations are a combination of white noise and first-order Gauss-Markov process noise.

Listing 1.9. MVO model properties

```
paramsVO = struct with fields:
        scale: 2
       sigmaN: 0.1390
```

```
        tau: 232
     sigmaB: 1.1576
  driftBias: [0 0 0]
```

The function ***supportVisualOdometryModel***, shown in Listing 1.10, is used to calculate simulated measurements produced by MVO. When it is called, this function extracts the model parameters from the ***paramsVO*** structure, calculates the deviations, which are then used to correct the vehicle position and orientation data. A more detailed description of the monocular visual odometry algorithm can be found in [19–21].

Listing 1.10. MVO model function (Matlab)

```
function [posVO, orientVO, paramsVO] ...
  = supportVisualOdometryModel(pos, orient, paramsVO)

% Extraction of model parameters
scaleVO = paramsVO.scale;
sigmaN = paramsVO.sigmaN;
tau = paramsVO.tau;
sigmaB = paramsVO.sigmaB;
sigmaA = sqrt((2/tau) + 1/(tau * tau)) * sigmaB;
b = paramsVO.driftBias;

% Calculation of deviations
b = (1 - 1/tau). * b + randn(1,3) * sigmaA;
drift = randn(1,3) * sigmaN + b;
paramsVO.driftBias = b;

% Calculation of output data from MVO
posVO = scaleVO * pos + drift;
orientVO = orient;
end
```

5 Simulation Setup

As can be seen in Listing 1.11, several variables are defined as pre-fusion sensor noise parameters before the simulation loop, sampling frequencies for GPS and NGO sensor models, with a sampling frequency for camera data of 25 Hz, and a sampling frequency for GPS data of 10 Hz. The simulation duration of 70 s is also defined in this section together with the ***supportPreallocateData*** function shown in Listing 1.12. This function is responsible for allocating data space to all variables in which the simulation results will be recorded for subsequent interpretation, whether graphical or numerical.

Listing 1.11. Simulation preparation and setup (Matlab)

```
% Setting the simulation run time
numSecondsToSimulate = 70;
numIMUSamples = numSecondsToSimulate * sampleRate;

% Definition of sampling frequency of data from MVO
imuSamplesPerCamera = 4;
numCameraSamples = ceil(numIMUSamples/imuSamplesPerCamera);

% Definition of sampling frequency of data from GPS
imuSamplesPerGPS = (sampleRate/gpsSampleRate);
assert(imuSamplesPerGPS == fix(imuSamplesPerGPS) ...
```

```
'GPS sampling rate must be an integer factor of IMU sampling rate.'
numGPSSamples = (numIMUSamples/imuSamplesPerGPS);

% Definition of data space reallocation for results interpretation
[pos, orient, vel, acc, angvel, posVO, orientVO, posGPS, velGPS, posEst,
orientEst, velEst] ...
  = supportPreallocateData(numIMUSamples, numCameraSamples, numGPSSamples);

% Setting the measurement noise parameter for data fusion from MVO
RposVO = 0.1;
RorientVO = 0.1;

% Setting the measurement noise parameter for data fusion from GPS
RvelGPS = gps.VelocityAccuracy.^2;
RposGPS = gps.HorizontalPositionAccuracy.^2;
```

Listing 1.12. Allocation of variables to record results (Matlab)

```
function [pos, orient, vel, acc, angvel, posVO, orientVO, posGPS, velGPS,
posEst, orientEst, velEst] ...
  = supportPreallocateData(numIMUSamples, numCameraSamples, numGPSSamples)

% Own trajectory of the vehicle (Ground Truth)
pos = zeros(numIMUSamples, 3);
orient = quaterion.zeros(numIMUSamples, 1);
vel = zeros(numIMUSamples, 3);
acc = zeros(numIMUSamples, 3);
angvel = zeros(numIMUSamples, 3);


% MVO output
posVO = zeros(numCameraSamples, 3);
orientVO = quaterion.zeros(numCameraSamples, 1);

% GPS output
posGPS = zeros(numGPSSamples, 3);
velGPS = zeros(numGPSSamples, 3);

% Filter output
posEst = zeros(numIMUSamples, 3);
orientEst = quaterion.zeros(numIMUSamples, 1);
velEst = zeros(numIMUSamples, 3);
end
```

5.1 Carrying Out the Simulation

The simulation itself works as a multi-step cycle. The cycle is created using the overlay command ***for***. This main loop is indexed by the number of samples from the IMU, therefore the core of this function is executed exactly as many times as the signal from the IMU sensor will be sampled during the simulation. Since the IMU sensor operates in this simulation at a frequency of 100 Hz, and the duration of the simulation is exactly 70 s, the number of samples from the IMU is exactly 7,000. For fusion purposes, samples from GPS sensors and cameras are cameras sampled exactly 1,750 times and GPS data will be sampled exactly 700 times, i.e. one sample of GPS data per 10 samples of IMU data.

At the very beginning of the cycle, data modeling the actual characteristics of the vehicle (Ground Truth) are always generated one step in advance, namely its position, orientation, speed, acceleration, and angular velocity. All this data is generated from the object of the actual trajectory of the vehicle. Using these actual vehicle properties, measurements of individual sensors are subsequently generated. In the next step, data from the IMU are sampled, which are then used as parameters of the prediction function, which estimates the values of the filter states based on these accelerometer and gyroscope measurements.

The steps described above are performed one after the other, but the following parts of the cycle are performed conditionally, depending on whether we want to apply data from all sensors in the simulation. This means all IMU and GPS and cameras, or if we want to use only GPS or only data from MVO as an assistance sensor for IMU. This is solved in the program using variables ***useGPS*** or ***useVO***, which acquire only ***true*** or ***false***, and are then conditionally checked in the ***if*** commands in the loop. This allows us to record and compare the results obtained in situations where data from all sensors are available, or vice versa, when one of them is unavailable.

The fusion of MVO and GPS data is performed in two separate conditional pieces of code, which are performed only when two conditions are met. The first condition is described above, i.e. the ***useVO*** variable or ***useGPS*** to be true and then the sensor is applied. The second condition for performing data fusion is the availability of a new sample of data from a given sensor. When an MVO sensor is applied and new data samples from this sensor are available, these samples are preprocessed and weighted by the Kalman gain in the ***supportVisualOdometryModel*** helper function, then the ***fusemvo*** function is called to perform the data fusion itself tracking the position and orientation of the vehicle.

The same procedure occurs even if a GPS sensor is applied and new data samples from this sensor are available, but no auxiliary function is used to process these samples, the measurement samples are generated directly using the GPS sensor model. The data fusion is performed by the called ***fusegps*** function, which updates the values of the filter states tracking the position and speed of the vehicle based on new GPS measurements.

The filter output and the actual vehicle values are then projected in real time onto the properties of the vehicle objects in the driving scenario using the ***supportUpdatePose*** function. In this way, the actual movement of the vehicle (blue cube) and the movement of the vehicle estimated by the fusion filter (green cube) are continuously visualized, as seen in Listing 1.13. The Matlab filter loop code and ***supportUpdatePose*** function are shown in Listing 1.13.

Listing 1.13. Creating a simulation loop – the core of the fusion filter (Matlab)

```
% Variables for recording the maximum deviation
orientErrMax = 0;
posErrMax = [0, 0, 0];

% Indexation
gpsIdx = 1;
cameraIdx = 1;

for i = 1:numIMUSamples
  % Generation data of real trajectory
  [pos(i,:), orient(i,:), vel(i,:), acc(i,:), angvel(i,:)] = traj();

  % Generation of accelerometer and gyroscope data from Ground True trajectory
  [accelMeas,gyroMeas]=imu(acc(i,:),angvel(i,:),orient(i));
```

```
  % State prediction based on IMU measurements
  predict(filt, accelMeas, gyroMeas);

  if (i == mod(i, imuSamplesPerCamera)) && useVo
    % Generating a sample of data from the MVO module
    [posVO(cameraIdx,:),orientVO(cameraIdx,:),paramsVO]= ...
    supportVisualOdometryModel(pos(i,:), orient(i,:), paramsVO);

    % Statuses correction / updating based on new samples from MVO
    fusemvo(filt,posVO(cameraIdx,:), RposVO, orientVO(cameraIdx), RorientVO);
    cameraIdx = cameraIdx + 1;
  end

  if (0 == mod(i, imuSamplesPerGPS)) && useGPS
    % Generating a sample of data from the GPS module
    [posGPS(gpsIdx,:), velGPS(gpsIdx,:)] = gps(pos(i,:), vel(i,:));

    % Statuses correction / updating based on new samples from GPS
    fusegps(filt,posGPS(gpsIdx,:),RposGPS,velGPS(gpsIdx,:), RvelGPS);
    gpsIdx = gpsIdx + 1;
  end

  [posEst(i,:), orientEst(i,:), velEst(i,:)] = pose(filt);

  % Vehicle visualization update in the driving scenario
  supportUpdatePose(estVehicle, posEst(i,:) velEst(i,:), orientEst(i));
  supportUpdatePose(groundTruthVehicle, pos(i,:) vel(i,:), orient(i));

    % Calculation of deviations and errors
    if useErrScope
      orientErr = rad2deg( ..., dist(orientEst(i,:), orient(i,:)));
      posErr = posEst(i,:) - pos(i,:);
      errscope(orientErr, posErr(1), posErr(2), posErr(3);
        if orientErr > orientErrMax
          orientErrMax = orientErr;
        end
        if abs(posErr(1)) > abs(posErrMax(1))
          posErrMax(1) = posErr(1);
        end
        if abs(posErr(2)) > abs(posErrMax(2))
          posErrMax(2) = posErr(2);
        end
        if abs(posErr(3)) > abs(posErrMax(3))
          posErrMax(3) = posErr(3);
        end
    end
    % Location viewer update
    if usePoseView
       viewer(posEst(i,:), orientEst(i,:), ... pos(i,:), orientEst(i,:));
    end
  % Update graphs and driver scenario visuals
  updatePlots(scene);
  drawnow limitrate;
end
```

5.2 Interpretation and Analysis of Results

The behavior of the filter and the outputs it produces are interpreted through several visualization elements and graphs during the simulation. The simulation results are described for four model situations, which show us whether the use of a camera sensor implementing the MVO algorithm is suitable for locating a land vehicle.

As already mentioned, during the simulation run, the movement of the vehicle in the driving scenario is monitored in real time by means of several visualizations, as can be seen e.g. in Figs. 1 and 2. The visualization in Fig. 3 interprets the observable difference in position, orientation and speed between a real vehicle (blue cube) and an estimated vehicle (green cube), the properties of which are the result of the work of the fusion filter. Figure 6 shows a visualization of the vehicle axes (actual and estimated) which, during the simulation, interpret the rotations of these axes due to the movement of the vehicle, the difference between the actual orientation and the orientation estimated by the filter, which is being quantified in the following sections. The Matlab system function called "***HelperOrientationViewer – Orientation visualization***" was used to visualize the rotations during the simulation, which is available as an example.

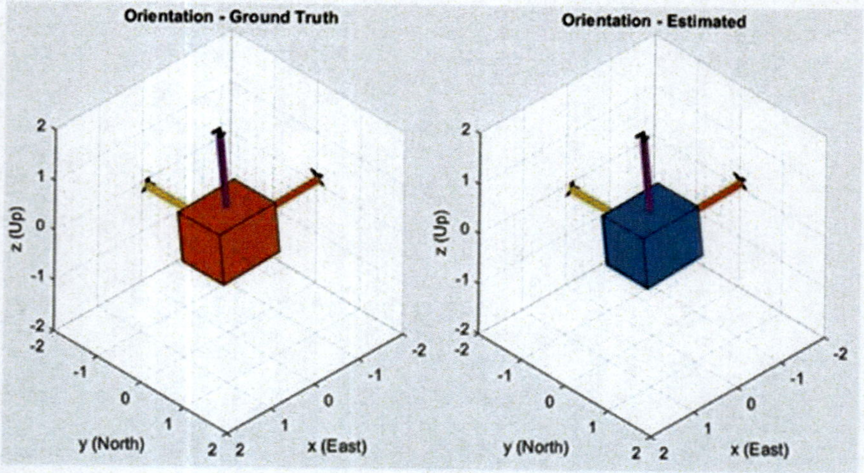

Fig. 6. Actual vehicle orientation and estimated vehicle orientation

In Figs. 7, 8, 9 and 10 four subdivided graphs are shown which, in real time, each after one cycle of the simulation loop, record the difference in the orientation and position of the estimated vehicle from the actual one. The first graph shows the orientation deviation in degrees. The second, third and fourth graph show the deviation in position in meters for x, y, and z-coordinate axis, i.e. in the space in which the vehicle moves. The creation of these graphs is based on the Matlab system function "***HelperScrollingPlotter – Streaming plotting***".

5.3 Localization Based on Data from the IMU

In a situation where data is not available from any sensor (neither GPS nor MVO) responsible for updating the predicted states, and the filter predicts states only on the basis of data obtained from the IMU, information about position and

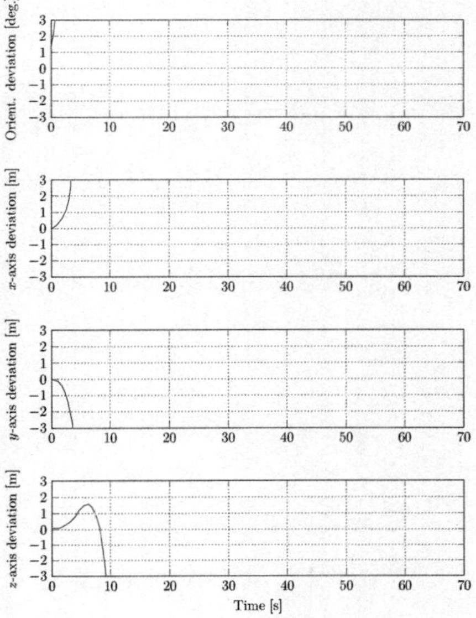

Fig. 7. Deviations in position and orientation during simulation using data from the IMU only

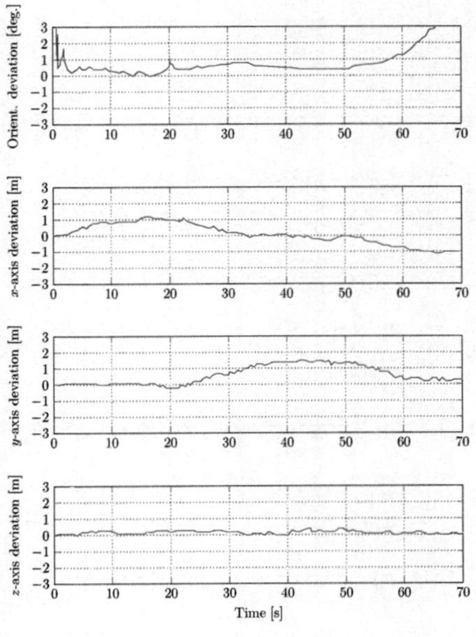

Fig. 8. Deviations in position and orientations during the simulation – fusion of IMU and GPS

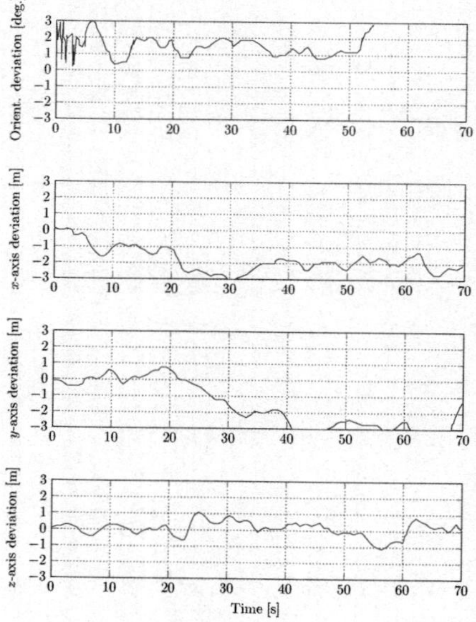

Fig. 9. Deviations in position and orientations during the simulation – fusion of IMU and MVO

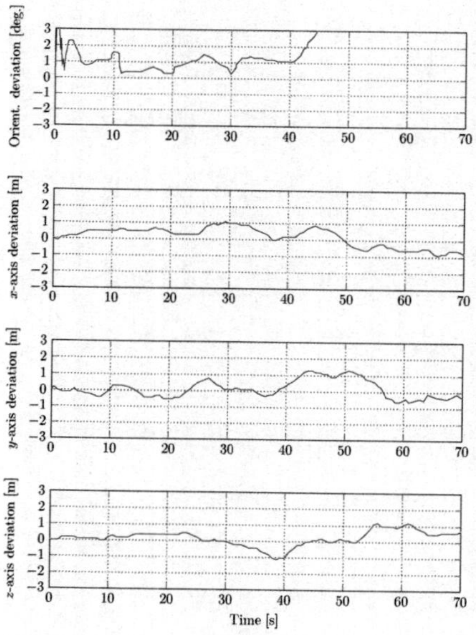

Fig. 10. Deviations in position and orientations during the simulation – fusion of IMU, MVO and GPS

orientation of the vehicle is completely lost within a few seconds after initialization. As can be seen in Fig. 6, the orientation gets out of the monitored range, i.e. ± 3°C from the initial value after one second from the start of the simulation. With regard to the deviation in position, a range of ± 3 m from the actual position of the vehicle was determined. The position deviation in the x-axis and y-axis directions has increased almost exponentially since the start of the simulation and left the specified range in both cases after approximately 3 s. The position deviation in the z-axis direction left the range within 9 s. Simulations of such a situation, when only data from the IMU unit are available, illustrate the consequences of a GPS signal failure in the case when the system does not have a redundant sensor that could substitute the GPS. As a result, we conclude that affordable IMUs cannot provide reliable vehicle information for more than a few seconds on their own. In Fig. 11 we see a trajectory shown in red, which represents the output of the filter with data on the position of the vehicle in case the filter does not have the data responsible for updating the predictions.

5.4 Localization Based on IMU + GPS

In the next simulated situation measurement data from the IMU is available, as before, but this time GPS data is also available. At the mentioned sampling frequencies, after every 10 predictions based on the measurement samples from the IMU, the fusion filter updates the states based on the measurement data from the GPS. The modeled GPS receiver works with the features of a standard, affordable GPS receiver. As can be seen in Fig. 12, the yellow line, which represents the output of the vehicle position filter, almost copied a blue line that indicated the actual position of the vehicle over time.

The very accurate results in the localization based on the fusion of IMU and GPS are also confirmed by the graphs in Fig. 12, where it can be clearly seen that the deviation in position was within the observed range of ± 3 m throughout the simulation, for each axis of the 3-D space. The smallest deviation in position occurred in the z-axis direction, due to the fact that the vehicle did not perform any part of its uphill or downhill movement, but only moved along the plane. The maximum deviation in position in this case was 1.29 m, along the y-axis. The orientation was mostly within the observed range during the simulation, at the beginning the orientation deviation peaked at higher values, which was caused by a higher acceleration of the vehicle in the beginning. The maximum deviation in orientation reached the value of 6.57°C at the peak at the beginning of the simulation, which was the best result in determining the orientation among the four simulated cases.

5.5 Localization Based on IMU + MVO

The third simulated situation was crucial because it showed us whether the fusion of the sensor data from the IMU unit together with data from the monocular visual odometry system is able to relevantly substitute the IMU + GPS system in the absence of a GPS signal. In order for the camera to be able to apply image

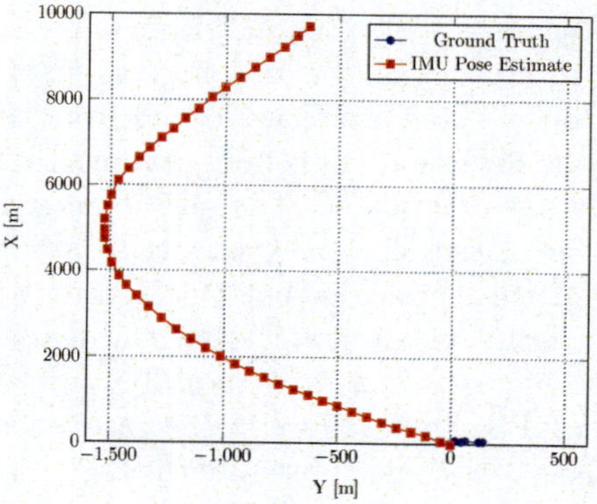

Fig. 11. IMU based localization

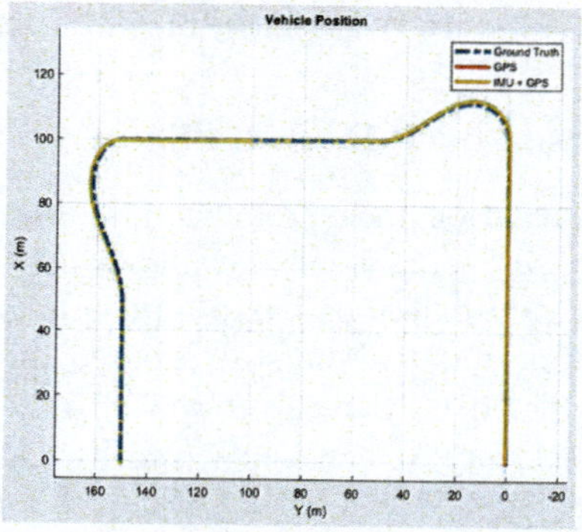

Fig. 12. Localization based on the fusion of IMU and GPS

processing algorithms to the recorded image and derive data about the position and orientation of the vehicle, it needs a higher sampling frequency than the GPS receiver, in our simulation the MVO data were sampled at 25 Hz, i.e. the update of the filter states was performed after every 4 predictions based on samples from the IMU unit. The red line in Fig. 13 shows that the position estimation based on MVO data achieves very good accuracy in recording changes in vehicle movement, thus the shape of the estimated trajectory relatively accurately copies the actual trajectory of the vehicle.

After the fusion of MVO data with IMU data, the scale ambiguity of the range, which is a factor influencing the camera's ability to recognize distances, was removed. The result after the fusion is shown by the yellow line, which quite well copied the actual trajectory indicated by the blue dashed line. More detailed results are shown in Fig. 9, where it can be seen that the deviation in orientation again reached the best values at the beginning of the simulation, but for most of the simulation time it was also at acceptable values, but subsequently at the end of the simulation it continuously got out of range again. The position deviation was higher in this case than when integrated with the GPS. The maximum deviation in position reached −4.067 m for the movement in the direction of the y-axis. The lowest value of the deviation was again in the direction of the z-axis (1.169 m), and the highest deviation was in the direction of the x-axis, which reached −3.159 m.

5.6 Localization Based on Fusion of IMU + GPS + MVO

In the last simulated situation, the vehicle was located on the basis of a fusion of data from all available sensors, i.e. from the IMU unit, and both GPS and camera sensors. This model situation has shown whether it is relevant to integrate data from MVO into a fusion filter even if a GPS signal is available. The fusion filter works in the same way here, but in this case the update is performed not only when each new GPS data sample is available, but also for each new MVO data sample. This means that the prediction was still performed based on data from the IMU at a sampling frequency of 100 Hz, but after every 4 predictions, the filter states were updated based on MVO data, and after every 10 predictions, the states were updated based on GPS data. The result of such a fusion is shown by the purple line in Fig. 14, which at first glance almost exactly copied the actual trajectory of the vehicle, indicated by the blue dashed line.

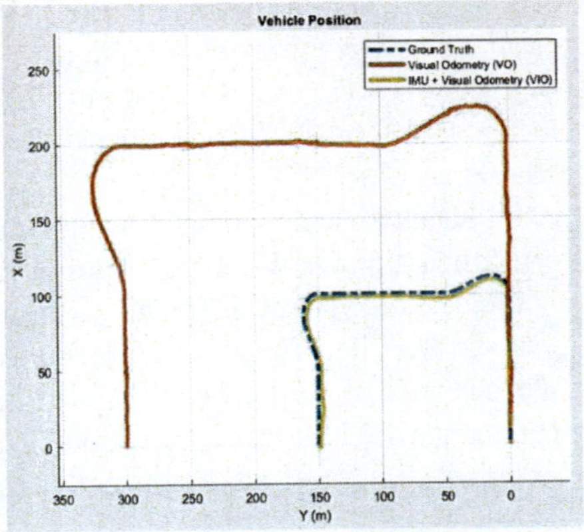

Fig. 13. Localization based on fusion of data from IMU and MVO

The numerical expression of accuracy is shown in Fig. 10, which has shown us that such a fusion method achieves excellent positioning results. The maximum deviations in position reached the following values for the individual axes – X: 1.13, Y: 1.23 and Z: 1.21 m, which represents higher accuracy than fusion of data from the IMU and the GPS system. However, the orientation deviation in this case was the worst of the previous cases, except for the localization based only on data from the IMU itself, which means that the system gained higher positioning accuracy with this fusion, but at the expense of orientation accuracy.

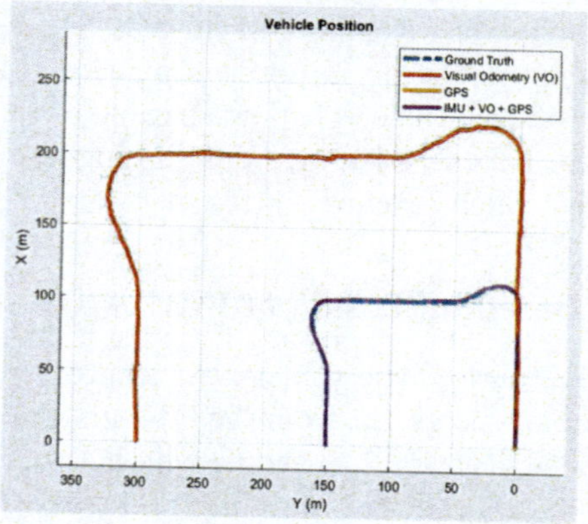

Fig. 14. Localization based on fusion of IMU, GPS and MVO

5.7 Discussion of Simulation Results

The results from all four simulated situations are summarized in Table 1, from which can be seen that the values of deviations in position and orientation are the worst in the case where the filter estimates the states of the vehicle only on the basis of predictions from the obtained IMU data. These values are very different from the reality after a few seconds, because the IMU sensor was uncorrected and accumulated more and more error over time. The results of this simulation confirmed that standard, lowcost IMU sensors based on MEMS technology, which are unable to calibrate themselves, need to be integrated with other sensors in order to obtain more reliable results.

In the second simulated situation, in which data was merged from the IMU and the GPS system, very consistent results were obtained throughout the simulation, the deviation in vehicle orientation was in this case significantly the lowest of all four simulations performed. The results of the fusion of data from the IMU and the GPS system, together with the low obtaining cost of GPS receivers, suggest that this is rightly the most widely used localization system used in practice.

The results from the third simulated situation show that the use of MVO data, in fusion with other sensors, to locate the vehicle makes sense and is applicable. The deviations during the simulation were mostly at acceptable values. For assistance systems affecting surrounding vehicles and objects, or for assistants influencing driving, such values are insufficient, but for the purpose of navigating the vehicle in areas without GPS signal availability, these localization data are very supportive, for example when localization takes place at underground or covered intersections.

The simulated situation, in which data from all IMU, GPS and camera sensors were fused, as expected, performed best in estimating vehicle position, with maximum position deviations being the lowest of the four simulations. However, the deviation in orientation reached the highest value. These results show that in applications demanding higher positional accuracy, such a fusion has its merits, but this is at the expense of the accuracy of the vehicle orientation.

Table 1. Maximum position deviations in the x, y and z axis and in the vehicle orientation

Technology	Maximum deviation in position in the direction of the axis [m]			Max. orientation deviation [deg.]
	x	y	z	
IMU	9853.98	1664.92	−4205.38	171.25
IMU + GPS	−1.26	1.29	−0.28	6.57
IMU + MVO	−3.16	−4.07	1.17	28.22
IMU + GPS + MVO	1.13	1.23	1.21	45.35

6 Conclusion

The paper describes the design of a fusion filter, which can determine the position, orientation and velocity of the land vehicle based on data from IMU, GPS and camera. The operation of this fusion filter was tested using several simulations performed in the Matlab environment. The simulated situations showed that in road traffic situations where no GPS signal is available, and without redundancy, the system loses information about the position and orientation of the vehicle after only a few seconds. The proposed fusion filter for the integration of data from all available sensors, i.e. the IMU, GPS and camera achieved the highest accuracy in determining the position, so the simulations confirmed the suitability of using a camera sensor implementing the algorithm of monocular visual odometry to locate the vehicle. The results of the work can be used in the implementation of a localization system in an autonomous robotic vehicle that will run in conditions challenging the GPS signal reception.

Acknowledgment. The paper was partially supported by the Slovak Research and Development Agency, grant No. APVV-17-0190 and the Slovak Cultural Educational Grant Agency, grant No.039STU-4/2021. The research presented in this paper was also partially supported by Technology Agency of the Czech Republic via grant reg. no. TK02030013.

References

1. Fujdiak, R., Mlynek, P., Misurec, J., Slacik, J.: Simulation of intelligent public light system in smart city. In: 2017 Progress In Electromagnetics Research Symposium-Spring (PIERS). IEEE, pp. 2515–2519 (2017)
2. Kromer, P., Hasal, M., Nowakova, J., Heckenbergerova, J., Musilek, P.: Statistical and nature-inspired modeling of vehicle flows by using finite mixtures of simple circular normal distributions. IEEE Intell. Transp. Syst. Mag. **12**(4), 182–194 (2020)
3. Fujdiak, R., Masek, P., Hosek, J., Mlynek, P., Misurec, J.: Efficiency evaluation of different types of cryptography curves on low-power devices. In: 2015 7th International Congress on Ultra Modern Telecommunications and Control Systems and Workshops (ICUMT), pp. 269–274. IEEE (2015)

4. Raso, O., Mlynek, P., Fujdiak, R., Pospichal, L., Kubicek, P.: Implementation of elliptic curve Diffie Hellman in ultra-low power microcontroller. In :2015 38th International Conference on Telecommunications and Signal Processing (TSP), pp. 662–666. IEEE (2015)
5. Fujdiak, R., Misurec, J., Mlynek, P., Janer, L.: Cryptograph key distribution with elliptic curve Diffie-Hellman algorithm in low-power devices for power grids. Revue Roum. des Sci. Techn. 84–88 (2016)
6. Krivtsova, I., Lebedev, I., Sukhoparov, M., Bazhayev, N., Zikratov, I., Ometov, A., Andreev, S., Masek, P., Fujdiak, R., Hosek, J.: Implementing a broadcast storm attack on a mission-critical wireless sensor network. In: International Conference on Wired/Wireless Internet Communication, pp. 297–308. Springer, Berlin
7. Fujdiak, R., Mlynek, P., Misurec, J., Masek, P.: Design of low-power random number generator using signal quantization error in smart grid. In: 2016 39th International Conference on Telecommunications and Signal Processing (TSP), pp. 7–10. IEEE (2016)
8. Mlynek, P., Misurec, J., Koutny, M., Fujdiak, R., Jedlicka, T.: Analysis and experimental evaluation of power line transmission parameters for power line communication. Meas. Sci. Rev. **15**(2), 64 (2015)
9. Mlynek, P., Misurec, J., Kolka, Z., Slacik, J., Fujdiak, R.: Narrowband power line communication for smart metering and street lighting control. IFAC-PapersOnLine **48**(4), 215–219 (2015)
10. Stava, M.: On precise fault localization and identification in noc architectures. In: 2019 22nd Euromicro Conference on Digital System Design (DSD), pp. 478–484 (2019)
11. Stava, M.: Efficient error recovery scheme in fault-tolerant NoC architectures. In: 2019 IEEE 22nd International Symposium on Design and Diagnostics of Electronic Circuits & Systems (DDECS), pp. 1–4 (2019)
12. Fujdiak, R., Masek, P., Mlynek, P., Misurec, J., Olshannikova, E.: Using genetic algorithm for advanced municipal waste collection in smart city. In: 2016 10th International Symposium on Communication Systems, Networks and Digital Signal Processing (CSNDSP), pp. 1–6 . IEEE (2016)
13. Krömer, P., Nowaková, J., Hasal, M.: Towards a new evolutionary algorithm for the minimum tollbooth problem. In International Conference on Computational Social Networks, pp. 116–125. Springer, Berlin (2018)
14. Fujdiak, R., Masek, P., Mlynek, P., Misurec, J., Muthanna, A.: Advanced optimization method for improving the urban traffic management. In: 2016 18th Conference of Open Innovations Association and Seminar on Information Security and Protection of Information Technology (FRUCT-ISPIT), pp. 48–53. IEEE (2016) 48–53
15. Masek, P., Fujdiak, R., Zeman, K., Hosek, J., Muthanna, A.: Remote networking technology for IoT: cloud-based access for Alljoyn-enabled devices. In: 2016 18th Conference of Open Innovations Association and Seminar on Information Security and Protection of Information Technology (FRUCT-ISPIT), pp. 200–205. IEEE (2016)
16. Stava, M.: A synthesis of reversible digital circuits to solve the Boolean satisfiability. In: Proceedings of the Fourth International Conference on Microelectronics, Computing and Communication Systems, pp. 233–246. Springer Singapore, Singapore (2021)
17. Falco, G., Einicke, G.A., Malos, J.T., Dovis, F.: Performance analysis of constrained loosely coupled GPS/INS integration solutions. Sensors **12**(11), 15,983–16,007 (2012)

18. Nguyen, V.H., et al.: Loosely coupled GPS/INS integration with Kalman filtering for land vehicle applications. In: 2012 International Conference on Control, Automation and Information Sciences (ICCAIS), pp. 90–95. IEEE (2012)
19. Corke, P., Lobo, J., Dias, J.: An introduction to inertial and visual sensing, pp. 519–535 (2007)
20. Li , M., Mourikis, A.I.: Improving the accuracy of EKF-based visual-inertial odometry. In: 2012 IEEE International Conference on Robotics and Automation, pp. 828–835. IEEE (2012)
21. Li , M., Mourikis, A.I.: High-precision, consistent EKF-based visual-inertial odometry. Int. J. Robot. Res. **32**(6), 690–711 (2013)

An AI-Based Platform Architecture with Situational Awareness for Travel Plans

Tae Moon Jung[1(✉)] and Inwhee Joe[2]

[1] PEOPULLEY Inc., 11, Jandari-Ro 3an-Gil, Mapo-Gu, Seoul, South Korea
moiiom@hanyang.ac.kr

[2] Hanyang University, Seoul 04763, South Korea
iwjoe@hanyang.ac.kr

Abstract. Demand for non-face-to-face services is increasing in all industries due to the impact of the global pandemic caused by COVID-19 over the past three years. This phenomenon also occurs in the travel industry without exception. Travel consumption patterns are changing in the form of FIT (Foreign Independent Tour) rather than package tours provided by existing travel agencies. Now, we are entering the era of travel tech using smartphones. To this end, technological efforts are being made to simultaneously address the two characteristics of individual free travel and non-face-to-face situations. In this paper, we propose an AI-based platform architecture with situation awareness for travel plans. It provides a personalized service using the ontology-based mobile service so that tourists can make travel plans according to their own circumstances, after considering the situation of individual tourists through sensor collection data while analyzing the congestion situation of tourist destinations through public data. The mobile platform consists of a total of four layers (Infrastructure, Cross-Cutting Infrastructure, Domain, Application), and has been implemented in an environment of Android 8.0 or higher and iOS 11.0 or higher through the AWS website.

Keywords: Congestion · Situational awareness · Ontology · Public data analysis · Personalized customization services · Mobile travel platforms

1 Introduction

Previously, users input destinations, travel periods, departure dates, and companion information into the system. Based on this, the system provided information to users through meta-information of travel destinations. The AI-Based Context-Awareness mobile platform introduced in this paper is as follows. It recognizes the user's situation through sensors by mobile devices and analyzes public open data that informs the current situation of the travel destination at the same time. Based on this, places with high travel satisfaction are selectively recommended. It also provides responsive service by modifying recommendations to reflect travel variables that may occur in the field. In this way, the system requires immediate and wide-ranging contextual awareness, and the big data collected for this requires systematic and efficient management. Accordingly, an ontology system architecture was applied to this platform [1].

R. Silhavy et al. (Eds.): CoMeSySo 2022, LNNS 596, pp. 376–384, 2023.
https://doi.org/10.1007/978-3-031-21435-6_32

2 Related Research

2.1 Situation Recognition Technology

Each data for situational recognition is collected through individual crawling or institutional APIs. Data collection is not much different from the existing methods for travel recommendation. However, the data collected in this study shows big differences in size and quality in terms of power, gas consumption data, real-time SNS and mobile devices data, and travel review data of web portals that were not previously considered. Accordingly, ontology-based data refined data attributes is necessary, through which situational awareness and congestion analysis can be performed. The legacy data collected is used to predict future time point, and real-time data such as telcos and financial companies are designed to respond immediately to field variables [2] (Fig. 1).

Table 1. Data set for situational awareness

Data set	Method
POI (Point of Interest) data by travel destination	–
Electricity / gas consumption data by commercial area and travel area	RESTful
POI-based weather information data	RESTful
User data by viewing type by mobile exhibition audio guide platform	RESTful
SNS (Instagram, Facebook) photos, hashtag data	Crawling
Review data of travel package sold on big portals	Crawling
Legacy and real-time data of mobile users	RESTful
Legacy data of credit card users from financial companies	RESTful
Legacy and real-time data of navigation users	RESTful
Sensor data (GPS, Accelerometer etc.) of smartphone users	RESTful

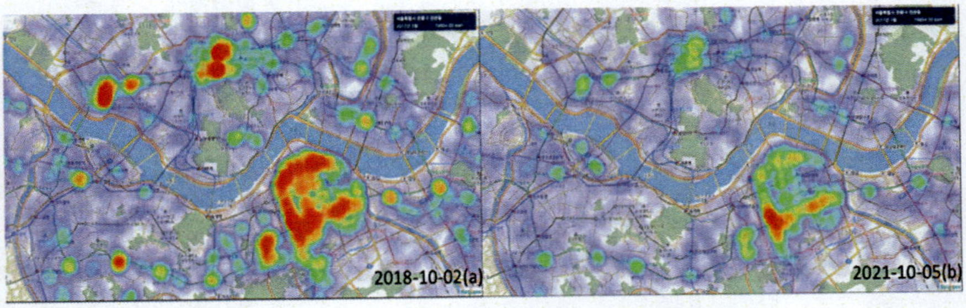

Fig. 1. Visualization of road ground activity analysis using monthly cumulative electricity and gas usage data (https://www.thingspire.com/)

2.2 Utilization of Ontology

There is a physical time difference between the time of making a travel plan and the time of actually departing the trip. By systematizing large-capacity data in an ontology refining method, the platform can accurately recognize the situation of each different point and provide additional alternatives by considering the situational variables occurring in the travel destination.

Table 2. Ontology classification of situational awareness information

Name	Definition	Description
Real-time traffic information when using a vehicle	Traffic information on the traffic link section	Information generation time (MCGT), link identification number (LIN), link speed (SR), link traffic volume (VR), link density (DR), link traffic time (TTQ), detector occupancy rate (DQ), section prediction Average speed (PALSR), estimated section transit time (PTTQ), estimated time (PTGT), additional input of communication information (AIT)
Traffic control information when using a vehicle	Schedule information on road control due to planned events or construction	Information generation time (MCGT), event position node (LNIN), event position link (LLIN), blocked lane type (LBOCTC), number of blocked lanes (LBOCN), event type classification (LBNIN), alternative route guidance (RART), Event start time (TSGT), Event end time (TEGT)
Road condition information when using a vehicle	Road surface abnormalities related to driving	Information generation time (MCGT), link identification number (LIN), link length (LQ), road surface condition (SOC), road blockability (LBOCC), number of blocked roads (LBOCN), alternative route guidance (RART), rainfall / snow level (DSPD), road surface temperature (DSTQ)

(*continued*)

Table 2. (*continued*)

Name	Definition	Description
Weather information	Provide weather information	Weather information generation time (CGT), weather information provision position (LC), weather forecast presence / absence (FOC), temperature (TQ), maximum temperature (HTQ), minimum temperature (LTQ), weather forecast probability (PP), weather specific status (SC), Visible Distance (VQ), Wind Speed (WSQ), Wind Direction (WDC), Humidity (HQ), Atmospheric Pressure (PQ), Sunrise Time (STG), Sunset Time (STGT)
Power consumption information	Regional power consumption information	Power consumption generation time (PCGT), link identification number (LIN), power consumption position (LP), power consumption (AP)
Gas consumption information	Information of city gas consumption by region	Information generation time (MCGT), link identification number (LIN)
Specific dates information	Provide information about a specific day	Monday (MD), Tuesday (TD), Wednesday (WD), Thursday (TTD), Friday (FD), Saturday (SD), Sunday (SS), Holidays (FRD), Border Days (ND), Extra Holidays (EFRD)
Tourism big data congestion information	Regional congestion information of public big data	Information generation time (MCGT), congestion degree information provision position (CLC), link identification number (LIN), congestion degree (COD), congestion intensity (CODS)

(*continued*)

Table 2. (*continued*)

Name	Definition	Description
Telcos information	Real-time location information for mobile smartphone subscribers of telcos	Communication information generation time (TCGT), communication information provision position (TLC), communication situation information (TID)
Financial companies	Large scale real-time location information for credit card subscribers of financial companies	Financial information generation time (FCGT), financial information provision location (FLC), financial information (FID)
Users feedback information	Providing user questionnaire archive information by travel location	Information generation time (MCGT), link identification number (LIN), information provision position (LBD), questionnaire affirmative information (APDB), questionnaire fraudulent information (ANDB), questionnaire generation medium (AFGJ)

2.3 Utilizing the Ontology of Situational Awareness Technology

The traveler decides the final route based on his/her personal situation (experience, personal preference, psychological conditions at the time) and various complex conditions of the destination (travel situation, access to other destinations, accommodation, restaurant, fun, latest trends, etc.). Therefore, the data set Table 1 is extensively collected and analyzed to comprehensively determine the situation of travelers and destinations. The data set in Table 1 collected for situational awareness can be into an ontology as shown in Fig. 2. For each situational guess, ontology inference is made, and the modeling of situational information, that is, schema creation work is performed. In Fig. 2, only the has XXX relationship and the subclass relationship are expressed, which are related to the contents of Table 2. The derived status information is generated by utilizing the basic status information in Table 2 [3].

3 System Architecture

3.1 Overview

The S/W architecture of this system is optimized to collect and process large amounts of data required for situation recognition while faithful to the basic structure. For the scalability of the entire travel recommendation application and flexible management of

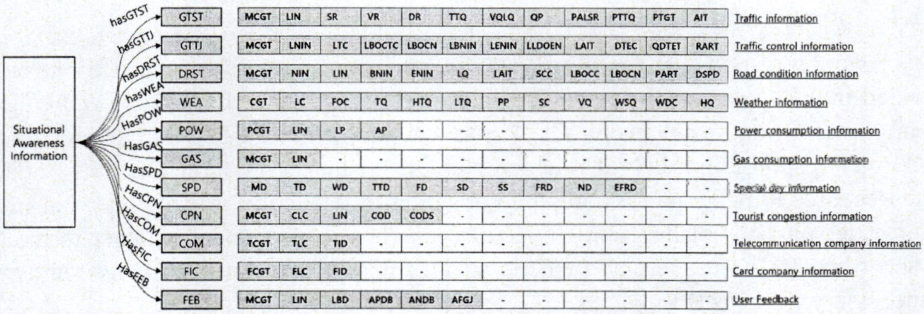

Fig. 2. Travel recommendation situational awareness information ontology

the service management team, it is composed of a modular integrated framework of a plug-in architecture method (Fig. 3).

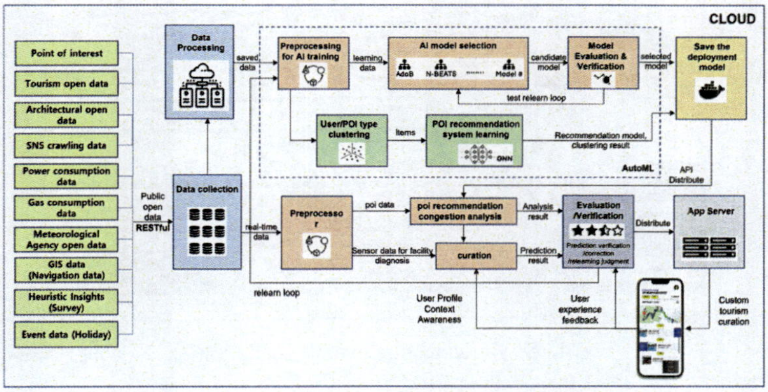

Fig. 3. Overall configuration chart of system architecture

By introducing a NoSQL-based large-capacity processing engine, large-capacity data accumulated during resource monitoring can be quickly processed in real time, and performance can be improved with multiple general servers (scale out) without using a single high-performance server (scale up). Guaranteed cost savings. Changes to the data model and extension of additional fields are simple, and in terms of readability, the 'JSON structure' has the advantage of intuitively grasping the data compared to the RDBMS table structure [4].

3.2 S/W Architecture Configuration

Based on the physical separation of one application server, the S/W architecture diagram consists of four properties: User (Client), External System, Application Server, Repository, and collection target. The S/W architecture expresses the organic relationship between the application's operating environment, function modules, and external libraries to be used, and through this, the overall configuration of the application server can be checked.

3.2.1 Infrastructure Layer

This layer is responsible for normalization and storage of raw data. This layer is divided into Repository, Resource Provider and is responsible for storage and collection/normalization, respectively. The Repository part provides RDBMS for relational and quick response in a hybrid way and NoSQL for managing large amounts of data. The Resource Provider part is composed of a plug-in-type collection module, and since it provides plug-ins of different Collectors, it is possible to select whether to mount it according to the usage environment, so it is possible to optimize it by removing unnecessary plug-ins (Fig. 4).

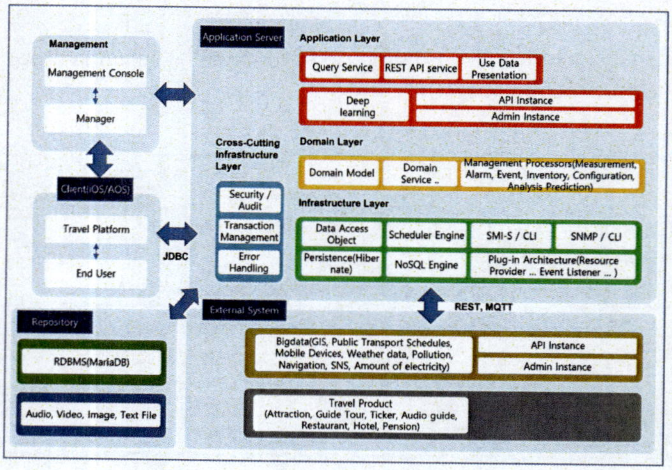

Fig. 4. S/W architecture overall configuration

3.2.2 Cross-Cutting Infrastructure Layer

Security (ACL, Audit) and transaction management functions that should be applied to all layers are implemented using AOP (Aspect Oriented Programming) technology. Using this AOP technology, it is possible to add, modify, and remove functions that can be commonly applied without modifying existing layers.

3.2.3 Domain Layer

Alarm, Analytics, and Prediction information is created based on the Inventory, Measurement data, and Event data collected from the collection target. After that, save the Repository to create data to be provided to the Application Layer and Presentation Layer. It corresponds to the 'heart' in the application layer and includes domain information and the state of business objects. This layer generates analysis, prediction, event, alarm, and inventory data with normalized data collected from the infrastructure layer and delivered to the presentation layer through the application layer.

3.2.4 Application Layer

It is a layer that coordinates application activities and does not contain business logic. It does not keep the state of the business object, but the processing state of the application.

4 Implementation

By simplifying the information input step in the form of an icon by reflecting the Gamification factor, user convenience was improved, the recommended itinerary was briefly indicated on the map, and elements necessary for the trip, such as restaurants and accommodation, were implemented so that it could be selected and changed.

In addition, the UX was configured so that the established itinerary could be easily modified through the swipe function assuming a situation in which the schedule and route were changed during the trip in the recommended place. Algorithms considering situational awareness are reflected in each recommendation element through the ontology data refining method, and through this, a recommendation service is provided based on more accurate information to users (Fig. 5).

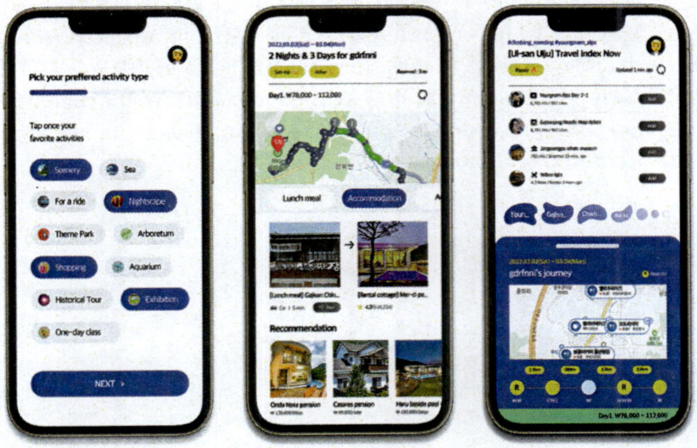

Fig. 5. Example of intelligent dynamic travel route recommendation and visualization technology development GUI

5 Conclusion

This paper proposes a system configuration for building a mobile platform to provide optimal recommendation factors by comprehensively considering the variables that may occur during the travel process from the travel planning process to the traveler. To this end, a situational awareness method that considers users and travel destinations simultaneously was applied by including public big data for situational awareness and collectable sensor data and managing them in an ontology method. In the future, research on artificial intelligence algorithms based on situational awareness and multiple recommendation methods based on users' situational awareness will be conducted. In addition, further research on the platform architecture that can be expanded to global services will be conducted by additionally applying the itinerary recommendation function linked to public transportation system and flight information in addition to the travel itinerary centered on private or rental cars based on free travel.

Acknowledgement. This research was supported by Culture, Sports and Tourism R&D Program through the Korea Creative Content Agency grant funded by the Ministry of Culture, Sports and Tourism in 2022(Project name: Customized tourism through AI-based tourist situation recognition and tourism information curation development of itinerary recommendation platform technology, Project Number: R2022020116, Contribution Rate: 100%).

References

1. Park, J., Jeong, H., Ru, H.: Personalized travel schedule creation system based on recommendation system. Korea Comput. Inf. Soc. Winter Acad. Conf. Proc. **25**(1), 105–108 (2017)
2. Kim, H.: Context-awareness technology for location based-service for ubiquitous learning. J. Korea Acad.-Ind. Cooperation Soc. **12**(11), 4869–4874 (2011)
3. Sim, J., Kwon, O., Kang, J.: Driver preference based traffic information recommender using context-aware technology. Knowl. Manag. Res. **11**(2), 75–93 (2010)
4. Kim, S., Kim, T.: Proposing an efficient system OS architecture for running artificial intelligence accelerators. In: Proceedings of the Korea Computer Comprehensive Academic Conference, pp. 1055–1057 (2019)

Cyber-Physical Test Facility for Certification of Robotic Unmanned Aerial Systems

Oleg Balyk[1], Mariya Zolotaeva[2], Alexey Bogomolov[3](✉), and Alexey Soldatov[4]

[1] Moscow Aviation Institute (National Research University), 4 Volokolamskoye highway, Moscow 125993, Russia

[2] National University of Oil and Gas "Gubkin University", 65 building 1 Leninsky avenue, Moscow 119991, Russia

[3] Federal Research Center "Computer Science and Control" of the Russian Academy of Sciences, 44 building 2 Vavylova str., Moscow 119333, Russia
a.v.bogomolov@gmail.com

[4] St. Petersburg Federal Research Center of the Russian Academy of Sciences, 39, 14th Line V.O., St. Petersburg 199178, Russia

Abstract. Based on the analysis of the needs of aviation equipment testing practice, the necessity of creating a cyber-physical testing complex for certification of robotic unmanned aerial systems is substantiated as a priority direction for the development of a flight model testing system that implements the optimal combination of full-scale flight experiments and tests using specialized simulation, mathematical and semi-natural modeling tools. The results of the theoretical substantiation of the structure and tasks of the cyber-physical test complex for certification of robotic unmanned aerial systems are presented. The functional structure of the certification system has been developed and the formal statement of the task of creating a test complex for robotic unmanned aerial systems has been completed. A unified methodology for choosing a rational image of the complex has been created, taking into account the general principles of its preparation for ensuring the certification of unmanned aerial systems. This technique makes it possible to objectively substantiate its composition and the minimum required volume of measures to bring the complex to a state in which it is able to ensure the planned flight tests in the shortest possible time with proper quality. It is shown that the use of the developed test complex for certification of robotic unmanned aerial systems will reduce the time spent on the test program by 2–2.5 times. To use the developed test complex as part of a flight model system for testing aviation equipment, it is necessary to develop and certify the means of semi-natural and simulation modeling and a complex mathematical model of an unmanned aerial system.

Keywords: Unmanned system · Cyber-physical system · Robotic system · Certification of unmanned systems · Testing of unmanned systems · Test complex

1 Introduction

The modern concept of certification of unmanned aerial systems (UAS) provides for a cycle of tests in order to obtain a reliable assessment of the ability of UAS to solve the

R. Silhavy et al. (Eds.): CoMeSySo 2022, LNNS 596, pp. 385–396, 2023.
https://doi.org/10.1007/978-3-031-21435-6_33

assigned target tasks in the given operating conditions and ensure flight safety [1–3]. The increase in the complexity of the created UAS, due to the requirements to increase their functionality, the introduction of on-board systems that operate using various physical principles (radar, laser location, infrared radiation, measurement of microparticle fluxes, television, radio control, stellar radiation, digital signal conversion, etc.) makes conflicting demands on test technology. On the one hand, taking into account the requirements for the reliability of quality assessment, the determination of characteristics should be carried out experimentally under conditions identical to the real application or close to them, on the other hand, UAS should be evaluated in the shortest possible time under resource constraints [4–6].

Qualitative and reliable assessment of UAS in the process of their certification for compliance with the specified requirements is possible only if there is a test complex (TC) that meets modern requirements for the technical feasibility of testing in the conditions of polygons and test bases equipped with testing and measuring equipment that fully meets the requirements for the examination of UAS samples, as well as the rational organization of the certification process and methodological support [7–9].

UAS test methods and, accordingly, the required structure and composition of ICs can differ significantly depending on various factors:

appointment of UAS;
type of UAS (aircraft, helicopter, etc.);
mass and dimensions of UAS, features of target loads;
type of propulsion system;
type of takeoff and landing;
features of the system and methods of control.

The IC should ensure that UAS tests are carried out with their evaluation in terms of spatial characteristics, accuracy, the number of measurement channels, the size of the measurement zone, and the composition of the air and ground target environment. Taking into account the wide variety of available types of UAS, studied in the process of testing characteristics, as well as the high rate of evolution of robotic systems, the task of determining the promising composition of IC for UAS certification is quite difficult.

The modern development of the theory and practice of testing aviation equipment is based on the widespread use of the flight model test system (LMTS), which implements the optimal combination of full-scale flight experiments and tests using specialized simulation, mathematical and semi-natural modeling tools [10–14]. This allows you to significantly reduce the time and cost of testing and testing aircraft.

2 Functional Analysis of the Certification System for Unmanned Aerial Systems

From the point of view of the development of IC certification of UAS, the procedure for conducting flight experiments (LE) performed in the process of factory tests and certification state tests (SGCT) is of practical interest, which is summarized in [1, 2, 15].

The scheme for conducting and analyzing the results of LE is universal for all types of tests. The certification testing program provides for a set of works, including a list of LEs, the main purpose of which is to evaluate all UAS characteristics defined in the terms of reference and certification basis [16–18]. For each item of the certification testing program, methods for conducting LE are developed, which determine the purpose, conditions for carrying out work, the composition of the funds involved, requirements, including those for IC, the procedure for conducting LE, algorithms for assessing characteristics. The most complex are the LEs, which provide for the assessment of the possibility of using UAS for the main intended purpose. The most important characteristic of UAS, flight safety, including in case of system failures, requires representative statistics and, as a rule, is determined by the results of simulation modeling, which are confirmed by the results of full-scale LE [19–22].

UAS characteristics are technical, operational, economic and other qualities that are essential from the point of view of its purpose and application. They can have quantitative or qualitative indicators, i.e. measures of characteristics that determine their compliance with the elements of measuring scales (symbols, numbers or numbers). These characteristics are interconnected and have a hierarchical structure. There are several main types of characteristics determined during the certification of UAS:

(a) Characteristics of the target efficiency, allowing to evaluate the specified quality and reliability of the performance of the functional tasks of the UAS. These include the probabilistic characteristics of detection, recognition, tracking of targets, the zone and range of action, the accuracy of aircraft navigation and determining the coordinates of objects, the timing of obtaining information, etc.
(b) Technical characteristics that determine the technical capabilities of the equipment to ensure the specified efficiency of the UAS and are divided into general technical and special.
(c) Operational characteristics that allow us to evaluate the convenience, time and cost of UAS maintenance.
(d) Ergonomic characteristics that evaluate the ease of use of the UAS by the crew in flight and in preparation for it.
(e) Economic characteristics that determine the cost of creating and operating UAS and the economic effect achieved from its use.

The division of characteristics into types is not strict, but only indicates their relationship and the need for a systematic approach in their assessment.

Thus, in the process of performing flight and ground tests, a database of a certified UAS should be formed, which is characterized by a large dimension, heterogeneity and dynamism. Maintaining such databases manually or with the fragmented use of computer technology is very difficult. In this regard, the relevance of creating an automated system for control (monitoring) of UAS certification is obvious.

The system of applied software and information models of technical test equipment and the control system of test equipment forms a virtual cyber-physical test complex, with the help of which the testers get the opportunity to remotely influence the test equipment in real time.

At the present stage of information technology development, based on the principles of building systems with an open architecture, distributed information and computing systems, Web technologies, RAD technologies (rapid prototyping technologies for applied information systems) [22, 24]. There has been a trend towards standardization and typification of virtual test complexes (Fig. 1) by moving application software modules from virtual test complexes into a network distributed hierarchical structure that allows you to quickly and easily, at the user level, adapt the modules for maintaining test programs, modules for maintaining test procedures and modules for maintaining these tests to specific test tasks (Fig. 2).

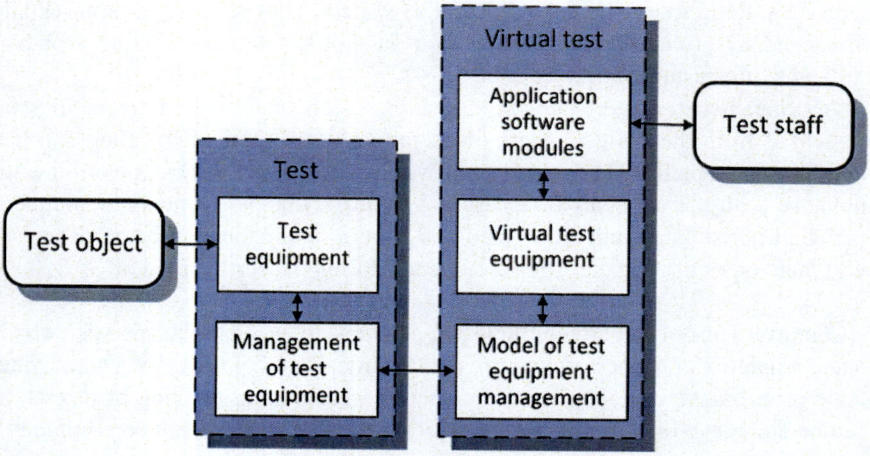

Fig. 1. Virtual facilities of test equipment

From the point of view of building systems with an open architecture, the UAS testing (certification) system can be represented as a seven-level hierarchical network structure.

At its first, lowest level, there are test tools (measurements of physical quantities) and their carriers. Data on measurements of the parameters of the tested UAS with the help of a complex of communication and data transmission tools, which forms the second hierarchical network level of the system, are transmitted to the third level in a single information space for collection, processing, storage and distribution to consumers. Based on these data, the fourth level of the system monitors and manages the UAS testing process in real time. As a result of processing measurements and evaluating UAS parameters (fifth level), using various analytical data processing systems, an assessment of the characteristics of test objects (sixth level) is obtained, on the basis of which a generalized assessment of UAS is formed with its reflection in the test report (seventh level).

The construction of an open architecture of the test system (Fig. 3) makes it possible to organically integrate various automated systems into it, which ensure the automated execution of certain tasks in the process of UAS certification.

3 Formal Statement of the Problem of Certification of Unmanned Aerial Systems

The solution of the problem of substantiating the sufficiency of IC to ensure the certification of UAS from a formal point of view should be carried out on a certain set.

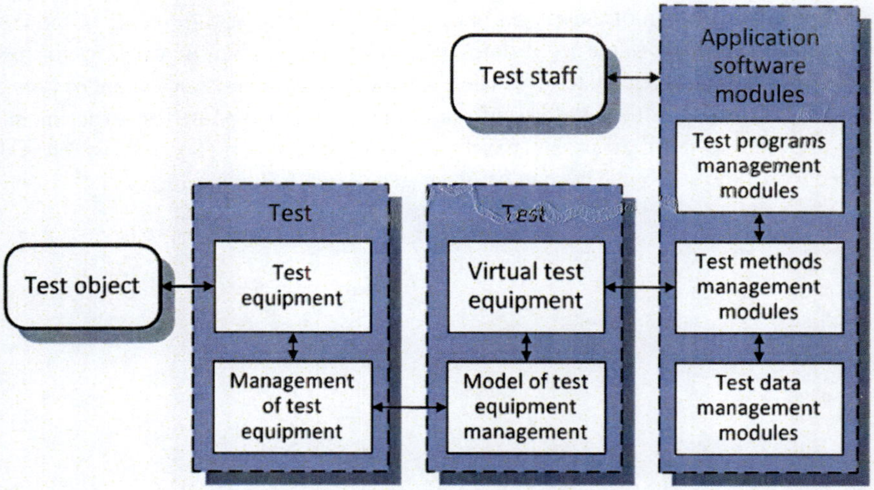

Fig. 2. Structure of test system

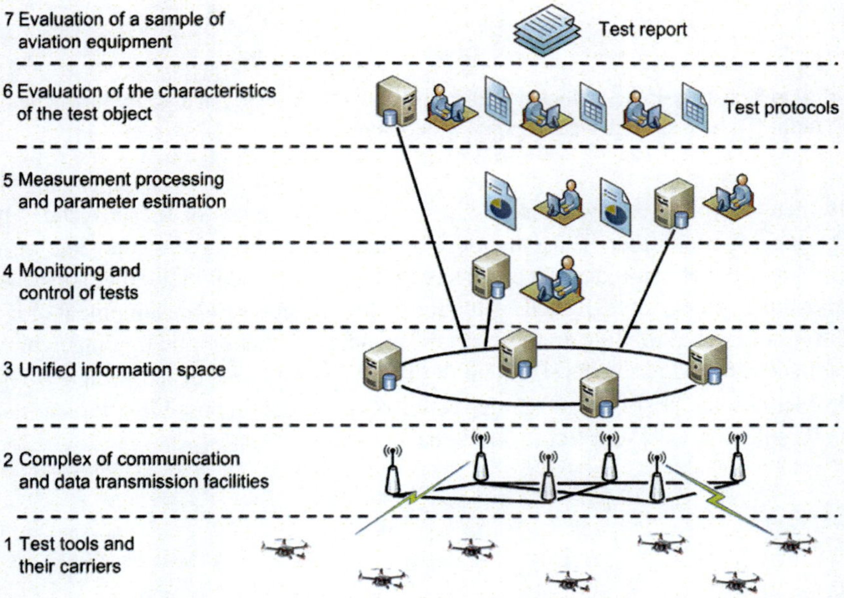

Fig. 3. The construction of an open architecture of the test system

The structure of the IC and the technical characteristics of its subsystems should be considered as a variable factor that allows the complex to be adapted to the requirements arising from the program and test methods for a specific UAS prototype.

Let's formalize the above. If resources, all possible current and potentially achievable states of each ith subsystem of the IC ($i = 1, 2, \ldots, N$) are represented by a certain set W_{0i} of the corresponding power, then all possible states of the complex as a whole W_0

are defined as the union of these sets. In accordance with the requirements of the UAS prototype test program, in the set of states of each IC subsystem, it is necessary to single out the set of its sufficient states W_i. Then the choice of a rational sufficient option for the formation of IC in the interests of providing a certain type of test experiment is carried out on the set of all possible sufficient states of the complex - W, which is the intersection of the sets Wi. This can be illustrated by the following Fig. 4.

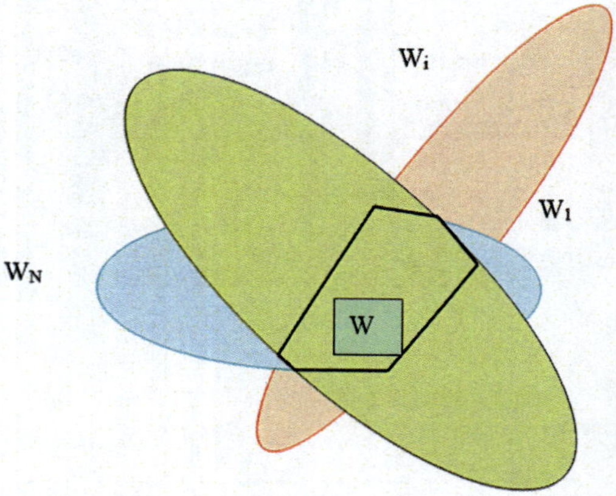

Fig. 4. Formalization of the definition of the set W, on which the problem of forming a rational image of the TF for a specific type of experiment is solved

In fact, the set W can be interpreted as a set of such states of IC subsystems that do not exclude each other. Those the development or maintenance of the state of one subsystem does not contradict the prospects for the development of another subsystem. The most obvious example of such a situation is the formation and development of IC in conditions of limited funding. In this case, the planned amount of total funding, which is divided and directed to the development of different subsystems of the complex. Thus, the problem should be solved at the intersection of the sets of all possible sufficient states of the IC subsystems (1) subject to additional constraints (2)

$$W = \bigcap_{i=1}^{N} W_i \tag{1}$$

$$W = \bigcup_{i=1}^{N} W_{0i} \tag{2}$$

The task of determining the sufficiency (rationalization of the appearance) of IC is inappropriate to directly solve for the entire UAS certification process as a whole. In accordance with regulatory documents, the certification process actually consists of a sequence of LEs, in each of which the compliance of the actual characteristics of the prototype with the requirements of the terms of reference and (or) aviation regulations is checked. Therefore, at the first stage, it is expedient to investigate the sufficiency of

IC for LE of the j-th type ($j = 1, 2,\ldots, k$), which are provided for by the certification program. Then the sufficiency of the entire IC as a whole will exist under the following condition (3)

$$\forall i, j(e_{ij} \geq e_{ij0}), \tag{3}$$

where

e_{ij} is the actual value of the sufficiency index of the i-th subsystem for carrying out the j-type LE, determined in accordance with the developed methodology;
e_{ij0} determined in accordance with the developed methodology, the required value of the sufficiency index of the ith subsystem for conducting LE of the j-th type.

It is obvious that the actual and required values of the IC sufficiency index, structured when the system is decomposed into N subsystems, to determine the sufficiency of the complex in the interests of providing certification, formally representing the technological chain consisting of LE k types, form, respectively, matrices E and E0 with dimensions ($N \times k$).

Due to the complexity of the IC and, accordingly, the impossibility of a sufficiently adequate formalization of the problem at the level of relatively simple analytical expressions (L. Zadeh's principle), one should use not optimization, but rationalization of the choice of the IC variant [5, 7, 22].

In practice, when controlling complex systems, it is practically optimal to consider such a control that keeps the controlled system in the attractor, in which the extremal of the objective functional corresponding to the current state of the system is located. It is precisely due to this that a reduction in the level of entropy of the system, sufficient from a practical point of view, should be ensured. We can say that optimal management means managing in accordance with the criterion of optimal goal setting for the system in its current state with the minimum possible deviation from the rational trend. Those. For complex systems, practice is the rational choice, not the optimal choice. The rationality of the choice makes it possible to abandon the use of the principle of strict optimality, which increases the practical applicability of the developed methodology for determining the sufficiency of the IC level. If this option is denoted by w, then the rationale for the preferred option for the formation of the IC can be performed using expression (4).

$$\exists w \forall i, j((w \in W) \cap (e_{ij} \geq e_{ij0}) \cap (EC_w = \sup A)), \tag{4}$$

where

A is the set of values of IC sufficiency indicators;
EC_w is the value of the IC sufficiency indicator in the state $w \in W$, $EC_w \in A$.

It is expedient to form the set W empirically, since This process may be based on poorly or not formalized considerations. It is in this that the rationality of the procedure for forming the appearance of the IC is manifested.

It should be noted that the methodical acceptance of the principle of rationality does not mean the rejection of other principles of optimization. So, for example, given the complexity of IC as a control object, which manifests itself in the fact that at the initial stage the sufficiency criteria form a certain vector with the likely presence of conflicts between particular performance indicators, the Pareto optimality principle occupies a special place.

4 Flight Model System for Testing and Certification of Unmanned Aerial Systems

To obtain a reliable assessment of the characteristics of UAS, with a limited amount of LE, it is necessary to widely use modeling methods. The methodological basis of the flight model test system (LMTS) of UAS is the technology of flight model tests of aircraft, developed in the early 80s of the XX century, which is based on the provisions of the methodology for the complete identification of aircraft using a priori information and flight results. Experiments and is a closed process of forming adequate models and obtaining reliable results with their help, confirmed by full-scale (flight) experiments [1, 2, 7, 20, 24].

The essence of the LMSI for UAS certification is to fulfill the following provisions:

(1) Obtaining for the same conditions a sufficient number of realizations of UAS performance indicators in flight tests and on the model.
(2) Carrying out a parametric refinement of the UAS model based on a comparison of the results of full-scale experiments and modeling, if the structure of the model is satisfactory.
(3) Carrying out a structural restructuring of the model, additional consideration of individual factors, addition of links in the presence of a residual difference between the output characteristics after an attempt at parametric refinement.
(4) Checking the statistical compatibility of the model and UAS in a number of purposefully chosen points of the factor space.
(5) Based on the calibration of the model (points 1–4), distribution of the UAS test results using simulation to the entire area of the factor space.

Thus, at first the isomorphism of the model and UAS is achieved, and then the assessment of this UAS on the model in all possible operating conditions.

The formalization of the process of substantiating the LMSI for UAS certification, taking into account the limitations on the cost (C) and time (T_i) of tests, is presented below:

$$M: \{K_s^*(A_{ki})\}_{i=1}^{3} = \arg\max E_{LMSI}(A_{ki}, K_{sj}, U_k),$$
$$C(K_{sj}) \le C^l(A_{ki}),\ T_u(K_{sj}) \le T^l(A_{ki})$$
$$\forall i = \overline{1, 3},\ j = \overline{1, n_v},\ k = \overline{1, n_y}, \tag{5}$$

where

A_{ki} is the tested variant ($i = \overline{1, 3}$) of the UAS product $A_{ki} \in \Omega_1$;
C^l, T^l limit values of the cost and time of the LMSI, respectively;
K_{sj} jth studied variant of LMSI belonging to the original set of alternatives Ω_2:

$$\begin{cases} K_{S_1} = \{N_{le1}, S_{pm1}, S_{im1}\}; \\ \ldots \\ K_{S_j} = \{N_{lej}, S_{pmj}, S_{imj}\}; \\ \ldots \\ K_{S_{nv}} = \{N_{len_v}, S_{pmn_v}, S_{imn_v}\}. \end{cases} \tag{6}$$

Here, the arguments determine the appearance of the LMSI, taking into account the choice of a different combination, respectively, according to the number of flight experiments, alternative options for the bench base during testing, namely, simulation benches and mathematical models with reasonable software and algorithmic support that implement a highly accurate assessment of the tactical and technical characteristics of the UAS sample;

U_k a vector that sets the external conditions during testing, including the parameters of the target and interference environment in the UAS coverage area;

E_{LMSI} is an indicator of the effectiveness of the created LMSI.

It is convenient to use the indicator of the maximum number (N_{ix}) of estimated performance characteristics of a UAS sample per unit of time as a generalized indicator of the effectiveness of the LMS E_{LMSI}, i.e. definition as

$$E_{LMSI} = \frac{N_{ix}(K_s(A_{ki}), T_u(K_s))}{T_u(K_s)} \tag{7}$$

Moreover, the realizable scope of tests of the UAS product in terms of the maximum number of evaluated characteristics N_{ix} for a given duration will depend ΔT_u on the capabilities of modeling tools (rigs and models) for assessing performance characteristics with a given accuracy and reliability of estimates:

$$N_{ix}(\Delta T_u) = N_{ile}(\Delta T_u) + N_{iim}(\Delta T_u) + N_{imm}(\Delta T_u) \tag{8}$$

This approach makes it possible not only to take into account the realizable volume in terms of the number of evaluated characteristics in the LMSI, but also the dynamic component of the evaluation process, i.e. time spent on characterization during testing.

Formalization of the process of high-precision evaluation of characteristics in the dynamics of the development and testing of UAS is represented by the following system of equations:

$$M: \begin{cases} V_{xo1}^{*}(t_1) = \underset{R_{LS},\Omega_{\xi},i=\overline{1,k}}{\arg\min} S_{vx}(R_{LS}, V_{xo1}(A_{oi}, t_1), V_e(t_1), \xi, D_{e1}(t_1)); \\ \vdots \\ V_{xon}^{*}(t_n) = \underset{R_{LS},\Omega_{\xi},i=\overline{1,k}}{\arg\min} S_{vx}(R_{LS}, V_{xon}(A_{oi}, t_n), V_e(t_n), \xi, D_{en}(t_n)); \end{cases} \quad (9)$$

where S_{vx} is the used measure of compliance V_{xo} of the estimated UAS performance characteristics vector with the reference values V_{xe} of the UAS characteristics;

R_{LS} many options for appearance (m), modes of operation, structure and parameters of UAS;

Ω_{ζ} a set of parameters of external conditions (ξ) and test modes for evaluating the characteristics of systems of this class formed in the process of testing based on them;

$D_{e1}(t_1)$, ..., $D_{en}(t_n)$ vectors of experimental data on the characteristics of UAS and other information sensors obtained in different periods of time t_1, ..., t_n;

A_{oi} the algorithm used (from a discrete set of k-algorithms) for evaluating the characteristics of the UAS product, in accordance with which the presented method M was developed,

$V_{xe}(t_1)$, ..., $V_{xe}(t_n)$ values of characteristics obtained during field tests of UAS.

To determine the measure of conformity (S_{VX}) of the estimated characteristics V_{XO}, a model for the formation of reference values of performance characteristics is used, as well as subtracting devices that perform the functions of comparing similar characteristics $V_{XO}^{(e)} - V_{XO}^{(m)}$. The results of flight or half-life experiments are considered reference V_{XO}.

5 Conclusion

The test complex for UAS certification is a cyber-physical system, therefore, when developing the IC sufficiency indicator, the relevant principles of complex systems theory and system analysis are taken into account. This made it possible to create a unified methodology for choosing a rational image of the complex, reflecting the general principles of its preparation for ensuring the certification of UAS.

As a result of the research, the expediency of developing and practical application of the methodology for choosing and substantiating the rational appearance of IC has been shown. This will make it possible to objectively substantiate the composition and the minimum required volume of measures to improve the complex in order to bring it to a state in which it is able to ensure the planned flight tests.

The use of LMSI in the interests of implementing the concept of accelerating UAS tests will significantly, by 2–2.5 times, reduce the time spent on the test program. For the application of LMSI, it is necessary to develop and certify the means of semi-natural and simulation modeling and a complex mathematical model of UAS.

Acknowledgements. This research was supported by the grant of the President of the Russian Federation for leading scientific schools of the Russian Federation NSh-122.2022.1.6.

References

1. Korsun, O.N., Stulovskii, A.V., Gulevich, S.P.: Comparative study of pontryagin maximum principle and direct method in aircraft optimal control design. In: IOP Conference Series: Materials Science and Engineering: Workshop on Materials and Engineering in Aeronautics, p. 012011 (2020)
2. Kharkov, V.P., Korsun, O.N., Pravidlo, M.N., Htang Om, M.: Nonlinear stochastic estimation of altitude and vertical flight speed in real time. In: IOP Conference Series: Materials Science and Engineering: Workshop on Materials and Engineering in Aeronautics, p. 012009 (2021)
3. Kulik, A.: Artificial intelligence-based aircraft accident threat parrying method. Proc. Telecommun. Univ. **7**(4), 110–117 (2021)
4. Bolshakov, A.A., Kulik, A.A.: Algorithm of a device designed to support decision making to counter the threat of an aviation accident. Herald of the Bauman Moscow State Technical University. Series Instrument Engineering, no. 3 (136), pp. 46–59 (2021)
5. Markov, N., Bogomolov, A., Shishov, A., Dvornikov M.: Information technology concept of integration of computing resources and physical processes in cyber-physical systems for personalized information about the potential danger of an emergency situation in high-altitude flight. Stud. Syst. Decis. Control **338**, 205-214 (2021)
6. Kuleshov, S.V., Zaytseva, A.A., Ronzhin, A.L.: The development of soft defined distributed infocommunication systems architecture based on the active data technology. Lecture Notes in Networks and Systems, vol. 95, pp. 257–265 (2020)
7. Soldatov, A.S.: The concept of creating a promising information-measuring system for flight tests of aviation equipment using Industry 4.0 technologies. Aerospace forces. Theor. Pract. **20**, 167–177 (2021)
8. Suyatinov, S.I., Buldakova, T.I., Vishnevskaya, Y.A.: Synergistic model of situational awareness of the human operator. Society 5.0: Human-Centered Society Challenges and Solutions. Cham, pp. 331–340 (2022)
9. Tobin, D., Bogomolov, A., Golosovskiy, M.: Model of organization of software testing for cyber-physical systems. Stud. Syst. Decis. Control **418**, 51–60 (2022)
10. Bychkov, E.V., Bogomolov, A.V., Kotlovanov, K.Y.: Stochastic mathematical model of internal waves. Bulletin of the South Ural State University. Series: Mathematical Modelling, Programming and Computer Software, vol. 13, issue no. 2, pp. 33–42 (2020)
11. Drozdov, S.V., Dragan, S.P., Bogomolov, A.V., Suleymanov, A.E.: A method to determine aircraft speed from ground acoustic measurements. Mosc. Univ. Phys. Bull. **75**(5), 459–464 (2020)
12. Kravets, A.G., Bolshakov, A.A., Shcherbakov, M.V.: Cyber-physical systems: Industry 4.0 challenges. Lecture Notes in Energy, p. 340 (2019)
13. Veshneva, I., Bolshakov, A., Kulik, A.: Increasing the safety of flights with the use of mathematical model based on status functions. Stud. Syst. Decis. Control 199:608–621 (2019)
14. Larkin, E.V., Nguyen, V.S., Privalov, A.N.: Simulation of digital control systems by nonlinear objects. Lect. Notes Data Eng. Commun. Technol. **124**, 711–721 (2022)
15. Kazakov, I.D., Shcherbakova, N.L., Shcherbakov, M.V., Brebels, A.: Accelerometer data based cyber-physical system for training intensity estimation. Stud. Syst. Decis. Control **259**, 325–335 (2020)
16. Soldatenko, S., Bogomolov, A., Ronzhin, A.: Mathematical modelling of climate change and variability in the context of outdoor ergonomics. Mathematics **9**(22) (2021)
17. Viet, N.T., Kravets, A., Duong Quoc Hoang, T.: Data mining methods for analysis and forecast of an emerging technology trend: a systematic mapping study from Scopus papers. Lecture Notes in Computer Science, vol. 12948 LNAI, pp. 81–101 (2021)

18. Larkin, E., Bogomolov, A., Akimenko, T., Privalov, A. Digital control of continuous production with dry friction at actuators. Smart Innov. Syst. Technol. **232**, 427–436 (2022)
19. Suyatinov, S.I., Buldakova, T.I., Vishnevskaya, J.A.: Identification of situations based on synergetic model. In: Proceedings—2021 3rd International Conference on Control Systems, Mathematical Modeling, Automation and Energy Efficiency, SUMMA 2021, pp. 509–514 (2021)
20. Kulik, A.A.: Development of a method for computation of aircraft safety control signal. Mechatron. Autom. Control **21**(11), 656–662 (2020)
21. Larkin, E.V., Akimenko, T.A., Bogomolov, A.V.: Modeling the reliability of the onboard equipment of a mobile robot. Izv. Saratov University. New Ser. Ser.: Math. Mech. Informl. **21**(3), 390–399 (2021)
22. Larkin, E.V., Bogomolov, A.V., Privalov, A.N., Dobrovolsky, N.N.: Relay races along a pair of selectable routes. Bull. South Ural State Univ. Ser.: Math. Mol. Program. Comput. Softw. **11**(1), 15–26 (2018)
23. Shcherbakov, M.V., Glotov, A.V., Cheremisinov, S.V.: Proactive and predictive maintenance of cyber-physical systems. Stud. Syst. Decis. Control **259**, 263–278 (2020)
24. Larkin, E.V., Bogomolov, A.V., Privalov, A.N., Dobrovolsky, N.N.: Discrete model of paired relay-race. Bull. South Ural State Univ. Ser.: Math. Mol. Program. Comput. Softw. **11**(3), 72–84 (2018)
25. Kravets, A.G., Seelantiev, A., Salnikova, N., Medintseva, I.: Redmine-based approach for automatic tasks distribution in the industrial automation projects. Stud. Syst. Decis. Control **418**, 261–273 (2022)

Information Models of UAVs Group Control

Anatoliy Gaiduk[1] and Sergey Kapustyan[2(✉)]

[1] Southern Federal University, Taganrog, Russian Federation
[2] Southern Scientific Centre, Russian Academy of Sciences, Rostov-On-Don, Russian Federation

Abstract. The control of unmanned aerial vehicles (UAVs) and their groups is a complex task. This task is especially complicated in the presence of uncertainty due to the incompleteness or unreliability of data coming from on-board information-measuring systems, as well as the inaccuracy of mathematical models, UAVs used by their control systems. When managing groups of UAVs, uncertainty may arise in the exchange of information between UAVs. The article deals with informational mathematical models of movements of UAVs and their groups, as well as signals from sensors used in control systems. An original approach is proposed to the formation of an estimate of the rate of change of signals used in checking the reliability of information from sensors, as well as a forecast model for information breaks.

Keywords: Control system of UAV · Sensor · Information · Reliability · Forecast model

1 Introduction

Unmanned aerial vehicles are increasingly being used in various spheres of human activity of industrial, social and defense orientation. In the English-language literature, these devices are called similarly - Pilotless Flying Devices (PLFD) or Unmanned Aerial Vehicle (UAV) [1–4]. Of course, many tasks such as the delivery of individual orders, monitoring of individual automobile or railway routes, the state of power lines, pipelines, ice conditions on rivers, inspection of the state border and other similar tasks are successfully solved by single UAVs [2, 3]. At the same time, the solution of such tasks as monitoring large forest areas, mapping territories, surveying marine areas, and solving many defense tasks with single UAVs is very difficult and most often turns out to be economically inefficient. Therefore, in these and many other similar cases, groups consisting of several UAVs are used [4–8].

Usually, a group of UAVs is understood as a set of a certain number of UAVs of the same type or different types that solve a common target task. Often, groups combine UAVs that complement the functionality of each other, so the capabilities of a heterogeneous UAV group are much wider than the capabilities of a single UAV of any type or a homogeneous group of UAVs. In this case, the group may include from several UAVs to several tens or even hundreds of UAVs of various types. This is especially true for small UAVs.

R. Silhavy et al. (Eds.): CoMeSySo 2022, LNNS 596, pp. 397–410, 2023.
https://doi.org/10.1007/978-3-031-21435-6_34

As is known, the solution of any problems in an automatic mode (without direct human participation) requires the development of analytical methods and corresponding algorithms implemented by computational means. In fact, analytical methods, algorithms, and programs that implement them for solving the above-mentioned UAV control tasks constitute the information support for UAV group actions. At the same time, UAVs must have some intelligence, that is, they must have on board a certain computer complex, a system of sensors and actuators that allow them to quickly adapt to changes in the current situation. Group control systems can be centralized (concentrated) and located on one of the UAVs, for example, on the leader, respectively, of the group and cluster. However, they can also be distributed, i.e. to combine the on-board computing devices (BVU) of the UAV group with a communication system.

Information support for UAV group control includes, first of all, mathematical models for describing and controlling UAV movements, models for planning UAV flight trajectories, taking into account the tasks of flying around moving and fixed obstacles, as well as algorithms for solving group control problems. In this case, models of signals coming from sensors and methods for their processing are of great importance. A separate task is the task of information interaction of UAVs both with each other and with the ground control center. Since UAVs are always under the influence of various external influences, the sensor signals are accompanied by interference, therefore, methods and models for processing information coming from sensors include methods and algorithms for filtering, evaluating the reliability and predicting signals [9]. As a rule, these methods are largely determined by the types and models of signals.

The purpose of this work is a comparative analysis of UAV mathematical models and signals used in control systems to describe movements and signal processing.

Currently, the most common are UAVs of the aircraft type and multicopters, therefore, UAVs of such types are considered in this paper.

2 Mathematical Models of UAVs

Models of almost all types of UAVs are developed on the basis of a model of motion of an absolutely rigid body in three-dimensional space [5, 10, 11]. In this case, as a rule, several coordinate systems are used, the formalism of Lagrange-Euler, Newton-Euler or generalized d'Alembert equations. These approaches generally lead to a system of 12 nonlinear differential equations of the 1st order. Taking into account the dynamics of the drives of the control elements, the order of this system is even more increased. Therefore, in practice, various simplifying assumptions are always introduced, which lead to various mathematical models of UAVs. Consider some of the applied UAV models.

The equations of motion of the UAV as a rigid body include six equations of kinematics and six equations of dynamics [5, 9, 11]. The equations of kinematics determine the connection of a moving coordinate system with a fixed, terrestrial coordinate system. The equations of dynamics for constant values of the mass and moments of inertia of the UAV in vector-matrix form, in the moving coordinate system associated with the UAV, have the form:

$$\dot{\mathbf{V}} = \Omega \mathbf{V} + m^{-1}(\mathbf{G} + \mathbf{A}_p + \mathbf{U} + \mathbf{R}_a), \tag{1}$$

$$\dot{\omega} = \mathbf{J}^{-1}[\mathbf{M}_A + \mathbf{M}_U + \mathbf{M}_{Ra} - \boldsymbol{\omega} \times \mathbf{J}\omega], \quad (2)$$

where

$$\mathbf{V} = \begin{pmatrix} V_x \\ V_y \\ V_z \end{pmatrix}, \mathbf{\&} = \begin{pmatrix} \omega_x \\ \omega_y \\ \omega_z \end{pmatrix}, \mathbf{r} = \begin{pmatrix} x \\ y \\ z \end{pmatrix}, \mathbf{R}_a = \begin{pmatrix} R_{ax} \\ R_{ay} \\ R_{az} \end{pmatrix}, \mathbf{M}_{Ra} = \begin{pmatrix} M_{Rax} \\ M_{Ray} \\ M_{Raz} \end{pmatrix}, \mathbf{U} = \begin{pmatrix} U_x \\ U_y \\ U_z \end{pmatrix},$$

$$\mathbf{M}_U = \begin{pmatrix} M_{Ux} \\ M_{Uy} \\ M_{Uz} \end{pmatrix}, \mathbf{\Omega} = \begin{pmatrix} 0 & \omega_z & -\omega_y \\ -\omega_z & 0 & \omega_x \\ \omega_y & -\omega_x & 0 \end{pmatrix}, \mathbf{A}_p = \begin{pmatrix} \rho g O \sin\vartheta \\ \rho g O \cos\vartheta \cos\gamma \\ -\rho g O \cos\vartheta \sin\gamma \end{pmatrix},$$

$$\mathbf{G} = \begin{pmatrix} -mg \sin\vartheta \\ -mg \cos\vartheta \cos\gamma \\ mg \cos\vartheta \sin\gamma \end{pmatrix} = \begin{pmatrix} G_x \\ G_y \\ G_z \end{pmatrix}, \mathbf{M}_A = \begin{pmatrix} 0 & G_z & -G_y \\ -G_z & 0 & G_x \\ G_y & -G_x & 0 \end{pmatrix} \begin{pmatrix} x \\ y \\ z \end{pmatrix},$$

$$\mathbf{J} = \begin{pmatrix} J_{xx} & -J_{xy} & -J_{xz} \\ -J_{xy} & J_{yy} & -J_{yz} \\ -J_{xz} & -J_{yz} & J_{zz} \end{pmatrix}.$$

Here **V** and **ω** – are the linear and angular velocity vectors of the UAV; $\mathbf{r}_c$ – is the radius vector of the point of application of the Archimedes force $\mathbf{A}_p$; **G** is the gravity vector; $\mathbf{M}_A$ – is the vector of the moment of the Archimedes force relative to the axes of the coordinate system associated with the UAV; **U** and $\mathbf{M}_U$ are vectors of control forces and moments formed by control planes; $\mathbf{R}_a$ and $\mathbf{M}_{Ra}$ – vectors of aerodynamic forces and moments due to air flow around the body; **J** is the matrix of moments of inertia; ψ, ϑ, γ are yaw, pitch and roll angles; ρ – air density; g is the free fall acceleration; × – vector product designation; O – UAV volume.

In [12], the UAV motion equations using the quaternion theory are represented by thirteenth-order equations, but also without taking into account the dynamics of control drives.

Note that Eq. (1) describes the dynamics of the translational, and Eq. (2) of the rotational motion of the UAV. The addition to Eqs. (1), (2) of the equations of kinematics and drives of controls - actuators (EA), obviously, significantly complicates the task of synthesizing UAV motion control systems.

In order to simplify the solution of this problem, first of all, the drives are assumed to be inertialess.

Taking into account the equations of kinematics, the equations are written directly in the terrestrial coordinate system [13]:

$$\dot{V} = g(n_x - \sin\theta), \dot{\theta} = \frac{g}{V}\left(n_y - \cos\theta\right), \dot{\psi} = -\frac{g}{V}\frac{n_z}{\cos\theta}; \quad (3)$$

$$\dot{x}_m = V\ \cos\theta\ \cos\psi, \dot{y}_m = V\ \sin\theta, \dot{z}_m = -V\ \cos\theta\ \sin\psi; \quad (4)$$

$$\dot{n}_y = u_1, \dot{n}_z = u_2, \quad (5)$$

where $V, \theta, \psi, x_{\mathrm{m}}, y_{\mathrm{m}}, z_{\mathrm{m}}$ – UAV speed, angles of inclination and rotation of the trajectory, coordinates of the UAV center of mass in space; n_x, n_y, n_z – longitudinal, normal and lateral components of overload, respectively; $q = \rho V^2/2$ – high-speed pressure; m is the mass; S is the given parameter; $n_x = (R(V, \rho, u_1) - A_x)/mg$; $\alpha = n_y/n_y^{\alpha}$; $\beta = n_z/n_z^{\beta}$; $n_y^{\alpha} = c_y^{\alpha} qS/\mathrm{mg}$; $n_z^{\beta} = c_z^{\beta} qS/mg$; $R(V, \rho, u_1)$ – engine thrust; $A_x = c_x qS$ $c_x = c_x(V, h, \alpha, \beta)$, $c_y^{\alpha} = c_y^{\alpha}(V, h)$, $c_z^{\beta} = c_z^{\beta}(V, h)$ – are aerodynamic characteristics; $h = y_{\mathrm{m}}$ – is the flight altitude, u_i – is the UAV control, $i = 1, 2, 3$.

In contrast to Eqs. (1), in model (3)–(5), the longitudinal motion is characterized only by the velocity V and its projections on the axes of the earth's coordinate system, and the rotational motion is characterized by the angles of inclination and rotation of the UAV trajectory. Therefore, the actual movement of the UAV is described by six differential Eqs. (3), (4). Equations (5) describe the actuators of the UAV, through which changes are made to its overloads. In this case, these devices are assumed to be inertialess. The input signals of these devices are controls. Despite the significant simplification of the UAV model, Eqs. (3)–(5) make it possible to ensure the movement of the UAV along rather complex curvilinear spatial trajectories with appropriate controls.

When solving some UAV control problems, even simpler mathematical models are used. For example, when solving the problem of controlling the exit of a group of UAVs to the required points, taking into account the mutual influence of UAVs on each other, the UAV model is described by the equations:

$$\dot{\varphi}_{hi} = \omega_{hi} + u_{hi}/\dot{D}_i, \, \dot{\omega}_{hi} = -2(\dot{D}_i \omega_{hi}/D_i) - u_{hi}/\dot{D}_i, \tag{6}$$

where φ and φ_g – are the required and current bearings of targets in the horizontal plane; u_{hi} – is acceleration (instantaneous lateral control); D_i and $\dot{D}_i$ – the distance to the required point of arrival of the ith UAV and the rate of its change; ω_i and ω_{hi} are the required and current angular velocities of the line of sight of the target point of the ith UAV [14]. When solving this problem, Eq. (6) are written in a vector-matrix form and the relations of the theory of optimal estimation and control are applied.

UAV models represented by Eqs. (1)–(6) are non-linear. This greatly complicates the solution of some UAV motion control tasks. Therefore, in many works, linear models are often used in the form of differential equations "input-output", transfer functions or in state variables [8, 15, 16]. In the latter case, the model is written in vector-matrix form and usually has the following form:

$$\dot{\mathbf{x}} = \mathbf{A}\mathbf{x} + \mathbf{B}\mathbf{u} + \boldsymbol{\zeta}_1, \, \mathbf{y} = \mathbf{C}\mathbf{x} + \mathbf{D}\mathbf{u} + \boldsymbol{\zeta}_2, \, \mathbf{v} = \mathbf{H}\mathbf{x} + \boldsymbol{\zeta}_3. \tag{7}$$

Here **x, u, y, v** and $\boldsymbol{\zeta}_i$ are the vectors of state variables, controls, controlled, observed UAV variables and external disturbances $i = 1, 2$; **A, B, C, D, H** – are matrices of constant coefficients corresponding to the dimensions. The first of Eqs. (7) describes changes in the state of the UAV; the state variables are usually the same variables as in Eqs. (1)–(4). The second of Eqs. (7) is the output equation, i.e. controlled variables, and the third equation is the observation (measurement) equation. It describes a measurement system that measures the values of some UAV variables necessary to control it. Most often, disturbances $\zeta_i = \zeta_i(t)$ are presented as random, such as white noise with known statistical characteristics, or are assumed to be absent. If real disturbances contain deterministic components, then their models are included in matrices **A, B, C, D**.

The dimension of the vector **x**, i.e. the number of state variables taken into account in the UAV model is determined by the content of the problem being solved. Thus, in [17], the models of the pitch angle control channel (of the third order) and the steering gear (of the first order) were used, given by transfer functions. The corresponding UAV motion model in state variables has the form:

$$\dot{\tilde{\mathbf{x}}} = \begin{pmatrix} 0 & 1 & 0 & 0 \\ a_{21} & a_{22} & 0 & 1 \\ a_{31} & a_{32} & 0 & 0 \\ 0 & 0 & 0 & a_{44} \end{pmatrix} \tilde{\mathbf{x}} + \begin{pmatrix} 0 \\ 0 \\ 0 \\ b_4 \end{pmatrix} u + \begin{pmatrix} 0 \\ 0 \\ 1 \\ 0 \end{pmatrix} \zeta, \; \boldsymbol{\vartheta} = \begin{pmatrix} 0 & 0 & 1 & 0 \end{pmatrix} \tilde{\mathbf{x}}, \tag{8}$$

where $a_{21} = -1/T^2$, $a_{22} = -2d/T$, $a_{31} = k_c/T^2$, $a_{32} = k_c T_1/T^2$, $a_{44} = -1/T_{\text{sg}}$, $b_4 = 1/T_{\text{sg}}$; T, T_1 are the time constants, k_c and и d are the transmission coefficient and the damping coefficient for the pitch channel of the UAV, respectively; T_{sg} – time constant of the steering gear; ζ – wind load.

This fourth-order motion model makes it possible to ensure the movement of the UAV along a given route under conditions of control and wind load restrictions using a predictive artificial neural network and reference points.

When planning trajectories, the simplest equations for each of the axes of the attached or earth coordinate system are often used. For example, when monitoring a certain rectangular area, the UAV makes straight flights at a constant height and turns [18]. In this case, to describe its motion along given trajectories, the equations can be used

$$\dot{x} = V_x,\; \dot{V}_x = \frac{b_x}{m} u_x - \frac{\beta_x}{m} V_x^2 + \zeta_x, \tag{9}$$

$$\dot{z} = V_z,\; \dot{V}_z = \frac{b_z}{m} u_z - \frac{\beta_z}{m} V_z^2 + \zeta_z, \tag{10}$$

$$\dot{\psi} = \omega_\psi,\; \dot{\omega}_\psi = \frac{b_\psi}{J_y} u_\psi - \frac{\beta_\psi}{J_y} V_\psi^2 + \zeta_\psi, \tag{11}$$

where $b_x, b_z, b_\psi, \beta_x, \beta_z, \beta_\psi$ are aerodynamic coefficients; $\zeta_x, \zeta_z, \zeta_\psi$ are projections of external disturbances; u_x, u_z, u_ψ – control of the longitudinal, transverse movement and heading of the UAV.

As noted above, the group application of UAVs is more effective, which requires mathematical models of UAV groups. In these cases, both linear and non-linear models are used. Thus, in [19, 20] the model of the UAV group was used as a set of Eqs. (7) taking into account the mutual influence of the UAV. This model of a group of N UAVs with a leader is described by the equations:

$$\dot{\mathbf{x}}_i = \mathbf{A}_i \mathbf{x}_i + \sum_{j=1; j \neq i}^{N} \mathbf{M}_{ij} \mathbf{x}_j + \mathbf{B}_i \mathbf{u}_i,\; i = \overline{2,\ N}, \tag{12}$$

$$\dot{\mathbf{x}}_{ij} = \mathbf{A}_{ij} \mathbf{x}_{ij} + \sum_{\nu=1;\ \nu \neq i;\ \nu \neq j}^{N} \mathbf{x}_{i\nu}^T \mathbf{M}_{ij}^{\nu} \mathbf{x}_{\nu j},\; i = \overline{1,\ N-1}. \tag{13}$$

Here $\mathbf{x}_1$ and $\mathbf{u}_1$ – n_1-vector of the state and ν_1-vector of the control; $\mathbf{A}_1$, $\mathbf{B}_1$, $\mathbf{M}_{1j}$ – matrices of parameters of the leader (leading UAV); $\mathbf{x}_i$ and $\mathbf{u}_i$ – n_i-vector of the state and ν_i-vector of the control i-th UAV, $i \neq 1$; $\mathbf{A}_i$, $\mathbf{B}_i$ – system matrices of own dynamics of the i-th UAV; $\mathbf{M}_{ij}$ is a matrix describing the influence of the j-th UAV on the state of the i-th UAV; $\mathbf{x}_{ij}$ – μ_i-vector of the relative position of the i-th and j-th UAV; $\mathbf{A}_{ij}$– a matrix describing the influence of the i-th and j-th UAV on their relative position; $\mathbf{M}^{\nu}_{ij}$ – a matrix describing the influence of the -th UAV on the relative position of the i-th and j-th UAV; T – is the symbol of the transposition operation. It should be noted that the pairs of matrices ($\mathbf{A}_i$, $\mathbf{B}_i$)are completely controllable for all $i = \overline{1,\ N}$.

The movement of the group is controlled by decentralized controls, which are formed by the control system of each ith UAV, taking into account the coordinates of the leader. The leader's coordinates, as well as their own, are determined by the navigation system of each i-th UAV. In [20], conditions were obtained for the coefficients of the matrices $\mathbf{M}_{ij}$ and $\mathbf{M}^{\nu}_{ij}$ describing the mutual influence of UAVs on each other, under which the movement of the UAV group, determined by the trajectory of the leader, is asymptotically stable.

In [18], as a model of horizontal movements of a group of UAVs, a set of Eqs. (9)–(11) was used, written for each i-th UAV of the group.

3 Models and Forecasts of Information Signals

As information signals in UAV control systems, signals generated by master devices or sensors are used. Drivers usually form the required (given) UAV trajectories. If the given trajectory of some i-th UAV is represented as the movement of a point in the earth's coordinate system $O_3x_3y_3z_3$, then the equations of its movement can be represented by the following expressions [21]:

$$\begin{aligned} &\dot{V}_i = a_i, \dot{\theta}_i = \alpha_i/V_i, \dot{\psi}_i = \beta_i/V_i, \dot{x}_i = V_i \cos\,\theta_i\, \cos\,\psi_i, \\ &\dot{y}_i = V_i\, \sin\,\theta_i, \dot{z}_i = V_i \cos\,\theta_i\, \sin\,\psi_i \end{aligned} \tag{14}$$

where $V_i = V_i(t)$ is the given speed, a_i is the acceleration, $\theta_i = \theta_i(t)$ is the angle between the velocity vector V_i and the horizontal plane, $\psi_i = \psi_i(t)$ is the angle between the direction of the projection of the velocity vector $V_{\mathrm{T}i} = V_{\mathrm{T}i}(t)$ on the horizontal plane and the axis Ox_3, which determines the direction of the trajectory; $x_{\mathrm{T}i} = x_{\mathrm{T}i}(t)$, $y_{\mathrm{T}i} = y_{\mathrm{T}i}(t)$, $z_{\mathrm{T}i} = z_{\mathrm{T}i}(t)$ – are the coordinates of the ith point moving along the given trajectory at time t. Equations (14) are always accompanied by the initial conditions $V_{\mathrm{T}i/0}$, $\theta_{\mathrm{T}i/0}$, $\psi_{\mathrm{T}i/0}$, $x_{\mathrm{T}i/0}$, $y_{\mathrm{T}/i0}$, $z_{\mathrm{T}/i0}$, which determine the speed, angular and linear coordinates of that state of the ith UAV, which is taken as the initial one, i.e. at $t = 0$.

By choosing the laws of changing the parameters $\alpha_{\mathrm{T}i} = \alpha_{\mathrm{T}i}(t)$ and $\beta_{\mathrm{T}i} = \beta_{\mathrm{T}i}(t)$, which determine changes in the speed, inclination angles and direction of the trajectory in time, it is possible to set the trajectories of each i-th UAV necessary to complete the task assigned to it.

In a particular case, when it is required, for example, the movement of the ith UAV with a constant speed V^* in a horizontal plane, at a constant given height h_i^*, parallel to the axis Ox_3, Eqs. (14) and the initial conditions take the form:

$$\dot{V}_{\mathrm{T}i} = 0, \dot{\theta}_{\mathrm{T}i} = 0, \dot{\psi}_{\mathrm{T}i} = 0, \dot{x}_{\mathrm{T}i} = V_{\mathrm{T}i/0}, \dot{y}_{\mathrm{T}i} = 0, \dot{z}_{\mathrm{T}i} = 0,$$

$$V_{Ti/0} = V^*,\ \theta_{Ti/0} = 0,\ \psi_{Ti/0} = 0,\ x_{Ti/0} = x^*_{T/i},\ y_{Ti/0} = h^*_i,\ z_{Ti/0} = z^*_{T/i}$$

Here $x^*_{Ti},\ z^*_{Ti}$ are the coordinates of the projection onto the earth's surface (which is considered flat) of the point from which the required movement of the i-th UAV should begin at the time taken as the reference point $t = 0$ [18].

Scalar perturbations $f(t)$ acting on UAVs are most often described by deterministic stepwise, linear or more complex polynomial functions of time, of the following form:

$$f_i(t) = (f_{0i} + \ldots + f_{\varsigma i} t^{\varsigma}) 1(t - t_{0i}), \tag{15}$$

where $f_{\nu i}$ are the parameters of the disturbance $f_i(t)$, acting on the i-th UAV,$\nu = [0,\ 1, \ldots,\ \varsigma]$; t_{0i} is the moment of the beginning of the influence of this perturbation; ς is the degree of perturbation (15). Parameters actually determine the values of the disturbance, its speed, second, third, etc. derivatives.

A more general perturbation model compared to (15) is the equations in state variables:

$$\dot{\mathbf{w}}_i = \mathbf{F}_i \mathbf{w}_i + \boldsymbol{\zeta}_{\tilde{f}i}, \tilde{f}_i = c^T_{\tilde{f}i} \mathbf{w}_i, t \geq t_{0i}. \tag{16}$$

where $\mathbf{w}_i$ are the state vectors of the perturbation generator model $\tilde{f}_i(t)$; $c_{\tilde{f}i}$ is the vector of constant coefficients; $\boldsymbol{\zeta}_{\tilde{f}i}$ is a vector of the white noise type, which is the input signal of the shaping filter of the random component of the disturbance $\tilde{f}_i(t)$. The nature of the deterministic and random components of this perturbation is determined by the type of matrix $\mathbf{F}_i$, and the numerical characteristics are determined by the initial conditions of the vector and the values of the coefficients of the vector $c_{\tilde{f}i}$. The filter matrix, which reflects the correlation properties of the random component of the perturbation $\tilde{f}_i(t)$, is contained in the matrix $\mathbf{F}_i$ [7, 8].

For example, if the wind has a gusty (oscillatory) character, then the model of its impact on a certain UAV can be described by the function

$$f(t) = f_0 + f_m \sin(\omega_f t + \varphi_f) + f(t), \tag{17}$$

where f_0 is the constant component; f_m – amplitude, ω_f and φ_f – frequency and phase of the oscillatory deterministic component; $f_{\text{cn}}(t)$ is the random component of the wind effect (17). Suppose this random component has a spectral density

$$S_{f_{\text{cn}}}(\omega) = \frac{\varphi_0^2}{\omega^2 + \varphi_1^2}, \tag{18}$$

where $\varphi_0,\ \ \phi_1$ are numerical coefficients characterizing the intensity and correlation properties of the random component. In this case, the model vector $\mathbf{w}_i$ in the form (16) of the wind disturbance (17), (18) has a dimension equal to four, and the matrix $\mathbf{F}_i$ and vector $c^T_{\tilde{f}i}$ have the form

$$\mathbf{F} = \begin{pmatrix} 0 & 0 & 0 & 0 \\ 0 & 0 & \omega_f^2 & 0 \\ 0 & 1 & 0 & 0 \\ 0 & 0 & 0 & -\phi_1 \end{pmatrix}, \boldsymbol{\zeta}_f = \begin{pmatrix} 0 \\ 0 \\ 0 \\ \zeta_{\text{cn}} \end{pmatrix}, c_f^T = \begin{pmatrix} 1 & 0 & 1 & 1 \end{pmatrix}. \tag{19}$$

The values f_0, f_m and φ_f from expression (17) are provided by the choice of the corresponding initial values of the vector $\mathbf{w}_0 = \left(w_{10}\ w_{20}\ w_{30}\ 0\right)^T$, and white noise ζ_{cn} has a spectral density $S_{\xi_f}(\varphi) = \varphi_0^2$, since the significant coefficients of the vector c_f are equal to one.

Usually, in UAV control systems, neither the setting actions (14), nor the variables of the objects, nor the disturbances, the models of which have the form (1)–(13), (15) or (16), can be used directly. The master influences are often formed by the leader of the group and come on board the UAV via a communication channel, and the variables or disturbances used in the control system are measured by sensors [15]. However, both the transmission process and the measurement process on board the UAV are always affected by random noise, so both the output signals of communication channels and sensors usually represent an additive mixture of the form

$$v(t) = \eta(t) + \xi(t), \tag{20}$$

where $\eta(t)$ is the useful information signal of the radio channel or measured by the sensor; $\xi(t)$ – interference close to white noise [6].

In order to isolate the useful component, the signal (20) is subjected to processing processes. These processes include: the least squares method, the Kalman-Bucy filter, filters of other types, the exponential smoothing method, the sliding smoothing method, and others. All these methods are covered in sufficient detail in numerous works, for example, in [6, 14, 22] and are not considered here.

As noted in [23], when developing information processing systems, special attention should be paid to the relevance and reliability of the information received from sensors. For this purpose, the sensors are usually checked for "plausibility". To solve this problem, an estimate of the rate of change of the signal coming from the sensor is usually used. This is due to the fact that the useful component of the sensor signal is usually narrowband and has a non-zero DC component. Since when the sensor fails, its output signal changes its value significantly, then a sharp jump in the rate of change of the sensor signal, in fact, is a sign of the unreliability of information coming from this sensor in the future.

Thus, the problem arises of determining the rate of change of signals of the type (20) with a non-zero useful component $\eta(t)$. At the same time, the use of differentiators of various types turns out to be practically impossible due to the presence of random components. Let us further consider a statistical approach to estimating the rate of change of sensor signals [24].

Suppose the sensor signal $v(t)$ (20) enters the processing system, where its dispersion, correlation function and correlation time are determined:

$$D_v = \int_0^\infty v(t)v(t)\,dt,\ R_{vv}(\tau) = \int_0^\infty v(t)v(t+\tau)\,dt,\ \tau_k = \frac{S_{Rv}}{D_v} = \frac{S_{Rv}}{R_{vv}(0)}. \tag{21}$$

Here, $R_{vv}(\tau)$ is the correlation function, $D_v = R_{vv}(0)$ is the signal dispersion $v(t)$; S_{Rv} is the area enclosed between the time axis τ and the graph of the function $R_{vv}(\tau)$; τ_k – correlation time (Fig. 1). As can be seen from (21), the value τ_k is determined so that the area of the rectangle is $OD_v\tau_k$ equal to the area S_{Rv}.

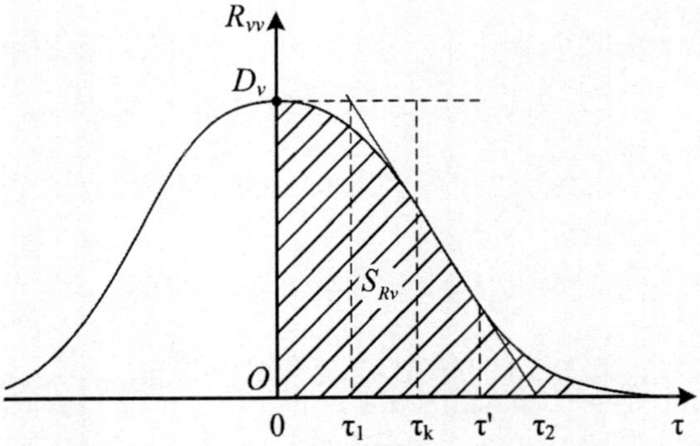

Fig. 1. Toward an estimate of the correlation time

τ.

It is also assumed that the signal $v(t)$ is characterized by a small value of the correlation time τ_k and a non-zero average value, and the average noise value $\xi(t)$ is zero, i.e.

$$\overline{v} = \int_0^\infty v(t)\,dt = \overline{\eta} \neq 0,\ \overline{\xi} = \int_0^\infty \xi(t)\,dt = 0. \tag{22}$$

The task is to estimate the rate of change of the useful signal $\eta(t)$, i.e. values $\dot{\eta}(t) = d\,\eta(t)/d\,t$, according to the results of its processing (21). Turning to its solution, first of all, we transform the expression for the correlation function $R_{vv}(\tau)$ (21) taking into account the connection of the output signal of the sensor with the measured information signal.

To this end, we substitute (20) into the second equality (21) and divide the integral of the sum into several integrals. As a result, we get

$$R_{vv}(\tau) = R_{\eta\eta}(\tau) + R_{\eta\xi}(\tau) + R_{\xi\eta}(\tau) + R_{\xi\xi}(\tau). \tag{23}$$

The information signal $\eta(t)$ and noise $\xi(t)$ are obviously uncorrelated, since they are generated by different factors, and the correlation functions of random signals close to white noise are very small at small values τ, therefore $R_{\eta\xi}(\tau) = R_{\xi\eta}(\tau) = 0$. Neglecting the value $R_{\xi\xi}(\tau) \approx 0$, from (23) we obtain the almost obvious expression

$$R_{vv}(\tau) = R_{\eta\eta}(\tau) = \int_0^\infty \eta(t)\eta(t+\tau)\,dt. \tag{24}$$

Since the signal $\eta(t)$ is a continuous function of time, for small τ we can assume that

$$\eta(t+\tau) \approx \eta(t) + \tau\,\dot{\eta}(t+\tau/2). \tag{25}$$

As above, from (24) and (25) we obtain

$$R_{vv}(\tau) \approx \int_0^\infty \eta(t)\eta(t)\,dt + \tau \int_0^\infty \eta(t)\,\dot{\eta}(t+\tau/2)dt$$

or

$$R_{\nu\nu}(\tau) \approx D_\eta + \tau\, R_{\eta\dot{\eta}}(\tau/2) \approx D_\nu + \tau\, R_{\eta\dot{\eta}}(\tau/2), \tag{26}$$

where

$$R_{\eta\dot{\eta}}(\tau/2) = \int\limits_0^\infty \eta(t)\, \dot{\eta}(t+\tau/2)dt \tag{27}$$

As is known, $R_{\nu\nu}(0) = D_\nu$, $R_{\eta\eta}(0) = D_\eta$, therefore, from (26) and the graph $R_{\nu\nu}(\tau)$ shown in Fig. 1, it follows that the correlation function $R_{\eta\dot{\eta}}(\tau/2)$ is negative and has a minimum $R_{\nu\dot{\nu},\min} = -D_\nu/\tau_{\max}$, at $\tau = \tau_{\max}$, where $\tau_{\max}$ is the minimum value of τ at which $R_{\nu\nu}(\tau) \approx 0$. With a further increase in τ, this function increases monotonically, tending to zero.

Returning to solving the problem, we differentiate (24) with respect to τ. As a result, we will have

$$\frac{d\,R_{\nu\nu}(\tau)}{d\,\tau} = \int\limits_0^\infty \eta(t) \frac{d\,\eta(t+\tau)}{d\,\tau} d\,t. \tag{28}$$

In this expression, we replace the derivative $d\,\eta(t+\tau)/d\,\tau$ with the derivative $d\,\eta(t+\tau/2)/d\,\tau$. In fact, the assumption is made here that the signal $\eta(t)$ has such properties that for small τ the approximate equalities

$$\frac{d\,\eta(t)}{d\,t} \approx \frac{d\,\eta(t+\tau)}{d\tau} \approx \dot{\eta}(t+\tau/2). \tag{29}$$

Taking into account that both t and τ denote time in this case, we can assume that the approximate equalities (29) are satisfied quite accurately, especially for small τ. Taking into account (29), from (28) also follows the approximate equality

$$\frac{d\,R_{\nu\nu}(\tau)}{d\,\tau} \approx R_{\eta\dot{\eta}}(\tau/2). \tag{30}$$

Thus, both in (26) and in (30) the same correlation function $R_{\eta\dot{\eta}}(\tau/2)$ appears between the variable $\eta(t)$ and its first time derivative $\dot{\eta}(t)$. To obtain the desired result, we find the estimate $R_{\eta\dot{\eta}}(\tau/2)$. Based on (30) and the graph in Fig. 1 we conclude that the function $R_{\eta\dot{\eta}}(\tau/2)$ is indeed negative for $\tau > 0$. In this regard, let us replace $\dot{\eta}(t+\tau/2)$ by $-|\dot{\eta}_{\max}|$ in (26) for $R_{\eta\dot{\eta}}(\tau/2)$. As a result, we will have

$$R_{\eta\dot{\eta}}\left(\frac{\tau}{2}\right) > -\int\limits_0^\infty \eta(t)\, |\dot{\eta}_{\max}| dt = -\,|\dot{\eta}_{\max}|\overline{\eta}.$$

Substituting this expression into (27), we obtain.
$R_{\nu\nu}(\tau) < D_\nu - \tau\, |\dot{\eta}_{\max}|\overline{\eta}$.

In accordance with the equality $D_v = R_{vv}(0)$ and (21) - $\overline{\eta} = \overline{v}$, from here at $\tau = \tau_k$ follows the required estimate:

$$|\dot{\eta}| \leq \frac{R_{vv}(0) - R_{vv}(\tau_k)}{\tau_k \, \overline{v}}. \tag{31}$$

Turning to the analysis of the resulting inequality (31), first of all, we note that the dimension of the quantity on its right side corresponds to the dimension of the first time derivative of the signals both $\eta(t)$, and $v(t)$.

The correctness of (31) is to some extent confirmed by the relationship that follows directly from (25) and (29) when the correlation function $R_{\eta\dot{\eta}}(\tau/2)$ is excluded from them:

$$R_{vv}(\tau') \approx R_{vv}(0) + (\tau' - \tau_1) \left.\frac{d\,R_{vv}(\tau)}{d\,\tau}\right|_{\tau=\tau_k}, \tau_1 \leq \tau' \leq \tau_2. \tag{32}$$

Time points τ', τ_1 and τ_2 are shown in Figs. 1. The validity of (32) in the case of signals with monotonic damped correlation functions (Fig. 1) is quite obvious.

Thus, if along with the autocorrelation function of the sensor signal, the correlation time and the average value of the sensor signal are determined, then it is possible to estimate the maximum rate of change of the useful sensor signal. However, this possibility can be realized if only the average value of the sensor signal, i.e. its useful component has an average value not equal to zero.

The resulting expression (32) can be given a fairly clear probabilistic interpretation, in the sense of assessing the reliability of the signal coming from the sensor. Indeed, if the sensor fails at time t_P, then its output signal at $t > t_P$ loses correlation with the previous correct values. Therefore, the correlation function $R_{vv}(\tau)$ (21), starting from this moment, sharply decreases, approaching zero. In this case, the estimate of the rate of change of the information signal, as can be seen from (32), increases just as sharply, which indicates the unreliability of the information coming from the sensor, starting from the time moment t_P.

If the signal $v(t)$ coming from any sensor turns out to be unreliable, then it cannot be used in the control system. If there is a backup sensor in the UAV sensor system, the faulty sensor is replaced by a backup one. However, very often the signal from the sensor is unreliable only in a certain small time interval $t_B - t_P$, where t_B is the moment when the sensor is restored. In other words, there is a "signal loss", "information break" for a finite time interval [25]. In these cases, especially in the presence of weight and size restrictions, it is advisable to use the predictive values of the information signal in the control system. These values can be found on the basis of reliable values obtained at time points preceding the time point t_P. When constructing a predictive model, it is usually assumed that the information signal $\eta(t)$ of the sensor before the loss of the signal and on the forecast interval changes according to a quadratic or linear law.

Let us assume that the sensor, the signal of which became unreliable at $t = t_P$, is discrete, and in the controller memory there are reliable values of the signal $\eta(t_P - kT_o)$, $k \in [1,\ 2, \ldots,\ m_o]$, where $m_o \in [5 \div 10]$ is the number of these values, T_o is the polling period of this sensor. Using these values, the least squares method (LSM) for the time $t \in [t_P,\ t_P + \mu\, T_o]$, $\mu < 1$ it is easy to find the coefficients $\eta_0,\ \eta_1,\ \eta_2$ of the quadratic regression equation $\eta(t_k) = \eta_0 + \eta_1 kT_o + \eta_2 (kT_o)^2$ [26]. The prediction equation has the form

$$\eta(t_\pi + (k-1)T_o) = \eta(t_\pi - T_o) + [\eta_1 + \eta_2(2\,(t_\pi - T_o) + kT_o)]\,kT_o, \qquad (33)$$

where $\eta(t_\pi - T_0) = \eta_0$ is the last reliable value (it is used in the control system at $t = t_\pi$ instead of $\eta(t_\pi)$). The variable k in (33) takes the values $k = 2, 3, \ldots, m_B$, where $m_B = (t_B - t_\pi - T_o)/T_o$, i.e. the forecast is carried out up to the point in time $t_B - T_o$, starting from the moment t_B, the signal of the restored sensor is used in the control system.

Note that when using the linear regression equation with the help of LSM, only η_0, η_1 and coefficient $\eta_2 = 0$, and the forecast procedure is simplified. However, the admissible time interval of the forecast becomes much smaller in this case.

The above relations in conjunction with the expressions of the accepted information processing method are usually used in the creation of information support for control systems for both single and groups of UAVs. The use of certain ratios from the above is determined by the type of UAV and the content of the tasks the group is focused on solving.

4 Conclusion

The paper considers the main types of mathematical models of UAVs and their groups, which are used in the development of motion control systems for these vehicles. It is shown that these problems can be solved using nonlinear or linear models of varying complexity - from models of the twelfth order and higher to models of the second order. The choice of model order is determined by the purpose of the system being developed. However, UAV models of the fourth or sixth order are most often used. Mathematical models of UAV groups are a combination of models of single robots supplemented with equations that take into account the mutual influence of UAVs on each other. When solving the problem of group control, the approach "with the leader" is more often used. One of the important tasks of information processing used in the motion control systems of both single UAVs and groups of UAVs is the assessment of the reliability of information coming from each sensor of the UAV sensor system. This is usually done by estimating the rate of change of the sensor output signal. The paper proposes a new approach to estimating the rate of change of these signals from sensors operating under conditions of stationary random noise such as white noise.

Acknowledgement. This article was prepared within the framework of the implementation of the State Budget of the SSC RAS, № gr. Project 122020100270-3.

References

1. Merkulov, V.I.: Group control of unmanned aerial vehicles in the tasks of monitoring the earth's surface. Inf. Measuring Control Syst. **1**(14), 30–38 (2016)

2. de Moraes, R.S., de Freitas, E.P.: distributed control for groups of unmanned aerial vehicles performing surveillance missions and providing relay communication network services. J. Intell. Rob. Syst. **92**(3–4), 645–656 (2017). https://doi.org/10.1007/s10846-017-0726-z
3. Ta Duc, A., Fantoni, I., Lozano, R.: Modeling and control of a convertible mini-UAV. Preprints of the 18th IFAC World Congress Milano (Italy) August 28–September 2, pp. 1492–1497 (2011)
4. Kapustyan, S.G., Gaiduk, A.R., Merkulov, V.I., Besklubova, K.V.: Group control of the UAV. Inf. Measuring Control Syst. **11**, 24–36 (2014)
5. Intelligent technologies for planning the movement of moving objects in three-dimensional non-deterministic environments. In: Pshikhopov V.K. (eds.) Moscow, Nauka, 232 p (2017)
6. Complexes with unmanned aerial vehicles. In: Verba, V.S., Tatarsky B.G. (eds.) 2 books: Book. 1. Principles and features of the construction of complexes with UAVs. Moscow, Radiotehnika, 512 p (2016)
7. Gaiduk, A.R., Kapustyan, S.G.: The concept of building systems for collective control of unmanned aerial vehicles. Radio control systems. In: Merkulov, V.I. (eds.) Optimization of control algorithms, vol. 4, pp. 39–46. Moscow, Radiotehnika (2018)
8. Kalyaev, I.A., Gaiduk, A.R., Kapustyan, S.G.: Models and algorithms of collective control in groups of robots, 280 p. Moscow, Fizmatlit (2009)
9. Pshikhopov, V., Medvedev, M., Gaiduk, A.R., et al.: Position-trajectory control system for robot on base of airship. In: Proceedings of IEEE Conference on Decision and Control, pp. 3590–3595 (2013)
10. Ahmed, B., Pota, H.R.: Flight Control of a Rotary Wing UAV Including Flapping Dynamics. Preprints of the 18th IFAC World Congress Milano (Italy), August 28–September 2, pp. 10,373–10,378 (2011)
11. Izadi, H.A., Zhang Y., Gordon, B.W.: Fault Tolerant Model Predictive Control of Quad-Rotor Helicopters with Actuator Fault Estimation. Preprints of the 18th IFAC World Congress Milano (Italy), August 28–September 2, pp. 6343–6346 (2011)
12. Roberts, A., Tayebi, A.: Position Control of VTOL UAVs Using Inertial Vector Measurements. Preprints of the 18th IFAC World Congress, Milano (Italy), August 28–September 2, pp. 2614–2619 (2011)
13. Anisimov, V.N., Kabanov, S.A.: Control of the trajectory of an aircraft when flying over obstacles using methods of analytical mechanics. Autom. Telemechanics **3**, 3–10 (2005)
14. Complexes with unmanned aerial vehicles. In 2 books: (Book. 2). In: Verba, V.S., Tatarsky, B.G. (eds.) Principles and features of the construction of complexes with UAVs, 824 p. Moscow, Radiotehnika. (2016)
15. Verba, V.S., Merkulov, V.I., Sadovsky, P.A., et al.: Radio control systems (Book 3). In: Merkulov V.I. (eds.) Theoretical foundations for the synthesis and analysis of radio control, 394 p. Moscow, Radiotehnika (2015)
16. Gaiduk, A.R., Kapustyan, S.G., Plaksienko, E.A., Kolokolova, K.V.: Multi-agent control of the parallel structure mechanism based on the decomposition approach. Sci. Bull. NSTU **1**(70), 51–66 (2018)
17. Abramov, N.S., Khachumov, M.V.: Simulation of pilotless aircraft route guidance as a goal pursuit task. Aerosp. Instrum. **9**, 9–22 (2013)
18. Gaiduk, A.R., Dyachenko, A.A., Kapustyan, S.G., Plaksienko, E.A.: Algorithms for autonomous group control of UAV horizontal movements. Sci. Bull. NGTU **2**(67), 120–134 (2017)
19. Mashkov, O.A., Samborsky, I.I.: Features of taking into account synergistic processes in the atmosphere when controlling the flight of groups of remotely piloted aircraft. Control Syst. Navig. Commun. **3**, 97–100 (2007)
20. Gaiduk, A.R.: Control of a group of UAVs with a restriction on control and state variables. Mechatronics Autom. Control **7**, 52–57 (2012)

21. Amelina, N.O.: Achieving consensus by an autonomous group of unmanned aircraft. Stoch. Optim. Inform. **6**, 127–132 (2010)
22. Merkulov, V.I., Kharkov, V.P., Shamarov, N.N.: Optimization of collective control of a group of unmanned aerial vehicles. Radio control systems (Book 4). In: Merkulov V.I. (eds.) Optimization of control algorithms, pp. 217–220. Moscow, Radiotehnika (2018)
23. Holovaty, I.N., Ustinov, A.I.: Application of active radio frequency identification technology to create systems for local positioning of mobile objects. In: Proceedings of the Scientific and Technical Conference "New Technologies in Advanced Detection, Navigation and Control Systems", pp. 127–132 (2012, May 24)
24. Gaiduk, A.R.: On the assessment of the reliability of sensor signals. Science and education at the turn of the millennium, pp. 81–85. Kislovodsk, Publishing House of KSTI (2018)
25. Miloserdov, I.V., Khalimov, R.A.: Algorithm for controlling an anti-radar missile during information breaks. Radio control systems (Book 4). In: Merkulov, V.I. (eds.) Optimization of Control Algorithms, 312 p. Moscow, Radiotehnika (2018)
26. Ljung, L.: System Identification: Theory for User, 429 p. Prentice–Hall (1987)

A Computational Approach to Calculate the Hydrographic Zero Along the Moroccan Atlantic Coast (Port of Mohammedia) Using Matlab

Mokhtar Abdenour(✉), Mohamed Lahmama, Assia Abdenour, Laila Mouakkir, and Mohamed Chagdali

Faculty of Sciences Ben M'sik, University Hassan II, PO BOX 7955, Casablanca, Morocco
mokhtar.abdenour-etu@etu.univh2c.ma

Abstract. The Hydrographic Zero (HZ) should be, according to the recommendations of the International Hydrographic Organization (IHO), as close as possible to the lowest astronomical sea level. It is the common reference level for nautical charts and tidal directories. The calculation of this reference level has been the subject of several previous studies, because of its utmost importance. Hence, the main goal of this study in the coastline of Mohammedia, using a database, sea state measurements and algorithmic analysis is to characterize the information in the recorded tide wave and then calculate the chart datum or Hydrographic Zero for the establishment of marine charts for navigation. The available tide gauge records covers the period from 31/03/2006 to 27/09/2009, and a Butterworth digital low pass filter in Matlab was used to obtain a filtered signal. Through a harmonic analysis of the tide gauge records done by means of a Discrete Fourier Transformation (DFT), the semi-diurnal character of the tides is proved. The frequency spectrum of the already filtered signal allowed us to identify the frequency content and the components of the tide. Then the Hydrographic Zero in the port of Mohammedia is calculated, it is around 2.081 m above the General Levelling of Morocco (NGM).

Keywords: Hydrographic zero · Tides · Sea level · Port of Mohammedia

1 Introduction

Hydrographic Zero (HZ) or chart datum is a reference level that is generally chosen as the level of the lowest astronomical seas. The "astronomical" characteristic

R. Silhavy et al. (Eds.): CoMeSySo 2022, LNNS 596, pp. 411–417, 2023.
https://doi.org/10.1007/978-3-031-21435-6_35

indicates that it is not a directly observed level but calculated from the generating force of the tide due to the gravitational actions of the Moon and the Sun [1]. During the past years, many definitions have been stated to approach this very important notion. Warburg in 1925 defined the Hydrographic Zero as "the level to which depths shown on the chart are referred and which should be used as the zero of the tidal predictions, or other tidal information" [2]. According to Wöppelmann "The Hydrographic Zero or chart datum is the reference level from which are counted, positively towards the nadir, the soundings carried on the charts and, positively towards the zenith, the water level" [3]. In the same context, the problem of determining the Hydrographic Zero is not recent .The problem of a datum for soundings is very old it goes back to the earliest days of hydrographic surveying [4]. Indeed, there are different and diverse criteria to establish a chart datum [3,4].

In general terms, defining the chart datum is of major importance, in fact water level measurements are made to serve the needs of the mariner, the engineer, the scientist, and the general public [5]. Thus, this work aims to establish a uniform level, or datum plane, to which observed water depths could be referred near to Mohammedia. In order to update the HZ at the level of the facade of Mohammedia, we exploit tidal records from which we extracted raw data. To extract the data of the astronomical tide and to be able to identify the frequency content of the signal of this one, we must proceed to the cleaning and filtering of the raw data via a low-pass digital filter. In addition to these regular movements, there are meteorological variations in water level, called heights and lows, which can be studied using statistical methods. Thus, a correction of the content of the cleaned signal is requested. A special attention is paid to the problems of poorly separated components and the proper conditioning of the harmonic analysis by the method of least squares [1]. The harmonic formula of the tide gives us:

$$h(t) = Z_0 + \sum_{i=1}^{n} A_i \cos(\omega_i t) + B_i \tag{1}$$

The application of the method of least squares allows to write:
The amplitude:

$$C_I = \sqrt{(A_i^2 + B_i^2)} \tag{2}$$

The phase:

$$\omega_i = arctan(\frac{A_i}{B_i}) \tag{3}$$

As for the line spectrum, it is obtained using a fast algorithm of the discrete Fourier transform (DFT) [6]. The spectrum of the tide is a mode of representation of the amplitude, or energy, as a function of frequency or period of the signal analyzed that allows to identify the frequency content of the tide signal. In addition, to properly observe and calculate the tides, the period called Saros [7] must be taken into account.

2 Methods

In order to achieve the prescribed objectives, we used Matlab software and T-Tide Harmonic Analysis Toolbox for Matlab, which is a program that provides for each tidal wave the amplitude and phase.

2.1 Study Area

The perimeter of this study covers the region of Mohammedia-Casablanca on the northern Atlantic coast of Morocco. Located on the Atlantic coast, 23 km from the port of Casablanca, Mohammedia, is situated at 7∘ 23.40′ west longitude and 33∘ 42.80′ north latitude. It is the main oil port of the Kingdom of Morocco, located in the southern part of Mohammedia Bay. The port includes two port facilities:

- The oil terminal;
- The inner harbor.

As for the local meteorology, the swell from the open sea bypasses the end of the main breakwater and enters the bay by propagating from north to south. The average tidal range is about 2.3 m during spring tides and 1.3 m during neap tides with an average level of 2.14 m. The prevailing coastal currents flow towards the WSW or ENE, parallel to the coast.

2.2 Tide Gauge and Data

The tide gauge installed in the port of Mohammedia is a coastal digital tide gauge Radar (VEGAPLUS 65) whose uncertainty at 95 of a measurement integrated over 2 min is about 1 cm. It is recommended by the International Hydrographic Organization to filter the extracted data in order to eliminate oscillations of short period. The position of the tide gauge of Mohammedia is 33° 42′ 54′ N, 7° 23′ 36′′ W. The study period is from 31/03/2006 to 27/09/2009, with a sampling interval of 6 min.

2.3 Data Processing Tools

In this work, predefined functions were used to clean, filter and identify the frequency content and components of the tide through a Discrete Fourier Transform (DFT). Then, a simulation via the T-tide toolbox under Matlab allowed to recover the amplitude and phase of each of the main harmonics of the ocean tide. T-Tide Harmonic Analysis Toolbox for Matlab is a box containing several programs written in Matlab format and allowing to perform harmonic analysis and tidal predictions, it was developed by Rich Pawlowicz, the first version being released in December 2001 [8]. The program takes a raw tidal time series as input and produces the amplitude and phase of the harmonic constituents for the time series, as well as error estimates and provides for each tidal wave the amplitude and the phase [8].

3 Results

This work has allowed to conclude a number of relevant results.

3.1 Signal Filtering

The tidal signal (Fig. 1) contains 289651 points and has high frequency components that correspond to disturbances in the shortwave level. Therefore, it was necessary to filter the obtained signal in order to localize the high frequencies. The filtered signal is obtained through a Butterworth digital filter designed using Matlab, this low-pass filter is with infinite impulse response of order 4 (Fig. 2).

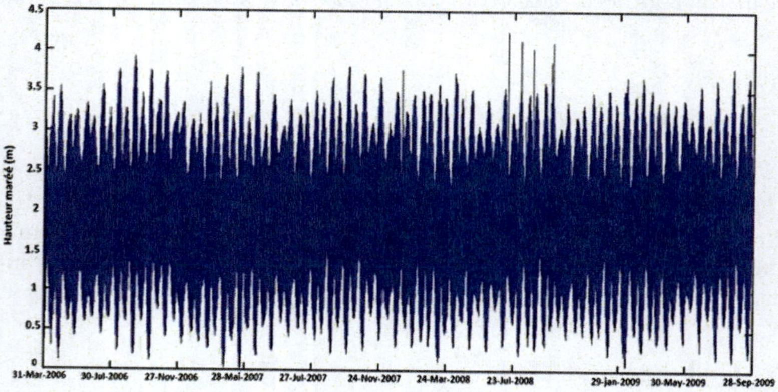

Fig. 1. Raw signal

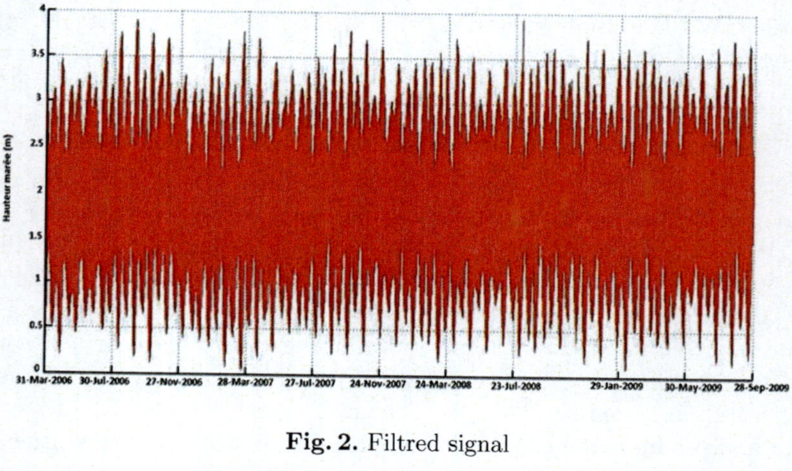

Fig. 2. Filtred signal

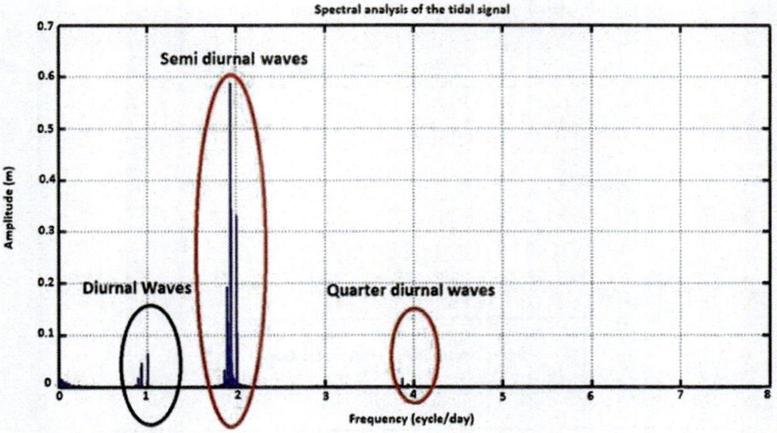

Fig. 3. Dominant waves

3.2 Spectral Analysis

A Discrete Fourier Transform (DFT), allowed to identify the frequency content of the already filtered signal and the components of the tide. In addition, a correction of the tidal signal is suggested, adding to the cleaned tidal signal the pressure signal and the one resulting from the wind. The results of the spectral analysis show the presence of diurnal, semi-diurnal and quarter diurnal waves. However, the semi-diurnal waves are the most dominant, which proves the semi-diurnal character of the tide in the port of Mohammedia (Fig. 3).

3.3 Predictions

For a 20-year prediction, and taking into consideration the so-called Saros period, which is exactly of the order of 18 years and 11 d and 7 h and 50 s [7], T-Tide allowed to retain for each tidal wave the amplitude and phase, the frequency is already known. Moreover, in order to distinguish between true deterministic (linear) frequencies and broad-spectrum variability, confidence intervals for the estimated tidal parameters are calculated using a selected algorithm. The semi diurnal wave M2 is the most dominant at Mohammedia harbor with an amplitude exceeding 0.5 m (Table 1).

3.4 Redefine the Hydrographic Zero

In order to redefine the Hydrographic Zero from the calculated tidal component base coefficients the approach shown in the Fig. 4 is adopted.

The steps followed for the updating of the HZ near to the port of Mohammedia have enabled to retain a level considered as the lowest possible astronomical tides, which is in the order of 2.081 m/NGM (Fig. 5). This level reappears normally and clearly in a periodic manner, after the period known as Saros [7].

Table 1. Synthesis of data on tidal types

Wave	Frequency	Amplitude	Phase
*NU2	0,0792016	0,0447	15,59
*H1	0,0803973	0,1675	337,22
*M2	0,0805114	0,6134	56,12
*H2	0,0806255	0,1698	287,69
*MKS2	0,0807396	0,0628	68,72

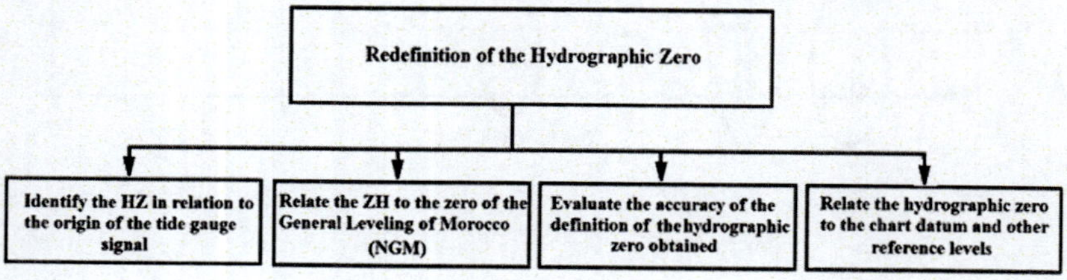

Fig. 4. Approach to redefine the hydrographic zero

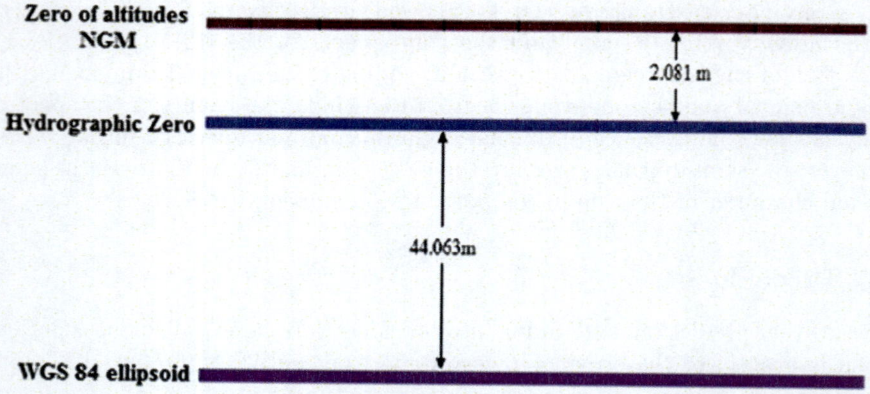

Fig. 5. Situation of the Hydrographic Zero in relation to reference levels

4 Conclusion

The purpose of this work is to calculate the Hydrographic Zero (HZ) at the level of the Atlantic facade of Mohammedia and thus to establish a uniform level, to which observed water depths could be referred. While the updating of the Hydrographic Zero is of capital importance for various fields including oceanography, port facilities, drilling platforms, marine and renewable energy, little attention is given to approach this crucial subject [9]. In order to update the Hydrographic Zero in the port of Mohammedia, we used tidal records from which we extracted raw data. Then, a Butterworth low-pass digital filter designed using Matlab has allowed to obtain the filtered signal. Moreover, a Discrete Fourier Transform (DFT) allowed to identify the frequency content of the filtered signal and the components of the tide.The results of the spectral analysis prove the semi-diurnal character of the tide in the port of Mohammedia. Using T-Tide Harmonic Analysis Toolbox for a 20-year prediction, and taking into consideration the so-called Saros period, we retained for each tidal wave the amplitude and phase. The steps followed for the updating of the Hydrographic Zero near to the port of Mohammedia have enabled to retain a level of the order of 2.081 m/NGM. This work contribute to understand the trend of sea state measurements over the study area and constitutes a study platform that will surely contribute to the resolution of problems related to sea level variations and climate change. Similarly, it could support future research on Hydrographic Zero and sea level changes in the Moroccan Atlantic coast.

Acknowledgements. This research did not receive any specific grant from funding agencies in the public, commercial, or not-for-profit sectors. We would like to thank the editor and anonymous reviewers for their comments that help improve the quality of this work.

References

1. Simon, B.: La marée. Institut Océanographique, Paris (2007)
2. Warburg, H.D.: Coordination of chart datums. Int. Hydrogr. Rev. (1925)
3. Wöppelmann, G., Allain, S., Bahurel, P., Lannuzel, S., Simon, B.: Hydrographic zero: towards a global determination. Ann. Hydrographiques **8**(777) (2011)
4. Gordon, D.L.: Rationalisation of Chart Datum in the British Isles. Int. Hydrographic Rev. **45**(1), (2015)
5. Gill, S.K., Schultz, J.R.: Tidal datums and their applications. NOAA, NOS Center for Operational Oceanographic Products and Services (2001)
6. Giannakopoulos, T., Pikrakis, A.: Introduction to Audio Analysis: A MATLAB Approach. Academic Press (2014)
7. Causeret, P.: A propos du Saros, de la rotation de la ligne des nœuds et du cycle de Méton. Cahi. Clairault 33–34 (2008)
8. Pawlowicz, R., Beardsley, B., Lentz, S.: Classical tidal harmonic analysis including error estimates in MATLAB using T_TIDE. Comput. Geosci. **28**(8), 929–937 (2002)
9. Mouakkir, L., Smaoui, H., Mordane, S., Chagdali, M.: Three dimensional modeling of hydrodynamics with wave-current interaction: Application to the Casablanca-Mohammedia coast. J. Phy. Chem. News. **55**, 74–80 (2010)

Nanotechnology on Perspective Computer Science

Mahyuddin K. M. Nasution[1](✉), Rahmad Syah[1], and Marischa Elveny[2]

[1] Data Science & Computational Intelligence Research Group, Excellent Center of Innovation and New Science, Universitas Medan Area, Jl. Kolam, Gedung PBSI, Medan, Sumatera Utara, Indonesia
mahyuddin@uma.ac.id, mahyuddin@usu.ac.id, rahmadsyah@uma.ac.id

[2] Data Science & Computational Intelligence Research Group, Universitas Sumatera Utara, Padang Bulan 20144 Medan, Indonesia
marischaelveny@usu.ac.id

Abstract. Computer science has long been an established science, but changes continue to occur both in terms of technology and application. Of course, the change comes from challenges that come from different technologies and knowledge. The challenge is a phenomenon, nanotechnology, which affects many other scientific fields indirectly. A science remains on its scientific principles, framing data and based on appropriate methods. Although the basic principles of computer science are not affected by changes in nanotechnology, the challenges of changing different disciplines due to nanotechnology need to be expressed from the point of view of computer science, which fall into two categories as is the relationship between computer science and nanotechnology.

Keywords: Algorithm · Paradigm · Innovation · Covid-19 · New science

1 Introduction

Inevitably, computer science is constantly changing to address the weaknesses of its own numerical technology [1]. The change is not only a result of a shift in insight into data from the point of view of algorithmic complexity in which software plays a role, but also a way of looking at material limitations and their atomic interactions in which hardware plays a role [2,3]. In addition, the pressure from the use of computers that has accumulated in life activities and the application of computers in various scientific fields has caused computer science not only to need to reinterpret its scope of work, but to rearrange strategies in its organization [4,5].

In this regard, the phenomenon of change in technology lately has raised new challenges from nanotechnology to all scientific fields or all subject areas of study, both from the education side and from the research side, as well as from the service side, where many nanotechnology products require socialization of

R. Silhavy et al. (Eds.): CoMeSySo 2022, LNNS 596, pp. 418–429, 2023.
https://doi.org/10.1007/978-3-031-21435-6_36

its use [6,7]. Likewise, the presence of nanotechnology has had a huge impact on the development of technologies that have features that users like [8]. This paper presents two different sides of the relationship between nanotechnology and computer science.

2 A Review to Perspective

Computers as a scientific field have long been establishing as a science that studies computer-related phenomena [9], especially about algorithms [10] and the structure of information [11], specifically related to the management of complexity [12]. In other words, computer science is the study of data, information and knowledge at different levels of complexity or abstraction, and from which other fields of scientific can be developed [13], or the technology that deals with the development and use of certain types of human-created artifacts [14]. The last two concepts reveal that computer science is needed in matters related to something extraordinary based on scientific fields such as physics, chemistry, biology as basic science to applied sciences such as materials science, pharmacy, engineering, medicine, but also social sciences [15,16].

From the point of view of large and small size considerations. Regardless of the capabilities of the space telescope, like Hubble [17] and James Webb [18], the activities associated with it still involve the computer directly, from controlling the telescope itself, recording the flow of information from the telescope's capture of space and celestial objects, and then interpreting that information by mining data [19]. Computationally speaking, it is a universe of discourse, a space that is still estimated at its breadth, but has the potential to be condensed into big data [20]. In contrast, the ability of the (electron)-microscope to capture atoms, in the form of molecules or microorganisms, and their interactions are also use computers to sense them so that the behavior of each can be known in detail [21]. Computationally, it is also the universe of discourse, a systematic of facts accumulated into big data [22]. In fact, computers play a role in describing things from all sides of life, including human thought and mathematically abstracting them into metadata, namely variables and their rules [23].

In many cases there are contracts between computer science and other disciplines, which are held through education or research [24]. The contract also picks up the promise of nanotechnology to respond to natural challenges and complete various assignments [25]. This assignment is as defined in the term "*nanotechnology*", namely as "*a field of applied science that focuses on the design, synthesis, characterization, and application of materials and devices at the nanometric scale*" [26]. The term, which became the title of an article published in 1987, states that nanotechnology concentrates on manufacturing with dimensions or tolerances in the range 0.1–100 nm, where $1\,\text{nm} = 1 \times 10^{-9}$ m [27]. Therefore, if the diameter of the atomic nucleus ranges from 10^{-15} to 10^{-14} meters, then nanotechnology is directly related to the atomic and molecular domain, which is widely discussed by scientific fields such as physics, chemistry, biology, pharmaceutical, agricultural, ecological, medical, engineering, materials science, geosciences, humanities, and industrial [28–33]. Thus, nanotechnology is a "*multidisciplinary field of discovery*". A scientific field that is currently growing rapidly has enormous uses in science and technology [34]. As a fellow technology, nanotechnology is inevitably closely related to computers both as science and as a system that organizes scientific activities, and because nanotechnology is always at a crossroads from computer science to other fields from a certain point of view, nanotechnology permeates scientific development activities from each of those fields [35].

As computer science derives from several other scientific fields - such as information technology, computer systems, information systems, data science, and others - nanotechnology after the discovery period has the potential to contribute to areas that are specifically discussed according to the interests of these different scientific fields [36]. Thus, based on the principles of epistemology and ethics, various ways of pushing that potential into appropriate scientific stages, starting from building concepts, providing limits on various related matters, establishing axioms, reasoning assumptions, testing hypotheses, examining theory, and conducting experiments so that the scientific system satisfies its interests naturally [37]. However, the scientific system is not just enough, what are the obstacles, disadvantages, or advantages of what is extracted from computer science so that it maintains its existence, or converges into its objectivity, inevitably nanotechnology follows the implementation of the scientific system [38]. Mathematics, according to Galileo's reasoning, i.e. "*Philosophy is written in this grand book, the universe, which stands continually open to our gaze. But the book cannot be understood unless one first learns to comprehend the language and read the letters in which it is composed. It is written in the language of mathematics.*" [39] - has provided an deep insight on the sciences on which the scientific system is built [40]. Therefore, in a philosophical context it is necessary to build a dialogue between the various interests of all reasoning - Promethean, Platonic, Cartesian, Newtonian, materialist - for dealing with success, order, common sense, certainty, uncertainty, complexity, well-being, and subjectivity both from individual, community, and globally pressures including human rights [37]. Such as the implementation of a study program in the mechatronics field at a study center, or the emergence of a data science study program at this time, indirectly illustrates the need for excellent study centers to increase the acceleration of a nation's development through innovations that not only produce useful products but especially new science [41,42]. Of course, all of that will be maximally obtained through multidisciplinary cooperation and between countries, where many new procedures/methods are optimally developed [43].

3 Phenomenon and Paradigm in Perspective

Nanotechnology is one of the recent phenomena [44,45]. As a science, however, nanotechnology is unwittingly by the organizers - academics and researchers - has framed the data for a nanoscience [46]. However, the framework is mixed with the interests of other scientific frameworks [36]. Likewise, data is still data that from a modeling point of view is useful for all sciences, and to then frame it according to the scientific interests of that science [47,48]. Therefore, this phenomenon is supported by a reality that naturally objects that become sources of data or as study subjects have the potential to support the improvement of human welfare. Imagine, how the data representation signal in the information channel moves singly, where electrons move from one atom to another, and the atom itself becomes a signal channel, indirectly makes it possible to save energy [49]. Mathematically, graph theory, follows natural reasoning, that in a straight line (as road and transportation) the arrangement of single-channel atoms forms a path without branching so that there is no other choice than to transfer electrons from one atom to another [50,51].

The idea of understanding data and their interactions in a single channel allows naturally to give birth to a model to frame the data in a linear conductor [52], then the framework arranges the components associated with it in a systematic arrangement where the variables are interrelated with each other to form the completion steps, which are also known as methods or in a different order as procedures [53]. Taking into account the nano size, the components are atoms, the atoms are the letters that make up words that convey meaning in sentences where language has grammar that organizes the word order (grammar) [54,55]. Natural language is used as a model to build a communication between humans and machines such as computers, and then becomes a programmable array based on nanotechnology [56]. The relationship between one model and another is a way of adapting a solution from a problem to a different problem. On that basis, the data model can be transformed into a base model or a shape model as described by Thales for calculating area, and the proof is carried out through dialogue in philosophy [40]. All of these are briefly expressed in one word as a paradigm.

4 Nano and Computer as Perspective

In general, technology has its own structure as the organization of its components. The structure of nanotechnology consists of a molecular structure and a molecular structure following the atomic structure of [57]. Atom is the basic unit of matter, with its component structure consisting of a number of positively charged protons, a number of negatively charged electrons, and neutrally charged neutrons. So the atomic structure consists of an atomic nucleus (neutrons and protons) and a cloud of electrons around nucleus [58]. With this structure introduce atoms interact with each other to form molecules. Atoms, molecules and their interactions in nano-sized space [59]. Thus, indirectly as an object from a computer, nano-sized space only allows for analysis outside the theory of chemistry or atomic physics using a computer by involving sensing from a device that is capable of enlarging imaging, in a simulation to capture atoms and their interaction processes and then describe them according to their dimensions of importance. On the other hand, nanotechnology is subjectively changing the capacity and appearance of computer machines in general [25]. Therefore, nanotechnology in a computer science perspective occurs in two categories, as an object or as a subject.

4.1 From Nano to Technology

With the component structure of nanotechnology, any technology that may contain components of nanotechnology will be associated with the similar structure. The structure of nanotechnology makes up every material that is implemented into different technologies, including computer technology. Before that, as a precursor there were various scientific fields which were then integrated into computer science and then implemented into computer technology. Science-based computers involve input devices, processing devices, and output devices. Imagine a cluster of vacuum tubes that make up a computer but with a smaller capacity than a cluster of transistors organized in a computer. Each transistor is a molecule that is not small, then every computer requires millions of transistors, and thus a computer is quite large and heavy. Nanotechnology, then, changed the representation of transistors to be smaller both in size and larger in capacity. Similar machines - entering information and data, processing it and presenting it - such as computers, now come with promising features besides energy saving. That way, the same machines have one tool with different status, as input as well as output.

Machines with nanotechnology interact with each other in connection and communication, both involving direct channels (using media) or not (using frequencies). In nanostructures, fibers are extremely versatile materials used in many technical applications as well as clothing and fashion, by which computers are linked to one another [60]. Along with that, various tools have emerged and are increasingly adapting and forming information channels that are able to control certain objects and their environment remotely. Through this technology, the Internet of Things (IoT) [61], carries out the delivery of all activities that were originally done by humans to machines so that extraordinary things that humans cannot do will be carried out by the machine (Micro-Nano-Robotics) directly or indirectly [62]. So, nanotechnology forms materials that are beneficial for the development of technology to be better in every way.

That way, when the storage area has a larger size in data storage capacity, then the data processor (microprocessor) works even faster, the complexity of the algorithm can begin to be decomposed one by one into modules that are able to work alone [63]. A large collection of data, or big data, collectively allows it to be processed by different machines, and integrate the results into a single unit of information that is easily presented [57]. So, nanotechnology has changed some of the principles of computer science where data literacy and computing are shifting towards cloud computing. In this regard, materials science is becoming an interdisciplinary field that examines the structure-property relationships in materials for application to many fields of science and engineering [64]. On that basis, the development of materials science cannot be separated from the support of computer science, even if it is to reveal metallurgical and mechanical properties [65–69].

4.2 From Science to Nano

Although different materials, components related to the molecular composition of sapiens, animals, and plants, including microorganisms such as Covid-19 [70], consist of atoms that specifically form G-C and A-T base pairing as nano-architectural motifs or gene structures, namely materials for designing life, and using computer science (algorithms) developed nanofabrication methods [71]. The structure of the gene, which consists of bases, contains messages from nature for life. In the spatial dimension, DNA is the natural building block for multi-dimensional construction, where dimensions can be synthesized rapidly with high yields, but lower dimension, from a single tereoisomer, and also their triangulated architecture imparts structurally stability [72].

In chemical metaphors, the computational intelligence of computer science makes it possible to parse genes by processing DNA-RNA-protein information and involves nanotechnology in cutting or adding pieces of gene structure [38]. This is the deepest idea of modern medicine for various diseases caused by proteins [73,74]. The structure of DNA-RNA-protein in general forms its own characteristics invariantly, and this can be recognized through knot theory or closed braids in mathematics [75,76]. Therefore, although the polymeric nature of genes in materials and technological applications is often overlooked due to the impression that RNA appears to be unstable, where ribonucleic acid (RNA) is a polynucleic acid [77], computational biology has thus become an interdisciplinary field applying computer science techniques, functional mathematics, and statistics to tackle problems inspired by biology [78,79]. In nanospace, genes are beautiful programs, so are computer programs in nanomemory. Computer pro-

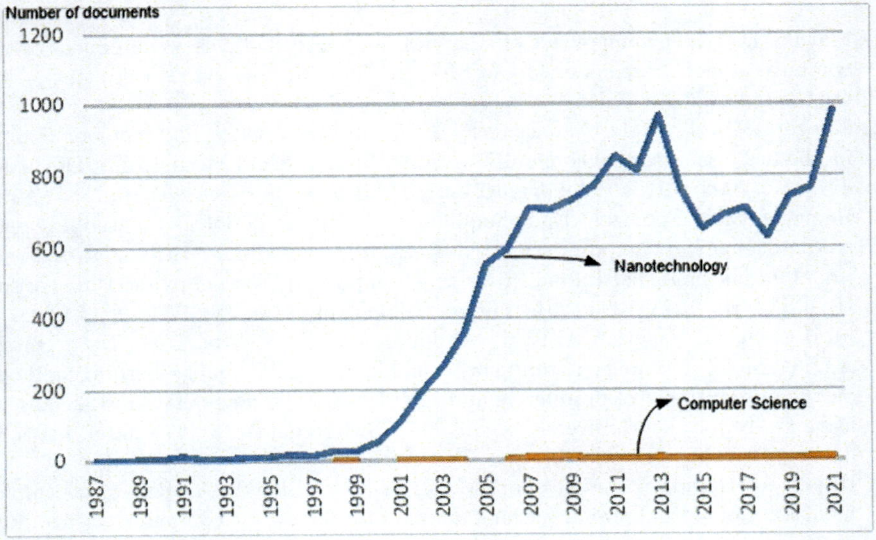

Fig. 1. Number of document per year about nanotechnology in Scopus database (July 2022)

grams are human efforts to facilitate their activities, these efforts also involve nanotechnology in all scientific aspects, and naturally develop in its perspective. Figure 1 shows the human effort to understand one of the world's phenomena, nanotechnology, along with its paradigm. With that, many ideas were expressed in other fields of study, but they were increasingly inspired by computer science.

As an applied science, following the complexity of nanotechnology, the world of medicine cannot stand alone, without the support of basic science and computer science. For example, by means of which it is collectively possible to provide patient-centred healthcare through the means of advanced technologies [80]: nanotechnology and computer science. Therefore, to build a complete understanding of something, multi-disciplinary collaboration is needed so that innovation embraces the birth of a new science.

5 Conclusion

Computer science and nanotechnology have a close relationship, although at the beginning they seemed separated. The development of nanotechnology is actually inspired by the computer itself as a technology, and changes the perspective of computer science towards various other scientific fields. From that position, computer science views nanotechnology from two sides, as an object or a subject. Based on that perspective, computer science is increasingly having capabilities and capacities, as well as the field of nanotechnology has received assistance to answer challenges from different sciences.

References

1. Matula, D.W.: The emergence of computational arithmetic as a component of the computer science curriculum. ACM SIGCSE Bull. **2**(3), 41–44 (1970). https://doi.org/10.1145/873641.873650
2. Bungartz, H.-J., Mehl, M., Zenger, C.: Computer science and numerical fluid mechanics - An essential cooperation. Notes Numer. Fluid Mech. Multi. Des. **100**, 437–450 (2009). https://doi.org/10.1007/978-3-540-70805-6_34
3. Requena, C.G., Gomez, M.E., Sahuquillo, J.: A new methodology for studying realistic processors in computer science degrees. In: Proceedings—IEEE 27th International Parallel and Distributed Processing Symposium Workshops and PhD Forum, IPDPSW, pp. 1283–1290 (2013). https://doi.org/10.1109/IPDPSW.2013.13
4. Mortazavi, M., Nasution, M.K.M., Abdolahzadeh, F., Behroozi, M., Davarpanah, A.: Sustainable learning environment by mobile-assisted language learning methods on the improvement of productive and receptive foreign language skills: a comparative study for asian universities. Sustain. (Switzerland) **13**(11) (2021). https://doi.org/10.3390/su13116328
5. Rajab, K., Hamdi, M., Al Reshan, M. S., Asiri, Y., Shaikh, A., Rajab, A.: Implementation of virtual training: the example of a faculty of computer science during COVID-19 for sustainable development in engineering education. Electron. (Switzerland) 11(5): 2022. https://doi.org/10.3390/electronics11050694

6. Syah, R., Elveny, M., Nasution, M.K.M., Ponkratov, V.V., Kuznetsova, M.Y., Poltarykhin, A.L., Babanezhad, M.: Numerical investigation of nanofluid flow using CFD and fuzzy-based particle swarm optimization. Sci. Rep. **11**. https://doi.org/10.1038/s41598-021-00279-6
7. Egwu, C.N., Babalola, R., Udoh, T. H., Esio, O.O.: Nanotechnology: applications, challenges, and prospects. Green Energy Technol. 3–15 (2022). https://doi.org/10.1007/978-3-030-95820-6_1
8. Makarucha, A.J., Todorova, N., Yarovsky, I.: Nanomaterials in biological environment: A review of computer modelling studies. Eur. Biophys. J. **40**(2), 103–115 (2011). https://doi.org/10.1007/s00249-010-0651-6
9. Newell, A., Perlis, A.J., Simon, H.A.: Computer science. Science **157**(3795), 1373–1374 (1967). https://doi.org/10.1126/science.157.3795.1373-b
10. Knuth, D.E.: Fundamental algorithm. In: The Art of Computer Programming, Vol. 1. Addison-Wesley (1973)
11. Wegner, P.: Programming languages. McGraw-Hill, Information Structures and Machine Organization (1968)
12. Djikstra, E.W.: Notes on structured programming. In: Structured Programming, Dahl. Academic Press, Dijkstra and Hoare (1972)
13. Sabin, R.E., Sabin, E.P.: Collaborative learning in an introductory computer science course. ACM SIGCSE Bull. **26**(1), 304–308 (1994). https://doi.org/10.1145/191033.191156
14. Fouh, E., Akbar, M., Shaffer, C.A.: The role of visualization in computer science education. Comput. Sch. **29**(1–2), 95–117 (2012). https://doi.org/10.1080/07380569.2012.651422
15. Syah R., Davarpanah, A., Elveny, M., Karmaker, A.K., Nasution, M.K.M., Hossain, M.A.: Forecasting daily electricity price by hybrid model of fractional wavelet transform, feature selection, support vector machine and optimization algorith. Electron. (Switzerland) **10**(18): 2021. https://doi.org/10.3390/electronics10182214
16. Syah, R., Alizadeh, S.M., Nurgalieva, K.S., Guerrero, J.W.G., Nasution, M.K.M., Davarpanah, A., Ramdan, D., Metwally, A.S.M.: A laboratory approach to measure enhanced gas recoveray from a tight gas reservoir during supercritical carbon dioxide injection. Sustain. (Switzerland) 13(21): 2021. https://doi.org/10.3390/su132111606
17. Shamir, L.: Automatic identification of outliers in Hubble Space Telescope galaxy images. Monthly Not. R. Astron. Soc. **501**(4), 5229–5238 (2021). https://doi.org/10.1093/mnras/staa4036
18. Carlisle, R.E., Acton, D.S.: Demonstration of extended capture range for James Webb Space Telescope phase retrieval. Appl. Opt. **54**(21), 6454–6460 (2015). https://doi.org/10.1364/AO.54.006454
19. Klotz, A.: Protocols for robotic telescope networks. Adv. Astron. **2010** (2010). https://doi.org/10.1155/2010/496765
20. Spasova, M., Stoeva, P., Stoev, A.: Archaeoastronomy in the big data age: origin and peculiarities of obtaining data on objects and artifacts. In: Proceedings of the International Astronomical Union, Vol. 15, issue no. S367, pp. 55–457 (2019). https://doi.org/10.1017/S174392132100017X
21. Schauer, P.: Comparison of photon transport efficiency in simple scintillation electron detector configurations for scanning electron microscope. Microsc. Res. Tech. **85**(5), 1870–1883 (2022). https://doi.org/10.1002/jemt.24048
22. Holmi, J.T., Lipsanen, H.: WITio: a MATLAB data evaluation toolbox to script broader insights into big data from WITec microscopes. SoftwareX **18** (2022). https://doi.org/10.1016/j.softx.2022.101009

23. Belcastro, L., Cantini, R., Marozzo, F., Orsino, A., Talia, D., Trunfio, P.: Programming big data analysis: principles and solutions. J. Big Data **9**(1), 1–50 (2022). https://doi.org/10.1186/s40537-021-00555-2
24. Richards, B.A.: Model curriculum for a liberal arts degree in computer science. ACM J. Educ. Resour. Comput. **7**(2), 2007 (2007). https://doi.org/10.1145/1240200.1240202
25. Vandenbunder, B.: New objects, new actors, new ethical issues: the interfaces between biology, nanotechnologies and computer sciences. Ethique et Sante **7**(1), 18–23 (2010). https://doi.org/10.1016/j.etiqe.2009.12.001
26. Bea, D.B., Tejeda, A.P., Pardo, A.A., Cuador, J.C.: Nanomedicine: general aspects of a promissory future. Rev. Habanera de Ciencias Med. **10**(3), 410–421 (2011)
27. Franks, A.: Nanotechnology. J. Phys. E: Sci. Instrum. **20**(12), 1442–1451 (1987). https://doi.org/10.1088/0022.3735/20/001
28. Rajneesh. Pathak, J., Singh, V., Kumar, D., Singh, S.P., Sinha, R.P.: DNA in nanotechnology: approaches and developments. Nanomater. Plants Algae Microorganisms 1: 79–102, 2018
29. Maher, K.O.: Nanomedicine and nanotechnology for heart failure research, diagnosis, and treatment. Heart Failure in the Child and Young Adult: From Bench to Bedside, pp. 779–784 (2018)
30. Afsana, F., Asif-Ur-Rahman, M., Ahmed, M.R., Mahmud, M., Shamim Kaiser, M.: An energy conserving routing scheme for wireless body sensor nanonetwork communication. IEEE Access **6**, 9186–9200 (2018). https://doi.org/10.1109/ACCESS.2018.2789437
31. Jadhav, S.S., Jadhav, S.V.: Application of nanotechnology in modern computers. In: 2018 International Conference on Circuits and Systems in Digital Enterprise Technology, ICCSDET (2018). https://doi.org/10.1109/ICCSDET.2018.8821086
32. Bayda, S., Adeel, M., Tuccinardi, T., Cordani, M., Rizzolio, F.: The history of nanoscience and nanotechnology: from chemical-physical applications to nanomedicine. Molecules **25**(1) (2020). https://doi.org/10.3390/molecules25010112
33. Azulay, H., Lutaty, A., Qvit, N.: How similar are proteins and Origami? Biomolecules **12**(5): 2022. https://doi.org/10.3390/biom12050622
34. Pathak, Y.V., Koomer, A.: Graduate programs on nanotechnology in pharmaceutical sciences. In: Technical Proceedings of the 2009 NSTI Nanotechnology Conference and Expo, NSTI-Nanotech 2009, vol. 3, pp. 410–413 (2009)
35. Pugliese, R., Regondi, S., Marini, R.: Machine learning-based approach: global trends, research directions, and regulatory standpoints. Data Sci. and Manag. **4**, 19–29 (2021). https://doi.org/10.1016/j.dsm.2021.12.002
36. Nasution, M.K.M., Hidayat, R., Syah, R.: Computer science. Int. J. Adv. Sci. Eng. Inf. Technol. **12**(3), 1142–1159 (2022). https://doi.org/10.18517/ijaseit.12.3.14832
37. Hinchcliff, J.: The future of the university: some ethico-epistemological explorations. Horizon **14**(2), 77–83 (2006). https://doi.org/10.1108/10748120610674049
38. Banzhaf, W., Dittrich, P., Rauhe, H.: Emergent computation by catalytic reactions. Nanotechnology **7**(4), 307–314 (1996). https://doi.org/10.1088/0957-4484/7/4/001
39. Remmert, V.R.: Galileo, god, and mathematics. In: Koetsier, T., Bergmans, L. (eds.) Mathematics and the Divine: A Historical Study, pp. 347–360. Tokyo, Elsevier, Amsterdam, Boston, Heidelberg, London, New York, Oxford, Paris, San Diego, San Francisco, Singapore, Sydney (2005)
40. Nasution, M.K.M.: Mathematical philosophy. J. Res. Math. Trends Technol. **2**(2), 45–60 (2020). https://doi.org/10.32734/jormtt.v2i2.4678

41. Nasution, M.K.M., Noah, S.A. M., Sitompul, O. S.: The excellent center: center for artificial intelligence technology. J. Phys. Conf. Ser. **1235** (2018)
42. Nasution, M.K.M.: The birth of a science. Hist. Sci. Technol. **10**(2) (2020)
43. Kaewkamnerdpong, B., Bentley, P., Bhalla, N.: Programming nanotechnology: learning from nature. Adv. Comput. **71**, 1–37 (2007). https://doi.org/10.1016/S0065-2458(06)71001-4
44. Karp, R.M.: WI'07/IAT'07/BIBM'07/GrC'07 joint keynote: computer science as a lens on the sciences: the example of computational molecular biology. In: Proceedings of the IEEE/WIC/ACM International Conference on Intelligent Agent Technology, IAT **2007**, xxi (2007). https://doi.org/10.1109/IAT.2007.8
45. Sanghera, P.: Quantum Physics for Scientists and Technologists: Fundamental Principles and Applications for Biologists, Chemists, Computer Scientists, and Nanotechnologists. Wiley (2011). https://doi.org/10.1002/9780470917121
46. Vilela, N., Omar, P.: Intelligent computational nanotechnology: the role of computational intelligence in the development of nanoscience and nanotechnology. J. Comput. Theoret. Nanosci. **11**(4), 928–944 (2014). https://doi.org/10.1166/jctn.2014.3446
47. Nasution, M.K.M., Aulia, I., Elveny, M.: Data. J. Phys. Conf. Ser. **1245**(1) (2019)
48. Nasution, M.K.M.: Understanding data toward going to data science. Cybern. Perspect. Syst. Lecture Notes in Networks and System (LNNS) **503**, 478–489 (2022). https://doi.org/10.1007/978-3-031-09073-8_42
49. Hoekstra, J.: Introduction to Nanoelectronic Single-Electron Circuit Design, 2nd edn. pp. 1–328 (2016). https://doi.org/10.1201/9781315364483
50. Nasution, M.K.M.: Aplikasi hipergraf berarah terhadap sistem transit kota. Epsilon Jurnal Matematika dan Terapannya **1**(2), 1–10 (2000). https://doi.org/10.13140/RG.2.2.21042.63684
51. Syah, R., Nasution, M.K.M., Shol, V.V., Kireeva, N., Jalil, A.T., Chen, T.-C., Aravindhan, S., Abood, E.S., Alkaim, A.F.: Designing a green supply chain transportation system for an automotive company based on bi-objective optimization. Found. Comput. Decis. Sci. **47**(2), 194–207 (2022). https://doi.org/10.2478/fcds-2022-0011
52. Wang, R.-J., Xu, T.-P., She, C., Nasution, Mahyuddin.: A new design for programmable logic array based on QCA-based nanotechnology. Optik **253** (2022). https://doi.org/10.1016/j.ijleo.2022.168581
53. Nasution, M.K.M.: Methodology. J. Phys.: Conf. Ser. **1566**(13) (2020). https://doi.org/10.1088/1742-6596/1566/1/012031
54. Maleki, F.K., Nasution, M.K.M., Gok, M.S., Arab Maleki, V.: An experimental investigation on mechanical properties of Fe2O3 microparticles reinforced polypropylene. J. Mater. Res. Technol. **16**, 229–237 (2022). https://doi.org/10.1016/j.jmrt.2021.11.104
55. Paidar, M., Bokov, D., Mehrez, S., Nasution, M.K.M., Ojo, O.O., Zain, A.M.: The influence of the backing plate materials on microstructure and mechanical properties of friction spot extrusion brazing of AA2024-T3 aluminum alloy and Brass sheets. J. Manuf. Process. **74**, 28–39 (2022). https://doi.org/10.1016/j.jmapro.2021.12.002
56. Azmi, Z., Nasution, M.K.M., Mawengkang, H.: Natural language processing to waterwheel with Jawa language. J. Phys.: Conf. Ser. **1116**(2) (2018). https://doi.org/10.1088/1742-6596/1116/2/022007
57. Paul, P.K., Dey, J.L.: Nanotechnology Vis–Vis Computing: With special references to cloud computing, big data management-A techno managerial knowledge study.

In: 2017 Innovations in Power and Advanced Computing Technologies, i-PACT 2017, Vol. 2017 , pp. 1–5 (2017, January). https://doi.org/10.1109/IPACT.2017.8245151
58. Centoducatte, R., de Oliveira, A. Z., Jorge, F. E., Camiletti, G. G.: ZORA double zeta basis sets for fifth row elements: Application in studies of electronic structure of atoms and molecules. Comput. Theor. Chem. **1207** (2022). https://doi.org/10.1016/j.comptc.2021.113511
59. Shodja, H.M., Ghafarollahi, A., Enzevaee, C.: Surface/interface effect on the scattering of Love waves by a nano-size surface-breaking crack within an ultra-thin layer bonded to an elastic half-space. Int. J. Solids Struct. **108**, 63–73 (2017). https://doi.org/10.1016/j.ijsolstr.2016.11.003
60. Araújo, M.: Fibre Science: Understanding How It Works and Speculating on Its Future. RILEM Bookseries **12**, 3–17. https://doi.org/10.1007/978-94-017-7515-1_1
61. Xu, G., Ding, Y., Zhao, J., Hu, L., Fu, X.: Research on the Internet of things (IoT). Sens. Transducers **160**(12), 463–471 (2013)
62. Gheorghe, G.I., Ilie, I., Istriteanu, S.: Research in micro-nano-robotics. Rom. Rev. Precis. Mecha. Opt. Mechatron. **46**, 83–89 (2014)
63. Cheng, Z., Xiao, J.: Algorithmic tile self-assembly model for the minimum set cover problem. J. Bionanoscience **6**(2), 69–77 (2012). https://doi.org/10.1166/jbns.2012.1079
64. Benjamin, N.D., Flores, C., Matlock, T., Lilian, P.D.: The emergence of immersive low-cost 3D virtual reality environments for interactive learning in materials science and engineering. MRS Online Proc. Libr. **1320**(1), 1–7 (2011). https://doi.org/10.1557/opl.2011.636
65. Paidar, M., Kazemi, A., Mehrez, S., Oladimeji Ojo, O., Matyuso Nasution, M.K., Mironov, S.N.: Investigation of modified friction stir clinching-brazing process of AA2024 Al/AZ31 Mg: metallurgical and mechanical properties. Arch. Civ. Mech. Eng. **21**(3), 1–17 (2021). https://doi.org/10.1007/s43452-021-00267-7
66. Paidar, M., Bokov, D., Nasution, M.K.M., Mehrez, S., Oladimeji, O.O., Omar Cooke, K.: Diffusion brazing of IN738 to SiC ceramic with Ag-Cu-Ti powder: effect of bonding time on metallurgical and mechanical properties. Results Phys. **31** 2021. https://doi.org/10.1016/j.rinp.2021.104956
67. Zhao, L., Nasution, M.K.M., Hekmatifar, M., Sabetyand, R., Kamenskov, P., Toghraie, D., Alizadeh, A., Iran, T.G.: The improvement of mechanical properties of conventional concretes using carbon nanoparticles using molecular dynamics simulation. Sci. Rep. **11**(1) (2021). https://doi.org/10.1038/s41598-021-99616-y
68. Paidar, M., Nasution, M.K.M., Mehrez, S., Bokov, D., Ramalingam, V.V., Zain, A.M.: The feasibility of friction stir spot extrusion-brazing of AA5083-H112 aluminum alloy to brass sheets with Zn interlayer. Mater. Lett. **3081** (2022). https://doi.org/10.1016/j.matlet.2021.131084
69. Bai, J., Paidar, M., Mehrez, S., Ojo, O. O., Nasution, M. K. M., Zain, A. M.: Friction spot extrusion brazing of copper to AISI 304 stainless steel with Zn interlayer: effect of shoulder surface modification. Arch. Civ. Mech. Eng. **22**(2) (2022). https://doi.org/10.1007/s43452-022-00386-9
70. Weiss, C., Carriere, M., Fusco, L., Capua, I., Regla-Nava, J.A., Pasquali, M., Scott, J.A., Vitale, F., Vitale, F., Unal, M.A., Mattevi, C., Bedognetti, D.: Toward nanotechnology-enabled approaches against the COVID-19 pandemic. ACS Nano **14**(6), 6383–6406 (2020). https://doi.org/10.1021/acsnano.0c03697
71. Lund, K., Williams, B., Ke, Y., Liu, Y., Yan, H.: DNA nanotechnology: a rapidly evolving field. Curr. Nanosci. **2**(2), 113–122 (2006). https://doi.org/10.2174/157341306776875811

72. Goodman, R.P., Heilemann, M., Doose, S., Erben, C.M., Kapanidis, A.N., Turberfield, A.J.: Reconfigurable, braced, three-dimensional DNA nanostructures. Nat. Nanotechnol. **3**(2), 93–96 (2008). https://doi.org/10.1038/nnano.2008.3
73. Razeghian, E., Nasution, M.K. M., Rahman, H.S., Gardanova, Z.R., Abdelbasset, W.K., Aravindhan, S., Bokov, D.O., Suksatan, W., Nakhaei, P., Shariatzadeh, S., Marofi, F., Yazdanifar, M.: A deep insight into CRISPR/Cas9 application in CAR-T cell-based tumor immunotherapies. Stem Cell Res. Ther. **12**(1) (2021). https://doi.org/10.1186/s13287-021-02510-7
74. Chupradit, S., Nasution, M.K.M., Rahman, H.S., Suksatan, W., Turki Jalil, A., Abdelbasset, W.K., Bokov, D., Markov, A., Fardeeva, I.N., Widjaja, G., Shalaby, M.N., Saleh, M.M.: Various types of electrochemical biosensors for leukemia detection and therapeutic approaches. Anal. Biochem. (2022). https://doi.org/10.1016/j.ab.2022.114736
75. Nasution, M.K.M., Harahap, A.S., Nasution, S., Herawati, E.: Triangle number of laurent polynomial for the closed braids σ^{2j}. In: IOP Conference Series: Materials Science and Engineering, vol. 300, Issue No. 1 (2018). IOP Conference Series: Materials Science and EngineeringOpen Access, vol. 300, Issue 115 (2018, February)
76. Nasution, M.K.M.: A computation in the braid group. IOP Conf. Ser.: Mater. Sci. Eng. **725**(1) (2020). https://doi.org/10.1088/1757-899X/725/1/012101
77. Li, H., Lee, T., Dziubla, T., Pi, F., Guo, S., Xu, J., Li, C., Haque, F., Liang, X.-J., Guo, P.: RNA as a stable polymer to build controllable and defined nanostructures for material and biomedical applications. Nano Today **10**(5), 631–655 (2015). https://doi.org/10.1016/j.nantod.2015.09.003
78. Soundararajan, H.C., Sivanandham, M.: Quantitative analysis of hepatitis B virion stability upon single-site amino acid mutation based on computational nanotechnology. Mol. Simul. **35**(6), 456–465 (2009). https://doi.org/10.1080/08927020802616699
79. Eshaghian-Wilner, M.M.: DNA sequence matching at nanoscale level. Bio-Inspired Nanoscale Integr. Comput. 1–555 (2009). https://doi.org/10.1002/9780470429983
80. Angelidis, P.A.: Uptake of Mobile ICT Health Services: Has the Time Come to become Commodity? In: Granelli, F., Skianis, C., Chatzimisios, P., Xiao, Y., Redana, S. (eds.) Mobilight 2009. LNICSSITE, vol. 13, pp. 292–302. Springer, Heidelberg (2009). https://doi.org/10.1007/978-3-642-03819-8_28

A 3D Visualization of Zlín in the Eighteen–Nineties in the Unreal Engine Environment

Pavel Pokorný(✉)

Department of Computer and Communication Systems, Faculty of Applied Informatics, Tomas Bata University in Zlín, Nad Stráněmi 4511, 760 05 Zlín, Czech Republic
pokorny@utb.cz

Abstract. This paper describes a 3D interactive desktop application that visualizes the appearance of the town of Zlín in the Eighteen–nineties. The Blender software suite was used to create a complex 3D model of Zlín and its municipality in previous years. Now, this model has been revisited, improved and extended. This was followed by work in order to export it into the Unreal Engine environment. Using this software, the terrain has been further improved to be more realistic and additional assets (grass, trees, bridges, fences) were added to achieve more reality. Then, a user interface with a basic menu was created, user interaction was implemented and initial animation has been added. Then everything was compiled into a fully functional application.

Keywords: 3D visualization · Modelling · Application · Historical visualization

1 Introduction

Nowadays, visualization is very often associated with data obtained from various fields. Visualization research at SCI [1] has focused on applications spanning computational fluid dynamics, medical imaging and analysis, healthcare data analysis, poetry, weather data analysis, biomedical data analysis, financial data analysis, network and graph analysis, etc. These data have generally numerical representation in nature – they have a value and meaning when the data are processed and analyzed. These processes nowadays are performed by computers [2].

The Computer Graphics field is a typical example for creating successful and amazing visualizations. In addition, as computer hardware evolves and performance improves, the possibilities and limits of computer graphics are increasing. The core system of 3D graphics (rendering pipeline) moved from software to raster graphics hardware via the development of geometry– and pixel–processing chips and sophisticated computing operations successively took over the graphics card (GPU) [3].

Animated visualizations offer a powerful way of showing your data and results. In the case of rendered output, it is a sequence of individual rendered frames that are displayed in sequence. Given the fact that the rendering process can select the location and direction

R. Silhavy et al. (Eds.): CoMeSySo 2022, LNNS 596, pp. 430–441, 2023.
https://doi.org/10.1007/978-3-031-21435-6_37

of the view and the quality of the computed frames, animated visualizations allow the creation of highly realistic visual outputs according to predefined user requirements [4].

It is an application that adds another level of visualization by allowing interactivity with the user. Based on user input, it can render arbitrary images and compose them into real–time animation. Such an application was developed at our faculty and is described in this paper.

2 A Short History of Zlín

Although it may seem that Zlín is a new and modern town, even today it is possible to find several buildings in this town whose history goes back deep into the past. This fact is evidenced by two of the most impressive buildings in the town - the church with a high tower and the chateau on which we can recognize the construction details of its early Renaissance style [5].

The first historical record of Zlín's existence comes from the year 1322 when it came into the possession of Queen Eliška, the widow of Wenceslas II, the king of Bohemia. Already at that time, Zlín was a township and the center of quite a widespread feudal estate. Zlín was the seat of the owners of the manor and lived through the gradual development of municipal government. The change in this peaceful evolution was brought about by the Industrial Revolution (Fig. 1).

Fig. 1. Zlín in year 1846 (painted image) [7]

An important milestone in the history of Zlín was the year 1894, when the shoe company of the Bat'a family was founded here. Thanks to the exceptional growth and prosperity of the company and the influence of Tomas Bata, who was also the mayor of the town, Zlín became a modern urban center in the interwar period, with an excellent architectural style and pace of life [6].

The activities of Tomas Bata expanded into many other fields of business: mechanical engineering (1903), tanning (1915), the energy industry (1917), agriculture (1917), etc. Tomas Bata ranked among distinguished personalities of the 20th century both in his home country and abroad. The scope of his activities was impressive, and his ideas, acts and initiatives appeal to us even today.

The visualization application we have created tries to show Zlín just before Tomas Bata started his business activities. At this time, Zlín had a population of less than three thousand inhabitants.

3 Used Resources and Software

The first phases that needed to be performed were to collect all suitable and available historical materials of Zlín in above mentioned period and to select adequate software for visualization application creation.

3.1 Acquiring Resources

The overall progress of this work was initiated by the collation of available historic materials and information about Zlín from the end of the 19th Century or the beginning of the 20th Century. The main resources were found in the Moravian Land Archive in Brno [8], State District Archive in Zlín - Klečůvka [9], the webpage [7], and the book on this subject [10]. Attention was mainly focused on sketches, drawings and photos.

The cadastral maps of that period are a very important part of the visualization process. In the Moravian Land Archive in Brno, there are two of these maps of Zlín available – from 1877, and 1897. Both maps were acquired and, because they were divided into several parts, they were loaded and combined in the GIMP software [11] in order to obtain the entire maps. These maps are colorized and in them, it is easy to distinguish wooden or brick buildings, gardens, roads, rivers, fields, etc. A part of the cadastral map of the center of Zlín, dating from 1877, is shown in Fig. 2. These maps were used for the creation of the model's terrain (e.g. the proper placement of edges and faces), terrain texture drawings, ground plans of buildings and the correct placements and orientations of the 3D models of buildings and accessories in the final scene.

Period photographs are the next important resources. Approximately 280 photographs were obtained from the above–mentioned resources, which captured the appearance of the town between the years 1870 – 1930.

3.2 Software Used

In order to select applications that we wanted to use, we compiled the requisite works that are required for creating high-quality visual outputs. The whole process includes 3d

Fig. 2. An original part of the Cadastral map of the Zlín center from 1877 [8]

model creation, correct texture mapping, and then to create a visualisation application that imports prepared models with their textures. All 3D models with appropriate UV maps and textures were made in Blender. The final application was developed in Unreal Engine.

Blender [12] is the free and open source 3D creation suite. It supports the entirety of the 3D pipeline—modelling, rigging, animation, simulation, rendering, compositing and motion tracking, video editing and 2D animation pipeline. This software is available under GNU GPL License and user can use it on the most common platforms – Windows, Mac OS, Linux, etc.

Blender opens a world of creativity, which has been traditionally exclusive to those able to afford high-end graphics software. Its GUI is the arrangement of windows and panels containing the controls for operating the program. Blender performs many tools that allow one to create and animate 3D models and characters in scenes combined with amazing visual effects. Scenes may be rendered using several different rendering algorithms into photo realistic images and animations (video clips). Video may be edited and compiled into movies [13].

Unreal Engine [14] is the world's most open and advanced real-time 3D creation tool. Continuously evolving to serve not only its original purpose as a state-of-the-art game engine, today it gives creators across industries the freedom and control to deliver cutting-edge content, interactive experiences, and immersive virtual worlds.

Unreal Engine is a complete suite of development tools for anyone working with real-time technology. From design visualizations and cinematic experiences to high-quality games across PC, mobile, console, Virtual Reality, and Advanced Reality, Unreal Engine gives one everything he need to create very high-quality applications. This engine with its own editor, offer complex tools for photorealistic rendering, dynamic physics and effects, robust data translation, life-like animation and much more. License for creators is free-to-use and 100% royalty-free; user can use it to create internal or free projects, or to develop linear content or custom projects for clients, but not for publishing off-the-shelf offerings [15].

Fig. 3. The rendered image of the interactive application from the previous work [16]

4 Modelling and Exporting

The 3D graphics of the model of Zlín from the end of the 19th century did not need to be created from the beginning. This model was created earlier, including an interactive web application that was directly generated in Blender (Fig. 3) [16] using the Blend4Web module [17]. However, such output had several disadvantages, so this work needed to be reworked.

The first disadvantage of this visualization method was the output itself. This web application was created directly in Blender using the interactive output for the web. However, the development of this rendering engine has been closed. In addition, its capabilities are limited, large scenes (which our visualization is) are not displayed continuously and it would be very difficult to make any improvements in quality, speed or control. Similarly, it would be very difficult (almost impossible) to further upgrade and extend this application. Therefore, it was decided to redesign the entire application so that only the model itself, including material properties and textures, is created in Blender and all these assets are then converted to the Unreal Engine, where the new visualization application will be created.

No new models were created in Blender, but it was necessary to make such modifications that allowed the whole scene to be seamlessly converted to the Unreal Engine. The first modifications were made at the level of the geometry of the 3d objects. Some parts of the map contained "holes" or were modelled in a very unrealistic way. Some building models were not placed correctly on the map surface or overlapped each other.

The riverbed model contained a large number of multiple vertices and the map surface extended into the river and was not properly connected to the surrounding area. All of these model issues were corrected in the Blender environment.

Next was to make changes to all material settings and textures. The reason for this was the above–mentioned changes of the rendering engine. The current rendering engine in Blender requires a different way of defining material properties that was not compatible with the previous version. So, it was necessary to create new materials with the appropriate parameters, manually assign them to each object and set the corresponding textures.

Finally, the uv texture maps, which often overlapped, were adjusted. These overlaps were removed as they would cause problems when converting the entire model to Unreal Engine. Also, the uv maps were transformed so that they were linked to the corresponding areas (Fig. 4). Finally, a script was used to remove the excess uv maps that were empty and causing poor display and consequently creating a incorrect light map in Unreal Engine.

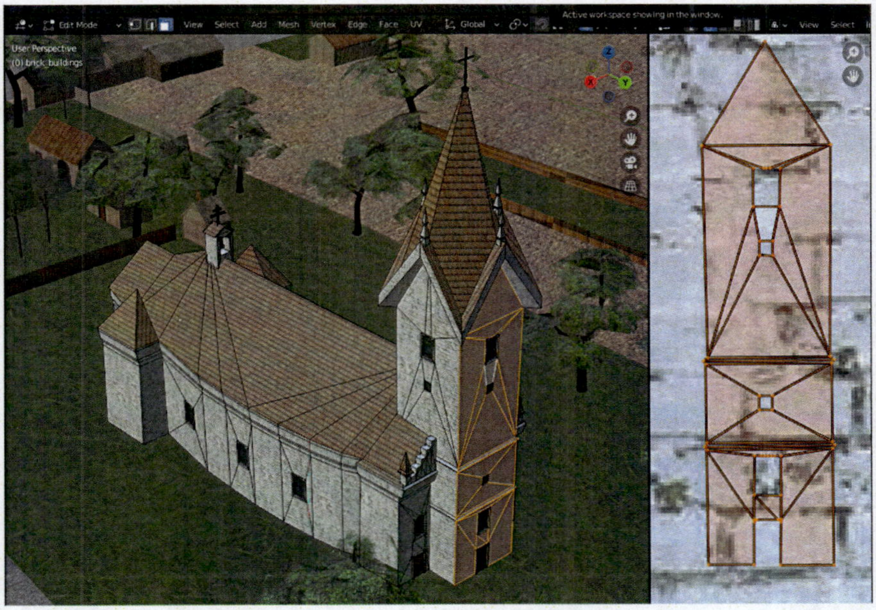

Fig. 4. Corrected UV map of the front surface of the church tower

In order to convert the model to Unreal Engine correctly, some additional modifications were made on the 3D model. The whole scene was placed in the origin of the global coordinate system and all modifiers and transformations were applied to all objects. Furthermore, the center points of the objects were set to meaningful positions. The normal vectors for all object faces were also recalculated to apply the textures on the correct side of the face.

The fbx format was used for the conversion process. This format was designed for single-step inter-operability between the different 3D software, easier data exchange, more efficient workflow as well as customizability [18]. Due to the large size of the scene, this conversion was made in parts.

5 Creation of the Application

5.1 Creating a Scene

A new project was created in the Unreal Engine with a third-person template without raytracing and with setting the display quality to maximum. This template provides the basic elements for creating the entire application – scene lighting, a character with basic animations and controls, and certain objects in the scene.

The basic element of the whole scene is the landscape model on which the other objects are placed. Although the landscape model was previously created in Blender, a new one was created in Unreal Engine using the internal Landscape Editor. The main reason for this was the impossibility of making sufficient adjustments for the imported landscape. The advantage of the landscape created in Unreal Engine is also the use of Landscape Splines and the overall improvement of the graphical quality of the whole scene with the use of multiple layers of materials.

The landscape model was created on the basis of an elevation map, i.e. on the basis of a black and white image that modifies the position of individual vertices of the object in space according to the brightness value of each pixel of this image. The elevation map was obtained from the source [19], which provides geographic data for more than 80% of the total surface of the Earth. This map was then applied to the terrain object in Unreal Engine to create its basic shape. Additionally, this shape was further modified manually using various graphical tools to make it easier to place objects on it and also to make it more realistic. Splines and Sculpt tools were used to create the riverbed.

Fig. 5. Redrawn surfaces on the map

A pack called Brushify - Environmental pack [20] was used as the basis for the terrain material, which contains basic textures and materials for the terrain – grass, mud, forest, snow, sand. The Quixel Megascans library [21] was used as the source for all other textures. According to the contemporary cadastral maps described in Sect. 3.1, the different materials and textures were redrawn to the correct locations of the terrain model (Fig. 5).

The next step was to import the buildings created in Blender. The buildings were inserted with the original materials without automatically generating collisions that would prevented the creation of a player on the map. In order to create the correct building collisions, the Complex object collision was used – a property that can be set for each mesh object. Imported buildings also had a problem with incorrect lighting due to wrongly created light maps. This problem was solved by increasing the resolution of the generated light map and additional modification of the material properties.

5.2 Scene Accessories

Three approaches were used to place additional 3D objects in the scene. The first approach is the Foliage mode, which is used for mass insertion of multiple objects by "drawing". This approach was used, for example, to create a forest at the edges of a scene. It is based on importing relevant objects into the Unreal Engine, then assigning these objects to the tool, and then "drawing" with these objects in the scene. In doing so, a number of parameters can be set, such as drawing intensity or application angle. The imported trees used in our scene have three levels of detail created, and when rendering, the level of detail displayed is selected based on the distance from the camera - the most detailed trees are rendered close to the camera. This approach allows one to insert a large number of trees into the scene and create forests without much impact on computer performance. In addition, all trees contain animations of branches due to wind, which is only applied in close view.

The second approach is to use the Procedural Foliage Spawner tool. Here one selects the layer (areas on the surface of the object) where the selected objects are to be applied. In this tool it is again possible to set a number of parameters such as the amount (density) of deposition, size, orientation, etc. In this way, the fields and all the grass containing also the different flowers were created. In case it happened that some objects applied automatically in this way overlapped into the surrounding areas where they should not have been placed, they were manually deleted.

The third option is the manual application of individual objects. Although this method of application is the most work-intensive, the advantage is the precise placement of each object. For example, trees along roadsides were placed in this way.

The lighting of the entire environment was created using the Directional Light tool. Furthermore, the Sky Atmosphere tool was used, which takes care of the overall appearance of the sky and also allows for alternating cycles of days and nights. In addition, a fog (the Exponential Height Fog tool) was added, with density varying according to height, and volumetric clouds that change dynamically were included to increase the realism of the whole scene.

The overall appearance of the scene in the Unreal Engine after inserting the buildings and all the other objects described in the previous paragraphs is shown in Fig. 6.

Fig. 6. The 3D scene with buildings and accessories in the Unreal Engine environment

5.3 Control and Application Menu

As mentioned in Sect. 5.1, the application template already had a predefined character object, with basic animations and controls. To this, additional controls were added such as changing the first– or third–person view, pausing the application or speeding up the time to move faster through the entire map. Also, a table was added with a list of keyboard shortcuts for controls that can be displayed/hidden in the app. This table is shown in Fig. 7 on the top right.

The Widget blueprint and new empty level were used to create the main menu. Then menu elements started to be added. The background image (screen) from our application was inserted with 75% transparency. Under that image there is a dark background layer to highlight that image. Next, Start, Authors and End buttons were added and placed in the middle of the screen using anchor points. Our faculty logo was added to the bottom right corner of the screen and the project name was inserted above the created buttons.

To create the button functionality, it was necessary to create nodes for the buttons in the event graph. Clicking on the Start button opens the main map, activates the control and allows users to browse the map immediately. A new widget was created for the Authors button containing basic project information, details about the authors and a button to return to the main menu. Free licensed music was added to the menu background. The sounds of the buttons when they are pressed were also set. These sounds are part of the editor.

6 Results

To create the final version and the subsequent executable file of the visualization application, the project had to be packaged. Unreal Engine includes an algorithm to easily pack the entire project. In the File menu the Package Project and the target platform are selected, in our case it is Windows (64-bit). Then, just select the folder on disk where

Fig. 7. View of the Zlín church in the created application with control table

Fig. 8. View of Zlín castle in the created application

the final package will be saved. Then all the scene shaders are compiled and the Unreal Engine will prepare the entire final package without any unused files.

The finished project is 3.07 GB in size and the initial packaging took almost half an hour. The package contains an executable file with.exe extension, a folder containing the game engine files, and a folder containing the entire packaged project in a single.pak file. Selected screenshots of the output from this application are shown in Figs. 7 and 8.

7 Conclusion

This paper describes a 3D interactive desktop application that visualizes the appearance of the town of Zlín in the Eighteen-nineties. The appearance of the town was created on the basis of historical materials – drawings, photographs and maps.

All models of buildings, their UV maps and textures were created in the Blender software suite. When all these objects were completed, they were then imported to Unreal Engine, where the visualization application was made as mentioned above.

The future goal is to improve and expand this application. This process will include modelling more detailed parts of the buildings and new natural assets in order to make the scene more realistic. The project could be further extended by adding a tutorial mode that would describe the objects in the scene with additional information, such as when the building was built or who the owner was at the time. We also plan to convert this application into virtual reality.

Acknowledgments. This work was supported by the Faculty of Applied Informatics, Tomas Bata University in Zlin, under the project RVO/FAI/2021/002.

References

1. Visualization, http://www.sci.utah.edu/research/visualization.html. Last accessed 18 July 2022
2. Srinivasa, K.G., Siddesh, G.M., Srinidhi, H.: Introduction to data visualization. In: Network Data Analytics—A Hands-On Approach for Application Development, pp. 321–331. Springer International Publishing, AG (2018).
3. Hughes, J.F., Van Dam, A., Foley, J.D., McGuire, M., Feiner, S.K., Sklar, D.F.: Computer Graphics: Principles and Practice, 3rd edn. Addison-Wesley, USA (2014). ISBN 978-0-321-39952-6
4. Rendering in 3D animation production. https://dreamfarmstudios.com/blog/the-final-step-in-3d-animation-production-3d-rendering/. Last accessed 18 July 2022
5. History of Zlín. http://www.ic-zlin.com/25153-history-of-zlin. Last accessed 18 July 2022
6. Personality Tomas Bata 1876–1932. http://en.tomasbata.org/personality. Last accessed 18 July 2022
7. Zlínské ulice. https://staryzlin.cz/zlinske-ulice.php. Last accessed 18 July 2022
8. Moravský zemský archív v Brně, http://www.mza.cz/. Last accessed 18 July 2022
9. Státní okresní archív Zlín. http://www.mza.cz/zlin/. Last accessed 18 July 2022
10. Pokluda, Z., Nováček, S.: Zlín ve fotografii (1890–1950). Zlín: ESPRINT Zlín, s. r. o. a Nadace Tomáše Bati, 2008. ISBN 978-80-254-3144-3
11. GIMP—GNU Image Manipulation Program. http://www.gimp.org. Last accessed 18 July 2022
12. Blender.org—Home of the Blender project. https://www.blender.org/. Last accessed 18 January 2021

13. Blain, J.M.: The Complete Guide to Blender Graphics, 3rd edn. CRC Press (2016). ISBN 978-1-4987-4647-2
14. The most powerful real-time 3D creation tool—unreal Engine. https://www.unrealengine.com/en-US/. Last accessed 18 July 2022
15. Uzayr, S.: Mastering Unreal Engine, 1st edn. CRC Press (2022). ISBN 978-1-03210-313-6
16. Pokorný, P., Dočkalová, P.: A 3D Visualization application of Zlín in the eighteen-nineties. In: Šilhavý, R. (ed.) Advances in Intelligent Systems and Computing 2018, LNCS, vol. 765, pp. 223–232. Springer, Heidelberg (2018). ISBN 978-331991191-5
17. Unleashing the Power of 3D Internet/Blend4Web. https://www.blend4web.com/en/. Last accessed 18 July 2022
18. FBX—Adaptable File Formats for 3D Animation Software. https://www.autodesk.com/products/fbx/overview. Last accessed 18 July 2022
19. CGIAR-CSI SRTM—SRTM 90m DEM Digital Elevation Database. https://srtm.csi.cgiar.org/. Last accessed 18 July 2022
20. Brushify—Environment Shaders Pack. https://www.unrealengine.com/marketplace/en-US/product/brushify-shaders-pack. Last accessed 18 July 2022
21. Quixel Megascans. https://quixel.com/megascans/home/. Last accessed 18 July 2022

Reference Method for the Care of Health Consultations of Older Adult Users Based on Chatbot, in a Context of Pandemic

Eleuterio Quispe Poma, Elbio Labra Chino, Fredy Abel Huanca Torres(✉), and Abel Angel Sullon Macalupu

Faculty of Engineering and Architecture, Professional School of Systems Engineering, Peruvian Union University, Juliaca, Perú
{eleuterioquispe,elbiolabra,abel.huanca,angeli}@upeu.edu.pe

Abstract. The COVID-19 pandemic has marked a considerable event in the history of all countries, causing a high degree of mortality rate in older adults, in Peru health care for this group of users has become relevant. In this sense, the EsSalud Center for the Elderly (Social Health Insurance) in the town of Sicuani (CAM - Sicuani) needs to protect its population. This research consisted of proposing a reference method for health consultations based on chatbot, avoiding face-to-face consultations; for which a reference method was developed to design and develop the chatbot, in order to allow it to answer user queries online and immediately. For this purpose, information was obtained from CAM - Sicuani regarding the daily queries made by users. Information was also collected on the topics, questions and answers related to COVID-19, this from the websites of the World Health Organization (WHO), EsSalud, Ministry of Health (MINSA). On the other hand, there are not many research and projects related to the chatbot. A chatbot development method was developed based on Amir Shevat's book consisting of 4 phases; The first phase called use case definition and exploration, the second conversation scripts, the third design and testing and finally the development of the chatbot. Regarding the results, through the analysis of the metrics it is shown that the chatbot is efficient in response time with an average of 0.8 ms in the confusion rate, the chatbot responds in real time; for the performance of the chatbot, it was verified that it is flexible and optimal, since the capacity of the chatbot is adaptable; for the last metric, which is the rating of the chatbot, it was evaluated based on star rating; through a visual scale, where an average of users was randomly selected for the chatbot tests in this first initial version, who were satisfied with a rating of 3 - 5 stars; With the chatbot, it contributes to having a significant improvement in the process of attending to queries, generating a positive attitude in users; In addition, CAM - Sicuani and its users will benefit from the chatbot in relation to minimizing costs and optimizing time, which is not subject to any alteration. We conclude that the chatbot in its initial version works efficiently, in addition this innovation will prevent people from attending these health consultation services in person, on the other hand the digital transformation is promoted and with this a competitive advantage is obtained.

Keywords: Cloud computing · Chatbots · COVID-19 · Health · Consultations · Older adults · CAM - Sicuani

R. Silhavy et al. (Eds.): CoMeSySo 2022, LNNS 596, pp. 442–455, 2023.
https://doi.org/10.1007/978-3-031-21435-6_38

1 Introduction

The COVID-19 pandemic has marked a milestone in all countries, causing a serious situation in society, economy and health, where Perú has not been immune to this evil. The WHO declared COVID-19 on March 11, 2020 as a new global pandemic [1], During the pandemic, a large number of mortality rates, economic repercussions, and increased poverty have been recorded throughout the world [2], all caused by the severe acute respiratory syndrome coronavirus 2 (SARS-CoV-2) virus facing to humanity [3], faced with the threat of this disease, several countries were very rigorous with the imposition of preventive measures, given that the entry of people from countries with a high degree of contagion is restricted [4], also in order to counteract the advance of the disease, public health policies were established [5], in the case of the Peruvian state, government actions were taken; where the state of national emergency was established with Supreme Decree No. 0044–2020-PCM and mandatory social isolation (quarantine) was ordered to control the virus [6], despite these efforts the outbreak of COVID-19, still it leads to a health problem, even more so, having more fatalities in older adults [7].

Taking into account the mortality rate of COVID-19, health care for this group of users has become relevant in Peru, since they are more affected compared to the rest of the people [8], Likewise, Supreme Decree No. 001–2021-MIMP [9]; where he mentions about statistics that show that the populations most vulnerable to the effects of COVID-19 are older adults, being them the main fatalities from COVID-19 [10], in reality the state of the older adult population in Peru is precarious, due to insufficient accessibility to health, education and pension services [11]. In this sense, the CAM - Sicuani, which is a public institution in Peru dedicated to the monitoring and care of the elderly; requires protecting its population in the context of the pandemic [12]. On the other hand, the CAM - Sicuani for the attention of the queries, does it manually and in person; generating crowds in the service process, causing long queues, discomfort and discomfort in users, also putting health at risk due to the COVID-19 pandemic, as well as the cost and time factor. The CAM - Sicuani in an agreement decided to look for alternatives through the use of information technologies; In order to avoid the exposure of this vulnerable population, it is also known that the confinement of people was an alternative to mitigate the possible infections caused by the virus [13].

Different existing technologies have been evaluated, such as: Chatbots, Google Docs forms, web page, video calls, telephone line and others; of which, chatbots are adapted for this type of attention; considering that they have come a long way from their beginnings as Eliza was the first chatbot developed in 1966 that was programmed to act as a psychotherapist [14], in the same way Parry was in the 70s, they were implemented intelligent personal assistants such as Siri in the 2010s, J. Herrera mentions that today the chatbot is in a stage of maturity in different areas, whether on the web, social networks and others [15], they are also reflected in various fields such as: means of transportation, hotels, education, veterinary consultations, medication consultations, electronic commerce, among others; since a customer support chatbot was designed and implemented for the Venice airport [16], now a personalized intelligent chatbot was designed for hotel services in Indonesia [17], likewise an educational chatbot was developed to answer queries [18], in a similar way a chatbot was implemented to change the lifestyle in education [19], on the other hand a chatbot prototype was designed for veterinary consultations [20], in the same way MediBot was implemented, a chatbot to consult information about drugs and their risks [21], in addition an e-commerce sales chatbot was developed to provide customer support and increase sales [22]. As can be seen, the use of conversational agents (chatbot) is convincing and evident, it is also considered that it will continue to grow in the following years, as shown in the Google Trends search engine [23]; The interest in the search for the term "chatbot" throughout the planet has been notorious in recent years.

Along with the start of the COVID-19 pandemic, many companies and businesses have chosen to implement the use of the chatbot, as well as entrepreneurs in order to reduce costs and generate higher profits. In addition, with the rise of the exponential contagion of COVID-19, the development of chatbots has been promoted in various technological investigations as a means of non-face-to-face care and thus counteract the advance of the virus. A chatbot was also developed as a communication tool during the COVID-19 health crisis in Spain [24], on the other hand [15], a chatbot was designed as a complementary tool to answer relevant civil queries from Spanish-speaking migrants who go to an NGO dedicated to their integration in Chile. The WHO did the same by implementing the structured WhatsApp chat to answer questions about COVID-19, and provide instant, reliable information 24 h a day, 7 days a week throughout the planet [25].

Thus, there are many chatbots, as an alternative in its development and implementation which had beneficial results; as shown in the investigations already carried out. However, there is no reference method for the design and development of the chatbot for the queries of older adult users with the characteristics of this research, which is why there was a motivation to design and develop the chatbot in this institution, aimed at improving the aging process with an active lifestyle in older adults; therefore, the chatbot will help prevent the COVID-19 breach. The developed chatbot is based on the proposed method; It consists of 4 phases, which are detailed in the method part. For its development with the C# programming language in.Net 5 and the Bot Framework emulator V4, cloud computing was also used with Microsoft azure services, QnaMaker and Luis AI cognitive services were used, and later it will be integrated with Facebook Messenger. For the tests, metrics or indicators were used, where the responsiveness of the chatbot in interaction with the user was evaluated, in addition to detecting performance anomalies, Live Metrics Stream. Due to the above, the objective of this research was to develop a reference method for health consultations based on chatbot, as a preventive measure against the COVID-19 pandemic; for which the chatbot was designed and developed, in order to allow users to answer queries online and immediately in the CAM - Sicuani.

2 Method

The proposed reference method of the chatbot was adapted from the method proposed by Amir Shevat [26]. This research consists of 4 phases. The first phase called case and exploration, the second conversation scripts, the third design and test and finally the development of the chatbot. All these phases were of vital importance in the investigation described below. In addition, it is shown in a general way in the following diagram Fig. 1.

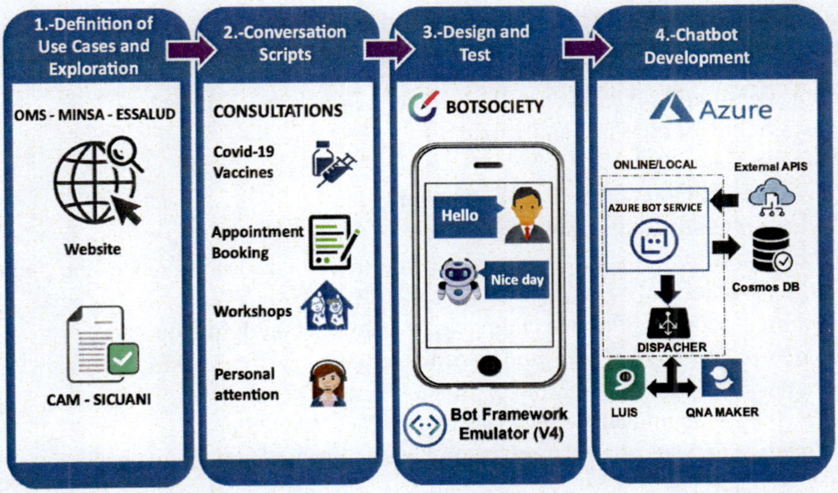

Fig. 1. Phases of the reference method for the care of health consultations based on chatbot, suitable from [26].

2.1 Definition of Use Case and Exploration

Obtaining Information. In this first phase, the use of the chatbot for user queries in the CAM - Sicuani, as a starting point, information was collected in meetings with the professionals responsible for the health service about frequent queries received by phone and in person from users. In the second instance, information on the topics, questions and answers was also obtained; this from the websites related to the COVID-19 pandemic such as: WHO, ESSALUD and MINSA. With the information collected, the topics, the repetitive questions and the respective answers were defined. For each topic a submenu opens with the related questions. on the other hand, the following basic requirements had to be identified for the chatbot:

- The Chatbot will be available to all queries from CAM - Sicuani users, related to the topic.
- The chatbot will generate a personalized appointment for users interested in having specialized attention.
- The Chatbot will save the user's data, once the appointment has been generated.
- The Chatbot will allow users to rate the bot.

The purpose of the chatbot for inquiries is to help users answer their questions in the shortest possible time. The chatbot as a conversational assistant facilitates and simulates an unstructured and structured conversation, since the interaction of the conversation is by questions and answers, finally CAM - Sicuani is very interested in clearing up any doubts of its users and they are pleased with the attention delivered by the chatbot.

Choice of Platform. Facebook Messenger was chosen for the choice of platform. due to the fact that it is the channel most used by cAM - Sicuani, it is for this reason that this medium will be given greater priority, since Facebook Messenger is the most used social network and its value is the mass of users it gathers; This facilitates the user experience when making queries.

Chatbot Personality Definition. The chatbot as a conversational assistant, in its responses to the user has a friendly tone.

Name: EsSalud Bot

Environment: Users of the CAM - Sicuani

Audience: Older adults aged 60 and over.

Task in question: Answer the repetitive and personalized questions to the users of the CAM - Sicuani.

Runtime Variations: Provides more casual conversation during queries.

Locally relevant social acceptance: Suitable for users of the CAM - Sicuani.

Service mark: Professional, fun but not too casual.

Values: Provide immediate response to users of CAM - Sicuani.

Derivative Personality: Friendly but not overly humorous.

2.2 Conversation Scripts

General Scheme. After the information consulted and compiled, the contents that the chatbot would contemplate were chosen, these were previously consulted and validated by the professional responsible for the CAM - Sicuani, the following topics being chosen:

- Appointments
  - Create appointment
  - See citation
- Consultations and contact
  - About vaccines
  - COVID-19 advice
  - About the workshops
  - Frequently asked questions
  - CAM Contact Center – Sicuani
- Follow us on our social networks
  - Website
- Chatbot Rating Unrelated
- Topics

Mapping of Intentions. The intentions represent the actions and tasks that the user will carry out, in this section is where a topic is subject with an intention; that is, the themes that have already been defined previously, are assigned to what the user wants to express. So, if the user wants to consult about the vaccines; has to trigger a defined expression. Since each intention must be predetermined according to the theme as keywords, important words that the Luis AI cognitive service detects what the user determines. Then the intentions are defined, as can be seen in Table 2.

Table 1. QnAMaker knowledge base.

Order	Questions	Answers
1	Information and contact	About address and number
2	Vaccines	About vaccines
3	Workshops	About workshops
4	Frequent questions	About frequently asked questions

Table 2. Mapping of intentions in Luis AI.

Order	Topic	Intention in Luis AI
1	Appointments	CreateAppointment,ViewAppointment
2	Information and contact	Say hello, See CenterContact, Dismiss, Thank you, See Options
3	Chatbot rating	Rate
4	Unrelated topics	None

Write Scripts. According to the Chatcompose official page [27], it defines scripts as a collection of archived messages. Also, these phrases or sentences determine the direction of a particular conversation. Scripts can contain conversation starters, answers to questions, and ideas for merging conversations. The scripts can even help the chatbot change topics accordingly and ask questions. For this research, the following scripts have been defined:

Attention to Queries. This script has the purpose that the chatbot can attend to the queries of the users; thus simulating a conversation with fast and scalable responses that improve the user experience, since it is available 24/7.

Appointments. This script determines a personalized attention to the user, for which the user had to book his appointment.

Bot Qualification. This script has the purpose of evaluating the chatbot by users, based on the star rating with a 5-point matrix, to classify the attributes on a satisfaction scale, since it determines with a rating where "1" star is low and "5" stars is excellent [28].

2.3 Design and Test

Design. The design of the chatbot prototype was carried out with the tool called Botsociety, since it has many of the functions for bots; To start designing, log in with the

Botsociety account after registering, since it supports functionalities on different social platforms such as: Facebook Messenger, WhatsApp, Skype, Twitter, Telegram and others [29]. For the design of this chatbot, Facebook messenger was chosen.

Tests. For tests Bot Framework Emulator V4 was used, doing conversation tests with the chatbot, inspecting the messages it sends and receives. The emulator shows the messages as they appear in a user interface, registers the requests and responses where the queries were made, appointment reservations and the chatbot qualification [30].

2.4 Development of the Chatbot

Choice of the Tool. The tools that were chosen are those mentioned above, which are defined in their respective sites and/or official pages as in: Microsoft Azure [31], Luis Cognitive Service (AI) [32], QnAMaker Cognitive Service [33], Microsoft Azure Cosmos DB [34], Visual Studio 2019 [35], Bot Framework Emulator (V4) [30], C#.

General Architecture. The general graph of the chatbot architecture and the respective links between the web application, the QnAMaker and Luis AI cognitive services, and the integration with Cosmos DB, which is typical of Microsoft Azure, was made. Next, the architecture scheme is presented in the following Fig. 2.

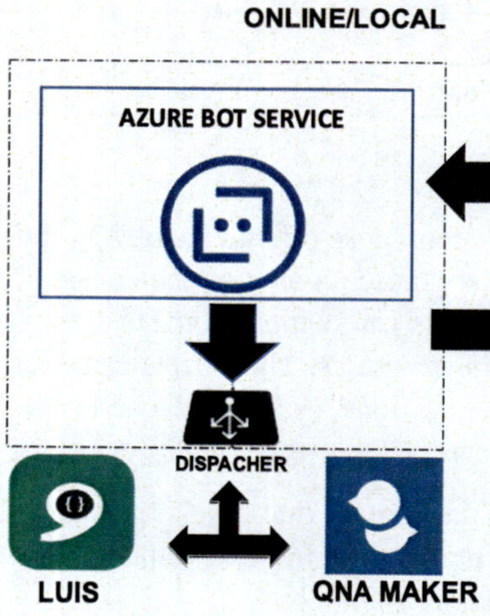

Fig. 2. General scheme of the operation of the chatbot [16].

Development. This stage basically starts by installing Visual Studio and Bot Framework version 4; once installed and configured, we open Visual Studio and create a new project, then we install .Net in the project, we also install Microsoft bot builder and create our resource group in the azure portal and configure the bot state with azure blob storage, we synchronize with our project with the API Keys and inside the project we install Microsoft bot builder of azure blob, once configured we add the images inside azure blob storage and create the cognitive service of Luis AI; where intentions are created, they get to be trained and tested. To later integrate it with our project, once integrated we create the carousel of options for it we create the hero cards; where the images that were uploaded to the azure blob storage are added, once the hero cards are configured, the cosmos DB database is created and configured in the azure portal, to store the information of the appointments and the rating of the bot by of the user. The project also downloads and installs the Entity Framework Core. We configure in Cosmos DB and develop the contact center flow, such as the rating flow and storage in the database. Next we create the citation flow, add the Luis AI entities, and also create the QnAMaker service. This knowledge base was created with the important data and EsSalud CAM - Sicuani, MINSA, WHO was used as a base source to later integrate it with the Bot Framework, for the tests metrics or indicators were used, where the response capacity was evaluated. Of the chatbot in interaction with the user, also to detect performance anomalies in real time, Live Metrics Stream was used.

Metrics to Evaluate the Chatbot
Confusion rate (CR): It is the number of messages made by the user, which are not understood by the chatbot.
Response time (RT): It is the response time of the chatbot to answer a user query in the shortest possible time.
Performance (R): It is the evaluation of the performance of the services offered by the azure platform.
User satisfaction (US): It is the user's rating regarding the experience with the chatbot, in which 1 represents the lowest rating and 5 the highest.

3 Results

For the results, runs were made which are related to the metrics, so that they are detailed below:

Run 01. The confusion rate (CR) metric was evaluated; where in the conversation flow the chatbot was confused with 50 expressions in the Luis AI cognitive service with topics not related to queries such as: <<Tell me a joke>> <<Do they sell pizza?>> <<Who are you?> > <<What's your name?>> and among others, which are detailed in Table 1. The chatbot, not understanding and/or understanding the user's expression, emits a response with the following phrase <<Try again with another term>> according to the Luis AI cognitive service, the chatbot recognizes the expressions that are already defined within the None intent; and those that are not within the mentioned intention will be trained and later added to the Luis AI cognitive service.

Of all the executed queries, the chatbot recognizes all the expressions that are defined within the None intent, with an immediate response that is optimal, since Luis AI's cognitive service shows a time score of 0.7 - 0.9 ms.

Run 02. In this run, the Response time (RT) metric was evaluated; where in the conversation flow the chatbot responds immediately to queries related to the topic, and which are defined in the intentions within the Luis IA cognitive service as: <<Hello>>

Table 3. User interaction with the chatbot according to the confusion rate (CR) metric.

Order	Queries not related to the topic (intention None)	Response	Score
1	Tell me a joke	Try again with another term	0.904
2	Do they sell pizza?	Try again with another term	0.861
3	Who are you?	Try again with another term	0.825
4	What's your name?	Try again with another term	0.891

<<Quote>> <<View appointment>> <<Thank you>> <<Qualify>> among others, which are detailed in Table 4. Where the chatbot issues a defined response according to the expression and intention made by the user, and those that are not within the aforementioned intention they will be trained and later added to the Luis AI cognitive service.

Table 4. Chatbot response time according to the Response time (RT) metric.

Order	Queries related to the topic according to each intention	Score
1	Hello Bot	0.744
2	Quote	0.982
3	See quote	0.767
4	Thank you	0.921
5	Rate	0.980

For response time (RT), the response time is optimal at an average of 0.75 - 0.9 ms, the chatbot recognizes the expression according to the intention, if it is not recognized, it will emit the following phrase < < Try again with another term > > according to the Luis AI cognitive service. It is worth mentioning that the chatbot responds to the intentions in real time.

Another way to evaluate the response time (RT) is through Azure's own Live Metrics Stream; where you can have an overview of input requests, requests, monitor performance, perform filters, observe errors and among others in real time. As shown in the following Fig. 3.

Run 03. For this performance metric (P), Live Metrics Stream was used; This metric is very important since with it we see the performance of the chatbot according to the capacity, even if there is a lot of interaction, its capacity can be increased as well as load balancing, which is why it is flexible, as can be seen in Fig. 4.

Run 04. For the user satisfaction metric (US); The star rating was used, where in Cosmos DB, the JSON data of the user's appointment and the corresponding rating are stored, as can be seen in the following Table 5.

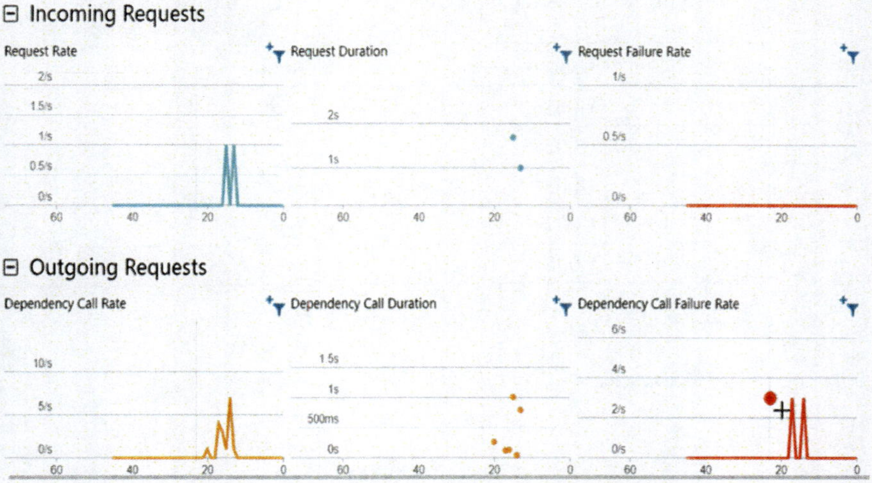

Fig. 3. Chatbot response in live metrics transmission.

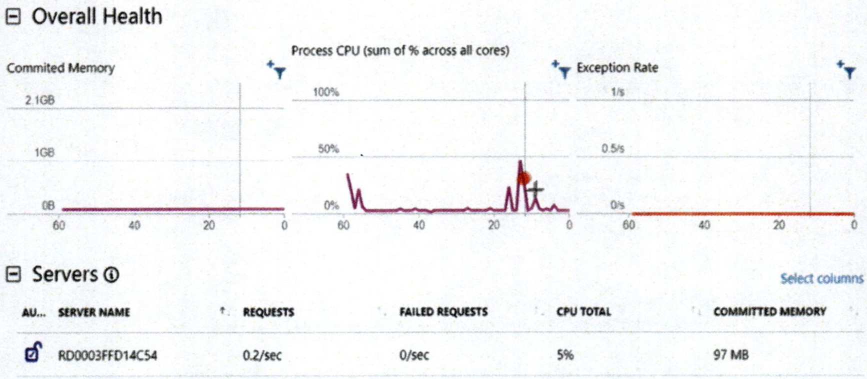

Fig. 4. The CPU process performance of the chatbot.

The final result of this research responds to the stated objective, given the most relevant image of this research is illustrated below. As can be seen in Fig. 5, where the user's interaction with the chatbot on Facebook messenger is observed.

4 Conclusions

This research consisted of proposing a reference method for health consultations based on chatbot, as a preventive measure against the COVID-19 pandemic, in order to allow frequent consultations to be answered and appointments to be booked at the CAM - Sicuani. The developed chatbot works efficiently in its initial version in interaction with the 50 randomly selected users, where in the user satisfaction (US) they were satisfied since it resolved their doubts immediately, as can be seen in Table 5. In the

Table 5. Star rating in the user satisfaction (US) metric.

Order	User	Rating
1	U001	☆☆☆☆
2	U002	☆☆☆☆☆
3	U003	☆☆☆
4	U004	☆☆☆☆
5	U005	☆☆☆☆☆

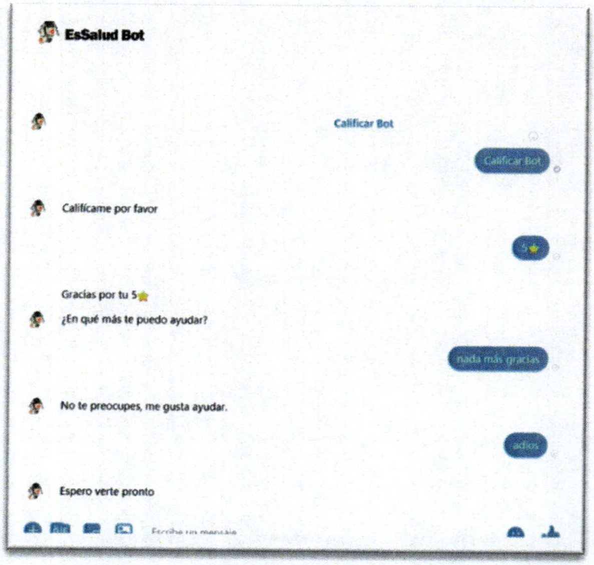

Fig. 5. Final result of the chatbot in Facebook messenger.

metric of response time (RT) the chatbot evaluates the response defined according to the expression and intention made by the user, it is worth highlighting those that are not within the mentioned intention will be trained and added later to the Luis AI cognitive service, a time of response in real time, with an average of 0.75 to 0.9 ms, which is detailed in Table 4. On the other hand, in the confusion rate (CR) metric, the chatbot was confused with 25 expressions or/and phrases with topics unrelated to the consultations in the cognitive service Luis AI; the chatbot, not understanding the expression, issues a response with the following phrase < < Try again with another term > > which is observed in Table 3. Also mention the finding of Live Metrics Stream; where you can see in real time, a whole panorama of metrics, such as the performance (P) and response time (RT) metrics, as can be seen in Figs. 4 and 3.

Which means that the chatbot solves consultations in real time, it also saves time, lowers costs and is available 24/7. Also highlight the limitations for the development of this research that were; the situation of COVID-19, budget to acquire Azure due to lack of financing. Despite these limitations, the chatbot-based referral method for healthcare consultations represents a significant contribution to the digital transformation of the insurance company; chatbots, together with the arrival of the pandemic, had a stage of maturity, where companies and/or entities have come to change the care process, as is the case of the WHO by implementing the structured WhatsApp chat to answer questions about the COVID-19, and provide information instantly, reliably 24 h a day, 7 days a week throughout the planet [26], with this innovation it is intended to contribute to future research such as: a guide for the development of chatbots, analysis of feelings, data mining, among others. Generating alternatives for other researchers to promote the development of new trends in research, formulating work studies that reduce the impact of a future crisis.

References

1. Organizacion Mundial de la Salud: La OMS caracteriza a COVID-19 como una pandemia—OPS/OMS | Organización Panamericana de la Salud, 11 de marzo de 2020. https://www.paho.org/es/noticias/11-3-2020-oms-caracteriza-covid-19-como-pandemia. Accedido 15 de agosto de 2021
2. Ciotti, M., Ciccozzi, M., Terrinoni, A., Jiang, W.-C., Wang, C.-B., Bernardini, S.: The COVID-19 pandemic. Crit. Rev. Clin. Lab. Sci. **57**(6), 365–388 (2020. https://doi.org/10.1080/10408363.2020.1783198
3. Quiroz Carrillo, C.G., et al.: Un nuevo coronavirus, una nueva enfermedad: COVID-19. Horiz. Méd. Lima **20**(2) (2020). https://doi.org/10.24265/horizmed.2020.v20n2.11
4. Mejia, C.R., et al.: Percepción de las medidas de salud pública en Perú para frenar el avance de la COVID-19. Rev. Cuba. Investig. Bioméd. **40**(1) (2021). Accedido 5 de septiembre de 2021. [En línea]. Disponible en http://scielo.sld.cu/scielo.php?script=sci_abstract&pid=S0864-03002021000100005&lng=es&nrm=iso&tlng=pt
5. Gonzales-Castillo, J.R., Varona-Castillo, L., Dominguez-Morante, M.G., Ocaña-Gutierrez, V.R.: Pandemia de la COVID-19 y las Políticas de Salud Pública en el Perú: marzo-mayo 2020. Rev. Salud Pública **22**(2) (2020). https://doi.org/10.15446/rsap.v22n2.87373
6. Decreto Supremo N° 044-2020-PCM. https://www.gob.pe/institucion/pcm/normas-legales/460472-044-2020-pcm. Accedido 5 de septiembre de 2021
7. Vega Rivero, J.A. et al.: La salud de las personas adultas mayores durante la pandemia de COVID-19. J. Negat. No Posit. Results **5**(7), 726–739 (2020). https://doi.org/10.19230/jonnpr.3772
8. Aquino-Canchari, C.R., del Quispe-Arrieta, R.C., Huaman Castillon, K.M., Aquino-Canchari, C.R.. del Quispe-Arrieta, R.C., Huaman Castillon, K.M.: COVID-19 y su relación con poblaciones vulnerables. Rev. Habanera Cienc. Médicas **19** (2020). Accedido 15 de agosto de 2021. [En línea]. Disponible en: http://scielo.sld.cu/scielo.php?script=sci_abstract&pid=S1729-519X2020000400005&lng=es&nrm=iso&tlng=es
9. Decreto Supremo que aprueba la Estrategia Preventiva para Centros de Atención para Personas Adultas Mayores - "YANAPAY60+", en el marco de la Emergencia Sanitaria por la COVID-19-DECRETO SUPREMO-N° 001–2021-MIMP. http://busquedas.elperuano.pe/normaslegales/decreto-supremo-que-aprueba-la-estrategia-preventiva-para-ce-decreto-supremo-n-001-2021-mimp-1929774-5/. Accedido 25 de julio de 2021
10. Dirección General de Epidemiología—Ministerio de Salud del Perú. Adultos mayores siguen siendo las principales víctimas mortales por COVID-19. CDC MINSA. https://www.dge.

gob.pe/portalnuevo/covid-19/adultos-mayores-siguen-siendo-las-principales-victimas-mortales-por-covid-19/. Accedido 15 de agosto de 2021
11. La situación de la población adulta mayor en el Perú, hallazgos de una investigación del Idehpucp para mejorar las políticas públicas. IDEHPUCP, 20 de noviembre de 2018. https://idehpucp.pucp.edu.pe/notas-informativas/la-situacion-de-la-poblacion-adulta-mayor-en-el-peru-hallazgos-de-una-investigacion-del-idehpucp-para-mejorar-las-politicas-publicas/. Accedido 15 de agosto de 2021
12. Resolución de Gerencia General N° 480-Essalud-2020. http://www.essalud.gob.pe/wp-content/uploads/RGG_480_GG_ESSALUD_2020.pdf?fbclid=IwAR3MGkip_iKy410Wq1LWfEwbYN1EYhFRL2uFDv0jkesROh2hJ34eOAy_6yw. Accedido 15 de agosto de 2021
13. Pinazo-Hernandis, S.: Impacto psicosocial de la COVID-19 en las personas mayores: problemas y retos. Rev. Esp. Geriatría Gerontol. **55**(5), 249–252 (2020, September). https://doi.org/10.1016/j.regg.2020.05.006
14. McNeal, M.L., Newyear, D.: Chapter 1: Introducing Chatbots in Libraries. Libr. Technol. Rep. **49**(8) (2013, December). Accedido 16 de septiembre de 2021. [En línea]. Disponible en https://www.journals.ala.org/index.php/ltr/article/view/4504
15. Herrera Toro, J.I.: Diseño de un Chatbot como herramienta complementaria para responder consultas civiles relevantes de migrantes hispanohablantes que acuden a una ONG dedicada a su integración en Chile. 2021, Accedido: 16 de septiembre de 2021. [En línea]. Disponible en: http://repositorio.uchile.cl/handle/2250/180182
16. Carisi, M., Albarelli, A., Luccio, F.L.: Design and implementation of an airport chatbot. In: Proceedings of the 5th EAI International Conference on Smart Objects and Technologies for Social Good, New York, NY, USA, pp. 49–54 (2019, September). https://doi.org/10.1145/3342428.3342664
17. Putri, F.P., Meidia, H., Gunawan, D.: Designing intelligent personalized Chatbot for hotel services. In: Proceedings of the 2019 2nd International Conference on Algorithms, Computing and Artificial Intelligence, New York, NY, USA, pp. 468–472 (2019, December). https://doi.org/10.1145/3377713.3377791
18. Sinha, S., Basak, S., Dey, Y., Mondal, A.: An educational Chatbot for answering queries. In: Emerging Technology in Modelling and Graphics, Singapore, pp. 55–60 (2020). https://doi.org/10.1007/978-981-13-7403-6_7
19. Kasthuri, E., Balaji, S.: A Chatbot for Changing Lifestyle in Education. In: 2021 Third International Conference on Intelligent Communication Technologies and Virtual Mobile Networks (ICICV) (2021, February), pp. 1317–1322. https://doi.org/10.1109/ICICV50876.2021.9388633
20. Huang, D.-H., Chueh, H.-E.: Chatbot usage intention analysis: veterinary consultation. J. Innov. Knowl. **6**(3), 135–144 (2021, July). https://doi.org/10.1016/j.jik.2020.09.002
21. Avila, C.V.S., Rolim, T.V., da Silva, J.W.F., Vidal, V.M.P.: MediBot: Um chatbot para consulta de riscos e informações sobre medicamentos. In: Anais Estendidos do Simpósio Brasileiro de Computação Aplicada à Saúde (SBCAS), pp. 1–6 (2019, June). https://doi.org/10.5753/sbcas.2019.6275
22. Khan, M.M.: Development of an e-commerce sales Chatbot. In: 2020 IEEE 17th International Conference on Smart Communities: Improving Quality of Life Using ICT, IoT and AI (HONET) (2020, December), pp. 173–176. https://doi.org/10.1109/HONET50430.2020.9322667
23. «Google Trends», Google Trends. https://trends.google.es/trends/explore?date=today%205-y&q=chatbot. Accedido 16 de septiembre de 2021
24. Bonales, G., Pradilla, N., Martínez, E.: Chatbot como herramienta comunicativa durante la crisis sanitaria COVID-19 en España. ComHumanitas Rev. Científica Comun. **11**(3) (2020). https://doi.org/10.31207/rch.v11i3.270

25. B. virtual: OMS habilita chat de WhatsApp para responder dudas sobre el Covid-19. https://bibliotecavirtual.insnsb.gob.pe/oms-habilita-chat-de-whatsapp-para-responder-dudas-sobre-el-covid-19/. Accedido 17 de septiembre de 2021
26. Shevat, A.: Designing Bots Creating Conversational Experiences, 1 edn. United States of America: O'Reilly (2017)
27. Cómo crear guiones de conversación para un chatbot. https://www.chatcompose.com/crear-guiones-conversacionales.html. Accedido 11 de enero de 2022
28. Pregunta de valoración por estrellas. ¿Cómo funcionan? https://www.questionpro.com/es/features/pregunta-deslizante. Accedido 7 de febrero de 2022
29. Botsociety: Design, preview and prototype your next chatbot or voice assistant. https://botsociety.io. accedido 14 de enero de 2022
30. JonathanFingold: Test and debug bots using the Bot Framework Emulator—Bot Service. https://docs.microsoft.com/en-us/azure/bot-service/bot-service-debug-emulator. Accedido 11 de enero de 2022
31. Qué es Azure: Servicios en la nube de Microsoft | Microsoft Azure. https://azure.microsoft.com/es-es/overview/what-is-azure/. Accedido 11 de enero de 2022
32. Introducción a Language Understanding (LUIS)—Azure Cognitive Services. https://docs.microsoft.com/es-es/azure/cognitive-services/luis/what-is-luis. Accedido 11 de enero de 2022
33. ¿Qué es el servicio QnA Maker?—Azure Cognitive Services. https://docs.microsoft.com/es-es/azure/cognitive-services/qnamaker/overview/overview. Accedido 11 de enero de 2022
34. Azure Cosmos DB: base de datos NoSQL | Microsoft Azure. https://azure.microsoft.com/es-es/services/cosmos-db/. Accedido 11 de enero de 2022
35. Visual Studio 2022 Community Edition: descargar la versión gratuita más reciente. https://visualstudio.microsoft.com/es/vs/community/. Accedido 11 de enero de 2022

Prototype Technology Decision Support System for the EBW Process

Vladislav Kukartsev[1,2,3](✉), Nazar Saidov[2], Artem Stupin[1,4], Olga Shagaeva[1], Olga Antamoshkina[5], and Larisa Korpacheva[1]

[1] Siberian Federal University, Krasnoyarsk, Russia
vlad_saa_2000@mail.ru
[2] Reshetnev Siberian State University of Science and Technology, Krasnoyarsk, Russia
[3] Bauman Moscow State Technical University, Moscow, Russia
[4] University of Cadiz, Cadiz, Spain
[5] Krasnoyarsk State Agrarian University, Krasnoyarsk, Russia

Abstract. Today, the main method of working out the welding modes and selecting the optimal parameters of EBW is welding of product samples. The aim of the study is to reduce the time of selection of optimal technological parameters of the process of electron-beam welding of products and to improve the quality of welded joints. The prototype consists of three program parts: Technologist's software, EBW simulation software and database. This prototype gives the user the opportunity not only to record the technological process of welding, but also to choose ready-made technological solutions, to simulate the EBW process according to specified technological parameters. The use of the proposed prototype makes it possible to minimize the costs of the enterprise for working out the technological process of electron-beam welding, as well as to create an information base on the conducted welding of products and ready-made technological solutions of the EBW process.

Keywords: Electron beam · Industrial and Innovation · Manufacturing value · Process simulation · Prototype · Software · Technological parameters of welding

1 Introduction

Electron-beam welding is one of the promising and high-tech methods of welding of various metals and alloys, primarily refractory, chemically active and heterogeneous quality steels, high-strength alloys based on aluminum and titanium. The most widely used type of welding is in the automotive, aeronautical and space industries, shipbuilding and nuclear power industry, where the quality of welded joints are the highest requirements [1–3].

Electron-beam welding has wide technological capabilities, such as the implementation of pulse operation mode, the movement of an electronic beam over the surface of the welded product with the help of a rapid deflecting system on any complex circuit, dynamic focus, multi-beam welding. Moreover, the technological capabilities of EBW

R. Silhavy et al. (Eds.): CoMeSySo 2022, LNNS 596, pp. 456–466, 2023.
https://doi.org/10.1007/978-3-031-21435-6_39

are significantly expanded as the process automation and the EBW process control [2, 4, 5] are improved.

To obtain a high-quality seam it is necessary to maintain at the required level a certain set of technological and energy parameters [6, 7]. Failure of optimal EBW modes results in various weld seam defects, even on well-welded materials. A number of defects, known as forming defects, are encountered in any melting welding. These are non-provares, pruning, sagging seam. In addition, there are specific EBW defects: root defects, extended cavities in the seam volume, «median» cracks and deviation of the seam from the joint [11–13].

In order to improve the quality of the process of electron-beam welding, many researchers carry out mathematical modeling of this process with the assignment of different process modes and materials [7–10]. Based on the experimental data and results of the simulation of the geometry of weld seams of stainless-steel products under various parameters of the EBW mode, multi-criteria analysis and optimization of the EBW process was carried out [13–17]. The work [18, 19] describes a mathematical model of the processes of evaporation, condensation, and diffusion of the AMg6 alloy during electron-beam welding. In the work [20] the dependence of electron-beam welding modes on the properties of the material is investigated. Until now one of the main methods of working out the welding modes and the choice of optimal parameters is welding samples of the product. For example, in the work [21] the technique of transition from argon-arc welding to EBW was developed. Electron-beam welding modes were selected on samples simulating weld joint.

The work [22–24] develops a technique that evaluates the efficiency of the technological process on the basis of geometric and thermophysical parameters of welded constructions.

In the work [25, 26] the samples are used to study the process of EBW and improve the quality of welding, by eliminating the seam defects.

2 Mathematical Support of the Software System

The determination of optimal modes of electron-beam welding is based on the mathematical model of the temperature field of the article in the case of electron-beam welding. This mathematical model represents the superposition of two heat sources of constant power, which move straight with constant speed. As a single source, the sum (1) of the instantaneous point source on the surface of the semi-infinite body and the linear heat source [27, 28]:

$$T(x, y, z, q, v, t) = T_1(x, y, z, q, v, t) + T_2(x, y, z, q, v, t), \tag{1}$$

where x, y, z – coordinates of the point under consideration in space;

q – effective electron beam power;

v – welding speed;

t – time calculated from the moment the source passed through the section in which the point in question is located;

T_1 – point instant heat source;

T_2 – linear instant heat source.

The algorithm for finding optimal welding parameters is presented in Fig. 1.

The search for technological parameters of electron-beam welding begins with initialization of data. The power of sources, material properties, welding speed and welding time are determined. Most of the data are taken from the database where EBW materials and process parameters are stored. The temperature field of the article is calculated according to Eq. (1). This process is iterative, as the calculation process clarifies the time of welding and the speed of welding.

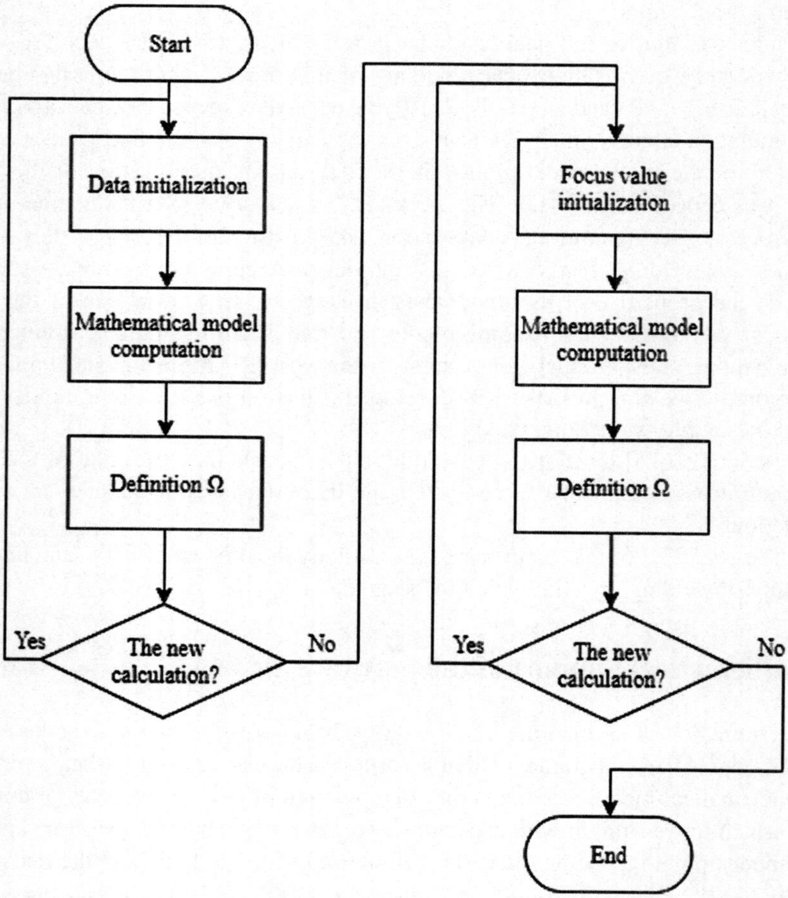

Fig. 1. Algorithm for determining optimum welding parameters.

The Ω function is used to determine the optimum parameters obtained. The minimum Ω value serves as a marker that the selected parameters of the welding process are optimal. Therefore, for the selected product (material, product thickness) calculation can be carried out many times, at different parameters of the welding process (power, welding speed).

The second part of the algorithm serves to determine the optimal focus of electron-beam welding. In this case, the mathematical model is calculated again for different focus positions (from the part above the surface to the seam root). The optimal focus value is also determined by the Ω function.

3 Software Design

Technology Decision Support System prototype for the EBW process is developed in C++ and consists of three main software components:

Technologist software.
EBW process modeling software.
Database.

EBW process modeling software and technologist software is a Windows application that is possible in the Windows 7/8/10 operating system environment. These components can function separately and run according to the tasks of technologists and scientists. The schematic diagram of the EBW process decision support system prototype is shown in Fig. 2.

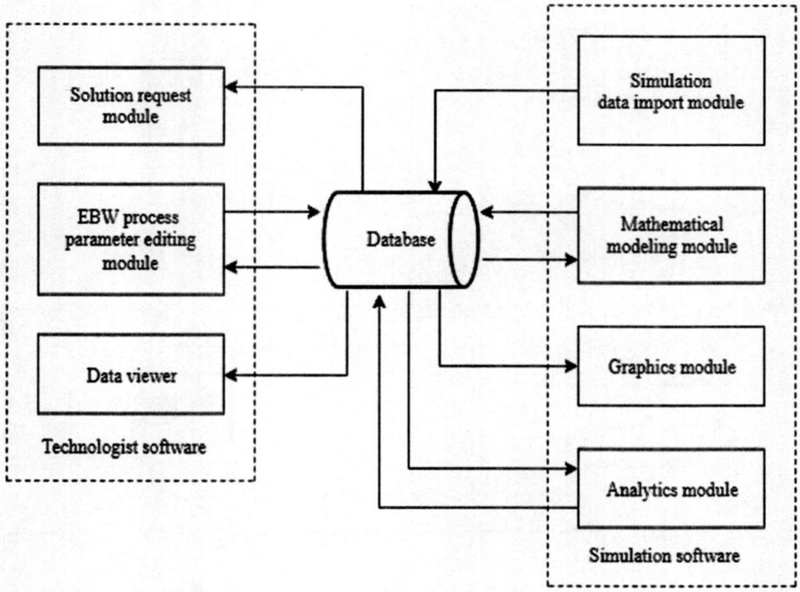

Fig. 2. Structural diagram of the prototype technology decision support system for the EBW process.

The flow chart of the prototype technology decision support system is shown in Fig. 3.

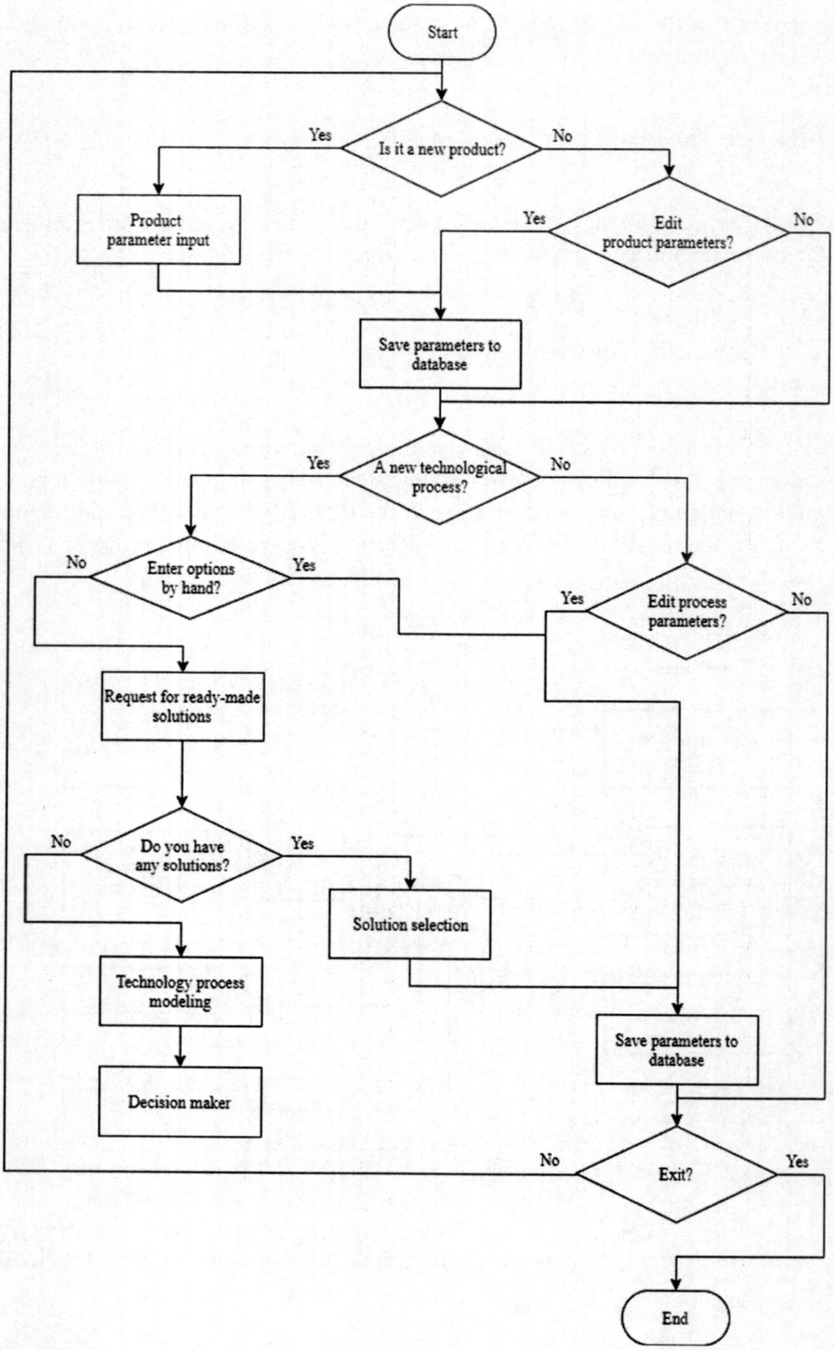

Fig. 3. Flow diagram of the prototype decision support system for the EBW process.

Technology software consists of three modules:

- The EBW Process Parameter Editing Module introduces and edits the parameters of the EBW process.
- The Solution Request Module selects from the Technology Solutions Database to perform the EBW process and achieve the required seam parameters.
- The data viewer displays graphs of the parameters of the welding processes held in the database.
- The simulation software consists of four modules providing the following functions:
- The mathematical modeling module implements the process of electron beam welding.
- Simulation simulation data import module performs data input and modeling graphs performed in third-party simulation software products.
- The graphics module performs graphical construction of results of mathematical modeling of the EBW process.
- The analytics module processes mathematical modeling data and develops technological solutions for the electron beam welding process.

The work begins with the software technologist. The technologist enters or edits the data of the welded product, and then goes to the input of the parameters of the technological process: electron beam current, accelerating voltage, focus current, welding speed. You can enter parameters manually or use already finished technical processes used previously. The prototype of the system also allows you to view recommended solutions and choose suitable parameters for welding the product. If they do not exist, the process is modelled and technological solutions for the EBW process are developed on the basis of the simulation results. Responsible for the last two steps of EBW process simulation.

4 Software System Informational Support

One of the main components of the decision support system for the EBW process and the program component of the system is the database, which contains data on the EBW process, modelling results, model parameters as well as obtained recommendations for the EBW process. MySQL [29, 30] is used as a database. The structural schema of the database is shown in Figs. 4 and 5.

The Decision Support System prototype database for the EBW process consists of two parts. The tables of product, techprocess, datawelding and samples (Fig. 3) provide the work of the technologist software and are intended for storing information about welded products and set process parameters.

In addition, the welding parameters values (electron beam current accelerating voltage, focus current, filament current, welding speed, vacuum chamber pressure) are stored in the database (datawelding table) during the EBW process. Also, in the database (sample table) there are links to the files with the results of quality control of welded products in the form of photos of polishes and expert opinions.

The second part of the database (Fig. 4) is intended for the work of the EBW Process Modelling Software and stores parameters and modelling results as well as proposed

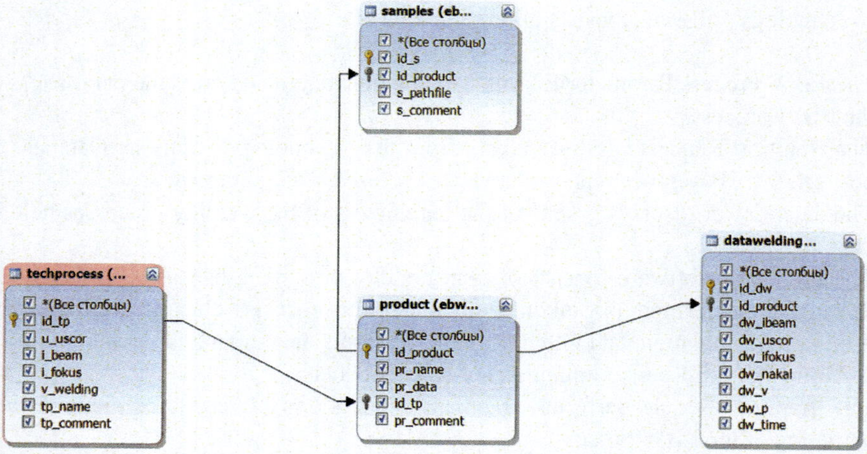

Fig. 4. Diagram of the prototype decision support database for the EBW process. Technological part.

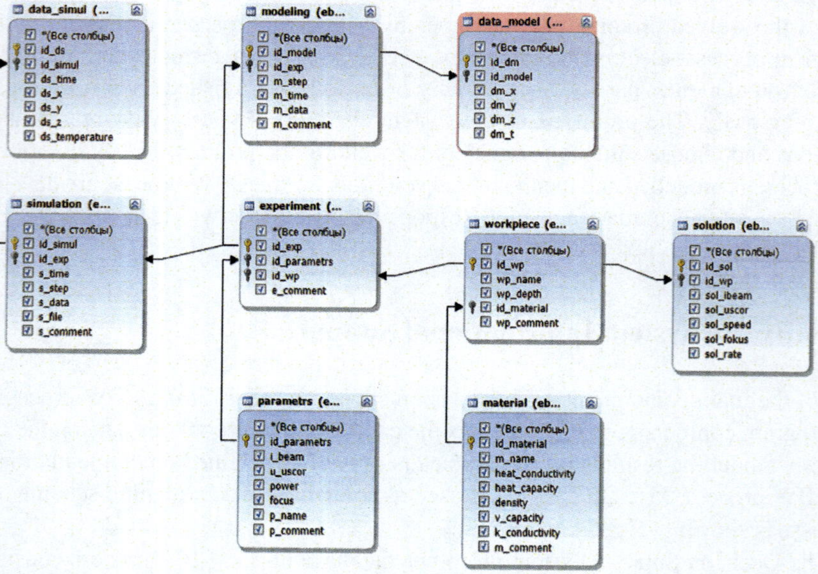

Fig. 5. Decision support prototype database schematic for EBW process. Modeling.

solutions for the EBW process. The central table for this database fragment is the workpiece table, which contains the geometric parameters of the workpiece. The material table contains information about the material properties of the blank, and the solution table contains ready-made technological solutions. These solutions are obtained by analyzing the modelling results contained in the other tables. The experiment and parametrs tables contain input data for mathematical modeling of the EBW process. The modeling and data_modeling tables are designed to store the results of mathematical modeling of the EBW process. And tables simulation and data_simulation store the results of simulation performed in third-party software products.

5 Description of the Programmer System

Figure 6 shows the basic form of the technologist's software.

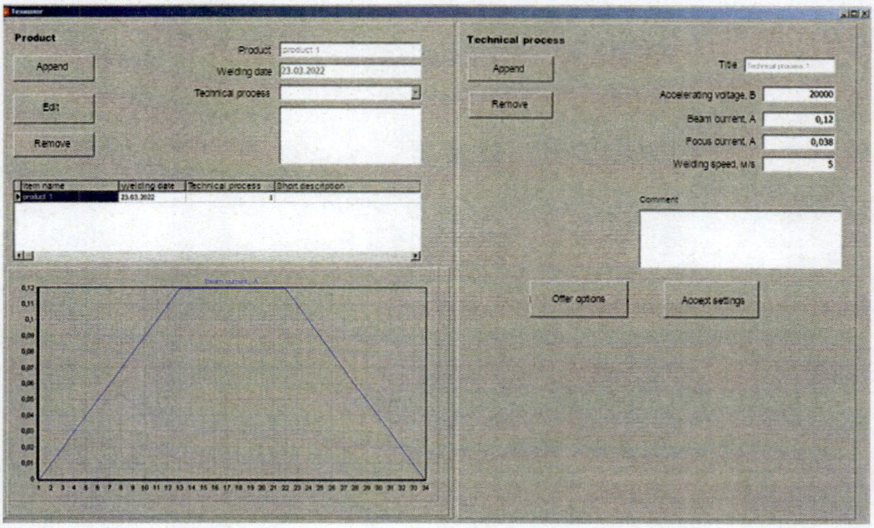

Fig. 6. The basic form of the software technologist.

The shape is divided into two parts. The data on the welded part is presented in the left part of the shape. With the buttons you can add a new product, edit the data on the already entered product or remove the product from the database. In addition, you can view the graphics of the technological parameters of the welded. On the right side are data on process parameters. And you can also add a new process to the database and its parameters. You can do this manually or choose the proposed solution. The list of ready-made solutions is formed dynamically with the help of the button «Offer parameters».

Figure 7 presents the main form of the EBW process simulation software.

The EBW process modeling system is based on the software system of mathematical modeling of the EBW process [23]. The basic form consists of four tabs: experimental planning, modeling, simulation modeling, analytics. The EBW process simulation work begins with the Experiment Planning tab. This tab allows you to manage experiments (create new ones, edit old ones) and experiment parameters. The tab is divided into three segments: materials, technical process, products. These segments introduce material parameters and process parameters of the EBW process, which will be used in the modeling of the EBW process. The remaining tabs are responsible for modeling the EBW process (Modeling), input and display of modeling results in third-party software products (Simulation modeling) and formation of ready-made technological solutions (Analytics).

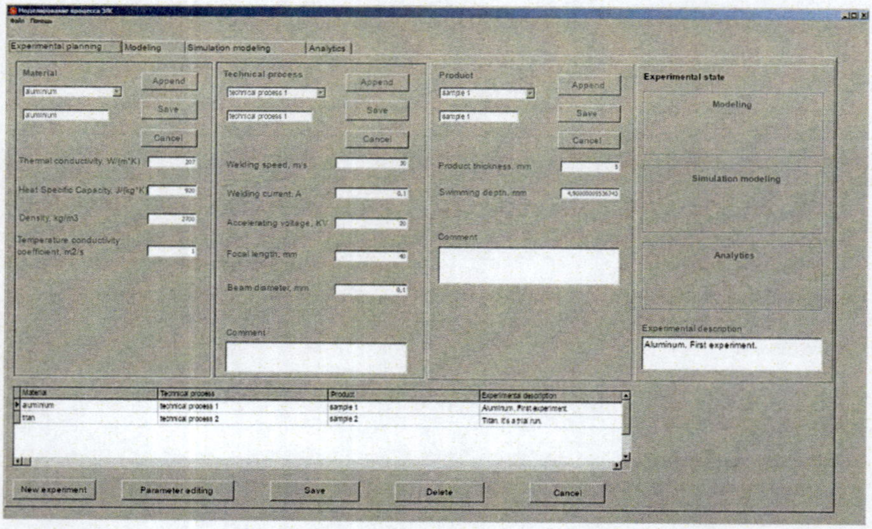

Fig. 7. Basic form of EBW process simulation software.

6 Conclusion

This paper proposes a prototype system for supporting technological decision-making for the EBW process. The prototype consists of three components. The link of the system is a database that allows to store and process information on the results of mathematical modeling, simulation modeling and a set of ready-made technological solutions.

The mathematical support of finding solutions allows carrying out several cycles of mathematical modeling of the EBW process, creating a set of solutions for the corresponding details and distinguishing from them the optimal solution. Application of the proposed software system makes it possible not only to minimize the costs of the enterprise for working out the technological parameters of the electronic process radial welding, as well as to create an information base on the conducted welding of products and finished technological solutions of the EBW process.

Further development of the decision support system involves the introduction of intelligent process simulation and welding data processing algorithms for more accurate selection of EBW process parameters.

Acknowledgments. This work was supported by the Ministry of Science and Higher Education of the Russian Federation (Grant No.075–15-2022–1121).

References

1. Salomatova, E.S.: Electron beam welding—from the invention to the present day. Bull. Perm Natl. Univ. Res. Polytech. un-ta. Mech. Eng. Mater. Sci. **1**, 74–87 (2013)
2. Bashenko, V.V.: State and prospects of development of electron beam welding. In: Materials of the first St. Petersburg International scientific and Technical conference Technologies and equipment of electron beam welding, vol. 2, pp. 5–21 (2008)
3. Li, Z., Chen, W.: Application progress of power beam processing technology in aeronautical industry. Aeronautica et Astronautica Sinica 2, **43**(4), 526882 (in Chinese). https://doi.org/10.7527/S1000-6893 (2022)
4. Laptenok. D.: Control of electron beam welding [control of the electron beam welding process]/Laptenok, V.D., Murygin, A.V., Seregin, Y.N., Braverman V.Ya.: (eds.) by Laptenok, V.D., Krasnoyarsk: CAA 234 (2000)
5. Kosenko, E.A., Nelyub, V.A.: Evaluation of the stress-strain state of a polymer composition material with a hybrid matrix. Polym. Sci. - Ser. D. **15**, 240–244 (2022). https://doi.org/10.1134/S1995421222020137
6. Sandhya, V., Naga Phani Sastry, M., HemaChandra Reddy, K.: Influence of welding speed, voltage, and beam current on the microstructure and mechanical properties of electron beam-welded titanium radial joints, materials today: Proceedings. https://doi.org/10.1016/j.matpr.2022.04.808
7. Kurashkin, S., Bukhtoyarov, V., Myrugin, A., Bocharov, A., Seregin, Y., Rogova, D.: Simulation model of the electron beam welding process for the formation of the effective technological parameters during the welding of aerospace structures. In: 2022 21st International Symposium INFOTEH-JAHORINA (INFOTEH). pp. 1–5. Institute of Electrical and Electronics Engineers Inc., East Sarajevo (2022). https://doi.org/10.1109/INFOTEH53737.2022.9751280
8. Nelyub, V.A., Fedorov, S.Y., Malysheva, G.V.: The study of the structure and properties of elementary carbon fibers with metal coatings. Inorg. Mater. Appl. Res. **12**, 1037–1041 (2021). https://doi.org/10.1134/S2075113321040316
9. Nelub, V., Gantimurov, A., Borodulin, A.: Economic analysis of data protection in systems with complex architecture using neural network methods. Econ. Ann. **185**, 178–188 (2020). https://doi.org/10.21003/EA.V185-17
10. Stoyanova, M.V., Novikov, A.D., Borodulin, A.S.: Evaluation construction made of polymer composite materials by molding using reusable flexible punches production profitability. IOP Conf. Ser. Mater. Sci. Eng. **934**, 012068 (2020). https://doi.org/10.1088/1757-899X/934/1/012068
11. Putilova, A.O.: Prevention of specific defects of ELS when welding products of large thickness. Collection of materials of the International scientific and practical conference dedicated to the Day of cosmonautics Actual problems of aviation and cosmonautics. NYE-monautics; Sibgau named after Reshetnev, M.F. 553–555 (2019)
12. Mladenov, G.M.: Electron beam welding: monograph [electron beam welding]. Perm: Publishing house Perm. NAC. FROM-sledge. polytech. un-ta, **374** (2014)
13. Grigoriev, V.: Investigation of the occurrence of specific defects in electron beam welding (ELS) of titanium alloys [investigation of the occurrence of specific defects in electron beam welding (Els) of titanium alloys/Grigoriev, V.V., Muravyev, V.I., Bahma-tov, P.V.//Welding production 36–42 (2019)
14. Mladenov, G., Koleva, E., Belenky, V.Y., Trushnikov, D.N.: Modeling and optimization of electron beam welding of steels. Bull. Perm Natl. Res. Polytech. Univ. Mech. Eng. Mater. Sci. **16**(4), 7–21 (2014)

15. Kanigalpula, P.K.C., Jaypuria, S., Pratihar, D.K., Jha, M.N.: Experimental investigations, inputoutput modeling, and optimization of spiking phenomenon in electron beam welding of ETP copper plates. Measurement **129**(1), 302–318 (2018)
16. Luo, M., Hu, R., Liu, T., Wu, B., Pang, S.: Optimization possibility of beam scanning for electron beam welding. Physics understanding and parameters selection criteria. Int. J. Heat Mass Trans. **127**(1), 1313–1326 (2018)
17. Koleva, E., Koleva, L., Trushnikov, D., Kolev, G., Petrova, Z.: Multicriterial optimization strategies for electron beam welding processes/VEIT-2021. J. Phys. Conf. Ser. **2240**, 012038. https://doi.org/10.1088/1742-6596/2240/1/012038. (https://iopscience.iop.org/article/https://doi.org/10.1088/1742-6596/2240/1/012038) (2022)
18. Salamatova, E.S.: Application of parallel computing for the study of evaporation during electron beam welding. Bull. Perm Natl. Res. Polytech. Univ. Aerosp. Eng. **39**, 96–108 (2014)
19. Salomatova, E.S., Trushnikov, D.N., Kaplin, A.I.: Modeling of evaporation processes in electron beam welding with dynamic positioning of the electron beam. Proceedings of Tula State University. un-ta. Technical sci. 6(2), 124–133 (2015)
20. Gribkov, V.S.: The influence of thermophysical properties of materials on the choice of the ELS. Electrometallurgy **9**, 30–35 (2015)
21. Zakharov, A.S.: Translation of the manufacturing process of products with automatic argon arc welding or electron beam welding. Selection of the ELS mode. Translation of the manufacturing process of the product from automatic argon arc welding to electron beam welding. Materials of the International scientific and practical conference of young scientists, postgraduates and students Electrophysical processing methods in modern industry, pp 401–404 Perm National Research Polytechnic University. Perm (2022)
22. Kurashkin, S., et al.: Mathematical modelling of waveguide paths by electron-beam welding. Proc. Comput. Sci. **200**, 83–90 (2022)
23. Brom, A.E., Stoyanova, M.V., Yazev, M.V., Korolev, S.A.: Assessment of technological resources for the production of composite products based on mathematical methods. IOP Conf. Ser. Mater. Sci. Eng. **934**, 012003 (2020). https://doi.org/10.1088/1757-899X/934/1/012003
24. Kukartsev, V., Mikhalev, A., Stashkevich, A., Moiseeva, K., Kauts, I.: Analysis of Data in solving the problem of reducing the accident rate through the use of special means on public roads. In: 2022 IEEE International IOT, Electronics and Mechatronics Conference (IEMTRONICS), pp. 1–4. Institute of Electrical and Electronics Engineers (IEEE), Toronto (2022). https://doi.org/10.1109/IEMTRONICS55184.2022.9795842
25. Sviridov, A.V.: Electron beam welding of gears made of heat–resistant steel EI—415. In: Proceedings of the IV International Conference Electron beam welding and funny technologies, pp. 92—208. Research University MEI, Moscow (2021)
26. Isaev, S.L.: Improving the quality of the weld of thin-walled brazed shells made of corrosion-resistant steel. In: Materials of the 5th International Scientific and practical conference Modern materials, equipment and technologies.—Kursk: CJSC, pp. 61–64 University Book (2015)
27. Konovalov, A.V.: Theory of welding processes. Publishing House of Bauman Moscow State Technical University (2007)
28. Murygin, A.V., Tynchenko, V.S., Kurashkin, S.O., Bocharov, A.N., Petrenko, V.E.: Software system for mathematical modeling of the electron beam welding process. Siberian Aerosp. J. **2**(22), 261–274 (2021)
29. MySQL [Electronic resource]. URL: https://www.mysql.com/. Accessed 01 May 2021
30. MySQL for Big Data [MySQL for Big Data]/Sh. Challawala, Dzh. Lakhatariya, Ch. Mechta, K. Patel'. M.: DMK Press 226 (2018)

Modeling the Temperature Distribution in an Electron Beam Welded Bimetallic Ring with COMSOL Multiphysics

Sergei Kurashkin[1,2,3(✉)], Elizaveta Shutkina[2], Yadviga Tynchenko[1,2], Vladimir Bukhtoyarov[1,2,3], Ivan Makarchuk[1], and Aleksey Bocharov[2]

[1] Siberian Federal University, Krasnoyarsk, Russia
scorpion_ser@mail.ru
[2] Reshetnev Siberian State University of Science and Technology, Krasnoyarsk, Russia
[3] Bauman Moscow State Technical University, Moscow, Russia

Abstract. Electric motors are used to control the movable table in electron beam welding. However, today, electric motors are widely used in the fuel industry. In particular, various types of electric motors are used for the injection and pumping of biofuel. In order to increase wear resistance and strength, it is proposed to connect the bimetallic ring to the rotor of the electric motor by electron beam welding. Since bimetal is understood as a coating on the surface of the base metal of another, for example, titanium and aluminum, it is accordingly necessary to work out the technological process, namely, to find the technological parameters to obtain a stable quality of the weld. On the basis of a mathematical model previously developed by the authors, in this work, simulation modeling of the electron-beam welding process in the manufacture of electric motors with a bimetallic ring in the COMSOL Multiphysics environment is carried out.

Keywords: Electron beam · Innovation · Mathematical modeling · Oil and gas industry · Process innovation · Simulation

1 Introduction

Numerical simulation [1] shall be used to describe the electron-beam welding processes. In the work [2–4] mathematical modeling of operation modes of welding electron gun and study of electron beam characteristics on different modes of operation of gun, such as focus mounting, angle of convergence of beam, diameter of beam is carried out. The study [5, 6] describes the wording adopted for numerical modelling of the electron-beam welding process, as well as the experimental work carried out for its validation and validation. The authors of the study [7, 8] consider the problem of mathematical modeling of the process of electron-beam welding on this-pad of the input and output of the electronic beam, also proposed a conceptual model, developed using the mathematical apparatus of the theory of thermal processes. The authors of [9] have developed a mathematical model for the study of the effect of electron beam de-focusing on the weld seam quality. The study [10] focuses on the problem of finding effective values for the parameters

R. Silhavy et al. (Eds.): CoMeSySo 2022, LNNS 596, pp. 467–474, 2023.
https://doi.org/10.1007/978-3-031-21435-6_40

of the EBW process, so it proposes a mathematical model to assess the quality of the process under consideration. The results of the study [11, 12] showed that the shape of the weld tub obtained by the numerical model corresponds to the experimental analysis of the cross section of the weld, also, the numerical model was also able to predict the receptivity to the formation of elixic fractures in the thermal exposure zone by studying the temporal gradient at the solidification stage.

Electric motors used in the oil and gas industry come in several types [13–15]:

- DC models.
- Synchronous motors.
- Asynchronous motors.

The latter group consists primarily of pumps, compressor rigs, drilling systems and equipment supplemented by electric centrifugal pumping mechanisms.

Asynchronous motors today make up the majority of electrical machines, are mainly used as electric motors and are the main converters of electrical energy to mechanical energy.

When comparing asynchronous and synchronous models, the latter have a number of important advantages [16]:

- Possibility of adjusting the reactive power value and correction.
- Ability to operate at rated power 1.
- The efficiency of synchronous models is about 1–3% higher than that of asynchronous models with the same power. This is explained as follows: the loss of electrical energy in the housing and in the winding of the stator for synchronous models is lower than for asynchronous models.

The advantages of asynchronous engines include the following features [17]:

- Constructive simplicity. Less qualified service.
- Easy to control even when used in emergency mode, connected with power surges in the network. No additional field extinguishing agent needed.
- Possibility of working in rooms with high humidity.
- Simplified system for self-startup.

However, the connection of the bimetal ring is a difficult task, as it requires precision, high repeatability and minimization of the weld seam. Electron-beam welding is suitable for the above requirements, which means that to create a technological process it is necessary to carry out research. Thus, within the framework of this article, the authors will conduct simulation modelling on the basis of previously developed and proven mathematical models, in order to test the welding modes and search for technological parameters, allowing welds to be made with the required weld dimensions.

2 Material and Methods

Simulation modelling is one of the most convenient and distributed ways to model complex systems and processes, such as manufacturing enterprises, information networks, mass service systems, world dynamics in the economy, environment, politics and so on. In continuous models, quantities are continuous functions of time. In accordance with this, the progress over time, that is, the recalculation of the variables during the model time, is carried out in their-matching model on the principle of «small delta-t» the next state of the system is determined by its previous state with a short interval between these states. In discrete models, changes occur intermittently and the condition of the elements remains constant [18].

The Specialist, when embarking on a task, should know the basis of dynamic processes, the approaches and methods of solving complex processes and systems, including analytical and simulation, as well as specific informationSimulation systems and their programming languages [19]. Among the many analytical modelling environments, the main ones are Maple, MathCAD, COMSOL Multiphysics and others. COMSOL Multiphysics® software is an integrated environment for numerically modeling and optimizing virtually any physical or physical-chemical system. The main solution algorithm used is the finite element method. Formulations based on the boundary element method and particle and ray tracing are available for a number of tasks. In the package you can create models for solving the problems of mechanics and acoustics, hydrodynamics and heat and mass transfer, electrodynamics and optics, chemistry and elec chemistry. Special value lies in the ability to simulate processes related to multi-physics and solve interdisciplinary problems [20].

The software was highly flexible and provided users with tools to create and customize descriptions of their own non-standard models based on mathematical equations. The platform includes a group of mathematical interfaces for setting and solving partial derivative equations in coefficient, general or weak form, as well as ordinary differential equations [21].

COMSOL Multiphysics has the ability to freely control all aspecs of the model, unlike other modeling packages, which usually have strict limitations, in addition, one of the key advantages of COMSOL Multiphysics pa-keta is multi-physics - the ability to solve a connection interdisciplinary tasks in a single environment. Working in the same development environment saves time and effort and focuses on the task and its solution. You can creatively model and solve problems that are difficult or impossible in a conventional approach by combining an arbitrary number of physical phenomena and defining user descriptions of physical phenomena, equations and expressions through a user interface (UI) [22]. Therefore, the COMSOL Multiphysics package is used in almost all industries: food, medical applications, aeronautics and space.

Figure 1 shows an example of a project in a COMSOL environment, namely a physical heat transfer model.

In early works [23–25], the authors developed mathematical models, carried out simulation modeling and verified the results obtained through field experiments. Thus, to solve the problem, the authors decided to use the COMSOL Multiphysics software.

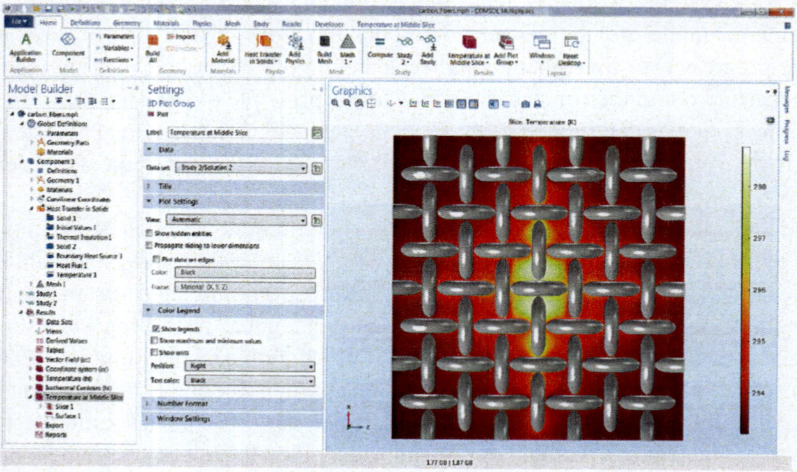

Fig. 1. Program interface COMSOL.

3 Result and Discussion

Figure 2 shows the temperature distribution when simulating the process of electron beam welding of a bimetallic ring on the basis of previously developed mathematical models and the technological process used to join products from titanium and aluminum alloy.

The results presented in Fig. 2 give an idea of the applicability of the previously developed mathematical models to a more complex design.

Temperature distribution on the surface of the product at higher power (the beam current increases, whereas in both cases the acceleration voltage was 30 kV) and the same welding speed is shown in Fig. 3.

As can be seen from Figs. 2 and 3, with an increase in power, the maximum temperature increases (Fig. 3), in this case it is higher than the melting point of titanium. The area with the maximum temperature is marked in red, since this is the place where the electron beam enters and exits. The next stage of this study will be the verification of the obtained technological parameters by means of field experiments.

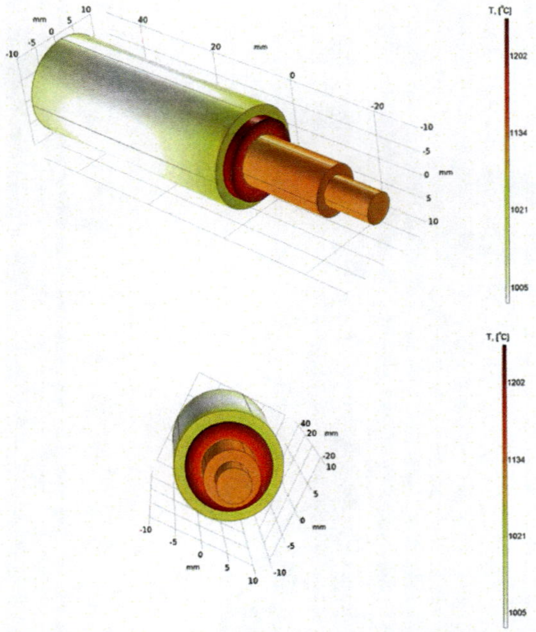

Fig. 2. Temperature distribution during electron beam welding when joining a bimetallic ring.

4 Conclusion

In this study, we simulate the heating of a bimetallic ring welded using an electron beam. Initial technological parameters were set (beam current—60 mA; skid voltage—30 kV; sharp focus and welding speed—1.8 cm/s) to obtain the required weld seam dimensions. Further varying the power, a series of experiments was carried out, showing how the energy is distributed on the surface of the product at all stages of welding (beam input; established mode; beam output), which allowed to identify such values of parameters, which have the required weld dimensions (seam width 1 mm; seam height 2 mm).

The simulation results in the COMSOL Multiphysics system show that the considered welding method is well applicable to solving the problem of forming a permanent joint. In addition, the created model makes it possible to select effective modes of welding bimetallic rings not only from titanium and aluminum alloys, but also from other materials.

Further implementation of the proposed computational model in the form of an independent module of the electron beam welding automation system will allow supporting the adoption of technological decisions for specialists in the machine-building industry.

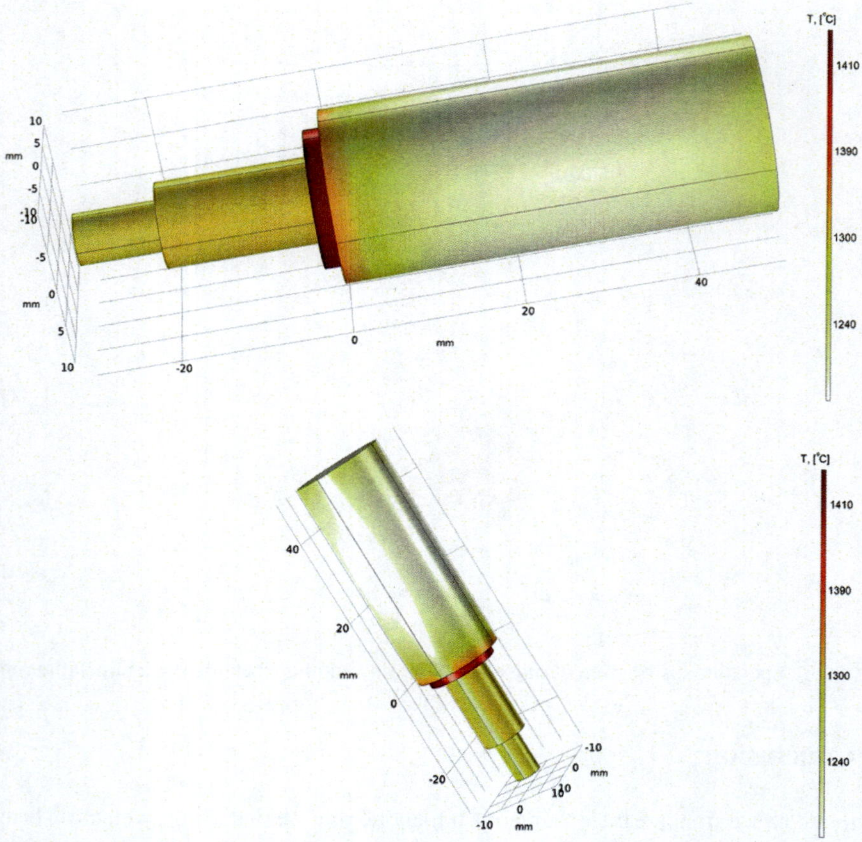

Fig. 3. Temperature distribution on the surface of the product when the bimetallic ring is connected to the rotor.

Acknowledgements. The studies were carried out according to the state assignment for the project "Development of a set of scientific and technical solutions in the field of creating biofuels and optimal biofuel compositions, providing the possibility of transforming consumed types of energy in accordance with trends in energy efficiency, reducing the carbon footprint of products and using alternative fuels to fossil fuels" (Contract FSRZ- 2021-0012) in the scientific laboratory of biofuel compositions of the Siberian Federal University, created as part of the activities of the Scientific and Educational Center "Yenisei Siberia".

References

1. Sitnikov, I.V., Belenky, V.Ya., Permyakov, G.L.: Development of a numerical model of electron beam welding of a gear wheel using one and three heat sources. Chem. Ecol. Urban Stud. Mater. Sci. Mech. Eng. **2**, 550–556 (2019)
2. Chulkov, I.S., Goncharov, A.L.: Mathematical modeling of the modes of operation of the welding electron gun. Electron beam welding and related technologies. Mater. Sci. Mech. Eng. **1**, 105–121 (2015)

3. Kosenko, E.A., Nelyub, V.A.: Evaluation of the stress-strain state of a polymer composition material with a hybrid matrix. Polym. Sci. Ser. D. **15**, 240–244 (2022). https://doi.org/10.1134/S1995421222020137
4. Brom, A.E., Stoyanova, M.V., Yazev, M.V., Korolev, S.A.: Assessment of technological resources for the production of composite products based on mathematical methods. IOP Conf. Ser. Mater. Sci. Eng. **934**, 012003 (2020). https://doi.org/10.1088/1757-899X/934/1/012003
5. Chimenti, M., Servera, M., Dialarmi, N., Wu, B., Jinwei, L.: Numerical modeling of the electron beam welding and its experimental validation. Finite Elem. Anal. Des. **121**, 118–133 (2016)
6. Nelyub, V.A., Fedorov, S.Y., Malysheva, G.V.: The study of the structure and properties of elementary carbon fibers with metal coatings. Inorg. Mater. Appl. Res. **12**, 1037–1041 (2021). https://doi.org/10.1134/S2075113321040316
7. Kurashkin, S.O., Tynchenko, V.S., Seregin, Yu.N., Petrenko, V.E., Milov, A.V., Murygin, A.V.: Mathematical models of beam input and output in the process of electron beam welding of thin-walled structures. J. Phys. Conf. Ser. **1515** (2020)
8. Nelub, V., Gantimurov, A., Borodulin, A.: Economic analysis of data protection in systems with complex architecture using neural network methods. Econ. Ann. **185**, 178–188 (2020). https://doi.org/10.21003/EA.V185-17
9. Kurashkin, S.O., Tynchenko, V.S., Seregin, Y.N., Petrenko, V.E., Murygin, A.V.: Mathematical functional for thermal distribution calculating during the electron-beam welding process. J. Phys. Conf. Ser. 1515 (2020)
10. Murygin, A.V., Kurashkin, S.O., Tynchenko, V.S., Rogova, D.V.: The use of ANSYS for modelling the energy distribution in steady mode with electron beam welding. J. Phys. Conf. Ser. **1889**, 042061 (2021). https://doi.org/10.1088/1742-6596/1889/4/042061
11. Stoyanova, M.V., Novikov, A.D., Borodulin, A.S.: Evaluation construction made of polymer composite materials by molding using reusable flexible punches production profitability. IOP Conf. Ser. Mater. Sci. Eng. **934**, 012068 (2020). https://doi.org/10.1088/1757-899X/934/1/012068
12. Salehi-Shabestari, A., Khakzadshahandashti, A., Rahimipour, M.: Numerical modeling of electron beam welding (EBW) of Zhs6u superalloy and its experimental validation. Mater. High Temp. **39**, 12–20 (2022)
13. Yakunin, A.N.: The use of modern series of Asynchronous Motors at oil industry facilities. Achievements, Probl. Prospect. Develop. Oil Gas Indus. **5**, 302
14. Knyazeva, Z.V., Yudin, P.Y., Petrov, S.S., Maksimuk, A.V.: Application of metallization coatings to protect submersible electric motors of pumping equipment from the effects of complicating factors in oil wells. Izvestiya Vuzov Powder Metall. Funct. Coat. **1**, 75–86 (2020)
15. Glazyrin, A.S., Kladiev, S.N., Afanasyev, K.S., Timoshkin, V.V., Slepnev, I.G., Polishchuk, V.I.: Development of a full-order observer with on-line monitoring of the moment of resistance for submersible asynchronous electric motors. Bull. Tomsk Polytech. Univ. Eng. Georesour. **329**(2), 118–126 (2018)
16. Odnokopylov, G.I., Bukreev, V.G., Rozaev, I.A.: Research of a fail-safe valve-inductor electric motor of a pump for oil production. Bull. Tomsk Polytech. Univ. Eng. Georesour. **330**(10), 69–81 (2019)
17. Martyanov, A.S.: Determination of the optimal energy of a storage device to reduce the sensitivity of submersible electric motors for oil production to voltage dips. Dyn. Syst. Mech. Mach. **6**(3), 55–61 (2018)
18. Kurashkin, S., Bukhtoyarov, V., Myrugin, A., Bocharov, A., Seregin, Y., Rogova, D.: Simulation model of the electron beam welding process for the formation of the effective technological parameters during the welding of aerospace structures. In: 2022 21st International

Symposium INFOTEH-JAHORINA (INFOTEH). Institute of Electrical and Electronics Engineers Inc., East Sarajevo, pp. 1–5, (2022). https://doi.org/10.1109/INFOTEH53737.2022.9751280
19. Simulation modeling: studies. manual/Elberg, M.S., Tsygankov, N.S.—Krasnoyarsk: Sib. feder. unit, 128 (2017)
20. Dickinson, E.J., Ekström, H., Fontes, E.: COMSOL multiphysics®: finite element software for electrochemical analysis. Mini Rev. Electrochem. Commun. **40**, 71–74 (2014)
21. Vajdi, M., Moghanlou, F.S., Sharifianjazi, F., Asl, M.S., Shokouhimehr, M.: A review on the Comsol Multiphysics studies of heat transfer in advanced ceramics. J. Comp. Compd. **2**(2), 35–43 (2020)
22. Sun, X., Luo, H., Soga, K.: A coupled thermal–hydraulic–mechanical–chemical (THMC) model for methane hydrate bearing sediments using COMSOL multiphysics. J. Zhejiang Univ. Sci. **19**(8), 600–623 (2018)
23. Tynchenko, V.S., Kurashkin, S.O., Murygin, A.V., Tynchenko, Y.A.: Energy distribution modelling in the weld zone for various electron beam current values in COMSOL multiphysics. J. Phys. Conf. Ser. **1889**(4), 042058 (2021)
24. Kurashkin, S., Rogova, D., Tynchenko, V., Petrenko, V., Milov, A.: Modeling of product heating at the stage of beam input in the process of electron beam welding using the COMSOL multiphysics system In. Proceedings of the Computational Methods in Systems and Software, pp 905–912 (2020)
25. Kurashkin, S.O., Tynchenko, V.S., Seregin, Y.N., Murygin, A.V., Kukartsev, V.V., Tynchenko, V.V.: The model of energy distribution during electron beam input in welding process. J. Phys. Conf. Ser. **1679**(4), 042036 (2020)

Voice Recording Setup in Clinical Practice

Jakub Steinbach[1(✉)], Jan Vrba[1(✉)], and Zuzana Urbániová[2]

[1] Department of Computing and Control Engineering, University of Chemistry and Technology in Prague, Technická 1905/5, 166 28, Praha 6, Prague, Czech Republic
jakub.steinbach@vscht.cz, jan.vrba@vscht.cz

[2] 3rd Faculty of Medicine Department of Otorhinolaryngology University Hospital Královské Vinohrady, Charles University, Staré Město, Czechia

Abstract. In this article, we introduce a novel voice recording setup in clinical practice. The aim of this setup is to take recordings of patients who underwent a surgery on vocal cords. The setup utilizes only one microphone for recording so no sound level meter measurement and calibration are required, as we want to emulate conditions where patients are taking their recording at home using smartphones or headsets. We also introduce a software for clinical practice that allows physicians to manage simple patient information and take recordings of their voice. Lastly, we propose the sequence of sounds and sentences that patients should say.

Keywords: Hardware · Voice recording · Voice processing · Otorhinolaryngology · Voice cords

1 Introduction

Nowadays, the patients who underwent surgery that affected their vocal cords are examined by physicians at regular intervals. During examination, physicians perform the perceptual diagnosis of the voice. During perceptual evaluation of the voice, the attending physician listens to the voice recording or directly to the patient and evaluates the patient's voice according to the UEP (Union of European Phoniatricians) and GRBAS scales [11]. Currently, there are machine diagnostic methods in which the accuracy of the diagnosis (compared to expert opinion) is 70% respectively 90% in terms of classification according to UEP or GRBAS and from the point of view of correct diagnostics [3,10,12].

There are several solutions to voice recording and processing such as so-called phonetographs (Vox Phonetography, Dr. Speech, Vocalab [8]). However, these are usually commercial programs, meaning they are not affordable for the general public.

During our complex project, our aim is to build a smartphone-enabled system that allows patients to record their voice at home. This recording will be evaluated and the patient will be able to see their progress in recovery from the

R. Silhavy et al. (Eds.): CoMeSySo 2022, LNNS 596, pp. 475–483, 2023.
https://doi.org/10.1007/978-3-031-21435-6_41

operation of the vocal cords. The part of this project is to build a recording set-up in clinical practice that allows high-quality sound recording. By gathering samples in a high quality, it is possible to train a model that would be utilizable even with recordings containing higher amount of noise, as various published results suggest [1,2]. Contrary to the previous setup [4], this setup is also designed without the need for microphone calibration, since it is expected that the finalized model will classify recordings taken mostly by mobile phone microphone.

In this article, we present the software and setup for clinical practice to record the patient's voice. Then, a specific text for extraction of potential features is proposed.

2 Recording Setup

The microphone in the setup is an AKG C 3000. It is a large diaphragm condenser microphone with 20–20,000 Hz bandwidth and cardioid polar pattern; phantom power is required. There are possibilities to use 10 dB switchable preattenuation pads and bass cut filter (with slope 6 dB/octave 500 Hz and below). For recording, we do not use these features. The AKG C 3000 is mounted in a microphone shock mount that is attached to the Konig & Meyer 23,840 microphone arm that is attached to a table. The microphone is connected with an cable to the RME BabyFace Pro FS sound card. This external sound card, connected to a computer via USB, allows us to record voice as a single channel with a sampling frequency of 192 kHz. All recordings are saved in the Waveform audio file format with 24 bit depth, which is a maximum bit depth supported by the used sound card.

The patient's voice recordings are taken in a soundproof room (depicted in Fig. 3) where the patient sits in a chair approximately 30 cm away from the microphone [7]. The patients are instructed through an intercom by a physician, who is sitting in an adjacent room, to read the given text and the sequence of vowels. Then, the physician starts recording the patient's voice via the Voice Analyst software, which is introduced in the following chapter. For the comfort of patients while maintaining the recommended distance, the room is equipped with an adjustable chair that can be moved in the room and whose seat can be raised or lowered.

3 Software Voice Analyst

The software for managing patient information and recording voice is written in Python 3.9 [9]. The graphical user interface (GUI) is based on the PyQt5 library and patient information is stored in a SqlLite3 database.

The database contains two tables, namely *Patients* and *Recordings*. The database scheme is shown in Fig. 2. The table *Recordings* stores information about the patient ID, the name of the recording file, the length of recovery at which the recording was recorded, and information about the state of the

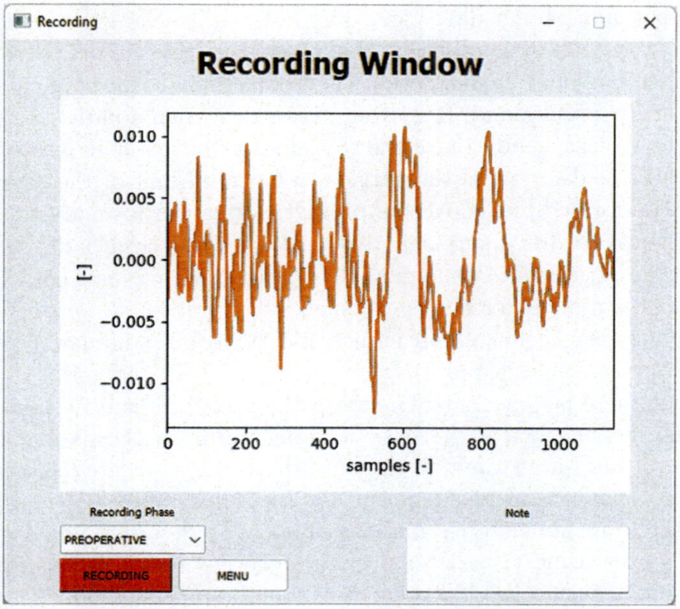

Fig. 1. Recording window of the voice analyst

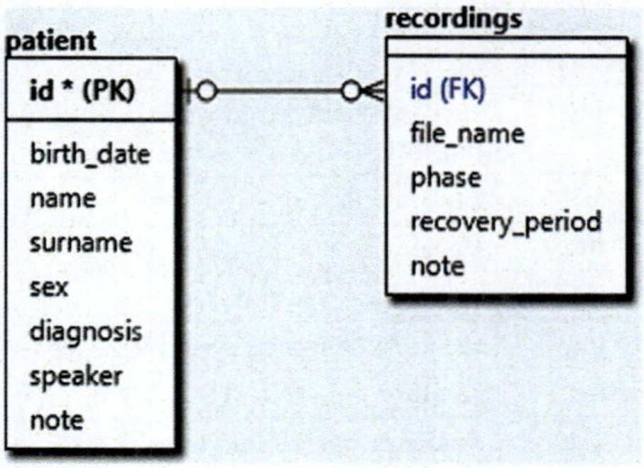

Fig. 2. Overview of database tables.

patient in the sense whether they have been subject to an operation (preoperative/postoperative), and the note made by the physician.

Table *patient* contains information about the patient. Namely the patient ID, name, surname, birth date, sex, diagnosis, native language, and note. So far, it is possible to assign 4 different diagnoses. Healthy subject, thyroidectomy, hemithyroidectomy and medialization. Thyroidectomy is the surgical removal of all or part of a thyroid gland. It is often performed when a patient has a thyroid cancer or some other condition of the thyroid gland such as hyperthyroidism or goiter. Hemithyroidectomy is the surgical removal of half of the thyroid gland. It is mostly performed to remove symptomatic or cancerous nodules. Medialization of the vocal cord is a surgical treatment for vocal cord paralysis when the injection of autologous fat tissue moves a nonfunctioning vocal cord toward the functioning one to allow for better voicing.

Both tables and their column names and types are summarized in the following tables (see Tables 1 and 2).

Note that there is dependency between the tables, as it is desired to remove all recording information, if the patient is deleted from the patient table. The files with recordings are stored on the hard drive in a separate folder, under the name that corresponds to *file_name* in the recording table.

The recording/playing is implemented via the sounddevice/soundfile libraries. The real-time visualization of the recorded signal is implemented via the Matplotlib library. The GUI consists of the Main menu window, New patient window, Load patient window, Recording window (see Fig. 1) and Show recording window. The recording is processed in a separate thread that utilizes signals to send data for real-time signal visualization. This is an important feature, as physicians are used to using this visual feedback to observe the recording process in real time. It also helps them to recognize that the patient is done because the physician cannot hear the patient during the recording (Figs. 4 and 5).

Table 1. Overview of *patient* table

Column name	Data type	Stored information
Id	String	ID of patient as acronyme of name and surname, datebirth, diagnosis and random integer
Birth_date	date	Patient date of birth
Name	String	Patient first name(s)
Surname	String	Patient last name
Sex	String	Patients assigned sex (biological sex)
Diagnosis	String	Healhty subject, HTE, TTE, another diagnosis
Speaker	String	Patients native language
Note	String	Note written by phisician with additional information

Table 2. Overview of *recordings* table

Column name	Data type	Stored information
Id	String	Foreign key referencing the primary key of table patient
File_name	String	File name that consists of user ID, date and time of recording
Phase	String	Preoperative/postoperation
Recovery_period	String	Number of days since patient came with a problem
Note	String	Note written by phisician with additional information

Fig. 3. Soundproof room

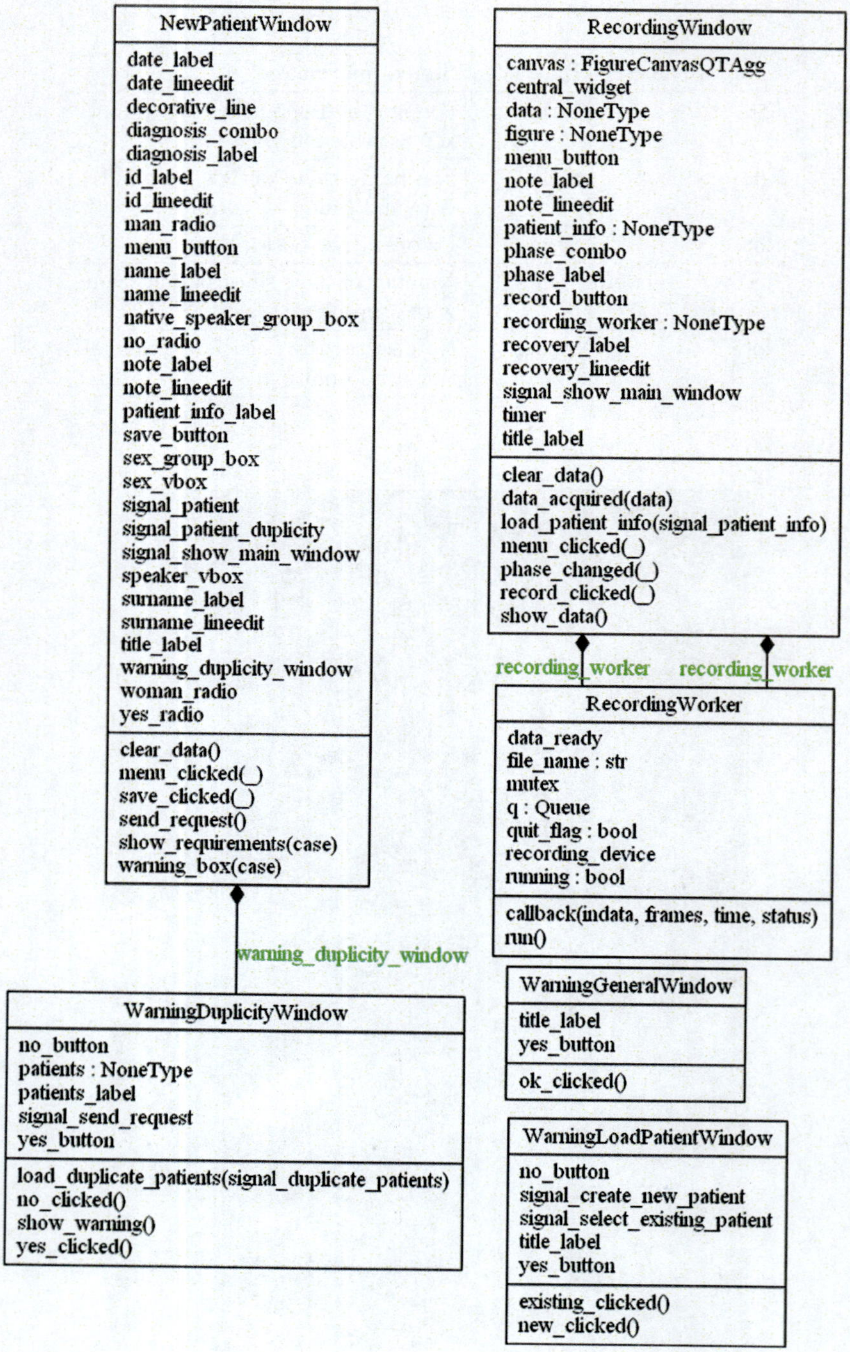

Fig. 4. UML diagram of voice analyst software

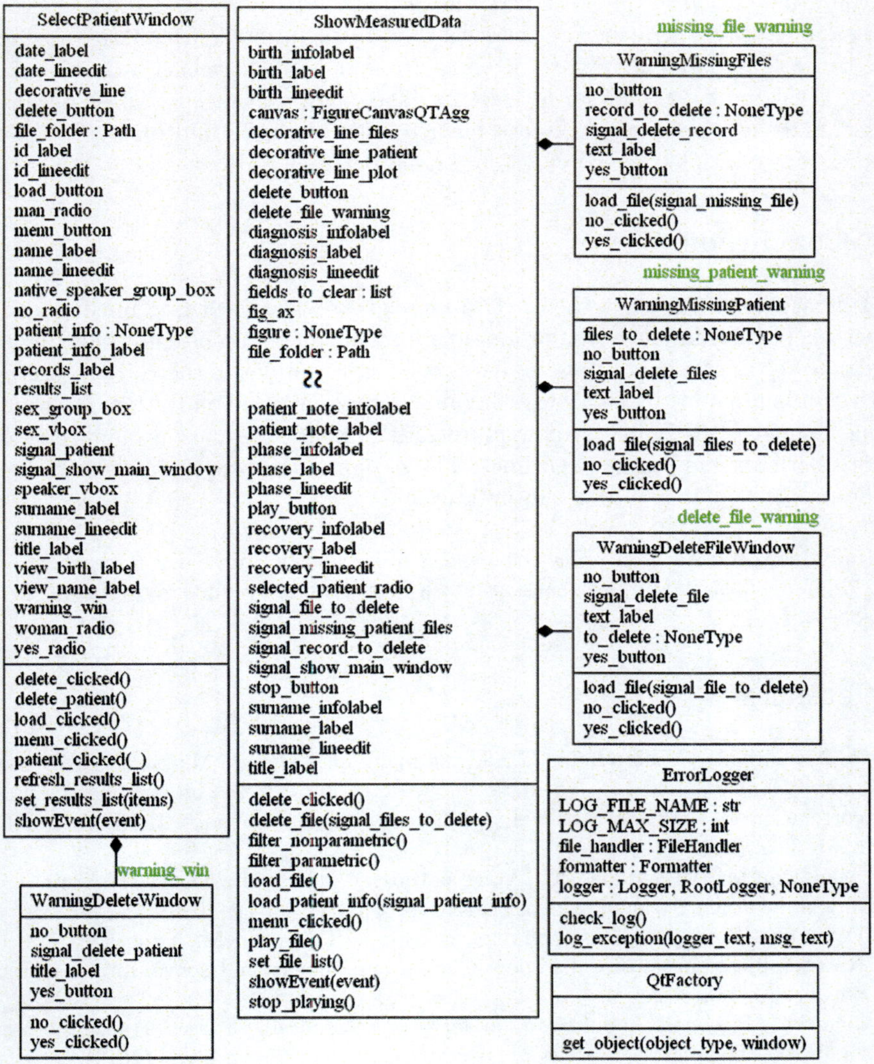

Fig. 5. UML diagram of voice analyst software

4 On Text Selection and Measurement

The text that patients read is chosen so that it contains all letters of the Czech alphabet, so the speaker has to pronounce stops, fricatives, nasals, laterals, thrills, ejectives, implosives, taps and flaps. This text was proposed in relation to the examination of the voice array [6]. After the patients read the text, they are asked to say a sequence of vowels a - e - i - o - u and repeat this sequence five times. The patients are instructed to pronounce the vowels at a comfortable sound level and hold them for at least 2 s. The vowel sequence is chosen because it seems to be a promising sound sequence that is used in multiple approaches to diagnose multiple specific vocal cord dysfunctions [2,5].

5 Conclusions

In this article, we present the novel approach to voice recording in diagnostic practice. The advantage of the proposed approach and setup lies in not having to calibrate the microphone. This emulates conditions similar to the voice recording with cellphones done by patients at home. A program based on Python programming language was built to allow physicians patient information management, record patients and replay recordings. The design of the programme was chosen to be intuitive and easy to use in practice.

Acknowledgements. Jan Vrba acknowledges his specific university research grant JIGA 445-85-2222. Jakub Steinbach acknowledges his specific university grant (IGA) 445-88-2202.

References

1. Al-Nasheri, A., Muhammad, G., Alsulaiman, M., Ali, Z., Malki, K.H., Mesallam, T.A., Ibrahim, M.F.: Voice pathology detection and classification using autocorrelation and entropy features in different frequency regions. IEEE Access **6**, 6961–6974 (2017)
2. Alhussein, M., Muhammad, G.: Voice pathology detection using deep learning on mobile healthcare framework. IEEE Access *6*, 41034–41041 (2018)
3. Fujimura, S., Kojima, T., Okanoue, Y., Shoji, K., Inoue, M., Omori, K., Hori, R.: Classification of voice disorders using a one-dimensional convolutional neural network. J. Voice (2020)
4. Jičínský, M., Mareš, J.: Software tool for voice disorder diagnostics. Transactions of the VŠB—Technical University of Ostrava. Mech. Ser. **65**(1), 11–18 (2019). http://transactions.fs.vsb.cz/2019-1/2051.pdf
5. Muhammad, G., Alhamid, M.F., Alsulaiman, M., Gupta, B.: Edge computing with cloud for voice disorder assessment and treatment. IEEE Commun. Mag. **56**(4), 60–65 (2018). https://ieeexplore.ieee.org/document/8337897/
6. NOVý, P., PEŠTA, J.: Fonetogram 4.5–lite (2008)
7. Schutte, H.K., Seidner, W.: Recommendation by the union of european phoniatricians (uep): standardizing voice area measurement/phonetography. Folia Phoniatr (Basel) **35**(6), 286–8 (1983)

8. Sicard, E., Menin-Sicard, A., Ruffle, L.: Vocalab: A user-friendly software for voice analysis and therapy. London (on line), United Kingdom (2021). https://hal.archives-ouvertes.fr/hal-03353490
9. Van Rossum, G., Drake, F.L.: Python 3 Reference Manual. CreateSpace, Scotts Valley, CA (2009)
10. Wang, Z., Yu, P., Yan, N., Wang, L., Ng, M.L.: Automatic assessment of pathological voice quality using multidimensional acoustic analysis based on the grbas scale. J. Signal Process. Syst. **82**(2), 241–251 (2016)
11. Wood, J.M., Athanasiadis, T., Allen, J.: Laryngitis. BMJ **349** (2014)
12. Xie, S., Yan, N., Yu, P., Ng, M.L., Wang, L., Ji, Z., et al.: Deep neural networks for voice quality assessment based on the grbas scale. In: Interspeech. pp. 2656–2660 (2016)

Development of Automated Control System of Electron-Beam Welding Process

Vadim Tynchenko[1,2,3](✉), Victor Kukartsev[1], Daria Shalaeva[2], Svetlana Zdrestova-Zaharenkova[1], Natalja Dzhioeva[1], and Kristina Moiseeva[2]

[1] Siberian Federal University, Krasnoyarsk, Russia
vadimond@mail.ru
[2] Reshetnev Siberian State University of Science and Technology, Krasnoyarsk, Russia
[3] Bauman Moscow State Technical University, Moscow, Russia

Abstract. In this study, the authors develop an automated system to control the process of electron-beam welding, in which the possibility of planning the technological process in the form of indication of a set of controlled and recorded parameters, as well as target values and tolerable ranges for them. Today, when welding products with a highly concentrated energy source, defects occur in the weld seam area, which leads to a large number of defects. The aim of the study is to increase the efficiency of the process of electron-beam welding of thin-walled structures due to the development and program implementation of a set of methods of automated planning and control of the technological process, as well as the statistical and intellectual processing of data obtained during the technological process. The program is developed for the installation of electron-beam welding, which is designed for the implementation of the technological process of electron-beam welding.

Keywords: Aerospace design · Automated control · Electron beam welding · Innovation · Modeling · Process innovation · Technology · Technological process

1 Introduction

Electron beam welding (EBW) is one of the most high-tech and high-performance welding methods. The authors of the work [1] consider the technological features of the EBW, as well as the main parameters and advantages of its application. The authors of [2–4] performed a simulation similar to electron beam welding processes, which showed that the dynamic positioning of the electron beam when creating several sources provides the possibility of introducing heat into several sections of the welded joint, which in turn makes it possible to obtain high–quality compounds of alloys of different metals. In the study of processes with EBW, pulse modulation is investigated [5], which leads to a change in heat input. Technological requirements for the quality of the EBW [6] directly affect the effectiveness and correctness of the implementation of the automated control system of the EBW. For example, the author of [7] presents his own control system for the EBW installation, where he highlights the most important task of positioning the

R. Silhavy et al. (Eds.): CoMeSySo 2022, LNNS 596, pp. 484–490, 2023.
https://doi.org/10.1007/978-3-031-21435-6_42

parts to be welded, but the technological characteristics do not allow you to know the time spent in advance to achieve the welding speed. Another system [8] is a program for controlling vacuum equipment specifically. The results of the practical application of EBW control models are presented in articles [9–11], where the basic principles of the construction and modeling of EBW systems are displayed. Sample Heading (Third Level). Only two levels of headings should be numbered. Lower-level headings remain unnumbered; they are formatted as run-in headings.

In order for the EBW automation system to function in accordance with the requirements imposed on it and be relevant, it is necessary to review various methods of electron beam welding. The author of the work [12] considers manipulators of EBP and SSD for effective work. When studying the effect of electron beam welding on metals and their alloys, it is very important to conduct a study of the formation of welded joints made by an electron beam using intermediate inserts [13]. Welding modes are determined (beam current, welding speed, sweep parameters), as well as the dimensions of the insert and the gaps between the insert itself and the base material. In order to create a complex automated EBW system in [16], the dependence of temperature on time on the surface of the roller is considered, which allows more precisely adjusting the setting effects of the shaft movement.

In the course of the work, methods of planning, monitoring and processing the results of the technological process of electron beam welding were investigated and implemented in the software system "Welding, Regulation, Correction, Analysis" (hereinafter referred to as WELDING). The functional purpose of PS WELDING is to control the technological process of electron beam welding, obtaining and processing information about the parameters of the equipment under test for mathematical and algorithmic processing of information in order to study the possibilities of modeling and forecasting the parameters of the technological process of electron beam welding. To study and implement the methods, modules for obtaining and processing information about the parameters of the welding process for mathematical and algorithmic processing of information were formed and implemented in the program code in order to study the possibilities of modeling and predicting the development of situations in the process of welding products.

2 Applied Mathematical Methods

The following mathematical methods are used in the software system:

1. In terms of statistical analysis: correlation analysis, regression analysis,
2. In terms of modeling: artificial neural networks, nonparametric regression.

2.1 Correlation Analysis

Correlation or correlation dependence is a statistical relationship between two or more random variables (or quantities that can be considered as such with some acceptable degree of accuracy). At the same time, changes in the values of one or more of these quantities accompany a systematic change in the values of another or other quantities [1].

As an estimate of the correlation dependence between the input variable parameters of the technological process and the output values in the WELDING system, the coefficients of pair correlation are used, which is the basic statistical approach used in statistical data processing packages.

2.2 Regression Analysis

Regression analysis is a statistical method of investigating the influence of one or more independent variables on a dependent variable. Independent variables are otherwise called regressors or predictors, and dependent variables are called criteria variables.

The method of nonparametric regression estimation is used as a basic regression estimation in WELDING PS (the description is given below). As estimates of the effectiveness of the regression model in the WELDING PS, the following are used: the error estimate of the regression model and the coefficient of determination, which are generally accepted basic statistical estimates of the quality of regression models in the theory and practice of regression analysis.

2.3 Artificial Neural Networks

The method of artificial neural networks is implemented in WELDING as an alternative method for restoring regression models. An artificial neural network is a computational structure that simulates biological processes, usually associated with the processes of the human brain [1]. Neural networks have a number of advantages over other methods of modeling and data analysis. These advantages are primarily due to the ability of neural networks to generalize, i.e., the ability to obtain an effective result on data that was not presented during network training.

2.4 Nonparametric Regression

Let there be some object. There is a sample of observations of the inputs and outputs of the object. The function describing the relationship between the input and output of an object is unknown. It is necessary, based on the available observations of the values of the inputs and outputs of the object, to restore the evaluation of the function describing the dependence between the input and output of the object. Regression is some averaged quantitative relationship between the input and output of an object.

3 Description of the Operation of the Program

The software system consists of eight interconnected logic modules:

1. The main module of the program implements the connection between the other seven modules of the software system, as well as the functionality of the main form of the project

2. The module for collecting information from the installation is designed to obtain parametric information during the welding process at the EBW installation. Implements interaction with data collection boards and monitoring process.
3. The process planning module is designed to generate process plans for the installation of EBW. Implements functions for calculating the values of variable parameters in accordance with user-defined constraints and sampling step.
4. The technological process execution module implements four types of the technological process being performed, providing information support during the technological process and automating the formation of protocols.
5. The results processing module automates the calculation of statistical criteria when processing process data. Implements, in particular, the calculation of the Kohren and Fischer criteria.
6. The graphical information display module is designed to build tables and graphs in order to structure the data provided to the user.
7. The modeling and forecasting module implement parametric and nonparametric approximation algorithms.
8. The database module is a service module designed to implement the interaction of the WELDING PS with the database stored on the MS SQL Server 2008 server.

4 Description of the Software System

All operations with the program are carried out through the window interface of the Windows application. To start working with the program, run the file svarka.exe.

The main application window shown in Fig. 1 will be opened.

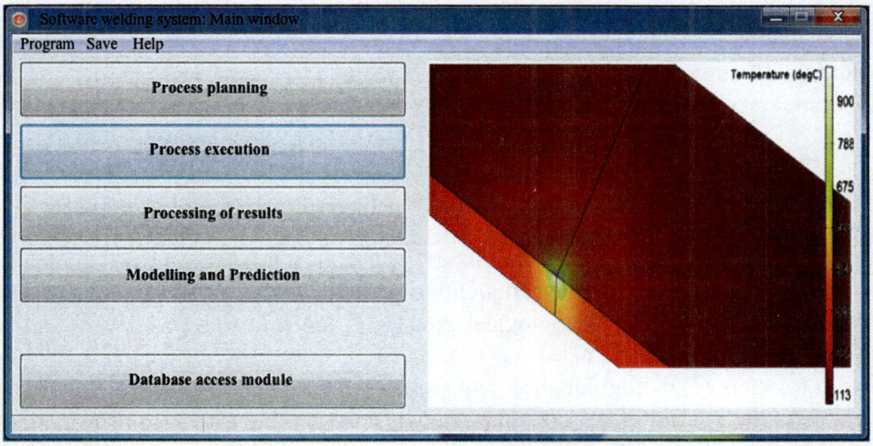

Fig. 1. Main window of PS WELDING.

The database module is a service module designed to implement the interaction of the WELDING PS with the database stored on the MS SQL Server 2008 server.

To adjust the database connection settings, you must:

Launch the main window of the WELDING PS. In the window, click on the "Database Access Module" button. The "Database Access Module" window will be displayed, the "Database Connection" tab in which there are four information input fields:

1. "Server Name/Address".
2. "User Name".
3. "Password".
4. "Database name".

Next, you need to enter the information in the input fields "Server name/Address", "User name", "Password", "Database name" and click "Connect". A message about a successful connection to the database will be displayed in the right part of the window;

When you click on the "Get a list of database tables" button, a list of the names of the database tables to which the connection was made will be displayed in the information field on the right.

Working with the database is possible only if there is a configured connection. Figure 2 shows the result of adding a new parameter to the database. To do this, you need:

1. Click the "Show table" button.
2. Fill in the input fields and click the "Add new" button.

Result of adding a new parameter shown in Fig. 2 will be opened.

The process planning module is designed to generate process execution plans for an electron beam welding installation. Implements functions for calculating the values of variable parameters in accordance with user-defined constraints and sampling step.

The technological process module implements four types of bench experiments, providing information support during the technological process and automating the formation of protocols.

5 Results

In this work, PS WELDING was developed, which is a prototype of an automated control system for the electron beam welding process. The program implements the possibility of planning the technological process in the form of specifying a set of controlled and recorded parameters, as well as target values and acceptable ranges of variation for them. The program allows you to track the execution of the technological process with the control of the correctness of its flow and the tolerance of deviations of the parameters of the technological process from the target. During the operation of the program, parameters are registered, the results of the planning and execution of the welding process are stored in the database. The program implements the possibility of processing the recorded data using a statistical assessment of the stability of the process and the method of restoring regression dependence after the end of the technological process or a series of technological processes.

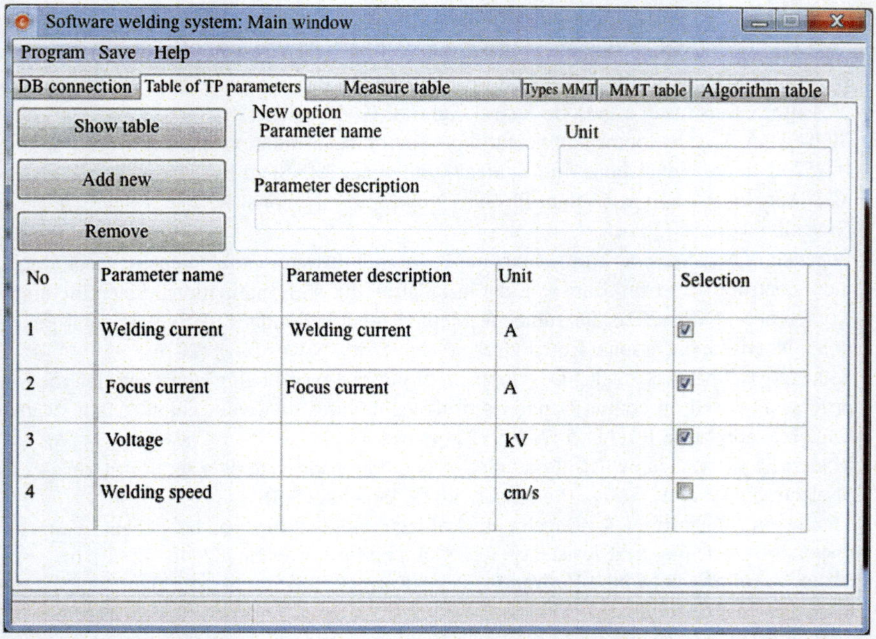

Fig. 2. Result of adding a new parameter, where: TP—technological parameters; DB—database; MMT—measuring and monitoring tools.

6 Conclusion

During the work on the project at this stage, the development of a set of practice-oriented approaches to planning, monitoring and processing the results of the implementation of the technological process of electron beam welding was carried out. The software code implements a software system for conducting a comprehensive study of the electron beam welding process. In addition, the program implements the possibility of processing the recorded data using a statistical assessment of the stability of the process and the method of restoring regression dependence after the end of the technological process or a series of technological processes.

In the future, the results of the study can be applied to create a full-fledged automated control system for the electron beam welding process, rather than its prototype.

References

1. Sidorov, V.P., Melzitdinova, A.V.: Electron beam welding. In: Technological Features and Equipment: Studies. Stipend. TSU Press, Togliatti (2013)
2. Mladenov, G.M., Trutnikov, D.N., Belenky, V.Y., Koleva, E.G.: Electron beam welding. Publishing House of Perm, Perm (2014)
3. Sudnik, V.A.: Fundamentals of welding modeling. In: The Handbook of ASM International Publishing House (USA): Technology of Mechanical Engineering, vol. 3, pp. 56–59 (2015)
4. Salomatova, E.S., Trushnikov, D.N., Belenky, V.Y.: Modeling of temperature in a steam-gas channel during electron beam welding. Therm. Process. Eng. **11**, 514–518 (2013)

5. Olshanskaya, T.V., Salomatova, E.S: Review of modern methods of electron beam control in electron beam welding. Bulletin PNRPU. Mech. Eng. Mater. Sci. **4**(18), 169–187 (2016)
6. Dragunov, V.K., et al.: Development of the technology of electron beam welding of high-pressure vessels made of 30XGSA steel. Oil Ind. **12**, 126–130 (2016)
7. Stelvaga, A.A.: Development of a control system for the installation of electron beam welding. In: 52nd International Scientific Student Conference MNSC-2014: Instrumental Methods and Techniques of Experimental Physics. Novosibirsk National Research State University, Novosibirsk, p. 56 (2014)
8. Popenko, Y.N., Medvedev, A.M., Starostenko, A.A.: Software development for vacuum equipment control of electron beam welding installation. In: 54th International Scientific Student Conference MNSC-2016: Instrumental Methods and Techniques of Experimental Physics, p. 35. Novosibirsk National Research State University, Novosibirsk (2016)
9. Kitaev, N.I., Speransky, S.K.: Development of optimal control algorithms for electron beam processing based on complex analysis of nanostructures of welded joints. Math. Methods Eng. Technol. MMTT **1**(11), 147–149 (2014)
10. Madera, A.G., Kandalov, P.I.: Principles of building a software package for thermal design of electronic systems. Softw. Prod. Syst. **3**(31), 435–438 (2018)
11. Marchenko, Y., Petrik, I.A.: Development of technology of electron beam heat treatment of dispersion-hardening heat-resistant steels. Compressor Power Eng. **3**, 40–42 (2016)
12. Brom, A.E., Stoyanova, M.V., Yazev, M.V., Korolev, S.A.: Assessment of technological resources for the production of composite products based on mathematical methods. IOP Conf. Ser. Mater. Sci. Eng. **934**, 012003 (2020). https://doi.org/10.1088/1757-899X/934/1/012003
13. Kosenko, E.A., Nelyub, V.A.: Evaluation of the stress-strain state of a polymer composition material with a hybrid matrix. Polym. Sci. Ser. D **15**, 240–244 (2022). https://doi.org/10.1134/S1995421222020137
14. Nelub, V., Gantimurov, A., Borodulin, A.: Economic analysis of data protection in systems with complex architecture using neural network methods. Econ. Ann. **185**, 178–188 (2020). https://doi.org/10.21003/EA.V185-17
15. Haikin, S.: Neural Networks a Complete Course, 2nd edn. Williams Publishing House, Moscow (2006)
16. Borisov, V.V., Kruglov, V.V.: Artificial neural networks. In: Theory and Practice, 2nd edn. Hotline-Telecom, Moscow (2006)
17. Stoyanova, M.V., Novikov, A.D., Borodulin, A.S.: Evaluation construction made of polymer composite materials by molding using reusable flexible punches production profitability. IOP Conf. Ser. Mater. Sci. Eng. **934**, 012068 (2020). https://doi.org/10.1088/1757-899X/934/1/012068
18. Kurashkin, S., Bukhtoyarov, V., Myrugin, A., Bocharov, A., Seregin, Y., Rogova, D.: Simulation model of the electron beam welding process for the formation of the effective technological parameters during the welding of aerospace structures. In: 2022 21st International Symposium INFOTEH-JAHORINA (INFOTEH), pp. 1–5. Institute of Electrical and Electronics Engineers Inc., East Sarajevo (2022). https://doi.org/10.1109/INFOTEH53737.2022.9751280
19. Pilinsky, M., Rutkovskaya, D., Rutkovsky, L.: Neural networks, Genetic Algorithms and Fuzzy Systems. Hotline-Telecom, Moscow (2006)
20. Nelyub, V.A., Fedorov, S.Y., Malysheva, G.V.: The study of the structure and properties of elementary carbon fibers with metal coatings. Inorg. Mater. Appl. Res. **12**, 1037–1041 (2021). https://doi.org/10.1134/S2075113321040316

IoT Device Using LoRaWAN for Data Transfer for Long Distances

Tibor Horák(✉), Peter Střelec, Szabolcs Kováč, Eduard Nemlaha, and Pavol Tanuška

Faculty of Materials Science and Technology in Trnava, Institute of Applied Informatics, Automation and Mechatronics, Slovak University of Technology in Bratislava, Trnava, Slovakia
{tibor.horak,peter.strelec,szabolcs.kovac,eduard.nemlaha, pavol.tanuska}@stuba.sk

Abstract. IoT devices can advantageously be used to monitor various variables. IoT communication is usually limited to shorter distances. The presented IoT device design allows use for areas where the data needs to be secured over longer distances and the frequency of sending is uncritical for control. It can be used for farmers or for cities where data sending can be done by multiple, must secure the communication of IoT device over long distances. LoRaWAN is one option to provide this type of communication. The demonstration is an IoT device that monitors soil moisture and provides data collection. The prepared IoT device can be used for designing IoT devices where data transmission over longer distances needs to be achieved. Collecting data using an IoT device, in this way, allow data to be obtained for analysis. The protocol can impact the power consumption of a device. The design of the IoT device is implemented in a way that it can be functional for as long as possible on battery power.

Keywords: IoT · LoRaWAN · LoRa transmits · Chirp spread spectrum

1 Introduction

The sustainability of water in the landscape and the ability to monitor its presence from rainfall, or the requirement to compare soil moisture and crop yields on non-irrigated, naturally grown crops and crops that are irrigated, is increasingly important to farmers. For this purpose, it is necessary to provide monitoring equipment that allows the moisture content of the environment to be monitored [1]. An IoT device that can measure soil moisture to a selected depth using a capacitive sensor appears advantageous. There is a need to provide communication over a longer distance. For small and medium farmers, this distance is in the order of kilometers or tens of kilometers. The data obtained from the measurements will allow crop planning and allow long term monitoring of soil moisture levels. Accordingly, irrigation or tillage operations can be efficiently managed and planned so that as much rainfall as possible can be in the soil [2].

R. Silhavy et al. (Eds.): CoMeSySo 2022, LNNS 596, pp. 491–500, 2023.
https://doi.org/10.1007/978-3-031-21435-6_43

Chirp Spread Spectrum (CSS) is a spread spectrum technique in radio communications. This type of modulation is used in nature by dolphins or bats. The first use of this technology was in 1940 by Professor Hüttman for radar equipment. Later, the chirp spread spectrum technology was modified by Sidney Darlington, whose research in 1940 produced the first pulse compression radar. It was used for covert military communications, which were intended to be secure, reliable, and resistant to jamming and eavesdropping [3].

A chirp signal has a constant amplitude that traverses the entire bandwidth in both linear and non-linear shapes from one end to the other in a certain amount of time. It uses broadband linear frequency modulated chirp pulses to encode the information. A chirp is a signal whose frequency decreases (down-chirp) or increases (up-chirp) over time; it is a relationship between time and frequency. It is a long-range wireless radio-frequency technology that is characterized by its high reliability, strong immunity to radio interference and low power consumption. This makes it particularly useful in large, busy environments and industrial factories. CSS as a technique is ideal for use in low-rate wireless networks [4–6].

2 Background

Long Range (LoRa) is a Low Power Wide Area Network (LPWAN) protocol. It is based on the Chirp Spread Spectrum (CSS) radio modulation technique. Distance is a major advantage and can present communication up to tens of kilometers, with low bit rates ranging from 0.3 to 50 kb/s. LoRa transmits over the licence-exempt megahertz radio frequency bands 433 MHz, 868 MHz (Europe) and 915 MHz (North America), 430 MHz, 920 MHz (Asia), 923 MHz (South Korea) [7].

LoRa enables transmission over very long distances. This can be greater than 16 km, which is approximately 10 miles in undeveloped or rural areas. The devices are also characterized by low power consumption, which is in the order of sleep mode μA and in the transmitting or receiving a signal over LoRaWAN requires less than 50 milliamps (mA) of the current. This means that the devices can have up to 10 years of battery life. Therefore, it is one of the suitable solutions for realizing the communication of IoT devices. The technology is presented in two parts of LoRa. LoRa physical layer and on the communication protocol, which is built on top of the underlying LoRa physical layer. The LoRaWAN communication protocol ensures reliable and secure communication. LoRaWAN network layer is shown in Fig. 1. The communication layer, Wide Area Network (WAN), can be an open source communication protocol defined by the LoRa Alliance consortium. LoRaWAN. This defines the communication protocol and system architecture for the network, while the LoRa physical layer enables long-range communication links [8–10].

LoRaWAN works well both indoors and outdoors or even in the presence of buildings, although it depends on the environment. The frequencies that are used to propagate the signal fall below 1 GHz and a very narrow frequency band is used, providing good coverage indoors. The LoRaWAN networks are useful and coverage can be obtained from a service provider with infrastructure or by building on its own. In this study, a custom LoRaWAN network was used to provide connectivity to the sensors [11, 12].

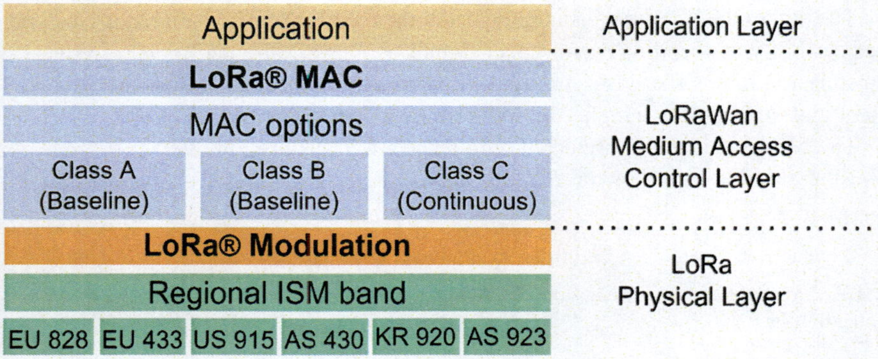

Fig. 1. LoRaWAN network layer

Physical Layer in Spectrum spreading in LoRa modulation is achieved by generating a chirp signal that changes frequency linearly in time. The advantages of such a modulation technique are that the timing and frequency spacing are identical, which eases both the complexity of the process and the design effort [13].

The spectral bandwidth is the same as the chirp bandwidth. The signal modulated using such a technique can be demodulated at the noise level. Lower transmit power is an advantage and can decrease the chance of multipath propagation and it's eliminating the Doppler Effect [14]. LoRa is characterized by three parameters Spreading Factor (SF), Bandwidth (BW) and Coding Rate (CR) [15] as it's shown in Table 1.

Table 1. Overview of technical specifications

Data rate (DR)	Spreading factors (SF)	Bandwidth (kHz)	Transfer speed (baud)	Receiver sensitivity (dBm)
0	12	125	292	– 135
1	11	125	537	– 133
2	10	125	976	– 132
3	9	125	1757	– 129
4	8	125	3125	– 126
5	7	125	5468	– 123
6	7	250	10937	– 120
7	FSK	–	50000	– 108

Spreading factor (*SF* Spreading factor)

Represents the number of chirps in one bit. LoRa allows up to seven configuration options ranging from 6 to 12. The length of the spread chirp sequence is defined as 2^{SF}. The Spreading Factor can thus be written as the relationship between the symbol

$$2^{SF} = \frac{R_c}{R_s}$$

The higher the SF value reaches, the better the receiver sensitivity and thus the higher the possibility of decoding even weaker signals. However, a higher SF value reduces the transmission rate. Otherwise, if a high SF value is not needed, we can choose a lower factor and thus achieve a higher bit rate [16].

The SF value is mainly selected with respect to transmission rate, range and signal robustness. Values are shown in Table 2.

Table 2. Available transfer speeds

Spreading factor (SF)	Spreading factor (chirp/symbol)
6	64
7	128
8	256
9	512
10	1024
11	2048
12	4096

The bandwidth is regulated by a standard for a given region or territory, which defines the parameters for broadcasting at given frequencies. Bandwidth mainly affects the transmission rate, receiver sensitivity and the part of the spectrum used [17, 18].

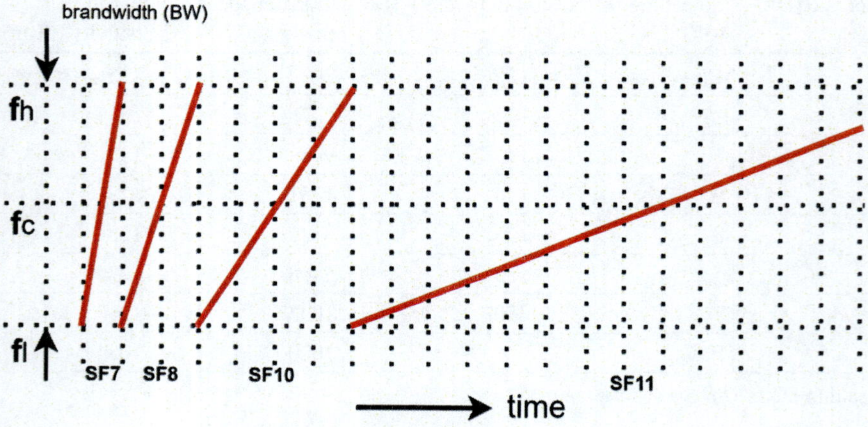

Fig. 2. Comparison of LoRa spreading factors

The bandwidth is configurable in a range of default values from 7.8 to 500 kHz. The bandwidth is equivalent to the chirp rate R_c. This means that BW = 400 kHz is the corresponding value of 400,000 cp/s. The bandwidth also affects the transmission rate, which increases in direct proportion with the bandwidth, as it is shown on Fig. 2. Based on the observed SF and BW parameters, we can define the symbol period Ts and the data rate (R_braw) [19].

$$\mathrm{Ts}(s) = \frac{2^{\mathrm{SF}}}{\mathrm{BW}}$$

$$R_{\mathrm{braw}}\ (b/s) = \frac{\mathrm{SF}}{\mathrm{Ts}}$$

CR is defined by Forward Error Correction (FEC). This allows us to correct the errors that occur during transmission. A message carrying some information is always 4 bits long and its length is adjustable between 5 and 8 bits. The CR determines the possible combination of the length of the data message to the code message, which means that the values of the code ratios can be 4/5, 4/6, 4/7, and 4/8. The higher the value, the more errors that occur can be corrected during backward decoding. The higher the CR, the more reliable the transmission. However, we must transmit more bits to send the same message as it is shown in Table 3. The CR value is configured according to the specific communication requirements. We can calculate the data transmission using the formula [20, 21].

$$\mathrm{bitrate} = \mathrm{SF} \cdot \frac{\mathrm{BW}}{2^{\mathrm{SF}}} \cdot \frac{4}{(4 + \mathrm{CR})}$$

Table 3. Ratio of CR and corrected bits

Coding ratio (CR)	Bits detected	Corrected bits
4/5	0	0
4/6	1	0
4/7	2	1
4/8	3	1

3 Materials and Methods

The experiment was based IoT device realized using a Raspberry Pi Pico RP2040, to which was connected a LoRa SX1262 chip from Semtech© and three capacitive Soil moisture sensor HW-390. The current of the LoRaWAN module is declared according to the specification at 4.2 mA and during transmission the current consumption is dependent on the radiated power. The SX1268 consumes less than 20 mA to transmit + 10 dBm signal at its antenna port and 120 mA at a maximum current for + 22 dBm [22].

Raspberry Pi Pico, the dev board of the RP2040, consumes about 1.3 mA in its sleep mode [23]. The device has LiFePO4 rechargeable battery and there is attached mini solar

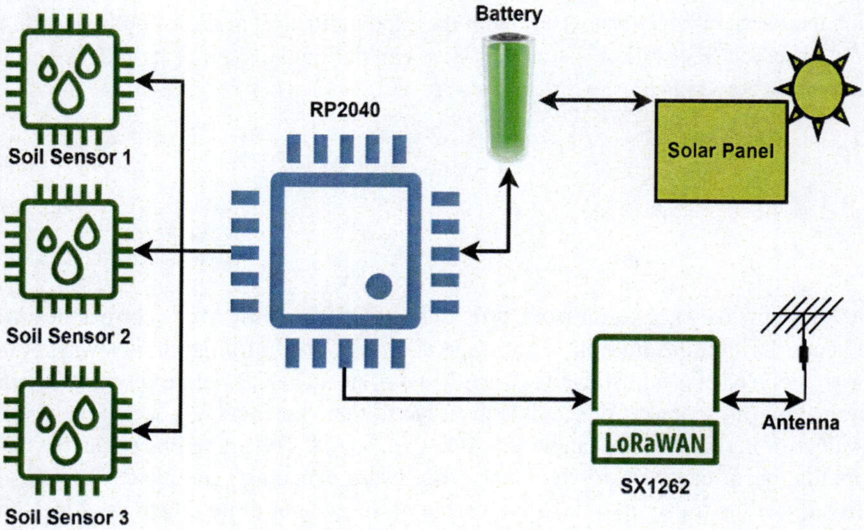

Fig. 3. The block scheme of IoT based on RP2040

panel. The battery has thermal protection and also a circuit to provide charging from the solar panel. Block scheme is on Fig. 3.

The soil sensor can be used as the HW-390 sensor [24]. It's a module consists of a fixed frequency oscillator, which generates a square wave and then returns it in the form of an analog signal in the range of 0–3 V. This signal provides a capacitive change at the receiving terminals. The higher the soil moisture value is, the greater the capacitance of the sensor and therefore, the higher the value read from the signal. A disadvantage is that an AD converter must be used. For the final solution, we used capacitive soil sensors that systems are built into the ATSAMD10 chip and using an I2C bus. The 3 sensors were placed in the soil at different depths, as shown in Fig. 4.

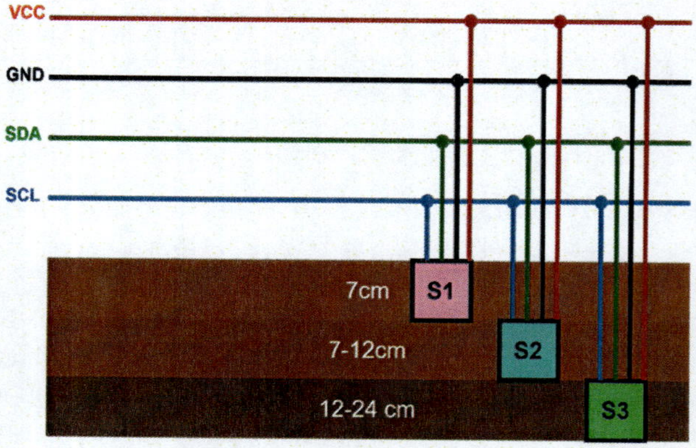

Fig. 4. Placing sensors in the soil

Data are read from the sensor using the I2C bus and sent according to the sensor identifier. For each sensor, a calibration of the values has been made and an address for the I2C bus has been assigned. The deployment in the soil was done over an area of $1m^2$. The RP2040 addresses each sensor. A 4-core shielded cable was used for connection. The power supply to the sensors was 3.3V. The sensed data is sent via messages to the LoRaWAN gateway. LoRaWAN Gateway is a LoRaWAN device built on RP with Gateway Shield SX1302 and this device is connected to the Internet. All data are acquired by Node Red tool and stored in a machine readable format in the database. Node-red is used to show values. Soil moisture can be analyzed and the results can be used for irrigation schedule or crop planting planning.

4 Experiment Implementation and Result

The communication of the connected IoT devices with LoRaWAN Gateway. The experiment measures the distance and the stability of connectivity between the IoT device and the gateway. The block diagram of communication and experimental setup is shown in Fig. 5.

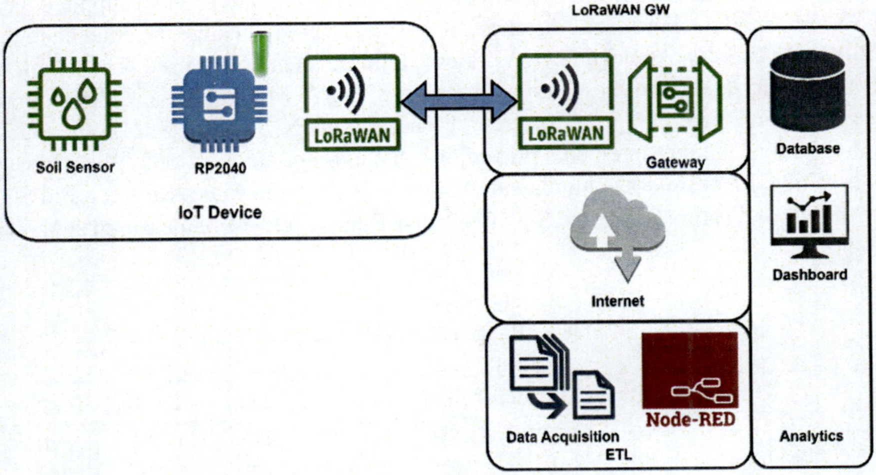

Fig. 5. IoT LoRaWAN block diagram

The IoT device sends data to Gateway. LoRaWAN gateway is connected to the internet and sends data to Node-RED. The gateway also has a function as a server where data are collected and stored in a MySQL relational database. The basic dashboard is implemented in Node-Red.

IoT devices have been deployed in the fields around the city of Trnava. Parameters were tested based on which stable operation was achieved. The aim was to verify over what distances the LoRaWAN protocol can be used to a custom LoRaWAN gateway. The experiment lasted for 3 months. The data transmissions that were realized using the created IoT device are to show the range, data rate and possibly the power consumption.

The sensors and measurement frequencies will be identical for each device. The impact on the battery life has a transmitting part. Therefore, it is possible to measure the soil moisture multiple times and then start the transmission and send the data once. The power consumption depends on the spreading factor used and the bandwidth used. Vendor parameters for spreading factors and time on air is in Table 4.

Table 4. LoRa spreading factors [25]

Spreading factor (for UL at 125 kHz)	Bit rate (bps)	Range (depends on terrain) (km)	Time on air for an 11-byte payload (ms)
SF10	980	8	371
SF9	1760	6	185
SF8	3125	4	103
SF7	5470	2	61

According to this assumption, the deployment in distance was realized. Communication over a distance of 3.7 km was sufficient to use the standard antenna supplied with the SX1262 module. For longer distances, it was necessary to change the antenna to the LoRaWAN Gateway. An 8dBi RAK Wireless Helium Antenna was used. The length of this antenna is 130 cm and it is designed for the 858–878 MHz frequency band, omnidirectional with 8 dB gain. The location of the LoRaWAN Gateway was on the tallest building and as the measurements proved, it is the appropriate location of the gateway that is critical for seamless communication. Over the days, the values for Received Signal Strength Indicator (RSSI) and Signal to Noise Ratio (SNR) were measured as shown in Table 5.

Table 5. Range experiment results

	Distance 1.1 km		Distance 3.7 km		Distance 5.2 km		Distance 10.3 km	
Antenna	RSSI [dB]	SNR [dB]	RSSI [dB]	SNR [dB]	RSSI [dB]	SNR [dB]	RSSI [dB]	SNR [dB]
5 dB	– 87.5	7.62	– 101.5	– 5.4	– 102.8	– 12.5	X	X
8 dB	– 91	7.97	– 101.5	– 8.66	102.8	– 13.8	– 104.8	– 13.9

The measurement of current and battery voltage measurements was so low that even without optimizing the code to solve the consumption, the IoT device can operate with the solar panel recharging without complications for several months, even in the winter, and year-round operation of such an IoT device is possible.

5 Conclusions

The IoT device with the integrated LoRa module can communicate by default for a distance of 3.7 km with a standard antenna setup for IoT with a combination of LoRaWAN Gateway. Applying higher gain antenna on Gateway, it's easy to get 5.2 km distance. Spreading factors are critical parameters for longer distances and our suggestion is to test this in the landscape. For the development of IoT is suggested to make this configurable for IoT devices.

The same is with the output power for the LoRaWAN module. Overall, the best option is an advanced setup based on a landscape. The theoretically higher spreading factor means a longer transmission time and logically higher current for the radio part of the LoRaWAN module. This should have an impact on the battery lifetime. With used setup it was possible to recharge battery even if the sun was not visible for more days and there was no outage in the communication.

Data collected from the IoT device are helpful and are used for planning of watering for farmers. This can be implemented almost everywhere and it's suitable for small or medium farmers, due to the fact that they have fields close to each other. The experiment shows that it's easy to implement and get stable data communication for distances of few kilometers. In the landscape, here farmer is suitable in the village and there is no direct visibility to all sensors, IoT device with LoRaWAN can transmit data to the Gateway.

Nevertheless disadvantage is the data rate limit. Such a setup is good enough in case that data are generated and collected every few minutes or better hours and they have small values. The plan is to keep this measurement running for a few years and we focus on collected data in the future.

Acknowledgments. This research was funded by the Scientific Grant Agency of the Ministry of Education, Science, Research, and Sport of the Slovak Republic and the Slovak Academy of Sciences, grant number VEGA 1/0176/22 Proactive control of hybrid production systems using simulation-based digital twin and VEGA 1/0193/22 Proposal of identification and monitoring of production equipment parameters for the needs of predictive maintenance in accordance with the concept of Industry 4.0 using Industrial IoT technologies. This article was also written thanks to the generous support under the Operational Program Integrated Infrastructure for the project: "Strategic research in the field of SMART monitoring, treatment and preventive protection against coronavirus (SARS-CoV-2)", Project no. 313011ASS8, co-financed by the European Regional Development Fund.

References

1. Kojima, Y., et al.: Low-cost soil moisture profile probe using thin-film capacitors and a capacitive touch sensor. Sensors **16**, 1292 (2016)
2. Bacco, M. et al.: Smart farming: opportunities, challenges and technology enablers. In: 2018 IoT Vertical and Topical Summit on Agriculture-Tuscany (IOT Tuscany), pp. 1–6 (2018)
3. Hur, H.: Chirp spread spectrum-based indoor localization (2009)
4. Naik, N.: LPWAN technologies for IoT systems: choice between ultra narrow band and spread spectrum. In: 2018 IEEE International Systems Engineering Symposium (ISSE). IEEE (2018)

5. Emmanuel, A.L. et al.: Optimization of spreading factor distribution in high density LoRa networks. In: 2020 IEEE 91st Vehicular Technology Conference (VTC2020-Spring). IEEE (2020)
6. Fusko, M. et al.: The importance of digitization and innovations for small and medium-sized enterprises. Proc. CBU Econ. Bus. **2**, 32–40 (2021)
7. Łukasik, Z., Ushakov, A.: Comparative analysis of data transmission technologies in industrial systems of the internet of things (IoT). J. Autom. Electron. Electr. Eng. **2**(1), 31–39 (2020)
8. Abbasi, M., Khorasanian, S., Yaghmaee, M.H.: Low-power wide area network (LPWAN) for smart grid: an in-depth study on LoRaWAN. In: 2019 5th Conference on Knowledge Based Engineering and Innovation (KBEI). IEEE (2019)
9. Jiang, X. et al.: Hybrid low-power wide-area mesh network for IoT applications. IEEE Internet Things J. **8**(2), 901–915 (2020)
10. Patel, D., Won, M.: Experimental study on low power wide area networks (LPWAN) for mobile internet of things. In: 2017 IEEE 85th Vehicular Technology Conference (VTC Spring). IEEE (2017)
11. Grübel, J. et al.: Dense indoor sensor networks: towards passively sensing human presence with LORAWAN. Pervasive Mob. Comput. **84**, 101640 (2022)
12. Froiz-Míguez, I. et al.: Design, implementation, and empirical validation of an IoT smart irrigation system for fog computing applications based on LoRa and LoRaWAN sensor nodes. Sensors **20**(23), 6865 (2020)
13. Sundaram, J.P., Du Shanmuga, W., Zhao, Z.: A survey on LoRa networking: research problems, current solutions, and open issues. IEEE Commun. Surv. Tutorials **22**(1), 371–388 (2019)
14. Ameloot, T. et al.: Characterizing the impact of Doppler effects on body-centric LoRa links with SDR. Sensors **21**(12), 4049 (2021)
15. Raychowdhury, A., Pramanik, A.: Survey on LoRa technology: solution for internet of things. In: Intelligent Systems, Technologies and Applications, pp. 259–271 (2020)
16. Cuomo, F. et al.: EXPLoRa: extending the performance of LoRa by suitable spreading factor allocations. In: 2017 IEEE 13th International Conference on Wireless and Mobile Computing, Networking and Communications (WiMob). IEEE (2017)
17. Mikhaylov, K., Petaejaejaervi J., Haenninen, T.: Analysis of capacity and scalability of the LoRa low power wide area network technology. In: European Wireless 2016; 22th European Wireless Conference. VDE (2016)
18. Šimon, M. et al.: Combined heuristic attack strategy on complex networks. Math. Prob. Eng. **2017** (2017)
19. Bor, M.C. et al.: Do LoRa low-power wide-area networks scale? In: Proceedings of the 19th ACM International Conference on Modeling, Analysis and Simulation of Wireless and Mobile Systems (2016)
20. Faber, M.J. et al.: A theoretical and experimental evaluation on the performance of LoRa technology. IEEE Sens. J. **20**(16), 9480–9489 (2020)
21. Bouguera, T. et al.: Energy consumption model for sensor nodes based on LoRa and LoRaWAN. Sensors **18**(7), 2104 (2018)
22. Semtech SX1262 https://www.semtech.com/products/wireless-rf/lora-core/sx1262 Accessed 1 Aug 2022
23. Raspberry Pi documentation. https://www.raspberrypi.com/documentation/microcontrollers/rp2040.html. Accessed 1 Aug 2022
24. Soil sensor HW-390. https://blog.zerokol.com/2021/01/hw-390-capacitive-soil-moisture-sensor.html. Accessed 3 Aug 2022
25. What are LoRa® and LoRaWAN®? https://lora-developers.semtech.com/documentation/tech-papers-and-guides/lora-and-lorawan. Accessed 3 Aug 2022

Using Technical and Structural Coefficients of Economic Statistics to Equalize Flows of Input-Output Table

Seyit Kerimkhulle(✉), Zhangeldi Aitkozha, Ainur Saliyeva, Zhandos Kerimkulov, Alibek Adalbek, and Roman Taberkhan

L.N. Gumilyov Eurasian National University, Nur-Sultan, Kazakhstan
kerimkhulle@gmail.com, aitkozha_zh@mail.ru, ainurka_ks@mail.ru, kerimkulovzh@gmail.com, q_al@mail.ru, rn_82@bk.ru

Abstract. In this article was attempts to solve the problem of (+) over evaluated and/or (−) under evaluated intersectoral flows in monetary terms, final demand budgets and value of employment, as well as technological and structural coefficients of the input-output table for statistics economics of Kazakhstan for 2018. Then using mathematical programming problems minimizing the objective function Euclidean distance between value of employment and final demand, total outlays, and total output, as well as using technical and structural coefficients of economic statistics was evaluation of equalization of flows of input-output table of flows and budgets. In particular, the value of the statistics of flows of agricultural sector, manufacturing, and service sector, respectively, was over evaluated by 0.1987, 1.2503 and under evaluated by 1.449 billion U.S. dollars in 2018. And the value employment flows, respectively, was under evaluated by 0.5513, 6.4375 and over evaluated by 6.9787 billion U.S. dollars.

Keywords: Technical coefficients · Structural coefficients · Economic statistics · Evaluation · Equalize · Flows of input-output table · Kazakhstan

1 Introduction

Known that the structural approach to the problems of the economic system, known as input-output analysis, developed by Leontief [1], laid the foundation for modern data processing technologies. In today, the input-output analysis methodology already covers a wide variety of fields, including economics, regional science, regional economics, city, regional and urban planning, environmental planning, public policy analysis, and public management [2].

Currently, the economy of different countries is very closely connected with each other. Therefore, the analysis and study of inter-production relations affects the effectiveness of economic and social policies of states. W. Leontiev's method of analysis of classical industrial relations was used to analyze the economic structures of different countries. The input-output table is a simple but very effective tool used in analyzing the structure of the economy and studying its structural changes.

R. Silhavy et al. (Eds.): CoMeSySo 2022, LNNS 596, pp. 501–??, 2023.
https://doi.org/10.1007/978-3-031-21435-6_44

Various factors influence the development of the country's economy. One of these problems is to determine the problems of development of social enterprises and their impact on the country's economy. For example, the problems of reconstruction to evaluate the development of shopping centers and increase their rationality were studied by using input-output analysis in [3]. In general, current shopping center research focuses only on the following issues: rent valuation, tenant location, retail agglomeration economics, retail demand externalities, and intangible asset value. And the advantage of this work is that shopping centers have an important role as city centers and catalysts for the development and reconstruction of territories. Its socio-economic contribution to the country's economy was revealed using the input-output model. The methodology of measuring their impact on the economy and employment was reviewed and analyzed. And the impact of social enterprise on the economy of the Czech Republic is considered in the work [4]. Basically, in most cases to date, the lack of proven impact of social enterprises on the economy is an obstacle to the development of this sector, not only from the point of view of investors and donors, but also from the point of view of politicians and even social enterprises themselves. The importance of this study is to identify gaps in research related to the impact of social enterprises (reflecting the local aspect). In the work, the efficiency measurement of social enterprises was carried out using the methods of local multiplier 3 (LM3) and analysis of costs and results. The research was conducted in 2015 based on the data of social enterprises (143 subjects) in the Czech Republic. And the results of the research show that the influence of social enterprises on the economy in certain regions of the Czech Republic is not small.

It is known that the Global Financial Crisis (GFC) of 2008, the crises related to the global COVID-19 disease had an impact on the economic development of the countries of the world. During the financial crisis, a lot of effort is spent to maintain the country's economy at a certain level. One such position was discussed using the input-output model in the work [5]. It shows that the GFC caused a 13% drop in GDP and a 16% drop in energy consumption. The stimulus packages are shown to have led to an increase in economic growth and energy consumption of 1.83% and 4.64%, respectively. So, in the work [6], using the global input-output database, the potential vulnerability of countries that are too much affected by the demand of the German automobile industry, namely Italy, is considered.

2 Methods, Models, Data

The methods, models and data for flow equalize of accounts and budgets of the input-output table will be built primarily using the statistical data of Kazakhstan for 2018 from OECD reports [7], then the mathematical programming problem for input-output will be used models [2], models [8] and projects [9] of open source system dynamics, as well as modern theories, methodologies and information technologies, the essence and content of which are detailed in [10–16].

Assume that by buying sector products **x**–the total output of one sector is distributed through sales to **Zi**–other sectors of the economy and consumers of **f**–final demand according to the system of equations [2]:

$$\mathbf{x} = \mathbf{Z}\mathbf{i} + \mathbf{f}, \tag{1}$$

and that by selling sector the value $\mathbf{x}'$–the total outlays of one sector are distributed through purchases to $\mathbf{i}'\mathbf{Z}$–other sectors of the economy and payments of $\boldsymbol{\varepsilon}'$–employment by sector according to the system of equations [2]:

$$\mathbf{x}' = \mathbf{i}'\mathbf{Z} + \boldsymbol{\varepsilon}', \tag{2}$$

where

$\mathbf{x} = \begin{bmatrix} x_1 \\ \vdots \\ x_n \end{bmatrix}$–total gross output (production) throughout the economy, x_i–the total output of sector i and $\mathbf{x}' = \left[x_1', \ldots, x_n'\right]$–all inputs are accounted for in the processing and payments sectors in monetary terms;

$\mathbf{Z} = \begin{bmatrix} z_{11} & \cdots & z_{1n} \\ \vdots & \ddots & \vdots \\ z_{n1} & \cdots & z_{nn} \end{bmatrix}$–interindustry flows, or transactions for a particular time period in monetary terms;

$\mathbf{i} = \begin{bmatrix} 1 \\ \vdots \\ 1 \end{bmatrix}$–vector $n \times 1$ column consisting of elements 1 and $\mathbf{i}' = [1, \ldots, 1]$–vector $1 \times n$ row consisting of n elements of 1;

$\mathbf{I} = \begin{bmatrix} 1 & \cdots & 0 \\ \vdots & \ddots & \vdots \\ 0 & \cdots & 1 \end{bmatrix}$–identity $n \times n$ matrix;

$\mathbf{f} = \begin{bmatrix} f_1 \\ \vdots \\ f_n \end{bmatrix}$–final demand in monetary terms and $\mathbf{f}_c = \begin{bmatrix} f_{c1} \\ \vdots \\ f_{cn} \end{bmatrix}$–final demand of structural coefficients;

$\mathbf{e}' = [e_1, \ldots, e_n]$–value of employment in monetary terms and $\mathbf{e}_c' = [e_{c1}, \ldots, e_{cn}]$–of employment of structural coefficients, then.

$\boldsymbol{\varepsilon} = \hat{\mathbf{e}}_c'$–produces a vector whose elements are the total labor income in each sector that accompanies the exogenous final demand.

Known that the $n \times n$ matrix of technical coefficients can be represented as [2]:

$$\mathbf{A} = \mathbf{Z}\hat{\mathbf{x}}^{-1} \tag{3}$$

where

$\mathbf{A} = \begin{bmatrix} a_{11} & \cdots & 0 \\ \vdots & \ddots & \vdots \\ 0 & \cdots & a_{nn} \end{bmatrix}$–the $n \times n$ matrix of technical coefficients;

$\hat{\mathbf{x}}^{-1} = \begin{bmatrix} 1/x_1 & \cdots & 0 \\ \vdots & \ddots & \vdots \\ 0 & \cdots & 1/x_n \end{bmatrix}$–diagonal matrix with vector $\mathbf{x}$ inversion elements along the main diagonal.

Further, we single out the list of input data parameters of economic system according to W. Leontiev Input-Output model (Eqs. 1–2) to equalize their values to obtain an equilibrium and productive growth of the country's economy, in particular:

- for blog of flows Processing Sectors are a $n \times n$ matrix of technical coefficients, if the number of sectors of the economy is n, then the number of technical coefficients to equalize will be n^2 and they are in the first quadrant of the Table 1;
- for blog of flows Final Demand are a vector $n \times 1$ column of final demand coefficients contains the base-year final demand flows in each sector divided by that sector's base-year gross output and the number of final demand coefficients to equalize will be n and they are in the quadrant of Final Demand of the Table 1;
- for blog of flows Value of Employment are a vector $n \times 1$ row of employment coefficients contains the base-year employment flows in each sector divided by that sector's base-year gross output and the number of employment coefficients to equalize will be n and they are in the quadrant of Value of Employment of the Table 1;

Table 1. Input data parameters for equalize input-output table accounts of economic system. Compiled by the author.

#	Sectors	1	...	n	$\mathbf{f}_c$	$\mathbf{x}_c$
		Processing Sectors			Final Demand	Total Output
1	Processing Sector 1	a_{11}	...	a_{1n}	f_{c1}	x_{c1}
...	...	...	$\ddots$	...	...	...
n	Processing Sector n	a_{n1}	...	a_{nn}	f_{cn}	x_{cn}
$\boldsymbol{\varepsilon}'_c$	Value of employment	e_{c1}	...	e_{cn}		
$\mathbf{x}'_c$	Total Outlays	x'_{c1}	...	x'_{cn}		

- information database input-output table to flows of transactions account for the Kazakhstan statistics, U.S. billion dollars, 2018 (see Table 2), which compiled from Source OECD Report [7];

- information database technical and structural coefficients for the Kazakhstan statistics, 2018 (see Table 3), which compiled from Source OECD Report [7].

Further, comparing the values of the initial data input-output fable to flows of transactions account for the Kazakhstan statistics, U.S. billion dollars, 2018 for the difference between monetary flows $\mathbf{f}' = [9.632, 66.730, 102.978]$–final demand (see Table 2, column $\mathbf{f}$) and flows $\boldsymbol{\varepsilon}' = [8.892, 59.042, 111.406]$–value of employment (see Table 2, row

Table 2. Input-output table to flows of transactions account for the Kazakhstan statistics, U.S. billion dollars, 2018, compiled from source OECD report [7].

Sectors	1	2	3	**f**	**x**
1	1.250	1.525	0.485	9.632	12.892
2	0.611	21.654	10.106	66.730	99.102
3	2.139	16.881	26.918	102.978	148.916
$\boldsymbol{\varepsilon}'$	8.892	59.042	111.406		
$\mathbf{x}'$	12.892	99.102	148.916		

Table 3. Technical and structural coefficients for the Kazakhstan statistics, 2018, compiled from source OECD report [7].

Sectors	1	2	3	$\mathbf{f}_c$	$\mathbf{x}_c$
1	0.0971	0.0154	0.0033	0.7470	0.8628
2	0.0474	0.2185	0.0679	0.6734	1.0071
3	0.1659	0.1703	0.1808	0.6915	1.2085
$\boldsymbol{\varepsilon}_c'$	0.6897	0.5958	0.7481		
$\mathbf{x}_c'$	1.0000	1.0000	1.0000		

$\boldsymbol{\varepsilon}'$) there is an imbalance of excess demand for $\Delta\mathbf{f} = \mathbf{f}' - \boldsymbol{\varepsilon}' = [0.740, 7.688, -8.428]$ billion U.S. dollars, in which it is observed that in the agricultural sector the demand for its products is too much by 0.740, also in the industrial sector–by 7.688 billion U.S. dollars, and in the third sector of services, the demand for its products is already short by 8.428 billion U.S. dollars, however, the funds in these aggregated accounts are balanced, which reveals the nature of the problem of this study.

Now, just as above, comparing the data values technical and structural coefficients for the Kazakhstan statistics, 2018 for the difference between the values $\mathbf{f}_c' = [0.7470, 0.6734, 0.6915]$–final demand of structural coefficients (see Table 3, column $\mathbf{f}_c$) and the values $\boldsymbol{\varepsilon}_c' = [0.6897, 0.5958, 0.7481]$–value of employment of structural coefficients (see Table 3, row $\boldsymbol{\varepsilon}_c'$) there is an (+) over evaluated of values on $\Delta\mathbf{f}_c = \mathbf{f}_c' - \boldsymbol{\varepsilon}_c' = [0.0574, 0.0776, -0.0566]$ is the share of the value of a unit of production of the entire economy, in which it is observed that in the agricultural sector the assessment of its products is (+) over evaluated by 0.0574, also in the industry sector–by 0.0776–the share of the value of a unit of production, respectively, of the agricultural sector and the industrial sector, and in the third sector services valuation on its products is already (−) under evaluated by 0.0566–a share of the value of a unit of services sector output, however, the sum of the valuations of these aggregate structural coefficients are equalized, which is the economic nature of the problems of this study research.

3 Results

3.1 Creating by Equalize Input-Output Table Accounts

To solve the problems of flows of transactions account disbalances and evaluations of technical and structural coefficients for the Kazakhstan economy statistics, we will use the equalize input-output model using:

- target function minimization $\rho(\boldsymbol{\varepsilon}', \mathbf{f})$–Euclidean distance between monetary flow vectors $\boldsymbol{\varepsilon}'$–value of employment and $\mathbf{f}$–final demand, and $\rho(\mathbf{x}', \mathbf{x})$–Euclidean distance between monetary flow vectors $\mathbf{x}'$–total outlays and $\mathbf{x}$–total output expressed in billion U.S. dollars:

$$\rho(\boldsymbol{\varepsilon}', \mathbf{f}) + \rho(\mathbf{x}', \mathbf{x}) \rightarrow \min, \tag{4}$$

- vector equality between monetary flows $\boldsymbol{\varepsilon}'$–value of employment and $\mathbf{f}$–final demand expressed in billion U.S. dollars:

$$\boldsymbol{\varepsilon}' = \mathbf{f}, \tag{5}$$

- vector equality between monetary flows $(\mathbf{x}')'$–total outlays and $\mathbf{x}$–total output expressed in billion U.S. dollars:

$$(\mathbf{x}')' = \mathbf{x}, \tag{6}$$

- equality of the current monetary budget $\mathbf{i}'\mathbf{f}$–final demand for the statistical value $\underset{2018}{f}$–final demand for 2018 in billion U.S. dollars:

$$\mathbf{i}'\mathbf{f} = \underset{2018}{f}, \tag{7}$$

- equality of the current monetary budget $\boldsymbol{\varepsilon}' i$–value of employment for the statistical value $\underset{2018}{e}$–value of employment for 2018 in billion U.S. dollars:

$$\boldsymbol{\varepsilon}' i = \underset{2018}{e}, \tag{8}$$

- equality of the current monetary budget $\mathbf{i}'\mathbf{x}$–total output for the statistical value $\underset{2018}{x}$ – total output for 2018 in billion U.S. dollars:

$$\mathbf{i}'\mathbf{x} = \underset{2018}{x}, \tag{9}$$

- equality of the current monetary budget $\mathbf{x}'\mathbf{i}$–total outlays for the statistical value $\underset{2018}{x'}$ – total outlays for 2018 in billion U.S. dollars:

$$\mathbf{x}'\mathbf{i} = \underset{2018}{x'}. \tag{10}$$

3.2 Realization by Equalize Input-Output Table Accounts

Let the data of the input parameters be given:

- statistical value of the final demand budget $\underset{2018}{f} = 179.34$ billion U.S. dollars in 2018;
- statistical value of the employment budget $\underset{2018}{e} = 179.34$ billion U.S. dollars in 2018;
- statistical value of the total output budget $\underset{2018}{x} = 260.91$ billion U.S. dollars in 2018;
- statistical value of the total outlays budget $\underset{2018}{x'} = 260.91$ billion U.S. dollars in 2018;
- values of simulation units at multipliers of technical coefficients $a^*_{ij} = 1_{ij} \times a_{ij}$, with final demand of structural coefficients $f^*_{cj} = 1_j \times f_{cj}$, at employment of structural coefficients $e^*_{cj} = 1_i \times e_{ci}$, $i, j = \overline{1, n}$.

Then the computer implementation of the equalize input-output mathematical model (Eqs. 4–10) makes it possible to obtain the results of the output data–evaluation and forecasting of the value, simulation and visualization of imbalances flows of transactions account and estimates of technical and structural coefficients for the Kazakhstan economy statistics:

- evaluation and equalization of intersectoral flows z_{ij}, $i, j = \overline{1, n}$, indeed, the increment in the value of the statistics of flows of agricultural sector, manufacturing, and service sector, respectively, will be (+) over evaluated and/or (−) under evaluated by:

$$\Delta \mathbf{Z} = \begin{bmatrix} +0.0117 & -0.1866 & -0.0238 \\ +0.0333 & +0.6920 & -1.9756 \\ +0.4963 & +5.9320 & -4.9793 \end{bmatrix},$$

  in billion US dollars (see Tables 2 and 4, respectively, matrix **Z**);

Table 4. Input-output table to flows of transactions account for the Kazakhstan statistics, U.S. billion dollars, 2018: compiled by the author based on the results of a computer experiment.

Sectors	1	2	3	f	x
1	1.240	1.712	0.509	9.433	12.893
2	0.578	20.962	12.082	65.480	99.102
3	1.642	10.948	31.898	104.427	148.916
$\boldsymbol{\varepsilon}'$	9.433	65.480	104.427		
$\mathbf{x}'$	12.893	99.102	148.916		

- evaluation and equalization of final demand flows f_j, $j = \overline{1, n}$, indeed, the increment in the value of the statistics of flows of agricultural sector, manufacturing, and service sector, respectively, will be (+) over evaluated and/or (−) under evaluated by:

$$\Delta \mathbf{f}' = \left[+0.1987 \; +1.2503 \; -1.4490 \right],$$

in billion US dollars (see Tables 2 and 4, respectively, column **f**);

- evaluation and equalization of value employment flows ε_i, $i = \overline{1, n}$, indeed, the increment in the value of the statistics of flows of agricultural sector, manufacturing, and service sector, respectively, will be (+) over evaluated and/or (−) under evaluated by:

$$\Delta \boldsymbol{\varepsilon}' = \left[-0.5413 \; -6.4375 \; +6.9787 \right],$$

in billion US dollars (see Tables 2 and 4, respectively, row $\boldsymbol{\varepsilon}'$);

- evaluation and equalization of value technical coefficients a_{ij}, $i, j = \overline{1, n}$, indeed, the increment in the value of the statistics of technical coefficients of intersectoral flows of agricultural sector, manufacturing, and service sector, respectively, will be (+) over evaluated and/or (−) under evaluated by:

$$\Delta \mathbf{A} = \begin{bmatrix} +0.0009 & -0.0019 & -0.0002 \\ +0.0026 & +0.0070 & -0.0133 \\ +0.0385 & +0.0599 & -0.0334 \end{bmatrix},$$

in physical volume indices (see Tables 3 and 5, respectively, matrix **A**);

- evaluation and equalization of value structural coefficients f_{cj}, $j = \overline{1, n}$, indeed, the increment in the value of the statistics of structural coefficients of final demand flows of agricultural sector, manufacturing, and service sector, respectively, will be (+) over evaluated and/or (−) under evaluated by:

$$\Delta \mathbf{f}_c' = \left[+0.0154 \; +0.0126 \; -0.0097 \right],$$

in physical volume indices (see Tables 3 and 5, respectively, column $\mathbf{f}_c$);

Table 5. Technical and structural coefficients for the Kazakhstan Statistics, 2018: compiled by the author based on the results of a computer experiment.

Sectors	1	2	3	$\mathbf{f}_c$	$\mathbf{x}_c$
1	0.0962	0.0173	0.0034	0.7316	0.8485
2	0.0448	0.2115	0.0811	0.6607	0.9982
3	0.1274	0.1105	0.2142	0.7013	1.1533
$\boldsymbol{\varepsilon}_c'$	0.7316	0.6607	0.7013		
$\mathbf{x}_c'$	1.0000	1.0000	1.0000		

- evaluation and equalization of value structural coefficients ε_{ci}, $i = \overline{1, n}$, indeed, the increment in the value of the statistics of structural coefficients of the value of employment flows of the agricultural sector, manufacturing, and the service sector, respectively, will be equal to

$$\Delta \boldsymbol{\varepsilon}_{\mathrm{c}}' = \left[0.0000\ 0.0000\ 0.0000\right],$$

equilibrium share of value added (see Tables 3 and 5, respectively, row $\boldsymbol{\varepsilon}_{\mathrm{c}}'$).

4 Discussions

The mathematical models, algorithms and information technologies developed in this paper can be the methodological basis for the mathematical support of decision-making using intelligent computing technology in the fields of human activity, social development, and business, and we present in following several different studies.

The following works [17–20] consider the effects of different economic sectors of countries on the national economy. The significant contribution of a certain regional project, sector to the country's economy is analyzed using the input-output table method. For example, the regional economic effects of the European Capital of Culture project were studied in [17]. Considering that Plzen in the Czech Republic will be the cultural capital of Europe in 2015, the paper determines what multiplier effects can be expected due to this event. Here, the main research work is arts, entertainment, and recreation, not the traditional sector, but also other sectors that can affect the growth of tourists, including "Wholesale and retail trade, vehicle repair", "Transportation and storage" and "Accommodation and catering" are translated. Thus, the regional input-output model was calculated, and their corresponding production, total added value, income, and employment multipliers were obtained. As a result, it is shown that these industries have the potential to influence the economy of the region.

In the world, real estate is considered one of the most important factors of economic activity. It includes manufacturing, retail, business services and more including, not only provides the necessary space for almost all types of economic activity, but also directly or indirectly contributes to economic activity and job creation. As such, the impact of the economy on a specific sector, namely commercial real estate, on different sectors of the South African economy is considered in [18]. The study examines the ability to consider the importance of commercial real estate to the economy using input-output analysis. The paper analyzes the direct, indirect, and induced effects of commercial real estate on economic activity, GDP, jobs created and taxes.

And the influence of the volume of production in Isfahan province of Iran and its other regions on each other is considered in the work [19]. The study was carried out using an input-output table. The table shows that there are 71 sectors for each region. Then, by expanding Leontiev's two-regional feedback matrix, the types of interregional effects are identified, including internal and external feedback and side effects. These effects during exogenous shock were calculated. Then, the sectors with the greatest influence in each region were identified and the intensity of cross-regional effects was analyzed using them. According to the authors, the side effects of sectors in a small region can be considered strong, so a small region becomes more dependent on a large region. However, for sectors with large comparative advantages and strong internal ties within a small region, the result may be the opposite. The impact of batik agricultural production on the Malaysian economy is considered in [20]. The input-output table approach was used in the study. Overall, the analysis shows that this industry, the batik industry, contributes significantly to the Malaysian economy through how the first-order effects lead to second- and higher-order effects for other sectors, as well as additional costs, especially the import of goods and labor.

5 Conclusion

Thus, in this article was attempts to solve the problem of (+) over evaluated and/or (−) under evaluated intersectoral flows in monetary terms, final demand budgets and value of employment, as well as technological and structural coefficients of the input-output table for statistics economics of Kazakhstan for 2018. Then using mathematical programming problems minimizing the objective function Euclidean distance between value of employment and final demand, total outlays, and total output, as well as using technical and structural coefficients of economic statistics was evaluation of equalization of flows of input-output table of flows and budgets. In particular, the value of the statistics of flows of agricultural sector, manufacturing, and service sector, respectively, was over evaluated by 0.1987, 1.2503 and under evaluated by 1.449 billion U.S. dollars in 2018. And the value employment flows, respectively, was under evaluated by 0.5513, 6.4375 and over evaluated by 6.9787 billion U.S. dollars.

Acknowledgments. This research was funded by the Science Committee of the Ministry of Science and Higher Education of the Republic of Kazakhstan (Grant No. AP09259435).

References

Leontief, W.: Quantitative Input and Output Relations in the Economic Systems of the United States. Rev. Econ. Stat. **18**(3), 105–125 (1936). https://doi.org/10.2307/1927837

Miller, R.E., Blair, P.D.: Input-Output Analysis: Foundations and Extensions, 2nd edn, pp. 1–750. Cambridge University Press (2009). https://doi.org/10.1017/CBO9780511626982

Musil, T.A.: Evaluating development and community benefits of shopping malls: A case study using input/output analysis. J. Financ. Manag. Prop. Constr. **16**(2), 111–125 (2011). https://doi.org/10.1108/13664381111153105

Tausl Prochazkova, P., Noskova, M.: An application of input-output analysis to social enterprises: a case of the Czech Republic. J. Entrepreneurship Emerg. Econ. **12**(4), 495–522 (2020). https://doi.org/10.1108/JEEE-08-2019-0114

Ali Bekhet, H., Yasmin, T.: Assessment of the global financial crisis effects on energy consumption and economic growth in Malaysia: an input-output analysis. Int. Econ. **140**, 49–70 (2014). https://doi.org/10.1016/j.inteco.2014.07.003

Sangalli, I.: The key role played by European partners in the German automotive value chain: a granular analysis based on the world input output database. Industria **41**(3), 439–478 (2020). https://doi.org/10.1430/98069

OECD homepage, https://stats.oecd.org. Accessed 17 May 2022

Sterman, J.D.: Business Dynamics: Systems Thinking and Modeling for a Complex World. McGraw-Hill Inc (2000)

Gnap, J., Konečný, V., Slávik, R., Beňová, D.: Possible impacts of regulating the weekly rest of road freight drivers on logistics in EU countries. Nase More **65**(4), 259–265 (2018). https://doi.org/10.17818/NM/2018/4SI.18

Low, S., Ullah, F., Shirowzhan, S., Sepasgozar, S.M.E., Lee, C.L.: Smart digital marketing capabilities for sustainable property development: a case of Malaysia. Sustainability **12**(13), 5402 (2020). https://doi.org/10.3390/su12135402

Balabanova, A., Petrova, S., Fomenko, V., Keschyan, N.: Labor potential of youth for the development of ecology in the digital economy. E3S Web Conf. **258**, 06048 (2021). https://doi.org/10.1051/e3sconf/202125806048

Balabanova, A., Keschyan, N., Borisova, T., Hachemizova, E.: Using digital platforms for environmental management. E3S Web Conf. **244**, 07006 (2021). https://doi.org/10.1051/e3sconf/202124407006

Stone, M.: Developments in B to B and B to C marketing and sales automation systems. J. Bus. Bus. Mark. **28**(2), 203–222 (2021). https://doi.org/10.1080/1051712X.2021.1920702

Kadyraliev, A., et al.: The impact of cargo turnover on the ecology of Kyrgyzstan. IOP Conf. Ser.: Earth Environ. Sci. **937**(2), 022086 (2021). https://doi.org/10.1088/1755-1315/937/2/022086

Yerzhanov, A., et al.: Building information security skills among young transport professionals. Transp. Res. Procedia **63**, 1481–1488 (2022). https://doi.org/10.1016/j.trpro.2022.06.159

Butsenko, E.: Development of information system for monitoring the rate of digital currency for investing. In: Lecture Notes in Networks and Systems, LNNS, vol. 502, pp. 176–185 (2022). https://doi.org/10.1007/978-3-031-09076-9_17

Nosková, M.: Regional economic effects of the European capital of culture project: the use of input-output analysis. Ekon. Manage. **19**(3), 57–74 (2016). https://doi.org/10.15240/tul/001/2016-3-005

Boshoff, D., Seymore, R.: Analysis of the South African input-output table to determine sector specific economic impacts: a study on real estate. S. Afr. J. Econ. Manage. Sci. **19**(4), 661–689 (2016). https://doi.org/10.17159/2222-3436/2016/v19n4a13

Abootalebi, M., Akbari, N., Khoshakhlagh, R.: Input-output analysis of interregional effects. J. Appl. Econ. Sci. **12**(1), 95–106 (2017)

Akhir, N.H.M., Ismail, N.W., Utit, C.: Malaysian batik industry contribution analysis using direct and indirect effects of input-output techniques. Int. J. Bus. Soc. **19**(1), 181–194 (2018)

Application of the Method of Expert Assessments in Determining the Level of Criticality of the Information Security Vulnerability of Smart Contracts

Anton Alekseevich Krivonogov(✉), Yuriy Nikolaevich Philippovich, and Sergey Aleksandrovich Kesel

Moscow Polytechnic University, Moscow 107023, Russian Federation
aakrivonogov97@yandex.ru

Abstract. The article focuses on the development of a methodology based on expert assessment to assess the level of vulnerability criticality for subsequent use in the analysis of smart contract code to make a decision on its placement in a blockchain system. The task includes a problem statement, basic approaches to threat and vulnerability analysis, a proposed methodology for determining the level of vulnerability criticality, its testing, and recommendations for future work. The issues of protecting smart contracts operating within the blockchain system from computer attacks and vulnerabilities are becoming increasingly important for both developers and participants in information interaction. One of the most important trends in this area is the development of methods and tools to detect, model, analyze and assess the initiation and execution of malicious attacks on the target component of the system. #COMESYSO1120.

Keywords: Computer attack · Method of collective expert assessment · Static and dynamic analysis · Code audit · Blockchain system · Criticality coefficient · Assessment methodology · Smart contract security

1 Introduction

Over the past few years, there has been an increasing trend in the number of incidents related to the impact of computer attacks on the main components of almost any blockchain system (BS)–smart contracts (SC). That is why security should be a priority criterion for a developer when developing SC code. Adhering to the rules of secure development, starting from the early design stage, it is possible to avoid the appearance of critical errors at all subsequent stages of SC life cycle [12]. To ensure security, first of all, it is necessary to take the most complex approach to internal code review to identify all potential information security (IS) vulnerabilities and attack vectors.

The security audit of SC code should be carried out without fail to identify IS vulnerabilities both at the code level and at BS level as a whole. However, at present, methods and models for identifying actual IS vulnerabilities both in SC and in BS as

R. Silhavy et al. (Eds.): CoMeSySo 2022, LNNS 596, pp. 512–530, 2023.
https://doi.org/10.1007/978-3-031-21435-6_45

a whole have not been sufficiently studied [7]. At the same time, the specificity of the occurrence of IS vulnerabilities in SC code requires the choice of adequate methods for analyzing and solving problems related to their protection. Currently, there are several methods aimed at finding and supporting decision-making when conducting a security audit of SC code (including static and dynamic methods of analysis) [6]. However, they, in turn, do not sufficiently take into account the specifics of the occurrence of IS vulnerabilities, and do not take into account such indicators as the possibility of exploiting a vulnerability, the degree of possible damage, the relevance, the labor intensity of its implementation, and other characteristics that are critical for assessment.

Based on this, in order to determine and implement the mathematical apparatus within the framework of the analysis system for conducting a security audit of the CS code, the first step is to investigate the area of occurrence and definition of IS vulnerabilities.

2 Basic Approaches to iS Threat and Vulnerability Analysis

The process of identification and assessment of IS vulnerabilities itself consists of analyzing IS vulnerability, including its identification and setting the value of IS vulnerability, as well as assessing the risk of IS vulnerability being realized. IS vulnerability identification is the process of finding, listing and describing IS vulnerability elements and includes the following steps [2]:

- Identification of IS vulnerability;
- Determination of the criticality of IS vulnerability;
- Determination of the consequences of implementing an IS vulnerability in SC code;
- Identification of existing control and management measures and tools.

At the same time, for an objective analysis of threats and vulnerabilities of IS SC it is necessary to resort to the use of assessment methods, among which the most popular are [4]:

1. Direct expert assessment, based on the knowledge of experts in the field of IS, who determine the list of threat parameters and their importance coefficients;
2. Statistical analysis based on the analysis of accumulated information about IS incidents (vulnerabilities) (e.g., frequency of threats, sources, causes, etc.);
3. Factor analysis, associated with the identification of factors leading to the implementation of threats with a certain probability (e.g., the availability of the means necessary for attackers to implement threats, the growth of malicious activity, vulnerabilities of information system, etc.).

The advantages of analyzing IS vulnerabilities by statistical methods include the possibility of adapting mathematical and statistical tools to the object, as well as universality, due to the lack of need for knowledge about possible attacks and the vulnerabilities used by them.

The method of expert assessments is advisable to apply in the absence of statistical data, while experts give answers about the current and future state of information objects characterized by certain parameters and properties. The forecasting methods based on expert assessments include the following [2, 4]:

1. The method of collective expert assessment, which involves summarizing the results of the work of a group of experts in the field of IS. In this case, in order to achieve an objective result and versatile analysis of the issue under study, individual, independent assessments of experts are processed;
2. The Delphi method, based on the "brainstorming" of a group of specialists. This method consists in organizing the collection and mathematical processing of expert assessments. This method is used when it is necessary to make a quick decision;
3. Forecasting by expert groups, which involves discussions on the issue under consideration, aimed at finding the only correct solution. This method is distinguished by the ability of experts not only to present their own ideas, but also to criticize others. The advantages of this method are ease of implementation and less chance of errors.

Due to the fact that the assessment of the criticality of IS vulnerability is carried out on the basis of a comprehensive study of the functioning of BS and an assessment of the influence of all factors on the security of both SC and digital assets (DA), then the most appropriate is the use of the method of collective expert assessments, which are the most effective in terms of maximum objectivity of the expert assessment [2]. At the same time, the objectivity of the assessment is determined by the consistency of expert opinion. The degree of agreement is assessed by the dispersion coefficient of concordance. To calculate the coefficient of concordance, expert assessments of the importance of parameters are ranked and the sum of the ranks for each IS vulnerability is calculated. The concordance coefficient (W) is calculated by the following formula [9]:

$$W = \frac{12 * S}{m^2\left(n^3 - n\right)}, \tag{1}$$

where

$$S = \sum_{i=1}^{n}\left(r_i - \frac{1}{n} * \sum_{i=1}^{n} r_i\right)^2; \tag{2}$$

m–number of experts; n–number of parameters; r_i–sum of ranks on each row of the matrix of the results of the ranking of n objects by the group of m experts.

The value W = 1 indicates the complete agreement of the opinions of experts, the value W = 0 indicates the complete inconsistency of the opinions of experts. A concordance coefficient of less than 0.5 indicates a lack of agreement among the opinions of the expert group so that a reliable forecast can be made based on the results of their survey.

However, the most important step in any peer review is the selection of experts. In [1], the following criteria for the selection of experts for conducting a complex analysis and audit of the security of the personal data processing information system (PDIS) are defined. Since BS is inherently a distributed PDIS, the criteria for selecting experts can be applied to it as well [3]. Thus, six main criteria can be distinguished:

1. Degree of the expert's competence (the availability of a scientific degree, academic title, work experience in the specialty, official position, number of published works);
2. Ability to navigate in the field of BS and IS;
3. Combination of narrow specialization and general outlook of the expert;
4. Ability to analyze and synthesize the studied problems;
5. Ability to process and assimilate qualitatively new information;
6. Combination of psychologically acceptable to each other in a group of experts of different ages, different scientific schools, etc.

3 Methodology for Determining the Criticality Level of iS Vulnerability

The assessment of the criticality of IS vulnerabilities in such a software component as a SC operating within BS is a rather non-trivial task and differs from the assessment of the criticality of IS vulnerabilities in a centralized system [7]. That is why it is necessary to most fully approach the definition of those specific characteristics that will be applicable to the current field of study, paying attention to the functional dependencies between them.

Based on this, the list of criteria for assessing criticality as a result of the implementation of IS vulnerability in relation to SC and DA stored inside it may look like this:

- Degree of initial security of BS-G_{INIT};
- Relevance of IS vulnerability implementation–R_{TH};
- Degree of possible damage from the implementation of an IS vulnerability–E_{DMG};
- Labor intensity of IS vulnerability implementation–D_{EXP}.

To more accurately determine the level of criticality of IS vulnerability, it is necessary to consider each of the characteristics separately, taking into consideration their specifics and the requirements imposed on them.

3.1 Degree of Initial Security of BS

One of the characteristics requiring detailed research is the degree of initial security of BS, within which the target SC performs its work. The initial security is understood as a generalized characteristic of the system, which depends on its technical features and operational characteristics [3, 10]. In this case, the level of initial security is determined by expert method on the basis of the characteristics given in Table 1.

The initial security level of BS is determined as follows:

- BS has a high level of initial security, if at least 70% of the characteristics correspond to the «high» level, and the rest to the average level of security, and at the same time, none of the characteristics should correspond to the «low» level;
- BS has a medium level of initial security, if at least 70% of characteristics correspond to the «medium» level or higher, and the rest should correspond to the «low» level of security;

Table 1. Operational and technical characteristics of BS.

Technical features and operating characteristics of the system	Level of security		
	Low	Medium	High
1. According to the type of BS used	Inclusive	Federative (consortium)	Exclusive
2. According to the consensus algorithm used	Proof-of-work, Proof-of-stake	Proof-of-authority, Raft etc	Proof-of-elapsed-time, Byzantine fault tolerant
3. According to BS development language used	Solidity, JavaScript, Python	Java, Rust	C++, Go, Simplicity
4. According to the number of nodes in BS	From 1 to 10	From 11 to 100	From 101 and more
5. According to the privacy mechanisms used	Not applicable within BS	Ability to use homomorphic encryption and/or code obfuscation	Ability to use interactive/non-interactive zero-knowledge proof (ZKP/NIZKP)
6. According to the programming language used by the SK	Solidity, Java	SolidityX, Vyper, Rust	Low-level programming languages (bytecode) (e.g. Michelson, etc.)
7. According to the use of the oracle mechanism in BS	Using untrusted external data sources	Using trusted and untrusted external data sources	Using a limited pool of trusted external data sources
8. The number of transactions processed per second (TPS)	Under 100 TPS	From 101 to 1000 TPS	From 1001 and more TPS
9. Using a certified cryptographic IS tool and GOST cryptography in BS (for example, CryptoPro CSP)	Not implemented	–	Implemented
10. Ability to dynamically update SC	Not implemented (SK are part of the platform)	Implemented by addressing a limited number of parties (centralized method)	Implemented

- BS has a low initial security level, if the corresponding conditions of items 1 and 2 are not met.

According to the results of the assessment, the initial degree of BS security can take one of three possible values (low, medium or high), which correspond to numerical indicators (from 1 to 3). The conditions for determining the initial level of security for the target BS are additionally given in Table 2.

Table 2. Conditions for determining the initial security level.

General level of initial security	Level of security according to characteristics		
	Low (%)	Medium (%)	High (%)
High level of initial security	0	$\sum \leq 30$	$\sum \geq 70$
Medium level of initial security	$\sum \leq 30$	$\sum \geq 70$	$\sum < 70$
Low level of initial security	$\sum < 70$	$\sum < 70$	$\sum > 0$

3.2 Relevance of iS Vulnerability Implementation

The relevance of the implementation of IS vulnerability is the most significant parameter in determining the criticality level of IS vulnerability, since depending on which of the numerical values it will take (0 or 1), the expediency of calculating the overall IS vulnerability criticality factor with the subsequent determination of its level depends on the specific BS implementation under consideration. To determine the relevance of the implementation of IS vulnerability in relation to SC operating within a particular BS, it is required to use the method of collective expert assessments, paying attention to such composite characteristics as possibility (frequency) of IS vulnerability implementation for similar information systems and risk of implementing an IS vulnerability.

The possibility of implementation of IS vulnerability is understood as an indicator, determined by expert means, which characterizes how possible it is to implement a specific IS vulnerability in relation to a specific BS implementation [3]. Three verbal gradations for this characteristic are introduced:

- low possibility of implementation–there are no objective prerequisites for the implementation of the target IS vulnerability (for example, there are no statistics on the implementation of a specific IS vulnerability in the target BS or the system is based on an exclusive BS with a limited number of nodes in the controlled area);
- medium possibility of implementation–there are objective prerequisites for the implementation of IS vulnerability, but the used parameters of BS significantly complicate its implementation (for example, there are statistics of implementation of this vulnerability in a similar BS, while within the target system appropriate protection measures are implemented);

- high possibility of implementation-there are objective prerequisites for the implementation of IS vulnerability and outdated BS settings or default settings are used.

The risk of IS vulnerability implementation is understood as an indicator, defined in an expert way, that characterizes how critical for the target BS is the implementation of a given IS vulnerability [1, 3]. This indicator also has three values:

- low risk–if the implementation of the threat can lead to insignificant negative consequences for the target BS, within which SC functions (for example, the termination of the correct functioning of the attacked SC without the possibility of stealing the DA);
- medium risk–if the implementation of the threat can lead to negative consequences for the target BS, within which SC functions (for example, the termination of the attacked SC and a number of other (secondary) SC with the possibility of theft of DA);
- high risk–if the implementation of the threat can lead to significant negative consequences for the target BS, within which SC functions (for example, the termination of the functioning of both the attacked SC and BS as a whole).

Separately, it should be noted that each characteristic gradation for determining the relevance of the implementation of IS vulnerability is placed in accordance with the numerical coefficient, namely:

- 1–for low probability of implementation/low risk;
- 2–for medium probability of implementation/medium risk;
- 3–for high probability of implementation/high risk.

3.3 Degree of Possible Damage from the Implementation of an iS Vulnerability

The degree of possible damage from the implementation of a specific IS vulnerability on SC is determined by the degree of negative consequences from the violation of the properties of privacy, integrity or availability in relation to SC, DA and BS infrastructure [3, 7]. The definition of the violation of each of the properties of IS is given in Table 3.

The indicator of the degree of possible damage is determined on the basis of expert approach using statistical methods of analysis for each of the assessed components (SC, DA, infrastructure BS) separately in accordance with Table 4, and then the general indicator of the degree of possible damage from implementation of IS vulnerability is calculated by the following formula:

$$E_{DMG} = \sum_{j=1}^{n} \max_{i}\left(E_j^i\right), \text{ where } i = P, I, A; j = SC, BS, DA. \quad (3)$$

3.4 Labor Intensity of iS Vulnerability Implementation

The final characteristic, which is also used in determining the level of criticality of IS vulnerability, is the labor intensity of IS vulnerability implementation. This indicator

Table 3. Determination of violation of IS properties.

IS property	Negative consequences from violation of IS properties
Privacy, E_{j1}^{P}	Violation of a property due to the ability of an external attacker's SC to access the attacked SC by manipulating a variable like tx.origin (or another variable for user authorization), as well as due to other software effects, through which you can get access to DA circulating in the target SC
Integrity, E_{j1}^{I}	Violation of a property by manipulating the logic of SC, which consists in its transition to an incorrect (alternative to the desired) state by influencing one or more target functions and/or variables
Availability, E_{j1}^{A}	Violation of the property due to multiple calls of functions (including recursive calling) of the attacked SC, as well as incorrect exception handling, which in turn may lead to additional load on the nodes of BS, and in some cases, partial or complete failure of the entire system

determines how many resources an attacker must spend to successfully exploit a IS vulnerability [7]. The labor intensity of IS vulnerability implementation depends primarily on the attacker's potential (P_{SPEC}), i.e., his specific knowledge and skills, as well as the attacker's software and hardware capabilities (SH_{CPB}) and the quantity of time spent on its implementation (T_{EXP}). Possible characteristics and their corresponding numerical values are defined in Table 5.

Based on the results of determining each of the characteristics, the indicator of labor intensity of IS vulnerability implementation is calculated by the following formula:

$$D_{EXP} = (SH_{SPB} + P_{SPEC}) * T_{EXP} \tag{4}$$

3.5 Calculation of iS Vulnerability Criticality Level

As a result of determining the listed indicators for the examined types of IS vulnerabilities, we can calculate a criticality score for each type of IS vulnerability using the following formula:

$$C_{SC} = \frac{(G_{INIT} + E_{DMG})}{D_{EXP}} * R_{TH} \tag{5}$$

At the same time, the indicator characterizing IS vulnerability criticality coefficient and the value corresponding to a certain level of IS vulnerability criticality correlate with each other in accordance with the previously proposed methodology in [7], as well as with the indicators given in GOST R 56545–2015, namely (Table 6).

This assessment shows how critical this or that IS vulnerability is both for the owner of SC and for BS as a whole. The obtained result can be used for further assessment of technological security of SC to make a decision on its placement in BS.

Table 4. Determining the degree of possible damage.

Qualitative indicator of the degree of damage	Quantitative indicator of the degree of damage	Negative consequences from violation of IS properties
High degree of damage	3	As a result of a violation of the three safety properties, significant negative consequences are possible for one component (e.g., loss of control over SC) The components of BS cannot perform all the functions assigned to them
Medium degree of damage	2	As a result of a violation of the two security properties, moderate negative consequences are possible for one component (e.g., loss of funds, temporary unavailability of SC) The components of BS cannot perform at least one of the functions assigned to them
Low degree of damage	1	As a result of violation of not more than one security property by one component, which at the same time is not critical for DA and BS as a whole, minor negative consequences are possible (for example, temporary unavailability of SC) BS components can perform all the functions assigned to them with a small delay and lag

4 Validation of the Proposed Methodology for Determining the Level of Criticality of iS Vulnerability

To prove the effectiveness of the proposed methodology, first of all, it is necessary to determine and calculate the criticality level for the most known and frequently implemented IS vulnerability–reentrancy (or recursive calling).

4.1 Determination of the Relevance of Reentrancy Implementation

This characteristic is determined by using the method of collective expert assessments, followed by the calculation of the concordance coefficient to determine the consistency

Table 5. Characteristics of determining the labor intensity of IS vulnerability implementation.

Indicator of an intruder's capabilities		Value in the implementation of IS vulnerability
Attacker's software and hardware capabilities, SH_{CPB}	Lack of auxiliary tools	0
	Use of standard software and hardware tools from publicly available sources (open-source)	1
	Use of specialized software and hardware tools	2
	Use of software and hardware tools, custom made for a particular BS or to implement a particular IS vulnerability	3
Attacker's potential, P_{SPEC}	Basic user (basic IS skills and knowledge, superficial knowledge of programming languages and disassembling features)	1
	Specialist (knowledge of general principles of BS construction, possession of sufficient practical and theoretical skills in the field of IS, knowledge of basic programming approaches and methods)	2
	Professional (knowledge of the principles of construction and operation of the attacked BS, knowledge of several programming languages, including SC, possession of specialized skills and knowledge in the field of IS)	3
Time taken to implement an IS vulnerability, T_{EXP}	Less than a minute	0,5
	From a few minutes to a few hours	1
	From a few hours to a few days	2

of expert opinions. The experiment was conducted by five different experts (or auditors) to achieve greater objectivity. A detailed calculation of the concordance coefficient

Table 6. Correlation of quantitative and qualitative assessments of criticality.

Qualitative assessment (level of criticality of IS vulnerability)	Quantitative assessment
Low	0.0–1.9
Medium	2.0–5.9
High	6.0–9.9
Critical	10

to determine the consistency of auditors in determining the relevance of reentrancy vulnerability is carried out according to formulas (1) and (2) is presented in Table 7.

Table 7. Determination of auditor consistency.

Parameters	Characteristic 1	Characteristic 2
Auditor 1	2	2
Auditor 2	3	2
Auditor 3	3	1
Auditor 4	2	2
Auditor 5	3	2
Sum	13	9
The square of the sum	169	81

The table shows which of reentrancy vulnerability characteristics, according to the auditors, are the most and least significant. The most critical characteristic is the characteristic for which the sum of ranks r_i is maximum, the least critical is the one for which the sum of ranks r_i is minimum. According to the formula for calculating the concordance coefficient, the coefficient $W = 0.64$ was determined, which indicates a sufficient consistency of opinions of a group of auditors so that, based on the results of a survey of auditors, it was possible to build a reliable forecast of the relevance of the implementation of reentrancy. Thus, the relevance indicator takes a value of 1. Since IS vulnerability under consideration is relevant in this case, it makes sense to proceed to the calculation of other characteristics.

4.2 Determination of the Degree of Initial Security of BS

To determine this indicator, it is necessary to refer to the list of operational and technical characteristics presented in Table 1, and apply them to the Ethereum, within which SC works. Characteristics for Ethereum are given in Table 8.

Based on the data in Table 2, we can confidently state that Ethereum has a low level of initial security and this indicator takes a value of 1.

Table 8. Operational and technical characteristics of Ethereum.

Technical features and operating characteristics of the system	Level of security
1. According to the type of BS used	Inclusive
2. According to the consensus algorithm used	Proof-of-work (with possible transition to Proof-of-stake)
3. According to BS development language used	C + +, Go
4. According to the number of nodes in BS	From 101 and more
5. According to the privacy mechanisms used	Not applicable within BS
6. According to the programming language used by the SK	Solidity
7. According to the use of the oracle mechanism in BS	Using untrusted external data sources
8. The number of transactions processed per second (TPS)	Under 100 TPS
9. Using a certified cryptographic IS tool and GOST cryptography in BS (for example, CryptoPro CSP)	Not implemented
10. Ability to dynamically update SC	Implemented

4.3 Determination of the Degree of Possible Damage from the Implementation of a Reentrancy

To determine this characteristic, it is necessary to assess the severity of consequences (possible damage) by determining the violations of the properties of privacy, integrity and accessibility to SC, DA and infrastructure of Ethereum according to Table 3:

- SC: Implementation of reentrancy leads to a violation of all three properties of information protection (loss of control over SC, change in its state and subsequent inaccessibility) [10]. In accordance with Table 4, for this component, the damage degree indicator takes a value of 3 (high degree of damage);
- DA: Implementation of reentrancy also leads to violation of all three properties of information protection (loss (or theft) of DA, blocking, purposeful change of balance state). In accordance with Table 4, for this component, the damage degree indicator takes the value of 3 (high degree of damage);
- Infrastructure of Ethereum: Implementation of reentrancy still leads to the violation of all three properties of information protection (to destruction or change of filling of transactions in individual control blocks, DoS) [10, 13]. In accordance with Table 4, for this component, the damage degree indicator takes the value of 3 (high degree of damage).

Therefore, according to formula (3), the degree of possible damage from the implementation of reentrancy takes a value of 9.

4.4 Determination of the Labor Intensity of Reentrancy Implementation

To calculate the general indicator of labor intensity of reentrancy implementation it is necessary to use expert assessments in terms of determining the values of intruder's capability indicators according to Table 5.

To implement this vulnerability, an attacker must first use specialized software and hardware tools to break into the Ethereum and search for a potentially vulnerable SC to exploit it (SH_{CPB} indicator takes a value of 2). For this he must have at least basic skills and knowledge in the field of IS and know the basics of object-oriented programming (to define a recursive function) (P_{SPEC} indicator takes a value of 1). The time it takes to exploit a recursive function in SC by calling reentrancy, according to statistics, can range from a few milliseconds to seconds (T_{EXP} indicator takes a value of 0.5) [5]. Therefore, the general indicator of labor intensity of reentrancy implementation is calculated by formula (4) and takes the value of 1.5.

4.5 Results and Analysis

According to the results of determining all the auxiliary characteristics, calculate the criticality indicator of reentrancy vulnerability according to the formula (5). The resulting criticality indicator this vulnerability quantitatively takes the value of 6.7. Referring to Table 6, we can conclude that reentrancy (recursive calling) has a high level of criticality when exploited by SC functioning in Ethereum.

By analogy with the determination of the criticality indicator of reentrancy vulnerability, calculations were also made for other most frequently implemented IS vulnerabilities in order to determine their levels of criticality. Figure 1 shows a graph with criticality coefficients of IS vulnerabilities to determine which of the analyzed vulnerabilities is the most or the least critical in the case of exploitation in SC functioning within the Ethereum.

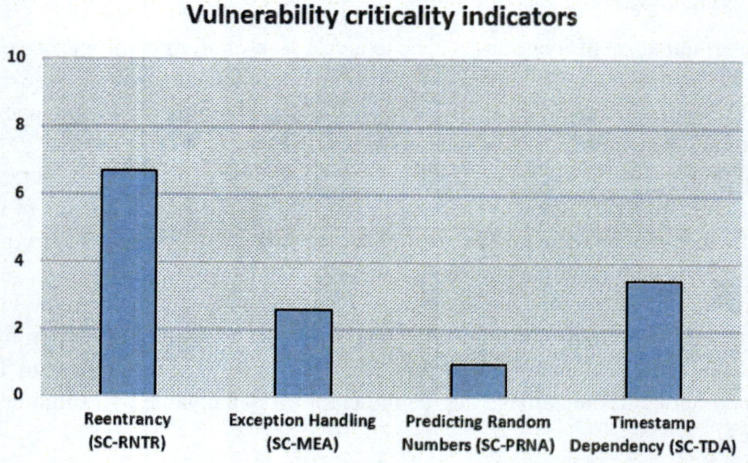

Fig. 1. Ethereum vulnerability criticality indicators.

The results of expert analysis and determination of the level and criticality factor for each of the considered IS vulnerabilities were used to form a list of IS vulnerabilities for Ethereum-based BS (see Table 9) [13]. When such a table is formed by an expert, it should be unique for each researched BS, and, at the same time, it should be periodically updated due to both changes in dynamic parameters of BS and/or SC, and emergence of new vectors of attacks and/or IS vulnerabilities.

Based on Table 9, we can conclude that reentrancy vulnerability is the most critical because of its low implementation complexity, medium probability of occurrence, and high degree of potential damage. The timestamp dependence vulnerability has a medium level of criticality because it is difficult to implement, unlikely to occur, and requires significant time and effort on the part of the attacker to exploit it successfully. At the same time, the exception handling vulnerability and predicting random numbers are the least critical, because they require high skill in the attackers and cause little damage to SC and the system within which they perform their work.

To validate the results and identified dependencies, a software prototype was developed, which includes modules of static and dynamic code analysis of SC with the possibility of forming recommendations to protect against the implementation of code threats when calling IS vulnerabilities. In order to simulate the impact of an attacker on the source code of SC, it is necessary to set such parameters as session timeout and maximum recursion depth (the value required for the number of the researched program code states) with the ability to further modify these parameters to determine the most complete and reliable analysis result. The interface of the software prototype of SC vulnerability analysis methodology is shown in Fig. 2.

When running this solution, it is necessary to load data with the sol extension containing SC code. After setting the initial parameters, the software prototype runs a static analysis block to detect malicious constructs in the code according to pre-defined test cases, and then a dynamic analysis block to detect false positive states of SC and BS. Based on the results of a complex analysis, the software prototype automatically generates a security report for SC code in pdf format for further analysis by an expert before being placed on BS node. An example of a security report with a detected SC-RNTR vulnerability is shown in Fig. 3.

5 Conclusion/Future Work

SC presents a new approach to concluding a contractual relationship between the parties to the information exchange. This technology makes it possible to replace traditional approaches in financial (or banking) operations, insurance, logistics, and medicine. That is why the main focus of this article is on identifying the most critical IS vulnerabilities in order to unambiguously establish the security level of this key component.

The developed methodology of IS vulnerability assessment will allow developers to most fully approach the research of possible vulnerabilities for a particular implementation of BS, within which SC perform their work. By classifying IS vulnerabilities and determining their criticality values, both the goal and the steps to protect the information circulating in SC can be properly formulated. The goal, in turn, will be to mitigate the threats created as a result of identifying programming errors and IS vulnerabilities.

Table 9. Operational and technical characteristics of Ethereum.

No.	Vulnerability	Description	Consequences of exploiting a vulnerability[1]	Vulnerability criticality coefficient	Vulnerability criticality level	IS requirements
1	Reentrancy (SC-RNTR)	The attacker creates a malicious SC, from which the original SC is called back before the first function call is completed. Exploitation of this vulnerability can lead both to the loss of DA, and to the failure of the entire BS, which would entail the branching of block chains into two independent streams (fork) [5, 13]	P, I, A	6.7	High	• Using send() or transfer() functions that do not allow recursive calls to withdraw DA; • Call selfdestruct() to stop functioning of the target SC in BS to investigate the incident further
2	Exception handling (SC-MEA)	Vulnerability can occur when calling an external SC method from the target SC, in which an exception occurs (e.g., due to exceeding the call stack limit). In this case, the exception in the external SC is not handled properly and is not passed to the calling SC [8, 11]	P, A	2.6	Medium	• When using low-level calling methods, make sure that the possibility of failure of this call is also handled; • Check the available gas limit in SC when performing external operations; • Using transfer() instead of send() to correctly handle exceptions when transferring DA

(*continued*)

[1] The consequences of exploitation of vulnerabilities are assessed in terms of violation of privacy (P), integrity (I), and accessibility (A) of DA, SC, and BS.

Table 9. (*continued*)

No.	Vulnerability	Description	Consequences of exploiting a vulnerability	Vulnerability criticality coefficient	Vulnerability criticality level	IS requirements
3	Predicting random numbers (SC-PRNA)	SC often use random numbers to determine the outcome of an execution. SC developers may use block headers or other unreliable source data to generate «random» values. Such a value can be predicted by an attacker to determine the outcome in their favor [8, 14]	P, I	1	Low	• Using a cryptographically pseudo-random number generator to determine the result of SC execution; • Using the block number or an external data source as a timestamp through the oracle mechanism
4	Timestamp dependency (SC-TDA)	SC can use a timestamp as an initial condition to perform some critical operations, such as transferring DA between wallets. The attacker, in turn, can predict a certain timestamp, which triggers a condition in their favor [7, 11]	P, I	3.5	Medium	• Using a cryptographically pseudo-random number generator to generate a block timestamp; • Using the block number or an external data source as a timestamp through the oracle mechanism

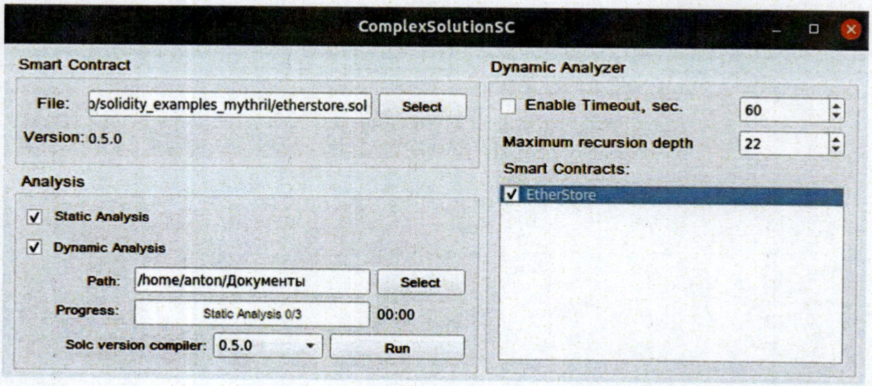

Fig. 2. Software prototype interface.

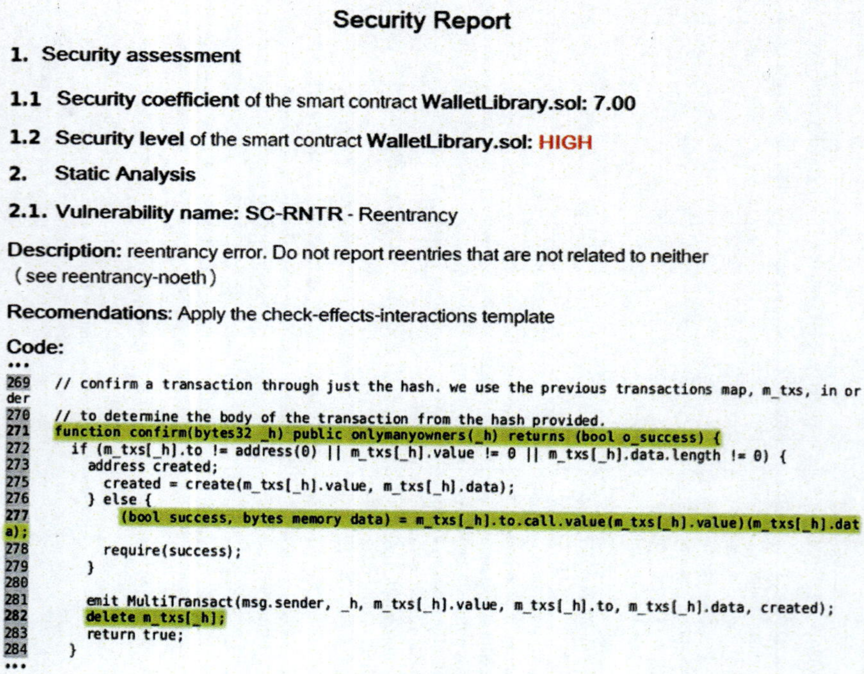

Security Report

1. **Security assessment**

1.1 **Security coefficient** of the smart contract **WalletLibrary.sol: 7.00**

1.2 **Security level** of the smart contract **WalletLibrary.sol: HIGH**

2. **Static Analysis**

2.1. **Vulnerability name: SC-RNTR** - Reentrancy

Description: reentrancy error. Do not report reentries that are not related to neither (see reentrancy-noeth)

Recomendations: Apply the check-effects-interactions template

Code:

```
...
// confirm a transaction through just the hash. we use the previous transactions map, m_txs, in or
der
// to determine the body of the transaction from the hash provided.
function confirm(bytes32 _h) public onlymanyowners(_h) returns (bool o_success) {
  if (m_txs[_h].to != address(0) || m_txs[_h].value != 0 || m_txs[_h].data.length != 0) {
    address created;
      created = create(m_txs[_h].value, m_txs[_h].data);
    } else {
        (bool success, bytes memory data) = m_txs[_h].to.call.value(m_txs[_h].value)(m_txs[_h].dat
a);
      require(success);
    }

    emit MultiTransact(msg.sender, _h, m_txs[_h].value, m_txs[_h].to, m_txs[_h].data, created);
    delete m_txs[_h];
    return true;
  }
...
```

Fig. 3. Example of a SC security report.

In our future work, we aim to refine the proposed IS vulnerability assessment methodology and software prototype, thereby developing a complex solution that allows to determine with sufficient accuracy the security level of SC code with the ability to generate a security report, which will help the developer in improving the target component and in eliminating flaws and potential vulnerabilities, for its placement in BS.

In the future, the following major tasks are expected to be initiated in order to achieve the goal, namely:

- Definition of additional dependencies for the most accurate determination of the level of security of SC code;
- Researching automated analysis tools to identify the most promising and comprehensive options in order to include them in a complex solution;
- Improving SC code security report to provide more details on identified weaknesses and to simplify the developer's understanding of the information presented;
- Optimization (in terms of time and memory) of the developed complex solution for its effective use in the analysis of SC code;
- Testing (approbation) of the complex solution on a set of active SC in BS to identify flaws and IS vulnerabilities in the code.

One of the main goals of SC is to fundamentally improve the conduct of business and transactions. Through their use, all transactions become transparent, traceable and irreversible in BS. Thus, the research conducted in the field of SC safety will bring this technology to a new level of its widespread use in various spheres and sectors of human activity.

References

1. Methodology for determining actual threats to the security of personal data during their processing in information systems of personal data. https://fstec.ru/. Accessed 20 July 2022
2. Kornienko, A.A., Nikitin, A.B., Diasamidze, S.V., Kuzmenkova, E.: Modeling computer attacks on a distributed information system. Izvestiya Peterburgskogo universiteta putej soobshcheniya **4**, 613–628 (2018)
3. Methodical document: Methodology for assessing information security threat. https://fstec.ru/. Accessed 20 July 2022
4. Zverev, E., Nikiforov, A.: Expert sampling: formation for a large population. MSFO i MSA v kreditnoj organizacii No. **2**(72), 68–74 (2019)
5. Atzei, N., Bartoletti, M., Cimoli, T.: A survey of attacks on ethereum smart contracts. In: Principles of Security and Trust, pp. 164–186. Springer, Berlin, Heidelberg (2017)
6. Borovik, V.S., Zenin, M.M., Gatchin, Y., Yuganson, A.N.: On the issue of security of smart contracts. Vestnik Chuvashskogo universiteta **1**, 79–87 (2018)
7. Repin, M.M., Krivonogov, A.A.: Problems of information security of smart contracts in systems based on distributed ledger technology. T.: OOO Izdatel'stvo TRIUMF, Moscow (2020)
8. Zhou, H., Milani Fard, A., Makanju, A.: The state of ethereum smart contracts security: vulnerabilities, countermeasures, and tool support. J. Cybersecur. Priv. 358–378. Department of Computer Science, New York Institute of Technology, Vancouver (2022)
9. Concordance coefficient. https://math.semestr.ru/. Accessed 24 July 2022
10. Andrianov, A.S., Vecherkin, V.B., Prokhorov, M.A., Tsvetkov, A.: Development of an approach to automating the process of primary processing of initial data for the analysis of the stability of automated systems for special purposes under destructive influences. Izvestiya TulGU. Tekhnicheskie nauki **10**, 463–472 (2018)
11. Mense, A., Flatscher, M.: Security vulnerabilities in ethereum smart contracts. In: Proceedings of the 20th International Conference on Information Integration and Web-based Applications and Services, pp. 375–380. ACM, Association for Computing Machinery, New York (2018)
12. Gupta, B.C.: Analysis of ethereum smart contracts—a security perspective. Indian Institute of Technology Kanpur, pp. 1–59 (2019)

13. Known attacks–ethereum smart contract best practices. https://consensys.github.io/smart-contract-best-practices/known_attacks/. Accessed 28 July 2022
14. Chen, J., Xia, X., Lo, D., Grundy, J., Luo, X., Chen, T.: Defining smart contract defects on ethereum. IEEE Trans. Software Eng. **48**, 327–345 (2020)

Stabilization of Second-Order Systems Using a P/I-Delayed Controller

Martin Strmiska[1(✉)], Libor Pekař[1], and José Mário Araújo[2]

[1] Tomas Bata University, Zlín, Czech Republic
strmiska@utb.cz

[2] Federal Institute of Bahia, Salvador, Brazil

Abstract. The present paper describes the stabilization and control of second-order systems using a proportional plus integral delayed controller (P/I-delayed controller). The stabilization of closed-loop control systems using a controller comprehends several techniques to increase the control circuit's stability. Even in some simple second-order plants, the classical PI controller cannot be applied to pursue the closed-loop stability, although the design requisites, for example, constant reference tracking capability. In this paper, the control will be analyzed using a P/I-delayed controller. The main goal of this article is to obtain the global stability region (the region of global stability) in the two-dimensional (k_p, k_i) plane for the delay-based P/I-delayed controller and to verify the control using the designed controller. The stabilization of linear-time invariant (LTI) systems with time delay in an integral part of the P/I-delayed controller will be studied in detail. Two typical examples, namely the control of an unstable plant and the control of a stable plant, are used as benchmarks to verify the proposal's effectiveness. The simulation results show that the presented stabilization method is effective and practically applicable in both analysis and control.

Keywords: Delay · PI controller · Stabilization · Second-order systems

1 Introduction

Stabilization is a method for achieving the controllability of a controlled system. A system is controllable if the output signal stabilizes at the desired value in a finite time [1–3, 10]. If the controlled and control system does not contain a delay, stability can be determined using several methods, such as the Hurwitz criterion, Routh's criterion, and others [1–3]. In most cases, purely analytical methods cannot determine stability if the regulated circuit contains a delay [7–9, 11]. In this case, we can use the methods of numerical mathematics.

R. Silhavy et al. (Eds.): CoMeSySo 2022, LNNS 596, pp. 531–539, 2023.
https://doi.org/10.1007/978-3-031-21435-6_46

We should deal with the stabilization of systems with delays since real situations contain delays, for example, transportation of materials [12], the transmission of signals [13], control of epidemics [14], traffic systems [15], network control systems [16], control of the RNA messenger in human cells [17], and others [18, 19]. This research aims to determine a stability region in the (k_p, k_i) plane for the P/I-delayed controller, to set the controller using the obtained parameters, and to demonstrate the simulation results. The focus in second-order plants is of particular interest in vibration control applications, a research topic of increasing interest [19].

The paper is organized as follows. The first part of the article is a theoretical background that describes the principle of the design method and the determination of the stabilization pair for the controller setting. The second intends to verify the theoretical findings by two numerical examples: the first one describes a stable system plant, whereas the second one represents an unstable system plant. For each example, the mathematical formulas are first derived, then the stability region is shown and described, and regulation with the received controller parameters is also demonstrated. The last part of this article is the Conclusion chapter, which briefly discusses all the results and possible solutions.

2 Theoretical Background

2.1 Principle of the Design Method

Consider the closed-loop control [1, 2, 7] in Fig. 1 with PI delay-based controller $C(s)$ and stable controlled system $G(s)$.

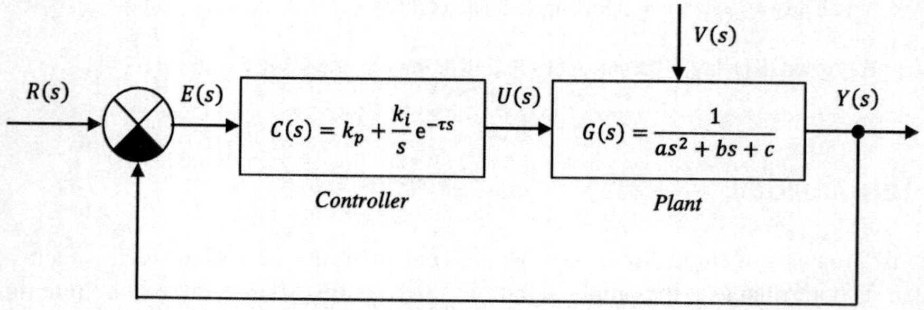

Fig. 1. Closed-loop control

The closed-loop control automatically regulates a system plant to maintain the desired magnitude. In the figure of the closed-loop control, $R(s)$ denotes the input reference signal, $E(s)$ the control error, $C(s)$ the controller, $U(s)$ the control action, $G(s)$ the plant, $V(s)$ the noise, and $Y(s)$ output signal.

Consider a controller given by transfer function

$$C(s) = k_p + \frac{k_i}{s}e^{-\tau s}, \tag{1}$$

where k_p represents a proportional gain, k_i an integral gain, $\tau \in \mathbb{R}_+$ is a delay, and s is a complex variable. Controllers that include delay in their structures are an interesting way to increase the degrees of freedom of classical controllers preserving the core and simplicity of that. Exempli gratia, delayed position feedback controllers are proposed in [19] for second-order dynamics of flexible systems.

Next, consider a controlled plant

$$G(s) = \frac{1}{as^2 + bs + c}, \tag{2}$$

where a, b, c are real coefficients, and s is a complex variable.

2.2 Determination of the Stabilization Pair (k_p, k_i)

Now we will determine a stabilizing pair (k_p, k_i) for which it will be possible to search for the region of stability of the closed-loop.

Let $Y(s)$ denote the output signal [1, 2] of the closed-loop control in Fig. 1 defined as

$$Y(s) = \frac{G(s)C(s)}{1 + G(s)C(s)} R(s). \tag{3}$$

The numerator of the $1 + G(s)C(s)$ is a characteristic quasi-polynomial $\Delta(s)$ [1, 4] of the closed-loop control

$$\Delta(s) = \text{num}[1 + G(s)C(s)]. \tag{4}$$

A regulated system is asymptotically stable if all roots of the characteristic equation lie in the left half-plane (LHP) of the complex plane [1–3]. To determine the characteristic quasi-polynomial $\Delta(s)$, we make a substitution of the expressions $G(s)$ and $C(s)$, i.e., (1) and (2) into an Eq. (4), and we get

$$\Delta(s) = as^3 + bs^2 + (c + k_p)s + k_i \mathrm{e}^{-\tau s}. \tag{5}$$

By substituting $s = o + \omega i$ into the characteristic quasi-polynomial (5), we obtain

$$\Delta(o + \omega i) = a(o + \omega i)^3 + b(o + \omega i)^2 + (c + k_p)(o + \omega i) + k_i \mathrm{e}^{-\tau(o+\omega i)}. \tag{6}$$

The characteristic quasi-polynomial (5) has at least one complex root $s = o + \omega i$, just as (6) is equal to zero.

Next, we use the formula $\mathrm{e}^{\alpha i} = \cos\alpha + i \sin\alpha$ and after some calculation, we get a mathematical expression divided into a real and an imaginary part

$$\begin{aligned} \Delta(o + \omega i) = {} & ao^3 - 3ao\omega^2 + bo^2 - b\omega^2 + co + k_p o + k_i e^{-\tau o} \cos\omega\tau \\ & + i\left(3ao^2\omega - a\omega^3 + 2bo\omega + c\omega + k_p\omega - k_i e^{-\tau o} \sin\omega\tau\right). \end{aligned} \tag{7}$$

Then, we put the real and imaginary parts equal to zero, and we get a system of two equations

$$\begin{array}{l} ao^3 - 3ao\omega^2 + bo^2 - b\omega^2 + co + k_p o + k_i e^{-\tau o} \cos\omega\tau = 0 \\ 3ao^2\omega - a\omega^3 + 2bo\omega + c\omega + k_p\omega - k_i e^{-\tau o} \sin\omega\tau = 0 \end{array}, \tag{8}$$

and finally, the solution of the system of two Eq. (8) is given by

$$\begin{aligned} k_i &= \tfrac{o^2+\omega^2}{\omega\cos\omega\tau+o\sin\omega\tau}\mathrm{e}^{\tau o}(2ao\omega+b\omega), \\ k_p &= -\left[a\left(o^2-\omega^2\right)+bo+c\right]+\tfrac{\omega\sin\omega\tau-o\cos\omega\tau}{\omega\cos\omega\tau+o\sin\omega\tau}(2ao\omega+b\omega). \end{aligned} \tag{9}$$

We must note that the characteristic quasi-polynomial (5) has at least one real root. By substituting $s = o$ into the characteristic quasi-polynomial, we obtain

$$\Delta(o) = ao^3 + bo^2 + \left(c + k_p\right)o + k_i\mathrm{e}^{-\tau o}. \tag{10}$$

Then, we put $\Delta(o) = 0$ and we get an implicit equation

$$\forall k_p \in \mathbb{R} : k_i = -o\left(ao^2 + bo + c + k_p\right)\mathrm{e}^{\tau o}. \tag{11}$$

We can now use the Eqs. (9) and (11) to find the stability areas [8, 9, 11].

3 Practical Part

3.1 Numerical Example I

Consider a fixed delay $\tau = 0.2\mathrm{s}$ and a of the following system

$$G(s) = \frac{1}{s^2 + 0.3s + 0.02}, \tag{12}$$

which has two distinct single real poles $s_1 = -0.2$ and $s_2 = -0.1$. Since both roots lie in the LHP, it is considered a stable open-loop plant system.

Now, considering the open-loop transfer function (12) and the controller (1), the characteristic quasi-polynomial $\Delta(s)$ of the closed-loop system can be computed as

$$\Delta(s) = s^3 + 0.3s^2 + \left(0.02 + k_p\right)s + k_i\mathrm{e}^{-0.2s}. \tag{13}$$

The main purpose is to get all the stabilizing PI delayed controllers for the open-loop system (12). In consideration of Eqs. (9) and (11), the stabilization pair $\left(k_p, k_i\right)$ is determined by

$$\begin{aligned} k_i &= \tfrac{o^2+\omega^2}{\omega\cos 0.2\omega+o\sin 0.2\omega}\mathrm{e}^{0.2o}(2o\omega+0.3\omega), \\ k_p &= -\left[\left(o^2-\omega^2\right)+0.3o+0.02\right]+\tfrac{\omega\sin 0.2\omega-o\cos 0.2\omega}{\omega\cos 0.2\omega+o\sin 0.2\omega}(2o\omega+0.3\omega), \end{aligned} \tag{14}$$

which represents a set of complex roots $\{\forall(s = o + \omega i) \in \mathbb{C}; o \in \mathbb{R} \wedge \omega \in \mathbb{R}_+\}$ and

$$k_i = -o\left(o^2 + 0.3o + 0.02 + k_p\right)\mathrm{e}^{0.2o}, \tag{15}$$

which represents a set of the real roots $\{\forall s = o; o \in \mathbb{R}\}$.

The stability region in the $\left(k_p, k_i\right)$ plane is shown in the Fig. 2.

Figure (Fig. 2) shows the stability of the closed-loop system (13). Red color represents the set of complex roots $\{\forall(s = o + \omega i) \in \mathbb{C}; o \in 0 \wedge \omega \in \mathbb{R}_+\}$, light red characterizes the set of complex roots $\{\forall(s = o + \omega i) \in \mathbb{C}; o \in \mathbb{R}_+ \wedge \omega \in \mathbb{R}_+\}$, blue illustrates

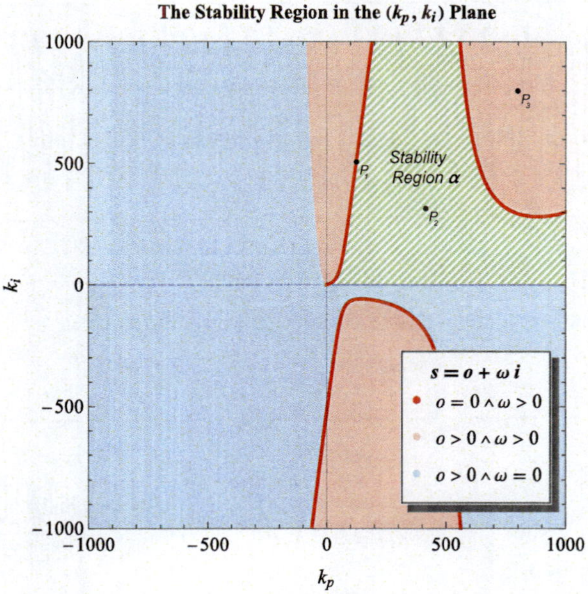

Fig. 2. The stability region in the (k_p, k_i) plane

the set of positive real roots $\{\forall s = o; o \in \mathbb{R}\}$ and green is a stability region α. As can be seen from the figure, the control system is marginally stable for the controller parameters $P_1 = (k_p, k_i) = (125, 509.6)$, asymptotically stable for the controller parameters $P_2 = (k_p, k_i) = (413, 318)$ in the stability region α and unstable for the controller parameters $P_3 = (k_p, k_i) = (800, 800)$. Lastly, we can verify points P_1, P_2, P_3 in the simulation (Figs. 3, 4 and 5).

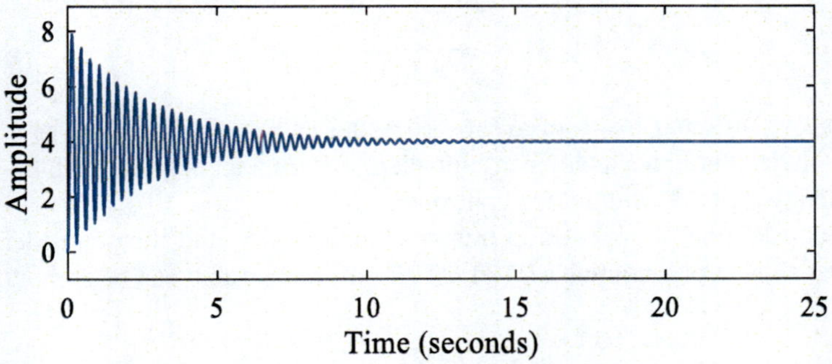

Fig. 3. The unit step response of the closed-loop system for the parameters $P_2 = (413, 318)$.–asymptotically stable case

The simulation results above show that the chosen design method for receiving the pair $P_n = (k_p, k_i)$ is correct. For the selected points P_1, P_2, P_3, it is possible to find the asymptotic stability region, the marginal stability area, and the unstable area.

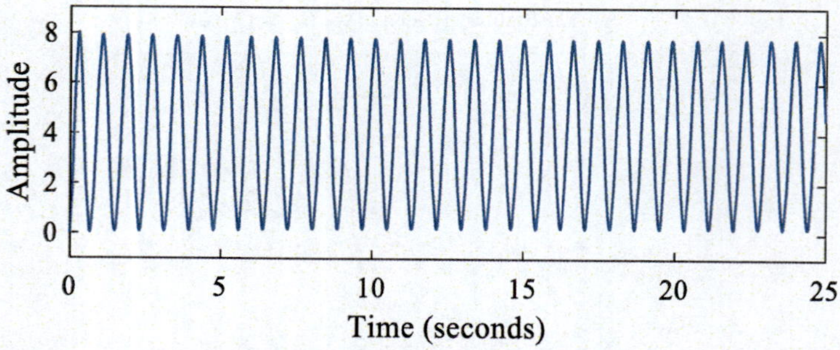

Fig. 4. The unit step response of the closed-loop system for the parameters $P_1 = (125, 509.6)$.–marginally stable case

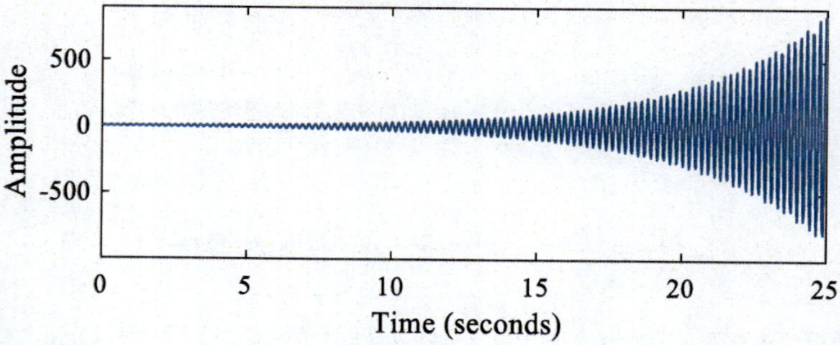

Fig. 5. The unit step response of the closed-loop system for the parameters $P_3 = (800, 800)$.–unstable case

3.2 Numerical Example II

Consider a fixed delay $\tau = 0.3$s and a controlled plant governed by the transfer function

$$G(s) = \frac{1}{s^2 - 0.3s - 0.28}, \tag{16}$$

which has two different real roots $s_1 = -0.4$ and $s_2 = 0.7$. Since one root lies in a right half-plane, it is an unstable open-loop plant system; i.e., open-loop plant system is asymptotically unstable without the controller.

Now, considering the open-loop transfer function (16) and the controller (1), the characteristic quasi-polynomial $\Delta(s)$ of the closed-loop system can be computed as

$$\Delta(s) = s^3 - 0.3s^2 + \left(k_p - 0.28\right)s + k_i e^{-0.3s}. \tag{17}$$

The main purpose is to get all the stabilizing PI delayed controllers for the open-loop system (16). By considering Eqs. (9) and (11), the stabilization pair $\left(k_p, k_i\right)$ is determined by

$$\begin{aligned} k_i &= \tfrac{o^2+\omega^2}{\omega\cos 0.3\omega + o\sin 0.3\omega} e^{0.3o}(2o\omega - 0.3\omega), \\ k_p &= -\left[\left(o^2 - \omega^2\right) - 0.3o - 0.28\right] + \tfrac{\omega\sin 0.3\omega - o\cos 0.3\omega}{\omega\cos 0.3\omega + o\sin 0.3\omega}(2o\omega - 0.3\omega), \end{aligned} \tag{18}$$

which represents a set of complex roots $\{\forall(s = o + \omega i) \in \mathbb{C}; o \in \mathbb{R} \wedge \omega \in \mathbb{R}_{+}\}$ and

$$k_i = -o\left(o^2 - 0.3o - 0.28 + k_p\right)e^{0.3o}, \tag{19}$$

which represents a set of positive real roots $\{\forall s = o; o \in \mathbb{R}\}$.

The stability region in the $\left(k_p, k_i\right)$ plane is shown in the Fig. 6.

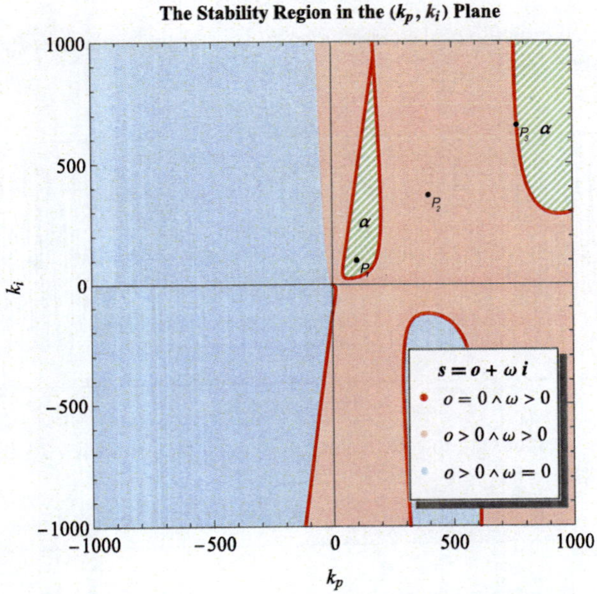

Fig. 6. The stability region in the $\left(k_p, k_i\right)$ plane

Picture (Fig. 6) shows the stability of the closed-loop system (17). Red color represents the set of complex roots $\{\forall(s = o + \omega i) \in \mathbb{C}; o \in 0 \wedge \omega \in \mathbb{R}_{+}\}$, light red characterizes the set of complex roots $\{\forall(s = o + \omega i) \in \mathbb{C}; o \in \mathbb{R}_{+} \wedge \omega \in \mathbb{R}_{+}\}$, blue illustrates the set of positive real roots $\{\forall s = o; o \in \mathbb{R}\}$ and stability regions α are represented by green color. As can be seen from the picture, the control system is asymptotically stable for the controller parameters $P_1 = \left(k_p, k_i\right) = (100, 100)$ in stability region α, unstable for the controller parameters $P_2 = \left(k_p, k_i\right) = (401, 370)$ and marginally stable for the controller parameters $P_3 = \left(k_p, k_i\right) = (772.5, 630)$. Finally, we can test points P_1, P_2, P_3 in the simulation (Figs. 7, 8 and 9).

The simulation results above show that the chosen design method for receiving the pair $P_n = \left(k_p, k_i\right)$ for a plant that is not stable in an open-loop is correct. For the selected points P_1, P_2, P_3, it is possible to find the asymptotic stability region, the marginal stability area, and the unstable area.

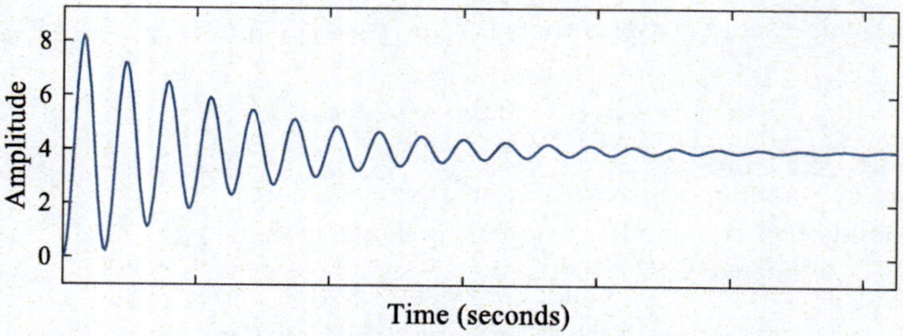

Fig. 7. The unit step response of the closed-loop system for the parameters $P_1 = (100, 100)$.–asymptotically stable case

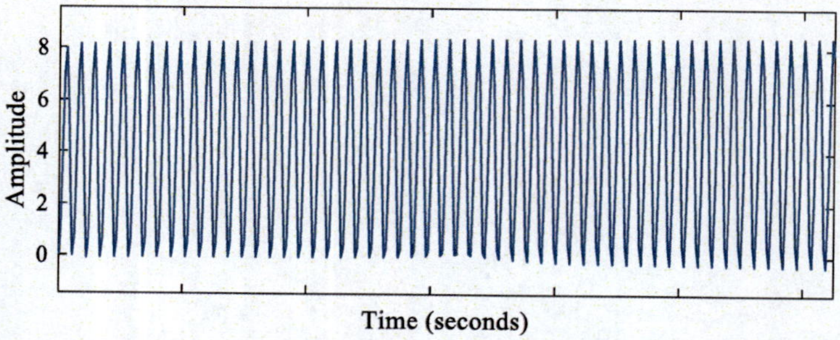

Fig. 8. The unit step response of the closed-loop system for the parameters $P_3 = (772.5, 630)$.–marginally stable case

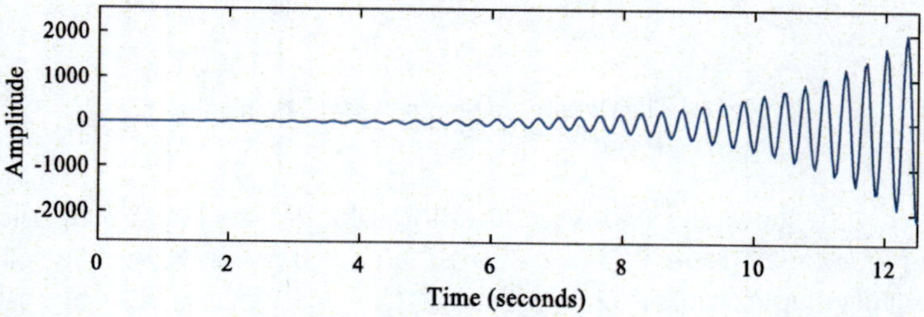

Fig. 9. The unit step response of the closed-loop system for the parameters $P_2 = (401, 370)$.–unstable case

4 Conclusion

The use of PI controllers is an intriguing topic for which the application can be found in many fields. This article focused on finding the stability regions for the second-order systems. Finding the stability regions is analytically complex if the closed-loop systems contain a delay; methods of numerical mathematics are used for such cases. The stability regions help to find the controller parameters and to understand in which cases the system is asymptotically stable, marginally stable, or unstable. In our case, the stability regions analysis is limited to using second-order plant systems. The stabilization pair, which corresponds to the graph in the global stability region, provides the designer with an easier way to accomplish the control requirements. This study can be extended to stabilization using higher-order polynomial plant systems or optimization methods where we can find optimal parameters for setting the controller.

Acknowledgements. The paper was supported by the Internal Grant Agency of Tomas Bata University in Zlín under the project number IGA/FAI/2022/008.

References

1. Hristuvarsakelis, D., Levine, W. S.: Handbook of Networked and Embedded Control Systems. Birkhäuser, Boston (2005)
2. Fridman, E.: Introduction to Time-Delay Systems: Analysis and Control. Springer, Cham (2014)
3. Xia, Y., Fu, M., Shi, P.: Analysis and Synthesis of Dynamical Systems with Time Delays. Springer, Berlin (2009)
4. Michiels, W., Engelborghs, K., Vansevevant, P., Roose, D.: Continuous pole placement for delay equations. Automatica **38**(5), 747–761 (2002). https://doi.org/10.1016/S0005-1098(01)00257-6,ISSN:0005-1098
5. Ogata, K.: Modern Control Engineering. Prentice Hall, New Jersey (2009)
6. Hamhalter, J., Tišer, J.: Funkce komplexní proměnné, 2nd vydání. České vysoké učení technické, Praze (2017)
7. Hamamci, S.E.: PI and PID stabilization of neutral and retarded systems with time delay. Turk. J. Electr. Eng. Comput. Sci. **20**
8. Richard, J.P.: Time-delay systems: an overview of some recent advances and open problems. Automatica **39**, 1667–1694 (2003)
9. Tan, N.: Computation of stabilizing PI and PID controllers for processes with time delay. ISA Trans. **44**, 213–223 (2005)
10. Zhang, X., Gao, X., Su, M.: Simplified stability criteria for delayed neutral systems. Math. Probl. Eng. (2014)
11. Gray, P.: Denial of shipment of radioactive material. Packag. Transp. Storage Secur. Radioact. Mater.
12. Wong, L., Lee, Y.C., Lee, H.W.: Optimal transmission of messages in computer networks—an optimal control problem involving control-dependent time-delayed arguments. J. Inequalities Appl. (2022)
13. Kumar, P., Erturk, V.S.: The analysis of a time delay fractional COVID-19 model via Caputo type fractional derivative. Math. Methods Appl. Sci. (2020)
14. Oertel, R., Wagner, P.: Delay-time actuated traffic signal control for an isolated intersection. In: Proceedings 90st Annual Meeting Transportation Research Board (TRB) (2011)
15. Joelianto, E.: Networked control systems: time delays and robust control design issues. In: Proceedings of 2011 2nd International Conference on Instrumentation Control and Automation, ICA 2011 (2011). https://doi.org/10.1109/ICA.2011.6130121
16. Kennell, D., Talkad, V.: Messenger RNA potential and the delay before exponential decay of messages. J. Mol. Biol. (1976)
17. Dollevoet, T., Huisman, D., Schmidt, M., Schöbel, A.: Delay propagation and delay management in transportation networks (2018)
18. Palpal-latoc, C., Bernardo, R.C., Vega, I.: Testing time-delayed cosmology. Preprint (2021)
19. Jnifene, A.: Active vibration control of flexible structures using delayed position feedback. Syst. Control Lett. **56**(3), 215–222 (2007)

Analysis of Data Gathering and Processing Modes During the Primary Identification of Radio Signals Through the Panoramic Spectral Analysis

Aleksey Studenikin[1,2], Anton Tokarev[1,2,3], Tatiana Demina[4], Serguei Pergamenshchikov[5], and Alexandra Salnikova[4,6](✉)

[1] JSC "IRCOS", 21 Zvozdnyy Avenue, Moscow 129626, Russia
[2] Department of Radio Engineering, Voronezh State Technical University, 14 Moscow Avenue, Voronezh 394026, Russia
[3] International Institute of Computer Technology, 29B Solnechnaya, Voronezh 394026, Russia
[4] Maikop State Technological University, 191 Pervomaiskaya, Maikop 385000, Russia
al.vl.salnikova@mail.ru
[5] Laboratoire de Mathematiques Raphael Salem, Universite de Rouen Normandie, BP12 Avenue de l'Universite, 76801 Rouen, France
[6] National Research University "MPEI", 14 Krasnokazarmennaya, Moscow 111250, Russia

Abstract. When evaluating the radio environment in wide frequency bands, an important task is to determine whether the signals belong to communication standards. To reliably check the compliance of the received emission with the standard, a detailed study of the "fine structure" of the signal is required, which involves significant computational costs, especially when updating the identification results at a high speed and in wide frequency ranges. To reduce the computational costs, in this case, the evaluation identification algorithms are used, and, for data gathering, a panoramic mode is applied, consisting in a cyclic retuning of the radio monitoring system receiver in the analyzed frequency range with the registration of short samples in the simultaneous analysis bandwidth. If the changes in the radio environment require a high speed periodic updating of the identification results, then the efficiency of monitoring depends not only on the complexity of the identification procedure, but also on how balanced the methods for initial data gathering and processing are. The aim of the present study is to compare the ways of implementing the processes of data gathering and processing during the primary signal identification as well as to work out the recommendations for optimizing promising approaches. The novelty of the proposed solution is in the modes introduced for the gathering and processing of spectral data and in the expressions for estimating their time parameters along with the recommendations for their optimization. #COMESYSO1120.

Keywords: Wideband radio monitoring · Signal identification · Accumulated spectrum · Spectral mask of emission · Panoramic spectral analysis

R. Silhavy et al. (Eds.): CoMeSySo 2022, LNNS 596, pp. 540–562, 2023.
https://doi.org/10.1007/978-3-031-21435-6_47

1 Introduction

A growing variety of standards for wireless information transmission, the introduction of a software-defined radio (SDR) technology, and the possibility of dynamic management of the radio frequency resource [1, 2] complicate the task of evaluating the radio environment in wide frequency bands. With a scarce amount of a priori information at hand, the determination of frequency boundaries and classes (standards) of radio emissions based on the reception and processing of the signals found in the controlled frequency band by the radio monitoring equipment (RME) appears to be a rather complex and time-consuming problem. Its solution involves retesting all possible hypotheses for each frequency section at which all kinds of emissions can be observed and thus requires significant computational resources.

At the same time, that the radio emissions occupy particular frequency range sections is established based on, foremost, a spectral analysis of the random processes received by RME [3], and when spectra are accumulated over long time intervals, different classes of signals will differ by a spectrum shape that makes them ready for a primary identification [4]. In this case, however, one should not count on obtaining a simple and error-free instrument, since the presence of noises and interference signal distortions during reception, as well as the differences in radio emission sources and the limited length of the received signal samples, and some other factors in practice give rise to noticeable deviations of the observed spectra from "references".

Nevertheless, the use of typical accumulated radio emission spectra belonging to different standards as the spectral masks for the primary signal identification makes it possible to develop quite effective algorithms for broadband radio monitoring [5]. The proliferation of short-range information transmission facilities based on SDR technology, as well as the spectrum sharing concept [6], will increase the need for radio monitoring under a partial and even complete absence of a priori information. Therefore, the problem of primary identification of signals has proven to remain relevant, and the success of its solution particularly depends on the management of data gathering and processing procedures.

2 Initial Data Generation for the Primary Identification of Signals

Typically, the analyzed frequency range (AFR) is many times greater than the bandwidth of simultaneous analysis (BSA) of the RME receiver that is B_{SA}, and the panoramic mode is used to collect radio traffic data. In this mode, the receiver is tuned to the starting AFR section with the bandwidth B_{SA}. . For this section, it generates the estimates of the spectral characteristics of active radio emissions (instantaneous or accumulated over the sampling slot t_s) and then retunes to the adjacent frequency section (which takes the time t_{ch}). After that, it estimates the specified spectral characteristics within this new section and, while continuing retuning with the step B_{SA}, , carries out scanning the entire controlled frequency range. At the next cycle of analysis, new spectral estimates can be combined with those previously obtained by averaging, maximizing, or otherwise. The panoramic spectrum (PS) thus generated then becomes the basis for the primary identification of radio emissions. The PS generation process is shown in Fig. 1.

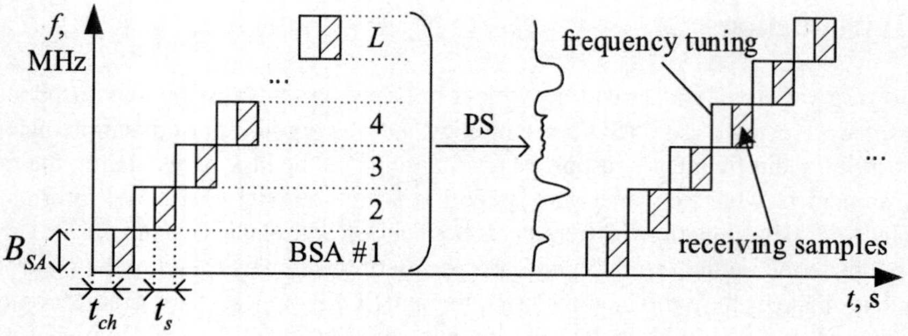

Fig. 1. Panoramic spectrum generation.

An effective mechanism for the generation of primary spectral estimates is the fast Fourier transform (FFT), according to which, a set of N_{FFT} samples of the complex spectrum is calculated by the samples of the complex envelope $\dot{u}_{CE\ l,r}(kT)$ of the signal received in the l-th frequency section of the AFR. The complex envelope is carried out with the sampling interval $T = 1/F_s$ where F_s is the sampling rate. Thus one gets:

$$\dot{c}_r(l, n_{CE}) = \sum_{k=0}^{N_{FFT}-1} w(k)\, \dot{u}_{CE\ l,r}(kT)\, \exp\left(-j2\pi n_{CE}k / N_{FFT}\right). \tag{1}$$

In (1), n_{CE} is the number of the spectral sample within the current band, k is the ordinal number of the time sample, $w(k)$ is the weighting function used to adjust the properties of the spectrum, $r = \overline{1, R}$ is the ordinal number of the time sample segment that is N_{FFT} by the sample size, R is the ratio of spectrum accumulation (RSA). Thus, the duration of the sample received in the BSA is $t_s = RN_{FFT}/F_s$.

The coefficients $\dot{c}_r(l, n_{CE})$ determined in (1) characterize the signal spectral components at the frequencies

$$f_{l,n_{CE}} = f_{l0} + n_{CE}F_s/N_{FFT}, \qquad -N_{FFT}/2 \leq n_{CE} < N_{FFT}/2,$$

where f_{l0} is the radio receiver tuning frequency when receiving data at the l-th band from the set of adjacent frequency bands in order. To generate the final spectral panorama, as a rule, not all N_{FFT} spectral samples are used, but a smaller number N of them, corresponding to the BSA and the receiver tuning frequency offset when retuning from the l-th band to the $(l + 1)$-th one:

$$B_{SA} = f_{(l+1)\,0} - f_{l0} = NF_s/N_{FFT}.$$

To reduce the effect of noise and the nature of the transmitted data on the shape of the spectrum, an averaging or the accumulation by the maximum of instantaneous spectral estimates over the sample recording interval is used. It should be noted that when using the accumulation, it is possible to analyze both the extended and the burst signals [7]. Therefore, as the initial data for analysis, one can recommend the set of the spectra maximally accumulated in all the BSAs corresponding to the analyzed frequency range:

$$S_{\max}(l, n_l) = 20\lg\left(\max_r |\, \dot{c}_r(l, n_{\min l} + n_l)\,|\right).$$

To obtain the PS from the L arrays made of N samples in each of them, the spectrum is generated for the frequency range having the size of $N_{all} = LN$ samples with the new indices $n = lN + n_l$. For such spectrum, the following equation is valid: $S(n) = S_{\max}(l, n_l)$.

3 The Concept of the Spectral Mask and Its Use for the Generation of Identification Decision Determining Statistics

Let the signals of J different classes (standards) be considered potentially possible in the analyzed frequency range. Then, in order to recognize the signal of the j-th standard, it is necessary to prepare in advance its "spectral mask" (SM), that is, an array of the samples $\mu_j(m)$ containing the typical spectral levels of this emission (for example, in decibels). The width M_j of this mask expressed in samples will depend on the width of the spectrum of the signals of the j-th class and on the frequency interval $\Delta f = F_s / N_{FFT}$ between the adjacent PS samples. With an increase in the spectral resolution, one can count on an increase in the reliability of recognition, however, at the same time, the computational complexity of the calculations and the sampling duration will increase, while the amplitude of burst signals will decrease. Therefore, it is desirable to choose the resolution of the spectrum representation in such a way that for all the SMs the width M is from several tens to several hundreds of samples. An example of SM for the UMTS base station (BS) signal superimposed on the accumulated spectrum S of this signal is shown in Fig. 2.

It is noteworthy that the above requirements explain the impracticality of identifying both broadband and narrowband signals by the spectrum with the same resolution. Indeed, the use of a wide frequency interval Δf does not allow representing the SM of narrowband signals with the proper spectral resolution, and the Δf reduction makes the SM of wideband signals too voluminous, which can lead to the negative effects described earlier.

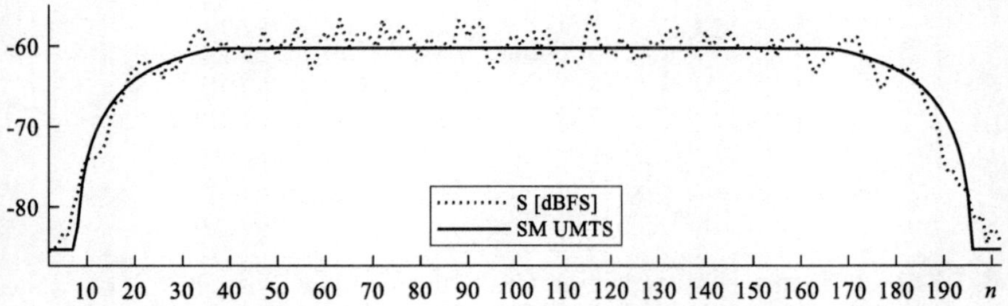

Fig. 2. SM for the UMTS BS signal superimposed on its accumulated spectrum.

Besides the spectral representation resolution influence, the shape of the accumulated spectrum is significantly affected by the radio signal propagation conditions. When a signal passes through real transmission channels, its spectral structure changes due to both the noise influence and the interference effects, including the ones caused by multipath propagation. The interference pattern largely depends on the mutual placement of the transmitting and receiving antennas that does not have time to noticeably change over a typical sampling duration of several milliseconds in the case of the low-moving (for example, wearable) sources.

At the same time, when data is accumulated over such intervals, the resulting spectrum of many signals turns out to be significantly dependent on a specific "portion" of data transmitted over the accumulation interval. Due to the features of distortions specified above, the main purpose of accumulation is to reduce the influence of broadband random processes such as noise and transmitted data on the shape of the spectrum, and that is why a long-term accumulation of hundreds and thousands of signal spectra would not make sense in this case. In practice, it has been found that, in the absence of pronounced pauses in signal emission (which is typical for BS signals), just dozens of accumulations are sufficient to stabilize its spectrum shape.

In the case of a burst signal, when tuning the receiver to its range, for guaranteed registration of at least one signal sample, a sampling duration is required that exceeds the emission pause time; to generate the required number of signal spectra, this value must be additionally increased, at least by $N_{FFT}R$ samples.

The spectrum accumulation factor R (and hence the sample duration) also affects the time efficiency of data gathering. Thus, if the time of data accumulation and the time of frequency retuning are equal by duration, only half of the data gathering time will fall on the sample recording. With twice the sampling duration, its recording will already account for 2/3 of the data gathering time. It means that the time efficiency of data gathering improves when the sampling duration grows.

4 A Generalized Signal Identification Algorithm

In a number of works on spectroscopy, the estimates of the magnitude of the similarity between the received and reference spectrum are proposed [8]. The essence of these estimates can be considered based on the example of one of the simplest of them–the spectral correlation coefficient (SCC) which is defined as follows:

$$SCC_j(n) = \mathrm{cov}_{S\,\mu_j}(n) \big/ \sigma_{S_j}(n)\, \sigma_{\mu_j}.$$

Here

$$\mathrm{cov}_{S\,\mu_j}(n) = \frac{1}{M_j} \sum_{m=0}^{M_j-1} \left(S(n+m) - \overline{S}_j(n)\right) \left(\mu_j(m) - \overline{\mu}_j\right)$$

stands for the covariance between the samples of the spectrum S and the SM μ_j, while

$$\sigma_{S_j}(n) = \sqrt{\frac{1}{M_j} \sum\nolimits_{i=n}^{n+M_j-1} \left(S(i) - \bar{S}_j(n)\right)^2}, \qquad \sigma_{\mu_j} = \sqrt{\frac{1}{M_j} \sum\nolimits_{m=0}^{M_j-1} \left(\mu_j(m) - \bar{\mu}_j\right)^2}$$

are the mean square values of the spectrum and the mask and

$$\bar{S}_j(n) = \frac{1}{M_j} \sum_{i=n}^{n+M_j-1} S(i), \qquad \bar{\mu}_j = \frac{1}{M_j} \sum_{m=0}^{M_j-1} \mu_j(m)$$

are the mean values of the spectrum and the mask, respectively.

The SCC values close to unity indicate that the shape of the spectrum segment starting from the n-th sample is close to the shape of SM $\mu_j(m)$. . The results of SCC calculation for the signals of GSM (BS) and DECT standards are shown in Fig. 3.

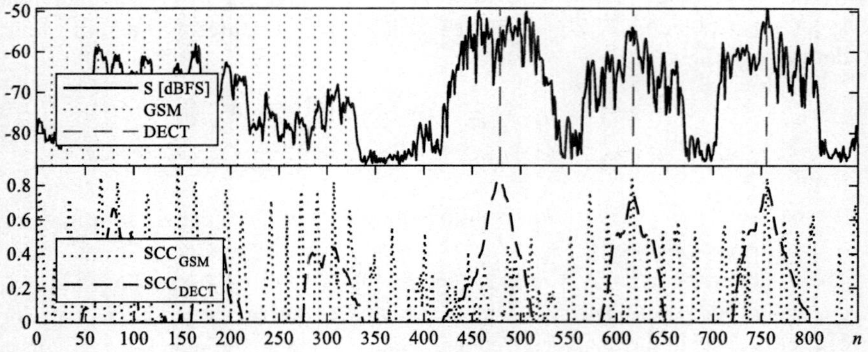

Fig. 3. Demonstration of the features of the spectral similarity estimates by the example of SCC.

As it follows from Fig. 3, for most signals, SCC takes on higher values when SM corresponds to the signal standard than when it does not. However, it can also be seen that in two cases DECT signals with a GSM SM produce SCC values which are higher than the SCC for real DECT SM. This case of false alarm is typical for the identification algorithms based on the spectrum shape, when a segment of the spectrum of the signal with a wider band turns out to be similar to the spectrum of the signal with a narrower band.

To reduce the probability of such errors, it is recommended to perform signal identification in order of decreasing the SM M_j width. That is, the complete signal identification algorithm should be built based on the "displacement" principle: the search for the positions of the signals of different standards should be performed starting from the widest SM, and, for each of the subsequent SMs, the search should be only for the totality of that sections of the frequency axis where no signals have been detected for the previous SM.

As the exception, one has to consider the signals whose spectrum widths differ little, for example, the LTE BS with a bandwidth of 5 MHz, and the UMTS BS, for which the local similarity maxima will correspond to both of these SMs, and when in a "displacement" mode, a standard with a larger spectrum width will more often be selected (in this case it is LTE). Therefore, for such sets of standards, out of several local similarity maxima, the largest one should be chosen as the identification result.

Taking into account the described features of the spectral similarity estimates, one can develop a generalized signal identification algorithm based on the following premises.

(1) The location of the spectra of the identified signals on the frequency axis is "tied" to their left boundaries, that is, to the starting SM sample.

(2) The measure of similarity initially chosen for identification $V_j(n)$ is transformed into the measure $X_j(n)$ that takes values in the range $\overline{0, 1}$, where $X_j(n) = 0$ means the minimum similarity, and $X_j(n) = 1$–the maximum one.
One denotes the maximum and the minimum possible values for the measure of similarity $V_j(n)$ as $V_{\max}$ and $V_{\min}$, respectively, while the magnitude of the measure of similarity $V_j(n)$ corresponding to the case of equality of the tested and reference spectra is denoted as V_{equal}. Then the similarity measure used for identification is defined as follows:

$$X_j(n) = \begin{cases} V_j(n) / (V_{\max} - V_{\min}), \; V_{equal} = V_{\max}, \\ 1 - V_j(n) / (V_{\max} - V_{\min}), \; V_{equal} = V_{\min}. \end{cases}$$

(3) As the possible signal location positions may be any PS samples from the range $n_1 = \overline{0, N_{all} - M_1 - 1}$.
(4) The masks are ordered by decreasing their width, that is, the processing starts with the widest SM $\mu_1(m)$, and only when the final decision on the presence or absence of such signals in the analyzed frequency range is made, the search starts with the SM $\mu_2(m)$, then, with $\mu_3(m)$ etc. The exceptions are the signals with a similar spectrum width; they are processed separately.

To determine, based on the PS $S(n)$ analysis, the positions of the signal spectra of different standards on the frequency axis, the following algorithm can be recommended.

(1) During the calculation of the array $X_j(i)$, among the possible SM shifts with respect to PS, a set of K continuous subsets $\mathbf{I}_k$ ($k = \overline{1, K}$) is selected, for all i values of which the condition $X_j(i) \geq t_{sim}$ is satisfied. Here t_{sim} is the minimum allowable spectral similarity value. Then, for each $\mathbf{I}_k$, a search for the local maximum of the estimate of the spectral similarity $X_j(i \in \mathbf{I}_k)$ is made. The shifts $i_{\max}$ corresponding to the local maxima are saved as the possible signal positions (PSP) corresponding to the j-th SM.
(2) If the new local maximum $X_j(i)$ is not detected during the M_1 steps after fixing the PSP $i_{\max}$, then the previously saved shift is assigned the status of the detected signal: $i_{\max} \rightarrow n_1^{(q)}$ (where q is the serial number of the detected signal of the first standard), and the part of the frequency axis related to the detected signal is marked as "the occupied" one and is not used in a subsequent signal search.
Another condition for stopping the search for alternative local maxima $X_j(i)$ is that the inequality

$$i > N_{all} - M_1 \tag{2}$$

holds meaning that the SM $\mu_1(m)$ no longer fits in the rest of the PS.
(3) If, however, a new local maximum $X_j(i)$ is found at the distance less than M_1 from $i_{\max}$, then of the two compared options, the PSP status remains with the one for which $X_j(i)$ is greater, while the option "conflicting" with it is discarded.
(4) After the completion of the PS $S(n)$ scanning due to the fulfillment of the condition (2), the SM $\mu_2(m)$ is loaded for processing, and the calculation along the frequency

axis of the statistics $X_j(i)$ starts again, but within the framework of the solution selection only such shifts i are considered allowed, for which the signals of the 2-nd standard will not fall on the sections occupied by the previously identified signals of the 1-st standard (except those comparable in SM width; for them the selection is based on the PSP value).

(5) Similarly, the signal search for all the subsequent SMs is performed.

5 Cyclic Signal Identification and the Data Update Problem

Carrying out signal identification according to the approaches described above presupposes that, by the time the calculation of the decision determining statistics begins, the PS generation has been completed, and its data is available for processing for an unlimited time. However, in practice, the situation with the availability of spectral data turns out to be ambiguous. The accumulation process is associated with noticeable time costs; therefore, with an AFR width of the order of several gigahertz, the spectral data for the entire PS is available only a few seconds after the start of the radio environment analysis. In addition, the very dynamism of the change in the modern radio environment suggests the need for cyclic re-identification. This means that PS needs to be periodically updated as fast as possible, but it is obviously not allowed to update PS until the data processing is completed, and thus the start of identification is always deferreded until PS data is ready.

Thus, the problem of optimizing the cyclic processes of spectral data gathering and radio signal identification becomes relevant. Recommendations for the implementation of data gathering and processing depend on the parameters of a panoramic analysis and the RME capabilities. Below are the results of a study of two approaches to the identification implementation based on the cyclically updated data. Further analysis suggests that:

(1) double buffering is used during the spectral data gathering and processing to improve the time use efficiency and to ensure the possibility of processing without gaps;
(2) a streaming hardware implementation of the identification algorithm is used, in which each frequency section analyzed at the same time is processed by all SMs, so that the processing time for this section does not depend on its occupancy. The choice between the "conflicting" PSPs of the same class is performed as they appear without saving their values in memory (except for the sets of SMs of comparable widths; for them, "displacement" occurs only for the first SM in the set, and the PSP values are saved for selecting solutions).
(3) at a data processing speed lower than the data accumulation rate, data gathering periodically stops awaiting for the calculation to be completed.

6 The Features of a Deferred Mode of Data Gathering and Processing

The deferred mode has been actually described above and it presupposes that the start of data processing is deferreded until the PS is fully ready. This mode has the following features:

(1) a relatively large deferred in the appearance of solutions, since information about the signals present in the scanned frequency is obtained only after the generation of PS and its processing by all SMs;
(2) for data gathering and processing, a memory buffer is required that provides the PS storage, the size of which in samples, taking into account double buffering, will be: $N_{mem\_def} = 2N_{all}$. The memory capacity required for the solution selection (i.e. the storage of PSP positions) is proportional to the value $N_{all} - \overline{M}$, where $\overline{M}$ is the average width of SM in samples, and its value depends on the PSP storage structure. For the sets of the SMs with comparable widths, it is also necessary to store PSP values. It means that there should be a memory capacity proportional to $N_{all} - \min\{\mathbf{M}_{eq}\}$, where $\mathbf{M}_{eq}$ is the array of widths of all SMs from such sets.
(3) a simply structured final stage of processing, presupposing that each of the SMs is loaded for processing only once, and the decisions on it are also made only once and finally in accordance with a fairly simple logic.

7 The Features of a Synchronous Mode of Data Gathering and Processing

The synchronous mode involves splitting the PS into N_{SF} spectral fragments (SF) consisting of W_{SF} BSA and representing a frequency band $B_{SF} = W_{SF}B_{SA}$ exceeding the widest SM M_1. In this context, the deferred mode can be considered as the synchronous one at $N_{SF} = 1$.

In this mode, the receiver sequentially collects data in the BSA until enough spectra are accumulated in memory to generate an SF, and then the processing of this SF is started as well as the simultaneous data gathering in the next SF, and so on. A diagram of such a process is given in the next section.

Now one can consider the features of the identification process implementation in this mode. The first of them is that the right boundary of the spectral region, for which the generation of the final identification results in SF is possible, may turn out to be dependent on the position of the signal found in it. This feature should be examined in more detail.

Let $i_{SF}(n)$ be the right boundary of SF with the number $n = \overline{1, N_{SF}}$. Since the information for the frequencies to the right of $i_{SF}(n)$ will become available only after data is collected in the next SF, it cannot be guaranteed that there will be no PSP to the right of the point $i_{SF}(n) - M_j$ when the processing approaches the boundary $i_{SF}(n)$. Instead, the area from this point to $i_{SF}(n)$ should be declared as an ambiguity domain of information (ADI) with the width of $N_{ADI_j} = M_j$ samples, and one's task is to continue calculating PSP for this ADI after data arrival for the frequencies to the right of $i_{SF}(n)$. Taking into account that the ADIs of all SMs are included in the ADI of the widest SM M_1, the generation of identification solutions for SF turns out to be possible only up to the sample $i_{SF}(n) - M_j$, while the samples lying to the right of it form an ambiguity domain of decisions (ADD).

Taking into account the features of the streaming hardware used for the implementation of the identification algorithm, it makes no sense to postpone the start of the decision selection procedure until the spectral data for the ADD is obtained: it should be carried out for the SF samples that are not included in the ADI, while the solutions obtained will be final only outside the ADD, and within its boundaries they will remain subject to further clarification.

The decision on the narrower SMs is made only after the completion of the "binding" to the frequency axis of the wide SMs (or after calculating the PSP for a set of SMs of a comparable width). Thus, when the spectrum of the found signal of the j-th class with the position i_{sig_j} obtained during the selection process intersects the boundary of the ADD for the same class, or the class with a comparable or greater SM width, this position cannot be considered final since this signal can be replaced by another one that will appear in the ADD after receiving the next SF. In this case, the ADD boundary will shift to the sample with the number i_{sig_j} while the ADD will expand to $N_{ADD} = i_{SF}(n) - i_{sig\_j}$ samples. The ratio of the SF boundaries, detected signals, and ambiguity domains for such a case is shown in Fig. 4.

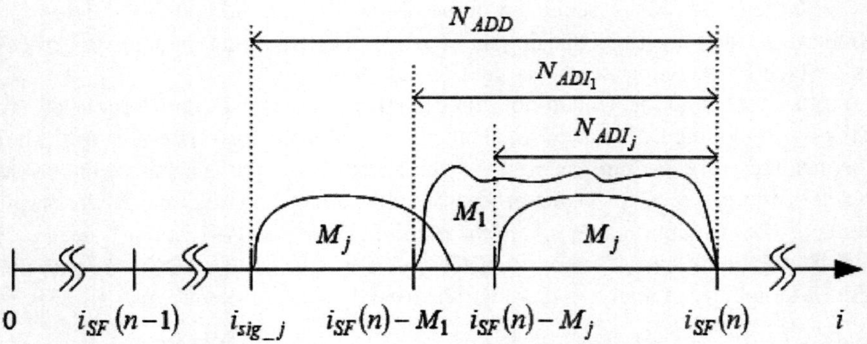

Fig. 4. The ratio of the SF boundaries and ambiguity domains in the synchronous mode.

After receiving data for the new SF, the following operations are to be carried out:

(1) the calculation of PSP (up to the ADI zone) together with the ADI of the previous SF;
(2) the pre-selection of solutions (up to the ADI zone) together with the ADD of the previous SF;
(3) the final selection of solutions (up to the ADD zone) together with the ADD of the previous SF.

Thus, when the SF with the number n is ready, the PSP is calculated for all SMs within $K_{SF} = W_{SF}N$ samples of the PS with numbers $\overline{(n-1)\,K_{SF} - M_j, nK_{SF} - M_j}$. The generation of the final solutions is possible for a different number of samples. Its minimum value corresponds to the absence of the ADD expansion in the $(n-1)$-th SF and its maximum expansion in the n-th SF, and, as a result, it can be determined as:

$$W_{SF\_id\_\min} = (nK_{SF} - 2M_1) - ((n-1)\,K_{SF} - M_1) = K_{SF} - M_1.$$

As for its maximum value, it corresponds to the maximum ADD expansion in the $(n-1)$-th SF and its absence in the n-th SF, and, as a result, it can be determined as

$$W_{SF\_id\_\max} = (nK_{SF} - M_1) - ((n-1)\,K_{SF} - 2M_1) = K_{SF} + M_1.$$

On average, after collecting data for the SF, it is possible to generate the final solutions for the number of samples equal to the SF width:

$$W_{SF\_id} = (K_{SF} - M_1 + K_{SF} + M_1)\big/2 = K_{SF}.$$

To simplify further analysis, it is assumed that the difference in the time needed for selecting solutions in the proposed modes can be neglected since the computational complexity of such procedures is low compared to the PSP calculation, and taking into account that ambiguity domains in synchronous mode requires only interrupting the decision selection process for the separate SF sections. During this interruption, data is collected and the PSP is calculated in the new SF, so that no additional deferred is introduced. At the same time, the amount of data processed during the selection process does not change in comparison with the deferred mode.

Another feature of the synchronous mode is that in order to ensure “seamless” processing when moving from one SF to another, at the moment the processing of the next SF begins, the processor must be in the state of completing the analysis of the last M_j samples of the previous SF. If the processor state cannot be fixed, and it works on a continuous data stream, then to obtain the required state, the processor will need to be reinitialized with the last M_j samples of the previous SF, and that will take some time t_{l_j}. For a simplicity of analysis, there will be used the value averaged over all SMs:

$$t_l = \frac{1}{J}\sum_{j=1}^{J} t_{l\_j}.$$

In the case when the processor structure provides for the possibility of fixing its state (for example, the state of the filter deferred elements), or when using the deferred mode, re-initialization is not required ($t_l = 0$). Also, it is not needed when processing the first SF in the PS, however, for a simplicity of further analysis, this exception is neglected.

The expressions for the time of data gathering and processing in SF (including the time needed for the selection of decisions), taking into account the time of re-initialization of the processor, take the form:

$$t_{SFg} = W_{SF}\,t_{1g}, \qquad t_{SFp} = W_{SF}\,t_{1p} + t_l,$$

where $t_{1g} = t_s + t_{cch}$, t_{1p} are the time of data gathering and processing in BSA, respectively.

Concluding the analysis of the synchronous mode, the following features of it should be mentioned.

(1) The average identification time of the signal since the moment of its airing is less in this mode than in the deferred mode, as it is shown below.

(2) The advantage of the mode is the reduction in the memory capacity required for collecting and processing data. To implement processing while taking into account AD and double buffering, the following number of spectral samples should be stored:

$$N_{mem\_sync} = 2\begin{cases} K_{SF}\,, \; t_l = 0\,, \\ K_{SF} + M_1\,, \; t_l > 0\,. \end{cases}$$

The memory capacity required for selecting solutions, i.e., for storing the PSP positions as well as the PSP values (for the sets of the SMs with comparable widths), is proportional to this value in the case when $t_l > 0$ and depends on the PSP storage structure.

(3) The disadvantage of the mode is the complication of the processing procedure due to taking into account the ambiguity domains and the processor structure, where, as new data arrives, it is necessary to ensure the correct transition between SFs for the PSP calculation and decision selection processes.

When choosing a data gathering and processing mode, its time parameters such as the real time coefficient (RTC) and the average time of identification (TID) are important. The next section provides estimates of their dependence on the main parameters of the identification process.

8 Estimation of the Time Parameters of the Proposed Modes

To estimate the time parameters of the proposed modes, one should consider in more detail the course of the process of cyclic signal identification in time. Its frequency-time diagrams corresponding to the synchronous mode are shown in Fig. 5. The diagram in the upper part corresponds to the case when $t_{1p} > t_{1g}$, and in the lower–to the case when $t_{1p} < t_{1g}$. The elements with the designation "*G*" represent the process of data gathering, while the elements with the designation "*P*"–their processing. The diagrams for the synchronous mode at $t_{1p} = t_{1g}$, as well as for the deferred mode (when $t_l = 0$) have a similar form.

Now it is necessary to obtain the expressions for the real-time coefficient RTC characterizing the proposed modes from the point of view of the continuity of the analysis. In general, this coefficient is defined as

$$RTC = t_{SFg} \big/ t_{SFp}.$$

The expression for RTC for the synchronous mode takes the form:

$$RTC_{sync} = W_{SF} t_{1g} \big/ \left(W_{SF} t_{1p} + t_{load} \right), \tag{3}$$

while for the deferred mode this coefficient can be defined as

$$RTC_{def} = L t_{1g} \big/ L t_{1p} = t_{1g} \big/ t_{1p}. \tag{4}$$

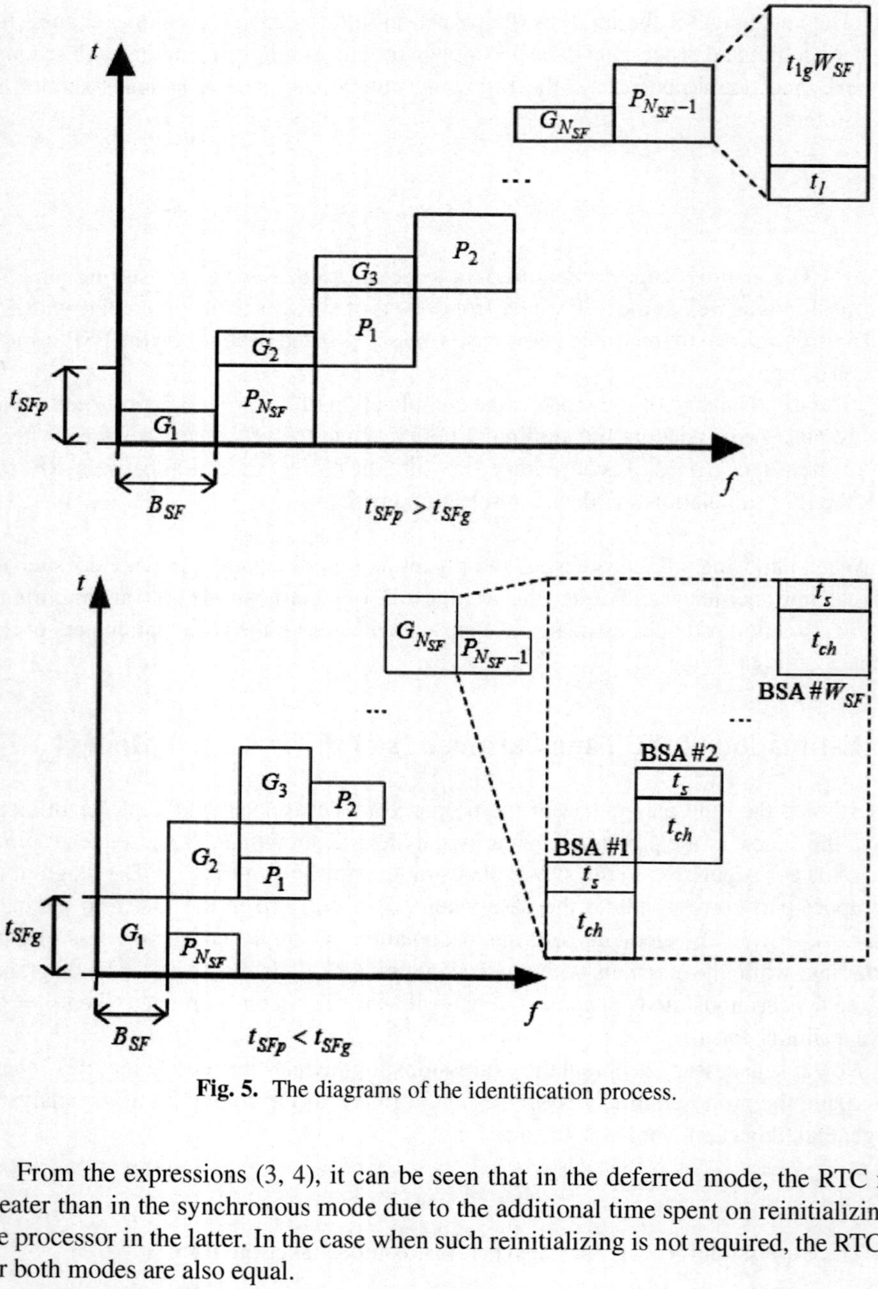

Fig. 5. The diagrams of the identification process.

From the expressions (3, 4), it can be seen that in the deferred mode, the RTC is greater than in the synchronous mode due to the additional time spent on reinitializing the processor in the latter. In the case when such reinitializing is not required, the RTCs for both modes are also equal.

To ensure the continuity of spectral data gathering in the modes considered, the condition $RTC \geq 1$ must be satisfied. Otherwise, when $t_{1p} > t_{1g}$, the data gathering process will be interrupted as this data processing is to be completed first. It is possible to ensure the continuity of data gathering at a low processing rate by increasing the ratio of spectrum accumulation RSA to such a minimum required value at which the data gathering time exceeds the processing time. The use of such optimization is characterized by the following features:

(1) the continuous data gathering, regardless of the rate of their processing;
(2) the average TID of the signal in comparison with the absence of optimization turns out to be less (which will be shown below);
(3) the optimization can be used both in the deferred and synchronous modes.

Now one defines the data gathering time in SF as the function of RSA that is R:

$$t_{SFg}(R) = W_{SF}\left(RN_{FFT}/F_s + t_{ch}\right).$$

Based on the data gathering continuity condition $t_{SFg}\left(R_{opt}\right) \geq t_{SFp}$, the optimal RSA R_{opt} satisfies the following inequality:

$$R_{opt} \geq F_s\left(t_{1p} + t_l/W_{SF} - t_{ch}\right)/N_{FFT}.$$

Then the RSA during optimization can be defined as

$$R_{opt} = \begin{cases} \left\lceil F_s\left(t_{1p} + t_l/W_{SF} - t_{ch}\right)/N_{FFT}\right\rceil, & t_{1p} > t_{1g}(R_{\min}), \\ R_{\min}, & t_{1p} < t_{1g}(R_{\min}), \end{cases}$$

where $\lceil\rceil$ is the ceiling operation. This expression is valid both for the synchronous and deferred modes (for the latter there should be $t_l = 0$).

In Fig. 6, one can see the diagrams of the dependencies of RTC and RSA R on the BSA processing time. The RTC diagrams are plotted on the left ordinate axis and are shown as dark lines, with heavy lines representing the absence of the RSA optimization, and normal thickness lines representing the application of this optimization. The RSA R diagrams are plotted along the right ordinate axis, and are shown as light lines. The vertical dash-dotted lines mark the points of equality of the time of data gathering and processing in the SF (for the synchronous mode) and in the BSA (for the deferred mode).

The calculated dependences shown in Fig. 6 and further correspond to the use of the RME ARGAMAK-D11 [9], for which $F_s = 25.6$ MHz, $B_{SA} = 24$ MHz, $t_{ch} = 2$ ms, and the typical analysis parameters are: the AFD width is 960 MHz, the SF width is $W_{SF} = 8$ and the number of SFs is $N_{SF} = 5$ in the synchronous mode, and $W_{SF} = 8.5$, $N_{FFT} = 1024$ in the deferred mode; the short samples are at $R_{\min} = 10$ ($t_{1g} = 2.4$ ms), $t_l = 768/F_s = 30$ μs, and $t_{1p} = 1$ ms when the number of SMs is $J = 5$ (or 5 ms at $J = 25$). These are the estimated values for the direct implementation of one of the integrated spectral similarity estimates.

According to Fig. 6, it can be seen that in the absence of RSA optimization, the excess of the BSA processing time of points of equality of the data gathering time leads to a decrease in RTC below one, which indicates stops in the data gathering process. When using optimization, RSA increases after the points of equality of the data gathering and processing time, and RTC does not fall below one, which means continuous data gathering.

Another practically significant time parameter of the data gathering and processing procedures is the average time of identification (TID) of the signal since the moment

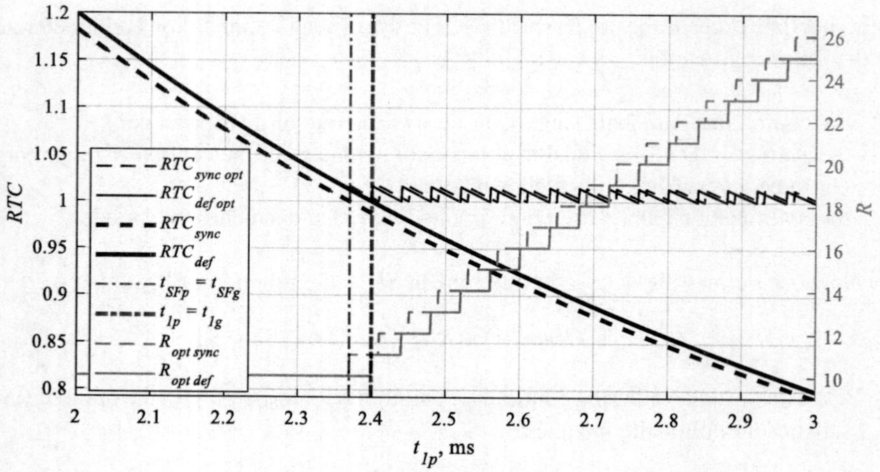

Fig. 6. The dependencies of RTC and RSA on the BSA processing time.

of its airing. To estimate TID for the proposed modes and to simplify the analysis, one agrees that:

(1) after going on the air, the signal is emitted continuously;
(2) the width of the signal spectrum is much less than BSA and can be neglected, that is, its identification occurs when the carrier frequency enters the BSA;
(3) the moment of the signal emission at the frequency f_{sig} coincides with the start of sampling in BSA with initial frequency f_{RC}.

In Fig. 7, one can see the marking of the frequency axis and the possible location of the initial BSA frequency of the receiver f_{RC} and the signal frequency $f_{sig\_1,2}$ in the synchronous mode. However, the following considerations are valid for both modes since deferred mode is the special case of the synchronous when $N_{SF} = 1$.

Taking into account that, for the portion of the current SF included in the ADI, the calculation of PSP is performed with a deferred–together with the next SF, a shift for the initial frequency f_1 of the analyzed range is introduced:

$$f_1' = f_1 - M_j F_s / N_{FFT}.$$

The marking of the analyzed range using this shift makes it possible not to take into account the AD processing deferred, so it will be further assumed that the frequency axis is shifted in the described way. The specific number of SM j does not matter for further analysis.

The right boundary of the p-th SF ($p = \overline{0, N_{SF}}$) is determined by the following expression:

$$f_{SF}(p) = f_1' + pB_{SF},$$

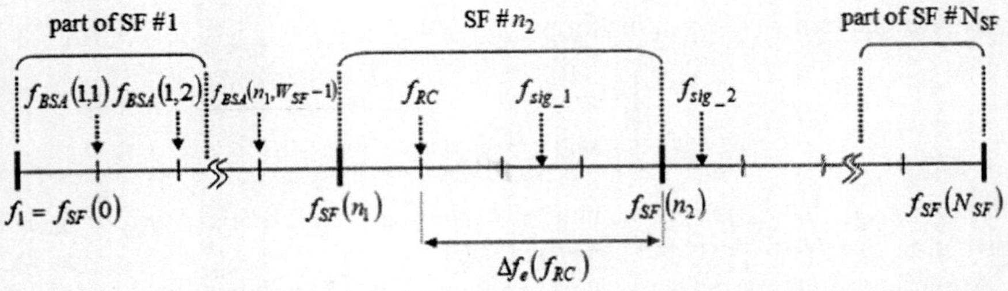

Fig. 7. The marking of the frequency axis for the signal identification in the synchronous mode.

while the right boundary of the q-th BSA belonging to it ($q = \overline{0, W_{SF}}$)–by the following one:

$$f_{BSA}(p, q) = f_{SF}(p - 1) + qB_{SA}, \qquad p > 0.$$

Then the distance along the frequency axis from the frequency f to the right boundary of the p-th SF (lying higher on the frequency axis) is determined by the following expression:

$$d(p, f) = f_{SF}(p) - f,$$

the distance from the frequency f to the right boundary of the SF that contains it is

$$\Delta f_e(f) = \min_{p=1,N_{SF};\ d(p,f)>0} \{d(p, f)\},$$

while the number of this SF is determined as

$$n_e(f) = \underset{p=1,N_{SF};\ d(p,f)>0}{\arg\min} \{d(p, f)\}.$$

Since, according to the condition, the moment of the signal emission at the frequency f_{sig} coincides with the start of sampling in BSA with the initial frequency f_{RC}, the identification time t_{idf} that depends on the signal frequency and the receiver tuning frequency, is the sum of the data gathering time t_{1f} in the remaining part of SF (from f_{RC} to the final frequency of SF) and the data gathering and processing time t_{2f} in the next SFs, the last of which contains the frequency f_{sig}:

$$t_{idf} = t_{1f} + t_{2f}. \tag{5}$$

If the time of SF data processing is longer than the time that takes to accumulate data in it, then, before starting data processing of SF containing the frequency f_{RC}, one should wait not only for the completion of the data gathering for the frequency section $\Delta f_e(f_{RC})$, but for the end of the previous SF data processing. With this in mind, the first term in (5) takes the form:

$$t_{1f} = \begin{cases} \left(\Delta f_e(f_{RC}) / B_{SA} W_{SF} - 1\right) t_{SFg} + t_{SFp}, & t_{SFp} \geq t_{SFg}, \\ \Delta f_e(f_{RC})\, t_{SFg} / B_{SA} W_{SF}, & t_{SFp} < t_{SFg}, \end{cases} \tag{6}$$

The second term in (5) is of the form:

$$t_{2f} = \begin{cases} (r+1)\, t_{SFp}\,, & t_{SFp} \geq t_{SFg}\,, \\ r t_{SFg} + t_{SFp}\,, & t_{SFp} < t_{SFg}\,. \end{cases} \tag{7}$$

where $r = n_e(f_{sig}) - n_e(f_{RC})$ is the number of the analyzed full SFs. At $f_{sig} < f_{RC}$, the substitution $r \to N_{SF} - r$ should be carried out.

According to the formulas (6, 7), it can be seen that when the distance along the frequency axis between f_{sig} and f_{RC} increases TID also increases (at $f_{sig} > f_{RC}$, otherwise it decreases), while the corresponding dependence has a piecewise linear (with the equal width of the linear segments) monotonic character. Therefore, the expression for the average TID is obtained by averaging the expressions for the minimum and maximum TIDs.

The minimum TID corresponds to the case when $f_{sig} = f_{RC} = f_{BSA}(p, W_{SF} - 1)$, where p that is the number of SF can be anything and does not depend on the signal and receiver frequencies. Then the first term in expression (5) takes the form:

$$t_{1\_\min} = \begin{cases} \left(1/W_{SF} - 1\right) t_{SFg} + t_{SFp}\,, & t_{SFp} \geq t_{SFg}\,, \\ t_{SFg}/W_{SF}, & t_{SFp} < t_{SFg}\,, \end{cases}$$

while the second term, taking into account that $r = n_e(f_{sig}) - n_e(f_{sig}) = 0$, is $t_{2\_\min} = t_{FSp}$.

Finally, the minimum TID is

$$t_{id\_\min} = \begin{cases} \left(1/W_{SF} - 1\right) t_{SFg} + 2t_{SFp}\,, & t_{SFp} \geq t_{SFg}\,, \\ t_{SFg}/W_{SF} + t_{SFp}, & t_{SFp} < t_{SFg}\,. \end{cases}$$

The maximum TID also does not depend on the signal and receiver frequencies and corresponds to the limiting case when, at the moment of the signal emission, the initial frequency of the receiver BSA turns out to be by an infinitesimal value greater than the signal frequency (and the signal has not gotten into the BSA), and these frequencies themselves turn out to be at the very beginning of the SF, that is, when

$$f_{RC} - f_{sig} \to +0, \qquad f_{sig} = f_{SF}(p),$$

where p, the number of the SF, can be anything.

Then, for the first term in the expression (5), one gets:

$$t_{1\_\max} = \begin{cases} t_{SFp}\,, & t_{SFp} \geq t_{SFg}\,, \\ t_{SFg}, & t_{SFp} < t_{SFg}\,. \end{cases}$$

The number of the analyzed full SFs, taking into account that $f_{sig} < f_{RC}$, is $r = N_{SF} - \left(n_e(f_{sig}) - n_e(f_{RC})\right) = N_{SF}$. Then the second term in the expression (5) can be defined as

$$t_{2\_\max} = \begin{cases} (1 + N_{SF})\, t_{SFp}\,, & t_{SFp} \geq t_{SFg}\,, \\ N_{SF} t_{SFg} + t_{SFp}\,, & t_{SFp} < t_{SFg}\,. \end{cases}$$

As a result, the maximum TID is

$$t_{id\_max} = \begin{cases} (2 + N_{SF})\, t_{SFp}\,, & t_{SFp} \geq t_{SFg}\,, \\ (1 + N_{SF})\, t_{SFg} + t_{SFp}\,, & t_{SFp} < t_{SFg}\,. \end{cases}$$

And the average TID is then

$$t_{id} = \frac{t_{id\_min} + t_{id\_max}}{2} = \frac{1}{2}\begin{cases} \left(1/W_{SF} - 1\right) t_{SFg} + (4 + N_{SF})\, t_{SFp}\,, & t_{SFp} \geq t_{SFg}\,, \\ \left(1/W_{SF} + 1 + N_{SF}\right) t_{SFg} + 2t_{SFp}\,, & t_{SFp} < t_{SFg}\,. \end{cases}$$

The resulting expression for the TID depends not only on the time spent for the data gathering and processing in the SF, but also on the SF width. To eliminate the last dependence, the expression should be rewritten in the following form:

$$t_{id} = \frac{1}{2}\begin{cases} (1 - W_{SF})\, t_{1g} + (4 + N_{SF})\left(W_{SF} t_{1p} + t_l\right), & W_{SF} t_{1p} + t_l \geq W_{SF} t_{1g}\,, \\ (1 + W_{SF}(1 + N_{SF}))\, t_{1g} + 2\left(W_{SF} t_{1p} + t_l\right), & W_{SF} t_{1p} + t_l < W_{SF} t_{1g}\,. \end{cases}$$

In the case when $t_l = 0$, the expression for the average TID simplifies:

$$t_{id} = \frac{1}{2}\begin{cases} (1 - W_{SF})\, t_{1g} + (4 + N_{SF})\, W_{SF} t_{1p}\,, & t_{1p} \geq t_{1g}\,, \\ (1 + W_{SF}(1 + N_{SF}))\, t_{1g} + 2W_{SF} t_{1p}\,, & t_{1p} < t_{1g}\,. \end{cases}$$

In Fig. 8, the diagrams of the dependencies of the average TID in the synchronous and deferred modes on the BSA processing time with and without RSA optimization are presented.

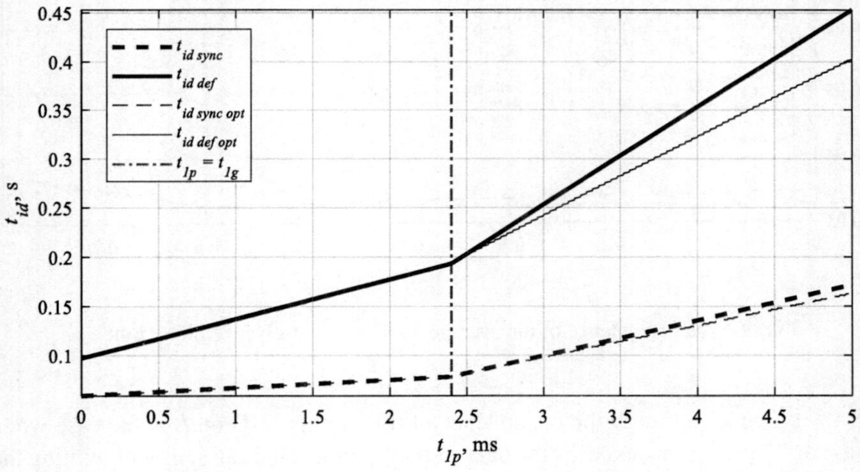

Fig. 8. The dependency of the average TID on the BSA processing time.

As Fig. 8 shows, for any processing time, the average TID in the synchronous mode is less than in the deferred one, and its value in both modes increases faster after the point of equality between the time of data gathering and the processing time. It can also be seen that the RSA optimization leads to a decrease in the average TID in both modes, and in the deferred mode the gain from it is greater than in the synchronous one.

Figure 9 shows the dependence of the average TID on the receiver retuning time in the proposed modes, as well as when applying the RSA optimization. As one can see in Fig. 9, without the RSA optimization, the minimum value of the average TID in both modes is achieved at such a receiver retuning time that ensures the equality of the data gathering time and the processing time. And when optimization is applied, the average TID ceases to depend on the retuning time.

In Fig. 10 the diagram of the dependence of the average TID on the number of the bandwidths of simultaneous analysis in the AFR L when the data processing time is less than the gathering time. Figure 10 demonstrates that the average TID is proportional to the width of SF W_{SF}. It is also noticeable that the average TID in the synchronous mode at any AFR width is less than in the deferred mode, and its minimum value corresponds to the case $W_{SF} = 1$ when the SF consists of only one BSA. For the same case, in Fig. 10, the following feature can be observed: the average TID with and without the accumulation coefficient optimization is the same. This feature is based on the initial assumption that the moment of the signal emission coincides with the start of sampling in the BSA.

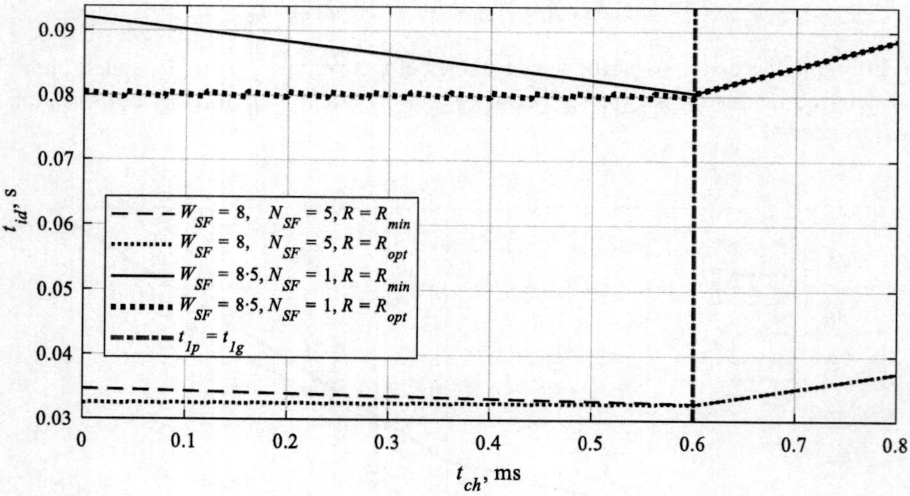

Fig. 9. The dependence of the average TID on the receiver retuning time.

Now it is time to turn to the dependence of the average TID on L in the case when the time of SF processing exceeds the data gathering time. One can start with writing the expression for the average TID $t_W(x)$ for the specified width of the SF x and the number of BSAs in the PS L:

$$t_W(x) = \frac{1}{2}\begin{cases} (1-x)\,t_{1g} + (4 + L/x)\left(xt_{1p} + t_l\right), & xt_{1p} + t_l \geq xt_{1g}, \\ \left(1 + x\left(1 + L/x\right)\right)t_{1g} + 2\left(xt_{1p} + t_l\right), & xt_{1p} + t_l < xt_{1g}. \end{cases}$$

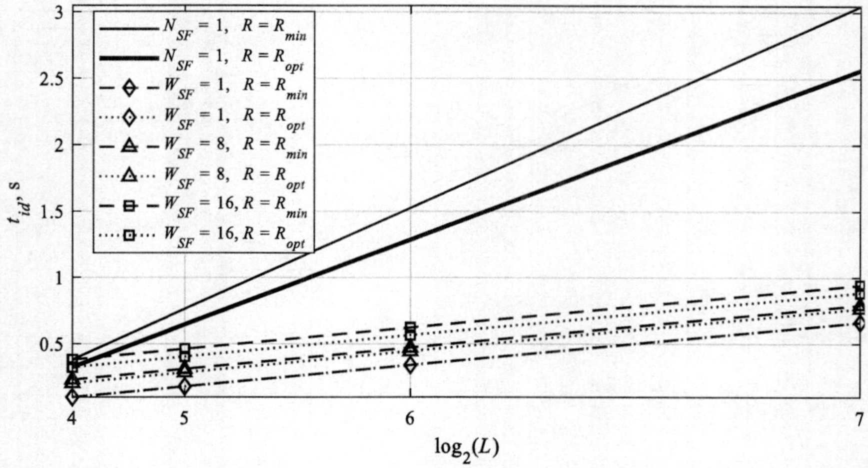

Fig. 10. The dependence of the average TID on the number of the bandwidths of simultaneous analysis in the AFR at $t_{SFp} < t_{SFg}$.

The next task is to determine the difference between the TID under the set SF width and under the single SF width:

$$\Delta t_W(x) = t_W(x) - t_W(1).$$

In Fig. 11, the diagram is presented that shows the dependence of $\Delta t_W(x)$ on L when the time of the SF processing is longer than the data gathering time. The smaller the value of $\Delta t_W(x)$ on the chart, the smaller is the average TID.

Figure 11 also informs us that the SF width providing the minimum value of the average TID differs depending on the number of the BSAs in the AFR L. When L decreases, the use of the SF with a narrower width is preferable, and when L increases, one prefers a wider SF width.

The next step is to obtain an expression for the optimal SF width providing the minimum value of the average TID depending on the number of the BSAs in AFR L and other analysis parameters. For that purpose, it should be agreed that L is a multiple of the SF width, and the array of its possible values (the divisors of L) is denoted as $\mathbf{W}_{arr}$. Then, for the optimal SF width W_{opt}, one can write:

$$W_{opt} = \underset{W_{SF} \in \mathbf{W}_{arr}}{\arg\min} \left[(1 - W_{SF})\, t_{1g} + \left(4 + L/W_{SF}\right)\left(W_{SF} t_{1p} + t_l\right) \right]. \quad (8)$$

To analyze the nature of the dependence of the optimal SF width on other factors, after equating the derivative of the expression (8) in brackets to zero, a real solution for it is obtained:

$$W_{opt\_R} = \sqrt{Lt_l / \left(4t_{1p} - t_{1g}\right)}. \quad (9)$$

It can be seen from (9) that, with an increase in the number of BSAs in AFR L, the time of re-initialization of the processor t_l, as well as the time of the data gathering in the

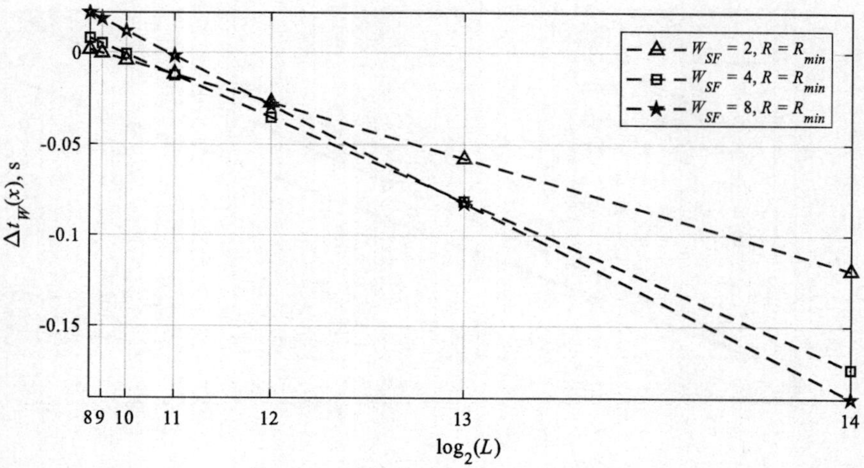

Fig. 11. The dependence of the difference of the average TID on the number of BSAs in the AFR at $t_{SFp} > t_{SFg}$.

BSA t_{1g}, the approximate (real) solution for the optimal SF width increases, while with an increase in the BSA processing time t_{1p}, it decreases. At zero time t_l, the optimal SF width becomes zero, which is consistent with the results of the analysis of the diagrams in Fig. 10.

It should be noted that when changing the SF width, it may turn out that the processing time no longer exceeds the data accumulation time. The same happens when using the RSA optimization. In this case, as it can be seen from the expression for the average TID, its minimum value is provided by the SF with the narrowest possible width.

At the end of the analysis, one compares the proposed approaches for the case when the RME "ARGAMAK-D11" [9], the typical values of the analysis parameters specified earlier, and $t_{1p} = 5$ ms are used. The task is to estimate the average TID gain from the use of the synchronous mode, as well as the RSA optimization, in comparison with the base case when using the deferred mode without optimization. The general expression for the gain is:

$$E = \left[\left(t_{id\_def} - t_{id\_x}\right) / t_{id\_def}\right] \cdot 100\,\%.$$

where t_{id_def} is the average TID for the deferred mode without the RSA optimization and t_{id_x} is the average TID in the compared cases. According to the results of the calculations, the gain when using the RSA optimization in deferred mode is 11% and when using synchronous mode without optimization it is 62% (at $t_{1p} = 2$ ms, it is 51%, but the RSA optimization cannot be applied in this case). With the RSA optimization the gain is 64%. These results confirm the pattern observed in the diagrams in Figs. 8, 9 and 10: switching from the deferred mode to the synchronous mode gives the most significant gain in TID, and the use of the RSA optimization allows it to be further increased, but by a relatively small magnitude.

9 Conclusion

The efficiency of the practical implementation of the identification algorithms significantly depends on the applied mode of data gathering and processing. The proposed modes, the deferred mode and the synchronous one, differ in the average identification time, the memory capacity required to store spectral data, and also in the complexity of the identification decision-making process.

The deferred mode, in which the identification is performed after the accumulation of the spectral data for the entire analyzed frequency range, is characterized by the least complexity of the identification decision-making process. However, the average identification time, as well as the required memory capacity for it, appear to be greater than for the synchronous mode.

The synchronous mode, in which identification starts after the completion of the data gathering for a fragment of the analyzed frequency range, in comparison with the deferred mode, is characterized by a lower average signal identification time (for example, it is 62% lower when using the RME "ARGAMAK-D11" and applying the typical analysis parameters), and it requires less memory capacity. Nonetheless, in this case, the process of making identification decisions is characterized by greater complexity, as it not only includes the algorithm for the refinement of these decisions, but it also presupposes the need to maintain a certain state of the processor at the start of the analysis of each of the spectral fragments.

Based on the results of the analysis, it can be concluded that if it is necessary to ensure the minimum possible average signal identification time and the memory capacity used, as well as if it is possible to implement an algorithm for refining the identification results, one should use the synchronous mode. Otherwise, in particular, when data gathering and processing are performed at a sufficiently high speed, the deferred mode is preferable.

The gain in average identification time from using the synchronous mode increases with the number of fragments into which the spectral panorama is divided for processing (and hence the increase in the width of the analyzed frequency range is also a positive factor). Therefore, with a small number of fragments in this mode (and a narrow width of the analyzed frequency range), using the deferred mode should be considered a better choice.

In both modes, when the time needed for processing the spectral fragment analyzed at a time exceeds the time needed for its accumulation, the data gathering process begins stopping. In this case, the application of the proposed optimization of the spectrum accumulation coefficient ensures a continuous data gathering process, and also reduces the average signal identification time (by 11% in the deferred mode, and by 2% in the synchronous mode when using RME "ARGAMAK-D11" and applying the typical analysis parameters).

Acknowledgements. This study was financially supported by the Ministry of Science and Higher Education of the Russian Federation (research project No. FSWF-2020–0022) as well as RFBR and CNRS (research project number 20–51-15001).

References

1. ITU radio regulations. https://www.itu.int/pub/R-REG-RR-2016. Accessed 01 Aug 2022
2. Holland, O., Bogucka, H., Medeisis, A.: Opportunistic Spectrum Sharing and White Space Access: The Practical Reality. Wiley, Hoboken (2015)
3. Rembovsky, A.M., Ashikhmin, A.V., Kozmin, V.A:. Radio Monitoring: Automated Systems and Their Components. Springer, New York (2018)
4. Recommendation ITU-R SM.1600-3: technical identification of digital signals. https://www.itu.int/rec/R-REC-SM.1600. Accessed 01 Aug 2022
5. Radiomonitoring and Radiolocation catalog 2016. https://cdn.rohde-schwarz.com/ru/downloads_45/common_library_45/brochures_and_datasheets_45/Radiomonitoring_and_Radiolocation_Catalog.pdf. Accessed 01 Aug 2022
6. Hoehn, J.R., Gallagher, J.C., Sayler, K.M.: Overview of department of defense use of the electromagnetic spectrum. Congressional Research Service, Washington (2021)
7. Mubarak Al-Sawafi, M., Ben-Yakov, U., Cutler, R. et al: Spectrum Monitoring Handbook. Radiocommunication Bureau, International Telecommunication Union, Geneva (2011)
8. Chang, C.: Hyperspectral Data Processing: Algorithm Design and Analysis. Wiley, Hoboken (2013)
9. ARGAMAK-D11 (Model 4): two-channel digital panoramic radio receiver. https://www.ircos.ru/en/rsv_d11-4.html. Accessed 01 Aug 2022

Business Model as a Basic Building Block of Business Architecture

Kamelia Shoilekova(✉)

Angel Kanchev University of Ruse, 8 Studentska Str., 7000 Ruse, Bulgaria
kshoylekova@uni-ruse.bg

Abstract. The accelerating rate of technology development and the market changes forced organizations to expand their focus on innovations following the aim to remain competitive in the face of shrinking product life cycle and growth of market uncertainty. Companies understood that they should work out their business model conception for strategy planning yet at the level of constructing the business architecture.

Keywords: Business architecture · Business model · Element

1 Introduction

In the very beginning of organization establishment, it is important to design an architecture allowing both the transparency at the business start-up and the probability for an effective and successful deployment of business strategy.

In spite of the natural progression from business models (representing business strategy, goal and conception) to business architecture (as an operating plan for implementing the conception, strategy and goal), these two ideas have been separately developed, neglecting that they are part of a whole.

2 Business Architecture

Business architecture as a term was adopted since the mid-nineties of twentieth century when many organizations tried to optimize their businesses.

Business architecture is a discipline that models business and it can be used as a bridge for converging the differences between business and technologies.

On the one side, business architecture can be defined as a manual worked out for software designers. The manual should describe a detailed route to be followed by the team of software architecture developers.

On the other side, business architecture can be presented as a set of business models, whereas a business model is a range of events linked to the business, which include various business functions, business processes, organization units, and activities of an enterprise.

R. Silhavy et al. (Eds.): CoMeSySo 2022, LNNS 596, pp. 563–567, 2023.
https://doi.org/10.1007/978-3-031-21435-6_48

3 Business Model

Business model is a sum of elements connected by a certain condition aiming to represent the functioning units of an enterprise. It is a combination of elements and factors showing the key interactions between the enterprise and its encirclement as a whole. The process of designing a business model of organization steps on the creation of a number of model prototypes, which describe in details both the transparency of the company development and its position on the competitive market. According to [3] the attention of business analyzers and software designers is focused on the business model due to the following reasons:

- Software developers need a description of all critical and complicated points of business activities in order to transform them into functionality that will further help the business development [5, 8];
- Time-to-market should be decreased as much as possible;
- Methods, aids and tools applied should allow fast transformation of any possibility into a new or additional functionality of the business [6].

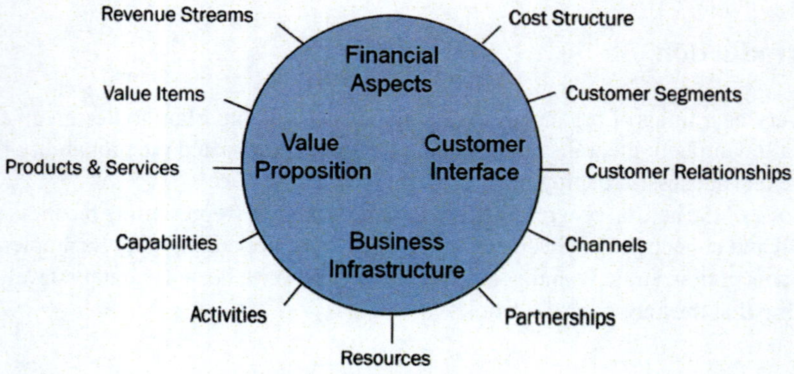

Fig. 1. Business model aspects [3]

Figure 1 shows a business model that is dependent and built on 11 main elements divided into 4 groups. It is noted that the elements used for the business model presented in this paper are the basic ones, but are not divided into groups. Figure 2 illustrates the nine basic elements, which influence the design of an enterprise business model.

4 Linking Business Model to Business Architecture

In substance, the main goal of a business model is to provide a realistic picture of the current-state organization. The presentation may change in the course of time, as it gives a clear view to what extent the organization schedule and strategies have been fulfilled, and whether it is necessary to re-think the possibilities for new, better strategies, goals and objectives. If these changes are included in the business architecture models, they can be more easily implemented within the software architecture, meaning that the organization goals shall be more rapidly achieved [4].

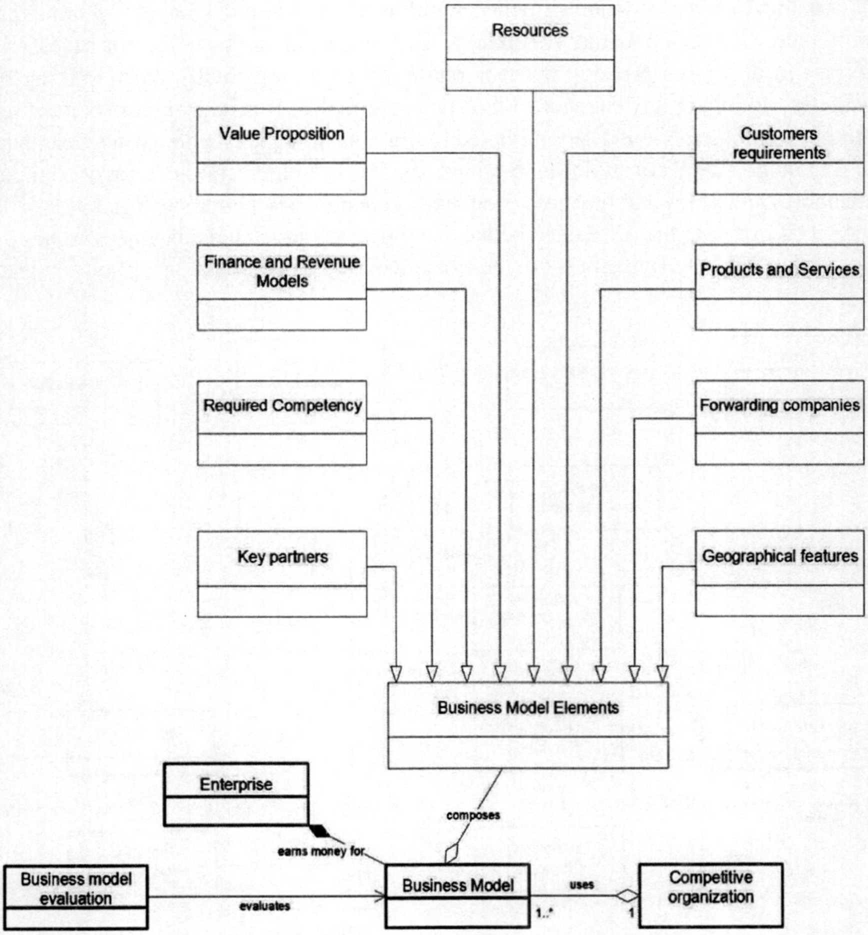

Fig. 2. Elements, which influence the design of business model

Freuently, the business models improve the communication among the business leaders providing them with a common perspective for the core business logic of the organization. The availability of a common perspective, structure, and understanding allows for working out a more attractive business design and a successful deployment of business models [2, 7, 9]. The only critical point in this situation is caused by the differences between the designing methods of IT developers and the business analyzers. However, the models are discussed on working meetings placing an accent on the concepts but not on the ways of their implementation. In these cases, participants show preferences for using UML as sketch, minding that it presumes non-compliance with the basic conceptions and rules.

Figure 3 shows an enterprise business model, and particularly, its links to other key elements, which participate in the construction of the business architecture. It is evident that for the business model design the organizations should guarantee that software developers have to design and construct all the elements of the business model in order to allow business agility blossoming in the conditions of the continual market dynamics and compliance with national state regulations. Besides, customers should not necessarily take part in designing process, but it is recommendable, especially when making the model description, or any changes of the current model, as it is important to contact real customers. Another essential point related to Fig. 3 is how the business model elements take part in the model construction. Business model is a unique composition of elements that enables analyzers and business executives generate new better strategies.

According to [1] the difference between a business model and a business strategy is the existence of more semantics, i.e. the existence of two different conceptions, at least.

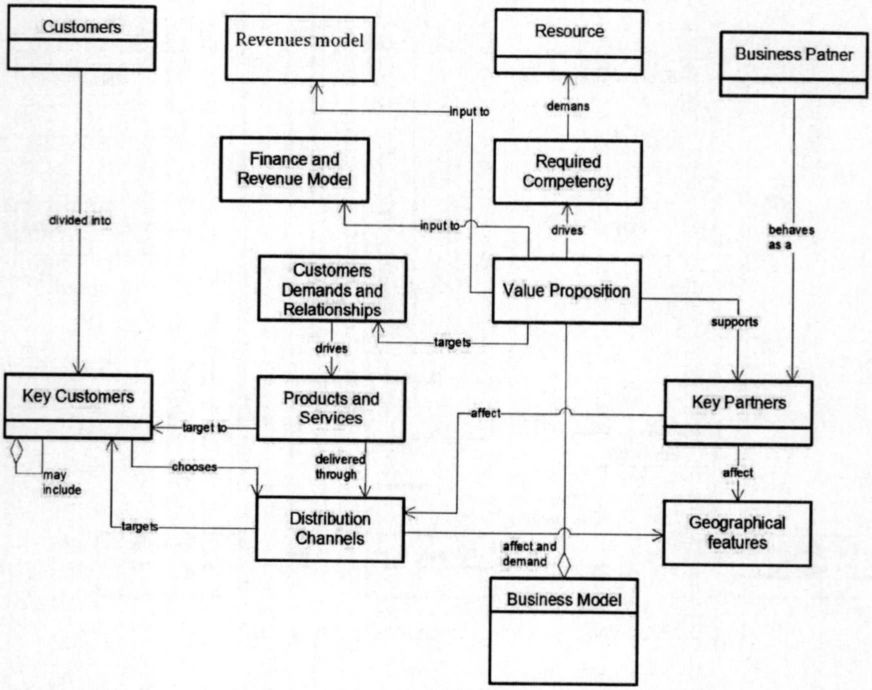

Fig. 3. Linking business model to other key elements taking part in business architecture

5 Conclusion

Business model is a basic part of business architecture. It appears as the main building block after outlining the goals, objectives, and perspectives of the business project. Business architecture, on the other side, should map in details all the particulars and functionalities of the business so that the software related to that business is properly developed in compliance with all the functionalities discussed and contained in the business architecture.

References

1. Osterwalder, A., Pigneur, Y.: Business model generation, 2010, Business Model Generation. 2010. Alexander Osterwalder, Yves… | by Daniel Good | Make Work Better | Medium
2. https://cdn.ymaws.com/www.businessarchitectureguild.org/resource/resmgr/docs/linkingbusinessmodelsb.arch.pdf
3. dupont.pdf
4. https://www.bptrends.com/essentials-of-business-architecture-design-the-business-phase/
5. Krastev, G., Voinohovska, V.: Smart mobile application for public transport schedules—logical model. TEM J. (2020). ISSN: 2217-8309. https://doi.org/10.18421/TEM92-15
6. Shoilekova, K.: Development of a business architecture model and its role in constructing an organizational architecture. In: Silhavy, R., Silhavy, P., Prokopova, Z. (eds.) CoMeSySo 2021. LNNS, vol. 232, pp. 216–223. Springer, Cham (2021). https://doi.org/10.1007/978-3-030-90318-3_21
7. Petrov, V.l., Rusev, R., Rusev, I.: Anaesthesiology in the big data era—are we ready for that?, Anaesthesiol. Intens. Care (2021) ISBN: 1310-4284
8. Krastev, G., Voinohovska, V.: Smart mobile application for public transport schedules—data organization and program realization 2020. Int. Congr. Hum. Comput. Interact. Optim. Robot. Appl. (HORA) 1–4 (2020). https://doi.org/10.1109/HORA49412.2020.9152884
9. Rusev, I., Rusev, R., Ivanova, B.: Training of medical staff for using data—usage of the full capability of smart devices In: EDULEARN21 Proceedings, pp. 8804–8810 (2021)

Facial Recognition Technologies Patent Landscape

A. S. Nikolaev[1(✉)], T. G. Maximova[2], I. E. Sakhno[1], A. A. Antipov[1], and S. V. Murashova[1]

[1] IP Management, Faculty of Technological Management and Innovations, ITMO University, 11 Tchaikovsky St., Saint-Petersburg, Russian Federation
{nikand,ip}@itmo.ru

[2] Faculty of Infocommunication Technologies, ITMO University, 11 Tchaikovsky St., Saint-Petersburg, Russian Federation

Abstract. This article presents a patent information analysis of patent activity in the field of face recognition technologies. The paper presents a view at the current state and prospects for the development of this technology. The study of patent activity accompanied by a thorough decomposition of the subject area, which includes the main areas of application of face recognition technology. The depth of the patent search was 20 years; more than 6 thousand patent documents studied using big data analysis methods. Because of the study, there is a decrease in patent activity in the field of face recognition technologies. This may be due to the consequences of the COVID-19 pandemic and the accompanying quarantine, remote forms of work and study, and various forms of restrictive measures. The authors give recommendations on the management of innovative activities in the field of face recognition for individual companies and organizations that regulate the global market.

Keywords: Facial recognition technologies · Intellectual property management · Patinformatics · Intellectual property analysis · Patent landscape

1 Introduction

Modern digital companies in the conditions of global uncertainty seek to reduce the number of high-risk innovative projects, giving preference to the modernization of existing technologies. The most popular format for innovative development is the export of technical solutions to related areas of human activity. This approach ensures a relatively rapid spread of technology due to its familiarity to the customer. Traditionally, successful technological expansion occurs in the areas of mass consumption of goods and services. Such technologies that have successfully come to the mass consumer from more local niches include a face recognition system.

Facial recognition is a way of identifying or confirming a person's identity by their face. The facial recognition system can be useful for identifying people in photos, videos or in real time. Facial recognition has a large number of applications, including security,

R. Silhavy et al. (Eds.): CoMeSySo 2022, LNNS 596, pp. 568–583, 2023.
https://doi.org/10.1007/978-3-031-21435-6_49

identity verification, Internet communication, and computer entertainment. The development of such systems began in the 1960s, beginning with a form of computer application. Since their inception, facial recognition systems have seen more recent adoption in smartphones and other forms of technology such as robotics. Since computerized face, recognition involves the measurement of a person's physiological characteristics, facial recognition systems classified as biometrics.

A face recognition method is a technology that can match a human face with a digital image or video frame from a face database, commonly used to authenticate users with an identity verification program. The system works by accurately identifying and measuring facial features from a given image. This method may consist both in a certain configuration of the elements that make up the face recognition system, and in the algorithm itself and in the method used in face recognition.

For the purposes of this study, we will indicate under the face recognition system not only the face recognition algorithm itself, but also the entire set of processes before and after it. At the first stage, the face detected and localized in the image. The system input receives an encoded video stream from the camera, which is a sequence of frames that can contain faces. The data enters the so-called decoder, which decodes the video stream into separate frames for further processing by the detector. Further, the detector on each frame determines the presence, in principle, of any objects in the frame that look like a human face. At the next stage, the alignment of the face image (geometric and brightness), the calculation of features and direct recognition by comparing the calculated features with the standards embedded in the database are performed. The main difference between all the presented algorithms will be the calculation of features and comparison of their sets with each other.

Thus, the algorithm is only part of the face recognition system. Most often, there are several algorithms in the system, and each is important for achieving a high-quality result. Let us say a good feature vector extraction algorithm and a bad detector algorithm will not give a good result. All high-quality algorithms together and allow you to get a high level of face recognition.

The face recognition system created as a security tool is now widely used as elements of the smart interface of various gadgets. This opens up new opportunities for manufacturers of smart devices. However, it is necessary to determine the patent purity of this technology in order to avoid problems with the copyright holders of patented solutions [5][1]. The purpose of this study is to develop a technological landscape for a face recognition system based on an analysis of the scientific, technical and patent-licensing activities of organizations involved in the development of face recognition methods and to identify development trends and promising areas of scientific, technical and commercial activities in this field of technology. The authors conducted a study of the technical level, novelty and development trends of the development object, and searched for patent and non-patent documents that reveal scientific, technical and engineering solutions similar to the research object.

[1] Patent portfolio analysis as a useful tool for identifying R&D and business opportunities - an empirical application in the nutrition and health industry [Text]/B. Fabry, H. Ernst, J. Langholz, M. Koster//WPI. - 2006. - Vol. 28, № 3. - P. 215–225.

This study focuses on a face recognition method based on an embedded algorithm for face detection on static images or real-time video sequences.

2 Discussion

The accuracy of face recognition systems as a biometric technology is lower than iris and fingerprint recognition face recognition system has become widespread due to the non-contact process and easy-to-use method. Therefore, it has become one of the three identification methods used in ePassports, as well as the biometric of choice for many other security applications [10][2]. Cognitive and perceptual models provide an important platform for interdisciplinary research, bringing scientists from seemingly disparate fields together and allowing them to share methodologies and results on a common problem. Data from neurobiological, psychological, perceptual, and cognitive experiments provide potentially useful insights into how our visual system encodes, stores, and recognizes faces. This knowledge can then linked to artificial solutions.

There are several reasons for the increased interest in facial recognition, including growing public concern about security, the need for identity verification in the digital world, and the need for facial analysis and modeling techniques in media management and computer entertainment. Therefore, recent advances in automated face analysis, pattern recognition, and machine learning have made it possible to develop automatic face recognition systems to solve these problems. The authors of the paper «Learning to resolve uncertainties for large-scale face recognition» consider the technology of eliminating uncertainties by ranking and "suppressing" low-quality images in the process of large-scale face recognition [1][3].

Also authors of the article «LocalFace: Learning significant local features for deep face recognition» propose the LocalFace recognition architecture, which increases the efficiency of extracting dynamic local features of the face. They also developed a technology that allows reconstructing high-resolution face images, which is comparable to advanced approaches based on super-resolution in the visual quality domain, and ensure that the corresponding low-resolution image [11][4].

The direction associated with the development and training of neural networks is already one of the most relevant and promising, along with the development of information transfer technologies, various kinds of cloud services and meaningful analysis of large amounts of data. Today, we can already say with confidence that in some application cases, systems based on neural networks and recognition are effective. The architecture and training methods of neural networks continue to develop actively and open up new

[2] Romanyuk O.N., Vyatkin S.I., Pavlov S.V., Mykhaylov P.I., Chekhmestruk R.Y., Perun I.V (2020) Face recognition techniques. Informatics, Control, Measurement in Economy and Environment Protection. 2020. Book. 10. № 1. pp 52–57.

[3] Abhijeet Boragule, Hamna Akram, Jeongbae Kim Ph.D. Learning to resolve uncertainties for large-scale face recognition.Pattern Recognition Letters Volume 160, August 2022, Pages 58–65 https://doi.org/10.1016/j.patrec.2022.06.004.

[4] Sisi Li, Zhonghual Liua, Di Wu. Low-resolution face recognition based on feature-mapping face hallucination. Computers and Electrical Engineering. Volume 101, July 2022, 108136. https://doi.org/10.1016/j.compeleceng.2022.1081360.

applications in practice. Answering the question of how neurons encode facial identity, during the experiments, the authors of the work «Face identity coding in the deep neural network and primate brain» concluded that artificial units have the same representation of facial features as primate neurons [15][5]. For example, the so-called recurrent neural networks already used today, which has significantly improved the quality of face recognition systems based on them. Still, the future of face recognition systems lies in interdisciplinary use, and in particular, in combining this technology with applications in the IoT space, such as automotive, smart home or smart city environments. A study by Sonia Poltoratski, Kendrick Kay, and Dawn Finzi shows that spatial processing in high-level visual areas dynamically used for recognition, and offers an efficient approach to link neural computation to behavior using receptive fields [12][6].

Kade Crockford and a number of other researchers at the same time draw attention to the ethical aspect of using this technology, which can restrict the rights of citizens, which requires the introduction of special regulations [7][7]. The legal regimes of many states have different ways of assessing the possibility of collecting photographs of citizens and their use for commercial purposes. Limiting the possibilities of using biometrics is one of the debatable issues about guaranteeing the rights and freedoms of citizens. The very discussion on this issue reduces consumer confidence in the technology and may negatively affect the speed of its introduction into new industries. The readiness of people to share data about their lives in order to improve the quality of service is one of the indicators of the development of an innovative culture of the population. This issue has already been the subject of our previous publications [2][8].

For this technology industry, the growing market for facial recognition is generally financially clean and simple. As investment in facial recognition technology increases and technology matures, we see that the use of facial recognition in some use cases, including new ones, is becoming more widespread. The COVID-19 pandemic has been a significant factor here. The COVID-19 pandemic is also leading to the emergence of facial recognition systems in combination with other biometric methods. Due to the pandemic, digitalization has accelerated in a number of areas, so the use of facial recognition technologies will inevitably increase in some of them.

[5] Jinge Wang, Runnan Cao, Nicholas J. Brandmeir. Face identity coding in the deep neural network and primate brain. Communications Biology volume 5, Article number: 611 (2022). https://doi.org/10.1038/s42003-022-03557-9.

[6] Sonia Poltoratski, Kendrick Kay, Dawn Finzi. Holistic face recognition is an emergent phenomenon of spatial processing in face-selective regions. Nature Communications volume 12, Article number: 4745 (2021). https://doi.org/10.1038/s41467-021-24806-1.

[7] Kade Crockford. How to regulate face recognition technology. Nature Human Behaviour volume 6, 476 (2022). https://doi.org/10.1038/s41562-022-01329-3.

[8] Antipov A.A., Maximova T.G., Nikolaev A.S., Agapova A.V., Murashova S.V. Philosophical Foundations of the Digital Culture//Proceedings of the 31st International Business Information Management Association Conference (IBIMA). Innovation Management and Education Excellence through Vision 2020, 2018, pp. 6668–66673.

3 Processing

3.1 The Current State of the Face Recognition Technology Market

Government and defense sector has the highest market share due to the growing demand for recognition solutions in law enforcement and security services (Fig. 1). These solutions used to identify and verify suspected criminals, as well as cross-border monitoring, among other things, in real time, while providing a secure environment. For example, in April 2018, the Indian Police in New Delhi implemented facial recognition systems to identify lost or abducted children. It applied to over 45,000 children, of which about 3000 children have found throughout the city.

This technology is widely used in the banking insurance and services sector to counter fraudulent transactions, so it has a significant market share. This technology used in the retail and e-commerce segment to improve retail customer service and sales and help retailers proactively prevent organized crime in retail. For example, in June 2019, Chinese e-commerce companies such as Alibaba.com, and JD.com, among others, integrated face authentication technology with sales information. This technology helps understanding the allocation of resources and targeted sales behavior. Thus, the market is gaining momentum in the retail and e-commerce sector [13][9].

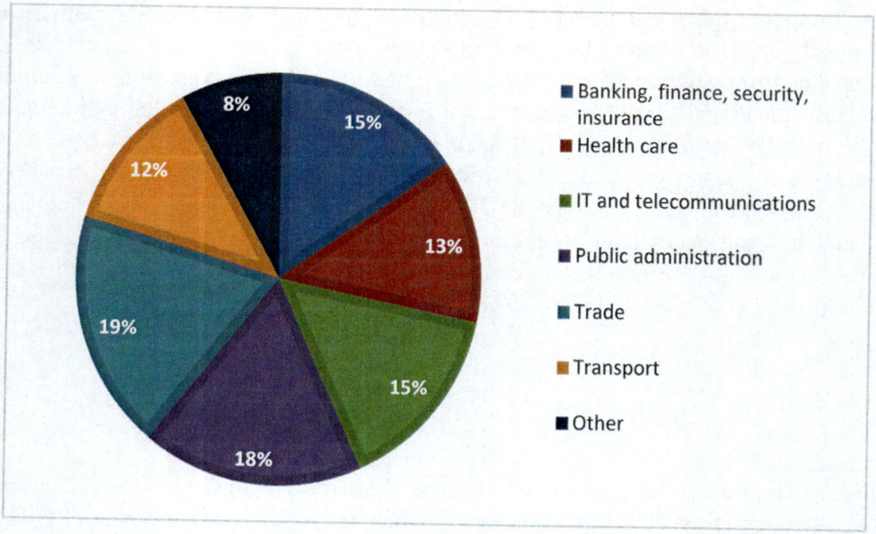

Fig. 1. Global facial recognition market share by end user, 2019–2021 [8][10]

As part of the study of market development prospects, we analyzed consumer expectations regarding facial recognition technology. The score formed based on Google

[9] The e-commerce companies are tackling slow sales growth with new technology based on shopping data.

[10] Market Research Report. Fortune Business Insights. 2019.

searches for the words "face recognition", "face recognition system" in English and Russian between January 1, 2004 and January 1, 2022. An assessment of demand dynamics based on the history of queries in Google Trends shows a rather low search activity for the queries "face recognition system" and "face recognition" compared to other technologies. We believe this is because this technology receives significantly less coverage in the media compared, for example, with a larger group of computer vision and machine learning technologies. Interestingly, this technology is twice as popular among users who make search queries in Russian. At the same time, the content generated by search algorithms for requests in Russian and English is almost identical and constitutes educational materials. You can also evaluate the popularity of requests in different countries of the world. It should note that the top five countries whose residents are interested in this technology do not include China and the United States, while there are low-income countries such as Sri Lanka, Ethiopia and Nepal. Look out for spikes in search queries in February 2010, June 2011, September 2017 and early 2020. This circumstance connected with the discussion in the media of the technology of smart cities that ensure the safety of citizens through a network of traffic cameras with the function of face recognition.

The current crisis caused by the ongoing pandemic has necessitated social distancing. Many organizations and governments have begun implementing work-from-home policies as part of the necessary coronavirus precautions. The ongoing crisis will force consumers and organizations to move to the online world and look for development drivers in a combination of traditional and remote approaches. The adoption of facial recognition solutions and other biometric technologies by end users continues to grow in response to the COVID-19 pandemic. Proper management and implementation of facial recognition technologies can help improve security and limit any physical contact. However, organizations must consider the interests of employees, as well as consumer privacy and security, due to the growing use of contactless biometric technologies and facial recognition software to address the challenges posed by the coronavirus. Currently, the ongoing COVID-19 pandemic is affecting more than 150 countries around the world and is increasing the importance of face authentication systems. For example, in 2020, Telepower Communication Co., Ltd., the world's leading provider of smart solutions, introduced facial recognition technology as an alternative to fingerprint scanners and to identify a person without any close contact [4][11].

The growing adoption of face recognition systems based on the artificial intelligence platform among enterprises expected to become one of the key market growth trends, which will increase the earnings of companies (Fig. 2).

This system uses artificial intelligence technology and machine learning algorithms to identify and verify identity from a digital image or video source. These technologies help increase user engagement, identify human expressions, manage security systems, analyze customer behavior, and enhance user security. For example, in June 2019, Vuzix, an American supplier of smart and augmented reality glasses, launched iFalcon Face Control Mobile, a fully autonomous AI-powered face recognition system. The device used to recognize faces in a crowd. In addition, large companies such as Amazon.com, Facebook, Inc., Microsoft Corporation, Google LLC, and others have used

[11] https://www.telpo.com.ru/face-recognition-machine/. Telpo.

artificial intelligence-based facial recognition applications to optimize user experience, performance, and security.

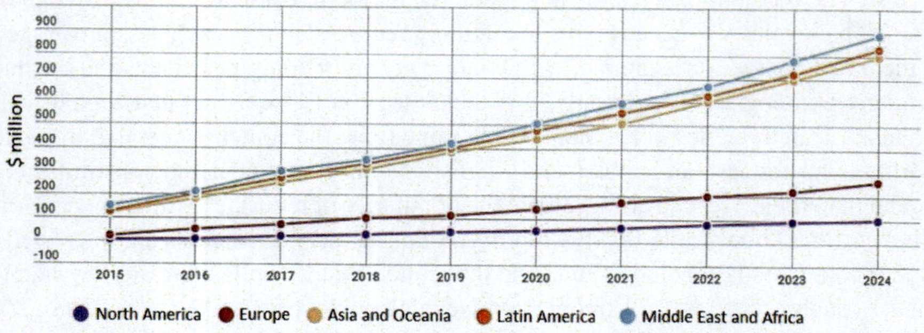

Fig. 2. Growth in revenues from the use of facial recognition technology [3][12]

Key market players are focusing on implementing various business strategies such as launching new products to meet the growing demand for facial recognition solutions. For example, in April 2020, Japan's THine Electronics, Inc., a semiconductor supplier, launched an artificial intelligence-based biometric facial recognition and fever detection system to provide a safe work environment. This system designed to detect body temperature and check people even when wearing facemasks. This system can also use in office buildings, shopping malls, and railway stations and other places where a large number of people gather [9][13].

Leading market players are focusing on expanding their product portfolio to strengthen their positions. Key market players such as FaceFirst, Inc. And Animetrics Inc., are focused on offering advanced facial recognition solutions and services to achieve a competitive advantage. In addition to this, various key giants such as NEC Corporation, Panasonic Corporation, Aware Inc. And others, focused on investing in research and development of facial recognition solutions to meet the growing demand from the industry.

Panasonic Corporation is one of the key market players that offers a variety of electronic solutions and technologies for end users in the consumer electronics, automotive, B2B and residential industries. The facial recognition systems offered by the company are easy to deploy and operate. In addition, they offer various advantages such as system extensibility, system cost reduction, high accuracy, face search and matching, etc.

List of leading profile companies: NEC Corporation (Japan), FaceFirst, Inc. (U.S.), IntelliVision (U.S.), Polaris Sensor Technologies, Inc. (U.S.), Aware, Inc. (U.S.), Ayonix Corporation (Japan), Cognitec Systems GmbH (Germany), Herta Security (Spain), Thales Group (Gemalto NV) (Netherlands), Animetrics (U.S.), Daon (Ireland), ID3 Technologies (France), Idemia (France). It should also be noted that the technology of

---

[12] Facial Recognition Market. Markets and Markets. 2020.

[13] New facial recognition and fever detection systems for safe work environment launched. Biometric Update. 2020.

the Russian company NtechLab with a project to detect pedestrians and cyclists on the roads [14].

3.2 Patent Activity in the Field of Face Recognition Technologies

For further analysis of the selected technological area, we used the patent publications of the World Intellectual Property Organization. After a patent search, a statistical sample formed, on which an analytical study carried out. The paper draws attention to the patent situation of face recognition systems, taking into account temporal dynamics and territorial characteristics. To classify innovative developments, we used sections of the International Patent Classification subclass G06K 9/00 from IPC 2021 that are relevant to the research topic. Figure 3 shows the dynamics of patent activity in the world in this area over the past 20 years.

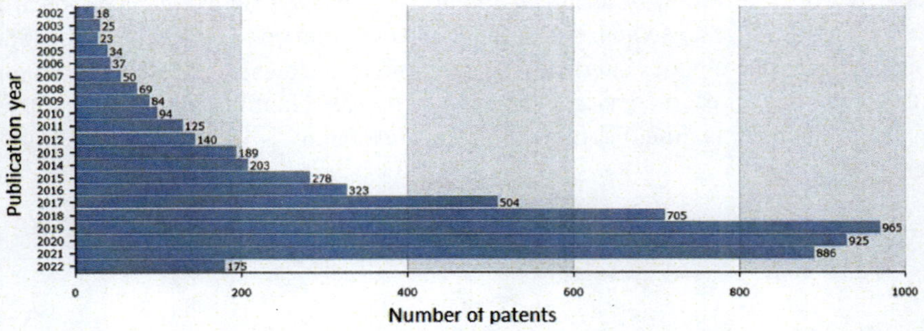

Fig. 3. Global patent activity in the field of face recognition systems [14]

The Chinese segment of patenting contains more than 5260 publications, which makes it the most representative of all studied. In order to make the selection more illustrative, we have focused on patent families containing an international publication. It is a set of patents (or applications) filed in several countries that related to each other by one or more common priority applications. This reduced the number of small families represented by only one national publication and lacking international coverage. Thus, the sample reduced to 320 patents. Further, each patent manually processed in order to weed out those that are not relevant and inappropriate for our subject area. Most of the patents excluded from the sample related only to the detection of a human face, and not to its recognition. That is, a face detected and other functions performed, for example, temperature measurement, and comparison of faces for cosmetic purposes, age determination, the use of digital masks, and others. I would like to note that face detection is part of the overall face recognition algorithm. After manual processing of patents, 142 patents obtained. The period of increased WIPO patent activity for facial recognition systems is 2015–2020 (Fig. 4).

Patent families demonstrate the relationship of a specific invention to patent documents issued based on it. An analysis of the size of patent families provides insight into the intentions of leading industry companies to develop domestically or target foreign markets. Small patent families contain 1–2 patent publications, which indicates

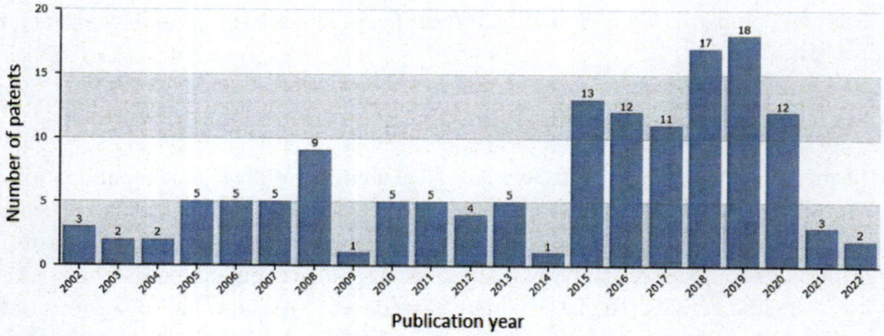

Fig. 4. Number of patents with international publication in the field of face recognition systems [14]

the degree of localization of technical solutions in selected subject areas. Typical for most technology areas. Medium patent families - represent up to 10 patent publications, which indicates the obvious international ambitions of the owners of patented solutions. In this technology area, there is a tendency to form patent families with a small number of patents, and the maximum number of patents in a family is 18, which is generally an average (Fig. 5.)

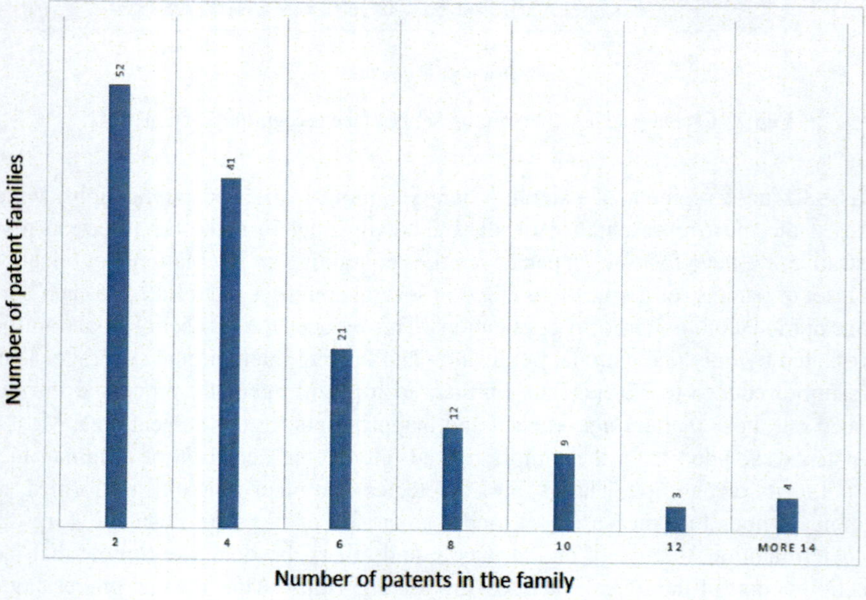

Fig. 5. Size of patent families in the field of face recognition technology [14]

Facial recognition technology in terms of patent maintenance, it is closer to regressing than to progressive, since the share of new patent applications is 23.24% of the total patent collection. Since there is a tendency to increase inactive patents in the total volume. As of May 2022, 41.5% of all patents in the sample were valid, and 27.46% of patent documents were no longer valid. Thus, there is a trend in the industry saturated with new solutions, in which right holders are still ready to use existing developments largely and implement them, rather than create something new. Figures 6 and 7 shows the number of patent families that companies form. It is important that the subsidiaries grouped under the parent company to increase the accuracy of this chart.

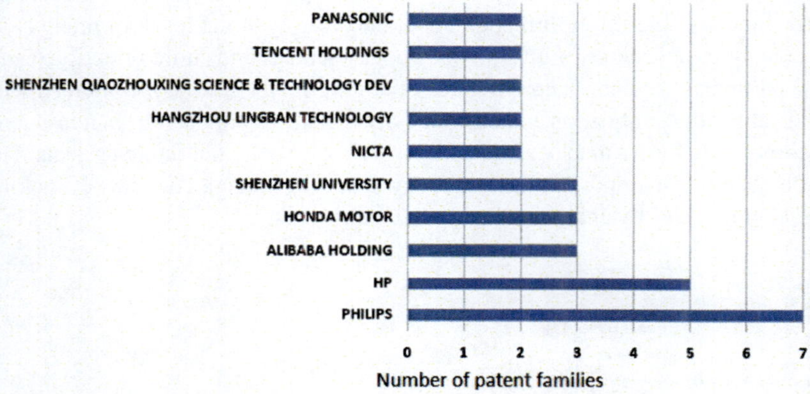

Fig. 6. Patent portfolios of market leaders of face recognition systems [14]

As part of the study, we analyzed the size of portfolios of applicants in the analyzed patent sample. As we can see, PHILIPS and HEWLETT PACKARD DEVELOPMENT (HP) are in the lead. About 80% of patents belong to companies that have only one patent, which indicates a large distribution of technology, while the position of a leader in this area is still quite vacant.

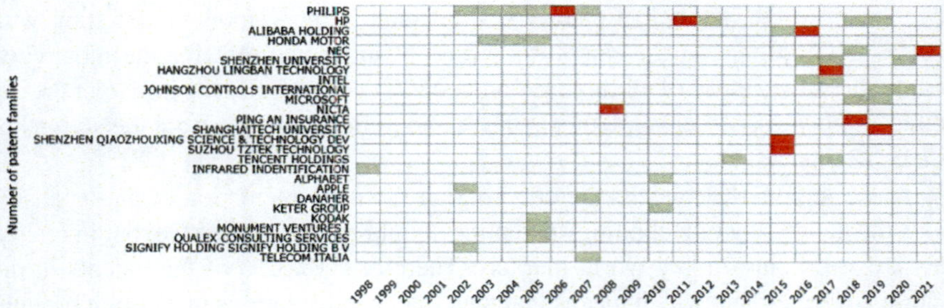

Fig. 7. Patent portfolios of market leaders of face recognition systems per years [14]

The leader and one of the pioneers in patents in the field of face recognition systems is PHILIPS, but at present, it is not actively developing developments in this technological area. The company now has three patents in force in Taiwan out of seven. HP obviously actively uses its developments of 2011–2012 and new patents appear already in 2018–2019, 3 out of 5 patents are valid in the US and Taiwan. At the same time, the most cited solutions in this area belong to IMAGE, which has a pronounced cluster of patent citations. Now its developments belong to the Apple. Among the industry connections in this segment, GN STORE NORD, MONUMENT VENTURES I, QUALEX CONSULTING SERVICES, which are subsidiaries of KODAK, should note. The patent citation profiles of these companies are consistent with each other.

An important element of the analysis of the corporate patent portfolio is the identification of key authors whose developments are of strategic importance for the development of the industry (Fig. 8). Information about the creators of the result of intellectual activity is an important source of information about scientists and entire scientific teams studying a particular problem. Knowledge of key figures in the industry, combined with information about the publication and other activity of authors obtained from non-patent sources, allows us to form a map of advanced researchers of potential future projects. The use of such analytics is important in technology scouting, when a company is looking for potential employees by analyzing their patent portfolios.

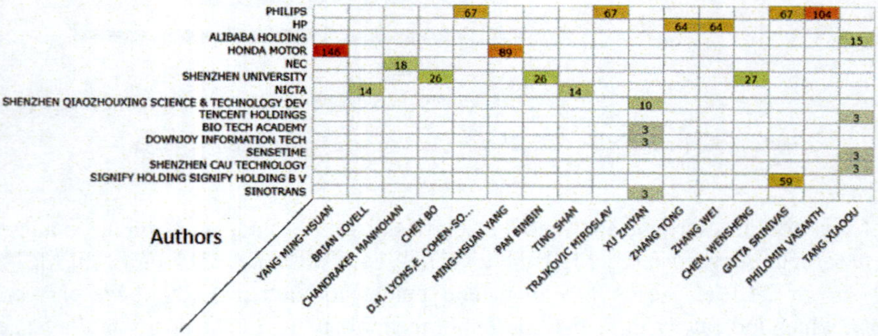

Fig. 8. Key authors in the field of face recognition systems [14]

The leader in the number of patents is the author Tang Xiaoou, cooperating with ALIBABA HOLDING. At the same time, Yang Ming-Hsuan remains the most cited author, as she is the author of all three HONDA MOTORS patents; she is also in the top five cited companies. Moreover, Gutta Srinivas and Philomin Vasanth are the authors of PHILIPS patents.

Since the key market players are counting on the implementation of an expansive strategy, the geography of technology patenting should be quite extensive, but in reality, patents are valid only in key world markets. Therefore, 18.22% of the patents in the collection are American, which means that these technical solutions most often brought to the US market. Chinese patents account for 14.49% of the collection, mainly from national companies. 9.35% of the patents in the collection belong to the region of the European Patent Office. However, as can be seen from the data in Fig. 9, far from all states remain covered within the European region.

Approximately 10% of patent applications in this area come from applicants from Japan, South Korea and Taiwan. There are also patents issued in Australia, Russia, Brazil, India, and South Africa. The states of first filing of patent applications with an

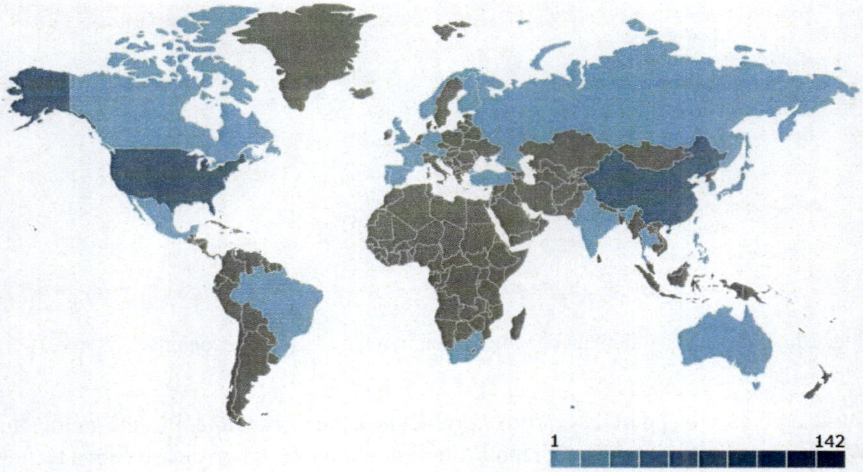

Fig. 9. Geography of patenting developments in the field of face recognition technologies [14]

overwhelming percentage are the United States (31.69%) and China (23.24%). At the same time, China is the absolute leader in the number of patents in the world for face recognition methods, but patents are domestic in more than 95% of cases and do not claim to enter the world stage. On the other hand, the US, which has about 8% of the total number of patents, tends to apply also to WIPO or initially chooses WIPO as the main application in order to globalize its technologies. WIPO as applicants in 21.13% of cases choose the Office of First Filing, while the European Patent Family only in 3.52%. Notable first filing offices also include South Korea (6.34%), Australia (2.82%) and the United Kingdom (2.11%). Traditionally, priority offices linked to the applicant's nationality or resident status, less frequently to patent strengths that are higher for US and European patents. Thus, the USA is the leading region for the development of this technology in the world market. China's patent activity in 2015–2020 is due, among other things, to the mass installation of CCTV cameras in cities in 2005–2015.

3.3 The Main Technological Niches in the Field of Face Recognition

Search by IPC[14] headings allows you to create collections of related patent documents that characterize one or another aspect of certain technologies [6]. That is why the assignment of technology to one or another classification code becomes important, since it allows for the simplest clustering of patent documents in the resulting collection. Thanks to the grouping of subclasses of the IPC, technological areas were identified, which presented in Figs. 10 and 11.

The patenting of facial recognition methods shows the relationship of this development to several technological areas. The development refers to computer technology, this group of the IPC dominated the rest throughout the entire period under review. This

[14] International Patent Classification (IPC), 2022. Geneva, Switzerland: World Intellectual Property Organization.

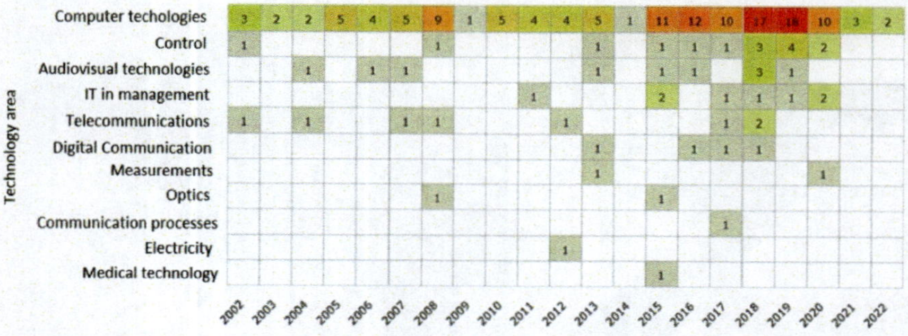

Fig. 10. Technological domain decomposition in relation to face recognition systems [14]

circumstance can considered because of problems in the work of the IPC, the architecture of which provides for a relatively small room for maneuver when classifying a technology to any classification group. At the same time, computer technology traditionally acts as a donor industry. This means that technical solutions find some final industrial application, and computer technologies become a tool for achieving other industrial tasks. This is how one can explain the development of such niches as Control, Audiovisual technologies, IT solutions for management and communication.

Subclass G06K - Image Data Reader - leads in patent applications. Also actively used codes are G06T - Processing or generation of image data, G06F - Processing digital data using electrical devices, H04N - Image transmission. Also popular are G06T 7/00 - Image analysis, e.g. bitmap to non-bitmap and G06F 17/00 - Data processing or digital computing equipment or methods specially designed for specific functions. All of these subgroups related to image and video processing.

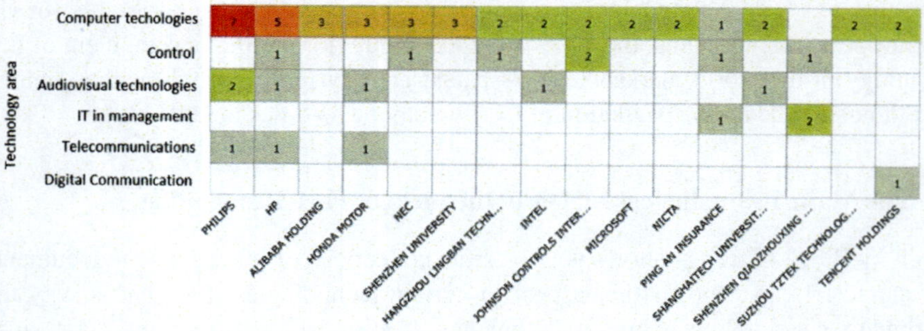

Fig. 11. Patent families for face recognition systems in technology areas by company [14]

The results of this study help to determine the diversity or specificity of the applicant's patent portfolio. The table in Fig. 11 allows users quickly identify the main technology area of the company studied. The least represented categories also serve as a means of identifying other potential uses for a given entity's patents. Thus, ALIBABA HOLDING and SHENZNEN UNIVERSITY turned out to be leaders in the field of computer technologies for face recognition systems, although their key competitors have patents applied in different areas, perhaps this is their competitive advantage. Another interesting argument is that scope specification correlates with patent novelty, since ALIBABA HOLDING and SHENZNEN UNIVERSITY have patent portfolios from new patents. The presence of universities among the major technological players shows the high business potential of educational organizations in terms of the release of relevant technological solutions. However, in terms of strength, university patents are by no means in the lead, as they rarely have international coverage.

If we consider the identification of national centers of competence in the field of face recognition, then we should note a balanced approach that applicants from different countries demonstrate when creating patent applications. So inventors from the USA, China and South Korea pay considerable attention to all the components of the face recognition system, from face detection and camera operation to image processing, facial features and emotions. For example, inventors from Turkey work only with face detection systems, that is, with technologies of the previous generation.

Based on patent applications, we can see a pattern that companies with a newer patent portfolio register their patents for more specific purposes, occupying small technological niches. This circumstance testifies to the growth of competition among market participants.

4 Conclusion

Government and commercial agencies are focusing on the introduction of mobile video surveillance systems to protect against criminal activity, and fraud is driving the growth of the market. In addition, the government and law enforcement agencies are implementing and promoting facial recognition systems to identify people in video, photographs or in real time, including help police find suspects or missing persons.

In recent years, patent activity in the field of face recognition technologies has falling, the reasons for this can mainly be COVID-19 and its accompanying quarantine, remote forms of work and study, and various forms of restrictive measures. At the same time, patents obtained based on existing ones, which describe methods of face recognition, for example, in a mask, as well as methods that measure temperature on the face, that is, innovations developed under the conditions of a pandemic. However, after the end of the COVID-19 pandemic, there will most likely be an increase in the number of patents, since, for example, the trend for contactless payment by a person was before the pandemic.

For companies that do not currently have leadership positions in this industry, it is important to study potential markets in which the presence of technological leaders not sufficiently developed both geographically and markets, taking into account government policies, which may include the development of modern security systems.

Another recommendation for the development of companies is to use the experience of China in terms of consolidation and complex setting of tasks for scientific organizations, universities, industrial companies and development institutions. This experience goes against the development of technology in this area in the United States, but a hybrid approach in this case will give a strong impetus. Companies should focus on the interdisciplinary use of facial recognition technologies, and also consider complex innovations and integrate with similar technologies, for example, recognition of a person by gait, recognition of emotions or deviant behavior.

When developing a new technology solution, organizations should study markets and companies. It is necessary to understand for what purposes and products the developed solution can used, so the patent collection can divided into two categories: scientific and industrial patents. A separate focus of attention is Chinese patents that go beyond China. In addition, an important area of work can be the study of patents that have entered the global market and their comparison with domestic patents. Understanding the reasons for the lack of some solutions on the world stage will be a key step in reap the benefits of US innovation in facial recognition.

Acknowledgments. The research described in this paper is partially supported by the ITMO University Research program (№619403).

References

1. Abhijeet, B., Hamna, A., Jeongbae, K.: Learning to resolve uncertainties for large-scale face recognition. Pattern Recogn. Lett. **160**, 58–65 (August 2022). https://doi.org/10.1016/j.patrec.2022.06.004(2022)
2. Antipov, A.A., Maximova, T.G., Nikolaev, A.S., Agapova, A.V., Murashova, S.V.: Philosophical foundations of the digital culture//proceedings of the 31st international business information management association conference (IBIMA). Innov. Manage. Educ. Excellence Vis. **2020**(2018), 6668–66673 (2018)
3. Facial Recognition Market. Markets and Markets. https://www.marketsandmarkets.com/PressReleases/facial-recognition.asp (2020)
4. Face recognition machine. https://www.telpo.com.ru/face-recognition-machine/Telpo. Last accessed 12 July 2022
5. Fabry, B., Ernst, H., Langholz, J., Koster, M.: Patent portfolio analysis as a useful tool for identifying R&D and business opportunities—an empirical application in the nutrition and health industry. WPI **28**(3), 215–225 (2006)
6. International Patent Classification (IPC), Geneva, Switzerland: World intellectual property organization (2022)
7. Crockford, K.: How to regulate face recognition technology. Nat. Hum. Behav. **6**, 476 (2022)
8. Market Research Report. Fortune Business Insights. 2019 https://www.fortunebusinessinsights.com/industry-reports/facial-recognition-market-101061 Last accessed 18 Aug 2022
9. New facial recognition and fever detection systems for safe work environment launched. Biometric Update 2020. https://www.biometricupdate.com/202004/new-facial-recognition-and-fever-detection-systems-for-safe-work-environment-launched (2020) Last accessed 18 Aug 2022
10. Romanyuk, O.N., Vyatkin, S.I., Pavlov, S.V., Mykhaylov, P.I., Chekhmestruk, R.Y., Perun I.V.: Face recognition techniques. Inf Control Meas. Econ. Environ. Prot. 10(1), 52–57 (2020)
11. Sisi, L., Zhonghual, L., Di, Wu.: Low-resolution face recognition based on feature-mapping face hallucination. Comput. Electr. Eng. **101**, 108136 (2022)
12. Poltoratski, S., Kendrick, K., Dawn, F.: Holistic face recognition is an emergent phenomenon of spatial processing in face-selective regions. Nat. Commun. **12**, Article number: 4745. https://doi.org/10.1038/s41467-021-24806-1(2021)

13. OECD Policy Responses to Coronavirus (COVID-19) E-commerce in the time of COVID-19. https://www.oecd.org/coronavirus/policy-responses/e-commerce-in-the-time-of-covid-19-3a2b78e8/. Last accessed 18 Aug 2022
14. Orbit Intelligence Homepage. https://www.orbit.com/ Last accessed 18 Aug 2022
15. Wang, J., Runnan, C., Nicholas, J.B.: Face identity coding in the deep neural network and primate brain. Commun. Biol. **5**, Article number: 611. https://doi.org/10.1038/s42003-022-03557-9 (2022) Last accessed 18 Aug 2022

Method for Distributed Mapping of Terrain by a Heterogeneous Group of Robots Based on Google Cartographer

Valeria Lebedeva(✉), Konstantin Kamynin, Igor Lebedev, Lev Kuznetsov, and Anton Saveliev

St. Petersburg Federal Research Center of the Russian Academy of Sciences (SPC RAS), St. Petersburg Institute for Informatics and Automation of the Russian Academy of Sciences, 39, 14th Line, 199178 St. Petersburg, Russia

lebedeva.v@iias.spb.su

Abstract. At present, the directions of robotic complexes for servicing vast territories are being actively developed. The use of these complexes makes it possible to carry out prospecting work, survey of agricultural plots, detection of fires, delivery of goods to hard-to-reach places and other work in a short time. To guarantee the performance of these works, it is necessary to solve the problem of group navigation in an unknown territory. In the context of this article, the task of navigation is understood as the construction of a map of the area, the availability and updating of the map for each agent of the group, and the ability of agents to plan a path and move around the area based on the available map. Existing algorithms allow solving the problem of navigating a group of robots, but they have many limitations: group homogeneity, high computational costs on group agents, low accuracy of building a map of the area, a highly loaded data transmission channel, and asynchrony of the received data. This paper presents a modification of the method of building a map by a group of robots based on a client-server architecture, which re-constructs a map of the area with high accuracy according to data from sensors installed on a group of robots. As a group of robotic devices, a group consisting of unmanned aerial vehicles and mobile robotic complexes in the form of four-wheeled ground platforms is considered. To test the proposed solution, experiments were carried out in a simulation environment. The purpose of the experiments is to prove the applicability of the developed solution based on client-server architecture on homogeneous and heterogeneous groups, to justify the use of unmanned aerial vehicles to improve the accuracy of the map, and to compare with the reference map the difference when building a map of the area with different composition of the group and with a set of sensors. The estimation parameter was the error calculated using ADNN metrics for SLAM algorithms relative to the ground truth. As a result, when using a heterogeneous group consisting of ground and air robots, the accuracy of the constructed map increases to 37% compared to using a homogeneous group. And comparing the values obtained using a heterogeneous group with respect to on-board sensors installed on robots, it was concluded that a map generated based on data from a depth camera is 1.7% more accurate than a map generated based on data with LiDAR.

R. Silhavy et al. (Eds.): CoMeSySo 2022, LNNS 596, pp. 584–597, 2023.
https://doi.org/10.1007/978-3-031-21435-6_50

Keywords: Heterogeneous group · Distributed SLAM · Mapping · Group of robots · Navigation · Data collection

1 Introduction

Nowadays, the direction of group robotics is actively developing. Groups of robots are widely used in various fields where it is required to distribute a set of tasks between different agents. The advantage of a group of robots lies in their high adaptability to various situations when they operate together, and an extended set of performed functions provides a high probability of successful implementation of the tasks.

Systems including heterogeneous robotic includes groups of unmanned aerial vehicles (UAVs) and groups of mobile ground robotic systems (RSs). Such systems can perform the tasks of protecting the territory, reconnaissance and monitoring the situation, and they can be used in the tasks of search and rescue operations, in forecasting or harvesting and detecting outbreaks of pests in agriculture [1]. A UAV group can provide monitoring of a given area; however, the detection of ground targets is limited by the speed of aircraft and flight altitude. A mobile robot can monitor certain areas of the terrain to determine the position of ground targets accurately, but its disadvantage is the restrictions on movement, considering the features of the terrain and obstacles [2]. Therefore, to obtain complete data on the surrounding space, data from UAVs and ground-based RSs are used: UAVs collect data to form a global terrain map and track the movement of ground objects [3], and ground-based RSs can perform more detailed construction of terrain maps in certain areas of the territory. By combining information from various groups of heterogeneous robotic we can get a high-precision map of the surveyed area with high detail in certain areas of increased interest [4, 5].

The aim of the work is to modify the method of simultaneous localization and mapping (SLAM) based on a client-server architecture to recreate a map based on data from sensors installed on a heterogeneous group of robots operating over a vast area. The novelty of the work lies in increasing the accuracy of the generated map of the area, due to the proposed algorithm for building a map based on a client-server architecture and data received from a group of UAVs and a group of ground-based RSs.

The paper considers the problem of movement of a group of robots in an environment with obstacles, in which a group of UAVs and a group of mobile robots must be connected through data received by sensors, including depth camera, or scanning laser rangefinders. Based on the data obtained, the sections are combined to reconstruct a general map of the area on which the group of robots operates.

The work contains the following sections: Sect. 1 presents an extended review and analysis of SLAM algorithms, and also contains a classification of these algorithms according to the approaches used to represent terrain maps and architecture for multi-agent systems, Sect. 2 describes the modification of the Cartographer package for the group; Sect. 3 contains a description of the implementation of the movement of UAVs and ground-based RSs in a simulation environment; Sect. 4 describes the procedure for conducting experiments and the results of comparing two combinations to identify the most effective method for constructing a map of the area by a group of robots.

2 Overview and Classification of SLAM Algorithms

In modern robotics, the SLAM navigation method is used to determine the location and orientation of autonomous robotic vehicles in an unknown environment. This method is the most promising and effective among other methods of navigation in an unknown space [6].

The principle of SLAM is based on the assessment of the map, recreated according to the received data from the sensors of the robotic vehicle, and the distance traveled. A robotic tool needs to know its location at every moment of time, as well as gradually scan the surrounding space, building a map of this space. The probabilistic definition of the SLAM method can be given as follows [7]:

$$u_{\{1:t\}} = \{u_1, u_2, u_3, \ldots, u_t\},$$

where u is the robot control signal at time t;

$$z_{\{1:t\}} = \{z_1, z_2, z_3, \ldots, z_t\},$$

where z is information about the environment being viewed by the robot at time t;

$$m,\ x_{\{1:t\}} = \{x_1, x_2, x_3, \ldots, x_t\},$$

where m is the rendered map and x is the obtained location of the robot at time t. Thus, to calculate the trajectory of the robot, everything can be represented in this form:

$$p(x_{\{1:t\}}, m|z_{\{1:t\}}, u_{\{1:t\}})$$

The expression above is illustrated in Fig. 1.

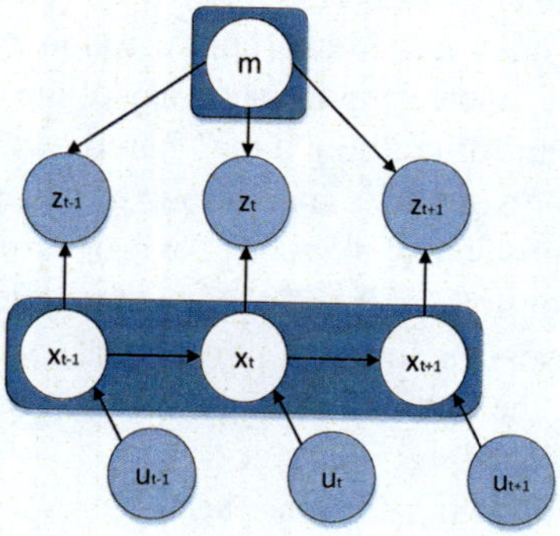

Fig. 1. Structure of the SLAM model

SLAM algorithms can be divided according to the approaches used to represent terrain maps. For example, grid-based SLAM, based on the definition of unique objects (feature-based), graph-based for sparser images, topological for recognizing types of locations and semantic for a deeper analysis of the environment [7].

Nowadays, SLAM navigation is widespread using ROS (Robotic Operation System) as packages. Widely known open-source packages are GMapping [8], Cartographer [9], Rtabmap [10].

Gmapping creates a map using LiDAR data. Usually a 2D LiDAR is used, capable of scanning the surrounding space 360 degrees around. Next, the finished map and data from the LiDAR are compared, as a result of which the location of the robot is estimated and corrected.

Google Cartographer is similar to GMapping in that it can build a map from 2D LiDAR and odometry data. Cartographer has a more complex gradation of points that make up maps – cells. As one and the same piece of terrain is explored, Cartographer smoothly changes the parameter of map points from "unknown" to "explored" with the flag "free" or "obstacle" [9]. As a result, Cartographer is less likely to miss any objects on the map. This is especially noticeable during sharp turns around the corner, when the robot can only scan a certain area of the terrain for a short period of time. The main feature of this method is that it consists of local SLAM and global SLAM. Local SLAM creates separate small parts of the map. Then, the global SLAM algorithm looks for matches between parts of the maps, as well as between sensor data and all maps created using local SLAM in parallel processes. As a result, the global SLAM forms a single map of the surrounding space [11]. In addition, Cartographer can work with 3D maps. This is an advantage for aircraft applications, as well as for mobile robots with large dimensions. On the other hand, this requires more computing power on the robot itself. Also, unlike the simpler GMapping, Cartographer can work with several leads at the same time (for example, at different heights to create a 3D map), as well as with RGBD and stereo cameras, thanks to which it is possible to get much more information about surrounding space.

Rtabmap is a SLAM method based on searching and matching visual sensor data using a memory mechanism where a database of visual images is stored in accordance with the robot's location data [12, 13]. The visual navigation mechanism works as follows. Rtabmap processes each new incoming 2D image, highlighting the main features that can be used to compare this image with others. Each image is compared with data on the location and orientation of the robot in space at the time of image acquisition. Thus, a database of visual images and the coordinates in which they were received is compiled. At the same time, each new image received is compared with the images already in the database according to a number of features. If the matching coefficient is high enough, the algorithm considers that the robot is currently located in an already previously explored place and, if necessary, corrects the odometry. It should be noted that such an approach strongly depends on the quality of the obtained images, as well as on the area where the mapping takes place. For example, Rtabmap will not work well in a simulation, where all walls and objects are the same, or in a real room, if there is practically nothing in it except walls, many images are identical. In parallel with this, a 2D map of the space in which the robot moves is built. Rtabmap can work with a wide range of sensors and their combination: LiDARs, stereo cameras, odometry, IMU. Both – LiDAR and RGBD cameras, or stereo cameras – can act as a data source for the map. However, it is necessary to choose one single type of sensors to be used.

Gmapping is a basic SLAM that does not require high computing power. It can be used for small simple robots with the need for mapping and localization using 2D LiDAR. At the same time, Cartographer and Rtabmap are already more resource-intensive packages and require appropriate computing power, as well as the presence of certain sensors (camera for Rtabmap). However, the capabilities of these algorithms are an order of magnitude higher – the quality of mapping on complex terrain – is superior to simple Gmapping.

Table 1 compares the described SLAM packages to select the most suitable one for solving the problem posed in this paper.

Table 1. Comparison of SLAM packages

Package name	Input data	The number of sensors connected at the same time	Output data	Accuracy of building a map on a test data set (Room 30 by 20 m)
Gmapping	2D LiDAR, odometry	1	2D map and robot position	Could not stitch several parts of the 2D map, so it is impossible to assess the accuracy of the built map
Cartographer	2D LiDAR, 3D lidar, RGBD and stereo cameras, odometry	Possibility to combine several sensors of different types	2D map, 3D map, and robot position	Up to 5 cm
Rtabmap	2D LiDAR, 3D lidar, RGBD and stereo cameras, odometry	1 sensor (LiDAR) and a monocular camera is required	2D map, 3D map, and robot position	Up to 5 cm

As a result of comparing modern SLAM algorithms such as Gmapping, Rtabmap and Google Cartographer, it was possible to establish that Google Cartographer is the most suitable algorithm for solving the problem of navigating a heterogeneous group of robotics over a vast area. Google Cartographer has more flexibility in terms of settings and choice of sensors. The ability to use not only LiDARs, but also depth cameras allow one to get more information. Also, Cartographer allows one to map large areas of premises and territories. Cartographer performs better in complex spaces, since using the map itself (rather than visual images, as in Rtabmap) to localize the robot allows for more efficient mapping. In addition, the system of division into local and global SLAM has a positive effect on the mapping process.

In addition to classification by the approaches used to represent terrain maps, there is a way to classify SLAM algorithms for use by multi-agent systems. Typically, distributed SLAMs are divided into two groups of algorithms: (1) algorithms that extend the single-agent algorithm, (2) algorithms that are initially developed as multi-agent algorithms. This division follows from the composition and type of agents in the system: are they all homogeneous and have equal capabilities or are there agents in the system that differ in computing power, functionality, and design.

Approaches with independent agents, such as in [14], contain several disadvantages. The main disadvantages of such approaches are the difficulty of tracking the relative orientation of neighboring RSs, merging maps in real time, and defining roles in a group. One of the most significant shortcomings is the failure of the entire system in case of loss of communication between the RS. The task of combining maps in a client-server architecture can be simple. One simple solution to the problem of merging maps in a client-server architecture is to have only one copy of the map that clients update.

The works [15–17] present a client-server architecture. All agents that explore an unknown environment transmit the data collected from the sensors to the server and do not perform any calculations themselves. The server must store the previous location of all agents in order to combine data from various agents into a single map. If the server does not contain information about the location of the robots, then several hypotheses about the location of the robots are formed to avoid incorrect map merging.

Advantages of the SLAM client-server architecture:

1. The system is resistant to the loss of client agents until all clients lose their connection to the server. With the active development of wireless communication technologies, it will be possible to reduce the waiting time and develop a system that can track and interact with multiple robots over a long distance in real time.
2. Agent-clients can be significantly cheaper and more energy efficient due to the use of less resource-demanding computing power.
3. RS can move at different speeds; there is no need for each PC to know and track the position of other PCs, which simplifies the communication architecture in a distributed system in software development.

As a result of the analysis and comparison of SLAM packages, this paper proposes a modification of the Cartographer package based on a client-server architecture for navigating a group of robots. The client robot collects data from sensors about the environment and transfers it to the server to build a global map.

3 The Developed Solution

The modification proposed in this article is intended for localization and navigation on the terrain of the group of robots. The modified Cartographer simultaneously operates on several agent robots that jointly analyze the external environment and build a map of the area. In the future, the unit of group will be called an agent, if it is necessary to emphasize its role in the executable algorithms, and a robot, if we are talking about physical equipment. The use of several agents that function together makes it possible to speed up the construction of a map of the area and increase the accuracy of the localization of each robot separately.

In the task set, the agents involved in solving problems of simultaneous localization and construction of terrain maps function autonomously and the only source of environmental data for them is a sensor presented in the form of LiDAR or a depth camera installed on a robot. Mobile platforms equipped with computing devices, as well as on-board sensors (LiDAR sensors or depth cameras) act as ground-based RSs, and quadcopter-type UAVs act as air RSs.

According to SLAM in the client-server architecture, each agent builds its own local map of the area and transfers the relevant information to the server, the server provides the construction and updating of a generalized global map of the territory [18]. Based on the results of generating an updated global map, the server sends each agent information about the changed part of the global map of the area.

To improve the accuracy of building a map by the RS group and the possibility of using such a map by robotic means, a modification of SLAM is proposed, implemented on the Robotic Operating System (ROS) using a client-server architecture.

The ROS system has a set of entities that describe the general structure of the software: node is an executable file that uses ROS to communicate with other nodes; topic is nodes can publish messages in a topic, as well as subscribe to a topic to receive messages within the ROS system; message is ROS data type used when subscribing and publishing topics.

Fig. 2. The structure of the proposed system

In the developed system that implements distributed SLAM navigation, three ROS nodes can be distinguished: robot, client, and server. In Fig. 2 the structure of the proposed system is showed. The robot can be equipped with an RGB-D or LiDAR depth camera. The client node is a ROS node hosted on an agent. The robot node publishes an RGB image and a depth image or LiDAR data to the client node. The client converts the received data into map points and publishes messages like "key frame and points of the created map in this key frame". Data transfer occurs through the connection of the client and server with one topic. Since topics in ROS can run in parallel, the SLAM system on the client can build a local map of the area, regardless of the state of the server. At the same time, the server, upon request from the agent, according to the developed solution, ensures the transfer of the current version of the generated global map and implements its integration with the local map of the agent.

The original implementation of the Cartographer package provides for the execution of three asynchronous flows on each agent: tracking, local mapping, and loop closing. In the tracking stream, data is retrieved from the sensor, followed by the selection of key frames. In the flow of local matching, the positions of key frames and map points are adjusted. In the loop closing thread, an optimization procedure is implemented to refine the existing local map of the area. The improvement proposed in the work consists in adding a separate stream to the Cartographer package, which transfers map data from the client to the server (Streaming).

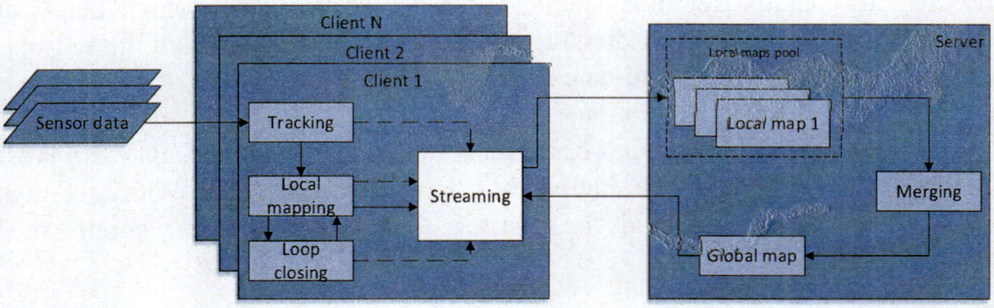

Fig. 3. Diagram of data flows as part of the modification of the Cartographer package

Figure 3 shows the updated data flow diagram within the Cartographer package, taking into account the proposed modification, which consists in adding a Streaming flow on the agent and new data flows indicated by dotted lines. According to the proposed modification of the distributed SLAM algorithm, the generated key frame and map points from the tracking stream, as well as data from the local matching and loop closing streams are transmitted directly to the server. After receiving the keyframe and map points from the tracking stream and the local mapping stream, the server-side one removes or corrects the keyframe information. The developed data transfer scheme makes it possible to increase the accuracy of generating a global map of the area, as well as to reduce delays in the process of communication between the server and client agents.

4 Description of UAV and Ground-Based Robot Models

This section describes the structure of an autonomous flight system for a multi-rotor UAV. The system was developed on the basis of ROS Navio [19]. The task of developing an autonomous flight system for UAVs is to adapt existing advanced technologies in an integrated flight system for a specific task. The use of the UAV allows increasing the accuracy of building a map of the area, as well as ensuring that each point on the map is linked to GPS coordinates, which can be used by ground platforms for navigating the terrain.

The UAV autonomous flight system was developed using a flight controller, an onboard computer, and a ground station. The flight controller and the on-board calculator are installed on the UAV. They control the UAV and receive information about the environment with the help of a depth camera. A ToF sensor is attached to the bottom of the UAV to determine the UAV's flight altitude and collect data that is converted into a point cloud about the terrain. The computational load on the onboard computer is high, therefore, a ground station is used to create and refine a map of the area from point cloud data, localization, path generation and navigation command generation.

The ground-based robot is a ground-based unmanned robot platform equipped with a sensor (LiDAR or depth camera), an IMU sensor, an odometer, and an on-board computer. A four-wheel differential drive is used as a kinematic movement model.

Data received from the sensor, IMU and odometer located on the platform, as well as data received from the ToF and sensor located on the UAV, are used by the on-board computer of the ground-based robot. The onboard computer on the ground-based robot processes this data and combines the incoming point clouds into a single height map in real time. Thus, the height map is updated as new data arrives from the sensors on the UAV and platform. An up-to-date elevation map is used to navigate the terrain and identify areas for further exploration.

To build trajectories for ground-based robot, the LHLRD-A* algorithm [20] is used, which builds the path on the current height map at the current time. With each update of the map, the algorithm rebuilds the trajectory to the target point, taking into account the updated data.

5 Experiments

In the experiments, the performance of the proposed solution for modifying the SLAM method based on the client-server architecture was evaluated. In addition, an analysis was carried out, a combination of robots and sensors providing the most accurate map of the area.

The server was implemented using an Intel Core i7-2640 k 2.80 GHz × 4 processor and an Intel Sandybridge Mobile graphics card. 3D LiDAR [21] in one series of experiments and an Intel RealSense D455 depth camera [22] in another series were installed as sensors on the models of robots.

In the course of the study, in order to identify the most accurate way to build a terrain map, two combinations were compared: LiDAR with the Cartographer ROS package and a depth camera with the Cartographer ROS package on a group of robots. The maximum viewing distance of the depth camera (IntelRealSense) is 6 m, and the accuracy of building a map reaches 3–5 mm. LiDAR has visibility up to 100 m, but the accuracy is 10–30 cm. To compare the results obtained, ROS bag files were written. ROS bag files are files in which data from ROS topics is written with appropriate time sampling.

For Cartographer in ROS, there is no difference between the data obtained from the recorded bag-file and the data of the robot obtained in real time. ROS bag files contain all data from ground robots that moved autonomously along the trajectory constructed by the LHLRD-A* algorithm and from the UAV that performed the mission of collecting terrain data. These bag files were then used in post-processing mode in Cartographer.

Experimental testing of the proposed solution was carried out in the Gazebo simulator. The group of agents consisted of four Pioneer3AT ground-based platforms. The iris is a model of multi-rotor UAV that was used as an air robot. The initial locality model is shown on Fig. 4a. The size of the area is 30 by 30 km. The results of the terrain map generation using the developed solution are shown on Fig. 4b, c.

As part of the experiment, each RS was assigned its own target coordinate. With the help of the LHLRD-A* traffic planning algorithm for a group of ground-based robots, each platform builds a movement route taking into account the terrain and obstacles on the ground. For the experiments, a complex terrain was modeled with a lot of hills, ditches, ridges, which directly affects the uniformity of the robot movement. When

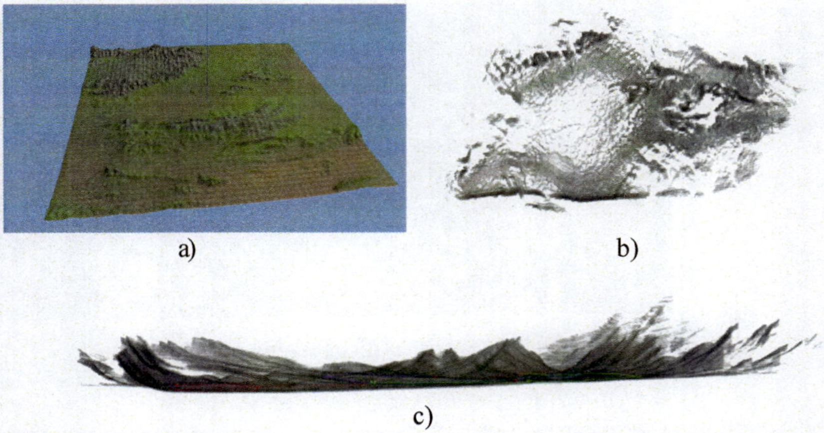

Fig. 4. Simulation results: a) the original map model b) the map built using the developed distributed SLAM algorithm and a group of four PCs c) the results of generating a global map of the territory using the developed distributed SLAM algorithm in projection onto the YOZ plane of the global reference frame.

moving the robot on such an uneven surface, excessive shaking occurs, which causes interference and errors in the operation of sensor systems. Therefore, the final built map has deviations from the reference map due to the type of sensor used, the difference in processing frequencies of data from the sensor and the inertial sensor (IMU) in conditions of excessive shaking.

To assess the accuracy of building a map, it is necessary to step by step compare the global maps obtained by surveying the area by the group of robots and the reference map of the area. To do this, the server calculates the error of building a global map by the group (Fig. 5). In each series of experiments, global maps of the area, recreated based on data from robot sensors, are converted into an Occupancy Grid and saved in pgm format using the map_saver from the map_server package. The occupancy grid is a two-dimensional representation of a map in space, which is a grid, each cell of which contains a certain state. A cell can take one of three states: busy, free, and unknown. In the experiments carried out, the occupancy grids of the saved maps were changed in such a way that the cells had only two values: occupied and free. This transformation is necessary to simplify the process of comparing maps. The next necessary steps for comparison are map alignment and calculation of difference metrics. Alignment was performed using the ICP (Iterative Nearest Point) algorithm [23], which allows finding such a transformation that minimizes the error of the sum of distances from each point of the map built by Cartographer to the nearest neighboring point on the reference map. In this work, the metric described in [24] was used, which calculates the average distance to the nearest neighbor (ADNN) as the sum of all distances divided by the number of occupied cells:

$$ADNN = \frac{\sum_{i=1}^{N} Nearest\_Neighbour(occupied\_grid\_cell(i))}{N},$$

where N is the number of occupied cells.

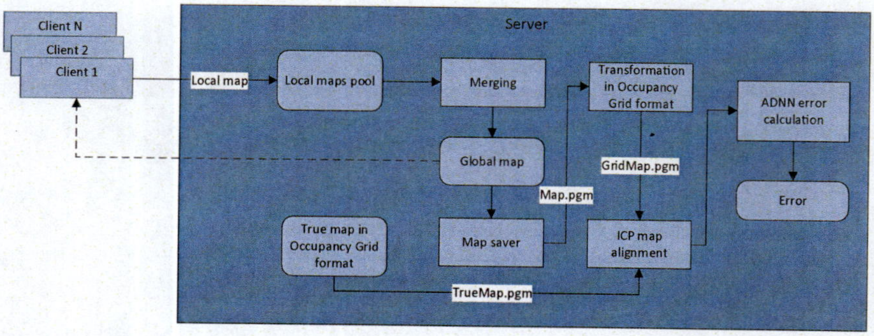

Fig. 5. Diagram of data flows for calculating the error of building a map

The experiments were carried out with two configurations of groups: only ground platforms, a group of ground platform and UAVs. An ADNN-based comparison of the mapping errors obtained by two configurations with two sets of sensors (LiDAR with Cartographer ROS package and depth camera with Cartographer ROS package) against ground truth is presented in Table 2.

Table 2. Map comparison: error calculated with ADNN metrics for SLAM methods relative to ground truth, in cm.

Group members	LiDAR	Depth camera
Only ground platforms	29.55	19.91
A group of ground platform and UAVs	15.01	14.76

Based on the obtained results, we can conclude that when we use a heterogeneous group consisting of ground and air robot, the accuracy of the constructed map increases by almost 50% in the case of the LiDAR and ROS package Cartographer, and by 26% higher in the case of with depth camera and ROS package Cartographer. Comparing the values obtained using a heterogeneous group with respect to the onboard sensors installed on the robots, we can say that the map generated based on data from the depth camera is 1.7% more accurate than the map generated based on data from LiDAR. Since depth cameras have a low visibility range, it is necessary to generate additional routes for the ground-based platforms in order to build a terrain map, which causes an increase in the time for collecting data on the work area. Although depth cameras are millimeter-precise, LiDAR has a greater range of view than depth cameras, allow one to collect a larger sample of terrain data in less time. For a heterogeneous group, including ground and air robots, the difference between the mapping error (1.7%) by the depth camera and LiDAR is insignificant, which proves the efficiency of using the proposed algorithm for constructing a terrain map. Using a depth camera as the primary data source for mapping also has advantages. The depth camera is able to identify more obstacles due to the fact that it scans 3D space and also acquires 2D images. Thus, all obstacles will be initially plotted on the map to the full extent of the space they occupy and saved. Since the objects will be mapped at the mapping stage, this will allow the robot to plan the optimal trajectory right from the starting point.

Thus, the conducted study shows that in order to create large-scale maps of high accuracy, a group of robots must be heterogeneous and include both ground and air robots. As sensors for collecting terrain data, LiDAR and a depth camera installed on ground and air robots are suitable. With this combination of sensors and group configuration, it is possible to achieve a small error (15.01 cm and 14.76 cm) calculated using ADNN metrics between the built and true maps. However, due to the low visibility range of the depth camera, each ground-based platform needs to build additional routes to collect as much terrain data as possible, which increases the duration of the work of the robot team on the ground.

6 Conclusion

As a result of the study, a modification of the SLAM algorithm was developed and tested based on the client-server architecture and the ROS package Cartographer. The proposed solution is resistant to loss of connection between the server and individual client agents and is able to function until all clients lose connection with the server. The resulting delays can be further optimized by organizing a high-speed communication environment between the server and clients.

As a result of a series of experiments comparing the accuracy of the obtained map using two configurations with two sets of sensors (LiDAR with ROS package Cartographer and a depth camera with ROS package Cartographer) with a reference map, it was possible to establish that, according to the ADNN metric, the use of LiDAR allows one to quickly re-create a map of the area than using a depth camera, due to the high visibility range of LiDAR. And the map matching error using LiDAR does not exceed 15.01 cm, which is acceptable when using a group of robots in an open area measuring 30 by 30 km.

In the future works we are going to test the developed software module in a critical situation when data is lost during sending a client data to the server. It will help to create a distributed SLAM system with low computing power on a client-server architecture for real-time operation.

Acknowledgements. The work is supported by the Russian Science Foundation project no. 20-79-10325.

References

1. Guérin, F., Guinand, F., Brethé, J.F., Pelvillain, H.: UAV-UGV cooperation for objects transportation in an industrial area. In 2015 IEEE International Conference on Industrial Technology (ICIT), pp. 547–552. IEEE (2015)
2. Fedorenko, R., Gabdullin, A., Fedorenko, A.: Global UGV path planning on point cloud maps created by UAV. In: 2018 3rd IEEE International Conference on Intelligent Transportation Engineering (ICITE), pp. 253–258. IEEE (2018)

3. Hood, S., Benson, K., Hamod, P., Madison, D., O'Kane, J.M., Rekleitis, I.: Bird's eye view: cooperative exploration by UGV and UAV. In: 2017 International Conference on Unmanned Aircraft Systems (ICUAS), pp. 247–255. IEEE (2017)
4. Williams, J., Jiang, S., O'Brien, M., Wagner, G., Hernandez, E., Cox, M., Pitt, A., Arkin, R., Hudson, N.: Online 3D frontier-based UGV and UAV exploration using direct point cloud visibility. In: 2020 IEEE International Conference on Multisensor Fusion and Integration for Intelligent Systems (MFI), pp. 263–270. IEEE (2020)
5. Aakash, C., Kumar, V.M.: Path planning of an UAV with the help of lidar for slam application. IOP Conf. Ser. Mater. Sci. Eng. **912**(6), 062013 (2020)
6. What is simultaneous localization and mapping (SLAM)? https://www.flyability.com/simultaneous-localization-and-mapping. Accessed 24 May 2022
7. Pedrosa, E., Reis, L., Silva, C.M.D., Ferreira, H.S.: Autonomous Navigation with Simultaneous Localization and Mapping in/Outdoor (2020)
8. Gmapping. http://wiki.ros.org/gmapping/. Accessed 24 May 2022
9. Google Cartogtapher: https://google-cartographer-ros.readthedocs.io/en/latest/#. Accessed 24 May 2022
10. Rtabmap: http://introlab.github.io/rtabmap/. Accessed 24 May 2022
11. Building Maps Using Google Cartographer and the OS1 Lidar Senso,Electronic resource. https://ouster.com/blog/building-maps-using-google-cartographer-and-the-os1-lidar-sensor/. Accessed 24 May 2022
12. Labbé, M., Michaud, F.: RTAB-Map as an open-source lidar and visual simultaneous localization and mapping library for large-scale and long-term online operation. J. Field Robot. **36**(2), 416–446 (2019)
13. da Silva, B.M., Xavier, R.S., Gonçalves, L.M.: Mapping and Navigation for Indoor Robots under ROS: An Experimental Analysis (2019)
14. Lazaro, M.T., Paz, L.M., Pinies, P., Castellanos, J.A., Grisetti, G.: Multi-robot SLAM using condensed measurements. In: 2013 IEEE/RSJ International Conference on Intelligent Robots and Systems, pp. 1069–1076. IEEE (2013)
15. Forster, C., Lynen, S., Kneip, L., Scaramuzza, D.: Collaborative monocular SLAM with multiple micro aerial vehicles. In: 2013 IEEE/RSJ International Conference on Intelligent Robots and Systems, pp. 3962–3970. IEEE 2013
16. Castro, D.A., Morales, C.F., De la Rosa, R.F.: Multi-robot slam on client-server architecture. In: 2012 Brazilian Robotics Symposium and Latin American Robotics Symposium, pp. 196–201. IEEE (2012). https://doi.org/10.1109/SBR-LARS.2012.39
17. Karrer, M., Schmuck, P., Chli, M.: CVI-SLAM—collaborative visual-inertial SLAM. IEEE Robot. Autom. Lett. **3**(4), 2762–2769 (2018)
18. Li, F., Yang, S., Yi, X., Yang, X.: CORB-SLAM: a collaborative visual SLAM system for multiple robots. In: International Conference on Collaborative Computing: Networking, Applications and Worksharing, pp. 480–490. Springer, Cham (2017)
19. ROS Introduction: http://wiki.ros.org/ja/ROS/Introduction. Accessed 24 May 2022
20. Zakharov, K., Saveliev, A., Sivchenko, O.: Energy-efficient path planning algorithm on three-dimensional large-scale terrain maps for mobile robots. In: International Conference on Interactive Collaborative Robotics, pp. 319–330. Springer, Cham (2020). https://doi.org/10.1007/978-3-030-60337-3_31
21. LiDAR integration with ROS: quickstart guide and projects ideas. https://www.generationrobots.com/blog/en/lidar-integration-with-ros-quickstart-guide-and-projects-ideas/. Accessed 24 May 2022
22. Intel RealSense: https://www.intelrealsense.com. Accessed 24 May 2022

23. Pomerleau, F., Colas, F., Siegwart, R.: A review of point cloud registration algorithms for mobile robotics. Found. Trends Robot. **4**(1), 1–104 (2015)
24. Santos, J.M., Portugal, D., Rocha, R.P.: An evaluation of 2D SLAM techniques available in robot operating system. In: 2013 IEEE International Symposium on Safety, Security, and Rescue Robotics (SSRR), pp. 1–6. IEEE (2013)

Geoportal Systems as an Access Point to Spatial Data Infrastructure

S. A. Yamashkin[1(✉)], A. A. Yamashkin[1], M. M. Radovanović[2], M. D. Petrović[2], and E. O. Yamashkina[1]

[1] National Research Mordovia State University, 68 Bolshevistskaya Street, 430005 Saransk, Russia
yamashkinsa@mail.ru
[2] Geographical Institute "Jovan Cvijić", Serbian Academy of Sciences and Arts (SASA), 9 Djure Jakšića, 11000 Belgrade, Serbia

Abstract. The study described in the article is aimed at developing the principles of forming the architecture of a geoportal system as access points to information arrays of spatial data infrastructure form a sufficient level of coordination and communication of participants in various industries to integrate, process and analyze spatial data, providing flexible access to spatial data based on the web-technologies. Two directions of development of the framework of the geoportal system are distinguished: deductive and inductive strategies. Their joint use will provide evolutionary improvement of geoportal systems. In the SDI system, the following bounded contexts for microservices can be distinguished: geoportals, systems for analyzing, processing and control of spatial data and distributed cloud storages. The division into subsystems of visualization, processing and integration of spatial information makes it possible, starting from the top level of decomposition of the system architecture, to achieve an increase in the cohesion of modules and a weakening of the link. The characteristic of the geoportal "The Natural and cultural heritage of the Republic of Mordovia", which performs the function of managing cultural and historical heritage systems is given.

Keywords: Spatial data infrastructure · Geoportal · Microsevices · Web service architecture · Human-web interaction · Cloud computing

1 Introduction

Spatial data play a significant role in the economies of countries and regions and serve as the basis for the digital transformation of economic sectors. This is due to the fact that digital information about the location of natural resources, mobile and infrastructural anthropogenic objects is used in most technological and information processes in the framework of solving problems of public administration, organizing commercial activities, creating connectivity and rationalizing the behavior of members of social and cultural communities [1].

R. Silhavy et al. (Eds.): CoMeSySo 2022, LNNS 596, pp. 598–612, 2023.
https://doi.org/10.1007/978-3-031-21435-6_51

The digitalization of the economy implies the need to transform technologies for collecting and integrating, processing and analyzing, disseminating and effectively using spatial data. As part of solving the problem of data collection, the development of technologies for aero- and space monitoring of land and the Internet of things is of current importance. The processing and analysis of spatial data today should largely rely on machine learning and artificial intelligence technologies, the introduction of cloud computing and distributed computing. Consolidated arrays of spatial data acquire practical value provided that they can be used flexibly by representatives of certain roles to solve the problem of providing an effective tool for making informed management decisions [2].

Geoportals are a technological basis that implements the ability and right of citizens, specialists, managers and scientists to receive and disseminate spatial information, become key components of sustainable environmental, socio-economic development of regions [3]. Systems of this class form a sufficient level of coordination and communication of participants in various industries for the integration, processing and analysis of spatial data, as well as the consistent overcoming of barriers to the innovative development of countries and regions, providing flexible access to spatial data based on web-technologies [4].

The purpose of the study presented in the article is the development of principles for the formation of the architecture of a geoportal system as an access point to the information arrays of the spatial data infrastructure.

2 Related Work

The geosystem approach to the analysis of the decisions made inevitably leads to the problem of territorial development, especially actualized by developments web-oriented spatial data infrastructure modules are required to provide communication between cloud spatial data storage centers, a geoportal system, and a system for analyzing and synthesizing spatial data. In this context, service-oriented model encapsulation strategies are relevant for sharing and integrating heterogeneous geospatial models in an open Internet environment (Fig. 1).

Today, geoportal systems are seen as a "spatial support" tool in disaster management, smart cities and e-government. Among foreign universities, the following scientific competitors can be distinguished: University of Melbourne, Lund University, Vancouver Island University, KU Leuven, Norwegian University of Science and Technology. It is impossible to pass by large commercial organizations, in which scientific work in the designated problem area is an integral component of business processes: Leica Geosystems, ESRI, ScanEx.

The geoportals implemented in the world are characterized by the property of project orientation: for example, the main driving force for the SDI in Japan was the need for information to support the emergency response infrastructure; in Malaysia and Indonesia, to support effective resource planning and development; in other countries – to improve the efficiency of regional and local authorities [5]. Over the years, the programs and priorities of the SDI have been refined to reflect the priorities of various governments and organizations. By order of the US White House [6], regarding the coordination of geographic data collection and access processes, the following key elements of the SDI are identified: technologies, policies, standards, and human resources needed to acquire, process, store, distribute, and improve the use of geospatial data.

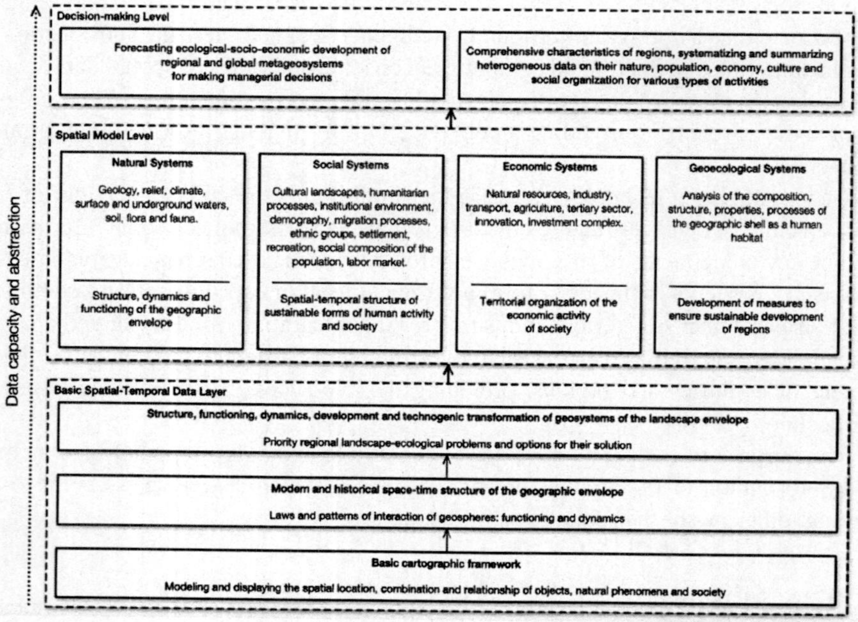

Fig. 1. General scheme of regional and global metageosystems

In the European Union [7], similar principles have been developed, and special emphasis is placed on the design and standardization of ascending and descending links between the SDI of states and the union as a whole. This principle was later developed in the INSPIRE directive [8]. As part of the Global Spatial Data Infrastructure [9] conference, it was proclaimed that the global spatial data infrastructure (GSDI) should cover the regulatory frameworks, organizational aspects, data, technologies, standards, mechanisms for providing spatial data, as well as financial and human resources necessary to ensure the achievement of goals.

Current trends are such that geoportal solutions are already a non-alternative basis for the work of state and municipal governments that ensure the functioning of strategic sectors of the economy [10]. They have firmly entered the practice of providing public services, interdepartmental interaction and publishing sectoral departmental information based on the principle of differentiating access to different categories of users. Most of the developed countries of the world have national geoportals: USA (data.gov, usgs.gov, drought.gov, phylo.org), Switzerland (geo.admin.ch), Finland (paikkatietoikkuna.fi), Czech Republic (cuzk.cz), China (chinageoss.org), France (geoportail.gouv.fr), Poland (geoportal.gov.pl), Spain (ceosspain.lpi.uv.es), Canada (open.canada.ca), Latvia (geoportal.lt). Some scientific organizations have geoportals, such as the universities of Manchester (manunicast.seaes.manchester.ac.uk) and Sydney (portal.gplates.org), the

Institute for Regional Development at the University of Castilla-La Mancha (http://ide.jccm.es).

Geoportals are the external part of the SDI, which acts as an access point to relevant spatial data, and are implemented as web portals, the core of which is the tools for searching and visualizing geospatial information and providing related services via the Internet [11]. Providers of geospatial data use geoportals to publish and distribute geospatial content, consumers - to gain access to information that is in demand in the economy and the social sphere.

Proper use of reliable spatial information in making scientifically based decisions helps to increase the efficiency of regional management in the areas of creating a favorable investment climate [12], creating a system of enterprises in the fields of agriculture and forestry [13], industry [14], transport [15], recreation and tourism [16], monitoring of natural ecological processes and elimination of their consequences [17].

The modern direction of solving the problem of introducing digital SDI and geoportal systems is the development and practical use of new automated methods and algorithms for interpreting remote sensing data and auxiliary spatial and temporal information using deep machine learning within the framework of SDI in order to constantly update cloud storages of large arrays of spatial and temporal data, distributed using geoportal systems.

3 Research Method

The software and hardware infrastructure of the geoportal system is a framework that includes physical and virtual resources and supports the functioning of the geoportal system as a web application accessible via the Internet. Due to the fact that the geoportal is available through the network to a weakly limited set of clients, it is necessary to emphasize the security and stability of the infrastructure to high loads, as well as the correctness of the parallel use of the system. The spatial DBMS and the geoportal database must provide storage of data that determine the state of the system and user settings. The architecture of the geoportal's spatial database should be designed based on the tasks the system is oriented towards. The geoportal framework as a pattern for designing and implementing a system that determines its architecture and structure should be divided into loosely coupled modules for managing data, operation logic, and appearance: web-based graphical geoportal user interfaces, business logic controllers, data models, and third-party libraries.

Universities are becoming an important link in the formation and development of geoportals, in the structure of which a multilevel system of knowledge about spatially distributed metageosystems is synthesized [18, 19]. The synthesis of digital spatial models is a multi-stage process in which spatial data is processed, analyzed and structured.

As part of the implementation of geoportal systems, it is important to solve the following specific tasks:

1. Visualization of spatial data within the digital map of the geoportal in order to ensure the possibility of their effective distribution and use.
2. Collection of spatial data received from external providers, including integration with IoT components.

3. Management space-distributed objects by sending control commands through the dispatcher's interfaces.
4. Implementation automation algorithms for solving the problems of building synthetic digital maps, forecasting the development of natural and natural-technogenic processes and phenomena, automatic or automated monitoring, decision-making based on consolidated statistical data and spatial processes.
5. Project-oriented use of intelligent algorithms based on the use of machine learning methods for the analysis of heterogeneous spatial data, including those based on deep neural network models.
6. Interactive generation of reporting documentation based on the spatial data of the geoportal, containing both the results of analytical processing of consolidated data and a system of automatically generated recommendations regarding the possibility of achieving specific parameters.
7. Functioning of the administration subsystem that provides functionality for manipulating the geoportal's spatial data with the necessary access rights.

The algorithm for developing a geoportal system as a tool for visualizing space-time processes occurring in complex territorial systems should include the following sequence of steps:

1. Definition of the task and scenarios for using the system. During this project stage, a list of problems is determined that the geoportal being developed is aimed at solving, and a register of key precedents is determined that forms the functionality space of the designed solution.
2. Design of the system. At this stage, the decomposition and refinement of the basic framework of the geoportal system for solving specific problems is carried out, spatial data is collected, the statistical characteristics of the system are evaluated, hypotheses are put forward to determine the level of abstraction of the territorial model, which ensures the acceptable accuracy of data visualization and the achievement of target effects.
3. Specification of the system architecture. Within the framework of this stage, the necessary list of algorithms is being developed that determines the functioning of a specific implementation of a geoportal system aimed at solving particular problems of analyzing natural and natural-technogenic processes.
4. Implementation of the graphical interface and visualization subsystem. At this stage, the interface component of the system is proposed and developed, which should determine the ways of interacting with data models, methods and tools for analyzing its functioning, optimizing and refining.
5. Iterative approbation of the geoportal system, which allows not only modeling specific spatial and temporal processes, but also calibrating the territorial model, refining the register of data used, and optimizing the quality of the result.

Two directions of development of the framework of the geoportal system should be distinguished. On the one hand, it is necessary to adhere to a deductive strategy, in which the geoportal is initially developed in order to achieve the possibility of covering solutions to the maximum number of design tasks. On the other hand, it is necessary to implement the inductive strategy as efficiently as possible, within which the positively proven improvements of private specific geoportals become the basis for the development of the system framework. The joint use of deductive and inductive strategies will provide an evolutionary improvement in the framework for building geoportal systems, as well as optimization of particular solutions developed on its basis.

Geoportals as an external component of digital spatial data infrastructures belong to the class of distributed web systems and, thus, inherit their characteristic features. In particular, this fact affects the non-functional requirements for geoportal systems. So, from the point of view of security, it is important to ensure the properties of confidentiality (protection against unauthorized access), availability (protection with parallel access to resources) and integrity (data safety) of the system. An urgent task in the design of geoportals is the problem of ensuring the quality of service, which consists in the need to fulfill a certain set of client requirements.

The qualitative requirement for scalability determines the ability of a geoportal system to meet the challenges of growing functionality and increasing workload. At the same time, it is advisable that the search time in geospatial databases does not grow exponentially with an increase in the amount of accumulated information about natural-social-production systems, and the amount of resource use does not increase uncontrollably with an increase in the number of users.

An analysis of the experience of building project-oriented digital SDI shows that the process of forming the architecture of systems of this class should be based on the position that in order to optimize the storage processes and efficient use of spatial data, the SDI should contain storage, analysis and synthesis subsystems connected through software interfaces, visualization and dissemination of spatial data (Fig. 2).

Microservice architecture involves organizing the interaction of isolated services, each of which performs its own and only its own specific task. In the SDI system, the following bounded contexts for microservices can be distinguished: geoportals (tools for visualization and management of spatial data), systems for analyzing, processing and control of spatial data, as well as distributed cloud storages. The division into subsystems of visualization, processing and integration of spatial information makes it possible, starting from the top level of decomposition of the system architecture, to achieve an increase in the cohesion of modules and a weakening of the coupling.

Geoportal systems are an access point to information consolidated in digital infrastructures of spatial data and are designed based on the tasks to be solved. Individual geoportals can be aimed at solving the problems of integrating information about objects of natural, historical and cultural heritage, managing land use systems (for example, agricultural enterprises), as well as industries and services whose activities are based on the implementation and use of Internet of Things components.

Based on the authentication module, the problem of role-based access control (RBAC) is solved, which involves the implementation of selective access control policies, while the access rights of system actors to use functionality and data are grouped taking into account the specifics of their application, forming roles in the system. The purpose of deploying a geoportal largely determines the services deployed within the framework of the CRUD spatial data management system and solving the problems of creating, reading, updating and deleting information. The central service in this context

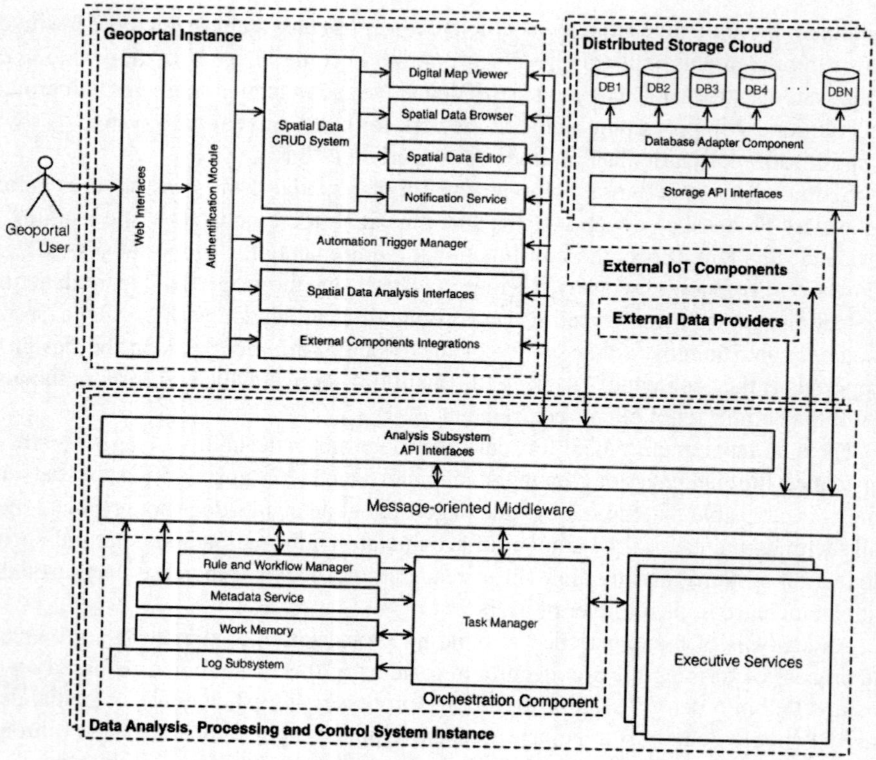

Fig. 2. Organization of interaction of services in the spatial data infrastructure

is the digital map visualization service, which allows you to control the layer-by-layer display of various thematic layers of spatial data and can be organized based on the Model-View-Presenter architectural pattern. The spatial data browser allows the user to interact with a system of filters that allow for an interactive search for spatial information. Spatial data editing services are developed and deployed in order to implement the ability to manage the spatial data storage, allow single and batch adding of spatial objects, import and export of information blocks, modify the spatial and attributive context. Notification services provide prompt receipt of important notifications, including those visualized within a digital geoportal map or spatial data browser.

Services for organizing interfaces for interaction with components of spatial data analysis represent a link that allows interaction with executive services of the subsystem for processing, analyzing and managing spatial data that are external to geoportal systems. On the other hand, the automation trigger management service allows you to solve the problem of optimizing routine tasks by launching certain computational calculations when certain conditions occur: temporary (onset of a certain time, expiration of a periodic interval) and event (fulfillment of spatial or attribute-oriented conditions). At the same time, the conditions can be focused both on the analysis of current parameters and on the use of statistical characteristics of the accumulated information.

Integration services with external components allow you to configure interaction with remote systems, including Internet of Things objects. In this case, integration services should be organized on the basis of the adapter pattern, which allows for the rapid and unified addition of equipment from new suppliers and modifications.

Distributed cloud storages of spatial data are formed on the basis of multi-model data storages. The link between geoportal systems and distributed cloud storage infrastructure of spatial data is the systems of analysis, processing and management of spatial data, based on technologies for automated information processing and data analysis. The core of systems of this type are executive services that encapsulate the software implementation of algorithms for processing and analyzing spatial data. To orchestrate the process of execution and interaction of individual services, it is necessary to deploy a task manager, the functioning of which is based on the subsystem for determining the rules for launching services and building the flow of works. It is advisable to build the middleware of the spatial data infrastructure as a set of programs for ensuring interaction between various services and components based on a focus on messaging in a distributed environment.

During the initial deployment of a geoportal, it is necessary to have guarantees regarding the stability of the system to loads of a certain level and at the same time ensure the necessary performance (task of the first type).

To achieve the designated goal, it is necessary to consistently solve the tasks aimed at: (1) determining the target performance P_t of the geoportal system based on a formalized set of criteria; (2) carrying out load testing, evaluating the actual performance P_l and comparing it with target indicators; (3) deploying a geoportal, evaluating performance under the condition of the work of real users P_r and comparing the obtained indicators with the target ones.

In the presence of a working geoportal solution, it may be necessary to solve a problem of the second type: achieving an increase in current performance by optimizing the current version of the geoportal system based on the sequential solution of the following tasks: (1) an integral assessment of the current system performance P_{r1}; (2) identification of bottlenecks of the geoportal system, its software and hardware components, the features of the functioning of which limit the ability to achieve target performance indicators P_t; (3) optimization of the system, followed by performance evaluation of the new version P_{r2} and comparison of the result with the target indicator P_t.

To solve the tasks outlined above, it is necessary to determine the methodology for generating a performance estimate P. This indicator should be applied (reflecting the real non-abstract performance of the geoportal system), integral (taking into account all key aspects of the system operation and options for its use); quantitative (measured in numbers to enable comparison of performance under different conditions for different versions of the system); interpretable (defined in terms of "good" - "bad").

Let us take as a set of key operations of the geoportal system Op a set of a finite number n of operations that are critical for the functioning of the spatial data infrastructure, the insufficient performance of which leads to losses of various types. In this case, operations performed in parallel by a large number of users deserve special attention.

$$Op = \{op_i | i \in \mathbb{N} \wedge i \leqslant N\} \tag{1}$$

Further, we will solve the problem of analyzing the performance of a geoportal system in order to optimize its functioning under high load conditions from the standpoint of studying the features of performing key operations.

Note that there are well-known methods and standards for evaluating the performance of corporate systems under high load conditions. Thus, the Apdex method (Application Performance Index, or application performance index) is an international open standard designed to solve the problem of objectively assessing the performance of information systems of various classes and allowing to reduce heterogeneous factors and statistical data on the performance of an information system to a simple value [20]. The positive features of the apparatus under consideration include ranking the analyzed operations by priority, taking into account objective factual data, and intuitive clarity of the assessment for the end users of the system and its owners. However, this approach also has disadvantages, consisting in insufficient flexibility of the apparatus and focusing mainly on system response indicators.

Let us give a sequence of steps to form a methodology for an objective assessment of the performance of a geoportal system. For each key operation op_i, we define the priority π_i, which is expressed as a natural number inversely proportional to the importance of the key operation and allows sorting the operations in order of the criticality of the impact of their performance quality on the operation of the geoportal system as a whole.

The next step is to collect performance data for each key operation op_i, which should be recorded as a set of performance indicators of the operation T_i with power m:

$$\mathrm{T}_i = \left\{\tau_{ij} | j \epsilon \mathbb{N} \wedge j \leqslant M\right\} \tag{2}$$

In this case, the performance indicator τ_{ij} must be expressed as a numerical value: it can be a measurement of the execution time of a key operation, a failure rate, the fact of launching a certain scenario and activating conditions, the number of errors, and so on. Also, each performance indicator τ_{ij} can be assigned a priority π_{ij}, indicating the importance of the criterion. For each performance indicator τ_{ij}, we define a finite number of statements obtained on the basis of the function $f_{ij} : \tau_{ij} \to X_{ij}$ that characterize the values of the set of performance indicators clustered according to a certain principle:

$$X_{ij} = \left\{x_{ij(k-1)} | k \epsilon \mathbb{N} \wedge k \leqslant K\right\} \tag{3}$$

Each statement x_{ijk} is associated with the value of the risk factor ω_{ijk}. The values of each performance indicator must be associated with at least two statements: x_{ij0}, for which $\omega_{ij0} = 0$ (for the case in which there is a negative impact on the operation of the geoportal system), and x_{ij1}, for which $\omega_{ij1} = 1$ (for a positive result). Additionally, it is possible to determine a finite number of statements x_{ijk} with risk factors ω_{ijk} characterizing states of varying degrees of moderation, and the outcome can be considered the more positive, the greater the value ω_{ijk} accepted for it.

Denote by N_{ijk} the number of outcomes for a series of trials that led to the receipt of the statement x_{ijk}. Thus, after conducting a series of L observations regarding the execution of the operation op_i in terms of estimating the performance indicator τ_{ij}, we can estimate the performance $P_{\Sigma ij}$ with respect to operation i from the point of view of the indicator j for the analyzed geoportal system:

$$P_{\Sigma ij} = \frac{\sum_{k=0}^{K-1} \omega_{ijk} N_{ijk}}{\sum_{k=0}^{K-1} N_{ijk}} : (\omega_{ij0} = 0) \wedge \left(\omega_{ij(K-1)} = 1\right)$$
$$\wedge \left(0 < \omega_{ij(k-1)} < 1 : k \in (1, K-1)\right) \tag{4}$$

In this case, the performance indicator $P_{\Sigma ij}$ will be measured in the range from zero (worst, unsatisfactory outcome) to one (ideal test). Of interest is an alternative indicator built according to the principle of $P_{\mathrm{H}ij}$ histograms and characterizing the distribution of the number of performance indicators in a series of L_{ij} tests by A_{ij} groups. To calculate it, we define through $\tau_{ij(L)}$ the vector formed on the basis of the values of the parameters τ_{ij}, obtained on the basis of L_{ij} tests.

$$P_{\mathrm{H}ij} = \left\{ h_{ija} \left| \begin{array}{l} h_{ija} = \sum_{l=1}^{L_{ij}} \tau_{ijl} \left[\begin{array}{c} \tau_{ijl} \geqslant \min\left(\tau_{ij(L_{ij})}\right) + (a-1)\sigma \\ \tau_{ijl} < \min\left(\tau_{ij(L_{ij})}\right) + a\sigma \end{array} \right] \\ \wedge \left(a \in \mathbb{N} \wedge a \leqslant A_{ij}\right) \\ \wedge \sigma = \frac{\left(\max\left(\tau_{ij(L_{ij})}\right) - \min\left(\tau_{ij(L_{ij})}\right)\right)}{A_{ij}} \end{array} \right. \right\} \tag{5}$$

In this case, the number of groups A can be selected manually in order to increase the information content of the metric or determined on the basis of a rule. So, according to the Sturges rule, the number of groups depends on the amount of data:

$$A_{ij} = 1 + \log_2 L_{ij} \tag{6}$$

However, this rule is not suitable for skewed datasets and can be modified to take into account the features of samples with a non-standard probability distribution by taking into account the skewness coefficient g_{ij} and the standard deviation $\sigma_{g_{ij}}$.

$$A_{ij} = 1 + \log_2 L_{ij} \left(1 + \frac{\left|g_{ij}\right|}{\sigma_{g_{ij}}}\right) \tag{7}$$

For an integral estimate $P_{\Sigma i}$ of the performance of the key operation op_i with respect to the entire set of performance indicators T_i, one can resort to using aggregate functions, for example, taking the minimum.

$$P_{\Sigma i} = \min_{1 \leqslant j \leq M} P_{\Sigma ij}\left[\pi_{ij} \geqslant \pi_{ij(min)}\right] \tag{8}$$

In this case, the value of $P_{\Sigma i}$ will be equal to the worst estimate of the performance of the key operation op_i relative to each criterion from the set T_i, which allows the metric to characterize the reliability of the geoportal system with respect to information about the value of the lowest performance indicator, characterizing the bottleneck in the system. Taking into account the values of $\pi_{ij(\min)}$ (threshold significant priority), it is possible to cut off operations of definitely low significance and not take into account their performance.

$$P = \langle P_{\Sigma}, P_{\mathrm{H}} \rangle = \left\langle \begin{bmatrix} P_{\Sigma 1} \\ P_{\Sigma 2} \\ \cdots \\ P_{\Sigma N} \end{bmatrix}, \begin{bmatrix} P_{\mathrm{H}1} \\ P_{\mathrm{H}2} \\ \cdots \\ P_{\mathrm{H}N} \end{bmatrix} \right\rangle. \tag{9}$$

Based on certain rules, the system performance values P can be compared to terms interpreted from the position of "good"-"bad".

4 Results and Discussions

Consolidation of data in digital SDI is of particular importance for solving the problem of information support for educational activities in the field of interdisciplinary research based on the use of spatial data. At the same time, the final artifacts of SDI data analysis become a key element that ensures the management of systems for disseminating knowledge about the key components of sustainable development of regions. Their use makes it possible to increase the efficiency of information processes that ensure the development of educational and socio-economic systems.

The geoportal "The Natural and cultural heritage of the Republic of Mordovia" performs the function of organizing the management of cultural and historical heritage systems. The framework of the project was initially designed and developed from scratch on the basis of a component-oriented approach to effectively solve the problem of disseminating information about the natural and cultural heritage of the region. The graphical interfaces of the information system were created using adaptive layout technologies to achieve the necessary UX indicators and ease of use from smartphones and desktop devices (Fig. 3). The geoportal database is designed based on the tasks to be solved in the project based on the relational approach.

Information in the geoportal is presented in 3 languages: Russian, English and Spanish. The data within the framework of the project are structured in 69 information blocks from 6 thematic sections: "Administrative structure, nature and resources", "History", "Population and social infrastructure", "Economy", "Ecology", "Municipal area heritage" and are provided with illustrative material: more than 800 photographs of objects of natural, historical and cultural heritage.

The basis of the content content is formed by more than 1240 information blocks about settlements with information about toponymy, geography, history are placed on the geoportal with reference to an electronic map, 830 objects of cultural and historical heritage and 95 specially protected natural areas are described with spatial reference to the digital map of the region. Based on the materials of the regional SDI, synthesized, among other things, using the new methods and algorithms for machine analysis of spatial data presented in the dissertation, 139 thematic scalable maps of various thematic directions were developed within the framework of the project. Maps of vegetation, landscapes and erosion processes have been developed and updated based on new methods and algorithms proposed in the dissertation.

Fig. 3. Geoportal «The Natural and Cultural Heritage of Mordovia»

In the paper [21], presented in the ISPRS International Journal of Geo-Information (the first quartile of Scopus), the experience of the conducted research is noted as another step towards the formation of tools for easy access to spatial data through web systems that use digital maps as a visual interface for geographic information. In the same scientific journal, in the article [22], when developing a web-based geoinformation system for heritage management and solving the problem of geovisualization of spatial data in the canton of Nabon (Ecuador), the results in the field of engineering of qualitative and functional requirements, including those related to the availability of adequate navigation tools for searching through geospatial data and providing quick access to attributes of spatial objects. The article [23] noted that the experience gained and the generated recommendations were used in the development of a tourist geoportal for the rural region of Extremadura (Spain).

The article [24], published by the journal GeoJournal of Tourism and Geosites and devoted to the restoration of cultural landscapes for planning cultural tourism in Bali, noted the experience of using geoportal systems to solve the problem of developing tourism based on the natural environment and culture, and, as a result, associated with significant risks. In the scientific work [25], in turn, the experience gained in the field of designing geoinformation systems was used to solve the problem of assessing the impact of spatial differences on the quality of urban life (comparative analytical study of three cities in Algeria). In particular, the results aimed at realizing the possibility of a quick comparative analysis of geographic information in space and time are taken into account. Finally, a study [26] was published in the scientific journal Discrete Dynamics in Nature and Society, based on the experience of using neural network algorithms in solving the problem of spatial analysis in the framework of the problem of predicting the processes of rural recovery and predicting the sustainable development of ecotourism based on big data.

5 Conclusion

Geoportals are a technological basis that implements the ability and right of citizens, specialists, managers and scientists to receive and disseminate spatial information, become key components of sustainable environmental, socio-economic development of regions; systems of this class form a sufficient level of coordination and communication between participants in various industries for the integration, processing and analysis of spatial data, providing flexible access to spatial data based on web-technologies.

It is necessary to single out two directions for the development of the framework of the geoportal system. On the one hand, it is necessary to adhere to a deductive strategy, in which the geoportal is initially developed in order to achieve the possibility of covering solutions to the maximum number of design tasks. On the other hand, it is necessary to implement the inductive strategy as efficiently as possible, within which the positively proven improvements of private specific geoportals become the basis for the development of the system framework. The joint use of deductive and inductive strategies will provide an evolutionary improvement in the framework for building geoportal systems, as well as optimization of particular solutions developed on its basis.

Microservice architecture involves organizing the interaction of isolated services, each of which performs its own and only its own specific task. In the SDI system, the following bounded contexts for microservices can be distinguished: geoportals (tools for visualization and management of spatial data), systems for analyzing, processing and control of spatial data, as well as distributed cloud storages. The division into subsystems of visualization, processing and integration of spatial information makes it possible, starting from the top level of decomposition of the system architecture, to achieve an increase in the cohesion of modules.

The article proposes a technique for generating a performance assessment in the form of applied, integral, quantitative, interpretable indicators that eliminate the problem of the Apdex methodology, which consists in insufficient flexibility of the apparatus and focus mainly on system response indicators. Based on certain rules, system performance values can be compared to terms interpreted from the position of "good" - "bad". On the basis of the proposed methods for the construction of geoportals, the geoportal system "The Natural and Cultural Heritage of the Republic of Mordovia" has been developed, which performs the function of organizing the management of cultural and historical heritage systems. The framework of the project was initially designed and developed from scratch on the basis of a component-oriented approach to effectively solve the problem of disseminating information about the natural and cultural heritage of the region.

Acknowledgments. The study was supported by the Russian Science Foundation, grant № 22-27-00651, https://rscf.ru/en/project/22-27-00651/.

References

1. Reetz, B., Riede, H., Fuchs, D., Jerg, M., Hagedorn, R.: DWD Geoportal—A Central Hub for Open Data, API and Communication (2022). https://doi.org/10.5194/ems2022-468

2. Dawidowicz, A., Nowak, M., Gross, M.: Land administration system and geoportal service for the need of a fit-for-purpose national urban greenery management system (UGMS). The concept for the EU member state of Poland. Acta Sci. Pol. Adm. Locorum **21**, 53–81 (2022). https://doi.org/10.31648/aspal.7454
3. Reddy, G.P.O.: Geoportal platforms for sustainable management of natural re-sources. In: Studies in Big Data, pp. 289–314. Springer, Singapore (2022)
4. Gkonos, C., Enescu, I.I., Hurni, L.: Spinning the wheel of design: evaluating geoportal graphical user interface adaptations in terms of human-centred design. Int. J. Cartogr. **5**, 23–43 (2019)
5. Masser, I.: A CRC title, part of the Taylor & Francis imprint, a member of the Taylor & Francis Group, the academic division of T&F Informa PLC. In: Geographic Information Systems to Spatial Data Infrastructure, pp. 265–280. CRC Press, Boca Raton (2019)
6. Executive Order of the White House: Coordinating Geographic Data Acquisition and Access: The National Spatial Data Infrastructure. Office of the Press Secretary: The White House, Washington, DC (1994)
7. GI2000-Towards a European Geographic Information Infrastructure (EGII)—A Discussion Document for Consultation with the European GI Community. European Commission, Brussel (2000)
8. Directive 2007/2/EC of the European Parliament and of the Council of 14 March 2007 Establishing an Infrastructure for Spatial Information in the European Community (INSPIRE). Published in the Official Journal (2007)
9. The Global Spatial Data Infrastructure Association Spatial Data Infrastructure Cookbook. Brussel (2012)
10. Purwanto, B. (2022) Data classification and access control in Indonesia one map policy geoportal
11. Carbonell-Carrera, C., Hess-Medler, S.: Interactive visualization software to im-prove relief interpretation skills: spatial data infrastructure geoportal versus augmented reality (2019)
12. Drebot, O., Komarova, N., Tarnavskyi, V., Komarov, D.: Geoportal of open data of the Bila Tserkva city territorial community as a component of the national spatial data infrastructure. Agrosvit **31**(3) (2022). https://doi.org/10.32702/2306-6792.2022.3.31
13. Barrera Vielma, M.G., Terrones, I.L.: Nodo integrador de la informaciốn ge-ográfica del Agro: Geoportal IDE MINAGRI de Chile. Rev. Cartogr. 185–191 (2019)
14. Sanchez-Gallegos, D.D., Sosa-Sosa, V.J., Marin-Castro, H.M., Tuxpan-Vargas, J.: An interoperable cloud based geoportal for discovery and management of earth observation products. Comput. Sci. Inf. Technol. (2018)
15. Cheng, X., Zhang, R., Zhou, J., Xu, W.: DeepTransport: learning spatial-temporal dependency for traffic condition forecasting. In: 2018 International Joint Conference on Neural Networks (IJCNN). IEEE (2018)
16. Yamashkin, A.A.: Cultural landscapes space-temporal systematization of in-formation in geoportals for the purposes of region tourist and recreational development. GeoJ. Tour. Geosites **29**, 440–449 (2020)
17. Barmpoutis, P., Papaioannou, P., Dimitropoulos, K., Grammalidis, N.: A review on early forest fire detection systems using optical remote sensing. Sensors (Basel) **20**, 6442 (2020). https://doi.org/10.3390/s20226442
18. Dareshiri, S., Farnaghi, M., Sahelgozin, M.: A recommender geoportal for geospatial resource discovery and recommendation. J. Spat. Sci. **64**, 49–71 (2019). https://doi.org/10.1080/14498596.2017.1397559
19. Gamez, M.R.: The geoportal as strategy for sustainable development. Int. J. Phys. Sci. Eng. (IJPSE) (2019). https://doi.org/10.29332/ijpse.v3n1.239
20. Sevcik, P.: Defining the application performance index. Bus. Commun. Rev. **20**(8) (2005)

21. Roose, M., Nylén, T., Tolvanen, H., Vesakoski, O.: User-centred design of multidisciplinary spatial data platforms for human-history research
22. Collado, A., Mora-Navarro, G., Heras, V., Lerma, J.L.: A web-based geoinformation system for heritage management and geovisualisation in Cantón Nabón (Ecuador). ISPRS Int. J. Geoinf. **11**, 4 (2021). https://doi.org/10.3390/ijgi11010004
23. Engelmo-Moriche, Á., Nieto-Masot, A., Horcajo-Romo, A.I.: Tourist geoportal of a rural region as a means of disseminating its offer and resources. Case study: extrema-dura. Rev. Estud. Andaluces **41**, 125–143 (2021)
24. Adhika, I.M., Putra, I.D.G.A.D.: Reinvigorating cultural landscapes for planning cultural tourism in Bali. GeoJ. Tour. Geosites **33**, 1462–1469 (2020). https://doi.org/10.30892/gtg.334spl03-594
25. Allal, A., Boudjemaa, K., et al.: The effect of spatial differences on the quality of urban life a comparative analytical study of three cities in the high plateaux region of Algeria. GeoJ. Tour. Geosites **40**, 181–190 (2022). https://doi.org/10.30892/gtg.40122-818
26. Tang, Y.: Discrete dynamic modeling analysis of rural revitalization and ecotourism sustainable prediction based on big data. Discrete Dyn. Nat. Soc. **2022**, 1–9 (2022)

Low Input Voltage DC-DC Converter for Harvesting Energy by Microbial Fuel Cell

Andrew Lazukin[1(✉)], Anton Saveliev[2], and Konstantin Krestovnikov[2]

[1] ITMO University, Kronverksky Pr. 49, bldg. A, St. Petersburg 197101, Russia
lazukin@scamt-itmo.ru

[2] St. Petersburg Federal Research Center of the Russian Academy of Sciences (SPC RAS), St. Petersburg Institute for Informatics and Automation of the Russian Academy of Sciences, 39, 14th Line, 199178 St. Petersburg, Russia
saveliev@iias.spb.su

Abstract. Marine robotics, environmental monitoring, control systems are becoming increasingly relevant nowadays. The necessity of using marine electronic is observed in all branches of human activity from the need to monitor the state of open-type aquaculture farms to the field of shelf exploration, Arctic exploration, and a number of military tasks. The solution to the issue of providing electrical energy to marine autonomous devices can be energy sources based on the metabolism of microorganisms, for example, based on a microbial fuel cell. However, low power and low generated voltage prevent the introduction of microbial fuel cell technologies into practice. To overcome this disadvantage in the current work, an ultra-low-voltage energy converter is proposed, which allows providing autonomous electronic devices with electrical energy from a microbial fuel cell. As part of the study, microbial fuel cells were manufactured, experimental verification was carried out with these protocols using the example of a temperature sensor with sending a signal to the subscriber using LoRa modulation. The application of the developed solutions confirmed the possibility of autonomous operation of the sensor, obtaining data on the ambient temperature, as well as transmitting this signal at a distance of up to 950 m.

Keywords: Autonomous power supply · Microbial fuel cell · Ultra-low-voltage converter

1 Introduction

At the moment, the use of environmental control systems based on electronic automated devices is becoming increasingly relevant. This mainly concerns autonomous sensors and environmental information collection systems. The main problem in the development of this direction is the search for energy sources that allow generating electric current for a long time, while it is necessary that these energy sources are easy to maintain. A special place is occupied by the problem of power supply of marine autonomous robotic complexes and environmental control complexes. As a rule, such complexes

R. Silhavy et al. (Eds.): CoMeSySo 2022, LNNS 596, pp. 613–627, 2023.
https://doi.org/10.1007/978-3-031-21435-6_52

require functioning for a long time at a great distance from the place of residence of a person [1].

The problem of finding power supply for autonomous complexes cannot be solved using some separate universal power generator. To solve this issue, one of the most common options is solar energy [2–4]. However, photoelectric converters are highly dependent on weather conditions which is a significant disadvantage. Another alternative option, which is also traditionally used to power autonomous complexes, is the use of wind energy [1]. One of the disadvantages of wind generation is the need to install this wind generator on a large mass carrier in order to withstand the wind generator mast, wind gusts, waves and storms that affect the monitoring complex.

Another option for providing electricity can be the conversion of kinetic energy of sea waves into electrical energy [5]. There are many options of converters, they can differ both in the principle of energy generation and in complexity. The most common option is generators operating based on electromagnetic induction. In such generators the interaction of a permanent magnet with a coil is used where, in the process of changing the magnetic flux density, electromagnetic induction occurs, resulting in an electric current that, after appropriate transformations, is either stored in batteries or directly feeds the load [6, 7].

In addition to the above options, it is also possible to use an electric energy converter from the kinetic energy of waves with a triboelectric effect. In [8], the authors describe the creation of a nanogenerator that generates an electric current under the influence of mechanical vibrations of the generator. The generator is made in the form of a block of flat plates on which aluminum electrodes are attached. These electrodes have a ribbed surface along which polytetrafluoroethylene (PTFE) balls move under different inclination of the plates. Due to the movements of the buoy, the falling wave causes the movement of the balls. A large difference in electronegativity leads to the loss of electrons for aluminum, since it comes into effective contact with PTFE. The authors demonstrated the possibility of powering LEDs and sensors of temperature and salinity of water from the energy of this generator. It is also possible to combine the use of triboelectric and induction generators [9, 10].

In addition to the triboelectric generator, piezoelectric converters are also used. The authors of [11] describe the method of converting mechanical wave energy into electrical energy using a piezoelectric converter built into a small-sized buoy. It is worth noting a triboelectric generator made according to the type of bottom algae [12]. The authors have developed a flexible triboelectric generator in the form of algae, which can be placed at the bottom of the reservoir, on the surface, attached to a carrier, such as a buoy, or to any other underwater surface where there are low-frequency mechanical vibrations of water.

From the above examples, it is important to note that in all the cases described above, for normal energy generation, either the removal of generators above water is necessary, as in the case of solar panels and wind generators, or the presence of generators on a floating base in the form of a buoy that allows one to capture mechanical vibrations with subsequent conversion of mechanical energy to the electric one. At the same time, the authors do not describe generators that function autonomously for a long time, being at the bottom of a reservoir, and basically all the proposed ideas are the development of previously created concepts.

Microbial fuel cells (MFC) were also implemented among the energy sources. The paper [13] demonstrates the world's first experience in using MFC as an energy source for powering sensors operating as part of a marine monitoring buoy. The authors note that the use of MFC in their installation can provide a conditionally infinite power supply time. According to the authors, this possibility is due to the properties of MFC. Such properties are resistance to corrosion and a constant influx of nutrient substrate. In this case, the nutrient substrate, being in a dissolved state in water, diffuses into a microbial fuel cell. This is possible since MFC is an open system with the possibility of free exchange of matter and energy. This allows both to supply bacteria with nutrition and to remove metabolites that accumulate in the working area of the energy source. A distinctive feature of all MFC is their low power and low voltage, which, as a rule, is not enough to power electronic devices. For this reason, to power the sensors for this buoy, an MFC operating voltage converter (about 0.35 V) was manufactured to a voltage of 3–6 V required for the operation of the sensors. The resulting energy is accumulated in a supercapacitor with a capacity of 1 F. The authors state that the efficiency of the converter was 85%. At the same time, the design of the harvester and its principle of operation are not disclosed by the authors. The paper notes that the average power consumption of the device when sending temperature humidity data was 18 mW. The advantage of this power supply is the complete absence of any maintenance in combination with the necessary unlimited operation time. At the same time, the authors note that one of the difficulties in the operation of this energy source is the need to convert energy and then store it in supercapacitors, which may prove to be a significant obstacle to the development of these fuel cells in real practice. Most commercially available DC-DC converters operate at a voltage of 0.7 V, which is higher than the operating voltage of the MFC. In addition, existing solutions require significant input currents for their operation, which does not allow the use of such devices with MFC. Ready-made converters, such as TPS61201 (Texas Instruments), are impractical for MFC, since the power consumption of such a circuit is 2 mW at an input voltage of 0.3 V. Thus, the use of a very high power MFC is required to start this converter. Thus, the work [14] demonstrates the development of an energy converter for microbial fuel elements. The converter is made based on an induction converter with a key based on a field-effect transistor switched by an external signal generator. However, the disadvantages of this system is the need to supply electrical energy to start the converter and for its subsequent operation. The authors note that stable operation with the possibility of powering sensors is possible when a voltage of 5 V. At the same time, the operation of the device is not stable and in the case of a temporary decrease in the power of the MFC, the entire operation of the electrical circuit will be stopped and will require manual restart of the master generators, such as LTC6906 or ALD1502, which are used in the described work as the master generators of the induction converter. In [15], a commercial BQ25504 converter is used as a DC-DC converter. The authors draw attention to its high efficiency of 80%, the presence of maximum power point control, as well as the possibility of programming the output voltage. However, the high current requirements of the converter led to rapid wear of the MFC operation, which required the authors to install several MFC by manually switching the operating modes between them every 2 days.

In [16], the features of creating an energy harvester are given. This work is aimed at providing autonomous power supply to environmental monitoring sensors for several months. In this paper, the authors used a two-stage conversion system. At the beginning, the voltage from the MFC accumulates in the supercapacitor until a certain level is reached. After that, the S-882Z power management chip supplies power to the TPS61200 series converter. This is necessary because the existing converters consume too much current, which the MFC cannot provide. At the same time, in order to power the sensors and the successful operation of the system considered by the authors, it is necessary to provide a power of at least 1.45 mW.

A similar option of the two-stage transformation was used by the authors of [17]. At the same time, the MFC power required for operation should have exceeded 3 mW.

Based on the analysis of the literature, it can be concluded that there is a problem of conversion, accumulation, and use of low-power energy from microbial fuel cells, as well as the organization of power management from these elements. This paper considers the possibility of using a blocking generator to power the payload from MFC, as well as its main parameters. In the following sections, the production of a MFC as a model energy source for this converter is described, two options for controlling the payload in automatic mode are considered. For experimental verification, the study considered the power supply of a temperature sensor with subsequent signal transmission via the Lo-Ra protocol.

2 Statement of Problem

The results of the analysis of literature showed that most of the payload elements powered by MFC have an electrical consumption of 10–100 mW with a voltage consumption of 3–5 V. The generated voltage of the MFC is in the range of 300–500 mV, the power of the MFC is from 1.5 mW and higher. Thus, in this study it is required to solve the following tasks:

1. development of an ultra-low-voltage converter with an operating voltage of 300–500 mV, with a starting power consumption of no more than 1.5 mW;
2. implementation of a microbial fuel cell with an operating voltage of 300–500 mV with a power of no more than 1.5 mW;
3. development of a load – demonstrator with an electric power consumption in the range of 10–100 mW in pulse mode.

The developed MFC is supposed to be delivered by means of an autonomous uninhabited underwater vehicle (AUV) with positive buoyancy, low power of the power supply system and light in weight. The AUV has four propellers, two of which are responsible for loading, and the other two for maneuvering in the horizontal plane. A camera is installed in the lower part of the AUV body to obtain images necessary for searching for the ArUco marker and positioning the device relative to it.

3 Ultra-low Voltage Energy Converter

The peculiarity of using MFC is that the generated voltage is very low and is in the range of 300–500 mV, depending on the operating parameters [18, 19]. Formally, this voltage may be higher, but such a voltage, as a rule, cannot be used to power DC-DC converters, since such converters need a high current to start, which leads to a temporary decrease in the voltage of the MFC and the inability of the converters to work. The serial connection of the MFC does not solve the problem since it leads to excessive polarization of the MFC electrodes and a significant increase in internal resistance [20].

From the results of the analysis of the literature data, it can be seen that existing solutions often find their application in energy collection systems generating relatively high currents (units and tens of mA). The creation of such currents is quite affordable for thermoelectric converters, but it requires large resources from the MFC [21]. Based on the results of the analysis, it can be noted that there are three main strategies for the construction of harvesters. In the first case, direct conversion by specialized microcircuits is used, which leads to high energy losses. In the second case, the strategy of an induction converter with an external supply of the master generator is considered. This is characterized by the complexity of the system and its low stability, dependence on external power supply and the presence of shutdown risks [14].

The third strategy assumes the presence of a two-stage conversion, where energy is accumulated in the supercapacitor at the beginning, followed by the power supply of specialized chips operating from this supercapacitor. At the same time, the capacitance of supercapacitor is usually measured in tens and hundreds of farads, which leads to significant leakage currents, while these circuits require external voltage monitoring systems to activate DC-DC converters [17].

In this paper, it is proposed to use a blocking oscillator as an ultra-low voltage voltage converter. The advantages of the blocking generator are simplicity of design, the possibility of obtaining a significant voltage gain, the availability of discrete components, low voltage of the converter start, low operating currents. The following scheme of the blocking generator is proposed for consideration (Fig. 1).

This oscillator uses a DSS60600 low saturation bipolar transistor. The transformer is made with enameled wire ($d = 0.1$ mm) on a ring ferrite core R $20 \times 10 \times 5$ PC40. The primary winding L1 has an inductance of 15 mH, the secondary winding L2 has an inductance of 135 mH. A 100 μF ceramic capacitor is installed at the input of the converter, and a low-dropout diode is installed at the output. The installation of a capacitor is necessary to increase the current supplied to the input of the converter at the moment when the L1 coil is fully connected to the load at the moment of saturation of the transistor. The output of the converter is a high-speed rectifier diode with low voltage drop and low leakage current. When transistor Q1 is turned off, the current in the primary winding of L1 is zero. The accumulated energy of the transformer gives through the winding L2. The decreasing magnetic flux in the core of the transformer induces a voltage on the winding L2. Therefore, the diode D1 opens, after which the energy stored by the transformer is spent on replenishing the energy of the supercapacitors C1–C2. The capacitance of the supercapacitor is selected based on the needs of the load. In our case, 2 VEC3R0105QG supercapacitors connected in series with a capacity of 1 F were selected.

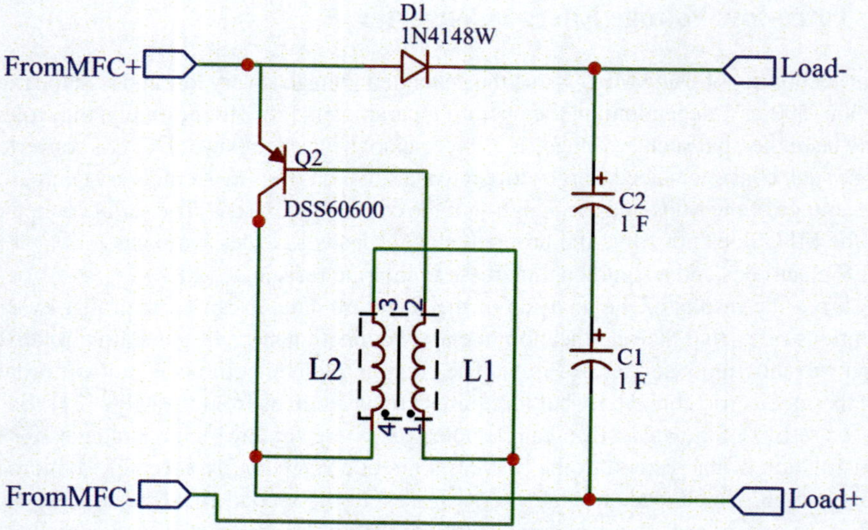

Fig. 1. Schematic diagram of blocking generator

4 Microbial Fuel Cell

MFC is a special technical device in which bacteria create conditions for changing the redox potential on the electrodes. The principle of operation is based on the ability of a number of facultative-anaerobic bacteria to use a solid electrode as an electron acceptor produced during the metabolic activity of these bacteria. The transfer of electrons by bacteria to the electrode (anode) occurs from their surface by means of terminal cytochromes, special conductive proteins, or carrier molecules (transducers) [22]. In parallel, with the production and transfer of electrons to the accumulator, bacteria get rid of protons that diffuse from the bacterial cell and are directed to the second electrode (cathode) MFC for their subsequent reaction by electron and oxygen, which results in water. Structurally, microbial fuel cells should have an anaerobic zone in which the anode is located, with bacteria attached to it, and an aerobic zone in which the cathode is located, on which protons are restored. At the same time, MFC can be prepared considering the further absence of their maintenance. Such MFC are usually installed in places with free access of organic substances and with a high level of moisture. The ideal option is the bottom of the reservoir. MFC installed at the bottom of the reservoir are called benthic MFC [23].

5 Manufacturing of Microbial Fuel Cell

A benthic microbial fuel cell was manufactured for experiments [24]. These fuel cells are distinguished by the absence of a proton-selective membrane and consist of two chemically inert electrodes, one of which (the anode) is located in the anaerobic layer of benthos or silt of a reservoir. The thickness of the benthos is often 4–6 cm. The second electrode (cathode) is placed above the anode in the aerobic zone, where the cathode freely communicates with oxygen dissolved in water. For ease of installation, such fuel cells can be placed in a chemically inert frame that holds the electrodes. If necessary, one of the electrodes can be installed in a container with sludge in advance. The parameters of the electrode area are selected based on the specific power of the electrodes in terms of power/area. In the literature there are various data on the specific power of MFC. The value can be at the level of hundreds of mW, which refers to electrodes with various modifications, and at the level of few mW [25, 26]. Based on the vastness of the data, we assume that the specific power of the MFC will be at the level of 15 mW/m^2.

Thus, for the manufacture of MFC with a power of up to 1.5 mW, it will be enough to use an anode with a size of up to 15 × 60 cm^2. To simplify the assembly, the anode was divided into 4 parts, each of which was used to design a separate MFC. Figure 2 shows the appearance of a separate MFC.

Fig. 2. Benthic microbial fuel cell

Carbon felt of the brand CARBOPON-22 is used as electrodes. Current-removing wires are connected to the anode. The cathode is a sheet of carbon felt with sides of 15 × 15 cm^2. The electrodes are placed in a 2.5 L plastic container with sides 16 × 18 × 12 cm. The anode is placed at the bottom of the tank. Then a layer of silt 4 cm thick is placed on the anode. The cathode is attached to the lid of the container using plastic puffs. Round holes with a diameter of 5 mm with increments of 1 cm are made along the perimeter of the container. The holes are necessary for the mass transfer of organic matter and oxygen dissolved in water. The current-removing tubes are connected to the electrodes and exit through a pre-made hole in the MFC housing.

6 Manufacturing of the Load Power Driver

The ATmega328P microcontroller is used as a load, which receives power from the DS18B20 temperature sensor and the LoRa 02 SX1278 radio module at 433MHz.

Considering that the load is powered by a supercapacitor battery, a control circuit is needed that would supply power to the load at a time when the energy stored in the supercapacitor will be sufficient to power this load. In this paper, two options for load management are considered.

The first option provides for the installation of a real-time clock based on the DS1307 chip. The block diagram of such a device is shown in Fig. 3.

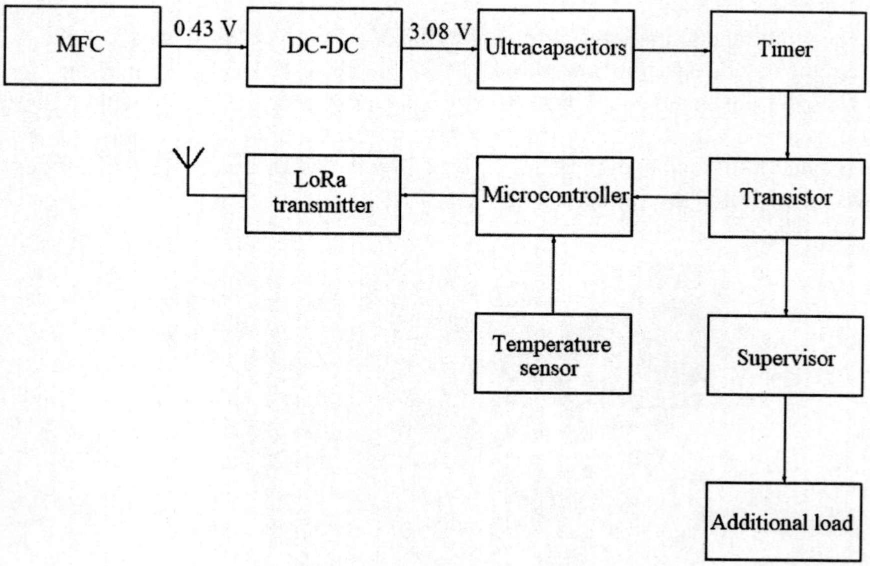

Fig. 3. Block diagram of a control device with a payload based on a timer

The real-time clock operates in timer mode and generates an interrupt signal once in 24 h, which lasts 6 s. By means of this interrupt signal, a key transistor is opened, which in turn supplies power to the load. The real-time clock itself is powered by supercapacitors. It is assumed that during the day MFC manage to charge the supercapacitor battery to the required level. The merit of such a solution is the high time accuracy of switching on the load, as well as the low current consumption. The disadvantage of such a solution is the danger of erasing volatile memory with loss of orientation in time if the blocking generator is disconnected from the energy source, and the supercapacitor energy is not enough to save memory. In addition, there is a danger of switching on the load when the supercapacitors are not sufficiently charged, which can lead to a power failure. If the supercapacitors are recharged, the device has an auxiliary load necessary to reset the excess charge accumulated in the supercapacitors.

The second option is power management with the help of a supervisor. In this case, the power supply to the load occurs only after the voltage of the supercapacitor battery reaches a certain threshold level at which the supervisor generates an interrupt. The block diagram is shown in Fig. 4.

The order of operation of this device is as follows:

1. current is supplied to the input of the voltage converter from an MFT or other power source with a low output voltage;

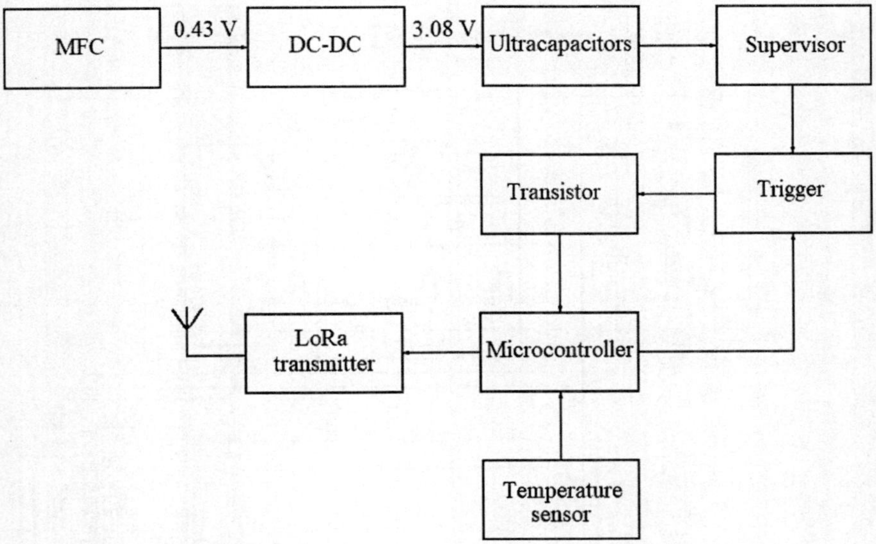

Fig. 4. A block diagram of a control device with a supervisor-based payload

2. the converter increases the voltage and transmits it to the ion battery;
3. supercapacitors accumulate a charge, which leads to a gradual increase in voltage on their plates;
4. the supervisor monitors the voltage. When a certain voltage is reached, the supervisor generates an interrupt applied to the trigger;
5. when an interrupt is received from the supervisor, the trigger generates a high static level at its output and simultaneously at the gate of the field-effect transistor operating in the key mode;
6. when the transistor key is opened, the load circuit is closed, which leads to its useful work;
7. when the load completes its work, an interrupt is formed by a microcontroller or other pulse shaper that is part of the load. The generated pulse is applied to the trigger;
8. when a new interrupt is received from the load, the trigger switches, rotating its output from a high logical state to a low one, which leads to the termination of the voltage supply to the gate of the transistor;
9. the transistor key closes, disconnecting the load power;
10. the charge of the supercapacitors continues until the next moment of the formation of the interrupt by the supervisor.

The schematic diagram of the device is shown in Fig. 5.

The control of this device is based on the MAX821 supervisor, the transistor key is controlled by the K561TM2 trigger. The trigger is switched via two pins P1 from the supervisor and S1 from the microcontroller. The trigger operates in the switch mode, the repeated supply of control signals to the same trigger input will not cause it to switch. The DS18B20 temperature sensor chip is used as a load. At the same time, it is possible

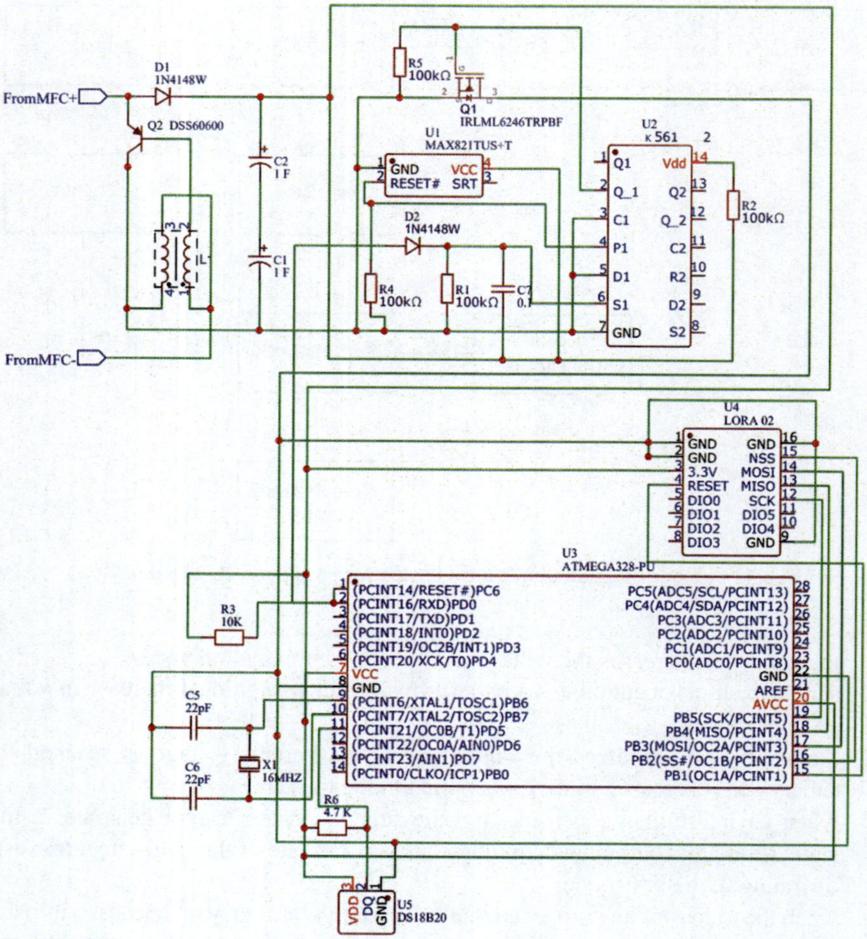

Fig. 5. Schematic diagram of a control device with a payload based on a supervisor

to use any other low-power sensor or load. The load is controlled by the ATmega328P microcontroller; however, the use of others is also allowed: STM, PIC, ESP32. The LoRa 02 transceiver is used to send data.

The disadvantage of this device is its large power consumption compared to the timer due to the constantly on super-visor. However, both options have their advantages, which will be discussed below.

7 Results

For the convenience of further work, we assume that we will represent the four MFC as U1, U2, U3, U4, which corresponds to four identically produced MFC. At the time of manufacture, each MFC created a voltage of about 150 mV. These results are presented in Table 1.

Table 1. Characteristics of microbial fuel cell operation

	U1	U2	U3	U4
No-load voltage after assembly, mV	151	144	135	155
No-load voltage after 14 days, mV	920	898	901	909
Voltage under load after 14 days, 1 kΩ	644	603	590	613
Power at load after 14 days, 1 kΩ, mV	0.41	0.363	0.348	0.365

Then all the MFC were connected to an external resistor with a nominal value of 1000 Ω. The MFC was in this state for two weeks until the moment of maximum power. Then the MFC data were tested in operation and connected to a blocking generator. At the same time, it should be noted that the voltage of the MFC without load was close to the maximum possible theoretical value.

7.1 Blocking Oscillator. Selection of Winding Inductance

To determine the optimal ratio, a series of experiments was carried out to select the optimal inductance of the transformer windings. The results are presented in Table 2.

Table 2. Parameters of the converter operation at different ratio of inductance winding-current

Inductance L2, mH	I in mA	U in mV	P in mW	I out uA	U out, V	P out mVT	η %
Coil option 1 L1 - 8 mH							
40	1	408	0.408	42	4.2	0.1764	43.23
55	0.97	423	0.410	44.9	4.49	0.2016	49.13
73	0.95	436	0.414	44.8	4.48	0.2007	48.45
95	0.94	442	0.415	44.7	4.47	0.1998	48.09
113	0.9	451	0.405	46	4.6	0.2116	52.13
Coil option 2 L1 - 15 mH							
46	1.1	377	0.414	40.7	4.07	0.1656	39.94
67	1.1	396	0.435	43.4	4.34	0.1883	43.24
90	0.99	410	0.405	43.7	4.37	0.1909	47.04
110	0.94	421	0.395	44.2	4.42	0.1953	49.36
136	0.95	433	0.411	45.6	4.56	0.2079	50.54

(continued)

Table 2. (*continued*)

Inductance L2, mH	I in mA	U in mV	P in mW	I out uA	U out, V	P out mVT	η %
Coil option 3 L1 - 30 mH							
48	0.82	520	0.426	1.2	0.12	0.1440	0.033
68	0.82	517	0.423	1.26	0.126	0.0032	0.037
90	0.85	522	0.443	1.22	0.122	0.0001	0.033
147	0.88	523	0.460	1.2	0.12	0.0001	0.031

In total, three inductance options for L1 were investigated, in each of which the inductance of L2 was consistently changed. The experiments were carried out in such a way that the power supply of the converter was in the range of 0.4–0.45 mW. The load for the converter was a 100 kΩ resistor with a 220 μF smoothing capacitor connected. Thus, it can be noted that the inductance options of the windings L1 between 3 and 15 mH practically do not affect the output efficiency of the device, however, in the case of a lower inductance L1, a slightly higher voltage is required for operation in general. In the third case, when the inductance of L1 was 30 mH, the converter showed extremely low performance, despite the wide level of variations in the inductance of L2. Based on this, we can conclude that for the conversion of voltage from MFC, the optimal option is the inductance of the coil L1 about 15 mH.

The next step was to determine the optimal supply voltage of the generator. The manufactured blocking generator showed the possibility of energy conversion in a wide range of supply voltage (see Fig. 6).

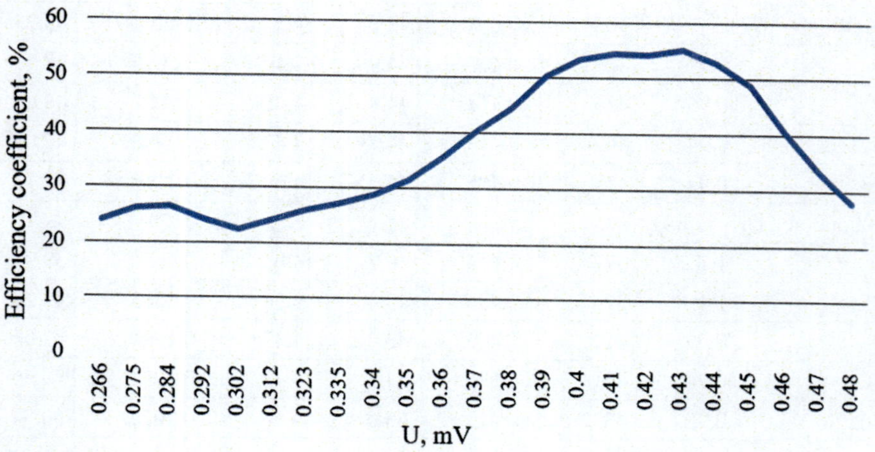

Fig. 6. Dependence of efficiency on the supply voltage

With a supply voltage from 266 to 480 mV, the optimal range for operation is 390–440 mV, where the efficiency of the device is 50%. The maximum efficiency of the device is 55.2%. It is worth noting that the efficiency depends on the power consumption. When replacing the resistor from 100 to 50 kΩ, the efficiency increased by an average of 2%, reaching 57.8%.

The converter starts at a voltage of 266 mV with a required current of 1 uA. At the same time, with a voltage of 370 mV and a current consumption of 56 uA, the converter was able to charge a capacitor connected to a 100 kΩ resistor to the level of 1 V. The maximum voltage to which the capacitor was charged was noted at an input voltage of 460 mV and a current consumption of 2.86 mA. This voltage was 7.3 V.

7.2 Demonstration of the Load Operation

The ATmega328P microcontroller, DS18B20 temperature sensor and LoRa SX1278 433MHz radio module were used as the load. In the first case of the load implementation, the timer operation is considered. When four MFC were connected to the load, an intensive accumulation of charge on the supercapacitors occurred. As a result, the total inclusion of four MFC turned out to be excessive and, in the future, just one MFC out of four was used in experiments in order to exclude overcharging of supercapacitors and their breakdown, as well as failure of the payload due to too high voltage supply. It is worth noting that to protect the device from over-voltage on charge, an auxiliary supervisor was used, configured to activate at a voltage above 3.08 V. When the load voltage is exceeded, the supervisor automatically turns on the additional load, which leads to an additional voltage drop, but not lower than 3.08 V. Thus, the possibility of powering the load from a single MFC with a power of 0.3–0.45 mW was shown.

In the second case, all 4 MFC were used to implement the load, that is, the power supply to the load is controlled by a supervisor. When the voltage reached 3.08 V, the load supply circuit was closed. The microcontroller interrogated the temperature sensor, and then sent the received data over the radio channel, followed by the formation of a delay, leading to the switching of the trigger and the opening of the power supply circuit with a transistor key. At the same time, 46 MJ was used for one iteration of sending data, the voltage drop on the supercapacitors was 0.03 V. In this configuration, the signal was sent with a period of 1 time in 10 s, which significantly exceeds the currently available results of the works presented in the literature review.

In both options of load control schemes, a useful operation took place, as a result of which the microcontroller interrogated the temperature sensor, and then sent data. The data transmission distance over the LoRa protocol was 950 m.

8 Conclusion

As a result of the work carried out, an ultra-low-voltage MFC energy converter was manufactured. Optimal parameters of the inductance of the transformer coils were selected. It is shown that the converter starts working at a voltage of 266 mV and a current consumption of 1 uA. The optimal supply voltage range is 390–440 mV. The maximum efficiency is 57.8%. Model MFCs with a total capacity of 1.8 mW were manufactured and the possibility of powering the converter with them was confirmed. Two options of load control drivers based on a timer with a real-time clock and based on a power monitor supervisor are shown. As a result of the demonstration, the actual possibility of receiving a useful signal about the ambient temperature and transmitting this signal at 950 m is shown. The results obtained make it possible to take the next step towards the practical implementation of microbial fuel cell technologies for powering autonomous marine monitoring systems.

Acknowledgements. Grant of the President of the Russian Federation MK-3094.2022.1.6.

References

1. Roy, A., et al.: Electrical power supply of remote maritime areas: a review of hybrid systems based on marine renewable energies. Energies **11**(7), 1904 (2018)
2. Albaladejo, C., et al.: A low-cost sensor buoy system for monitoring shallow marine environments. Sensors **12**(7), 9613–9634 (2012)
3. Knepp, T.N., et al.: Development of an autonomous sea ice tethered buoy for the study of ocean-atmosphere-sea ice-snow pack interactions: the O-buoy. Atmosp. Measur. Tech. **3**(1), 249–261 (2010)
4. Chen, J., et al.: Simulation and design of solar power system for ocean buoy. J. Phys. Conf. Ser. **1061**(1), 012018 (2018)
5. Xu, R., et al.: Recent progress on wave energy marine buoys. J. Mar. Sci. Eng. **10**(5), 566 (2022)
6. Wang, H., et al.: Design and experiment of a new wave power conversion device for self-powered sensor buoy. In: IECON 2021–47th Annual Conference of the IEEE Industrial Electronics Society, pp. 1–5. IEEE (2021). https://doi.org/10.1109/IECON48115.2021.9589049
7. Guerriero, P., et al.: Merged photovoltaic/wave system for the power supply of a marine buoy for harbour monitoring. In: 2019 International Conference on Clean Electrical Power (ICCEP), pp. 664–668. IEEE (2019). https://doi.org/10.1109/ICCEP.2019.8890205
8. Wang, H., et al.: A stackable triboelectric nanogenerator for wave-driven marine buoys. Nanomaterials **12**(4), 594 (2022)
9. Yu, D., et al.: A novel direct-driven triboelectric–electromagnetic hybridized wave energy converter for buoy power supply. Appl. Nanosci. **12**(5), 1697–1711 (2022)
10. Chen, X., et al.: A chaotic pendulum triboelectric-electromagnetic hybridized nanogenerator for wave energy scavenging and self-powered wireless sensing system. Nano Energy **69**, 104440 (2020)
11. Alizzio, D., Quattrocchi, A., Montanini, R.: Development and characterisation of a self-powered measurement buoy prototype by means of piezoelectric energy harvester for monitoring activities in a marine environment. ACTA IMEKO **10**(4), 201–208 (2021)
12. Wang, Y., et al.: Flexible seaweed-like triboelectric nanogenerator as a wave energy harvester powering marine internet of things. ACS Nano **15**(10), 15700–15709 (2021)
13. Tender, L.M., et al.: The first demonstration of a microbial fuel cell as a viable power supply: powering a meteorological buoy. J. Power Sources **179**(2), 571–575 (2008)
14. Wu, P.K., et al.: A low power DC/DC booster circuit designed for microbial fuel cells. Process Biochem. **47**(11), 1620–1626 (2012)
15. Khaled, F., Ondel, O., Allard, B.: Microbial fuel cells as power supply of a low-power temperature sensor. J. Power Sources **306**, 354–360 (2016)
16. Zhang, F., Tian, L., He, Z.: Powering a wireless temperature sensor using sediment microbial fuel cells with vertical arrangement of electrodes. J. Power Sources **196**(22), 9568–9573 (2011). https://doi.org/10.1016/j.jpowsour.2011.07.037

17. Donovan, C., et al.: Sediment microbial fuel cell powering a submersible ultrasonic receiver: new approach to remote monitoring. J. Power Sources **233**, 79–85 (2013)
18. Kumbar, S.S., et al.: Enhancement in cathodic redox reactions of single-chambered microbial fuel cells with Castor oil-emitted powder as cathode material. Materials **14**(16), 4454 (2021)
19. Yang, H., et al.: Performance optimization and microbial community evaluation for domestic wastewater treatment in a constructed wetland-microbial fuel cell. Environ. Res. **212**, 113249 (2022)
20. An, J., et al.: Understanding energy loss in parallelly connected microbial fuel cells: non-faradaic current. Biores. Technol. **203**, 280–286 (2016)
21. Khaled, F., et al.: Microbial fuel cells as power supply of a low-power temperature sensor. J. Power Sources **306**, 354–360 (2016)
22. Logan, B.E.: Exoelectrogenic bacteria that power microbial fuel cells. Nat. Rev. Microbiol. **7**(5), 375–381 (2009)
23. Ramírez-Vargas, C.A., et al.: Microbial electrochemical technologies for wastewater treatment: principles and evolution from microbial fuel cells to bioelectrochemical-based constructed wetlands. Water **10**(9), 1128 (2018)
24. Tavakolian, M., Taleghani, H.G., Khorshidian, M.: New design of benthic microbial fuel cell for bioelectricity generation: comparative study. Int. J. Hydrogen Energy **45**(43), 23533–23542 (2020)
25. Kabutey, F.T., Ding, J., Zhao, Q., Antwi, P., Quashie, F.K.: Electrical current generation from a continuous flow macrophyte biocathode sediment microbial fuel cell (mSMFC) during the degradation of pollutants in urban river sediment. Environ. Sci. Pollut. Res. **27**(28), 35364–35380 (2020). https://doi.org/10.1007/s11356-020-09812-y
26. Kamaraj, Y., Punamalai, G., Kandasamy, S., Kasinathan, K.: Influence of long-term organic and conventional fertilization on bacterial communities involved in bioelectricity production from paddy field-microbial fuel cells. Arch. Microbiol. **202**(8), 2279–2289 (2020). https://doi.org/10.1007/s00203-020-01947-3

Underwater Object Tracking with 2D Sonar Signals Preprocessed Using the Virtual High-Dynamic Range Enhancement Method

Jakub Grabek(✉) and Bogusław Cyganek

Department of Electronics, AGH University of Science and Technology, Al. Mickiewicza 30, 30-059 Kraków, Poland
jgrabek@agh.edu.pl, cyganek@agh.edu.pl
https://iet.agh.edu.pl/

Abstract. In this paper, a custom underwater 2D sonar signal acquisition, enhancement, and tracking system are presented. The system is based on a modern 2D multibeam sonar by Teledyne allowing data gathering and real-time operation. First, our custom setup allows 2D sonar signal acquisition for gathering underwater datasets, which are rare in the public domain. Second, two underwater object tracking methods are presented and tested, which are proposed to operate on the 2D sonar signal enhanced by the Virtual High-Dynamic Range (VHDR) method. It can modify sonar images by applying a number of tone-adjusting curves, potentially revealing hidden details. The tracking experiments were carried out on the original, as well as on the VHDR-processed data. Compared to original sonar images, the presented method decreased the tracking error by up to 26% and removed all tracking failures on preprocessed data. The presented system can also be easily customized, both from hardware and software points of view. Our created 2D sonar dataset is available free from the Internet and is planned to be expanded in the near future.

Keywords: Multibeam sonar · Underwater tracking · VHDR enhancement

1 Introduction

There is a problem with acquiring significant amounts of usable sonar data for digital signal processing tasks, such as denoising, object detection, and recognition. The main reasons are high signal acquisition costs, legal restrictions, which block the transfer of archived data to the public domain, as well as usability of available data [1,2]. These high demands come from the relatively high cost of a sonar device and the additional work during operation needed to achieve high-quality data. Researchers often try to reduce costs mainly through partnerships with privately held companies, specialized in underwater surveys or navy.

R. Silhavy et al. (Eds.): CoMeSySo 2022, LNNS 596, pp. 628–636, 2023.
https://doi.org/10.1007/978-3-031-21435-6_53

However, those entities often restrict access to the provided data only to their partners, which results in high-quality data being left in archives and not being beneficial for the scientific community. Also, there is no widely used standard for sonar recording files. The data available in the public domain is often shared in a form that can be difficult or impossible to read without proprietary software, making compiling records from different sources a very challenging task.

Hence, our first motivation was to prepare underwater data for future ML development focused both on sonar and visual data. The first step was to build an embedded platform capable of transporting and controlling different sensors/devices. The 2D multibeam sonar by Teledyne was selected since its data can be interpreted as underwater 2D images, where a wave (frequency of a few hundred kHz - few MHz) generated from transducers plays the role of light in visual images. Unlike side-scan sonars, it can provide usable images without the need for constant movement and precision information on sonar head position and orientation. Both sonar types are often used as a data source for various Machine Learning and Artificial Intelligence solutions [3–5]. As a result, the first contribution of our work is the proposed novel 2D sonar dataset, available free from the Internet [6].

The second motivation was to develop useful underwater tracking methods based on these gathered 2D sonar signals. The underwater detection and tracking problems are essential, especially for autonomous and unmanned operations [7,8] and there is a strong push towards ML/AI-based methods [9,10]. A significant part of solutions are acoustic-based and often use multibeam sonars, which provide data in a more usable form. For this purpose two tracking methods - MIL [11] and CSRT [12] - were investigated as well tested with robust performance. However, since they were designed for aerial images, their operation with 2D sonar signals with a low dynamic range was not sufficiently accurate. Therefore we propose to enhance the methods by preprocessing the input 2D sonar signals with our designed Virtual High-Dynamic Range (VHDR) method [13]. Thanks to this modification, the tracking quality in both methods were greatly improved, as shown in the experimental results. Although the proposed system was verified in the context of underwater tracking, both our novel 2D sonar dataset as well as the VHDR-based sonar signals preprocessing method can be used in other underwater tasks, such as object detection and obstacle avoidance, which we plan to investigate in the future. Also, in the future, we plan to enlarge our 2D sonar dataset and label the visible objects in order to be able to train a deep convolutional neural network for object detection and tracking.

2 Image Acquisition Platform

The sonar system is a part of the image acquisition platform, which was first developed to gather visual spectrum data for algorithm development and testing. It consists of an underwater scooter with steel wings used as a mounting base for additional equipment, shown in Fig. 1. It was used to gather data taken by single and multiple underwater cameras used with lighting and is now extended by a multibeam sonar head by Teledyne with a storage and control system.

(a) System preparation

(b) Jaworzno's post-excavation lagoon

Fig. 1. Images of the system during tests.

2.1 Hardware

As an image gathering device, the Teledyne M900-130 [14] 2D multibeam imaging sonar is used. It has a maximum range of up to 100m and operates using a 900kHz wave with an update rate up 25 Hz. The device is powered and controlled over an Ethernet cable fitted into a watertight container.

In the container, the onboard computer with needed equipment is fitted, using FDM (Fused Deposition Modeling) 3D printed chassis as presented in the schematic in Fig. 2. The Li-ion battery provides power to the computer and Inertial Measurement Unit (IMU) sensors through the step-down converter and directly powers sonar. Such configuration allows the system to work over 8 h without a recharge. Depending on the configuration selected for the job, the chassis has multiple mounting points to test different sensors. For the test on shallow water, there is also a possibility to use a screen that visualizes sonar data in real-time. The casing is intended to work up to 50m underwater and was trimmed to have slightly positive buoyancy.

For the onboard computer, the NVIDIA Jetson Nano is used. It combines an ARM 64bit processor (Quad-Core ARM A57 @ 1.43 GHz) with 128-Core Maxwell GPU. Additionally, it provides all standard communication interfaces such as USB or Ethernet and communication buses (I2C, SPI, UART) needed for additional sensor testing. Small dimensions were also advantageous while fitting the computer inside the container.

Data from the Inertial Navigation Unit is used for system position and orientation computation. Such information is saved alongside sonar data for further use. The Kalman Filter is used to minimize errors in resulting values.

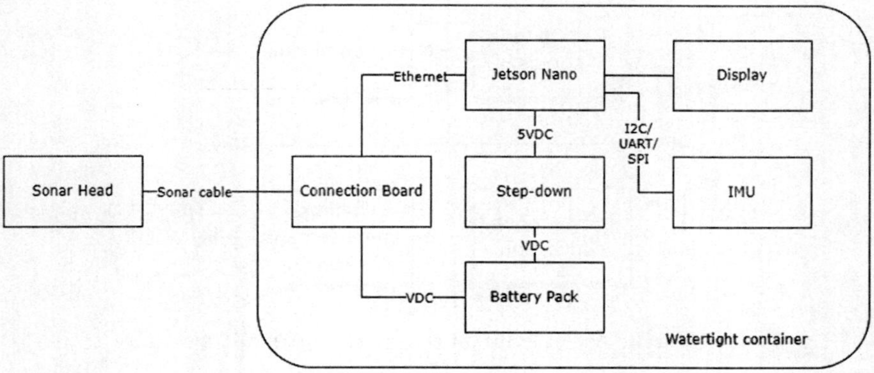

Fig. 2. System schematic.

2.2 Software

The software's core is responsible for sonar control and saving acquired data. It was compiled using C/C++ with libraries provided by Teledyne. The user can change sonar and post-processing parameters such as range, acquisition gain, Time-Varying Gain, noise threshold values, and speed of sound in water. During the process, the application can also visualize recorded data in real-time.

The acquired data can then be preprocessed according to needs in a specific experiment. Image processing modules for speckle noise removal [15], image sharpening, dynamic range correction [13] were implemented. Such processed data can be saved for future experiments or be used as inputs in ML/AI methods such as object detectors, trackers, or positioning algorithms (i.e., Simultaneous Localization And Mapping).

3 Image Enhancement with VHDR

Virtual High Dynamic Range [13] processes an input Low Dynamic Range (LDR) image - here the original 2D sonar frame - by several tone adjustment curves, resulting in a number of tone-converted images. The role of these is to expose different ranges of the input image, potentially revealing hidden details. These are then fused back to form one High Dynamic Range (HDR) image with range conversion and contrast enhancement, as presented in Fig. 3.

The luminance of the image is adjusted using a modified logistic function :

$$s_0(x, k, x_0) = \frac{A}{1 + e^{-Dk\frac{(x-x_0)}{R}}} \tag{1}$$

$$x_0 \in P_{min}, P_{mid}, P_{max} \tag{2}$$

where D is a domain range constant, k is a parameter, P_{max} and P_{min} are maximal and minimal pixel ranges of the input low dynamic range image. $A = P_{max}$ was selected for further calculations and $P_{mid} = (P_{max} - P_{min})/2$.

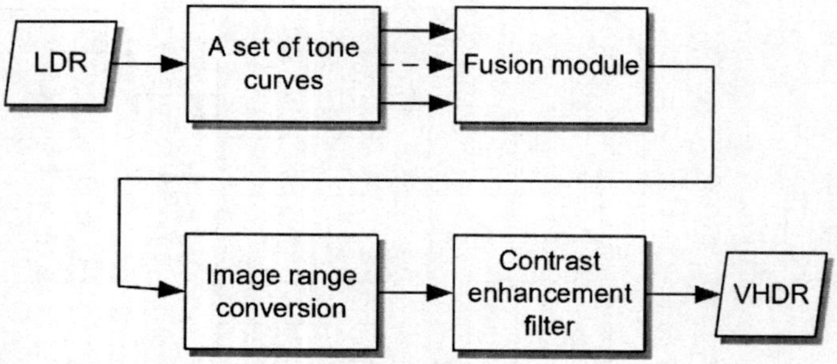

Fig. 3. Architecture of the VHDR method [13].

Function (1) was augmented by multiplicative and additive components to provide a concave, convex, and S-like shape to cover the full image range, i.e. [0, 255], and presented as follows:

$$s_1(x, k, x_0) = (\frac{|x_0 - P_{mid}|}{P_{mid}} + 1)s_0(x, k, x_0) + s_0(0, k, x_0) \tag{3}$$

where middle value of the LDR is $P_{mid} = (P_{max} - P_{min})/2$ end equals 127.5 for 8-bit input. Selection of the midpoint parameters leads to multiple curves differently affecting processed image:

1. Convex case - The midtone data of the input image is stretched toward the highlight of the output side, resulting in a bright-toned image.
2. S-curve - The highlight and shadow of the input signal are enhanced, resulting in a higher contrast image.
3. Concave case - The midtone data of the input is stretched toward the shadow of the output, resulting in a dark-toned image.

Parameter k values are chosen experimentally to achieve tone mapping adequate to meet the requirements of the current task. As mentioned, VHDR was shown useful to enhance underwater visible signals [13]. In this work, we show that enhancing noisy and low-contrast 2D sonar signals is also useful.

4 Tests

Every recorded sonar ping is converted into a reflection map, where image luminosity is proportional to the wave reflected from objects and stored in lossless proprietary Teledyne's format. All data gathered during tests in Jaworzno's post-excavation lagoon is converted into lossless PNG images (1100 × 600 pixels) from this format. The data is presented in the polar coordinate fan-shaped scan image as this approach minimizes objects' distortions.

The test data consists of 11 recordings ranging between 200 and 400 frames each. In total, 3200 frames were selected where man-made objects such as pipes or submerged quarry equipment are visible. Every frame center position of tracked object was recorded to be used as a reference point during the tracking test. As input for tested trackers, the ROI 51 by 51 pixels was selected, which center resides in the corresponding reference point in the first frame of the sequence. The ROI size was selected to test tracking quality on bigger objects which are not always fully visible on the image (i.e., log pipes or large quarry equipment such as excavators).

The second dataset was prepared by applying the VHDR scheme to the original data, as shown in Fig. 4. The goal was to highlight mid-tones as well as improve image contrast. Table 1 presents parameters used during dataset creation, which were selected by a trial-and-error process for a whole dataset. The tracking performance was tested on both datasets. The tracking methods operate as follows. However, no neural network-based solution was tested due to insufficient data.

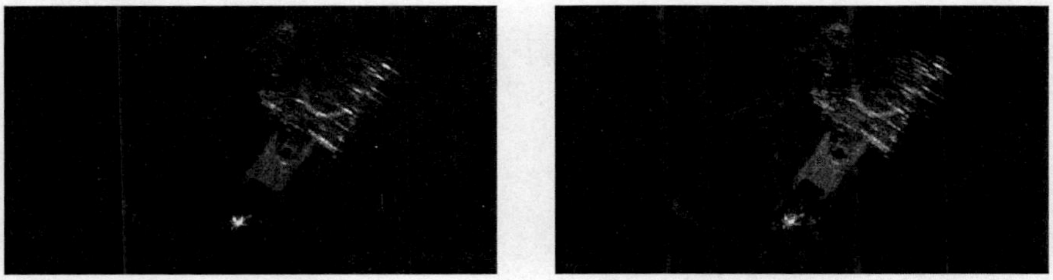

(a) Original data (b) VHDR processed data

Fig. 4. Images used during experiment.

Table 1. VHDR paramteres used during preprocessing.

x_0	0	127.5	255
k	0.3	0.7	1.4

MIL tracker [11] The system contains three components: image representation, appearance model, and motion model. Image representation consists of Haar-like features that are computed for each frame. The appearance model is composed of a discriminative classifier combined with a motion model returning a new object position.

CSRT tracker [12] CSRT tracker is a two-step algorithm. First, in the localization step, image features are extracted from the target estimated position from the previous step. New object position is estimated by summing correlation responses over all channels. In the update step, histograms are extracted, and spatial reliability maps are reconstructed. Based on that filter's values and channel reliability weights are updated.

A tracking error was defined as the averaged euclidean distance between the tracker response on the position of the selected object and the real position defined by a human observer. Obtained results are presented in Table 2 and in Figs. 5 and 6.

VHDR application on tested data resulted in tracking improvement for both used trackers. In the case of MIL, the mean error decreased by over 19%, and the CSRT tracker achieved a result of 26% better. There are frames where trackers failed to find a tracked object on the original dataset. There is no such problem for processed data, which indicates that VHDR highlighted the low value and, therefore, hidden features allowing more efficient tracker model updates leading to removal frames without tracking prediction.

Table 2. Results of the experiment.

Tracker type	Mean error [pixels]		Tracking fails count	
	Original data	Processed data	Original data	Processed data
MIL	31.665	25.526	54	0
CSRT	19.698	14.491	76	0

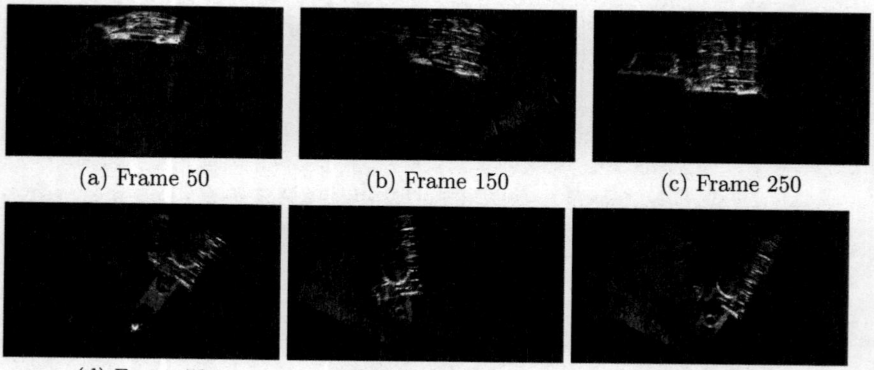

(a) Frame 50 (b) Frame 150 (c) Frame 250

(d) Frame 50 (e) Frame 150 (f) Frame 250

Fig. 5. CSRT Tracking. Blue square indicates tracker position, red dot indicates reference position.

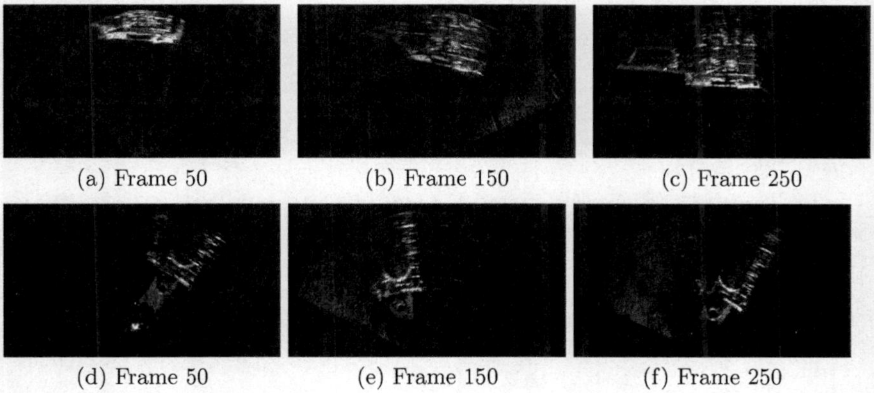

Fig. 6. MIL Tracking. Blue square indicates tracker position, red dot indicates reference position.

5 Conclusion and Future Works

In this paper, an underwater sonar acquisition system was presented. The system allows data gathering, as well as online data processing. Hence, the first contribution of our work is the proposed novel 2D sonar dataset, available free from the Internet [6]. As a showcase, we show that the presented sonar VHDR image processing scheme can positively impact tested tracking methods. Thanks to this, the trackers' distance error was decreased by 19% and 26% for MIL and CSRT trackers, respectively. The proposed approach indicates that the VHDR application can improve it for non-neural-based solutions by increasing the visible and usable feature count.

The system will be expanded in the future to allow simultaneous control over sonar systems and coupled underwater cameras to test and develop new multispectral algorithms as the sensor fusion provides significantly more useful information. The gathered sonar and visual data will be used to create new datasets, both for regular and multispectral use.

The data will be selected to address object detection and tracking problems and will be shared in open form. After enlarging our 2D sonar dataset and its labeling, we also plan to test the impact of sonar VHDR preprocessing on neural network-based methods. The methods will focus on object detection, tracking, positioning, and mapping algorithms.

Acknowledgements. This work has been supported by AGH University of Science and Technology under subvention funds no.16.16.230.434, and by the Teledyne Marine as a part of Teledyne Marine's Academic Product Grant.

References

1. Mitchell, H.B.: Image fusion: theories, techniques and applications. Springer Science & Business Media (2010)
2. Cyganek., B.: Object detection and recognition in digital images: theory and practice. Wiley (2013)
3. Yulin, T., Jin, S., Bian, G., Zhang, Y.: Shipwreck target recognition in side-scan sonar images by improved YOLOv3 model based on transfer learning. IEEE Access **8**, 173450–173460 (2020)
4. Zhang, W., Zhou, T., Xu, C., Liu, M.: A SIFT-Like feature detector and descriptor for multibeam sonar imaging. J. Sens. (2021)
5. Liu, Y., Ye, S., Zhang, A., Li, M.: Calibration system design for multi-beam sonar. J. Phys. Conf. Ser. **2121**(1), 012043 (2021)
6. Link to used data: https://home.agh.edu.pl/jgrabek/datasets/sonar2022.zip
7. Ghafoor, H., Noh, Y.: An overview of next-generation underwater target detection and tracking: an integrated underwater architecture. IEEE Access **7** (2019)
8. Luo, J., Han, Y., Fan, L.: Underwater acoustic target tracking: a review. Sensors **18**(1), 112 (2018)
9. Sejin, L., Byungjae, P., Ayoung, K.: Deep learning from shallow dives: sonar image generation and training for underwater object detection. ArXiv preprint arXiv:1810.07990 (2018)
10. Panetta, K., Kezebou, L., Oludare, V., Agaian, S.: Comprehensive underwater object tracking benchmark dataset and underwater image enhancement with GAN. IEEE J. Oceanic Eng. **47**(1), 59–75 (2021)
11. Babenko B., Yang, M-H., and Belongie, S.: Visual tracking with online multiple instance learning. In: CVPR 2009 IEEE Conference on Computer Vision and Pattern Recognition, pp. 983–990. IEEE (2009)
12. Lukezic, A., Vojir, T., ˇCehovin Zajc, L., Matas, J., and Kristan, M.: Discriminative correlation filter with channel and spatial reliability. In: Proceedings of the IEEE Conference on Computer Vision and Pattern Recognition, pp. 6309–6318. (2017)
13. Cyganek, B., and Wozniak, M.: Virtual high dynamic range imaging for underwater drone navigation. In: ICIAE2018 the 6th IIAE International Conference on Industrial Application Engineering (2018)
14. Seatronics LTD: Teledyne Blueview M900 series. https://seatronics-group.com/rov-sensors/teledyne-blueview-m900-series/. Last accessed 2 Jun 2022
15. Grabek, J., Cyganek, B.: Speckle noise filtering in side-scan sonar images based on the tucker tensor decomposition. Sensors **19**(13), 2903 (2019)

Error-Free 2d Pictogrammic Syntax in a Programming Learning Environment for Preschool Children

A. G. Kushnirenko[1(✉)], A. G. Leonov[1], and S. A. Polikarpov[2]

[1] Federal Research Center "Scientific Research Institute for System Analysis of the Russian Academy of Sciences", 36-1 Nakhimovsky Prospect, Moscow 117218, Russian Federation
agk_@mail.ru

[2] Steklov Mathematical Institute of Russian Academy of Sciences, Ulitsa Gubkina, 8, Moscow 117966, Russian Federation

Abstract. When mastering the basics of programming by preschoolers, serious difficulties are created by the need to diagnose and correct syntactic errors. With the traditional method of "on-screen" program editing, these difficulties can be avoided by blocking the child's actions that lead to syntactic violations. Today, the technique of compiling programs from material objects (tangible objects) with command icons printed on them is gaining popularity. When using this technique, no blocking of erroneous user actions is possible. In this situation, the authors propose to protect the child from syntactical errors by postulating the two-dimensionality of the program and defining the syntax and semantics of the pictogrammic programming language for beginners using indentation so that any placement of pictograms in the cells of a two-dimensional table gives a syntactically correct and executable program. This approach has been implemented and tested in the Russian educational environment "PiktoMir" of pictogrammic programming for preschoolers.

Keywords: Program · Pictogram · Textless programming environment · Screenless programming · PiktoMir · Preschooler · Robot · Structured programming · Syntactical errors

1 Introduction

In recent years, work has been carried out all over the world to reduce the age of children's familiarization with the basics of programming and the development of algorithmic (computational) thinking in children [1–10]. Similar work is underway in Russia. The article [11], summarizing the results of the work of the Russian Academy of Sciences on teaching programming to preschoolers and elementary school students, describes a system of scientific (according to L.S. Vygotsky) programming concepts intended for mastering by preschoolers, as well as a method for introducing children to these concepts in an annual systematic course of programming for preschoolers of graduation groups of Russian kindergartens [12]. This course has been successfully completed by

R. Silhavy et al. (Eds.): CoMeSySo 2022, LNNS 596, pp. 637–651, 2023.
https://doi.org/10.1007/978-3-031-21435-6_54

thousands of preschoolers in the city of Surgut over the past three years. The course uses the PiktoMir pictogrammic programming environment [13] developed by Federal Research Center SRISA/NIISI RAS.

PiktoMir is a textless, pictographic programming environment. PiktoMir allows children to create programs to control the movement of real robots-toys on the floor and virtual robots on the tablet screen (Fig. 1), using pictograms of robot commands and control structures of a textless programming language.

Fig. 1. Real and virtual robots of the PiktoMir environment.

Programs for controlling PiktoMir's robots can be made from pictures on the tablet screen. Alternatively, programs can also be assembled in the real world from pictograms printed on various tangible objects, such as wooden cubes or magnetic cards. For example, in the repertoire of PiktoMir robots there is the virtual robot Spinner, moving along a checkered platform and able to execute orders: TURN LEFT, TURN RIGHT, FORWARD, PAINT (Fig. 2). In the Sect. 3 of the article, we will designate 4 pictograms for these orders with symbols (L) (R) (F) (P).

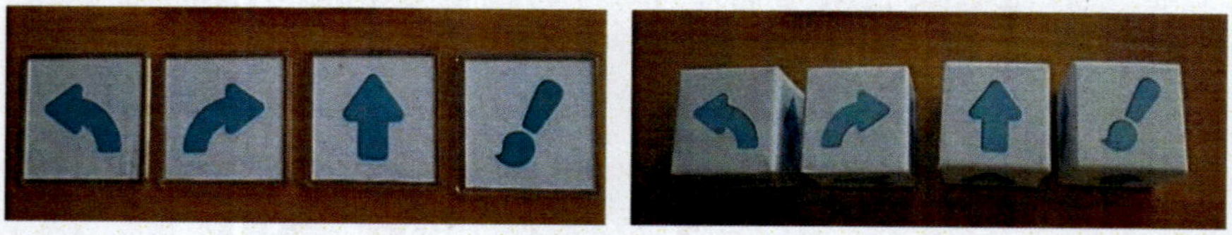

Fig. 2. Pictograms of Spinner's commands placed on magnetic cards and on wooden cubes.

Using these 4 commands, children can compose the simplest programs for controlling the Spinner robot, for example, a program using **repeat 4 times loop** (Fig. 3).

1.1 A Distinctive Feature of the Textless Programming Environment PiktoMir

The use of the textless style of programming for working with children aged 5–7 years and the use of tangible objects for compiling programs are quite common today [14–20]. Our approach is different in that we initially set the task of systematically mastering the basics of programming by children.

The word "systematically" used above needs clarification. Having started intensively programming, mankind in the late 1950s came to the need to use a small number of standardized structured control flow constructs. In 1968, this thesis was called "structured

Fig. 3. A simplest programs for the Spinner robot.

programming”. The list of these standard control flow constructs consists of a branch construct, two repetition constructs, a subroutine construct, and a block construct that allows nesting of constructs. The systematic development of the basics of programming requires familiarity with these constructions and their pairwise combinations. It is to support this systematic development of the basics of programming by preschoolers and other beginners that the textless programming environment PiktoMir was created in Russia. A distinctive feature of the PiktoMir environment is its algorithmic completeness. PiktoMir supports all control flow constructs of structured programming, including the construction of a subroutine without parameters. Moreover, the system allows limited use of numbers and counting in robot control programs using a stationary Magic Jar robot.

There are about 20 commonly used combinations of control structures and we have developed a methodology that allows children 6–7 years old to create more than a hundred programs during the course of a year, practicing the combination of all constructs of structured programming, including the construction of a subroutine without parameters. Preliminary familiarization of children with the basics of structural programming can be provided in the annual cycle of 30 half-hour lessons. Full development of the basics requires 50–60 half-hour lessons. The Pikto language [21], used in the PiktoMir system, makes it possible to compose programs in textless form for controlling real toy robots and virtual screen robots. The latter can be movable, like the Spinner, or motionless, like the Magic Jug with stones that acts as a counter, shown in Fig. 1 on the right. These robots are able to execute unconditional orders, and also know how to answer some queries, returning a logical or integer value, which makes it possible to include in the Pikto language standard control flow structures **loop while** <condition> **do** and **if** <condition> **do**, where <condition> is the answer of some robot to some command-question, for example, to *the Spinner*’s command-question CAN YOU TAKE A STEP? or *the Magic Jug’s* command-question JUG IS EMPTY? The language also supports the control structure **loop** <repeater> **times do**, where <repeater> can be specified by the icon of the command-question of some robot, for example, *the Magic Jug’s* command-question HOW MANY STONES ARE IN THE JUG?, or a pictogram depicting the face of a dice.

In the Sect. 3 of the article, we will designate these 6 pictograms with symbols **(1) (2) (3) (4) (5) (6)**. The Pikto language also provides for nesting control structures and splitting the program into sections with predefined one-letter names (Fig. 5). In the

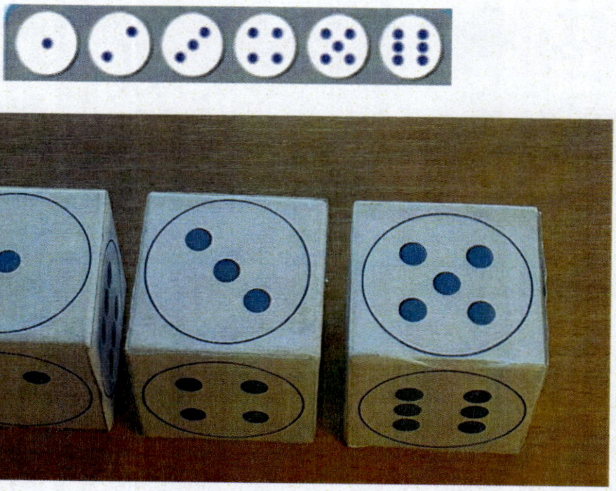

Fig. 4. Icons of repeaters on the screen of the tablet and on the faces of the dices.

Sect. 3 of the article, we will designate these 5 pictograms with symbols **(А) (Б) (В) (Г) (Д)**.

Fig. 5. Pictograms of one-letter names of sections on the screen of the tablet.

1.2 The Main Problem Solved in This Article

The PiktoMir environment, along with the traditional approach in which children compose a program on a tablet screen, supports an approach in which children compose programs from various tangible objects, on the faces of which pictograms of commands and control structures are applied (see Figs. 3, 4 and 5). The student manually, mechanically, places these pictogram-bearing objects in the cells of an imaginary two-dimensional table. To help the child line up the cubes evenly, the table can be graphed on cardboard or on a magnetic board.

Preschoolers like to build programs on the surface of the school table from wooden blocks. This activity for preschoolers is easier than moving icons around on a tablet screen and does not pose a risk of CTS. In our previous practice of working with preschoolers, using classical syntax, we assembled programs from material objects only in the first few lessons. As programs become more complex, children begin to make syntactical errors and the teacher does not have time to help all the children in the group to overcome these errors.

In addition, it should be noted that some rules of classical syntax do not look very natural in pictogrammic programming. When a teacher tries to explain some of the rules of classical syntax to children composing programs from wooden cubes, psychological stress can arise if both the tablet and the teacher refuse to accept some programs that the children think are obviously correct. For example, it is difficult to explain to children why of the two programs in Fig. 6 the tablet considers the program in the right half of the drawing correct, and the program in the left half as incorrect.

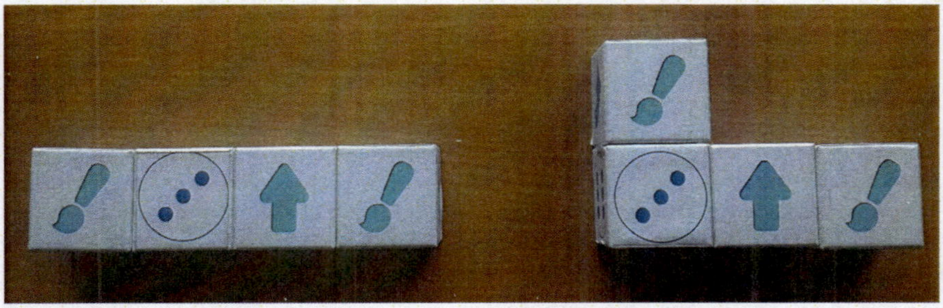

Fig. 6. Incorrect and correct program in terms of classical syntax.

Recent studies show [22] that the basics of programming are more successfully mastered in environments that protect the child from the possibility of making a syntactic error by blocking actions that lead to such errors. Therefore, the question arises **how to avoid syntax errors when composing programs from tangible object**.

2 Methodology

2.1 Proposed Way to Solve the Main Problem

When drawing up programs by manipulating tangible objects, it is practically impossible to control the work of the child and prevent any erroneous actions of the child since these actions are performed in the material world, over which the computer has no control. For example, when building from wooden blocks a program with nested loops for a KIBO robot [19], a child can mistakenly add an extra block with a third closing operator bracket END REPEAT (see the program in the left half of Fig. 7) and it is impossible to prevent him from performing this erroneous action. Later we will see that in PiktoMir (see the program in the right half of the Fig. 7) such errors are impossible, since there are simply no pictograms of closing brackets in PiktoMir.

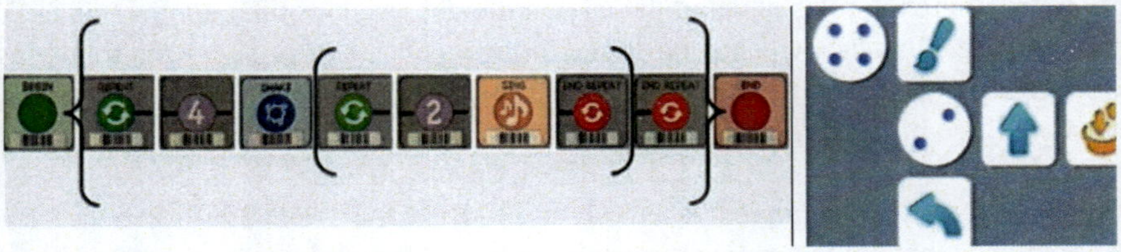

Fig. 7. Nested loops in KIBO environment and in PiktoMir environment.

A teacher working with one or two children making a program out of wooden blocks on a table can prevent children from making syntactical errors by giving advice to the children in real time. An educator working with a group of a dozen children, each of whom independently creates his own program from wooden blocks on his desk, is not able to do this. Therefore, you need to look for another method to avoid syntactic errors. Here is such a method based on the following simple idea:

- let's define a pictogrammic programming language so that any programs composed of a given set of pictograms with robot commands and control flow structure headers are syntactically correct and executable.

To implement this idea, we, following the example of the Python language, "endow" the program with a two-dimensional structure, and use indentation to set the block structure. Thus, we postulate that programs composed of pictograms must be organized a priori in a two-dimensional table. With the material method of programming, a violation of this postulate is described by a phrase easy understandable to children: "pictograms are unevenly spaced" and when programming on a touch screen, the impossibility of such a violation can be automatically ensured. Thus, when compiling programs from tangible objects, the only possible syntactic (or ruther 2D-syntactic) violation is *unevenness*, a violation of the two-dimensional structure. This *unevenness* is easily recognized and corrected by children, starting from the age of 3–4 years, and therefore in the future we assume that the pictograms are evenly spaced, that is, the program is endowed with a two-dimensional structure.

With any method of compiling a program, it is necessary to first explain to the children how the computer will execute the program and what syntactic rules for compiling the program will need to be observed in order for the computer to understand the program. The simpler the syntax rules are, the easier it will be for children to understand and follow them. The ideal in two-dimensional pictogrammic programming is the complete abolition of any syntactic restrictions, except for the violation of two-dimensionality. In this article, the ideal is realized both theoretically and practically. Namely:

- rules are formulated that allow to interpret any arrangement of icons in the cells of a two-dimensional table as a syntactically correct executable program,
- these rules are implemented in the Pikto language of the PiktoMir system.

It is important that these rules methodically turned out to be quite natural and effective in mastering the basics of programming by preschoolers and elementary school students. For example, according to these rules, both programs in Fig. 6 are correct. Before explaining the rules in detail, let us describe the result of their application to the program in Fig. 7 (right), which controls the Spinner robot. Our rules say the following. The program in question consists of an outer REPEAT 4 TIMES loop, the body of which consists of three blocks. The first block consists of a single FORWARD command, the second block consists of a REPEAT 2 TIMES loop, and the third block also consists of a single TURN LEFT command. The body of the REPEAT 2 TIMES loop consists of two commands that are located on the same line as the loop header.

2.2 The Advantage of Shifting the Emphasis from Syntax to Semantics

Although the use of the rules of syntactically unrestricted pictogrammic programming eliminates the need to explain syntactic restrictions to children, it does not eliminate the need to explain to children the rules for splitting a program into nested constructs, which determine the semantics of the program, the rules for its execution. The advantage of shifting the emphasis from syntax to semantics is that, unlike syntax, semantics is illustrative and its understanding is possible in practice, in experiment. For by successively pressing the step-by-step program execution button, the child sees clearly which program icon the computer is executing and what this execution leads to. Visualization is especially important for speeding up the first steps in programming. According to our methodology, these first steps are organized as follows. Having compiled a program from material objects, say, from cubes on a table, the child shows the program to a computer (tablet). The computer recognizes the received image, turns it into a program in its memory and executes this program. No matter how the child arranges the cubes in the cells of a two-dimensional table, this arrangement is recognized by the computer as some correct program, and the child gets the opportunity to see the process of executing this program, to see whether the behavior of the robot when executing this program corresponds to the ideas that the child tried to realize when compiling it. If the behavior does not match the original ideas, the child changes the program and has the opportunity to see how the modified program is executed. Thus, in any situation, the child can experiment and does not get into situations in which he must remember the rules for programming and try to speculatively understand which of these rules he has violated.

3 Results

We managed to put the idea expressed in Sect. 2.1 into practice. The rules in this section for interpreting programs can be applied to any pictographic programming environment. For definiteness, we will state these rules on the example of the PiktoMir environment.

3.1 General Scheme for Interpreting the PiktoMir Program Composed in Error-Free Syntax

There are no arithmetic or logical variables and constants in PiktoMir, which means there are no arithmetic or logical expressions. So, you don't have to worry about errors in expressions. Syntactically invalid program fragments in PiktoMir could potentially occur in two situations:

(a) when splitting the program into sections-subroutines with unique names,
(b) when nesting control flow structures.

Situation (a). We resolve this situation as follows: we ignore possible inconsistencies, allowing them to manifest themselves one way or another during the execution of the program. Namely,

- we consider the call of a subroutine with an indefined name to be valid and execute this call as a single-cycle command No Operation;
- a line of code that could be interpreted as a redefinition of a subroutine name, since that name has already been assigned higher in the code, is considered a continuation of the code of the previous section and all code lines adjacent to this line is also included into the previous section.

Fig. 8. A 2 × 5 rectangle and a program that fills it.

Let us explain these two rules with an example. Figure 8 shows a 2 × 5 rectangle and a correct program that fills it. This program consists of an unnamed main section and two subroutines named A and B. In Fig. 9 on the left shows an erroneous program. The programmer, instead of choosing the name B for the second subroutine, which is used in the main section, mistakenly reselected the name A. As a result of this error, the program specified a call to an undefined subroutine B and an attempt was made to define subroutine A twice. PiktoMir will understand this program as shown in Fig. 9 on the right, that is, it will assume that there is only one subroutine A in the program. When executing the main section of this program, a call to a non-existing subroutine B will be interpreted as a single-cycle No Operation command.

Situation (b). The correctness of nested control structures is ensured by the fact that, inspired by the example of the Python language, we set the block structure implicitly, using indentation. The Pikto language allows nested repetition and branch constructs. This means that within one section of the program, the balance of operator brackets must be maintained. Explicit violation of the balance in the Pikto language is impossible, since in the Pikto language there are no pictograms to end a construction. Information about where a control construct ends is implicitly specified in Pikto by placing initial indents on program lines. And these rules are set in such a way that no matter what are indentation of various lines of the program, you can always unambiguously determine where any

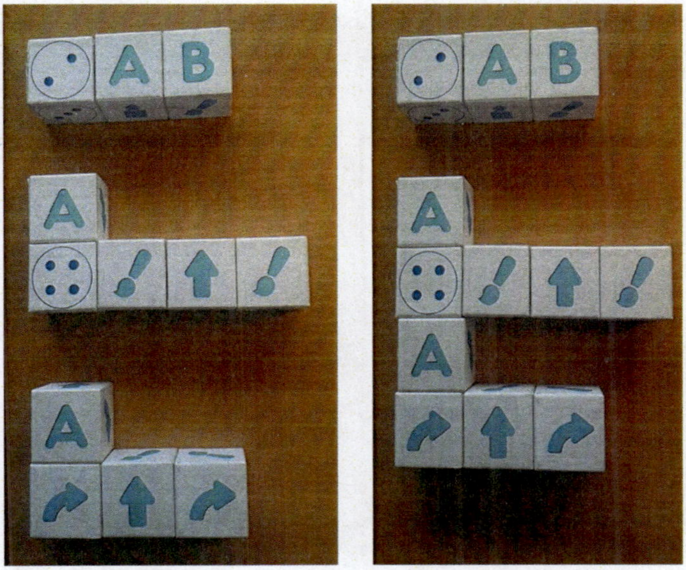

Fig. 9. A program with errors in the naming of subprograms.

control structure ends without disturbing the balance of control brackets. Thus you can unambiguously determine how the program will be executed.

3.2 Program Examples

Let's start with an example program consisting of one unnamed section.

Fig. 10. A program for drawing a capital Greek letter Π.

On Fig. 10, the "repeat 3 times" loop begins with an icon **(3)** placed in the first position of the first line with an indent of 0. The body of this loop begins on the same line, starting from the second position. **We always consider an indent greater than the indent of the beginning of the construction (see the next section) as a sign of the inclusion of the next line in the body of the construction**. The second line begins with such an indent, so the entire second line is included into the body of the loop. In the third line, the indent is zero, which means that the loop body ended on the second line and the command (**P**) located in the first position of the third line is not included in the loop body and forms a new block that will be executed after the loop is completed.

The embeddings of control structures are defined similarly.

Fig. 11. Setting nested loops with indents. The complete and compact forms of a programm drawing a capital letter E.

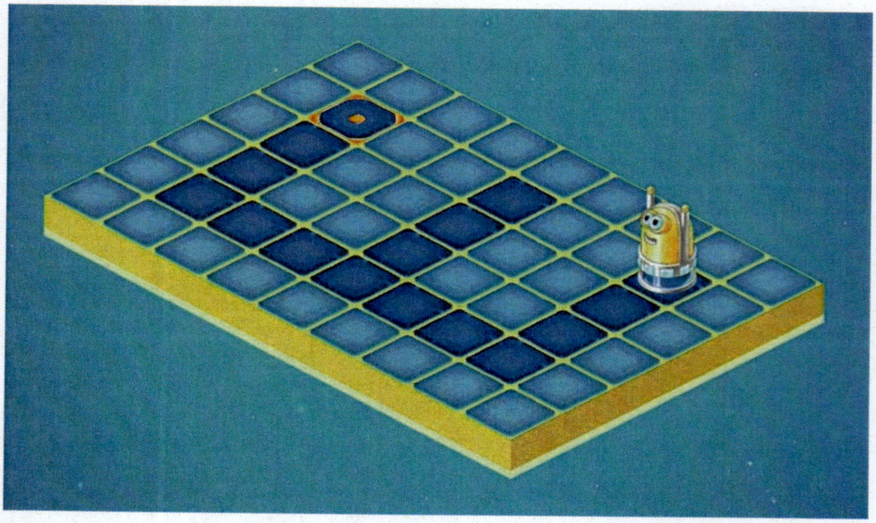

Fig. 12. The result of executing of the program on Fig. 11. The starting position of the robot is marked with a dot.

Please note that the programs in Fig. 11 consists of two sections separated by a blank line. The result of running this program is shown in Fig. 12.

3.3 Outline of a Formal Algorithm for Compiling the Body of a Single Subroutine in the Pikto Language

We consider the problem of compiling the body of one section of a Pikto program, which is a rectangular table T of fixed size, filled with pictograms from the alphabet

{( )} ∪ **A** ∪ **B**, where

( ) – background pictogram denoting an unfilled (empty) cell of table T,

$\mathbf{A} = \{\mathbf{a}_0, \mathbf{a}_1, \ldots, \mathbf{a}_{m-1}\}$ is a set of action icons (**a**ction), i.e. robot command-orders and section calls;

$\mathbf{B} = \{\mathbf{b}_0, \mathbf{b}_1, \ldots \mathbf{b}_{n-1}\}$ is set of pictograms of **be**ginnings of control structures. Cells filled with elements of sets **A** and **B** are called *non-empty*, a table row is called *non-empty* if it contains at least one non-empty cell. *The indent* of a non-empty line is the number of empty (background) cells preceding the first non-empty cell. The indent of a line consisting only of empty cells is considered equal to 0. The indent of any cell of a line is the number of cells in the line preceding the given cell.

The sets **A** and **B** depend on the set of robots that the program will have to control. The mechanism of formation of these sets is not important to us; by the time the section is compiled, these sets are assumed to be formed.

For example, when compiling the unnamed section of the program in Fig. 11 it is assumed that the set **A** consists of 9 pictograms

(L) (R) (F) (P) (A) (Б) (В) (Г) (Д),

i.e., in the compiled section, only calls to the commands-actions of the Spinner robot and calls to sections with one-letter names are used, and the set **B** consists of 6 pictograms

(1) (2) (3) (4) (5) (6),

that is, in the compiled section, only loops with an explicitly specified number of repetitions from 1 to 6 are used as control flow structures. Our task is to generate from the table T a sequence S of elements in the alphabet

$\{\mathbf{a}_0, \mathbf{a}_1, \ldots, \mathbf{a}_{m-1}\} \cup \{\mathbf{b}_0, \mathbf{b}_1, \ldots \mathbf{b}_{n-1}\} \cup \{(\mathbf{e})\}$, where

(e) is an explicit universal closing operator bracket (**e**nd) missing in the source program. The required number of instances of **(e)** must be placed in the compiled program in order to unambiguously specify the block structure of this program.

After compiling each section, a multi-section program, including calls (maybe recursive) of subroutines without parameters and calls of commands-actions and calls of commands-questions of the robots used, can be executed according to generally accepted rules.

The sequence S is constructed as follows. We create an empty sequence S and an empty stack of non-negative integers, which will store the indents of the headers of unclosed control structures and start scanning the table T line by line.

- each encountered pictogram ( ) is ignored;
- each encountered pictogram from **A** is added to the end of S;
- each encountered pictogram from B is added to the end of S, thereby opening a new operator bracket in the program S and, in addition, the indent of this pictogram from the beginning of the line is written to the stack;
- when moving to a new line, we calculate its indent I, and, while an indent greater than or equal to I is found at the top of the stack, we remove this indent from the stack and write the universal closing bracket (e) into S, thereby closing one of the previously opened operator brackets;
- the end of the section is processed as a line with an indent of 0.

Let us illustrate the described algorithm with two examples presented in Tables 1 and 2.

Table 1. The only section of the program in Fig. 10

(3)	**(P)**	**(F)**	**(P)**	**(F)**	( )
( )	**(P)**	**(F)**	**(P)**	**(F)**	**(R)**
(P)	( )	( )	( )	( )	( )

The result of compiling this section will be the sequence

(3) (P) (F) (P) (F) (P) (F) (R) (e) (P)

in which the opening bracket **(3)** in the first position matches the closing bracket **(e)** in the penultimate position.

Table 2. An unnamed section of the program in Fig. 11 (full form)

(2)	( )	( )	( )
( )	**(3)**	( )	( )
( )	( )	**(A)**	**(L)**
( )	**(P)**	**(L)**	( )

The outer loop of the section begins with an icon **(2)** with a zero indent. The next three lines have a non-zero indent, so we consider them to be part of the body of the outer loop. The second line of this body, indented 1, contains the heading **(3)** of the inner loop. According to our algorithm, the body of this inner loop should include the third line with an indent of 2, but should not include the fourth line, which will be executed upon each completion of the inner loop. The result of compiling this section will be the sequence

(2) (3) (A) (L) (e) (P) (L) (e)

in which the inner loop **(3) (A) (L) (e)** is nested in the outer loop **(2)… (e)**.

4 Discussion

The rules for interpreting programs composed in error-free syntax proposed in Sect. 3 are implemented in PiktoMir both for the mode of compiling a program from material objects and for the traditional on-screen mode of compiling a program. None of these regimes has yet carried out mass experiments on compiling sufficiently complex programs. At present, in our preschool programming course, tangible object programming is practiced only in the very first lessons, and we limit ourselves to single-section linear programs and programs with one numerical repeater. Our hypothesis is that the use of error-free syntax will make programming from material objects efficient in many more lessons. We plan to test this hypothesis in the 2022–2023 academic year.

5 Conclusion

It can be expected that in the coming years, the system of preschool education in Russia will introduce the basics of programming to children, starting from the age of 4–5 years, using textless, screenless environments. The Russian Academy of Sciences is now developing a methodology and software for such education in several hundred kindergartens in Russia (Fig. 13). This work is aimed at making teaching programming to preschoolers more effective. In our opinion, the use of the error-free syntax described in this article can help the teacher alone, without assistants, to cope with the organization of individual work on compiling programs from material objects in a group of preschoolers of 8–12 people.

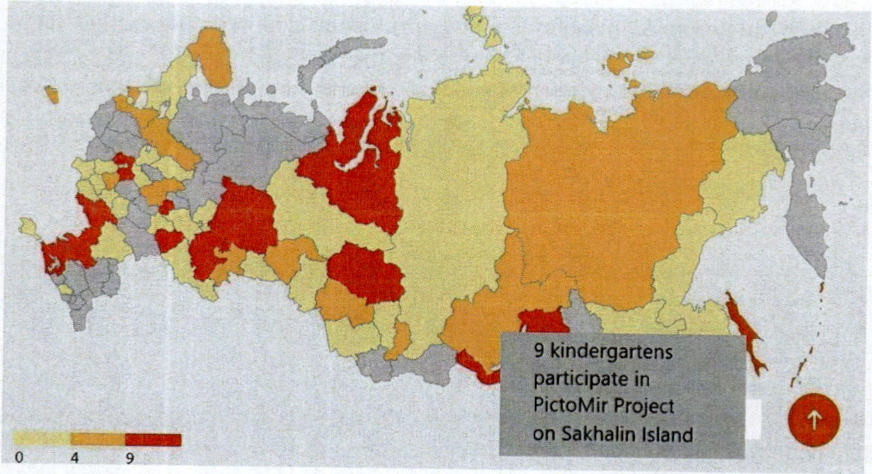

Fig. 13. Russian kindergartens participating in the PiktoMir project as of August 2022. An explanation of the legend is given on the example of Sakhalin Island.

Acknowledgments. The authors are grateful to N. O. Besshaposhnikov, K. A. Mashchenko and A. A. Maly for helpful discussions.

Sources of Financing The work was carried out jointly by a member of the Steklov Mathematical Institute RAS S.A. Polikarpov (analysis of world experience, statement of the problem) within the framework of the scientific project No. 19-29-14152 mk "Fundamental foundations for the formation of mathematical literacy for the digital society at the primary level of education" and employees of the FGU Federal Research Center NIISI RAS (development and implementation of algorithms) within the framework of the state task for topic FNEF-2022-0010.

References

1. Ershov, A.P.: Programming, the second literacy. In: Computer and education: Proceedings of the IFIP TC 3 3rd World Conference on Computer Education, Part 1, pp. 1–17. North-Holland Publishing Company, Amsterdam (1981)
2. Papert, S.: Mindstorms: Children, Computers, and Powerful Ideas, 242 p. Basic Books, NYC. https://mindstorms.media.mit.edu (1980). Last accessed 19 Sept 2022
3. Resnick, M., et al.: Scratch: programming for all. Commun. ACM **52**(11), 60–67 (2009). https://doi.org/10.1145/1592761.1592779
4. Flannery, L.P., Kazakoff, E.R., Bonta, et al.: Designing ScratchJr: support for early childhood learning through computer programming. In: Proceedings of the 12th International Conference on Interaction Design and Children (IDC'13), pp. 1–10. ACM, New York, NY, USA (2013). https://doi.org/10.1145/2485760.2485785
5. Kalaš, I.: Recognizing the Potential of ICT in Early Childhood Education. Analytical Survey, pp. 1–149. UNESCO Institute for Information Technologies in Education. https://iite.unesco.org/pics/publications/en/files/3214673.pdf (2010). Last accessed 19 Sept 2022
6. Richtel, M.: Reading, writing, arithmetic, and lately, coding. The New York Times, 10 May 2014. https://www.nytimes.com/2014/05/11/us/reading-writing-arithmetic-and-lately-coding.html. Last accessed 19 Sept 2022

7. Semenov, A.L.: Conceptual problems of teaching computer science, algorithm studies and programming at school. Herald Cybernet. **2**(22), 11–15 (2016) (in Russian)
8. Semenov, A.L.: Goals of general education in the digital world. In: Proceedings of the III International Conference Informatization of Education and Methods of e-Learning in 2 Parts. Part 2. Krasnoyarsk, pp. 383–388 (2019) (in Russian)
9. Betelin, V.B., Kushnirenko, A.G., Semenov, A.L., et al.: About digital literacy and environments for its development. Inform. Primen. **4**(14), 100–107 (2020) (in Russian). https://doi.org/10.14357/19922642004014
10. Agliamutdinova, D.B., Besshaposhnikov, N.O., Kushnirenko, A.G., et al.: Problems of early learning to program. How to bridge the gap between pictographic and textual programming styles. Int. J. Educ. Inform. Technol. (NAUN) **15**, 331–343 (2021). https://doi.org/10.46300/9109.2021.15.35
11. Betelin, V.B., Kushnirenko, A.G., Leonov, A.G., et al.: Basic programming concepts as explained for preschoolers. Int. J. Educ. Inform. Technol. (NAUN) **15**, 245–255 (2021). https://doi.org/10.46300/9109.2021.15.25
12. Besshaposhnikov, N.O., Kushnirenko, A.G., Leonov, A.G., et al.: Digital educational environment PiktoMir: experience of development and mass implementation of an annual programming course for preschoolers. Inform. Educ. **10**, 28–40 (2020) (in Russian). https://doi.org/10.32517/0234-0453-2020-35-10-28-40
13. Rogozhkina, I.B., Kushnirenko, A.G.: PiktoMir: teaching programming concepts to preschoolers with a new tutorial environment. Proc. Soc. Behav. Sci. **28**, 601–605 (2011). https://doi.org/10.1016/j.sbspro.2011.11.114
14. Bers M.U., Resnick, M.: The Official ScratchJr Book: Help Your Kids Learn to Code. No Starch Press (2015)
15. Polikarpov, S.A., Rudchenko, T.A.: Paper and digital textbooks in primary school. Advantages and disadvantages of the approaches. In: Proceedings of the III International Conference Informatization of Education and Methods of e-Learning in 2 Parts. Part 2. Krasnoyarsk, pp. 617–621 (2021) (in Russian)
16. Polikarpov, S.A.: Mathematical education in Russia: modern approaches to math teacher preparation. In: Current issues in mathematics education materials of the American-Russian workshop. COMAP (2017). https://www.comap.com/Free/C_I_IN_MAT_ED/index.html. Last accessed 19 Sept 2022
17. Meet Cubetto: https://www.primotoys.com. Last accessed 19 Sept 2022
18. Matatalab Coding Set: https://matatalab.com/en/coding-set. Last accessed 19 Sept 2022
19. Sullivan, A., Elkin, M., Bers, M.U.: KIBO robot demo: engaging young children in programming and engineering. In: Proceedings of the 14th International Conference on Interaction Design and Children (IDC'15). ACM, Boston, MA, USA (2015). https://doi.org/10.1145/2771839.2771868
20. Patent US20140297035A1. https://patents.google.com/patent/US20140297035. Last accessed 19 Sept 2022
21. Besshaposhnikov, N.O., Leonov, A.G.: Pictogramming programming language PIKTO. Herald Cybernet. **4**(28), 173–180 (2017) (in Russian)
22. Mladenović, M., Mladenović, S., Žanko, Ž: Impact of used programming language for K–12 students' understanding of the loop concept. Int. J. Technol. Enhanc. Learn. **1**(12), 79–98 (2020)

Comparison of Stabilization with P/I-Delayed Controllers for Second-Order Systems Using Built-In MATLAB Heuristic Optimization Methods

Martin Strmiska[1(✉)], José Mário Araújo[2], and Libor Pekař[1]

[1] Tomas Bata University, Zlín, Czech Republic
strmiska@utb.cz
[2] Federal Institute of Bahia, Salvador, Brazil

Abstract. In control theory, optimization is a process that makes some system behavior as likelihood, functional, or practical as possible (depending on the design requirements). From a mathematical point of view, optimization is a way to find the minimum or maximum of the mathematical function by which the system is described. Moreover, if there is a time delay in the feedback loops inside a controlled system, the system has an infinite spectrum – infinitely many roots (poles); this degrades the control's quality, especially in terms of stability, robustness, and oscillation response. For these reasons, it is necessary to optimize the system transient response. This paper will compare selected heuristic methods of optimization of linear-time invariant (LTI) systems with time delay in an integral part of the P/I-delayed controller. Some examples are chosen as benchmarks to verify the effectiveness of the optimization. In future research, we would like to focus on implementing selected modern optimization algorithms for solving non-smooth, non-convex, and non-Lipschitz optimization problems.

Keywords: Second-order systems · Stabilization · Heuristic optimalization

1 Introduction

Stabilization is a method for achieving the internal or BIBO-stability of a controlled system. The roots of stability analysis of dynamical systems, with particular focus in nonlinear and linear time invariant ones can be traced back to the work of Lyapunov and some essential books in systems theory [1–3]. If the control system does not contain time delay, stability can be determined using several methods, such as the Routh-Hurwitz criterion [4]. In most cases, purely analytical methods cannot determine stability if a system contains delay [5–8]. In this case, we can use the methods from numerical mathematics or frequency domain, as the Nyquist criterion [9].

Real world problems often include delays, for example, transportation of materials [10], the transmission of signals [11], control of epidemics [12], traffic systems [13], network control systems [14], traffic systems [15], control of the RNA messenger in

R. Silhavy et al. (Eds.): CoMeSySo 2022, LNNS 596, pp. 652–662, 2023.
https://doi.org/10.1007/978-3-031-21435-6_55

human cells [16], and others [16, 17]. The presence of delays in control systems is in general pervasive and can cause undesirable effects as loss of stability in closed loop and noise amplification [17]. However, in some situations, the designer can introduce a delay in the controller structure, in order to achieve some benefits, keeping the structure as simple as possible, as an instance, the output feedback with delays in the controller [18–20].

As an unwanted effect of using delays in controller design, the infinite dimensional spectrum rises due to the quasi-polynomial nature of the characteristic equation [21, 22]. To assure the dominance of stable roots of the system can be an intricate task to the designer. Stability charts as function of system parameters, as the delay itself, or the controller gains gives in most cases insightful in stability analysis or design of the system [20, 22–26].

This research aims to determine optimal gains in the plane for the P/I-delayed controller proposed in [26] using heuristic optimization approaches to tune the controllers with the goal to asset tracking error performance of the closed-loop system. By setting a convenient box-bounded region in the stability chart plotted in the plane, simulation-based optimization of a cost function (performance index), as integrals of the constant reference tracking error, can be conducted to tuning the gains for the controller. Notice that the central problem has some of the plagues for certain classes of optimization techniques, as the lack of an analytical objective function, non-smooth and non-convex nature of that function. Hence, heuristic optimization techniques, that are in most case derivative free, are the insightful ones to attack the optimization problem to tune the controller parameters [27]. The focus in second-order plants is of particular interest in vibration control applications, a research topic of increasing interest [28]. Moreover, a simple PI controller cannot be capable to stabilize certain unstable second-order plants in closed loop, because it limited degree of freedom to modify the coefficients of the characteristic polynomial. Unstable vibration can be found, for example, in friction-induced (or stick-slip) vibration systems [29].

The paper is organized as follows. The Sect. 2 of the article is a theoretical background that describes the principle of the design method and the determination of the stabilization pair for the controller setting. In the Sect. 3, a summary of the heuristic optimization techniques used is given. Section 4 offers the verification of the proposed approach in three numerical examples. For each example, the mathematical formulas are first derived, then the stability regions are shown and described. Lastly, regulation with the designed optimum controllers in IAE – Integral of Absolute Error - sense is illustrated. The last part of this article is the Conclusion chapter, which briefly discusses all the results and future directions.

2 Preliminaries and Statement of the Problem

2.1 Closed-Loop Control and Stabilization Pair (k_p, k_i)

Consider a closed-loop system [26] in Fig. 1 with P/I-delayed controller $C(s)$ and a controlled system $G(s)$.

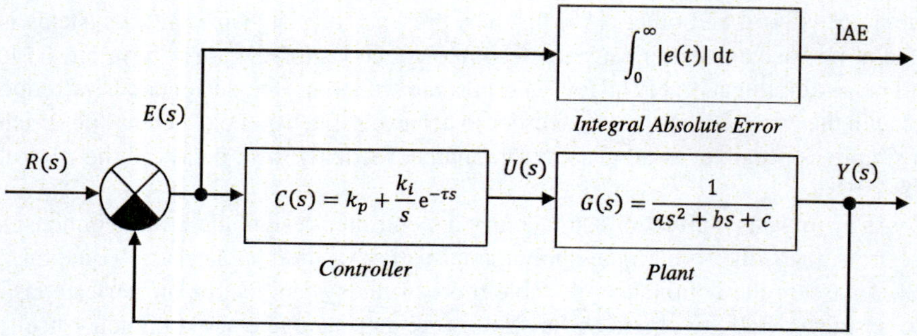

Fig. 1. Closed-loop Control

The closed-loop control automatically regulates a system plant to maintain the desired magnitude. In the figure of the closed-loop control, $R(s)$ denotes the input reference signal, $E(s)$ the control error, $C(s)$ the controller, $U(s)$ the control action, $G(s)$ the controlled plant, *IAE* the integral absolute error, and $Y(s)$ output signal.

Consider a controller $C(s)$ and a controlled plant $G(s)$ which are determined by the transfer functions

$$C(s) = k_p + \frac{k_i}{s}\mathrm{e}^{-\tau s}, \quad G(s) = \frac{1}{as^2 + bs + c}, \tag{1}$$

where k_p represents a proportional gain, k_i an integral gain, $\tau \in \mathbb{R}_+$ is a delay, s is a complex variable, and a, b, c are real coefficients.

To stabilize a given closed-loop, it is useful to find the stability regions in which to search for the optimal controller settings. To find the stability regions we use the stabilization pair (k_p, k_i), which is given by the equations

$$\begin{aligned} k_i &= \frac{o^2 + \omega^2}{\omega \cos \omega\tau + o \sin \omega\tau}\mathrm{e}^{\tau o}(2ao\omega + b\omega), \\ k_p &= -\left[a\left(o^2 - \omega^2\right) + bo + c\right] + \frac{\omega \sin \omega\tau - o \cos \omega\tau}{\omega \cos \omega\tau + o \sin \omega\tau}(2ao\omega + b\omega), \end{aligned} \tag{2}$$

where o is real part and ω is imaginary part of the complex number $s = o + \omega i$, for which the characteristic function equal zero when (2) is satisfied, it will be possible to search for the region of stability of the closed-loop system. The stabilization pair was derived in [26].

2.2 Error Index

Among the most useful measures to asset the constant reference tracking performance are the integrals error [31]. Given a set of the controller parameters with closed-loop stability guarantee, the designer can draw a solution with fast reference tracking, somewhat common since lower gains in this set have slow transient responses and the higher can lead the response to be excessively oscillatory. The integrals of the error, absolute or quadratic, weighted or not by the time, are an excellent cost function since that it has been proved to contains stationary solutions. In this study, the adopted error index is the integral of the absolute error – IAE:

$$\mathrm{IAE} = \int_0^\infty |e(t)|\mathrm{d}t. \tag{3}$$

Since that IAE is a function of the controller gains, the following optimization problem can be solved to the optimum IAE design

$$\begin{aligned} &\min_{k_p, k_i} \mathrm{IAE} \\ &s.t. \left(k_p, k_i\right) \in B \end{aligned}, \tag{4}$$

in which B is an adequate box region in the $\left(k_p, k_i\right)$ plane, that is, have a reasonable intersection with stability region in the charts. This optimization problem is not trivial because the IAE is not an explicit function of the gains; hence, derivative free, simulation base optimization techniques, also known as heuristics, are the most affordable way to compute the solution.

3 Heuristic Optimization Methods

To solve the optimization problem of Eq. (4), three heuristic approaches are explored, and the results are compared to illustrate their capabilities. A summary of these methods is offered in the next subsections. Details on these techniques can be found in [27].

3.1 Simulated Annealing

Simulated annealing (SA) is a numerical heuristic optimization technique based on the principles of thermodynamics. The technique's name is derived from the equation for calculating the probability, which corresponds to the course of controlled cooling of the material during annealing. The algorithm combines elements of local search with the Metropolis algorithm (the algorithm of accepting a worse solution).

At each iteration of the simulated annealing algorithm, a new point is randomly generated (candidate). If the generated candidate shows a better value of the criterion function, it replaces the original solution. Otherwise, the original solution is replaced by the generated (worse) candidate with only a certain probability.

3.2 Genetic Algorithm

In general, genetic algorithms (GA) are used to solve optimization problems. They are based on the principles of genetics and mechanisms of natural selection. The algorithm repeatedly modifies a population of individual solutions.

The genetic algorithm works as follows: at each iteration, individuals from the current population are randomly selected and used as parents to produce the children for the following generation. Over subsequent generations, the population converges toward an optimal solution.

3.3 Pattern Search

A pattern search (PS) algorithm is based on searching the neighborhood of the criterion function at discrete points that are calculated according to a given pattern. The template is a set of directional vectors defined over the space of optimized parameters (the definition domain of the criterion function).

At each iteration of the pattern search algorithm, the pattern either moves to the point that best minimizes its objective function or shrinks in size if no point is better than the current point, until the expected accuracy has been achieved or the algorithm reaches a maximum number of predetermined iterations.

4 Numerical Optimization Results

4.1 Considered Plant Models and Their Controllers

Let us consider three transfer functions G_1, G_2, G_3, for which the optimization will be performed:

$$G_1(s) = \frac{1}{s^2 - 0.1s + 10}, \quad G_2(s) = \frac{1}{s^2 - 2}, \quad G_3(s) = \frac{1}{s^2 - 1.1s + 0.28}, \tag{5}$$

where the P/I-delayed controller of each plant is

$$C_1(s) = k_p + \frac{k_i}{s}\mathrm{e}^{-0.44s}, \quad C_2(s) = k_p + \frac{k_i}{s}\mathrm{e}^{-0.23s}, \quad C_3(s) = k_p + \frac{k_i}{s}\mathrm{e}^{-0.345s}. \tag{6}$$

The results were obtained by the following steps:

1. Firstly, we determined the stability regions in the (k_p, k_i) plane for each controlled plant (see Subsect. 4.2),
2. Then using SA, GA, and PS methods we got the results which can be seen in the Table 1,
3. Finally, we simulated the control in MATLAB.

Table 1. Computed solutions with the heuristic methods

Plant	Start point $[k_p, k_i]$	Domain	Optimal solution			
			Method	k_p	k_i	IAE
G_1	[50, 50]	$k_p \in 0, 100$ $k_i \in 0, 400$	SA	35.61	90.23	0.595
			GA	33.10	80.06	0.597
			PS	35.32	89.71	0.595

(continued)

4.2 The Stability Region in the (k_p, k_i) Plane

See Figs. 2, 3 and 4.

Table 1. (*continued*)

Plant	Start point $[k_p, k_i]$	Domain	Optimal solution			
			Method	k_p	k_i	IAE
G_2	[300, 300]	$k_p \in 100, 400$ $k_i \in 50, 400$	SA	143.45	399.96	0.484
			GA	143.17	391.80	0.493
			PS	143.34	400.01	0.484
G_3	[115, 37]	$k_p \in 25, 130 k_i \in 50, 500$	SA	77.47	206.59	0.925
			GA	82.04	228.92	0.947
			PS	78.18	209.03	0.925

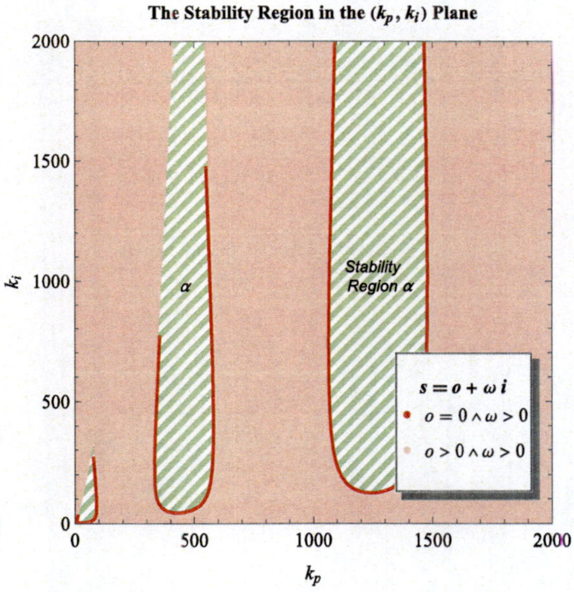

Fig. 2. The stability region of G_1

4.3 Main Results

The optimization problem was solved using the MATLAB Optimization Toolbox, with the three heuristic methods described, using the built-in functions for GA, PS and SA. A personal computer Acer Spin 3 (8 GB memory and Intel Core i3 processor) was used. The box domain B was chosed by taken from the stability charts a reasonable intersection with the stable regions α in the Figs. 2, 3, and 4. Table 1 gives the gains and objective function for each of the methods. All these methods have capability to search a global optimum, the consistence of the solutions displayed in Table 1 are illustrative of the effectiveness of the proposed approach. The unit step response for the three plants in closed loop are displayed at Figs. 5, 6, and 7. The non-optimized case, the reference, and

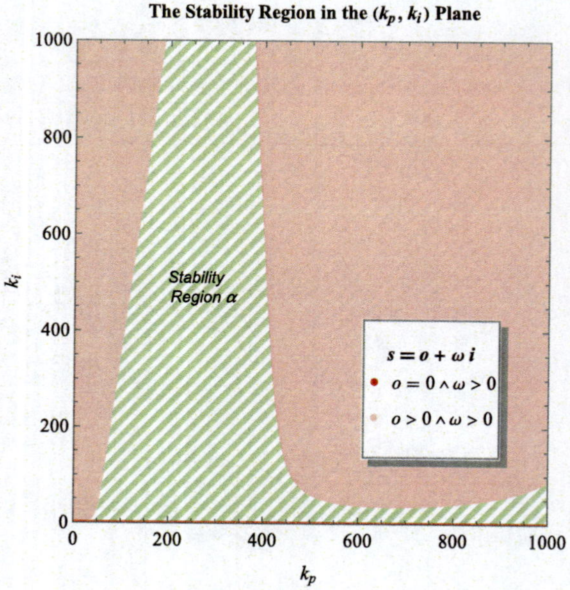

Fig. 3. The stability region of G_2

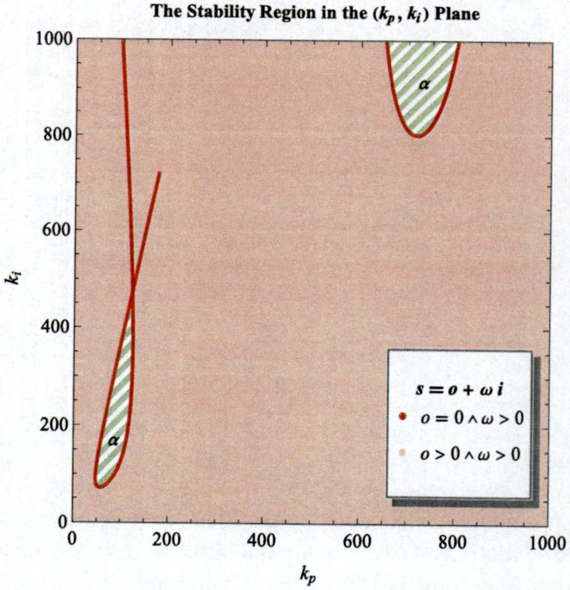

Fig. 4. The stability region of G_3

the optimized solutions for the three methods applied can be seen in each of the Figs. 5, 6, and 7. As we can observe from the figures, in all three cases (SA, GA, PS) there is a very fast output signal stabilization (and almost non-oscillating). There is a high probability of the obtained solutions is close to a global one, given that they are quite similar. Some of the responses are superimposed in the graphics. The supplied starting point (initial guess) was the same for all the methods, in order to give a fair comparison of the solution.

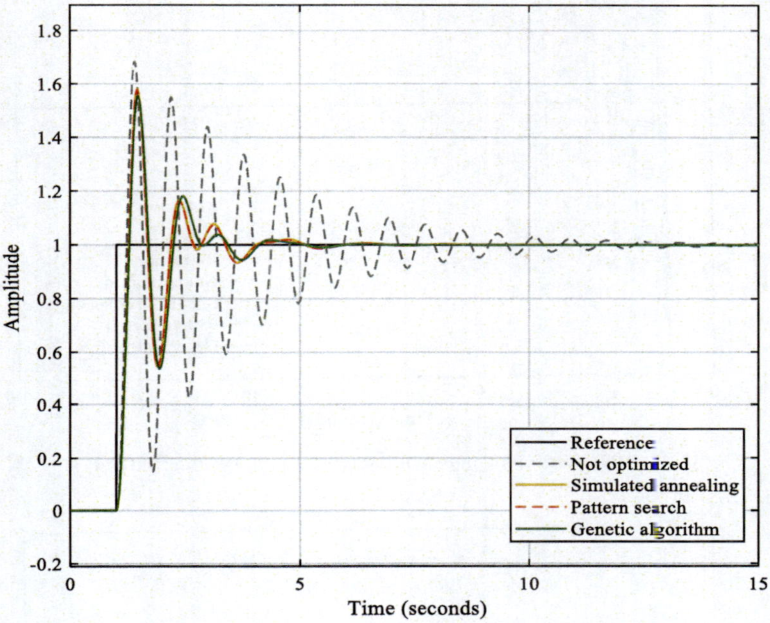

Fig. 5. The unit step response of the closed-loop system for the G_1

5 Conclusion

The use of PI controllers is an intriguing topic for which the application can be found in many fields. This article focused on finding the stability regions for the second-order systems. Finding the stability regions is analytically complex if the closed-loop systems contain a delay; methods of numerical mathematics are used for such cases. The stability regions help to find the controller parameters and to understand in which cases the system is asymptotically stable, marginally stable, or unstable. In our case, the stability regions analysis is limited to using second-order plant systems. The stabilization pair, which corresponds to the graph in the global stability region, provides the designer with an easier way to accomplish the control requirements. This study can be extended to stabilization using higher-order polynomial plant systems or optimization methods where we can find optimal parameters for setting the controller.

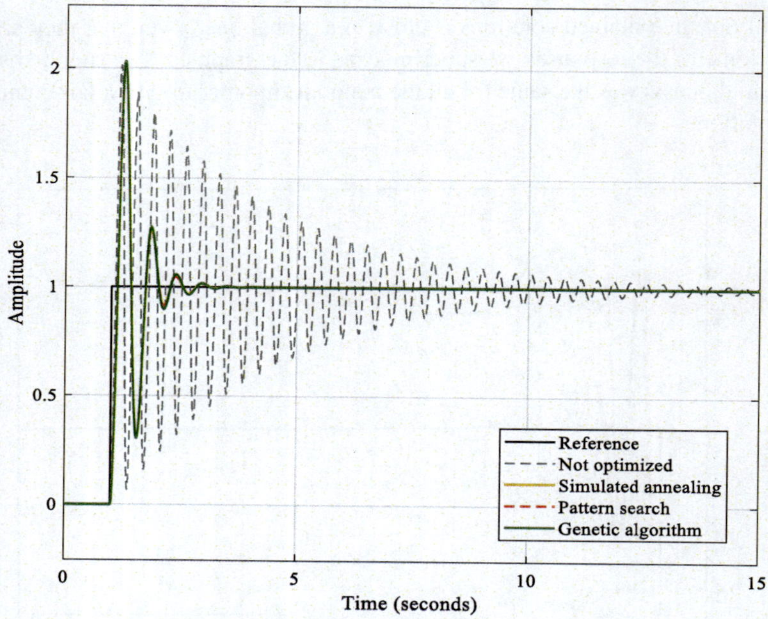

Fig. 6. The unit step response of the closed-loop system for the G_2

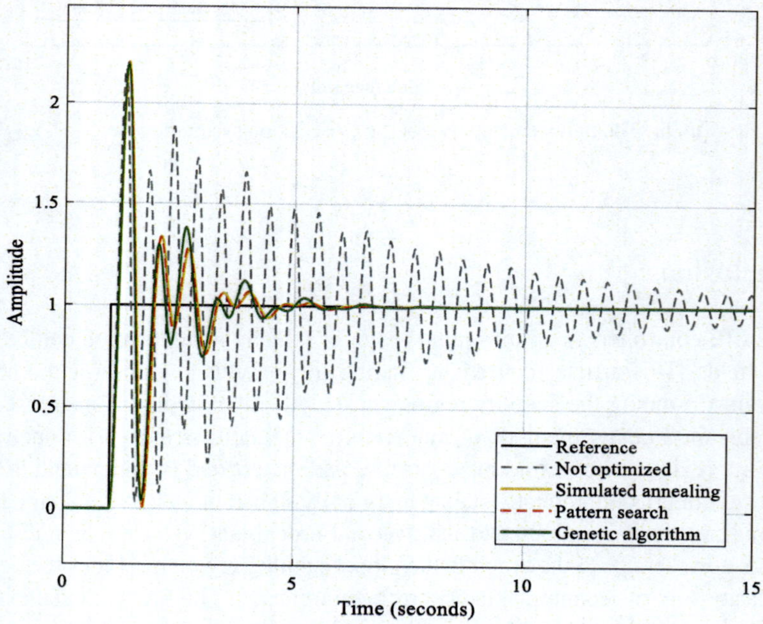

Fig. 7. The unit step response of the closed-loop system for the G_3

Acknowledgements. The paper was supported by the Internal Grant Agency of Tomas Bata University in Zlín under the project number IGA/FAI/2022/008.

References

1. Lyapunov, A.M.: The General Problem of the Stability of Motion, pp. 1–251. Kharkiv Mathematical Society, Kharkiv (1892)
2. Zadeh, L.A., Deoser, C.A.: Linear System Theory: The State Space Approach. Dover Publications (2008)
3. Khalil, H.K.: Nonlinear Systems. Prentice Hall (2014)
4. Barnett, S.: Polynomials and Linear Control Systems. Dekker (1983)
5. Hristuvarsakelis, D., Levine, W.S.: Handbook of Networked and Embedded Control Systems. Birkhäuser, Boston (2005)
6. Fridman, E.: Introduction to Time-Delay Systems: Analysis and Control. Springer, Cham (2014)
7. Xia, Y., Fu, M., Shi, P.: Analysis and synthesis of dynamical systems with time delays. Springer, Berlin (2009)
8. Michiels, W., Engelborghs, K., Vansevevant, P., Roose, D.: Continuous pole placement for delay equations. Automatica **38**(5), 747–761 (2002)
9. Araújo, J.M.: Discussion on 'state feedback control with time delay'. Mech. Syst. Sig. Process. **98** (2018). https://doi.org/10.1016/j.ymssp.2017.05.004
10. Wong, L., Lee, Y.E., Lee, H.J.: Optimal transmission of messages in computer networks—an optimal control problem involving control-dependent time-delayed arguments. J. Inequal. Appl. (2022)
11. Kumar, P., Erturk, V.S.: The analysis of a time delay fractional COVID-19 model via Caputo type fractional derivative. In: Mathematical Methods in the Applied Sciences (2020)
12. Oertel, R., Wagner, P.: Delay-time actuated traffic signal control for an isolated intersection. In: Proceedings 90st Annual Meeting Transportation Research Board (TRB)
13. Joelianto, E.: Networked control systems: time delays and robust control design issues. In: Proceedings of 2011 2nd International Conference on Instrumentation Control and Automation, ICA 2011. https://doi.org/10.1109/ICA.2011.6130121
14. Kennell, D., Talkad, V.: Messenger RNA potential and the delay before exponential decay of messages. J. Mol. Biol. (1976)
15. Dollevoet, T., Huisman, D., Schmidt, M., Schöbel, A.: Delay Propagation and Delay Management in Transportation Networks (2018)
16. Palpal-latoc, C., Bernardo, R.C., Vega, I.: Testing time-delayed cosmology. Preprint (2021)
17. Mirkin, L., Palmor, Z.J.: Control issues in systems with loop delays. In: Hristu-Varsakelis, D., Levine, W.S. (eds.) Handbook of Networked and Embedded Control Systems. Control Engineering. Birkhäuser, Boston (2005)
18. Jnifene, A.: Active vibration control of flexible structures using delayed position feedback. Syst. Control Lett. **56**(3), 215–222 (2007)
19. Peng, J., Xiang, M., Wang, L., Xie, W., Sun, H., Yu, J.: Nonlinear primary resonance in vibration control of cable-stayed beam with time delay feedback. Mech. Syst. Signal Process. **137**, 106488 (2020)
20. Elmadssia, S., Saadaoui, K., Zaguia, A., Ezzedine, T., Wang, Q.-G.: Stabilization domains for second order delay systems. IEEE Access **9**, 53518–53529 (2021)
21. Busłowicz, M.: Sufficient conditions for instability of delay differential systems. Int. J. Control **37**(6), 1311–1321 (1983)

22. Pekař, L., Strmiska, M., Song, M., Dostálek, P.: Numerical gridding stability charts estimation using quasi-polynomial approximation for TDS. In: Proceedings of the 23rd International Conference on Process Control (PC), pp. 290–295 (2021)
23. Insperger, T., Stépán, G.: Stability charts for fundamental delay-differential equations. In: Semi-Discretization for Time-Delay Systems. Applied Mathematical Sciences, vol. 178. Springer, New York, NY (2011). https://doi.org/10.1007/978-1-4614-0335-7_2
24. Szaksz, B., Stepan, G.: Stability charts of a delayed model of vehicle towing. IFAC Pap. Online **54**(18), 64–69 (2021)
25. Lv, Z.-H., Zhang, J.-F., Ouyang, H.: Receptance-based computation of stability crossing curves for single-input-multiple-output second-order linear systems with two time-delays. Int. J. Struct. Stab. Dyn. **22**(01) (2022). https://doi.org/10.1142/s021945542250002x
26. Strmiska, M., Pekař, L., Araújo, J.M.: Stabilization of second-order systems using a P/I-delayed controller. In: Proceedings of the 6th Computational Methods in Systems and Software (2022)
27. Floudas, C., Pardalos, P.: Encyclopedia of Optimization, 2nd edn. Springer. USA, 2009. ISBN: 978-0-387-74759-0
28. Araújo, J.M., Santos, T.L.M.: Special issue on control of second-order vibrating systems with time delay. Mech. Syst. Signal Process. **137**, 106527 (2020)
29. Vahid-Araghi, O., Golnaraghi, F.: Negative damping instability mechanism. In: Friction-Induced Vibration in Lead Screw Drives. Springer, New York, NY (2010)
30. Novella-Rodriguez, D.F., Muro-Cuellar, B.: Control of second order strictly proper unstable systems with time delay. Rev. Mex. Ing. Quím [Online] **10**(3), 551–559 (2011)
31. Smith, C.A., Corripio, A.B.: Principles and Practice of Automatic Process Control. Wiley (2006)

Using a Distributed Architecture of a Geographic Information System to Support Thin Clients

Maxim Polenov(✉), Daniil Ivanov, and Dmytry Bespalov

Department of Computer Engineering, Southern Federal University, Taganrog, Russia
{mypolenov,dabespalov}@sfedu.ru

Abstract. The paper discusses the features of using a distributed data processing architecture, based on the client-server model, as the basis for building a geographic information system that is actively used by clients with low-performance computing facilities. This problem is solved by using the proposed modification of the traditional distributed architecture which consists in transferring the GIS core to the server component. The purpose of the work is to consider the principles of distribution of the functionality of a geographic information system for server and client components and the organization of their interaction. The proposed approach will allow, when implemented in a software application, to reduce the requirements for the computing facilities of users. The proposed architecture can be used in other distributed systems as well. Especially in those where the task is to display three-dimensional information on thin clients.

Keywords: Geographic information system · Data processing · Modified distributed architecture · Thin clients

1 Introduction

The use of Geographic Information Systems (GIS) technologies makes it possible to significantly increase the efficiency and quality of work with spatially coordinated data compared to traditional mapping methods. Currently, almost everyone has their own pocket version of GIS, provided by various services such as Google maps [1], Microsoft Bing Maps [2] and others. They allow you to view satellite photos of the Earth, determine the user's current location, build routes, and much more.

However, the range of tasks to be solved by these services is limited. It is necessary to consider the computing power of the device on which the task will be solved when developing GIS applications for solving specific problems. Even high-performance computing systems can encounter performance issues with such applications. In this regard, the solution of these problems is transferred to the server part of the application or service [3]. This happens in almost all services and the user receives only a solution to this problem.

R. Silhavy et al. (Eds.): CoMeSySo 2022, LNNS 596, pp. 663–669, 2023.
https://doi.org/10.1007/978-3-031-21435-6_56

The imaging or rendering is performed by the GIS core in these services. The core receives all the necessary data: maps, routes, other graphic designations and other information. And based on this data, the user receives the finished image. For low-power computing devices, even such a task can be quite complex and time-consuming, since several maps can be loaded simultaneously in different projections [4]. Thus, we will get a system that will not be able to work efficiently in real time.

To solve this problem, you can do the following: increase the computing power of user devices or transfer the GIS core and, as a result, rendering, to the GIS server part.

Since it is not always possible to increase the computing power of devices, especially when the devices are made in the form of complete monoblocks or similar systems and mobile devices, the second option can be considered preferable, since it does not require device upgrades, but only a modification of the organization of the application itself. Also, the main advantage of this solution is to increase the performance of the GIS while maintaining its full functionality. The relevance of these studies lies in the fact that at present there are systems that can support the work of only thin clients, which often have little functionality and are not able to solve heavy computing tasks.

2 Geographic Information Systems Architecture

The most promising among the currently used GIS architectures is the distributed architecture, which provides centralized storage, analysis and provision of spatial data to the network for remote users. Distributed GIS [5, 6] has the advantages of a multi-user architecture without its disadvantages, since any change in geospatial data is carried out only through web server moderators.

Based on the information obtained from the analysis of popular GIS services, we will compile a generalized diagram of the distributed GIS architecture of these services, shown in Fig. 1.

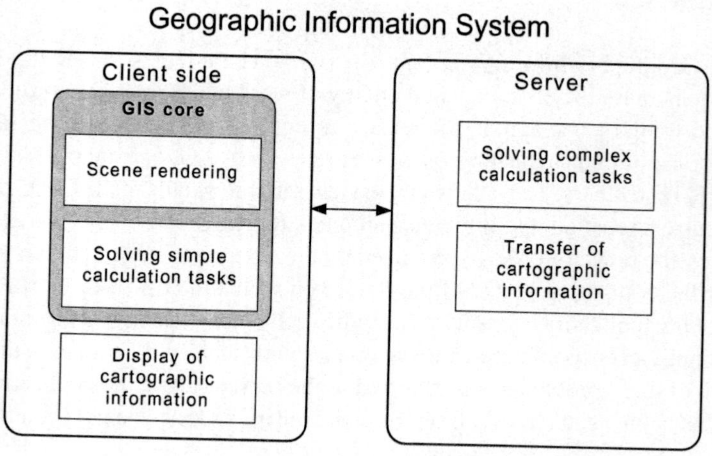

Fig. 1. The distributed GIS architecture

However, applications based on this architecture may encounter problems displaying real-time processed 3D data when such applications are used on low performance computing devices.

To solve this problem, a modified distributed architecture of geographic information systems is proposed [7]. This architecture is shown in Fig. 2.

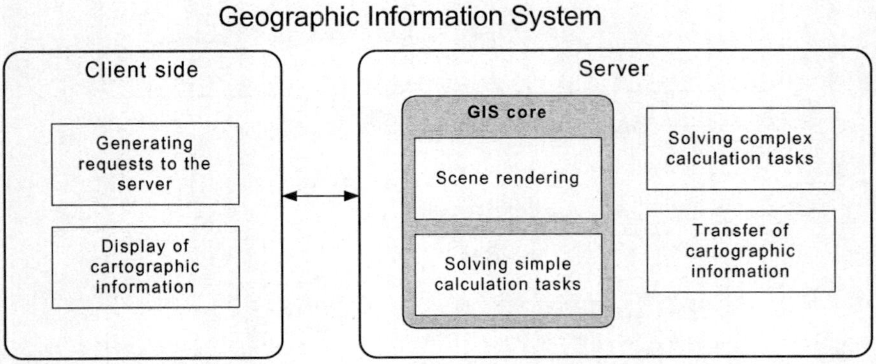

Fig. 2. The modified distributed GIS architecture

The main difference between the proposed approach and the classical distributed architecture is the transfer of the core of the geographic information system to the server component of the application. The interaction of the core and other parts of the GIS is carried out within a single system, which is a significant advantage, since possible losses of the initial data required to solve complex computational tasks are excluded.

Also, along with the core, the main computational load associated with rendering the scene is transferred to the server, which makes it possible to use computing devices with low performance as client computers (thin clients).

3 GIS Application

To determine the functionality of the modified distributed architecture, it was decided to develop a software application based on it. When developing the application, the fact was considered that in this architecture the core of the geographic information system is transferred to the server part, and it is necessary to implement two components: client and server.

The organization of the client component is shown in Fig. 3.

The organization of the server component is shown in Fig. 4.

These diagrams show the main libraries that were used in the development of the application. The connection between the graphical interface and the part of the application responsible for the implementation of the geoinformation component is carried out using signals and slots provided by the Qt framework [8].

The osgEarth geoinformation library [9] is an open project, a geospatial developer package and a shell simulating the Earth's surface.

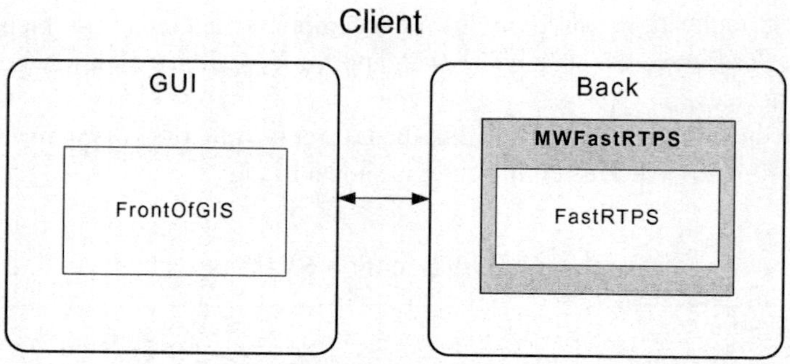

Fig. 3. The client component of application

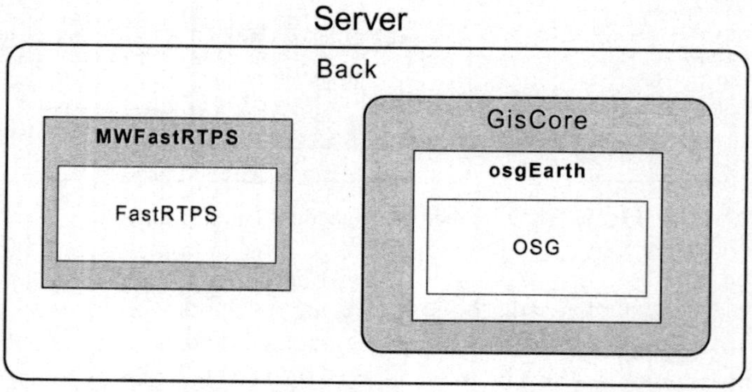

Fig. 4. The server component of application

The main component of the osgEarth geoinformation library is the OSG (OpenSceneGraph) graphics library [10]. OSG is a high-performance three-dimensional graphics toolkit with open-source code used by application developers in such areas as visual modeling, games, virtual reality, scientific visualization and modeling.

Further the GisCore wrapper library was created and based on the above two libraries. It extends the functionality of these libraries and provides a convenient interface for practical and fast work with geospatial information.

The FastRTPS library [11] implements the RTPS protocol (Real Time Publish Subscribe), which provides communication between the publisher and the subscriber via transport layer network protocols [12]. This library uses the publisher-subscriber design pattern, in which the senders of messages, called publishers, are not directly tied by the program code of sending messages to subscribers. Based on the FastRTPS library, the MWFastRTPS wrapper library (Middleware) was created to simplify the use of the FastRTPS library.

Two types of messages were used to transfer data between the server and client components: a string and an image of arbitrary size and format. For data processing, the best option would be to use a standard data format that could be serialized and deserialized either by standard C++ libraries or by libraries used in development. These conditions were satisfied with the JSON data format (JavaScript Object Notation) [13], since the selected Qt framework has a built-in ability to work with this data type and provides translation of the object to a string and back.

Also, a protocol of information and logical interaction between the application components was developed to determine the content of the JSON file.

The analysis of existing geographic information systems also made it possible to single out a general approach to organizing a graphical interface, which was tested on many users. In almost all systems, it is almost the same, so this approach was taken as the basis for developing the application. The scheme of the graphical interface of the application is shown in Fig. 5.

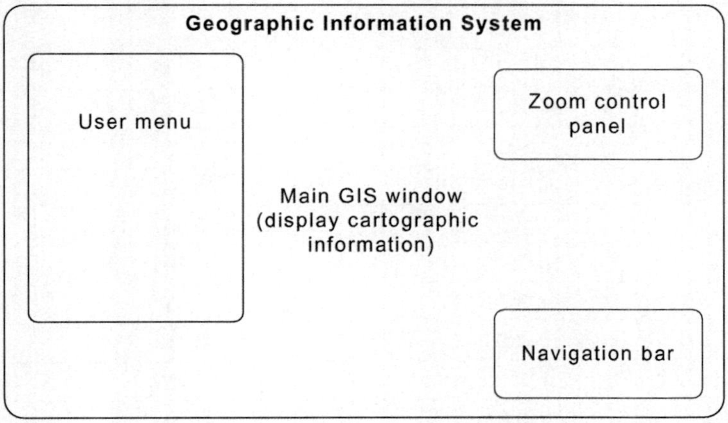

Fig. 5. The GUI of GIS application

The menu is a window for managing basic functions, such as switching between maps, displaying coordinates at a specified point, and others. The zoom control panel is presented as a slider to zoom in or out on the map. The navigation bar will allow you to quickly return to your current location, go to observe the entire map, or perform another function to move the observer over the map.

The QML language [14] and the Qt framework were chosen for the development of the graphical interface. This framework has a powerful QML graphics module that defines and implements the language infrastructure and provides an application programming interface (API) that allows application developers to extend the QML language. The graphical interface was also presented in the form of a library for further development and the possibility of improving the application.

The mechanism of signals and slots provided by the Qt framework was used also in the development of the library. This mechanism was used to quickly connect various elements of the graphical interface to the application functions.

4 Experiments

The study of the proposed architecture was carried out on the model of organization of interaction between the server and client components of GIS in a local network.

Several experiments were carried out with the test applications in conditions of high CPU usage and multiple measurements of various characteristics and time of working out the GIS functionality were carried out.

The characteristics of one of the classical GIS under equivalent load conditions were chosen as reference. The results obtained are presented in Table 1.

Table 1. Comparison of the execution time of basic operations

Operation type	"Classical" GIS, reference time, ms	GIS on modified distributed architecture, time, ms
panUp	1.865	1.32
panLeft	1.702	1.113
panRight	1.101	0.917
panDown	1.871	0.864
changeNormalizedScale	0.044	0.262
scalePlus	0.05	0.907
scaleMinus	0.054	0.705
press	1.007	1.437
move	0.062	0.318
setViewpointAboveMap	1.011	0.969
changeScaleInPoint	0.206	0.206

When using the classical GIS application to perform a calculation task, there is an increase in the processing time of operations such as pan, as well as setViewpointAboveMap, since these operations require a calculation associated with coordinate transformation. It should also be noted that when working with the application during the execution of the calculation task, the presence of delays in the execution of operations was noticeable even in the GUI of the program.

When working in the GIS application on modified distributed architecture, the calculation task was solved on the server side. Since this task cannot fully load the powerful server processor, this did not affect the operation of the application. However, it is worth noting that even in the absence of the impact of the lack of computing resources on the server part of the geographic information system, the average time to complete operations has increased. This is due to the fact that in this architecture there is an additional data transfer over the network.

5 Conclusion

The paper considers the proposed modified distributed architecture of the geoinformation system as one of the options for optimizing the operation of client-server GIS applications.

When developing a GIS software application, based on this architecture, a graphical interface, the necessary libraries were implemented, and a protocol for information and logical interaction between the server and client components was developed.

The conducted experiments demonstrated the maintaining of the necessary functionality and user characteristics of GIS. It should also be noted that the developed GIS application based on a modified distributed architecture in practice has shown the feasibility and convenience of its use on low-power client computers (thin clients), especially when solving performance-demanding computational tasks.

The proposed architecture can be used in other distributed systems as well. Especially in those where the task is to display three-dimensional information on thin clients.

Acknowledgments. The reported study was funded by Russian Foundation for Basic Research (RFBR) according to the research project No. 20-07-00559.

References

1. Google maps. https://www.google.com/maps
2. Microsoft Bing Maps. https://www.bing.com/maps
3. Longley, A., Goodchild, F., Maguire J., Rhind, J.: Geographical Information Systems and Science, 2nd edn. Wiley (2001)
4. Gimond, M.: Intro to GIS and spatial analysis. https://mgimond.github.io/Spatial/
5. Agrawal, S., Gupta, R.D.: Web GIS and its architecture: a review. Arab. J. Geosci. **10**(23), 1–13 (2017). https://doi.org/10.1007/s12517-017-3296-2
6. Coulouris, G., Dollimore, J., Kindberg, T., Blair, G.: Distributed Systems. Concepts and Design, Fifth ed., Addison-Wesley (2012)
7. Polenov, M., Ivanov, D., Bespalov, D.: Implementation of a geoinformation system based on distributed data processing architecture. Inf. Commun. **8**, 60–65 (2021) (in Russian)
8. Qt 5.15. https://doc.qt.io/qt-5/index.html
9. osgEarth. A 3D Mapping Engine & SDK for OpenSceneGraph. http://osgearth.org/
10. OpenSceneGraph. https://www.openscenegraph.com/
11. FastRTPS installation. Developer Guide. https://dev.px4.io/v1.11_noredirect/en/setup/fast-rtps-installation.html
12. eProsima. Fast DDS Documentation. https://fast-dds.docs.eprosima.com/en/latest/
13. Introducing JSON. https://www.json.org/json-en.html
14. Qt QML. https://doc.qt.io/qt-5/qtqml-index.html

Estimating the Approximate Necessity of Essential Medicines in Syria Using Machine Learning

Labeb Abood(✉)

Belgorod State University, Belgorod 308007, Russian Federation
marcabood028@gmail.com

Abstract. Essential Medicines are one of the very crucial parts of the health care needs. This article focuses on the estimation of the approximate necessity of essential medicines in the Syrian Arab Republic. Since 1977, when the idea of essential drugs or medicines was introduced by the WHO, it has been playing a vital role in society's betterment. Syria is no exception in that case. Since 2011, Syria's health care systems have been facing many problems. To make the situation better, it is necessary to focus on the modernization of the health care needs. After researching briefly about the usage, regulations, supply, and production of essential medicines in the Russian Federation, Germany, and the USA, the application of artificial intelligence seems like an ultimate necessity to make health care needs seamless and rapid. In this article, a machine learning approach is applied to predict the demand for different essential medicines in different regions of Syria. The task includes research methods, implementation, results, related work, and analysis for future progression. Random Forest Classifier is used to estimate the accuracy of the approach method.

Keywords: Essential medicine · Health care needs · Random forest classifier · Machine learning · Artificial intelligence in healthcare · Syrian healthcare system

1 Introduction

Currently, we are living in a time where there is no choice but to be frugal in every sector of life. It is necessary to choose the most up-to-date system and technology to ensure basic rights like medical care for every human being. Countries like America, Russia, and Germany have already taken their health system to a place where a patient can get almost any necessary health-related services by using just a smart phone. For many years, the primary goal of using technology in healthcare has been the rapid diagnosis and cure of diseases. With the addition of information technology, the biggest advantage in this medical field has been in data collection. Billions of people are getting sick all over the world and all their information is being stored online by internet and servers. As a result, health workers can easily store patient information for life. In Syria, the process is not that easy. Collecting data from different regions of Syria is challenging. Still, in recent times, many positive measures have been taken place related to health care needs.

R. Silhavy et al. (Eds.): CoMeSySo 2022, LNNS 596, pp. 670–679, 2023.
https://doi.org/10.1007/978-3-031-21435-6_57

The aim of this research is to predict an approximate demand of essential medicine from different regions of Syria, so that the distribution of medicine can be effective, and as per the demand of the people of that region.

The research followed several steps and is divided into two major parts. At first the proposed methodology of collecting data for essential medicine has been outlined, then the approximate estimation of essential medicine has taken place using machine learning.

Machine learning is swiftly infiltrating many areas within the healthcare industry, from diagnosis and prognosis to drug development and epidemiology, with significant potential to transform the medical landscape [1]. There is some research that has been carried out to understand the prediction of essential medicine demand using machine learning. Several countries are already getting the benefit of such an analysis.

2 Estimation of the Demand of Essential Medicines

Today, we stand on the threshold of the new medical revolution, just as big and far-reaching [2]. Earlier, either doctors or health care representatives wrote patient information on the paper for either collecting data to improve the overall health system or understanding the potential and existing health condition of a patient. And with the unimaginable advancement of digital media and information technology, the old paper document writing system has been eliminated. The combination of technology with the medical world has made it much easier to collect all these records. A highly available and well organized medical data for any country can help to update the essential medicine list in every two years by WHO more successfully. To make the research more effective, it is necessary to understand the underlying problems related to health care needs in Syria, essential medicine, and its production, supply, and availability in different regions. The article will outline these along with the usages of machine learning for predicting the demand of essential medicine in Syria.

2.1 Problem Statement

Medical data is more accessible in some countries than in others. Collecting data from Syria is comparatively difficult and time-consuming. The principal task of this research is based on the demand of the essential medicine for the Syrian health care system, and all the data collected using different methods are from various places of Syria, including Damascus, Halab (Aleppo), Hims (Homs), Tadmur (Palmyra), and so on. For collecting data, another struggle is that the patients don't feel very comfortable giving information for an independent research about what drugs they are being prescribed. But they are more comfortable to share their symptoms. The research named "Symptoms Based Disease Prediction Using Machine Learning Techniques" described the relative study of various Machine Learning algorithms for the detection of various diseases [3]. The system that is used to predict the diseases processes the symptoms provided by the user as input and gives the output as the probability of the disease [4]. Mostly, such systems used Big Data, and different Convolutional Neural Network algorithms. For type of data that have been collected, implemented using Machine Learning algorithm, i.e. K-nearest Neighbors, Decision Tree, Naïve Bayesian, random forest [4].

2.2 Research Methodology

Two major challenges were faced in the research's method. One of these is data collection, and the other is to estimate the future demand of the essential medicine using machine learning. From the collected data first prediction of disease was done, defined in Fig. 1.

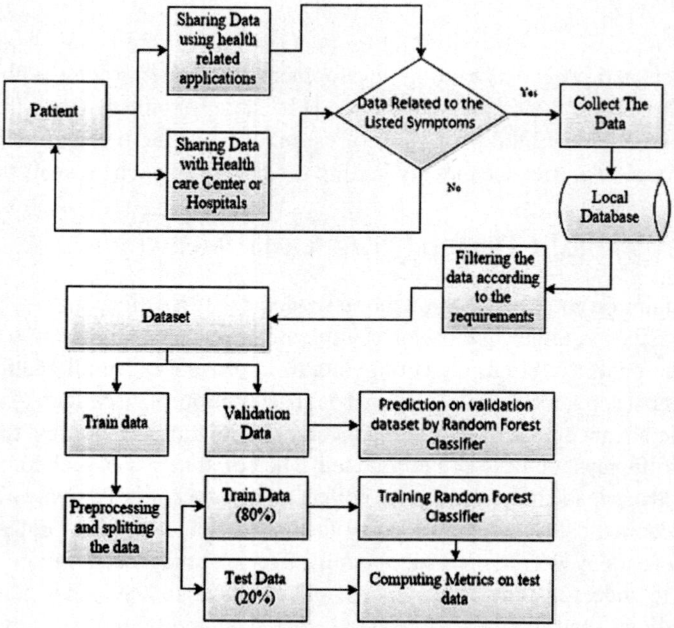

Fig. 1. The process of collecting data, and performing the data classification using random forest classifier.

Methods like document examination, observations, and questionnaires have been chosen for data collection. From the 2019 list of essential medicines by WHO, only twenty widely used drugs have been used in this study. Processes like, Data preparation, cleaning, and model building have taken place by studying various researches and existing machine learning approaches. To understand the quality of the model, a confusion matrix has been used. The prediction from the random forest classifier is applied and has been collected. The results will show the prediction of the disease, and these data will be utilized to predict the medicine for the disease defined in Fig. 2.

From that data, it is possible to predict the demand of the essential medicine by using a sorting algorithm. The proposed research methodology allows to apply the machine learning approach in different regions of Syria separately. After finding out the demand it will be compared to understand which regions demand what sort of essential medicines, and the result will show a count of maximum to minimum (Fig. 3).

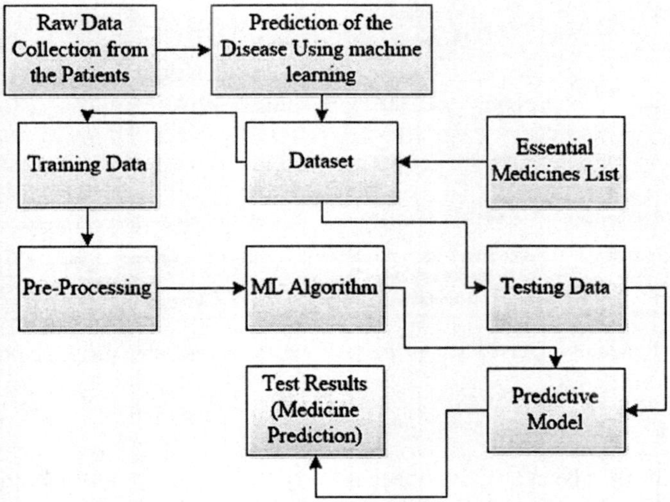

Fig. 2. The proposed framework for performing the prediction of the medicines for the diseases.

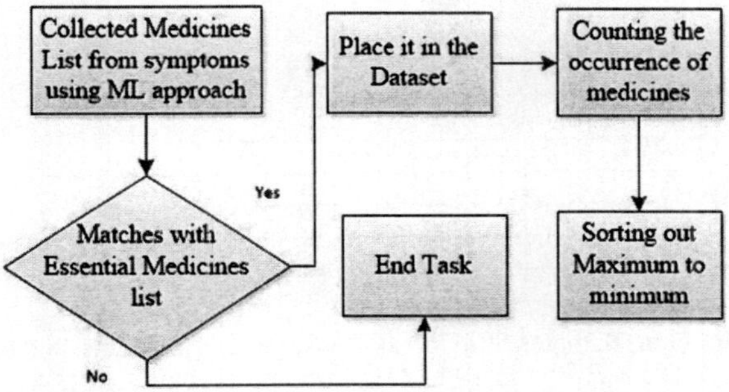

Fig. 3. The proposed approach to product the demand of essential medicines.

2.3 Understanding the Role of Essential Medicines

The World Health Organization (WHO) publishes a list of essential medicines every two years. Nearly 150 countries around the world follow the World Health Organization's list of essential medicines. These drugs are very important for public health. Contemporary morbidity, cost, accessibility, and treatment factors are considered in compiling this list for each country. The World Health Organization has recently not only planned a list of essential medicines but also published a list of essential diagnostics. The list focuses on global public health challenges, including cancer, with an emphasis on effective and timely solutions. A list of essential medicines produced by the World Health Organization (WHO) in 2019 was followed for this study (Fig. 4).

Item	Description	Comments
Paracetamol	Inj IM	Doctors have been advisedto avoid intramuscular injection of paracetamol in patients with fever.
Paracetamol	Injection:10mg/ml, 100ml v vial	It is an analgesic (it relieves pain) and an antipyretic (it lowers fever).
Tramadol	Injection: 50mg in 2ml amp	It is used to relieve moderate to moderately severe pain, including pain after surgery
Hyoscine Hydrobromide	Injection: 400 micrograms/ml	It is taken to prevent travel sickness (motion sickness).
Epinephrine (adrenaline)	Injection: 1 mg (as hydrochloride or hydrogen tartrate) in 1ml apoule	It is used for emergency treatment of severe allergic reactions

Fig. 4. Randomly selected five medicines from Syria's essential medicine list 2019

2.4 Dataset

The dataset that has been used is collected from a large scale online and offline survey about symptoms of various diseases in different regions of Syria. In the dataset symptoms like: abdominal pain, abnormal menstruation, acidity, acute liver failure, altered sensorium, anxiety, back pain, belly pain, blackheads, bladder discomfort, blister, etc. are used (Fig. 5).

Symptoms

abdominal_pain	acidity	altered_sensorium	anxiety	back_pain	...
1	0	1	0	0	Allergy
0	0	0	1	0	Common Cold
0	1	0	0	1	Diabetes
0	0	1	0	0	...

Diseases

Fig. 5. Random Samples of Medical Dataset with Symptoms, and diseases.

This dataset comprises either 0 and 1. It is a clean dataset with no null values [5]. If the patients have symptoms listed here, then it will be 1, but if it is absent, then it will be 0. Diseases like allergy, common cold, diabetes, and so on, are recorded to predict. After predicting the disease, the recorded data will be used for estimating the medicine. The prediction will be done using the suggestion of medical professionals using a machine learning approach. Then the dataset will be used to count the occurrences of the medicines. All the collected data from different regions will be implemented differently, so that it can be easier to compare, and find out the demand from maximum to minimum.

2.5 Random Forest Classifier

Random Forest is a popular machine learning algorithm that belongs to the supervised learning technique. It can be used for both classification and regression problems in Machine Learning [6]. Random Forest classifier works by constructing a multitude of decision trees at training time, and the output is the class selected by most trees [7, 8] (Fig. 6).

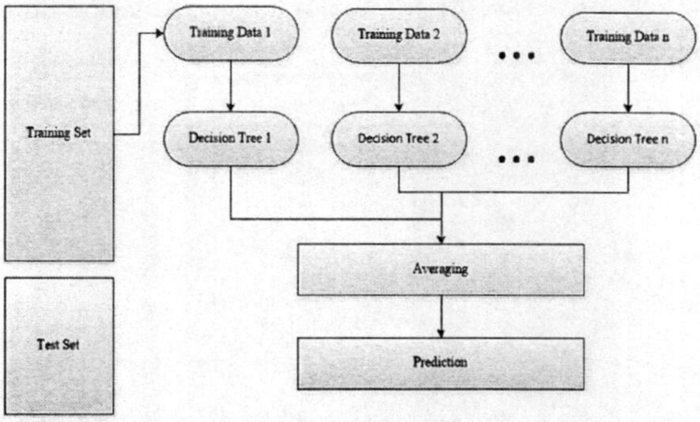

Fig. 6. A simple diagram of the random forest algorithm.

Random Forest Classifier during the time of classification often uses the Gini index. The formula is used to decide the nodes on a decision tree. It applies the class and probability to determine the Gini of each branch on a node, determining which of the branches is more likely to occur [9].

$$Gini = 1 - \sum_{i=1}^{C} (Pi)^2$$

Here, pi represents the relative frequency of the class you are observing in the dataset and c represents the number of classes [9].

2.6 Implementation

Predicting future trends provides an additional value for improved healthcare system management in today's global business trend and step forward technologies [10]. The proposed methodology requires implementation of two prediction algorithms using a random forest classifier. Python programming language is used to implement them. The program starts with importing libraries like numpy, pandas, seaborn, etc. Then these steps are taken place [11]:

- Reading the dataset;
- Splitting the data;
- Training and testing the model;
- Model building (Random Forest Classifier);
- Fitting the model;
- The creation of a function for predicting the disease.

It is necessary to check the accuracy of the model, and for that, the model will be trained on the train data, using a confusion matrix. Confusion matrix or error matrix is a widespread approach in machine learning to evaluate the quality of a classifier is to cross–classify predicted and actual decision classes [12] (Fig. 7).

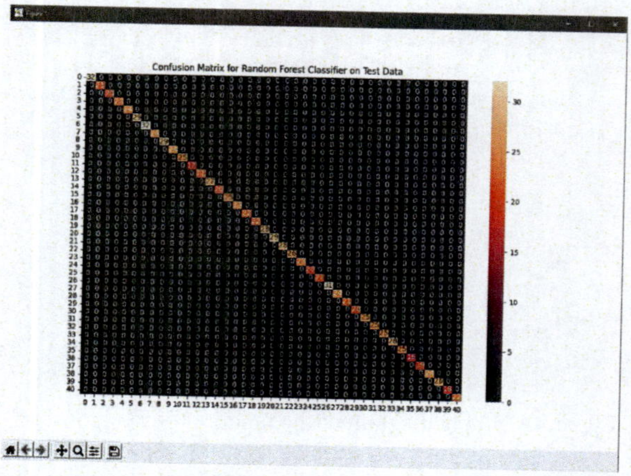

Fig. 7. Confusion matrix for random forest classifier on test data (disease prediction)

In the final part of the implementation, the function shows the prediction from the input data as 'Disease_prediction': 'Hypertension'. Then, with the collected data, prediction of medicine is implemented in a similar way (Fig. 8).

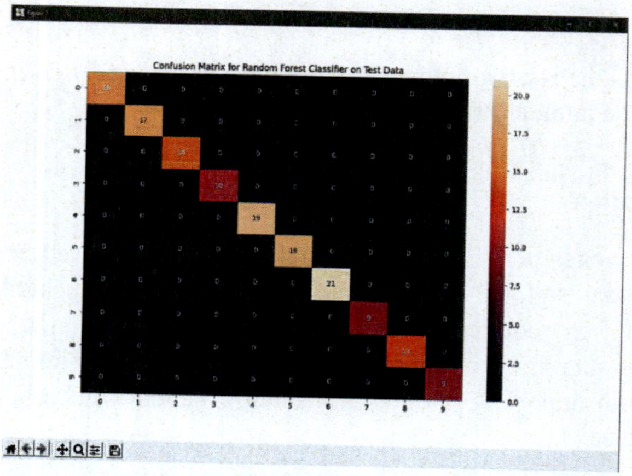

Fig. 8. Confusion matrix for random forest classifier on test data (medicine prediction)

Then, from the collected data, a function for counting medicines for different regions of Syria is implemented to find out maximum to minimum demands.

2.7 Results and Analysis

From various researches, it is evidenced that the data-driven predictive modeling with machine learning (ML) could become the cornerstone of health supply chain planning and operational management [13]. The results from the proposed models also show silver lining for those who are interested in the application of the machine learning approaches to health care needs, including the production, demand, and necessity of essential medicines.

The 2019 list of essential medicine in Syria by WHO contains a lot of medicines. For this research, ten essential medicines were selected from the implemented models based on the outcome of the predictions. From the counting occurrences function, Paracetamol shows the highest usages among the patients. Considering this in mind, the places, and the number of patients took part in this research, the following demands are predicted for different regions (Fig. 9).

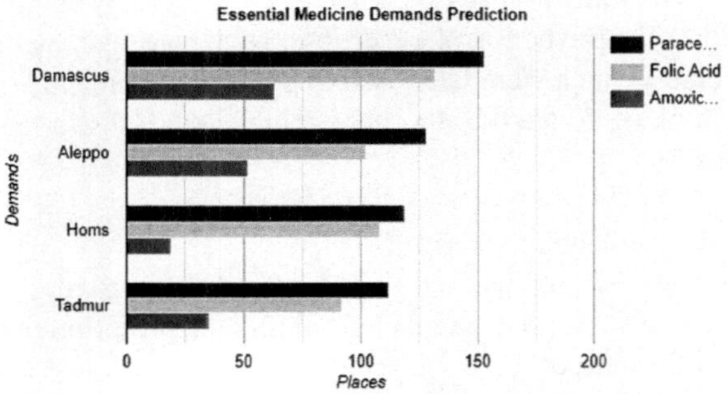

Fig. 9. Prediction of demands of some essential medicines in different regions of Syria using collected data.

To get an overall idea or predict the demand, it is necessary to be mindful about the fact that the population, weather, gender, education, and age play vital roles according to some researches. All the results and prediction shows:

- A more advanced dataset is necessary to implement the models more accurately.
- Patients should have given as much options of symptoms as possible during the time of data collection.
- More machine learning models should check the accuracy of the models, and a combined result could make the prediction additionally successful and accurate.
- The Dataset should be not only symptom based independently collected data but also with items like prescriptions, laboratory tests, details about important diseases, etc. [14].

The limitations for creating an up-to-date dataset in Syria are more complicated than in many other countries. However, there is no alternate rather than trying to make it possible for the improvement of the correctness of implementing different machine learning approaches. In a recent study, [15] there is a description of fundamental steps for preparing medical imaging data in artificial intelligence algorithm development, which explains current limitations to data correction and explores alternative approaches to address the problem of data availability. Such studies help to estimate the demand more accurately.

3 Conclusion/Future Work

When considering machine learning as a task, the task can be planned as a generalization of classical approximation tasks, since in the real-life of the applied projects, the data can be non-numeric, incomplete, and heterogeneous. [16, 17] Machine learning models are now well used in improving health care system worldwide.

The biggest hurdle in this research was data collection. If it were the case that a list of the prescriptions of patients for the past few years from different areas of Syria was available, then it would be easy to predict the demand. But the information collected was the signs and symptoms of diseases. First, it was necessary to find out the disease from the dataset, then the prediction of the drugs. From there, a dataset has been created of essential medicines, which were later used to predict the demand.

In the future, looking forward to thoroughgoing some tasks to make the proposed approach more efficient:

- Enhancing the data collecting approach,
- Understanding other research methodologies related to this topic
- Encouraging patients in Syria to share medical reports, prescriptions, and analysis in order to improve the dataset.
- Access to collecting data from all the regions of Syria,
- Using simultaneously three or four classifier models to predict more accurately,
- Developing health care applications to collect data, and so on.

The availability or right to essential drugs is not limited to those who can afford it, but is universal. Over 150 countries around the world use the World Health Organization's list of essential medicines. From which it is proved that essential medicines are very important for public health. And countries like Syria, Yemen, and other developing, and under-developed countries need to focus on more advanced, and effective health education to make sure the ultimate demands and needs of people.

If artificial intelligence (AI) technologies have a similar impact on healthcare as in other industries, then health care can become more effective and more efficient, improving the daily lives of millions of people [18]. It is a must understand that AI is not one technology, but a collection of them [19]. This is the new future, and the earlier researchers related to medicine, or any health care system, will understand this, the quicker and better the outcome will be.

References

1. Ascent of machine learning in medicine. Nat. Mater. **18**, 407 (2019). https://doi.org/10.1038/s41563-019-0360-1
2. Ahmed, Z., Mohamed, K., Zeeshan, S., Dong, X.: Artificial intelligence with multi-functional machine learning platform development for better healthcare and precision medicine. Database (Oxford). 2020 Jan 1;2020:baaa010. https://doi.org/10.1093/database/baaa010. PMID: 32185396; PMCID: PMC7078068

3. Hamsagayathri, P., Vigneshwaran, S.: Symptoms based disease prediction using machine learning techniques. In: 2021 Third International Conference on Intelligent Communication Technologies and Virtual Mobile Networks (ICICV), pp. 747–752 2021.https://doi.org/10.1109/ICICV50876.2021.9388603
4. Gomathy, C.K.: The prediction of disease using machine learning. Int. J. Sci. Res. Eng. Manage. (IJSREM) (2021)
5. Disease prediction using machine learning. https://www.geeksforgeeks.org/disease-prediction-using-machine-learning/. Last accessed 08 July 2022
6. Machine learning random forest algorithm. https://www.javatpoint.com/machine-learning-random-forest-algorithm. Last accessed 06 Aug 2022
7. Ho, T.K.: Random Decision Forests (PDF). In: Proceedings of the 3rd International Conference on Document Analysis and Recognition, Montreal, QC, 14–16. pp. 278–282
8. Ho, T.K.: The random subspace method for constructing decision forests (PDF). IEEE Trans. Pattern Anal. Mach. Intell. **20**(8):832–844. https://doi.org/10.1109/34.709601
9. Random forest algorithm for machine learning. https://medium.com/capital-one-tech/random-forest-algorithm-for-machine-learning-c4b2c8cc9feb. Last accessed 03 July 2022
10. Dash, S., Shakyawar, S.K., Sharma, M., Kaushik, S.: Big data in healthcare: management, analysis and future prospects. J. Big Data **6**, 54 (2019)
11. Disease prediction using machine learning. https://www.geeksforgeeks.org/disease-prediction-using-machine-learning/. Last accessed 05 June 2022
12. Düntsch, I., Gediga, G..: Confusion matrices and rough set data analysis. In: Published Under Licence by IOP Publishing Ltd Journal of Physics: Conference Series, (vol. 1229), 2019 3rd International Conference on Machine Vision and Information Technology (CMVIT 2019) 22–24 February 2019
13. Mbonyinshuti, F., Nkurunziza, J., Niyobuhungiro, J., Kayitare, E.: The prediction of essential medicines demand: a machine learning approach using consumption data in Rwanda. Processes **10**(1), 26. https://doi.org/10.3390/pr10010026
14. What is health data. https://datasaveslives.eu/what-is-health-data. Last accessed 05 Feb 2022
15. Willemink, M.J., Koszek, W.A., Hardell, C., Wu, J., Fleischmann, D., Harvey, H., Folio, L.R., Summers, R.M., Rubin, D.L., Lungren, M.P.: Preparing medical imaging data for machine learning. Radiology. **295**(1), 4–15 (2020). https://doi.org/10.1148/radiol.2020192224. Epub 2020 Feb 18. PMID: 32068507; PMCID: PMC7104701
16. Hastie, T., Tibshirani, R., Friedman, J.: The Elements of Statistical Learning: Data Mining, Inference, and Prediction. Springer, p 746
17. Prudius, A.A., Karpunin, A.A., Vlasov, A.I.: Analysis of machine learning methods to improve efficiency of big data processing in Industry 4.0. J. Phys. Conf. Ser. **1333**, 032065 (2019). https://doi.org/10.1088/1742-6596/1333/3/032065
18. Goldfarb, A., Teodoridis, F.: Why is AI adoption in health care lagging? Brookings 9 Mar 2022
19. Davenport, T., Kalakota, R.: The potential for artificial intelligence in healthcare. Future Healthc J. **6**(2), 94–98 (2019). https://doi.org/10.7861/futurehosp.6-2-94.PMID:31363513; PMCID:PMC6616181

Web Scanner: An Innovative Prototype for Checking Web Vulnerability

Atishay Jain, M. R. Aadithyanarayanan, Aniket Anand, Harsh Maheshwari, Sudhanshu Gonge(✉), Rahul Joshi, and Ketan Kotecha

Symbiosis Institute of Technology, Symbiosis International (Deemed) University, Pune, India
{atishay.jain.btech2020,sudhanshu.gonge,rahulj}@sitpune.edu.in, head@scaai.siu.edu.in

Abstract. With growing concerns about network security, various ways to protect your network from unauthorized access have been proposed. A new style has been proposed to reveal implicit discrepancies that can harm the network. Vulnerability analysis is the most commonly used method. Vulnerability is the hidden weakness of a system that makes it vulnerable to attack. By assessing the vulnerabilities of these systems, you can identify and develop new ways to protect your system from destruction. The impact of the vulnerability makes online security less humorous, so it is necessary to consider in advance how to mitigate its destructive impact. To overcome these issues, web security scanners are among the most essential tools for assessing the security of your website that are proposed in this research work. It allows regular stoners to identify vulnerabilities by detecting website vulnerabilities such as (i) Cross-site scripting (XSS), (ii) Cross-site request forgery (CSRF), (iii) Distributed denial of service (DDoS), and (iv) SQL injection. A tool that can be used to indicate. It also checks for various data manipulation viz. (i) URL manipulation, (ii) Session expiration, (iii) Privilege escalation, and (iv) Cookie-based attacks. It provides information about all open ports. This experiment carried out in this research work focuses on how our application can scan common and advanced vulnerabilities, that could be present on the website and show them to the user with minute detail of every attack. The user is given a security score displayed on the screen showing how trustworthy the website is with all the information the user might provide to the website.

Keywords: Vulnerabilities · Security · Web · Website · Scanner

1 Introduction

With advances in information technology, user's data safety has become a more urgent concern. Most software developers are unaware of the various security measures that need to be implemented in their systems. The aim of this software is used to consider the applications and find the bugs that programming languages may introduce into their systems. It is much more important to develop new strategies and methods. This will help to protect users from the risk of attacks from unauthorized access. Not only does

R. Silhavy et al. (Eds.): CoMeSySo 2022, LNNS 596, pp. 680–691, 2023.
https://doi.org/10.1007/978-3-031-21435-6_58

software designed with excrescencies render stoners exposed to assaults, but networks frequently become a critical component by jeopardizing the druggies' security.

In today's world, an automated scanner to check for website faults and vulnerabilities is critical. A web tester evaluates a website to find any flaws [1]. Black-box testing [2, 3] refers to testing performed by scanning tools without knowing the source code. The efficacy of each scanner should be tested by evaluating these scanners. The purpose of this testing is similar to an attacker's conduct based on the fazing technique; the code stored within the program is not one of the prerequisites for doing black-hat testing. Specific data is generated for each type of attack in this technique, and the relevant results are monitored for vulnerabilities, which is done automatically by our web scanner applications. The hackers search the web for vulnerabilities that they can exploit to launch any type of attack, such as taking control of the user account, gaining access to sensitive data, or injecting malicious code into the user input field, so this issue should be continuously evaluated in order to reduce its detrimental influence [4, 5]. Not only does software created with excrescencies render the user exposed to assaults, but continuously changing networks also play a role in jeopardizing stoners' security.

The knowledge and a comprehensive grasp of the vulnerabilities are required for assessing and removing them [6, 7]. The basic idea knows behind these vulnerabilities. There may be lots of problems that need to be solved such as, (i) The appearance of the data in the system, (ii) Rectify the vulnerabilities that need to be corrected to make the system free, and (iii) Need to find alternatives solution for these vulnerabilities [6–9].

A variety of methodologies have been used to detect these vulnerabilities, and necessary measures have been adopted. Only a few examples include static analysis, attack graph development and analysis, and the usage of vulnerability scanners. Today, however, vulnerability scanners are widely used to discover flaws [10–12]. They are extremely important in the creation of attack graphs.

As cyberattacks are increasing in the modern era, we have built a website that will check the vulnerabilities of other websites so that the user can ensure the security of their valuable data. Our web-scanner will take the URL of other websites and will detect different vulnerabilities to which the website is vulnerable, and it will also give a score to the website based on their securities. By using, scanner users can make a better judgment about whether to trust the website with their crucial and valuable information or not.

2 Motivation

Due to the constant growth of Internet services, usage, speed, user-friendly browsing, worldwide access, and so on, almost every aspect of human existence is impacted by the Internet." On the other hand, hackers are using new strategies and ways to assault the digital world by exploiting web application weaknesses [13]. Many apps are going online, but how safe these goods are a problem since it is tied to the security of the user who will be utilizing the app. As a result, it becomes vital to identify vulnerabilities in websites that may pose a serious threat to the user's security [8].

The examination of these vulnerabilities is critical in the aim of securing the social digital environment. It may be accomplished in two ways. "First, manual analysis is error-prone because of the human nature of forgiveness, dynamic change in technology, and fraudulence attack tactics. Second, conventional web application vulnerability scanners may generate false alert rates from time to time. As a result, a framework to identify several layers of vulnerabilities, spanning from client-side vulnerabilities to communication-side vulnerabilities to server-side vulnerabilities, is required" [18]. As a result, we designed a web application that checks for these vulnerabilities on the website and informs the user.

3 Related Work

Several studies on the crawling coverage and detection performance of web application vulnerability scanners have been published [3]. More than 10 scanners were examined, with the major findings being that good crawler coverage is critical for spotting any vulnerabilities and that free scanners work just as well as commercial scanners [9]. Which covers over 50 free and commercial scanners, confirms the same.

Chen's [11], which covers roughly 20 scanners and is updated regularly, and El Idrissi et al. [12] which evaluates 11 scanners, both come to the same conclusion that scanners can perform as well as commercial scanners. Suto states in [5] that while properly training or configuring a scanner improves detection performance, it also significantly increases the complexity and time effort required to operate it. Bau et al. also show that the eight scanners they utilized in their study have varied strengths, i.e., they find distinct vulnerabilities [6].

Vega et al. [7], who compare the vulnerabilities found by the four scanners in their review with alarms generated by the intrusion detection system (IDS) Snort [13], validate this. According to Qasaimeh et al., the five scanners they tested perform differently not just in terms of the number of findings discovered, but also in terms of the frequency of false positives [8]. Smaller investigations with two and three scanners were published in [9, 10], respectively, confirming that various scanners have variable detection strengths. Other projects focused on expanding the scope of vulnerability scanning. Crawling and therefore scanning speed may be enhanced by taking into account the state changes of a web application as it processes requests, as illustrated in [14]. Van Deursen et al. introduce Crawljax, a Selenium WebDriver-based crawler that enhances the Rawling of Ajax-based online applications in [15]. Pellegrino et al. accomplish the same result by dynamically evaluating JavaScript code in web pages [16]. In [17], Zulla explores ways to enhance web vulnerability detection in general, including methods for detecting login forms on websites automatically.

Figure 1 shows us at what percentages the fields contribute to data compromise and which fields are affected the most. It is no surprise to us that the computer science and engineering fields cover more than half of the pie with a whopping 58.9%. These fields are most affected by the compromise of important information through the vulnerabilities of the websites and other aspects to the hackers. These fields contain thresholds of

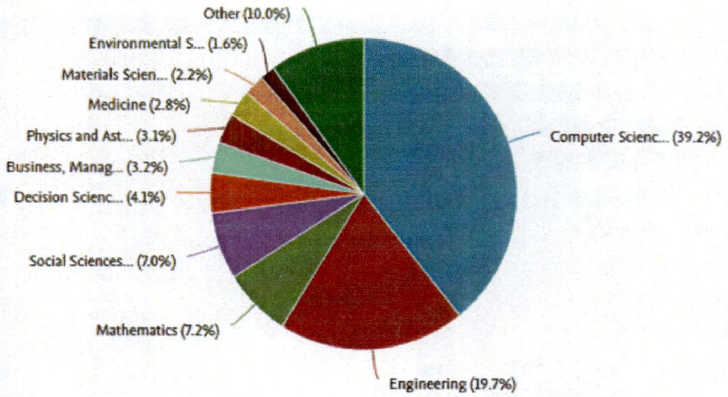

Fig. 1. Data compromise according to the different streams.

information that are no less than a gold mine and hence are widely targeted. Other than computer science and engineering fields there are mathematics, social science, and business and management fields which are some other hotspots for compromise of data. The above pie chart covers this distribution accurately [19, 20].

Our research expands on prior findings, such as the fact that free scanners function similarly to commercial scanners, that various scanners have varying strengths in terms of detection skills, and that crawling coverage is critical for detecting any vulnerabilities. However, our job extends much beyond what is now being done. To begin, the supplied solution - JARVIS – covers not just crawling coverage, but also the trustworthiness of authenticated scans has a big impact on the number of vulnerabilities that may be found. JARVIS is also scanner agnostic, meaning it may be used with nearly any vulnerability scanner currently available [21, 22].

We also conduct a thorough assessment using a number of scanners and test applications to demonstrate the benefits and feasibility of our technical solutions. Finally, to our knowledge, this is the first study to evaluate the advantages and disadvantages of combining several scanners. Fakhreldeen A. evaluates several web scanners based on the OWASP Top 15-2015 application security concerns; the developer uses this evaluation to select the optimal scanner for each application [1]. Sneha Parmar (2018) et al. in [2] employs a specialized scanning method using specific-source tools, which improves detection accuracy over utilizing only one scanning tool. Web vulnerability scanners may cover practically every area of support using several types of scanners.

The distribution of how much the state nations donate to web scanning and vulnerability scanning technologies is shown in Fig. 2. With its technical breakthroughs, the United States of America is the clear leader above the others, followed by China and India. These nations are heavily involved in the development and upkeep of the vast majority of online scanning technologies available to consumers. India and China are not far behind the United Kingdom and Germany.

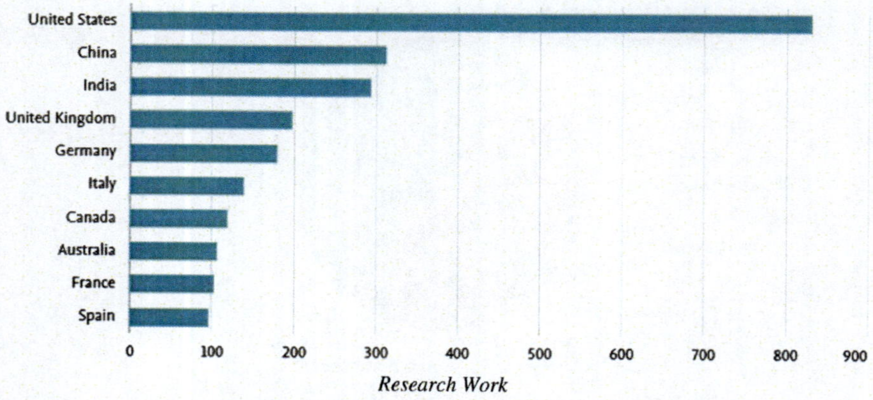

Fig. 2. Top country-by-country research on web vulnerability

David A. et al. (2014) [6] Shelly uses two testbeds, one with vulnerabilities and the other without, to assess the performance of commercialized and open platform web scanners, and propose a unique technique to decrease the number of false positives and false-negative attacks.

Yuliana M. et al. (2015) in [7] gives an overview of the results of running Qualys Guard WAS and Acunetix WVS against a specific testbed. WAVS can detect and assess the efficiency of various penetration testing methods by identifying the most difficult vulnerabilities.

Xiaowei Li and YuanXue et al. (2011) in [8] compare each request's answer often on the web to assess the behaviormodelsodel of the apps. This methodology includes two phases: the first phase involves identifying attacks and the second phase involves evaluating each request and response to look for deviations.

Mikko Vimpari et al. (2019) in [9] employs the Choosing by Advantage technique of qualitative research (CBA). It was based on typical advantages and disadvantages derived from an analytical investigation to assess each one.

4 Problem Definition and Objectives

A) *Problem Definition*

To check for common and advanced vulnerabilities through our web application Web scanner and provide users with all details about the vulnerability found.

To identify the vulnerabilities of the website for the purpose of improving the security features and creating a secure website. Web security is necessary to prevent hackers and cyber thieves from gaining access to sensitive material containing personal information. The attack is performed on the website which impacts both the company and the website users (clients, consumers, employees). As it relates to users themselves, risks are also numerous: Identity theft (both on the Internet and in the "real world") Personal data theft, including banking data and email.

B) *Scope and Objectives*

i) ***Scope***

- It gives information about the vulnerabilities by analyzing all the necessary security issues.
- It doesn't entirely replace the existing system but it mostly automizes the scanning process and all the data used.
- It gives the user all the necessary privileges to access and modify data intended for them.

ii) ***Objectives***

- It gives the user all the necessary privileges to access and modify data intended for them.
- Create awareness among users about whether to trust a website or not with its valuable information.
- To provide information about all the common and advanced attacks on the same platform.
- The organization hosting the Website can know where they can improve to better secure their website.

C) ***Hardware and Software Requirements***

i) ***Software Requirement***

1) Platform: Python.
2) Technology: PyQt5, Python 3.6, JSON, and Above.
3) IDE: PyCharm, QT designer.
4) Database: MYSQL 5.5.
5) Comm API (QtCore, QtGui, QtWidgets), BeautifulSoup, os, whois, and Bane Library.

ii) ***Hardware Requirement***

Hardware must be-
System of the following configuration

1) Processor: Dual Core of 2.2 GHz.
2) Windows: 7, 10, 11 and above.
3) Hard Disc: 201 GB.
4) RAM: 2GB.
5) Monitor, Keyboard, Mouse.
6) Internet connection.

5 Flowchart and Algorithm

A) **Flowchart** (Fig. 3)

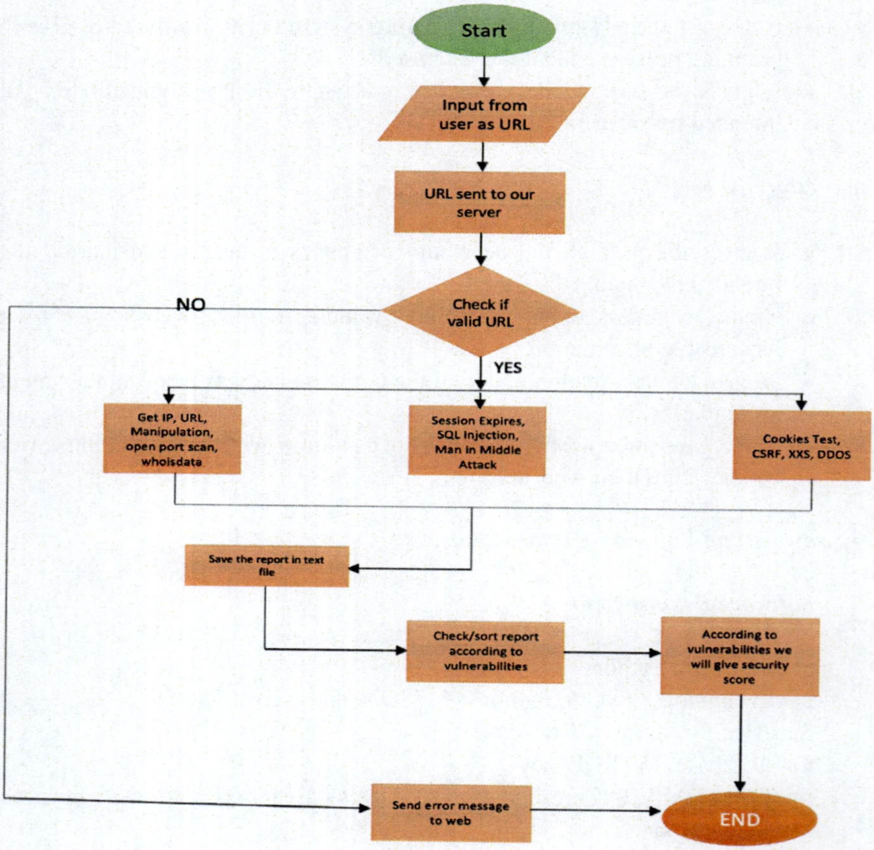

Fig. 3. Flowchart

B) **Algorithms**

Algorithm: WebScanner to scan vulnerabilities in websites
// Input: URL.
 // Output: Vulnerabilities of a website with security score.
 Step 1. Start.
 Step 2. Enter URL in the TextField and click on the scan website button.
 Step 3. The URL is sent to the server.

Step 4. Check if the URL is valid.

Step 5. If valid:

5.1 Get IP address, scan port, and whois data.

5.2 Scan for vulnerabilities like XXS, DDOS, CSRF, etc.

5.3 Save the report in a text file.

5.4 Report is sorted according to vulnerabilities find through the scan process i.e., 5.2.

5.5 Security scores and vulnerabilities are shown on the interface.

5.7 End.

Step 6. If invalid:

6.1 Sends the error message showing URL is not Valid.

6.2 End.

Step 7. Stop.

The proposed Methodology can be depicted below mentioned. Different libraries are imported into the main file which contains different types of vulnerability tests body part of the main.py file shows how the imported libraries are implemented. The user interface is created by our application Web scanner.

6 Results and Discussion

The proposed system is built on a python-based windows machine with NetBeans and PyCharm as IDE. The suggested approach is put through a series of tests to determine the system's legitimacy, as detailed below. The system is tested using both common and rare URLs. Different types of URL sites had been tested by our application as shown below.

Fig. 4. WebScanner Interface

The text field in the interface is typed with the URL "google.com" shown in Fig. 5. It is necessary for the user to enter the website name along with the extension for the Web

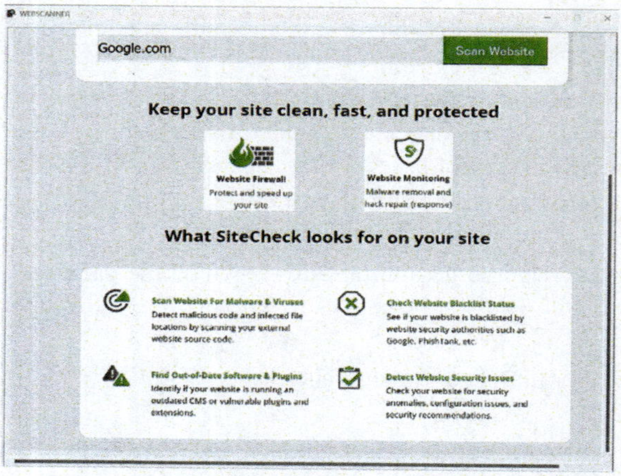

Fig. 5. WebScanner Interface

Scanner to work properly which is also mentioned in the interface. After entering the URL, the user is just a click away just a click away from getting all the security-related information about the website. As soon as the user clicks on the scan website button, a background process starts and after a while, a screen appears showing the website security score shown in Fig. 6 (a). At the top, the name of the website is displayed, and just below it shows information related to website monitoring, the website being blacklisted, and the website's firewall. Website IP address, domain name, and the server are also displayed. Figure 6 (b) gives details information on the vulnerabilities. It shows the links found after URL manipulation, all the open ports present, and details on the website. SQL injectable is also indicated. In our example google is not SQL injectable. Figure 6 (c) reveals the who's data, website blacklist status according to different websites, and website malware. The UI clearly indicates all the vulnerabilities and gives the user a clear picture of whether to trust the website or not giving all the specifies.

The Fig. 4 and Fig. 5 images show the interface, where the user can enter the URL. The Fig. 6 (a), The Fig. 6 (b) and The Fig. 6 (c) images are showing the security score and provides details about different vulnerabilities present on the website.

7 Conclusion

To list the vulnerabilities in online applications or remote hosts, many ways may be employed. To secure the network system, vulnerability assessment is crucial. Various scanners identify different sorts of vulnerabilities, according to our findings, but no one instrument can detect all types of vulnerabilities. Thus, we have created our web application which can perform almost all such types of vulnerabilities tested under a single application.

Due to their significant influence on the functional functioning of any website, this investigation and review of Web application vulnerability scanners focused on many common and complex assaults including SQL and XSS. It can be concluded from the

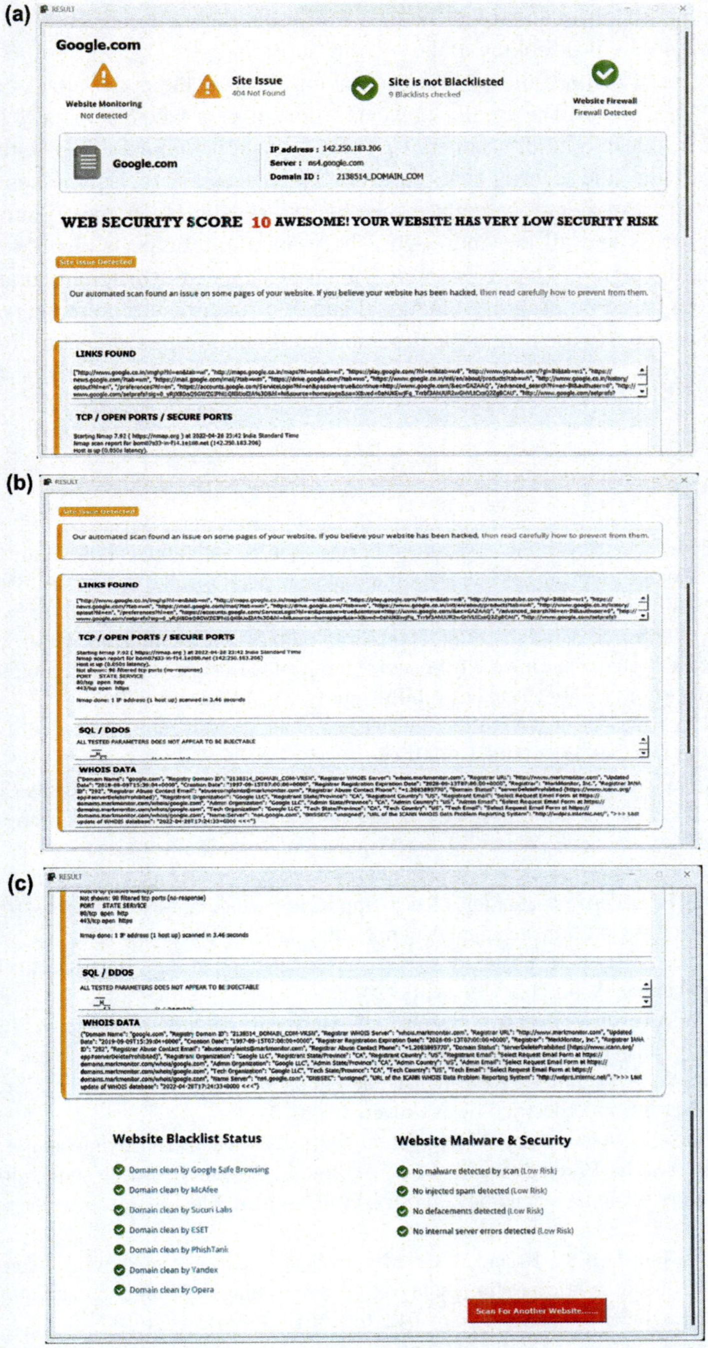

Fig. 6. Result with Web Security Score, b: Result with Web Security Score, c: Result with Web Security Score

results that our scanner can provide adequate information to the user on whether to trust a website or not by just looking at the security score.

In general, it can see the use case of having most of the cyber-attacks under one platform. As a result, if a developer wants to examine a website for any intentional vulnerability, our one in all scanners will check for every vulnerability without much hassle. It provides the security analysis with all the necessary security issues to prevent the hours. It doesn't entirely replace the existing system but it mostly atomizes the scanning process and all the data used. The website prototype is also developed by considering the eight golden rules of HCI. It also includes the different parameters for its evaluation based on usability functional and non-functional requirements for users. This will help users to use it multiple times to identify the various vulnerabilities coming from the website. This will prevent the malicious passive attack which is performed by hackers.

References

1. Fakhreldeen, A., Eltyeb E.: Assessment of open-source web application security scanners. In: College of Computer Science and Information Technology, KAU, Khulais, Saudi Arabia, March 2017
2. Pakorn, I.: A comparative study of security vulnerabilities in responsive web design framework. Malardalen University School of Innovation Design and Engineering, June (2018)
3. Sneha, P.: Vulnerability Checker for Info security. SRM University (2016)
4. Fakhreldeen, A.: Using WASSEC to evaluate commercial web application security scanners. Int. J. Soft Comput. Eng. (IJSCE) (2017)
5. Kinnaird, M.: Open-Source web vulnerability scanners. Marymount University (2016)
6. Shelly, D.A.: Using a web server test bed to analyze the limitations of web application vulnerability scanners. Faculty of the Virginia Polytechnic Institute and State University (2014)
7. Yuliana, M.: Security evaluation of web application vulnerability scanners' strengths and limitations using custom web application. California State University (2015)
8. Li, X., Xue, Y.: BLOCK: a black-box approach for detection of state violation attacks towards web applications. Vanderbilt University (2011)
9. Mikko, V.: An evaluation of free fuzzing tools. University of Oulu Department of Information Processing (2019)
10. Qianqian, W., Xiangjun, L.: Research and design on Web application vulnerability scanning service. North China Electric Power University (2014)
11. Mohammed, R.: Assessment of web scanner tools. Al-Mustansiriyah University (2018)
12. Wang, B., Liu, L., Li, F., Zhang, J., Chen, T., Zou, Z.: Research on web application security vulnerability scanning technology. In: State Grid Xinjiang Electric Power Research Institute (2020)
13. Shar, L.S. Tan, H.B.K., Briand, L.C.: Mining SQL injection and cross-site scripting vulnerabilities using hybrid program analysis. In: Proceedings of International Conference on Software Engineering (ICSE '13), pp 642–651. IEEE Press (2013)
14. Li, Y., Wang, Z., Guo, T.: Reflected XSS vulnerability analysis. Int. Res. J. Comput. Sci. Inf. Syst. (IRJCSIS) **2**, 25–33 (2013)
15. Shar, L.K., Tan, H.B.K.: Automated removal of cross-site scripting vulnerabilities in web applications. Inf. Softw. Technol. **54**, 467–478 (2012)

16. Haixia, Y., Zhihong, N.: A database security testing scheme of web application. In: 4th International Conference on Computer Science and Education, pp. 953–955. IEEE (2009)
17. Junjin, M.: An approach for SQL injection vulnerability detection. In: 2009 Sixth International Conference on Information Technology: New Generations IEEE, pp. 1411–1414
18. Marashdih Abdalla Wasef, ZaabaZarulFitri Cross-Site Scripting Detection Approaches in Web Application. Int. J. Adv. Comput. Sci. Appl. **7**(10), 155–160 (2016)
19. Park, Y.J., Park, J.C.: Web application intrusion detection system for input validation attack. In: Third 2008 International Conference on Convergence and Hybrid Information Technology, IEEE, PP 498–504
20. Avancini, A., Bruno, F.K.: Security testing of web applications: a research plan. In: IEEEICSE '12, Proceedings of the 34th International Conference on Software Engineering 2012, Zurich, Switzerland, pp. 1491–1494
21. Prokhorenko, V., Choo, K.-K.R., Ashman, H.: Web application protection techniques: a taxonomy. J. Netw. Comput. Appl. **60**, 95–112 (2016)
22. Panda, S, Ramani, S.: Protection of web application against SQL injection attacks. Int. J. Mod. Eng. Res. (IJMER) **3**(1), 166–168 (2013). www.ijmer.com. ISSN: 2249-6645

Analysis of the Performance of NewSQL Databases Compared to RDBMS Based on Linux OS

Sarah Myriam Lydia Hahn(✉), Ionela Chereja, and Oliviu Matei

Computers and Information Technology, Technical University Cluj Napoca, Cluj-Napoca, Romania
sarah-hahn@gmx.net , oliviu.matei@holisun.com

Abstract. Besides the well-known relational database management systems (RDBMS) a new kind of databases called NewSQL databases were developed. NewSQL databases aim to combine the advantages of RDBMS and NoSQL databases. In this article the performance of the most popular RDBMS and NewSQL databases is measured and compared to each other. The objective is to measure whether there is a improve of performance for NewSQL databases as promised. To get comparable results a data set was prepared as well as homogeneous test settings were defined. Furthermore KPIs for the measurement were specified. The outcome of the performance tests are displayed and explained. Conclusions as well as recommended actions are made.

Keywords: NewSQL · RDBMS · Performance · Linux · Cloud technology · Test series · Distributed databases · Big data · Latency · Execution time

1 Introduction

In the 1960s the first hierarchical and network database management systems (DBMS) were developed. Applications using these systems were much quicker and had cheaper costs because the data had not to be handled in a stand-alone application [6]. DBMS were followed by two models - the hierarchical model by an Information Management Systems (IMS) and the network model from Conference on Data Systems Languages (CODASYL) [11]. CODASYL was a community with the objective to develop a standard programming language for all systems [14]. The first approaches were not totally satisfying for everyone. From time to time new kinds of DBMS were developed to compensate the disadvantages of the existing databases. This article focuses in the following on NewSQL databases. Only DBMS which are relevant for the history of NewSQL databases are shortly described in the variation of time.

R. Silhavy et al. (Eds.): CoMeSySo 2022, LNNS 596, pp. 692–704, 2023.
https://doi.org/10.1007/978-3-031-21435-6_59

The first databases were overtaken by RDBMS. One of the most known article about relational databases is from the year 1970 “A Relational Model of Data for Large Shared Data Banks” written by. Edgar F. Codd. In this work the relational model of a database was first mentioned and defined [12]. A database is well-structured storage of data. There are two types of data in a database. The information itself and metadata - both have to be stored [32]. RDBMS are defined by having various tables with data in the storage which can be manipulated by updates [6]. DBMS is a software which supports the user e.g. in the case of maintenance, access and data queering [32].

Caused by new infrastructural developments, especially in the cloud environment, there are new needs and use cases concerning DBMS. To fulfill these new requirements NoSQL databases were developed. Their wording descends from “not only SQL”. It describes the missing attributes and SQL compared to RDBMS [23]. New main features of these kind of databases are data division and duplication for scaling in the field of cloud computing [25]. Furthermore disadvantages from RDBMS were eliminated such as a schema, difficulty with the handling of mass data and expensive hardware costs to have an adequate performance [23,31].

However also NoSQL databases have disadvantages like the missing consistency and the non existing Structured Query Language (SQL). That is the reason why NewSQL databases were developed. The intention of this kind of databases is the combination of the advantages of RDBMS and NoSQL databases [15,30]. It has to be questioned whether these kind of databases really only have advantages and if it is overall better choice than the well-known RDBMS and NoSQL databases.

Based on DB-Engine Ranking, a popular ranking for DBMS, relational ones are at the front. The first four databases are RDBMS (Oracle, MySQL, Microsoft SQL Server, PostgreSQL). These are followed by a few NoSQL databases. The first NewSQL database in the ranking is CockroachDB at place 58. There are six parameters for the calculation of the popularity - number of search results (Google, Bing), search frequency (Google Trends), frequency of technical discussions (Stack Overflow, DBA Stack Exchange), number of job offers (Indeed, Simply Hired), number of LinkedIn profiles and number of Twitter tweets [16]. The result can be interpreted that NewSQL are not widespread because they did not achieve their objective to combine only advantages. Nevertheless it could be possible that they are not so popular and there is still a big potential of optimization of software.

There are already articles available regarding the properties of NoSQL and NewSQL databases as well as the usage of these for transformation tools and analytical processes [9,10,19,21]. An important characteristic for a decision is the performance of a database. The performance of the most popular NewSQL databases was measured in the past [20]. The key question if the performance of NewSQL databases is really better than the performance of the traditional RDBMS is not answered yet. There is no article which compares the performance of the most popular NewSQL to the most popular RDBMS. Such a performance test and analysis is made in this article.

The performance tests are limited to Linux operating systems (OS). The most popular RDBMS and NewSQL databases in the field of Linux OS are introduced. It can be used as a market survey and supports he decision-making of a DBMS. This overview as well as important definitions are part of the chapter "related work and explanations". As the central part of the article a performance test is designed and executed. The design including the performance test and the test setting can be found in the section "methodology". In the following the results are shown in "performance test results". Last but not least the "conclusion" of the test is presented.

2 Related Work and Explanations

This section explains related work and explanations for the following performance test.

2.1 RDBMS

Edgar F. Codd introduces twelve rules which defines a RDBMS [13]:

1. Information rule: All information is only represented as values in tables. Also column or table names are values in metadata tables.
2. Guaranteed Access Rule: Every single value can be logically accessed by ordering table name, primary key value and column name.
3. Systematic treatment of missing values: Empty values, also called null values, are supported to store.
4. Dynamic on-line catalog based on the relational model: The database description is also stored as values in a table. This means that accessing data and database description can be done in the same way.
5. Comprehensive Data sublanguage Rule: There is at least one well-defined language for accessing data and view definitions, data manipulation, integrity constraints, authorization and transaction boundaries.
6. View Updating Rule: A view can be updated, deleted, transformed - in fact the data based on it is modified.
7. High-level Insert, Update, and Delete: Each of data retrieval, insertion, update and deletion has to be possibly made in a single operation.
8. Physical Data Independence: By a change of access or physical resources the RDBMS is still the same with its functions.
9. Logical Data Independence: The physical and the logical level are independent. There is no need to have a look at the hardware performance but on the logic.
10. Integrity Independence: Integrity rules are definable in the given language (rule 5) and are stored in the metadata.
11. Distribution Independence: The sublanguage can be used independently of possible distribution of data.
12. Non-Subversion Rule: A low-level language cannot fulfill all of the requirements of high-level language.

2.2 NewSQL Databases

Matthew Asslet mentioned NewSQL databases first in an article in 2011 [2]. These databases in particular integrate the scalability including high performance and availability with the ACID model [4,24,30]. The acronym ACID is put together from the properties atomicity, consistency, isolation, durability. A transaction is a request which adheres these four properties [18].

NewSQL databases are especially made for online transaction processing (OLTP) workloads like an electronic cash register. Therefore models which are known from NoSQL databases like column-oriented data model, distributed nodes or in-memory processing are used [29].

Last but not least the well known query language from RDBMS SQL is supported by NewSQL databases. This language is very common and it is not necessary to learn a new language with this new kind of database. A parallel execution of read and write transactions is ensured with a non-locking concurrency control mechanism [4,24].

2.3 Performance

During the decision making process regarding a DBMS the performance is one of the properties which should be included in the evaluation. The performance of a data integration process can be measured with different indicators such as [33]:

- Time efficiency: grade of low response and processing times
- Resource utilization: resources utilized by the process differentiate between the type of resource and the quantity
- Capacity: demand that can be executed without missing the other requirements
- Modes: modes of data processing (batch, real-time etc.).

To have a comparable result all the settings regarding hardware and data sets are the same. The only different indicator is time efficiency. In this environment time efficiency goes along with latency. The test focuses on the standard latency regarding a DBMS:

- Write latency: the time it takes to insert a number of records into the database
- Update latency: the time it takes to update a number of records into the database
- Read latency: the time it takes to read a number of records from the database.

3 Methodology

3.1 Databases

For the performance test the most popular databases were chosen. DB-Engine Ranking offers a ranking especially for RDBMS. The most popular ones are Oracle, MySQL, Microsoft SQL Server, PostgreSQL. The fifth rank has a popularity at about a quarter of the fourth rank and is not of note [16].

For NewSQL databases there is no separate ranking at DB-Engine Ranking. For that reason a research of existing scientific work (including, but not only [1,7,8,17,22,24,26,27]) was made. The authors mention the following NewSQL databases: Actor DB, Altibase, Apache Trafodion, Citus, CockroachDB, Drizzle, Faircomdb (former C-TreeACE), FaunaDB, GemFire XD, Google Spanner, GridGain IMDB, Harper DB, MariaDB Xpand (former ClusterixDB), NuoDB, Singlestore (former MemSQL), TIBCO Active Spaces, TiDB, TokuDB, Vitess and VoltDB.

For the relevance of the performance test only DBMS which are still operative are of note. Furthermore the performance test is limited to Linux OS and the systems have to fit into the test setting. Another requirement for the test setting is equal to Linux OS which is an open-source OS that the database also has at minimum a free version or is an open-source database. The databases Citus, CockroachDB, Faircomdb, GridGain IMDB, Harper DB, TIBCO Active Spaces, TiDB, Vitess and VoltDB comply with these requirements. Out of these the four most popular ones are CockroachDB, TiDB, VoltDB and GridGain IMDB [16]. From the RDBMS Oracle does not fit in the test setting because in the actual version there is no free version available.

There was no optimization of the DBMS regarding this use case - the standard configuration was used which is set by the producer for the first tests. But of course the databases can be optimized for real use cases and they should be. In the case of the performance test the overall performance is measured and not a real use case is set up. Furthermore the graphical user interface or terminal which came along with the installation was used.

3.2 Data Set

To benchmark the results of the performance test it is important that the setting and the data sets are equivalent. For all the test series the same data set was used. The data set contains weather data from 2008 to 2020 from different Romanian cities which was collected for predicting soil moisture [3,5,28]. Each city has its own observation station with different metrics - temperature in different heights, soil temperature, moisture and/or rainfall. Each observation station also tracks whether the coupler is attached and the status of the battery. The data is collected as .hobo or .dtf files. The files were converted to .csv files with the software HOBOware. These files were converted into SQL statements for the insert, update and select. For the different databases different SQL dialects were used.

Table 1 shows the content of the data set with its columns and the regarding data type 1.

The 13 collected years of data are together over 12 million of data rows.

3.3 Test Settings

For comparability reasons virtual machines (VMs) for the test settings were used to have identical conditions. The VMs were set up with the virtualization soft-

Table 1. Data set columns

Column name	Data type	Length
id	integer	
bad_battery	Text	100
batt_v	Decimal	
coupler_attached	Text	100
coupler_detached	Text	100
date	Text	100
end_of_file	Text	100
event_units	Decimal	
ext__line_event_units	Decimal	
good_battery	Text	100
host_connected	Text	100
index	Integer	
moisture_m_m	Decimal	
moisture_m_m_2	Decimal	
rainfall	Decimal	
stopped	Text	100
Temp_C	Decimal	
Temp_F	Decimal	
Temp_F_1	Decimal	
Temp_F_2	Decimal	
Temp_F_3	Decimal	
Temp_F_4	Decimal	
Temp_F_5	Decimal	
Temp_F_6	Decimal	
Temp_F_soil_1	Decimal	
Temp_F_soil_2	Decimal	
Temp_F_soil_3	Decimal	
time	Text	100
timestamp	Text	100
table_name	Text	255
city	Text	255

ware from Oracle VirtualBox 6.1.18. Each of them has the same configuration. A separate VM for each database was set up to avoid any dependencies. Each machine is a single node cluster. The VMs are based on a desktop PC for having the same physical hardware beneath:

- Processor: AMD Ryzen 3 2200U, 2500 MHz
- CPU: 2 physical cores
- RAM: 8 GB (4 GB to virtual machine)
- Disk: 50 GB to virtual machine.

The test is based on a system running on Linux OS. Nevertheless not every database supports the same OS. Table 2 shows which database goes along with which OS in tests. The objective was to use the same Linux OS whenever possible.

Table 2. Linux OS per NewSQL DBMS

DBMS	Linux OS
CockroachDB	Ubuntu 20.4 LTS
GridGain IMDB	RedHat 8.3
Microsoft SQL Server	Ubuntu 18.4 LTS
MySQL	Ubuntu 18.4 LTS
PostgreSQL	Ubuntu 18.4 LTS
TiDB	RedHat 7.9
VoltDB	Ubuntu 18.4 LTS

3.4 KPIs

The latency is measured for write, update and read operations. For each of them a SQL query is executed and the result time is stopped.

- Write operation: One separate SQL statement for the insertion of each of x records.
- Update operation: One separate SQL statement for the update of each of x records.
- Read operation: One separate SQL statement for the reading of each of x records.
- Read operation total: One SQL statement for the reading of all of x records.

Each test is set up as one SQL transaction and performed for at least five consecutive times. By taking the average execution time this ensures that outlier are removed. The number of records are defined in steps of 1.000. The starting point is set at 1.000 leading up to 10.000. In total this accumulates per database to four different SQL statements multiplied with ten different amount of records multiplied with five executions.

4 Performance Test Results

With the used setting it was not possible to execute any of the performance test series with TiDB. The database should have been part of the tests. Since there are no results it is shown with zero/zero in each diagram.

The first diagram Fig. 1 displays the results of the write latency. It can be seen clearly that CockroachDB has a bad latency compared to the other databases. In fact it was not possible to execute the tests above 4000 records.

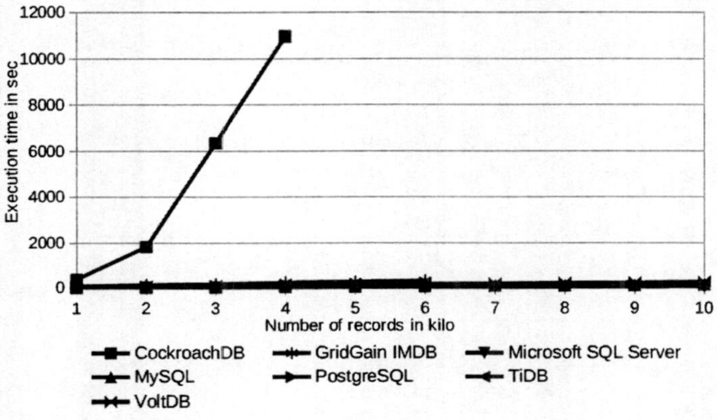

Fig. 1. Write latency

To have a closer look at the results of the other database a detail view is shown in Fig. 2. The NewSQL databases GridGain IMDB and VoltDB do poorly compared to RDBMS. Especially GridGain IMDB got no results over 6000 records. The RDBMS with the best write latency is Microsoft SQL Server followed by PostgreSQL and MySQL.

Figure 3 represents the performance of updates. Again CockroachDB has the poorest performance and it was not possible to execute the tests with more than 4000 records.

The detail view in Fig. 4 shows that the best performance has PostgreSQL. The next databases are VoltDB, Microsoft SQL Server and MySQL. GridGain IMDB is far behind and again it was not possible to execute the update statement with more than 6000 records.

Also the read latency in Fig. 5 shows as their slowest database CockroachDB with a execution stop at over 4000 records.

Taking a closer look at the other databases in Fig. 6 the same result can be seen for the update latency. PostgreSQL has the lowest execution time followed by VoltDB, Microsoft SQL Server and MySQL. Again GridGain IMDB is at the last place.

Last but not least the total read latency is displayed in Fig. 7. Compared to the other latencies MySQL and GridGain IMDB are the slowest databases. CockroachDB is in the middle.

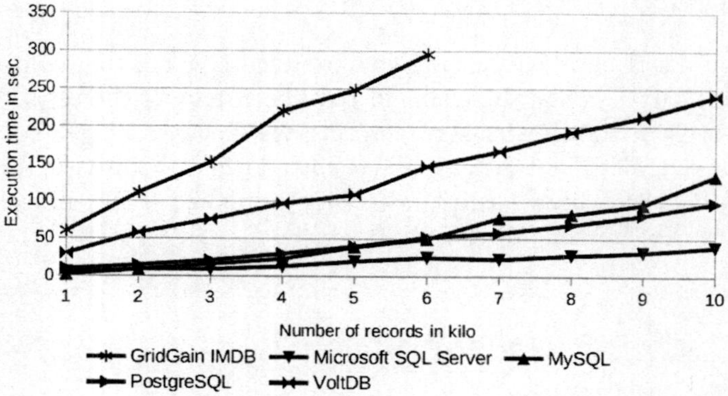

Fig. 2. Write latency in detail

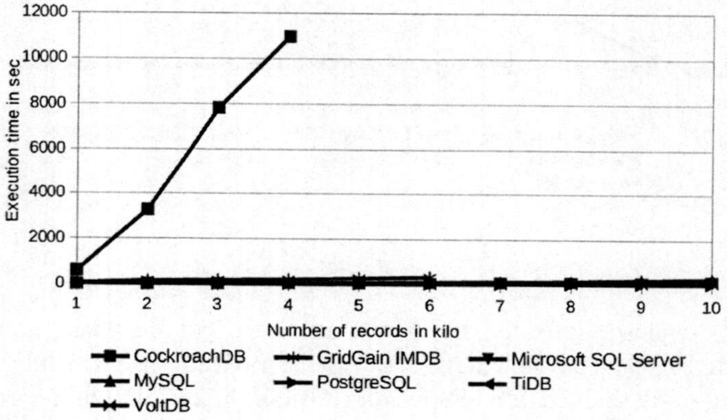

Fig. 3. Update latency

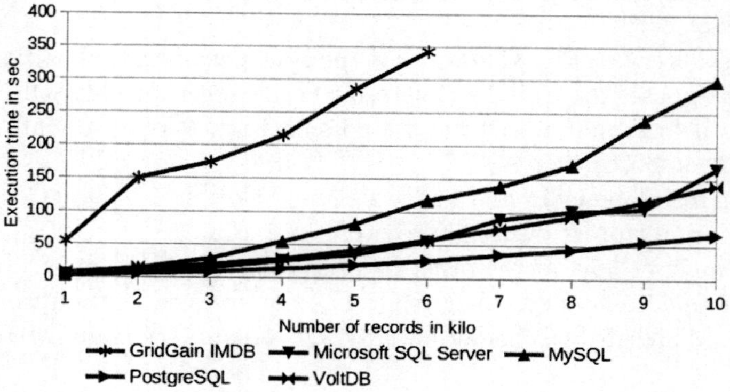

Fig. 4. Update latency in detail

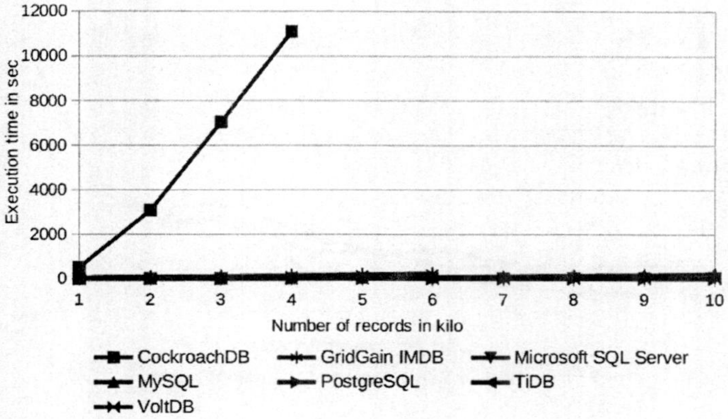

Fig. 5. Select row-by-row latency

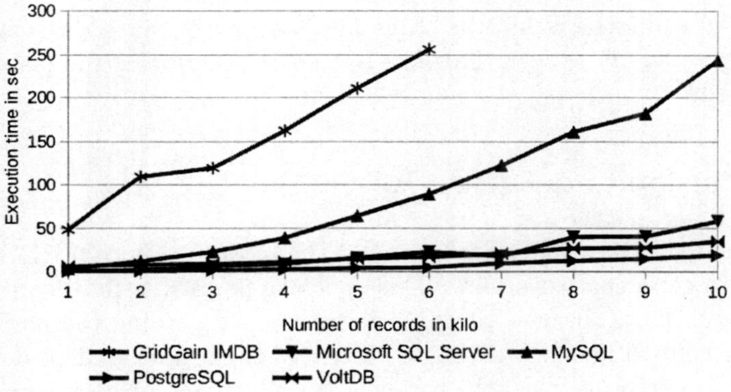

Fig. 6. Select row-by-row latency in detail

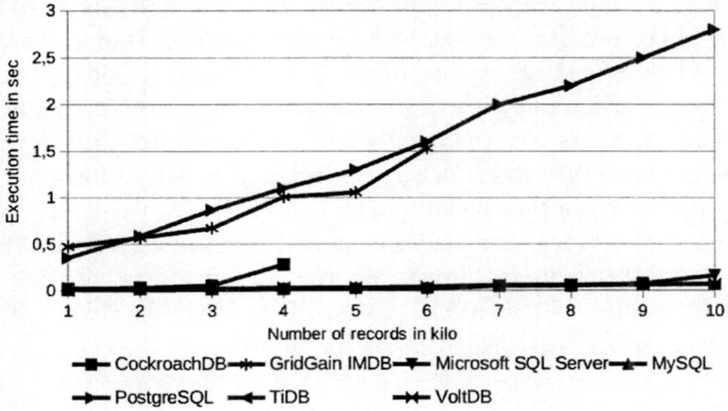

Fig. 7. Select latency

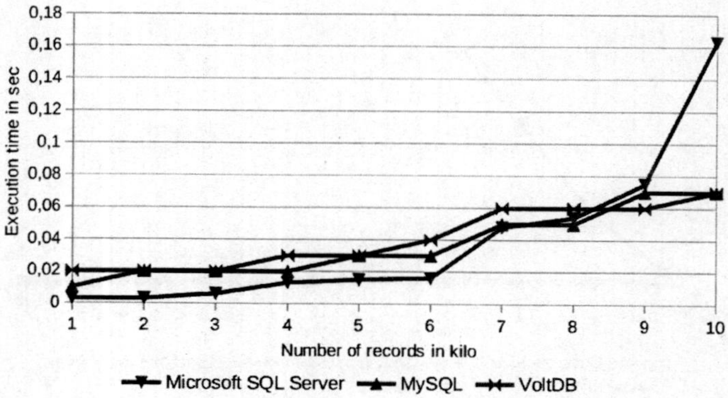

Fig. 8. Select latency in detail

In Fig. 8 it can be seen that the other databases have almost the same latency except Microsoft SQL Server. This one has an outlier during the last test with 10,000 records.

5 Conclusion and Future Research

The most popular NewSQL databases and the most well known RDBMS were mentioned to get a short overview. This article opposes the traditional RDBMS to a new kind of databases - NewSQL databases - regarding the performance. The results help to decide whether a switch to these new kind of database is reasonable. Other properties which can be an advantage or disadvantage for a database are not mentioned in this article. Nevertheless it has to be involved in a decision for or against a system.

If the focus is set on the results of the performance test the RDBMS PostgreSQL and Microsoft SQL Server had the best results. Under these circumstances it is not possible to switch to NewSQL databases. However the NewSQL database VoltDB has almost the same performance than PostgreSQL. It is important that these test results are only a snap-shot. Depending on the further development of these systems another execution of these performance tests in the future could have another result.

Also other settings for this evaluation should be mentioned, e.g. other OS or using more hardware resources. In the second case also tests with big data can be made. Also special use cases and optimization of the hardware or software for the execution could be tested.

References

1. Almassabi, A., Bawazeer, O., Adam, S.: Top NewSQL databases and features classification. Int. J. Database Manag. Syst. **10**, 11–31 (2018)
2. Anslett, M.: NoSQL, NewSQL and beyond: the drivers and use cases for database alternatives. The Drivers and Use Cases for Database Alternatives, NoSQL, NewSQL and Beyond (2011)
3. Anton, C.A., Matei, O., Avram, A.: Collaborative data mining in agriculture for prediction of soil moisture and temperature. In: Computer Science On-Line Conference, pp. 141–151. Springer (2019)
4. Aslett, M.: How will the database incumbents respond to NoSQL and NewSQL (2011)
5. Avram, A., Matei, O., Pintea, C.M., Pop, P.C., Anton, C.A.: Context-aware data mining vs classical data mining: case study on predicting soil moisture. In: International Workshop on Soft Computing Models in Industrial and Environmental Applications, pp. 199–208. Springer (2019)
6. Bakuya, T., Matsui, M.: Relational database management system. US Patent 5,680,614 (21 Oct 1997)
7. Binani, S., Gutti, A., Upadhyay, S.: SQL vs. NoSQL vs. NewSQL—a comparative study. Database 6(1), 1–4 (2016)
8. Chandra, U.: A comparative study on: NoSQL, NewSQL and Polygot persistence. Int. J. Soft Comput. Eng. (IJSE) 7 (2017)
9. Chereja, I., Hahn, S.M.L., Matei, O., Avram, A.: Multidimensional analysis of NewSQL database systems. In: Computer Science On-line Conference, pp. 221–236. Springer (2021)
10. Chereja, I., Hahn, S.M.L., Matei, O., Avram, A.: Operationalizing analytics with NewSQL. In: Computer Science On-line Conference, pp. 249–263. Springer (2021)
11. Chopra, R.: Database Management System (DBMS). A Practical Approach. S. Chand Publishing (2010)
12. Codd, E.F.: A relational model of data for large shared data banks. In: Software Pioneers, pp. 263–294. Springer (2002)
13. Codd, E.: An evaluation scheme for database management systems that are claimed to be relational. In: 1986 IEEE Second International Conference on Data Engineering, pp. 720–729. IEEE (1986)
14. Committee, C.S., et al.: CODASYL data base task group: April 1971 report. ACM, New York (1971)
15. Fatima, H., Wasnik, K.: Comparison of SQL, NoSQL and NewSQL databases for internet of things. In: 2016 IEEE Bombay Section Symposium (IBSS), pp. 1–6. IEEE (2016)
16. solidIT Consulting & Software Development GmbH: Db-engines (2021). db-engines.com/
17. Gurevich, Y.: Comparative survey of NoSQL/NewSQL DB systems. Department of Mathematics and Computer Science, The Open University of Israel (2015)
18. Haerder, T., Reuter, A.: Principles of transaction-oriented database recovery. ACM Comput. Surv. (CSUR) **15**(4), 287–317 (1983)
19. Hahn, S.M.L., Chereja, I., Matei, O.: Analysis of transformation tools applicable on NewSQL databases. In: Computer Science On-line Conference, pp. 180–195. Springer (2021)
20. Hahn, S.M.L., Chereja, I., Matei, O.: Comparison of the performance of NewSQL databases based on Linux OS (to appear). In: Proceedings of the Computational Methods in Systems and Software. Springer (2021)

21. Hahn, S.M.L., Chereja, I., Matei, O.: Evaluation of transformation tools in the context of NoSQL databases. In: Proceedings of SAI Intelligent Systems Conference, pp. 146–165. Springer (2021)
22. Hajoui, O., Dehbi, R., Talea, M., Batouta, Z.I.: An advanced comparative study of the most promising NoSQL and NewSQL databases with a multi-criteria analysis method. J. Theor. Appl. Inf. Technol. **81**(3), 579 (2015)
23. Han, J., Haihong, E., Le, G., Du, J.: Survey on NoSQL database. In: 2011 6th International Conference on Pervasive Computing and Applications, pp. 363–366 (2011)
24. Kaur, K., Sachdeva, M.: Performance evaluation of NewSQL databases. In: 2017 International Conference on Inventive Systems and Control (ICISC), pp. 1–5. IEEE (2017)
25. Kelly, A.: Making sense of NoSQL: a guide for managers and the rest of us by Ann Kelly and Dan McCreary (2013)
26. Khasawneh, T.N., AL-Sahlee, M.H., Safia, A.A.: SQL, NewSQL, and NoSQL databases: a comparative survey. In: 2020 11th International Conference on Information and Communication Systems (ICICS), pp. 013–021 (2020)
27. Kumar, R., Gupta, N., Maharwal, H., Charu, S., Yadav, K.: Critical analysis of database management using NewSQL. Int. J. Comput. Sci. Mob. Comput. **3**, 434–438 (2014)
28. Matei, O., Rusu, T., Petrovan, A., Mihuţ, G.: A data mining system for real time soil moisture prediction. Procedia Eng. **181**, 837–844 (2017)
29. Moniruzzaman, A.B.M.: NewSQL: towards next-generation scalable RDBMS for online transaction processing (OLTP) for big data management. Int. J. Database Theory Appl. **7** (2014)
30. Pavlo, A., Aslett, M.: What's really new with NewSQL? ACM Sigmod Rec. **45**(2), 45–55 (2016)
31. Sharma, V., Dave, M.: SQL and NoSQL databases. Int. J. Adv. Res. Comput. Sci. Softw. Eng. **2**(8) (2012)
32. Sumathi, S., Esakkirajan, S.: Fundamentals of Relational Database Management Systems, vol. 47. Springer (2007)
33. Theodorou, V., Abelló, A., Lehner, W.: Quality measures for ETL processes. In: International Conference on Data Warehousing and Knowledge Discovery, pp. 9–22. Springer (2014)

Improving the Performance of Effort Estimation in Terms of Function Point Analysis by Balancing Datasets

Huynh Thai Hoc(✉), Vo Van Hai, Ho Le Thi Kim Nhung, and Roman Jasek

Faculty of Applied Informatics, Tomas Bata University in Zlin, Nad Stranemi 4511, Zlin 76001, Czech Republic
{huynh_thai,vo_van,lho,jasek}@utb.cz

Abstract. This research proposes an approach to improve the performance of effort estimation based on the balancing of each group for categorical variables. The proposed model is based on function point analysis, Industry Sector, and deep learning. The Pytorch library is used to build the deep learning model with the dataset ISBSG (release 2020). The accuracy of our model is compared with that of the Adj-Effort approach. We adopt the prediction level at 0.3, Mean Absolute Error, Mean Balanced Relative Error, Mean Inverted Balanced Relative Error, and Standardised Accuracy as criteria for validation. The findings demonstrate that our proposed model outweighs the unbalanced and Adj-Effort approaches.

Keywords: Software effort estimation · Adj-Effort · Deep learning · Balance class weight

1 Introduction

The software application is almost a requirement for most activities worldwide. It is used in various industries, such as education, economic, or health care insurance systems. As a result, many rely on a high-quality application that yields precise outcomes. However, one of the crucial challenges in a software system is developing a successful software project. According to the Standish report 2020, 50% of software systems are troubled, and only 31% are successful [1]. The leading causes of software failure are a lack of knowledge of prerequisites and incorrect size estimation [2]. These might be able to persuade scientists to investigate software that could estimate size and effort.

R. Silhavy et al. (Eds.): CoMeSySo 2022, LNNS 596, pp. 705–714, 2023.
https://doi.org/10.1007/978-3-031-21435-6_60

Software effort estimation (SEE) is primarily required at the beginning of the project to adjust related resources [3]. Due to there might be no information available at the early stage of the project, the managers tend to focus on size measurements. They might estimate the likely effort with the help of these measurements. Several methods were developed prior to the development of the use case points [4] or the function point analysis [5] method.

However, there is an advantage in that many previous completed projects have been recorded as a reference dataset. This might aid researchers in surveying, comprehending, and forecasting SEE for unobserved data. For example, the International Software Benchmarking Standards Group (ISBSG) is a non-profit organisation. A group of national software metrics associations founded it in Australia in 1997 [6]. Their mission was to encourage the employment of data from the IT industry to optimise software processes and products.

For instance, the ISBSG repository (R1, release 2020) contains 9592 completed project records spanning from 1989 to 2020. They cover various activities, such as new development, enhancement, migration, re-development, etc. Furthermore, they cover fields such as banking, insurance, and finance. They also recorded most of the relevant attributes based on the activities of projects, such as size attributes, sizing, effort, etc. Over the years, this group of data has also been studied and used by scientists to build a predictive model that estimates software effort by function point. In 2020, Hoc et al. proposed the Adj-Effort approach based on FPA. Their model outperformed Capers Jones and Mean Effort [7].

$$Effort_{Adj} = 2.4639 \times AFP^{0.8813} \times PDR^{0.8433} \tag{1}$$

where AFP stands for the Adjusted Function Point and PDR is the abbreviation of the productivity of the project [8].

On the other hand, Silhavy et al. [9] studied a new software development effort estimation based on EI, EO, EQ, ILF, EIF, categorical variables (Industry Sector, Relative Size, Business Area) and stepwise regression. ISBSG (release 2015) was used as the historical dataset for their study. As a result, the performance obtained from their proposed approach might be better than baseline models (IFPUG, Regression Clustering-based, and Spectral Clustering-based Models).

However, the number of projects in each categorical variable might not balance. This might negatively impact the accuracy of effort estimation. In this paper, we will study SEE based on ISBSG (release 2020), where EI, EO, EQ, ILF, EIF and Industry Sector as input variables for our proposed model. We also consider enhancing the accuracy of SEE by deeply verifying the Industry Sector to ensure that all the different types in the Industry Sector will be involved in the model's training.

The rest of the study is organised as follows: Sect. 2 presents the Research Questions; Sect. 3 describes the Related Works; Sect. 4 depicts Dataset Description; Sect. 5 proposes Experimental Design; Sect. 6 discusses Criteria Validation; Sect. 7 presents Result and Discussion, and Sect. 8 concludes with Future Work.

2 Research Questions

Two research questions should be answered:

(1) RQ1: how accurate is the model between the deep learning approach compared with the Adj-Effort approach?
(2) RQ2: how accurate deep-learning-based model between balance and without balance in the number of projects based on the Industry Sector?

3 Related Works

Deep learning allows computational models made up of various processing layers to gain knowledge of varying levels of complexity. According to [10], these methods have significantly advanced in various fields, such as speech recognition, visual object recognition, or object detection. Deep learning investigates complicated systems in large data sets by employing back-propagation algorithms to demonstrate how a machine changes the internal parameters used to compute each class's representation based on the performance of the previous layer.

Several studies have used artificial networks and deep learning to improve effort estimation in software engineering. As mentioned in the publication [11], scholars used a diverse range of models, such as link artificial neural networks, or feed-forward neural networks, to build the effort estimation model. This publication demonstrates that many neural networks, higher order neural networks, and deep learning neural networks have been used for software effort estimation.

In 2019, Ming Qin et al. [12] proposed a BiLSTM-CRF framework based on a deep learning model to enhance the performance of effort estimation based on function points. This framework combined a bidirectional long short-term memory (BiLSTM) and conditional random fields (CRF). It includes forward and backwards long-term memory networks for learning from previous tokens.

In addition, M. Choetkiertikul et al. [13] proposed a deep learning-based, fully end-to-end predictive model to predict story points. It eradicates the requirement for users to individually design features from textual descriptions of problems. It integrates two effective deep-learning architectures, LSTM and Recurrent Highway Network. Their approach outperforms Random Guessing, Mean and Median methods.

Moreover, learning models from unbalanced training data is a complex and time-consuming task for machine learning. Current deep learning methods might focus on class-balanced or moderately unbalanced data when training models, ignoring the threat of understanding from considerably unbalanced training data. Anand et al. [14] investigated the impact of class imbalance on the back-propagation algorithm in shallow neural networks. The authors demonstrate that the majority class effectively overwhelms the net gradient, which is in charge of updating the model's weights in imbalanced class situations. During early iterations, this rapidly declines the fault of the majority group. However, it frequently increases the error of the minority group, causing the network to evolve stuck in a sluggish convergence mode. As mentioned in [15], there are several approaches to tackle this problem, such as random over-sample (replicates random samplings from the minority group) or random under-sampling (remove random samples from the majority group) [16].

This paper will adopt a random over-sample by replicating random projects from Defence, Electronics & Computers, Medical & Health Care, and Mining groups.

4 Dataset Description

We use the ISBSG repository (release 2020) as the historical dataset. This version has more than 9592 completed records. It has 251 attributes, including numeric attributes such as Summary Work Effort (SWE), EI, EO, EQ, EIF, ILF, and categorical variables such as Industry Sector (IS). The IS is a categorical attribute. It contains a total of fifteen kinds of the sector, including Banking, Communication, Construction, Defence, Electronics & Computers, Financial, Government, Insurance, Manufacturing, Medical & Health Care, Mining, Professional Services, Service Industry, Utilities, and Wholesale & Retail. This dataset will be pre-processed by filtering based on the following criteria:

1. The data quality attribute is tagged A or B.
2. EI, EO, EQ, EIF, ILF and Industry Sector are not empty.
3. The development type is tagged New Development.

Moreover, according to [1], deleting outliers is appropriate for estimating precise and accurate software effort based on previous datasets. The interquartile range (IQR) method is used in this study to eliminate outliers. The value SWE might be considered removed if they are out of the range from $Q1 - 1.5 \times IRQ$ to $Q3 + 1.5 \times IRQ$. $Q1$ is the lower quartile, $Q3$ is the upper quartile, and IRQ is presented in Eq. (1). Table 1 shows descriptive characteristics of the SWE between the training and testing datasets.

$$IQR = Q3 - Q1 \tag{2}$$

Furthermore, as shown in Fig. 1, the distribution of groups in the Industry Sector is unequal. Many projects are centred on Banking, Communications, Government, Insurance, Manufacturing, and the Service Industry, with less emphasis on other sectors such as Defense, Electronics & Computers, or Mining.

5 Experimental Design

Figure 2 describes the effort estimation system design based on balancing the amount of Industry Sector. This process has four primary phases. In the first step, we load the raw dataset from the ISBSG repository (Release R1, 2020). Following that, we clean the raw database by adopting the rules presented in Sect. 4. Next, we use the one-hot encoding technique to transfer categorical variables into numeric values. There are fifteen groups for IS category (see Fig. 1), meaning that we will have more than fifteen columns generated. Each row has only one attribute with the value '1' in the column corresponding to that group, and the rest of the other groups will be assigned

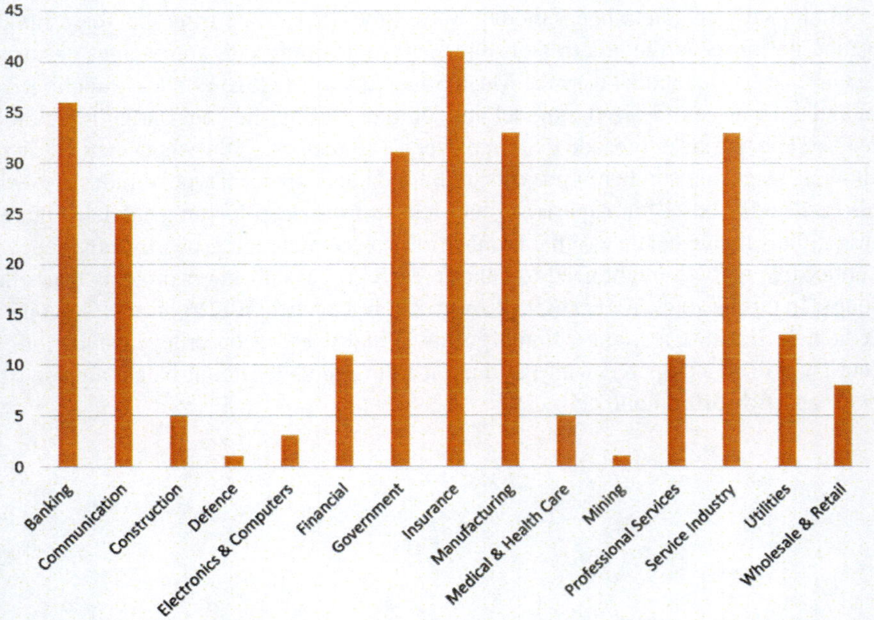

Fig. 1. The chart of the observational dataset based on the Industry Sector

Table 1. Descriptive observed SWE

Dataset	Mean	Standard Deviation	Minimum	25%	50%	75%	Maximum
Training	5726	7242	47	1470	2952	6944	46,101
Testing	8967	8229	362	3368	5078	11,824	34,947

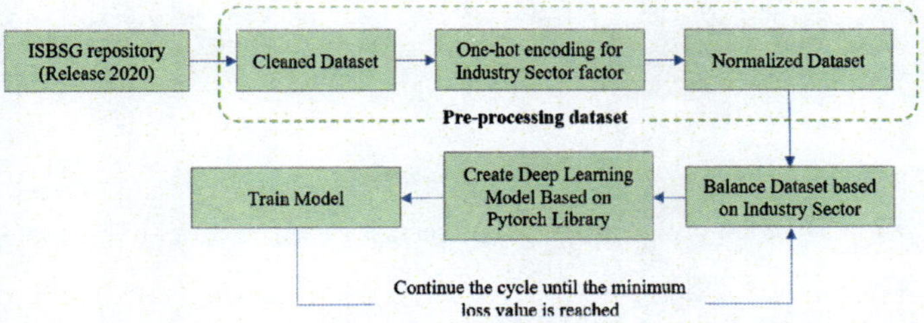

Fig. 2. The flow diagram of effort estimation based on balancing the number of projects based on the Industry Sector

'0'. Since there is a significantly different value between SWE, EI, EO, EQ, EIF, ILF, and categorical variables, we proceed to the next step of normalising the data for the values SWE, EI, EO, EQ, EIF, and ILF. In this paper, we choose the max-min normalisation method [17]. Second, we randomly split the cleaned dataset into two subsets, 80% of the projects used for training and the rest used for testing.

In order to generate a linear model, we use the nn package from the Torch library, and the optimiser is the Adam algorithm, whose learning rate and weight decay are 1e-3 and 1e-6. This model uses the Mean Squared Error (MSE) as a loss function. The gradients for all tensors in the model are computed using the backward() function on the loss. The optimiser is made to iterate over all tensors using the step() method. In the following stage, the number of projects used for balance for each type of Industry Sector will be decided based on experience and fed into the deep learning model. Figure 3 present the suitable balance of the number of projects for each group by experience. It is noticeable that we might need to balance each group with fewer projects than other groups. In this case, we replicate for Construction, Defence, Electronics & Computers, Medical & Health Care, and Mining groups. Finally, based on criteria validation, its accuracy will be compared with the deep learning model without balancing Industry Sector and Adj-Effort approach.

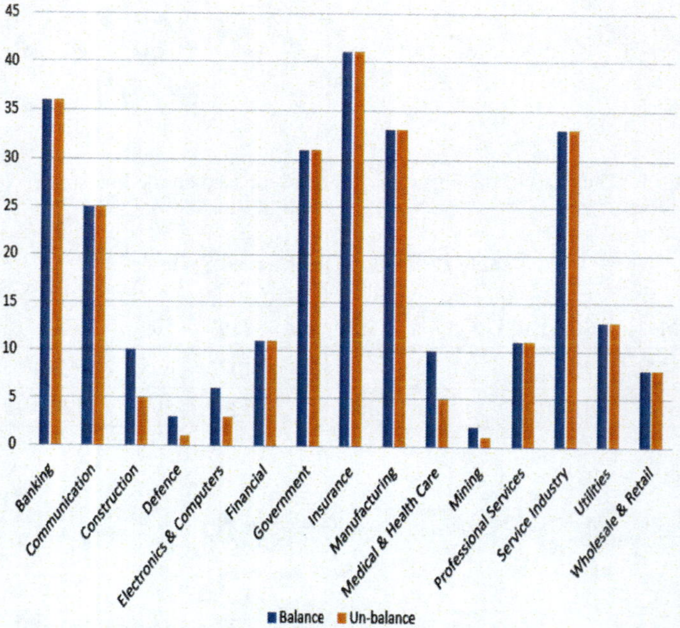

Fig. 3. The Flow Diagram of the number of projects for each category of the Industry Sector

6 Criteria Validations

Several well-known error measures will be used to validate the accuracy of SEE. They are essential factors of evaluation prediction performance. Prediction at level 0.3 (PRED(0.3)), Mean Absolute Error (MAE), Mean Balanced Relative Error (MBRE), Mean Inverted Balanced Relative Error (MIBRE), and Standardised Accuracy (SA).

The following are the given equations:

$$MAE = \frac{1}{N}\sum\nolimits_{i=1}^{N} |y_i - \hat{y}_i| \quad (1)$$

$$PRED(x) = \frac{1}{N}\sum\nolimits_{i=1}^{N} \begin{cases} 1, \ if \ \frac{|y_i - \hat{y}_i|}{y_i} \le x \\ 0, \ otherwise \end{cases} \quad (2)$$

$$MBRE = \frac{1}{N}\sum\nolimits_{i=1}^{N} \frac{|y_i - \hat{y}_i|}{\min(y_i, \hat{y}_i)} \quad (3)$$

$$MIBRE = \frac{1}{N}\sum\nolimits_{i=1}^{N} \frac{|y_i - \hat{y}_i|}{\max(y_i, \hat{y}_i)} \quad (4)$$

$$SA = \left(1 - \frac{MAE}{\overline{MAE_p}}\right) \quad (5)$$

where N is the number of projects, y_i is the actual effort for the i-th project, $\hat{y}_i$ is the predicted corresponding effort, and the x value is considered 0.3 - as many studies recommend. $\overline{MAE_p}$ is the mean value of a large number, typically 1000, runs of random guessing [18].

7 Results and Discussion

Table 2 and Fig. 4 present the accuracy of effort estimation obtained from deep learning by balancing the number of records based on each kind of group for the Industry Sector factor. Its result is compared with the un-balancing approach and Adj-Effort approach. As can be seen, in general, the MAE obtained from the deep learning method is minimal compared to Adj-Effort. Similarly, the values of MBRE and MIBRE obtained from the deep learning method are 0.48/0.21 and 0.48/0.26, respectively. These values are smaller than those obtained from Adj-Effort (0.53/0.31).

On the other hand, PRED(0.3) and SA obtained from the deep learning approach might be better than those obtained from Adj-Effort. The findings might answer the **RQ1** that the deep learning approach outperforms better than the Adj-Effort.

Table 2. The performance effort estimation based on deep learning with/without balance versus Adj-Effort

Method	PRED(0.3)	MAE	MBRE	MIBRE	SA
Balance	0.84	2338.80	0.48	0.21	0.60
Un-Balance	0.60	2442.32	0.48	0.26	0.44
Adj-Effort	0.40	2511.97	0.53	0.31	0.43

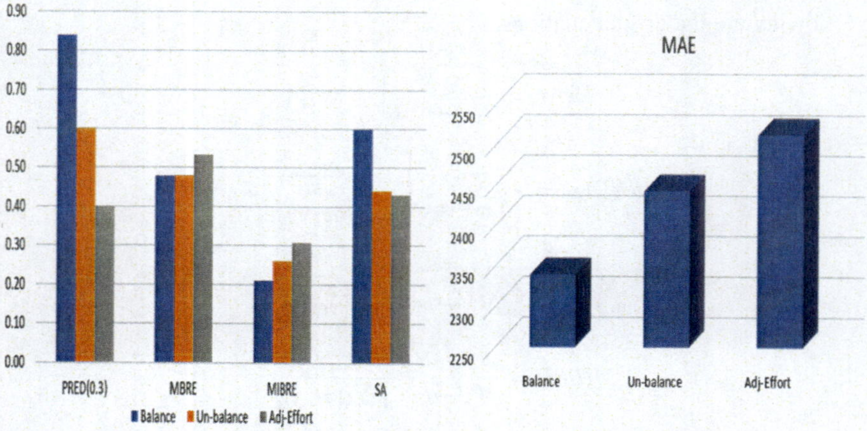

Fig. 4. The statistical charts compare the effectiveness of models

Furthermore, PRED(0.3) and SA obtained from the balanced approach are 84% and 60%, respectively. By contrast, values obtained from the un-balanced method are 60% and 44%. It demonstrates that the statistical criteria obtained from the balanced approach outperformed better than the un-balance approach. It is possible to conduct for **RQ2** that a balanced approach might properly approach to help the effort estimation more accurate than the others.

Moreover, Table 3 depicts the t-test result of the balance-based deep learning model compared with un-balance and Adj-Effort. The findings demonstrate that the proposed model differs significantly from the others in terms of SA (p-value < 0.05). Based on this result, we might conclude that our proposed model might be more accurate than the deep learning model without balancing the number of projects for each kind of IS.

Table 3. Hypothesis t-test results

	t-value	p-value
Un-balance	2.11	0.045
Adj-Effort	2.12	0.044

8 Conclusion and Future Work

This study compared the balance-based deep learning model with the unbalanced deep learning model and the Adj-effort. The results generally show that the deep learning method outperforms the comparison method in prediction. Furthermore, the balance of different sector types enhances the model's performance. This finding suggests that if the input value is a categorical variable, researchers should consider balancing the amount of dataset on each type to improve the model's efficiency.

On the other hand, balancing projects for each group might have several challenges. According to Johnson, J.M et al. [15], under-sampling may reduce the amount of data that the model should be learned. Oversampling lengthens training time due to the larger size of the training set and might have been linked to overwhelming. In the future, intelligent sampling methods [19] should be studied to retain helpful information for learning.

Acknowledgement. This work was supported by the Faculty of Applied Informatics, Tomas Bata University in Zlín, under project IGA/CebiaTech/2022/001.

References

1. Johnson, J.: Chaos 2020: Beyond Infinity. Standish Group (2020)
2. Curcio, K., Navarro, T., Malucelli, A., Reinehr, S.: Requirements engineering: a systematic mapping study in agile software development. J. Syst. Softw. **139**, 32–50 (2018). https://doi.org/10.1016/j.jss.2018.01.036
3. Ochodek, M., Nawrocki, J., Kwarciak, K.: Simplifying effort estimation based on use case points. Inf. Softw. Technol. **53**(3), 200–213 (2011). https://doi.org/10.1016/j.infsof.2010.10.005
4. Karner, G.: Metrics for objectory, LiTH-IDA-Ex-9344: 21. Diploma thesis, University of Linköping, Sweden (1993)
5. Albrecht, A.J.: Measuring application development productivity. (1979)
6. ISBSG: ISBSG International Software Benchmarking Standards Group, Release R1 (2020)
7. Hoc, H.T., van Hai, V., Nhung, H.L.T.K.: An approach to adjust effort estimation of function point analysis. In: Computer Science On-line Conference, pp. 522–537 (2021)
8. IFPUG. http://www.ifpug.org/. International Function Point Users Group
9. Silhavy, P., Silhavy, R., Prokopova, Z.: Categorical variable segmentation model for software development effort estimation. IEEE Access **7**, 9618–9626 (2019)
10. LeCun, Y., Bengio, Y., Hinton, G.: Deep learning. Nature **521**(7553), 436–444 (2015). https://doi.org/10.1038/nature14539
11. Kumar, P.S., Behera, H.S., Kumari, A., Nayak, J., Naik, B.: Advancement from neural networks to deep learning in software effort estimation: perspective of two decades. Comput. Sci. Rev. **38**, 100288 (2020)
12. Qin, M., Shen, L., Zhang, D., Zhao, L.: Deep learning model for function point based software cost estimation—an industry case study. In: 2019 International Conference on Intelligent Computing, Automation and Systems (ICICAS), pp. 768–772 (2019). https://doi.org/10.1109/ICICAS48597.2019.00165
13. Choetkiertikul, M., Dam, H.K., Tran, T., Pham, T., Ghose, A., Menzies, T.: A deep learning model for estimating story points. IEEE Trans. Software Eng. **45**(7), 637–656 (2019). https://doi.org/10.1109/TSE.2018.2792473
14. Anand, R., Mehrotra, K.G., Mohan, C.K., Ranka, S.: An improved algorithm for neural network classification of imbalanced training sets. IEEE Trans. Neural Netw. **4**(6), 962–969 (1993). https://doi.org/10.1109/72.286891
15. Johnson, J.M., Khoshgoftaar, T.M.: Survey on deep learning with class imbalance. J. Big Data **6**(1), 1–54 (2019). https://doi.org/10.1186/s40537-019-0192-5
16. van Hulse, J., Khoshgoftaar, T.M., Napolitano, A.: Experimental perspectives on learning from imbalanced data. In: Proceedings of the 24th International Conference on Machine Learning, pp. 935–942 (2007). https://doi.org/10.1145/1273496.1273614

17. Patro, S., Sahu, K.K.: Normalisation: a pre-processing stage. arXiv preprint arXiv:1503.06462 (2015)
18. Shepperd, M., MacDonell, S.: Evaluating prediction systems in software project estimation. Inf. Softw. Technol. **54**(8), 820–827 (2012)
19. Mani, I., Zhang, I.: kNN approach to unbalanced data distributions: a case study involving information extraction. In: Proceedings of the Workshop on Learning from Imbalanced Datasets, vol. 126, pp. 1–7 (2003)

The Model of Interface Structure for Technical Systems Based on a Communication of Logics

Sergei Kucherov(✉), Yulia Lipko, Alexander Sviridov, and Svetlana Belikova

Institute of Computer Technologies and Information Security, Sothern Federal University, 44 Nekrasovsky St., Taganrog, Russian Federation
skucherov@sfedu.ru

Abstract. The concept of an interface has many formulations and reflects the specifics of its implementation in different conditions. This leads to the fact that a software-implemented interface of technical systems in the existing development paradigm becomes a result of designers and developers' empirical activity, which is based on a generalized subjective idea of the algorithms and possible scenarios for using these systems. This result is next converted into information display logic. The information display logic determines a content of the interface, which is dictated by the data available at the corresponding stage of operation and the list of control actions. The interface itself is presented as a set of screens, the behavior and chronology of which is determined by the business logic. This paradigm contradicts the way of user's thinking based on the logic of working with a technical system, which is individual even for identical technical systems. As a result, the user must set the points of intersection for a work logic and an information display logic in the form of available screen forms on his own. To avoid this problem, we propose to re-define the interface model structure as a communication of a work logic and of an information display logic. This new methodological base should be based on the categories of user's thinking, and the interface itself should be based on a single model of logics communication. The article presents the concept and the model of interface structure for technical systems, based on the proposed methodological framework.

Keywords: Human-machine interface · Interface model · Interface structure · Polymodal interface · Logic of working with a technical system · Information display logic · Communication

1 Introduction

Existing approaches to building human-machine interfaces focus on visualizing the functioning of a technical system in such a way that the operator is given optimal access to as much information as possible [1]. The logic of displaying this amount of information is formed empirically at the stage of interface design and development and is subject to the developer's subjective idea of possible algorithms and scenarios for using the technical system [2]. In turn, each user has his own idea of the technical system, based

R. Silhavy et al. (Eds.): CoMeSySo 2022, LNNS 596, pp. 715–722, 2023.
https://doi.org/10.1007/978-3-031-21435-6_61

on which the rules for its application for solving professional problems are formed, or, in other words, the logic of working with the technical system.

A situation arises in which an interface that is conditionally perfect in terms of design and visualization laws is not perceived by the user, since it requires the latter to independently build the points of intersection of the logic of work with the logic of displaying information in the form of accessible screen forms [3]. This leads to the need to learn not only the principles of the functioning of the technical system, but also how and in what order the system displays the available information. As a result, in the short term, there is a decrease in the efficiency of human-machine systems, in the long term, the risk of errors by the operator remains due to possible contradictions between the logic of work and the logic of displaying information. If we take into account the fact that each user is able to effectively interact only with certain ways of presenting information (visual, textual, audio, etc.), then the problem of developing interfaces for technical systems becomes one of the fundamental ones in modern science.

The paper proposes the theoretical foundations for the development of interfaces for human-machine systems, devoid of the above disadvantages. The second section presents the results of the study of the problem of communication in the existing paradigm of interface development. In the third section, a model of the structure of the interface of technical systems based on the communication of logics is proposed, which forms the methodological basis for the overdue paradigm shift in the design and development of systems in general and interfaces of technical systems in particular.

2 The Problem of Communication in the Existing Paradigm of Developing Interfaces of Technical Systems

Existing approaches to the development of interfaces of technical systems consider communication as an exchange of knowledge between the participants in the development process (technical system developer, interface developer, user, if he is involved in the design process, and other stakeholders). An analysis of existing studies [4–7] showed that domain experts and developers are at different levels of subject knowledge about the process and method of interface development, which affects the logic of distribution of functions between the user and the system in the interaction process: the user often does not can quickly navigate the proposed functions, the set of which does not always correspond to the content of the current step of his professional activity.

In the existing paradigm of developing interfaces for technical systems, communication between experts and developers is carried out in a verbal form. The requirements stated by the experts and the description of the tasks are fixed and transformed. The information received from the domain expert is used in the future to build various models necessary for the design process.

The work [8] studies the field of analysis of the processes of designing HMI systems, which shows that today this process is not formally defined, communication models represent the communication between the developer and the user, which increases the possibility of semantic gaps. The need to formalize the process of designing an HMI system is determined. The work [9] presents a study of the convenience of using the graphical interface of the medical system for general practitioners. The communication model is the developer's observation of the work of users, the results of which determined the most appropriate type of system interface. During the study, users (physicians) acted as a "data source" for the development team - they demonstrated examples of their work with medical records and answered questions from the developers. The work [10] proposes a software tool for the user that allows him to communicate with the developer using the tools for designing the interface of the future system. As the authors of the work emphasize, the main feature of their proposed development is the implementation of the possibility of determining functional dependencies between database entities based on the analysis of relationships between interface elements. The software tool developed by the authors implements a communication model between users and developers, processing the interface assembled by users and, on its basis, creates a relational application database, and generates structural and behavioral models of the system (represented in the form of the UML language). The paper [11] studies the effectiveness of the interaction between the operator's work and the technical system. The task of developing an HMI system is defined in which users and developers use such communication methods as observations in the control room, interviews, surveys, and documentation reviews, in which users acted as "data sources" for developers and did not have the opportunity to way to influence the logic and process of building an HMI system. In [12], a method for designing the HMI system of a modern car is discussed. During the design process, a persona is introduced, and co-creation is used to generate design options for the HMI system. Thus, it can be determined that the communication model between the user and the developer is carried out by exchanging information about the preferred design option for the system. The works [13, 14] show that it is in the process of communication during the translation, perception, and transformation of information that a semantic gap arises first between the participants, then more widely between the goal and the result of the design (or, in other words, there are the phenomena of performance abyss and evaluation abyss according to [15]).

Despite the abundance of approaches to the design and development of human-machine system interfaces, all are united by one characteristic property - both the starting point and the end result are presented as objects. It does not matter whether at the initial stage the tasks solved by the user [16] or the goals facing him [17] are taken as the basis, the interface development process will be reduced to converting such an initial set of objects into a final one in the form of screen forms.

Based on the analysis carried out, it follows that in the existing paradigm of the development of interfaces of technical systems, the following picture has developed. The user and the developer implicitly and independently form their own logic: the user - the logic of solving the problem, the developer - the logic of displaying information. The developer also forms a set of interface windows based on the information display logic. As a result, the user finds himself in a situation in which he is forced to fill in the logic of solving the problem with content empirically created by the developer, and in case of inconsistencies, change the logic unilaterally.

Let's consider the problem in more detail. The logic operated by the user reflects the conditional sequence of steps in the method of solving a professional problem using a technical system. For example, for a CNC machine, such logic may be as follows: (1) enter the program into the machine control system; (2) indicate the type of part; (3)

indicate the type of material; (4) depending on the type of part and material, set the processing parameters; (5) initiate the launch of the program. The logic of solving problems is formed by the user because of training and subsequent accumulation of professional experience and reflects the logic of solving a problem using a technical system. The logic used by the interface developer, in turn, reflects some conditional sequence of steps in the method of displaying information and managing the process of the technical system functioning. For a CNC machine, it can be as follows: (1) display a dialog for selecting or loading a program; (2) display the dialog for setting program execution parameters; (3) display the dialog for launching and tracking program execution parameters. This logic is formed by the developer empirically, based on the available initial data on the technical system, and represents the logic of displaying information.

From a general standpoint, the problem-solving logic used by the user and the display logic generated by the developer are similar in the above example, however, the user implies a limitation of the parameters available for configuration depending on the selected type of part and material, while the developer of such a task does not put. As a result, in the case of using a technical system with such an interface, the user will be forced to simplify his logic for solving the problem by removing steps from it that allow avoiding errors in setting up the program for a specific material and type of part.

Considering the fact that the methods of interface design considered in this section are aimed at obtaining comprehensive information about the content of the interface, we can conclude that the existing development paradigm does not set the task of communicating logics. As a result, a contradiction is formed between how the user sees the application of the technical system and how the interface developer imagines the possibilities of using the technical system.

3 The Model of the Structure of the Interface of Technical Systems Based on the Communication of Logics

The user interface, reflecting the meaning of the system management process, should be a methodological tool that allows you to organize the form and content of communication between the logic of solving user problems and the logic of displaying information. Given the real situation, namely the fact that the complexity of changing interfaces for all previously created and operated technical systems is not commensurate with the benefits of its results, it is proposed to build such an interface over the existing mechanisms of user interaction with the technical system. This will make it possible to communicate logics without interfering with the already implemented interface of technical systems.

The interface of technical systems based on communication of logics is intended for ordering information and control signals in the context of tasks solved by the user in accordance with the developed skills for the implementation of professional activities. A model of the interface structure is proposed, in which the form and content of communication of logics has three levels of abstraction, built over the existing interface of a technical system (Fig. 1):

1. user's system image;
2. image of the system's capabilities for the user;
3. the image of the functioning of the system under the control of the user.

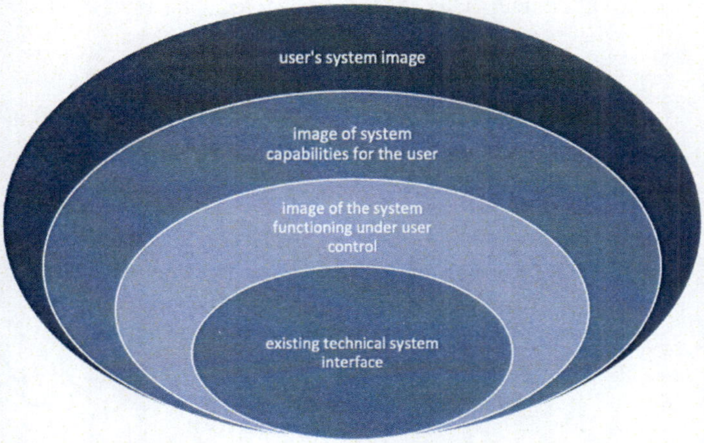

Fig. 1. The structure of the interface of technical systems based on the communication of logics

At the level of abstraction "user image of the system", the unity of the form and content of communication of logics is an integrated representation of the system through a set of concepts that reflect objects and functions that are significant to the user. A tag cloud can be used as a tool for representing a set of concepts (Fig. 2).

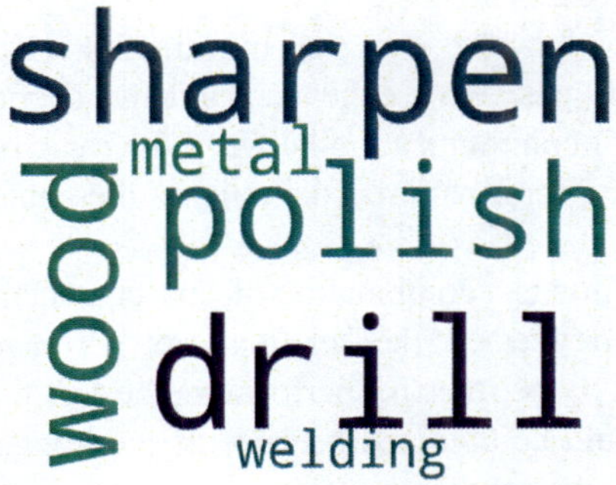

Fig. 2. User's system image

Such a representation serves for the initial acquaintance of the user with the capabilities of the technical system without the need to interact with it. To create such an image, a domain expert defines a set of concepts and functions relevant to user activity. Further, based on the relevance criterion (for example, frequency of use), the scale of the text for the concept changes up or down. As the image is used, the values of the criteria can be adjusted, changing the scale of representation of concepts.

At the level of abstraction "the image of the system's capabilities for the user", restrictions are introduced on the combination of functions and objects that can be implemented and created/used by the technical system. The capabilities of the system are represented by a graph with two types of vertices - an object and a function. The edges connecting the vertices are the rules for performing actions to implement the function on the object (Fig. 3). The content of the edges is considered at the third level of abstraction.

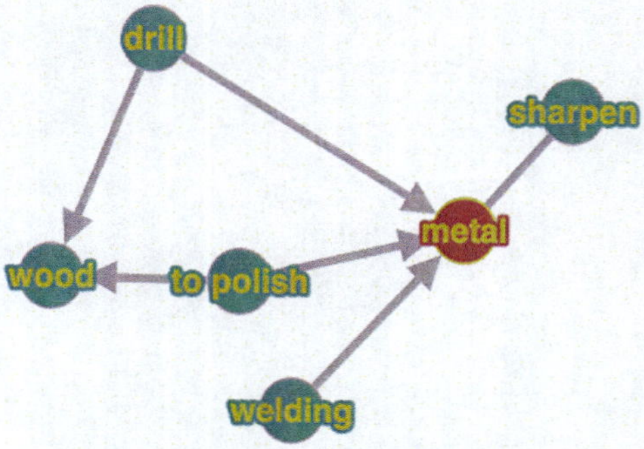

Fig. 3. Image of system capabilities for the user

Such a representation serves to acquaint the user with the space of possibilities provided by the system (types of objects and functions performed on them), which, in addition to the objects and functions themselves, reflects restrictions on possible combinations. The representation graph can be depicted in a three-dimensional space, the areas of which correspond to the groupings of functions and objects according to their thematic orientation (for complex systems). In this case, the user is provided with appropriate navigation mechanisms.

At the level of abstraction "the image of the functioning of the system under the control of the user", it is proposed to use the mechanisms of action [18] to form the unity of the form and content of communication between the logic of solving user problems and the logic of displaying information. This unity is the content of edges connecting objects and functions.

The mechanism of action is a combination of four characteristics (Fig. 4): elements on which the function is performed; the function to be performed; the tool with which the function is performed or the rules for performing the function in the case of manual labor; the result that should be obtained because of the implementation of the action according to the model of the mechanism.

Element	Function	Tool	Result

Fig. 4. Structure of the mechanism of action

Each mechanism contains the rules for the implementation of the user's action with the technical system in its interface. Each action is aimed at achieving a certain result known to the user, which serves as a criterion for the correctness of the composition of the rules and the implementation of actions according to them (Fig. 5).

Element	Function	Tool	Result
program file	copy	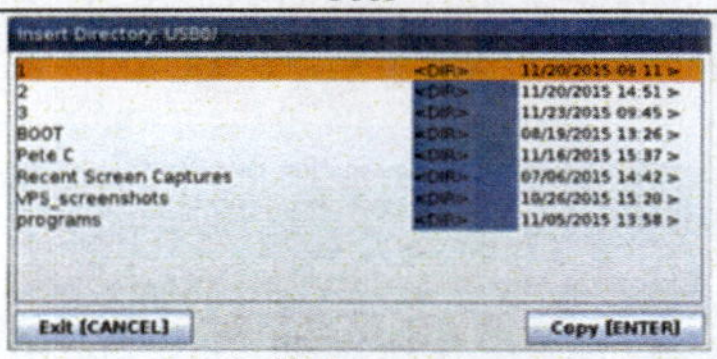 file browser	program file has been copied

Fig. 5. The image of the functioning of the system under the control of the user

In the considered example, the action mechanism is used to describe the logic of the user's action in the technical system. The composition and sequence of the mechanisms of action determines the form of the logic of solving the problem, and the content of the mechanisms includes elements of the logic of displaying information, which ultimately creates the necessary effect of communication of logics.

4 Conclusions

This article presents the results of a study of the problem of communication in the context of interaction between a user and a technical system through an interface and proposals for its solution. While studying the existing design and development paradigm, it was found that the existing interface model does not provide communication between the logic for performing professional tasks by the user and the logic for displaying information. This leads to several negative consequences, including a decrease in the efficiency of the functioning of the human-machine system and an increase in the likelihood of errors due to inconsistencies in logics.

The solution to the formulated problem can be found through a change in the interface development paradigm. The first step in the creation of a new paradigm is the proposed next-generation interface model, which ensures the communication of logics. This model differs from the known interface models by the introduction of three additional levels of abstraction aimed at eliminating contradictions between the logic of displaying information and the logic of solving problems by the user. Such an approach, in addition to the communication of logics, will ensure the polymodality of the interface since the ones proposed in Figure Forms (Figs.1, 2, 3 and 4) are one of the special cases of visual representation of information.

The interface structure model presented in the article based on the communication of logics is a conceptual representation of the image of solving the scientific problem of designing user interface design tools, reflecting the meaning of the system management process.

Acknowledgments. This work has been supported by the grants the Russian Science Foundation (project No. 22-19-00723).

References

1. Bachtiar, A.M., Dharmayanti, D., Ramadhan, E.G.: Analysis of interaction design model in content marketing domain using design sprint method. J. Eng. Sci. Technol. **15** (2020)
2. Singh, H.P., Kumar, P.: Developments in the human machine interface technologies and their applications: a review. J. Med. Eng. Technol. **45**, 552–573 (2021). https://doi.org/10.1080/03091902.2021.1936237
3. Khramov, V.V.: Development of a human-machine interface based on hybrid intelligence.(2020). https://doi.org/10.25559/SITITO.16.202004.893-900
4. Szekely P., Sukaviriya P., Castells P., Muthukumarasamy J., Salcher E. Declarative interface models for user interface construction tools: the MASTERMIND approach
5. Rasmussen, J: Skills, rules, and knowledge; signals, signs, and symbols, and other distinctions in human performance models System design for human interaction (pp. 291–300): IEEE Press (1987)
6. Vandervelpen, C., Vandenbergh, J., Luyten, K., Coninx, K.: D5.1 model-based user inter-face development methods for the design of context-driven. Adaptable Interact. Syst. 27 (2004)
7. Man, Y.: Human-Machine interface considerations for design and testing in distributed sociotechnical systems. Göteborg, Sweden (2015). https://core.ac.uk/download/pdf/198044413.pdf. Access date: 01 June 2022
8. Kaufmann, A., Maier, T.: Influencing parameters for the development of adaptive HMIs & requirements for the categorization of control functions generated via expert knowledge. In: Kantola, J.I., Nazir, S., Salminen, V. (eds), Advances in Human Factors, Business Management and Leadership. AHFE 2021. Lecture Notes in Networks and Systems, vol. 267. Springer, Cham (2021).https://doi.org/10.1007/978-3-030-80876-1_40
9. Rizvi, R.F., Marquard, J.L., Hultman, G.M., Adam, T.J., Harder, K.A., Melton, G.B.: Usability evaluation of electronic health record system around clinical notes usage–an ethnographic study. Appl. Clin. Inform. **8**(4), 1095 (2017)
10. Fischer, G., Giaccardi, E.: Meta-design: a framework for the future of end-user development. In: End user development, pp. 427–457. Springer, Dordrecht (2006)
11. Li, X., et al.: The control room operator: the forgotten element in mineral process control. Miner. Eng. **24**(8), 894–902 (2011)
12. Li, X., Ge, X., Wang, Y.: The psychological and statistical design method for co-creation HMI applications in the Chinese automotive market. SAE Technical Paper, 2017. №. 2017-01-0650
13. Matsunaga, H., Nakazawa, H.: Development of adaptive human-machine interface to match human satisfaction—comparison of satisfaction among different work styles. Seimitsu Kogaku Kaishi/J. Jpn. Soc. Precision Eng. **66**, 859–864 (2000). https://doi.org/10.2493/jjspe.66.859
14. Cross, E.S., Ramsey, R.: Mind meets machine: towards a cognitive science of human-machine interactions. Trends Cogn. Sci. (2021). https://doi.org/10.1016/j.tics.2020.11.009
15. Norman, D.: Design of Everyday Things: Revised and Expanded. MIT Press, London (2013)
16. Lewis, C., Rieman, J.: Task-centered user interface design: a practical introduction (2006)
17. Szczepansky, B.: User Goals (2020). https://medium.com/goal-directed-design/user-goals-f55c42be30de. Access date: 01 June 2022
18. Belikova, S.A., Rogozov, Y.I., Sviridov, A.S.: Method of user interface design based on semantic approach. Adv. Intell. Syst. Comput. **763**, 311–318 (2019).https://doi.org/10.1007/978-3-319-91186-1_32

Classification of Interaction Participants in the Formation of the Trajectory of Diagnosis and Treatment of Bronchopulmonary Diseases to Design Agents of a Multi-agent System

Nafisa Yusupova[1], Rustem Zulkarneev[2], Dmitry Rizvanov[1], Rashit Nasyrov[1], and Diana Bogdanova[1](✉)

[1] Ufa State Aviation Technical University, St. K.Marx, 12, 450008 Ufa, Russia
dianochka.bogdanova@gmail.com

[2] Bashkortostan State Medical University, St. Lenina, 3, 450008 Ufa, Russia

Abstract. The article deals with the formation of a rational trajectory of diagnosis and treatment of bronchopulmonary diseases of patients in hospital conditions. The description of the process of treatment taking into account various factors is given. The task of forming effective trajectory of diagnostics and treatment of bronchopulmonary diseases is considered in the context of clinical guidelines describing the process of treatment, individual characteristics of patient's health, and available diagnostic equipment and medicines. To solve this problem it is proposed to use a multi-agent approach. This approach will take into account clinical guidelines describing the treatment process, the individual characteristics of the patient's health, and the available diagnostic equipment. The interaction of real participants in the process can be modeled as the interaction of corresponding prototypes—the agents of a multi-agent system. To design the agents of a multi-agent system, a classification of the participants of such interaction for forming a diagnostic and treatment trajectory is proposed. At the first stage digital platform AnyLogic is proposed to be used as a tool for modeling the interaction processes. A formal statement of the problem of forming a trajectory for diagnosing and treating a patient, the results of primary data processing necessary for subsequent modeling, state diagrams of the main agents of the future model are presented, and directions for further research are determined #COMESYSO1120.

Keywords: Diagnostic and treatment trajectory · Agent · Multi-agent system · Agent-based modeling

1 Introduction

The problem of shaping a patient's diagnostic and treatment trajectory, taking into account his/her personal characteristics and the interests of all stakeholders in this process, using artificial intelligence technologies to improve the efficiency of medical data processing, reduce risks and improve the quality of life of the patient for implementing the transition to personalized medicine and high-tech health care, is topical today.

R. Silhavy et al. (Eds.): CoMeSySo 2022, LNNS 596, pp. 723–731, 2023.
https://doi.org/10.1007/978-3-031-21435-6_62

Research in the application of information technologies, including intelligent technologies, to medical data processing is being actively pursued by many researchers in various countries [1–4]. However, some questions remain unresolved, in particular the consideration of the interests of participants in the process of diagnosis, treatment, and patient follow-up, and have not been fully explored.

This article is devoted to the development of a model for shaping the trajectory of diagnosis and treatment of patients, taking into account all stakeholders. Section 1 describes the problem situation and presents the problem statement. Section 2 is devoted to the analysis of known approaches for solving such problems. Section 3 classifies the participants of interaction in the process of patient treatment, presents a multi-agent model for forming a rational trajectory of diagnosis and treatment of a patient and its implementation using the digital platform AnyLogic. Section 4 presents the results of primary processing of statistical data for subsequent modeling and their interpretation.

2 Problem Description and Problem Statement

Forming a diagnostic and treatment trajectory is a complex process, often poorly formalized, that can have many alternatives depending on a variety of factors. Such factors include the patient's health status, the presence of certain comorbidities, allergic reactions to certain medications, the resources available to the medical institution (availability of diagnostic equipment, available medications), and many others. The person who decides on the choice of a particular diagnostic and treatment trajectory is the treating physician-therapist.

There are clinical guidelines that describe the treatment process for various conditions. These guidelines may change (improve) over time. And undoubtedly, the treating physician needs information support when choosing one or another diagnostic and treatment trajectory.

The task is to use the available clinical guidelines describing how to treat a patient, the individual characteristics of the patient's health and the resources available to the medical institution in the form of diagnostic equipment and medications to form the most effective diagnostic and treatment trajectory designed to improve the patient's condition as much as possible. The aim of this work is to build a model reflecting the patient's diagnostic and treatment trajectory and to determine the parameters of this model.

The aim of the study is to build a multi-agent model reflecting the patient's diagnostic and treatment trajectory, to determine the parameters of this model and the primary processing of the data required for subsequent modeling.

3 Known Approaches to Solving Such Problems

Among domestic developments using a multi-agent approach in healthcare, we can highlight [5], which considers the implementation of a portal using multi-agent technology in healthcare. The multi-agent system developed and presented in [5, 6] simulates the interaction between general practitioners, the head doctor of a clinic, specialists of a hospital, ambulance service, medical university, managers of the regional Ministry of Health, territorial compulsory medical insurance fund, authorized pharmacological enterprise, patient of medical institutions. In [7] an overview of the application of multi-agent systems for various tasks in the field of health care is given. The authors highlight such areas of application of multiagent systems as: investigating the effectiveness of different mechanisms ensuring the interaction of agents, different scheduling heuristics, etc.; using a multiagent system for operational planning of the treatment process in real hospitals; the multiagent structure is used to build a simulation model of a particular hospital. Such a model reproduces as accurately as possible the organizational structure of the hospital (or its part), available resources (wards, beds, equipment, staff), the mechanism of interaction between units during the treatment process, and actual statistical characteristics of the patient flow. The simulation model makes it possible to improve the organization of the treatment process itself.

In [8], the authors return a multi-agent system for the formation of a schedule for the execution of a procedure in a sanatorium-and-health work.

Among foreign sources, it is possible to highlight the following areas of application of MAS in healthcare: decision-making support for healthcare managers, balancing and maximizing the use of available resources [9]; MAS-based DSS for managing hospital resources [10]. In [11–13] the main directions applications of MAS in the health care sphere.

4 Proposed Approach

4.1 Classification of Interaction Participants in the Formation of the PATient's Treatment Trajectory

The proposed classification of participants in the interaction of the process of the patient's treatment in a hospital, which includes not only the patient and the nursing staff, but also other stakeholders, with description of the main roles is presented in Table 1.

4.2 Choosing a Digital Platform for Multi-agent Modeling

To obtain the best trajectory of in-depth diagnosis and treatment, a multi-agent approach is proposed [14–16]. A multi-agent system (MAS) unit is planned for development, which simulates the interaction of the main participants: patient, diagnostic doctor, therapist, specialized doctors, pharmacist and others. The prototypes of the participants will be the corresponding agents of the MAS, which communicate with each other. As a result of the MAS functioning, the search for the most rational combination of therapeutic (medicinal) and physical influences on a specific person (treatment pathway) and the issuance of personalized recommendations taking into account such characteristics as the patient's health features, lifestyle, demographic data, contraindications to certain groups of drugs, development and severity of disease, associated diseases, personal preferences, characteristics of drugs, their side effects, interaction with drugs, etc., will take place. For this purpose, it is planned to give the appropriate knowledge and target functions to the MAS agents.

The interaction of real participants in the process can be modeled as the interaction of the corresponding prototypes—the agents of the multi-agent system.

Table 1. Classification of participants in the interaction of the process the patient's treatment in the hospital

No.	Participant	Role in the treatment process
1	Patient	Directly passes all diagnostic procedures and treatment
2	Attending doctor	Monitors the patient's health status and is responsible for the choice of methods and means of treatment
3	Diagnostic doctors	Provide diagnostic procedures and interpretation of results
4	Specialized doctors	Carry out a more detailed study and description of the patient's condition, can perform a specific appointment to the patient
5	Pharmacists	Participate in the provision of medicines for the process of diagnosis and treatment of the patient
6	Paramedical personnel	Ensure that the patient undergoes treatment and necessary diagnostic procedures
7	Head of department (administrative person)	Carries out general administrative management of the process of diagnosis and treatment of patients in the department, taking into account the available resources (material, financial, human, etc.)
8	Relatives, patient's surroundings	Participate in the process of interaction between the patient and the attending physician, which is necessary for making decisions in the process of diagnosis and treatment

At the initial stage it is supposed to use digital platform AnyLogic [17] as a simulation tool, which allows building an agent-based model describing the trajectory of patient diagnosis and treatment. Digital platform AnyLogic allows you to build models using three approaches: discrete event, agent-based and system dynamics. The capabilities of multi-approach simulation modeling allow you to develop complex models of phenomena, processes, systems within a single concept and single digital platform.

4.3 Multi-agent Model for the Formation of a Rational Trajectory for the Diagnosis and Treatment of a Patient

In accordance with the previously proposed methodology for solving resource management problems in complex socio-economic systems [18, 19], the multi-agent model for the formation of a rational trajectory for diagnosing and treating a patient can be represented as follows:

$MAS = \ < A, States, Acts, KB, Constraints, F >$, where:
$A = \{a^i{}_j\}$—the set of agents of class i;
$States = \{states^i{}_j\}$—the set of states of the agents of class i;
$Acts = \{acts^i{}_j\}$—the set of possible actions of the agents of class i;
$KB = \{KB^i{}_j\}$—the set of knowledge bases of the agents of class i;
Constraints—the set of domain constraints;
$F = \{F^i\}$—the set of performance evaluation criteria for each class of agents.

Classes of agents are defined on the basis of the above classification in Table 1. By the trajectory of diagnosis and treatment of the patient we will understand the sequence of transitions between the states of the agent «Patient» from the initial to the final one. The transitions between agents' states can depend not only on their own actions, but also on the actions of agents of other classes from the «Acts» set. The functions F allow us to numerically estimate the state the agent is in at a given moment.

To form a rational diagnostic and treatment trajectory, it is required to determine the sequence of actions that would eventually lead the agent «Patient» to one of the final states with the best values of performance criteria from F for each class of agents.

4.4 Development of a Multi-agent Model for the Formation of a Trajectory for the Diagnosis and Treatment of a Patient in the Digital Platform AnyLogic

A multi-agent model using the digital AnyLogic platform is a set of state diagrams for selected classes of agents that describe the behavior of agents and their interaction.

The developed simplified model includes the processes of selection and pre-sorting of patients coming to the hospital. Then the treatment process itself takes place, at the end of which a new triage takes place. Based on the results of this triage, the patient is either discharged or, if the treatment was ineffective, can go through all the stages of diagnosis and treatment again with new prescriptions, that is, according to a new trajectory.

The passage of all stages of diagnosis and treatment of patients is represented as a process of transition of an agent from one state to another. Thus, in order to complete the whole process of treatment in the hospital, it is necessary for the agent «Patient» to move to the final state. Such a representation is quite convenient for modeling, as it makes the model itself extensible in terms of greater detail and subsequent refinement and complication. In addition, it greatly simplifies the processes of correct description of the interactions of the main participants at any of the stages of treatment.

Figures 1 and 2 show, as an example, the developed simplified diagrams of the states of the agents «Patient» and «Attending doctor», describing the treatment process.

Interaction of agents «Patient» and «Attending doctor» is implemented by exchanging messages between the agents of the corresponding types. For example, the attending doctor, when a patient comes to him or her, has to examine the patient, confirm the pre-established diagnosis, and prescribe a set of diagnostic and therapeutic procedures in accordance with the current regulations. This interaction is implemented by sending the «Patient» agent sends a message to the «Attending doctor» agent. The «Attending doctor» agent, receives the corresponding message, carries out a set of necessary actions for prescribing a certain treatment.

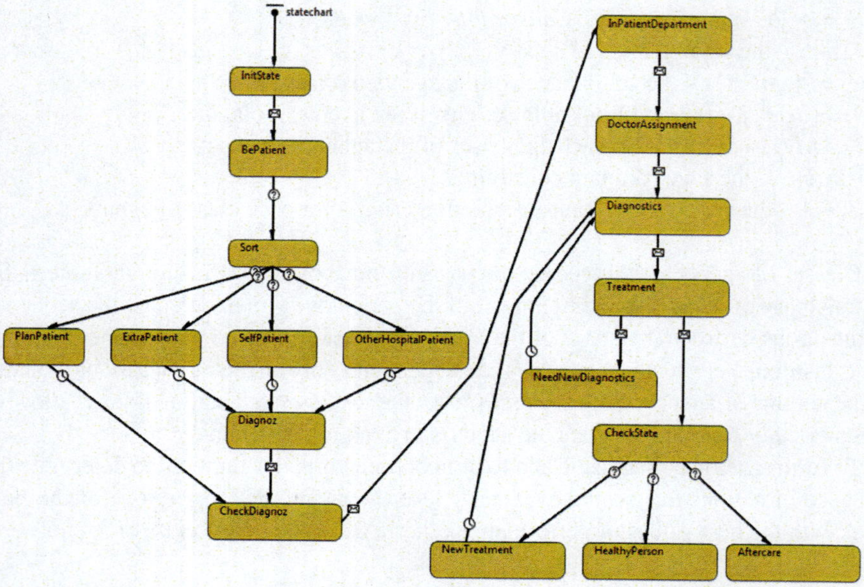

Fig. 1. Diagram of the «Patient» agent states

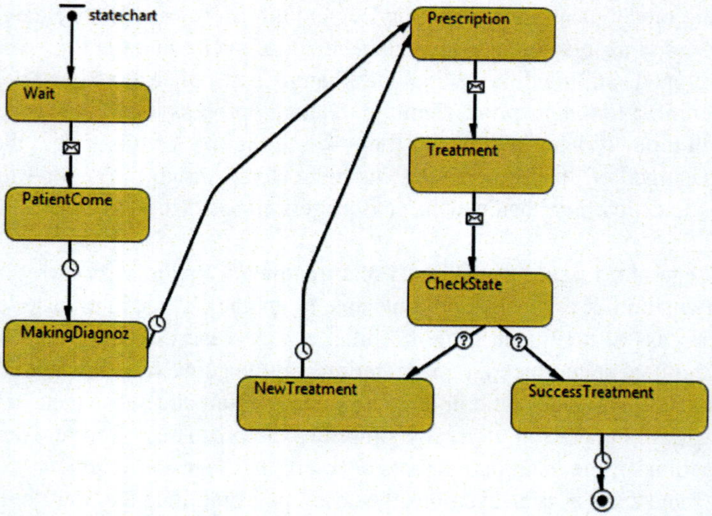

Fig. 2. Diagram of the states of the «Attending doctor» agent

5 Primary Data Processing for Subsequent Modeling

As parameters that characterize the process of treatment of bronchopulmonary diseases in patients hospitalized at the initial were the following:

The intensity of patients' admission to the hospital.

Parameters reflecting the distribution of patients admitted to the hospital (outpatient treatment or hospitalization).

Obtaining baseline data and their further processing is an important step for subsequent modeling. In our case, such data include the statistics of patients' admissions and their subsequent distribution.

As an example of initial data, we took real data on the admission of patients for 2021 to one of the clinics in the city of Ufa. Pre-processing consisted in collecting and processing data on inpatient admissions and subsequent sorting, which determines where the patient will be sent: to outpatient treatment or to one of the inpatient departments. We analyzed daily data reflecting patient admissions and their further distribution for the year 2021.

The results of the analysis are shown in Fig. 3.

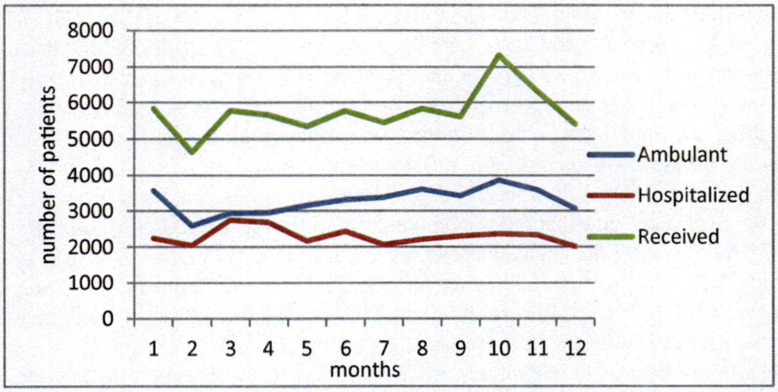

Fig. 3. Distribution of admitted patients.

The results of the analysis allow us to conclude that of the total number of inpatient admissions, most from 51 to 62% are sent for outpatient treatment, and about 32–48% of patients are hospitalized. At the same time, the variability of hospitalized patients is much higher, up to 1.5 times, while the variability of those patients who sent for outpatient treatment is only 1.2.

The analysis of admission schedules makes it possible to assume the presence of seasonality in this process. This circumstance requires a more thorough study and will be the subject of further research.

6 Conclusion

A classification of participants in the interaction in the formation of the trajectory of diagnostics and treatment of patients is proposed, which is used to design agents of a multi-agent system. It includes 8 main participants. A multi-agent model has been developed to form a rational trajectory for diagnosing and treating a patient. With the help of the AnyLogic digital platform, the state diagrams of the agent's "Patient" and "Attending doctor" were developed—prototypes of the main participants in the interaction. The primary processing of the data necessary for modeling the process of diagnosing and treating patients was carried out. It showed the need for a more thorough analysis of the statistics of patient admissions in order to identify seasonal factors.

In the future, it is planned to develop formalized target functions of the agents and the whole system, which will make it possible to numerically evaluate their performance and compare different alternative trajectories, as well as to pay more attention to agent-based modeling directly the process of diagnosis and treatment.

This study was funded by a Russian Science Foundation grant 22-19-00471.

References

1. Frolov, S.V., Kulikov, A.Y., Ostapenko, O.A., Strygina, E.V.: Medical decision support systems in medicine. Sci. J. Founders Olympus 9(32), 9–16 (2018). ISSN: 2413-7081
2. Kupeeva, I.A., Raznatovsky, K.I., Ravodin, R.A., Karelin, V.V., Bure, V.M., Gusarov, M.V.: Evaluation of the effectiveness of an intelligent system to support medical decision-making. Bull. St. Petersburg Univ. Appl. Math. Inf. Manag. Process. 2(62–68) (2016)
3. Ravodin, R.A.: Intelligent decision support system for medical decisions in dermatovenereology. Probl. Med. Mycol. 3(59–65) (2014). ISSN: 1999-6780
4. Doan, D.K., Kroshilin, A.V., Kroshilina, S.V.: Review of approaches to the problem of decision making in medical information systems under uncertainty. Tech. Sci. (05.02.00, 05.13.00, 05.17.00, 05.23.00) Fundamental research no. 12(26–30) (2015)
5. Wittikh, V.A., Gusarova, G.I., Kuznetsov, S.I., Pavlov, V.V., Skobelev, P.O., Surnin, O.L., Fedoseeva, L.S., Chernov, E.V., Shamashov, M.A.: Network multi-agent model of the health care management system of the region and the system for monitoring the efficiency and quality of the work of doctors in polyclinics. Informatization Health Compulsory Med. Insur. Syst. 11 (2004). www.idmz.ru
6. Wittikh, V.A., Ezhkov, V.N., Kotelnikov, G.P., Larionov, Y.K., Pavlov, V.V., Skobelev, P.O., Fedoseeva, L.S., Shamashov, M.A.: Development of an integrated multi-agent system for healthcare management in the region. In: Proceedings of the 4th International Conference on Problems of Control and Modeling of Complex Systems, Samara, June 17–24, SNC RAS, Samara, pp. 398–406 (2002)
7. Dorofeyuk, A.A., Dorofeyuk, Y.A., Mandel, A.S., Pokrovskaya, I.V., Chernyavsky, A.L.: Multi-agent systems in the problems of managing the medical process in large-scale complexes of large-scale research. In: VIII International Conference Control of Potentially Large-Scale Systems, MLSD'2015, Moscow, IPU RAS (2015)
8. Bogdanova, D.R., Rizvanov, D.A., Senkina, G.V., Popov, D.V.: Multi-agent system for scheduling procedures for vacationers in a sanatorium complex. In: Proceedings of the 8th International Conference on Computer Science and Information Technology (CSIT'2006), Karlsruhe, Germany, vol. 1 (S), pp. 118–124 (2006)
9. Benhuiji, N., Roy, D., Ansio, D.: Multi-agent system for patient-centered healthcare. IFAC-PapersOnLine **48**(3), 710–714 (2015)
10. Zoghlamy, N., Glaa, B., Rabah, S., Abed, M.: Health decision support tool: multi-agent system for hospital bed management. Int. J. Appl. Manag. Sci. **10**(1) (2018)
11. Shakshouki, E., Reed, M.: Application of multi-agent systems in healthcare: current technologies and plans for the future. In: 6th International Conference on Environmental Systems, Networks and Technologies (ANT 2015). Procedia Comput. Sci. **52**, 252–261 (2015)

12. Paranjape, R., Sadanand, A.: Multi-Agent Systems for Modeling and Modeling in the Healthcare Sector (2009)
13. Becker, K.A., Lorig, F., Timm, I.J.: An overview of multi-agent systems to improve home healthcare management. AIH@IJCAI (2018)
14. Weiss G.: Multiagent Systems: A Modern Approach to Distributed Artificial Intelligence, p. 619. MIT Press (1999)
15. Wooldridge, M.J., Jennings, N.R.: Intelligent agents: theory and practice. Knowl. Eng. Rev. 10(2), 115–152 (1995)
16. Skobelev, P.: Multi-agent systems for real time resource allocation, scheduling, optimization and controlling: industrial application. In: 10th International Conference on Industrial Applications of Holonic and Multi-Agent Systems (HoloMAS 2011), France, Toulouse, pp. 5–14. Springer (2011)
17. Mitrakov, A.A.: Fundamentals of Agent-Based Modeling System Design. Publishing house of PGNIU, Perm (2013)
18. Kovács, G., Yussupova, N., Rizvanov, D.: Resource management simulation using multi-agent approach and semantic constraints. Pollack Periodica **12**(1), 45–58 (2017). Published by the Akademiai Kiado, Budapest, Hungary
19. Yussupova, N., Rizvanov, D.: Decision-making support in resource management in manufacturing scheduling. In: Proceedings of 18th IFAC Conference on Technology, Culture and International Stability (TECIS 2018), vol. 51, Issue 30, pp. 544–547 (2018)

Intelligent Decision Support Based on Mental User Models: Research Design

A. V. Vicentiy[1,2](✉)

[1] Putilov Institute for Informatics and Mathematical Modeling—Subdivision of the Federal Research Centre, Kola Science Centre of the Russian Academy of Sciences, 24A, Fersman St., Apatity, Murmansk Region 184209, Russia
alx_2003@mail.ru

[2] Apatity Branch of Murmansk Arctic State University, Lesnaya St. 29, Apatity, Murmansk Region 184209, Russia

Abstract. This paper describes the research design in the field of user modeling for intelligent decision support. The relevance of the research, the novelty and the significance of the research results for the development of a new type of interfaces based on cognitive visualization of spatial data are described. This paper is the first step in a research project aimed at developing new approaches for creating cognitive user interfaces. The paper describes the overall research design, the research problem, the specific tasks within the problem, the scientific novelty of the research, and the expected results. In addition, we give a brief overview of the current state of research on the problem, describe the main directions of research in world science, as well as suggest some methods and approaches to implement the main stages of the project. The duration of the research is two years, the number of performers is three people. Thus, we present a brief overview in the field of intelligent decision support based on mental user models, describe our research strategy and discuss the problems of creating and using mental user models for decision-making purposes.

Keywords: Mental user model · Intelligent decision support systems · Information need · User modeling · Cognitive geovisualization · User interface · User experience · Human-computer interaction

1 Introduction

At present, intelligent information systems for decision support (DSS) are becoming more and more widespread. Such systems use information about the user to improve the efficiency of human-computer interaction and better meet the information needs of the user. Thus it is possible to increase the efficiency of user cooperation with the system, reduce the decision-making time and improve the quality of decisions [1].

R. Silhavy et al. (Eds.): CoMeSySo 2022, LNNS 596, pp. 732–740, 2023.
https://doi.org/10.1007/978-3-031-21435-6_63

Consideration of information about user preferences, his features and characteristics is carried out by creating a user model [2]. Various approaches to user modeling have been developing for quite a long time. However, to date, there is no common understanding among scientists and practitioners from various subject areas as to what exactly a user model should be. Also, there are currently no single, unified approaches to user modeling even within the same subject area.

Nevertheless, information systems using user models, as well as information technologies of user modeling, are actively developing despite the fact that there are a number of problems associated with the formal assessment of the adequacy of the models used and the uncertainty of the quality of the results obtained [3, 4]. Under these conditions, the task of creating a formalized approach to the synthesis of mental models of users to improve the efficiency of user interaction with the software system is relevant.

The range of scientific, technical and practical tasks for which user modeling is used is quite wide. At this stage, we limit our research to considering the problems of developing mental models and analyzing user information needs for intelligent decision support in geodata visualization and analysis. This limitation is due to two reasons. Firstly, we believe that, despite the rapid development in recent years of a variety of intelligent technologies for processing visual data, the analysis of geographic data is an area in which the role of a human being as the data interpreter and decision maker (DM), remains quite large [5]. Secondly, in previous years, with the direct participation of the author, a number of researches in the field of development of cognitive interfaces for spatial data processing have been carried out [6]. Thus, we plan to use the existing scientific groundwork to solve the problem of synthesizing mental models of users in the development of cognitive interfaces for spatial data processing.

The scientific significance and relevance of the problem of improving the effectiveness of human-computer interaction to provide decision support is related to improving the efficiency of perception and analysis of information by the decision maker and, consequently, increasing the speed and quality of decisions. In this context, the development and application of the user mental model in decision support systems can be the basis for the formation of recommended sources and ways of presenting information, which take into account not only the user's direct requests, but also his characteristics, typical behavior in decision-making and other characteristics. We proceed from the fact that the individual characteristics of a decision maker can significantly affect the process and the result of decision-making [7]. Thus, taking into account the characteristics of DM formalized by the user mental model allows us to influence the usefulness of the information issued to the user. That is, the use of the user mental model allows us to increase the pragmatic value of the information issued by the system to the user.

As a result of the research work, we expect to expand the frontiers of knowledge in the field of cognitive technologies of human-computer interaction. To this end, we propose to develop a formal approach to intelligent decision support for geodata visualization and analysis based on the synthesis of mental models and analysis of user information needs.

This paper is the first in a research project on this topic. It describes the general design of the research, the scientific problem of the research, the specific tasks within the problem, the scientific novelty of the research, as well as the expected results. The paper provides a brief overview of the current state of research on the problem, describes the main research areas in world science, as well as suggests methods and approaches for implementing the project.

2 Materials and Methods

A special feature of the research is the development of a new approach to intelligent decision support. This approach allows us to take into account the characteristics of the user, using a mental model. Thus, we can adapt the behavior of the decision support system to the specific user, to make the system personalized. The user model and the formalized description of information needs are oriented to decision support systems that are designed to visualize and analyze geodata by a human (decision maker). In the course of the project we will use various methods and approaches corresponding to the main objectives of the research.

To assess the current state of researches and to clarify the main directions of research in the world science we will perform a systematic review of scientific sources. The basis of the systematic review is the methodology proposed by B. Kitchenham and co-authors, which is the de facto standard for systematic literature reviews in computer science [8, 9]. In the framework of this methodology it is necessary to make a specification of the used sources of information and search queries, as well as to keep a research protocol, fixing in it all the intermediate results. In this way, it is possible to achieve correctness and reproducibility of the results of the systematic review.

To produce the principles and conceptual model of intelligent decision support we will rely on the methods of system analysis, conceptual modeling, as well as elements of functional-objective analysis [10].

To describe the development of the mental model and the representation of the user's information needs, methods of set theory and ontological design will be used, in particular, methods of automated formation and filling of thesaurus, ontology matching using the top level ontology, as well as methods of automated text analysis and extraction of named entities [11–13].

To solve the problem of integrating the obtained results within a unified approach to intellectual support of decision-making, methods of matching ontologies, ontological and conceptual modeling will be used. To implement a software prototype that demonstrates the features of the proposed approach, in addition to engineering approaches related to programming, known methods and technologies of data visualization, as well as elements of interactive visual analysis will be used.

3 Results

In order to develop the research design, it is necessary to have an understanding of the current state of research in the field of intelligent decision support systems, as well as to become familiar with the current approaches, methods and data sources that are used for user modeling. Therefore, in the first stage of research design we performed a brief review of scientific publications on the theme of the research. As a result of performing the review, we identified the main research areas in the world science and potential scientific competitors.

Modern intelligent information systems, including decision support and recommendation systems, are used in a variety of applications. Often such systems use information about users to generate personalized recommendations. Thus, users receive relevant resources from a variety of possible options, which makes the work of such information systems more effective. At the same time, user behavior is often quite complex. Therefore, the development of new approaches to create personalized user models is an important and urgent task nowadays [14, 15].

The main problem in developing human-computer systems is to create an information system that is designed to be used by many users, but which, from the perspective of a particular user, works as if the system had been created just for him. One way to solve this problem is to create user models that will influence the behavior of the human-computer system and provide the user with information resources appropriate to his specific goals, knowledge, and information perception abilities. Various works in the field of creating user modeling methods have been going on for several decades [16, 17], but there is still no common understanding of what exactly the user model should be to ensure the best functioning of the information system [18].

One of the main obstacles to creating effective user models in previous years was the insufficient amount of information about the user. It should be noted that significant progress has been made in recent years in obtaining and processing large volumes of heterogeneous data from various sources. Collecting and processing information about users of information systems for decision support (including recommendation systems) is no exception in this sense. Currently, there are information systems capable of collecting various data about the user, which in some works are also called digital user traces [19, 20]. Here is just a short list of user characteristics and features that are proposed to be taken into account to improve the efficiency of interaction between user and human-centered information system: age, gender, language, ethnicity, [21, 22] emotional reactions arising from interaction with information system, character, personality characteristics, [23–25] culture, socioeconomic status, educational level, technical skills, physical features, vital signs (pulse, blood pressure, temperature, etc.), [26–28]. Thus, it can be noted that information systems using user models, as well as information technologies of user modeling are actively developing despite the fact that there are a number of problems associated with the formal assessment of the adequacy of the models used and the uncertainty of the quality of the results obtained [29].

In the field of user modeling traditionally distinguish the following generalized types of models:

- static user models, based on immutable data sets collected during model creation;
- dynamic user models, which provide actualization of the user representation in the system due to the support of data update mechanisms;
- stereotype-based user models that allow classifying (referring to this or that stereotype) users on the basis of the collected information about the user;
- highly adaptive user models, allowing to build models of specific users rather than a class or group of users with similar characteristics.

As for approaches to adapt the behavior of the information system according to the chosen user model, there are usually two groups of methods:

- rule-based filtering;
- collaborative filtering.

Three main methods are used to collect data about the user:

- collection of data about the user at the first contact with the system, for example, at registration;
- collecting data about the user in the process of observing him during the work with the system;
- a hybrid approach that combines the advantages of the two above [30].

It is also possible to mention several names of scientists working in similar directions and close subject areas. Especially noteworthy are the works of such authors as: Alexis Tsoukias and colleagues from the University of Paris-Dauphine (France); David Contreras of the University of Arturo Prata (Chile); Linus Dietz of the Technical University of Munich (Germany).

Having identified the main areas of research in the development of intelligent decision support systems based on user modeling, we proceeded directly to the development of the research design. The research design is presented as a list of specific tasks and expected results. Also the sequence of tasks, necessary methods and materials are indicated, the significance of the main results of the research project is evaluated.

As part of the study, we distinguish three main tasks:

Task 1. Perform a systematic review of user modeling methods for intelligent decision support based on spatial and temporal data analysis. Despite the fact that it is possible to find a sufficient number of works devoted to user modeling, these works usually do not have a systemic nature and consider only narrow, problem-oriented tasks of user modeling when creating specific software systems. As part of our preliminary review of scientific publications and study of the subject area, we were unable to find a single modern systematic review on this theme. Therefore, we believe that performing a systematic review on the topic of research is an urgent task. The review will be published as a paper and can be used by other researchers in the future.

Task 2: Development of principles and conceptual model of intellectual decision support based on mental model and user information needs in the field of visualization and geodata analysis. The conceptual model is formed in accordance with the principles of intellectual decision support. The principles are the basis of the information environment of intelligent decision support based on the mental model and information needs of the user. It describes the information and functional level of the future conceptual model, defines the purpose of intellectual decision support, fixes limitations, and defines types of information sources.

Task 3. Development of a mental model and formalized description of user information needs for intelligent decision support based on analysis of spatio-temporal data. This task describes an approach for creating a mental model and a formalized description of user information needs in the context of the decision-making task based on the analysis of spatio-temporal data. A method based on taking into account the context of the tasks to be solved, as well as the analysis of the user's thesaurus, is proposed to create the models.

All tasks must be performed sequentially. None of the tasks can be started until the previous task is fully completed. The results of each task will either be published in a paper or presented at a conference in order to receive critical comments and recommendations, as well as to evaluate their significance and compliance with the world level of research in this field.

In addition to describing the objectives, we also included in the research design a description of the main expected results and their significance for the development of the research theme. The wording of the main results partially coincides with the wording of the tasks, but allows us to clarify it, as well as to give a more concrete idea of the results that we seek to obtain. The main results of the research are as follows:

(1) Systematic review of user modeling methods in the tasks of intelligent decision support based on analysis of spatial and temporal data. The significance of the result is to describe and systematize information about existing and currently most common approaches, methods and technologies for creating and using user models in intelligent information systems for decision support, as well as to identify the most promising areas of user modeling within the project tasks. The results of the systematic review will be used to fulfill Task 2 and Task 3 mentioned above.
(2) Principles and conceptual model of intellectual decision support based on the mental model and information needs of the user in the field of visualization and analysis of geodata. The significance of the result is that the principles define the purpose of intelligent decision support, fix the limitations, determine the types of information sources for the subsequent modeling of the user and the description of its information needs. The conceptual model allows to carry out decision support in the field of visualization and analysis of geodata on the basis of consideration of preferences, features and characteristics of the user, according to his mental model, and also the decision-making task expressed through the description of the information need of the user.
(3) The mental model and formalized description of the user's information need for intelligent decision support based on the analysis of spatial and temporal data. The mental model and the formalized description of the information need allow to determine the decision parameters, which most fully correspond to the main characteristics and preferences of the user when solving a specific decision-making problem. In addition, the results will be used to develop a software prototype of intelligent decision support information system.
(4) Integration of the obtained results within the framework of the developed approach to intellectual support of decision-making in the field of visualization and analysis of geodata on the basis of development of mental models and analysis of information needs of users. At this stage there is a coordination and harmonization of the results obtained earlier, the correctness of the proposed models and the possibility of their joint use are checked.

(5) Software prototype, demonstrating the features of intelligent decision support in the field of visualization and analysis of geodata based on the use of a user mental model and formalized description of the information needs of the user. The significance of the result is to test the proposed approach on the example of solving specific decision-making tasks, taking into account the use of spatial and temporal data by the decision maker. Also at this stage the speed of data processing on specific cases is tested in practice, as well as the possibility of practical use of different types of information sources for modeling the user and describing his information needs.

It can be noted that the result 4 and result 5 stand out from the general list and have a practical orientation. We believe that to confirm the results obtained in the previous stages of research, it is necessary to develop and test an information system that demonstrates the main results obtained in the field of user modeling and intelligent decision support based on the obtained models. The distribution model of this software is free/libre and open-source software.

4 Conclusions and Future Work

The paper describes the research design for developing an approach to intelligent decision support based on mental models of users. The description of the research design includes a description of the scientific problem the project is aimed to solve, its scientific importance and relevance. Also described the specific objectives of the study and the expected results. The assessment of the current state of research in the field of development of systems of intellectual support for decision-making and creation of user models is given, the main directions of research and scientific competitors are identified. Adequate methods and approaches to solve the problems of research have been selected.

As mentioned above, the research project is for two years and three main performers. We also plan to involve students from the Apatity branch of Murmansk Arctic State University, as well as undergraduate and graduate students from the Kola Science Center of the Russian Academy of Sciences, if funding permits. Currently we are working on the first task of the project. We plan to complete it by the end of 2022.

As a result of the project will form a scientific reserve in the field of development of intelligent decision support systems of a new type, focused on the user and allowing to take into account his characteristics in decision-making. Elements of practical implementation of the project (software prototype) will be used in the development of regional management support system in the form of individual modules.

References

1. Smirnov, A., Levashova, T., Petrov, M.: Scenario model of intelligent decision support based on user's digital life models. Inf. Control Syst. **4**, 47–60 (2021). https://doi.org/10.31799/1684-8853-2021-4-47-60
2. Smirnov, A., Ponomarev, A., Levashova, T., Shilov, N.: Conceptual framework of a human-machine collective intelligence environment for decision support. In: Proceedings of the Bulgarian Academy of Sciences, vol. 75, no 1, pp. 102–109 (2022). https://doi.org/10.7546/CRABS.2022.01.12

3. Araujo, T., Helberger, N., Kruikemeier, S., de Vreese, C.H.: In AI we trust? Perceptions about automated decision-making by artificial intelligence. AI Soc. **35**(3), 611–623 (2020). https://doi.org/10.1007/s00146-019-00931-w
4. Asniar Surendro, K.: Predictive analytics for predicting customer behavior. In: International Conference of Artificial Intelligence and Information Technology (ICAIIT), pp. 230–233. IEEE (2019). https://doi.org/10.1109/ICAIIT.2019.8834571
5. Vicentiy, A.V., Vicentiy, I.V.: The method of dynamic visualization of spatial data for cognitive interfaces of information systems supporting regional management. In: 19th International Multidisciplinary Scientific Geoconference SGEM 2019, pp. 667–672 (2019). https://doi.org/10.5593/sgem2019/2.1/S07.087
6. Vicentiy, A.V.: Development of methods and tools to support regional management in the Arctic zone of the Russian Federation based on cognitive interfaces. IOP Conf. Ser.: Earth Environ. Sci. **320**, 012139 (2019). IOP Publishing. https://doi.org/10.1088/1755-1315/302/1/012139
7. Pentland, B.T., Recker, J., Wolf, J., Wyner, G.: Bringing context inside process research with digital trace data. J. Assoc. Inf. Syst. **21**(5), 1214–1236 (2020). https://doi.org/10.17705/1jais.00635
8. Kitchenham, B., Brereton, P.: A systematic review of systematic review process research in software engineering. Inf. Softw. Technol. **55**(12), 2049–2075 (2013)
9. Kitchenham, B., Charters, S.: Guidelines for performing systematic literature reviews in software engineering. Technical report EBSE-2007-01, School of Computer Science and Mathematics, Keele University (2007)
10. Hevner, A.R., March, S.T., Park, J., Ram, S.: Design science in information systems research. MIS Q.: Manag. Inf. Syst. **28**(1), 75–105 (2004). https://doi.org/10.2307/25148625
11. Wulandari, I.A., Sensuse, D.I., Krisnadhi, A.A., Akmaliah, I.F., Rahayu, P.: Ontologies for decision support system: the study of focus and techniques. In: 10th International Conference on Information Technology and Electrical Engineering (ICITEE), pp. 609–614 (2018). https://doi.org/10.1109/ICITEED.2018.8534947
12. Kadima, H., Malek, M.: Toward ontology-based personalization of a recommender system in social network. Int. J. Comput. Inf. Syst. Ind. Manag. Appl. **5**, 499–508 (2013). https://doi.org/10.1109/SOCPAR.2010.5685957
13. Pilecki, B.M., Vicentiy, A.V.: Development of a method for extracting spatial data from texts for visualization and information decision-making support for territorial management. IOP Conf. Ser.: Earth Environ. Sci. Institute of Physics Publishing (2020). https://doi.org/10.1088/1755-1315/539/1/012087
14. Ben Hassen, A., Ben Ticha, S., Chaibi, A.H.: Deep learning for visual-features extraction based personalized user modeling. SN Comput. Sci. **3**, 261 (2022). https://doi.org/10.1007/s42979-022-01131-y
15. Wang, L., Huang, C., Lu, Y., Ma, W., Liu, R., Vosoughi, S.: Dynamic structural role node embedding for user modeling in evolving networks. ACM Trans. Inf. Syst. **40**(3), 21, Article 46 (July 2022) (2021). https://doi.org/10.1145/3472955
16. Fischer, G.: User modeling in human-computer interaction. User Model. User-Adap. Inter. **11**(11), 65–86 (2001). https://doi.org/10.1023/A:1011145532042
17. Hothi, J., Hall, W.: An evaluation of adapted hypermedia techniques using static user modelling. In: Proceedings of the 2nd Workshop on Adaptive Hypertext and Hypermedia, Southampton University, Electronics and Computer Science University Road, Southampton, Hampshire, UK (1998)
18. Piao, G., Breslin, J.G.: Inferring user interests in microblogging social networks: a survey. User Model. User-Adap. Inter. (UMUAI). **28**(3), 277–329 (2018) arXiv:1712.07691. https://doi.org/10.1007/s11257-018-9207-8. S2CID 3847937

19. Seeskin, Z.H., et al.: Uses of alternative data sources for public health statistics and policymaking: challenges and opportunities. In: Proceedings of 2018 Joint Statistical Meetings. American Statistical Association, pp. 1822–1861 (2018)
20. Han, M.L., Kwak, B.I., Kim, H.K.: CBR-based decision support methodology for cybercrime investigation: focused on the data-driven website defacement analysis. Secur. Commun. Netw. **2019** (2019)
21. Singh, H., Khalajzadeh, H., Paktinat, S., Graetsch, U.M., Grundy, J.: Modelling human-centric aspects of end-users with iStar. J. Comput. Lang. **68**, 101091 (2022). https://doi.org/10.1016/j.cola.2022.101091
22. Grundy, J., Khalajzadeh, H., McIntosh, J., Kanij, T., Mueller, I.: HumaniSE: approaches to achieve more human-centric software engineering. In: Ali, R., Kaindl, H., Maciaszek, L.A. (eds.) Evaluation of Novel Approaches to Software Engineering (ENASE 2020). Communications in Computer and Information Science (CCIS), vol. 1375, pp. 444–468. Springer, Cham (2021). https://doi.org/10.1007/978-3-030-70006-5_18
23. Curumsing, M.K., Fernando, N., Abdelrazek, M., Vasa, R., Mouzakis, K., Grundy, J.: Understanding the impact of emotions on software: a case study in requirements gathering and evaluation. J. Syst. Softw. **147**, 215–229 (2019). https://doi.org/10.1016/J.JSS.2018.06.077
24. Wirtz-Brückner, S., Jakobs, E.-M., Ziefle, M.: Age-specific usability issues of software interfaces. LandLeuchten (BMVI-funded) view project public acceptance and perception of carbon capture and utilization (CCU) view project. Proc. IEA. **17**, 1–10 (2009)
25. Stock, S.E., Davies, D.K., Wehmeyer, M.L., Palmer, S.B.: Evaluation of cognitively accessible software to increase independent access to cellphone technology for people with intellectual disability. J. Intellect. Disabil. Res. **52**, 1155–1164 (2008). https://doi.org/10.1111/j.1365-2788.2008.01099.x
26. Jim, A., Shim, H., Wang, J., Wijaya, L., Xu, R., Khalajzadeh, H., Grundy, J., Kanij, T.: Improving the modelling of human-centric aspects of software systems: a case study of modelling end user age in wirefame designs. In: Proceedings of the 16th International Conference on Evaluation of Novel Approaches to Software Engineering, pp. 68–79. SCITEPRESS—Science and Technology Publications (2021). https://doi.org/10.5220/0010403000680079
27. Grundy, J., Khalajzadeh, H., Mcintosh, J.: Towards human-centric model-driven software engineering. In: Proceedings of the 15th International Conference on Evaluation of Novel Approaches to Software Engineering, pp. 229–238. SCITEPRESS—Science and Technology Publications (2020). https://doi.org/10.5220/0009806002290238
28. Vicentiy, A.V.: The geoimage generation method for decision support systems based on natural language text analysis. In: Silhavy, R. (ed.) CSOC 2021. LNNS, vol. 230, pp. 609–619. Springer, Cham (2021). https://doi.org/10.1007/978-3-030-77442-4_51
29. Bahrainian, S.A., Crestani, F.: Tracking smartphone app usage for time-aware recommendation. In: Choemprayong, S., Crestani, F., Cunningham, S.J. (eds.) ICADL 2017. LNCS, vol. 10647, pp. 161–172. Springer, Cham (2017). https://doi.org/10.1007/978-3-319-70232-2_14
30. Jameson, A., Paris, C., Tasso, C. (eds.): User Modeling. ICMS, vol. 383. Springer, Vienna (1997). https://doi.org/10.1007/978-3-7091-2670-7

Comparison of Web Pagebuilders and Their Use in Teaching Web Development

Jiri Vojtesek(✉) and Jakub Josef Forman

Tomas Bata University in Zlin, Faculty of Applied Informatics, Nam. T.G. Masaryka 5555, 760 01 Zlin, Czech Republic
vojtesek@utb.cz, j_forman@utb.cz
http://www.utb.cz/fai

Abstract. Web development and web design are one of the main aims of informatics. The technology age allows the creation of websites with many frameworks, languages, and utilities. One helpful utility is Pagebuilders. These editors or applications are useful for designing prototyping, building landing pages, learning how web design UI/UX works, or creating webs and e-shops for customers in record time. The article describes block programming and comparison with Pagebuilders. The following pages compare five popular Pagebuilders and their usage with their pros and cons, style of their main aim, and functions mainly aimed at teaching web development.

Keywords: Visual Pagebuilders · Web design · Programming · Teaching

1 Introduction

The last article [1] introduces web development and its teaching. The report summarizes the main advantages and disadvantages of traditional web development and manual programming. It compares them with today's modern methods through Pagebuilders in web Contact Management systems (CMS). The article concludes by answering SEO questions about page builder solutions' advantages and disadvantages. This article describes block programming and Pagebuilders with their comparison.

Websites are becoming increasingly popular and are attracting young people to create them. Modern technologies and web development systems (web applications) allow young people to experience the other side of web development, from the design and background to the implementation and deployment on the server. Several resources on the Internet deal with tutorials on how to create web pages that the user writes themselves in HTML, CSS, and other languages, but pages created in this way are complicated for "beginner coders" to maintain. On the other hand, there are several solutions called Pagebuilders, also

R. Silhavy et al. (Eds.): CoMeSySo 2022, LNNS 596, pp. 741–753, 2023.
https://doi.org/10.1007/978-3-031-21435-6_64

known as Theme Builders, which aim to enable the creation of websites without the need to know a programming language and often without design and layout experience. There are many types of Pagebuilders, each suitable for different things.

Pagebuilders can also be found in some primary schools in the Czech Republic, where pupils are shown how to create their first simple website by dragging and dropping elements into a predetermined template. This method of demonstration, along with their design and practice, allows them to understand the design and layout of a website better than creating a basic Hello World page.

2 Teaching Websites in a Graphical Way

Creating websites using Pagebuilders could be categorized as block programming, as they are essentially blocks. Each block the student drags into the page is defined and has functions and settings similar to block programming.

2.1 Block-Base Coding

Block-based coding or programming utilizes a drag-and-drop learning environment, where programmers use code instruction "blocks" to construct animated stories and games – see Fig. 1. It's an entry-level activity where kids can gain a foundation in computational thinking through visuals instead of coding based on text [2].

When we mention blocks, it's a way to describe the "pieces" or "chunks" of instructions a user is putting together to tell their creation what to do. For example, Scratch programing language is block-based and is one of the most popular options for kids' coding languages. Also, we have many more options like Kodu Game Lab, Lego Mindstorms, and more [2]. Blocks are primarily defined as:

- Action blocks - Sound, Motion, ...
- Logic blocks - If statements and others
- Repetition blocks
- Property blocks - Variables
- Setter blocks - For changing variables.

Why is block coding for kids necessary? Block coding is a programming activity that allows children to understand how coding works to develop digital animation, games, or robots. It can also help get children used to computational thinking needed in programming [3]. Teaching children how to code at a young age will prepare them for the future. What's more, there are skills children can learn better through coding, such as the following:

- Mathematical skills
- Logic
- Logical thinking
- Writing
- Communication
- Creativity.

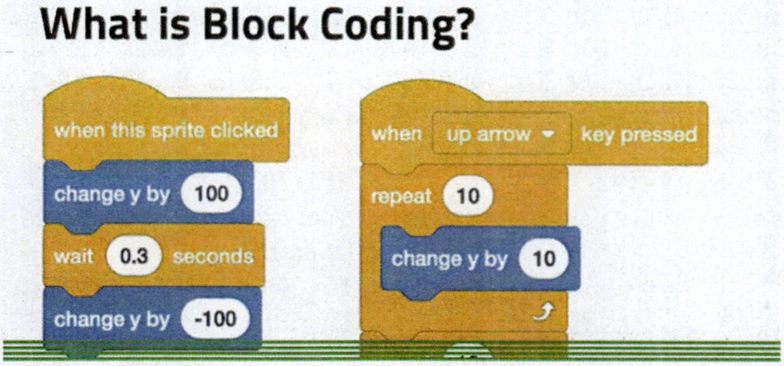

Fig. 1. What is block coding [3]

2.2 Block Programming or Websites

There is no such thing as block programming for websites, but as mentioned in previous section, web development using Pagebuilders can be partially included in this category. It does not mean that students will then be able to write code like a programmer, but they will understand the graphical layout of websites, how to adapt them for mobile phones and tablets, or the logic of the elements used on the page. If this work is viewed from a programming perspective, it is the front-end - what the user/visitor of the website sees and interacts with. In contrast, block programming for children tends to support the backend - the logic behind the application, the logical units, conditions, and actions.

However, a particular part of the logic also takes place on the front-end of the web pages, so Pagebuilders offer the possibility to customize what happens when a button is pressed, what action will be called, or what will happen in the background of the page, etc.

In our previous experiences, we've received feedback from students several times who have said that a form of graphic web design is more beneficial to them than cramming a code they don't understand or find difficult to know into a simple, functional design. However, this does not mean that students should not understand the basic knowledge of web programming and its application in practice. Pagebuilders help create a basic overview of how a graphical design of a website works, that it is made up of blocks (essentially HTML tags like `<span>`, `<div>`, ...), or that creating a website doesn't necessarily mean knowing the code completely.

3 Using Pagebuilders in Practice

Using Pagebuilders doesn't always mean the same thing. For example, we know companies and their websites/blogs using the WordPress content management system. The editorial system for several versions now included a Page Builder called Guttenberg instead of the original WYSIWYG editor. This Page Builder allows you to drag and drop blocks into a page, edit the content, and format it in any way you need [4].

Well-known companies that use WordPress (also some Page Builder) for blog posts, landing pages, or complete website content [5]:

- Sony Music
- PlayStation Blog
- TechCrunch
- The Next Web
- Time Magazine
- CNN Press Room
- Disney Books
- The Roling Stones
- Spotify Newsroom
- Wired
- Microsoft News
- TED Blog
- Angry Birds
- Usain Bolt
- The New York Times
- The White House
- Creative Commons
- 9to5Mac
- Apple
- Samsung
- and many more

Many large companies (not only IT-based) use their custom Page Builder solutions, which can be used to manage web content faster than manually. In addition, implementing a content management system or the Page above Builders allows them to reduce the costs spent on creating and managing web content using experienced programmers.

4 Known Pagebuilders

Several Pagebuilders are sometimes more, sometimes less helpful. The Pagebuilders below are chosen because of their different approaches and uses.

4.1 Bootstrap Studio

As the name suggests, Bootstrap Studio is a studio for creating websites based on the Bootstrap CSS framework. It offers a simple page layout similar to graphical editors Figma, Sketch, Photoshop, and others. Example Bootstrap Studio application is in Fig. 2.

This studio is straightforward to use, offering **advantages** such as [7]:

- Block dragging - which is included in the basic CSS Framework
- Defining custom blocks
- Custom CSS + customization options
- Importing CSS libraries
- Custom JavaScript + customization options
- Importing JavaScript libraries
- Work with external and internal libraries
- Ability to work with fonts
- FontAwesome Pro support
- Informative navigation bar showing HTML element embedding

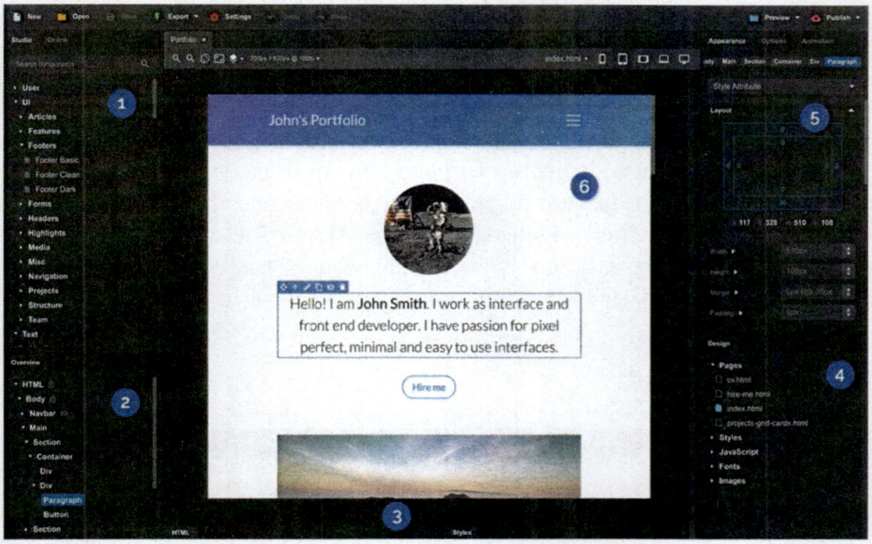

Fig. 2. Bootstrap studio application [6]

- Block styling using a simple web console-like editor
- Creating interlinking HTML pages
- Generate a production version
- Option of responsive design using settings or custom CSS
- Predefined templates
- Synchronizing blocks and design
- Online community library
- Realtime preview
- Code editing
- Animation
- Free Student License

But this studio comes with some **disadvantages** as well, including:

- Any possible PHP code generation default tool for the backend.
- Not being free.

Bootstrap Studio is a perfect start for high or elementary school students. It offers essential HTML elements, advanced animations, and the ability to export a production version of your site directly to the server. It is a studio that anyone can try, but those who have never tried the Bootstrap CSS framework may have trouble understanding how it works at first.

In addition to creating a site using blocks, it also offers to import templates that Bootstrap itself predefines, the studio's Online Author Library, and the community. You can even import only certain blocks from individual templates that can fit into a web presence.

This studio can be used while working with simple websites with no functions on the backend, i.e., static pages in the form of .html pages served by JavaScript. The end product is high-quality, well-organized HTML, CSS, and JS code and assets, including images, fonts, and more. The advantage of this studio is its strict interdependence with the Bootstrap CSS Framework.

4.2 Wix

Wix is a popular content management system (CMS) based on the basic Wix Page Builder editor, which allows you to interact with all elements of your website (see Fig. 3). Wix offers a basic layout for working with the editor using four panels. A top bar is dedicated to page view options, savings, and settings. The left menu is for adding new elements to the page and manipulating them if necessary. Right-click, then advanced element action menu was handling options and context menu. The last one is the actual view of the edited page, which is the identical view that the page user sees [8].

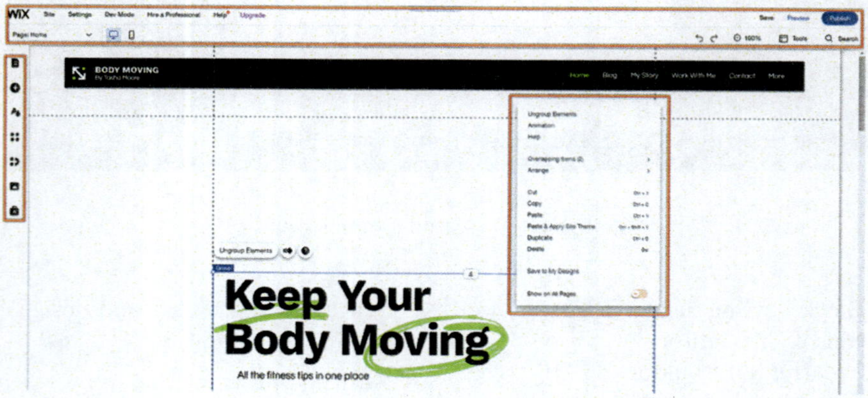

Fig. 3. Wix web application [8]

Wix offers several modern tools that can be used mainly to create a quick presentation, but there are also cases when they can be used to create a complex website. **Benefices** of Wix are:

- Basic Page Builder editor
- Allow use of more than 500 themes
- Third-party tool integration
- Wix ADI - Artificial Intelligence for design pages with a few simple questions
- Only paid plan
- Velo - open developer platform for custom code and design freedom
- Only Wix hosting with limits
- Wix live chat
- Wix eCommerce
- Wix Bookings
- Wix Blog
- Wix Music player
- Paid extended Editor X - Page Builder.

On the other side, Wix doesn't offer, and there are **disadvantages**:

- Does not have a Free or DEMO version
- Must use only WIX hosting
- Missing advanced SEO in the basic version
- It can't leave Wix with export data to the new website
- It does not support external 3rd plugins like WordPress.

Wix will have the primary use for simple websites that don't need extensive logic. Thanks to how the editor works and the possibility of inserting blocks, anyone can create a web page similar to the Word text editor.

However, this CMS is not free, and you need to buy additional parts for more advanced features - for example, the advanced Editor X. Apart from the not always reasonable price, it is not possible to move the data if the site owner decides to leave the platform and move to another one. Another disadvantage is its closed nature compared to classic and permanent content management systems.

4.3 Elementor

It is a plugin extending the basic functionality of WordPress CMS (Fig. 4). The editor offers advanced widgets in the paid version, but some are also available in the free version. This editor is the fastest-growing editor on the WordPress platform and offers a variety of 3rd party integrations and CMS extensions [9].

Elementor has built-in support for linking essential marketing tools, WordPress plugins, social networking, security features, and more [10]. If the default integration is insufficient for use case websites, it is possible to use additional 3rd party extensions. Elementor, an open source platform, allows you to create custom blocks (widgets), extend existing blocks with new features, or integration using 3rd party APIs.

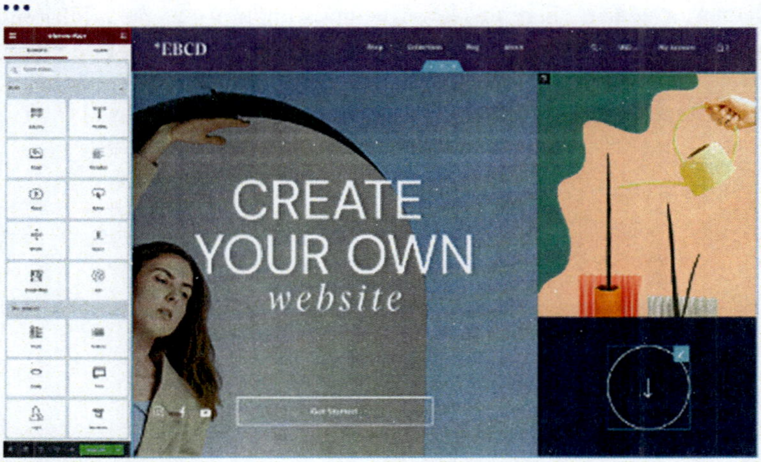

Fig. 4. Elementor web application [11]

Free version [12]:

- Free Widgets
- Custom theme builder
- Mobile design ready
- Global settings (colors, fonts, icons, etc.)
- Custom CSS (WordPress only)
- Basic animations
- Pro animations
- WooCommerce – eCommerce
- Forms Integration

Paid version (Pro) [12]:

- Dynamic tags
- Additional features for custom theme builder
- Global widgets – custom-defined widgets
- Custom CSS
- Headers and Footers
- 3rd party extensions & integrations

Elementor has gained extensive support from WordPress users through its Free version. It offers a sufficient number of features for testing and getting acquainted with the editor, creating the first landing pages or plugin creations for programmers.

The editor is helpful regarding a website or e-shop of larger dimensions and specific requirements for students and anyone who wants to work with the web and its sustainability. WordPress is easily extensible, and Elementor was created in a similar vein.

Elementor also offers a Pro paid version, which allows you to define better the content of pages called archives - pages with posts, products, product categories, search engine pages, and more. Then there are dynamic tags that allow you to automatically match URLs to their actual values or high-quality SEO optimization and page minification.

4.4 Weebly

Weebly is a similar type of Page Builder as Wix. The difference is mainly in the price and appearance of the editor. While the Weebly editor looks similarly to the Elementor editor, there are fewer widgets and options for additional customization. This editor is mainly for beginners who want to create a simple but functional website for free. Example application is in Fig. 5.

Basic features offered by Page Builder and main **advantages** [14]:

- Beginner-friendly templates
- Responsive themes
- eCommerce
- Video background
- Site search

Features that this editor does not offer or are **insufficient**:

- Worse support for mobile design
- It does not offer 3rd external plugins
- Bad support

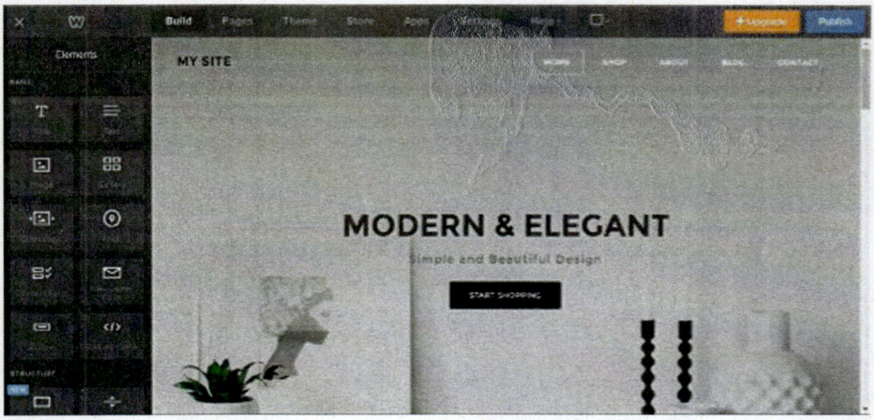

Fig. 5. Weebly web application [13]

If students are beginners and want to build a website through a simple free Page Builder, Weebly is a product worth checking out. Students have their site free of charge for as long as they want.

However, Weebly has weaknesses, such as the inability to customize mobile sites the way you want. Unfortunately, its templates aren't as good as their competitors.

4.5 Webnode

Editor Webnode is similar to the basic Gutenberg editor in WordPress but with the same principles as Wix. This editor doesn't offer the same extensive capabilities as Wix or Elementor, but it is sufficient for many users and meets all their web development requirements [15]. Its main advantage is the simplicity of page building, even for non-developers – see Fig. 6.

Advantages offered by this CMS [16]:

- User-friendly Page Builder
- Domain registration
- Webnode self-hosting
- Email accounts (under a domain)
- Free templates
- eCommerce
- Backups
- Free email address.

Features that this CMS does not offer or are insufficient:

- Basic eCommerce functions
- A basic blog without contributing
- It does not offer 3rd external plugins.

Fig. 6. Webnode web application [15]

5 Comparison of Pagebuilders

Table 1 compares Pagebuilders by price and focus type. Some offer a free version, and others do not. The table does not take into account student discounts and other special prices. Additional charge shows if you need to pay fees again to keep the site running after the 1st year. Fees include different extensions, more convenient features, and license renewals.

Table 1. Pagebuilders comparison [7,17–20]

	Bootstrap studio	Wix	Elementor	Weebly	Webnode
Price ($)	30–62	11–46	0–49	0–25	0–32
Type	HTML Editor	Single CMS	WP Plugin	Single CMS	Single CMS
Multiple sites	Yes	No	Yes	Yes	Yes
Additional charge	No	Yes	No	Yes	Yes

Users using Bootstrap Studio use this studio to create page designs. They then publish these pages as HTML files or enrich them with custom backend code to create another logical layer of pages. An example of such enrichment is a contact form and email submission. The significant advantage of this studio is the ability to trap custom JavaScript even with the formatting needed to optimize the site better. This studio deserves the attention of everyone who wants to create a simple yet functional design with perfect code organization.

Wix is one of the trendy platforms, but this editor/CMS does not impress with the price, which can climb to more than $ 500 per month. On the other hand, this editor cannot be denied the number of built-in and very useful features, such as the newsletter, which can be edited arbitrarily. In addition to editing, it offers detailed statistics of newsletter email campaigns, and, compared to other editors, everything is available already in the base [21]. This editor mainly aims at start-ups and users, offering features such as page generation, design, content or logo generation using AI. Still, each extension takes part of the finances, usually paid monthly.

Elementor appeals with its free version, which includes many features that other editors do not offer even in their basic paid versions. It is worth mentioning the versioning system and the offer of more than 40 blocks, over 50 ready-made pages, and more than 100 ready-made block templates, from which you can build a graphically very successful presentation with low operating costs. In the case of the paid version, everything is expanded by up to 4 times, and it is also possible to edit all the settings of the stand [22]. A perfect feature that users enjoy is the dynamic tags, which offer to insert content into the page from the administration without having to send it directly into the design.

Weebly is suitable for small businesses and beginners who want to create one of their first websites. Weebly is one of the few free editors, but in this version, it is constrained[23]. In using Weebly, we have encountered several obstacles to creating our creative designs without advanced knowledge of CSS. On the other hand, Weebly offers basic design customization, advanced eCommerce tools, and functional but low-detail mobile appearance settings. This editor does not provide custom functionality and logic, but this part can be partially replaced by the extensions that Weebly offers in its store.

Weebly and Webnode have a lot in common in style and usage. Both editors are aimed at beginners interested in creating websites. Webnode is too similar to Weebly in some parts. However, this editor offers modern page layouts. Unfortunately, however, they are not easy to edit [24]. The editing tool is highly problematic if a certain amount of customization is required for the design. Another major limitation is the color palette that Webnode offers. However, this studio has a very well and designed form of eCommerce that allows you to create and sell almost anything without specific requirements.

6 Conclusion

Pagebuilders are used on most modern web hosts to create simple websites. However, some advanced editors take their functionality and use to a much higher level than just dragging and dropping elements into a page. For use in secondary schools to support learning how to create web pages, it would be a good idea to use a basic student version of Bootstrap Studio, where students can try out the basics of Page Builder but also customize or extend individual parts with their code. Wix will serve more simple but professional sites, but one should keep in mind that the site will not be functional in the future without additional fees. Elementor, on the other hand, is a full-featured editor that builds on the foundation of WordPress and is easily extensible and easy to use. It is suitable for creating sustainable sites with complex logic that the WordPress CMS can handle, and no custom site code is required. The Weebly and Webnode editors are similar to the Wix CMS but offer free versions, which make these editors top-rated software for creating the first websites for small or start-up companies.

References

1. Vojtesek, J., Forman, J.J.: Using Pagebuilder to teach web development. In: Silhavy, R. (ed.) Software Engineering Perspectives in Systems. CSOC 2022. Lecture Notes in Networks and Systems, vol. 501. Springer, Cham (2022). https://doi.org/10.1007/978-3-031-09070-7_51
2. What is block-based coding [online]. Available at: www.idtech.com/blog/what-is-block-based-coding. Cited 11 July 2022
3. What is Block Coding for Kids [online]. Available at: funtech.co.uk/latest/what-is-block-coding-for-kids. Cited 11 July 2022
4. Say Hello to Gutenberg, the WordPress Editor [online]. Available at: wordpress.org/gutenberg/. Cited 11 July 2022
5. 40+ Most Notable Big Name Brands that are Using WordPress [online]. Available at: www.wpbeginner.com/showcase/40-most-notable-big-name-brands-that-are-using-wordpress/. Cited 11 July 2022
6. Bootstrap Studio Docs [online]. Available at: bootstrapstudio.io/docs/. Cited 11 July 2022
7. Bootstrap Studio [online]. Available at: bootstrapstudio.io/. Cited 11 July 2022
8. Wix Editor: Using the Editor Menus [online]. Available at: support.wix.com/en/article/wix-editor-using-the-editor-menus. Cited 11 July 2022
9. Browse All Elementor Features [online]. Available at: elementor.com/features/. Cited 11 July 2022
10. Connect Your Favorite Tools [online]. Available at: elementor.com/features/integrations/. Cited 11 July 2022
11. Build Intuitively with Our Drag & Drop Editor [online]. Available at: elementor.com/features/editor/. Cited 11 July 2022
12. Elementor Benefits - Pro VS Free [online]. Available at: elementor.com/help/elementor-pro-vs-free/. Cited 11 July 2022
13. Weebly Site Builder Review 2022 [online]. Available at: makeawebsitehub.com/review/weebly/. Cited 11 July 2022
14. Weebly Website Builder Review 2022 [online]. Available at: firstsiteguide.com/weebly-review/. Cited 11 July 2022
15. Webnode Recenze [online]. Available at: www.5nej.cz/webnode-recenze/. Cited 11 July 2022
16. Webnode features [online]. Available at: www.webnode.com/features/. Cited 11 July 2022
17. Wix Premium Plans [online]. Available at: wix.com/upgrade/website. Cited 11 July 2022
18. Explore Elementor Pro Plugin Plans [online]. Available at: elementor.com/pricing-plugin/. Cited 11 July 2022
19. Pricing [online]. Available at: www.weebly.com/pricing. Cited 11 July 2022
20. Website Builder Comparison Tool [online]. Available at: www.tooltester.com/en/website-builder-comparison/. Cited 11 July 2022
21. Wix Review: Top 7 Reasons to Use Wix in 2022 [online]. Available at: www.adamenfroy.com/best-website-builder/wix-review. Cited 11 July 2022

22. Elementor Review 2022: Does It Live Up to the Hype? [online]. Available at: www.websiteplanet.com/blog/elementor-review-can-really-build-wordpress-website-alone/. Cited 11 July 2022
23. Weebly Review 2022: Is It the Right Website Builder for You? [online]. Available at: www.websitebuilderexpert.com/website-builders/weebly-review/. Cited 11 July 2022
24. Reviews and Expert Opinion - Webnode [online]. Available at: www.websiteplanet.com/website-builders/webnode/. Cited 11 July 2022

Understanding the Labor Market from the Bottom Up with an Agent-Based Model

Samira Boulahbel-Bachari[1,2](✉) and Nadjia El Saadi[2]

[1] National Center of Research in Applied Economics for Development (CREAD), Bouzareah, Algiers, Algeria
boulahbelsamira@gmail.com
[2] Higher National School of Statistics and Applied Economics, LAMOPS, Kolea, Algeria

Abstract. We present an agent-based labor market model that incorporates worker internal migration. We examine at the labor markets of two regions: the origin region and the destination region. The characteristics of the workers, as well as the features of the regions in which they live, are taken into account. The model describes the individual's life cycle from birth to retirement. The model will so replicate births, deaths, education, labor market entry and exit. We evaluate worker mobility between the two regions when considering new labor market entry and exit. The proposed modeling is based on mathematical equations that describe the socioeconomic aspects of the actors and their behavior. The model differs from other labor market studies in that it considers worker internal migration, an analysis of the two labor markets involved (related to the regions of origin and destination), and the simulation tool used: "agent-based model."

Keywords: Agent-based model · Labor market · Internal migration · Mathematical equations · Heterogeneous agents · Selection

1 Introduction

Agent-based modeling is a branch of artificial intelligence inspired by complex systems, particularly those with sociological and biological metaphors. This relatively new branch of modeling was developed in the 1940s and became popular at the end of the 1980s as computer tools improved. It is founded on mathematical formulations and employs experimental methods to explore complex systems. Agent-based modeling enables the study of phenomena at the aggregate level that emerge from the interactions of agents who follow specific behavioral rules and may be constrained in their choices by a variety of factors.

Although its application is still limited, the utilization of such approach is expanding and affecting a wider variety of subjects, from computing to history and economics.

R. Silhavy et al. (Eds.): CoMeSySo 2022, LNNS 596, pp. 754–769, 2023.
https://doi.org/10.1007/978-3-031-21435-6_65

Given the similarity in the descriptions of the economic agent and the "computer" agent, using this kind of modeling in economics almost seems natural. In fact, the economic agent (consumer, household, company, government, etc.) functions similarly to the computer agent in that it develops within a context (the job market, the economy, a region, etc.), interacts with other agents in a social network or neighborhood, has skills (human capital), and partially or fully perceives its surroundings (asymmetry of information, perfect knowledge of the market). According to their abilities and resources, each economic agent has goals to reach, such as maximizing utility and optimizing resources.

Another advantage of employing economic agent-based models is the addition of other pertinent non-economic aspects, such as social, political, and psychological factors, allowing a more comprehensive study of the outcomes.

All of these advantages make agent-based economic modeling an effective tool for supplementing statistical and analytical methodologies. Furthermore, they allow us to assess the viability of some economic theories as well as the relaxation of two essential hypotheses of the new neoclassical economic theory. It is about the hypothesis of reasonable expectations and representative agents (Haldane and Turrell 2018). In fact, the representative agent is replaced by heterogeneous "agents" who follow behavioral rules defined by their characteristics, and rational expectations are replaced with limited rationality, which allows the agents to make decisions based on partial knowledge and a limited capacity for calculation. The combination of these two hypotheses allows for greater flexibility in analyzing the labor market, which explains the ongoing growth of research studies in this domain.

Agent-based models have a wide range of applications in economics, allowing for the modeling of economic actors of different types, including individuals, households, firms, and institutions. Thus, ABMs are used to examine the influence of new technologies on the electrical network (North et al. 2002), the banking sector and credit (Dosi et al. 2013), and previously unexplored linkages in the labor market. Labor market ABMs generate richer models that can be used to evaluate and evaluate economic policy (Neugart and Richiardi 2018).

Bergmann's (1974) modeling of the American economy that took into consideration the diverse activity sectors and Eliasson's (1977) modeling of the Swedish economy were among the first studies on the labor market which were based on agent-based models. The ARTEMIS model of the French labor market (Ballot 2002), the Richiardi model from 2006, and the Worksim model developed by Z. Lewkovicz in 2010 and refined by O. Goudet in 2015 are further examples of country specific models.

From the perspective of labor market analysis, we propose this work. Internal migrants are given special consideration in the evolution of the labor market in emerging countries. This group of workers is practically never included in the design and execution of employment policies in these nations, due to the absence of data. However, the fast urbanization that these countries are experiencing appears to be a significant obstacle. The evolution of modes of transportation, the advancement of information and communication technologies, and the unequal expansion in economic prospects have all contributed in the proliferation of this issue. Almost all of the population expansion in developing countries will occur in urban areas in the coming decades, boosting the rate of urbanization in these countries from 44 to 75% (UN Habitat 2008).

While rapid urbanization can contribute and ensure long-term economic progress, it can also place pressure on cities' infrastructure, ecology, and social fabric.

Taking internal migration into account in labor market analyses would thus allow for a better evaluation of national strategies and better adaption of public policies.

Economic models that analyze labor market dynamics often simplify the behavioral mechanisms and decision-making processes of agents and do not account for migrant workers. This description is usually constructed by a set of equations that supports a representative individual to whom one assigns entire rationality and flawless labor market knowledge.

The model provided in this paper depicts behaviors at the non-aggregated individual level to ensure agent heterogeneity and examines emergent macroeconomic behaviors.

The proposed modeling is based on mathematical equations that describe the socioeconomic aspects of the actors and their behavior. The model differs from other labor market studies in that it considers worker internal migration, an analysis of the two labor markets involved (related to the regions of origin and destination), and the simulation tool used: "agent-based model."

The paper is subdivided into three sections. Section 1 is a quick survey of the labor market research literature. The proposed labor market model is discussed in the Sect. 2. Finally, Sect. 3 will concentrate on labor market simulation and data analysis. The essential aspects of the work provided will be summarized in the conclusion.

2 The Labor Market: From Economic Modeling to Agent-Based Modeling

In general, labor market analysis is based on microeconomic approaches to comprehend individual interactions and the behavior of its agents, and macroeconomic techniques to investigate interactions influencing global variables. These considerations are critical for both researchers and economic actors.

While the first work on labor market analysis identified a lack of wage adjustment as the main cause of unemployment, new factors and conceptual frameworks have been introduced in recent studies such as human capital (Becker 1964), job search (Stigler 1962; Stigler and Friedland 1975), labor market segmentation (Piore and Doeringen 1971; Bulow and Summers 1986; Perrot and Zylberberg 1989), implicit contracts (Azariadis 1975; Baily 1974), or the impact of the efficiency of the wages (Leinbenstein 1966).

Even if these new theories go beyond the conventional theories' constraints, they are still difficult to be evaluated empirically and do not allow for a joint study at the micro and macroeconomic levels.

In the 1970s, new experimental methods for modeling the labor market as a dynamic and complex system emerged. These models are based on the interactions and behavior of diverse economic agents, allowing them to analyze newly emerging macroeconomic laws (bottom-up modelling).

These approaches enable more accurate modeling of individual behavior (unemployed, employers, employees) as well as labor market institutions (employment contracts). According to the research objectives, labor market multi-agent models are classified into two types: normative multi-agent models that aim to demonstrate the potential of the multi-agent approach in comparison to more traditional approaches, and multi-agent models calibrated to understand the functioning of a specific labor market and precisely quantify the effect of public policies. The latter labor market models can operate alone or as part of a larger macroeconomic model.

Bergmann (1974) and Eliasson (1977) were among the first to use agent-based models to analyze labor market. Bergmann's model uses predetermined rules to characterize the behavior of all actors (workers, companies, banks, financial intermediaries, government, and central bank). The author defines a single bank, a financial intermediary, and six firms from six distinct sectors. The model stands out for its complexity and the importance it places on rule-based decision-making, even if the assumption of representative firms is controversial for an ABM.

Eliasson (1977) models the Swedish economy by focusing on the labor market. The model's goal is to explain inflation at the microeconomic level and to investigate its relationship with earnings, investment, and business expansion. As a result, the labor market module focuses on enterprises and how they seek employees to meet their output targets. Manufacturers produce, sell their products, make investment decisions, and alter their expectations. Individuals divide their income between savings and consumption.

These two models, which are very similar to microsimulation models, simulate the entire population of a country.

Since the 2000s, new labor market models based on agents have been established. For example, the ARTEMIS model of the French labor market (Ballot 2002), which later evolved into WorkSim (Lewkovicz 2010; Goudet 2015), the EURACE model, Neugart's (2004), and Gemkow and Neugart (2011).

Ballot (2002) defines the French labor market as an institution undergoing endogenous change in his model. As a result, the behavior of agents and the situations they confront are influenced by their past. Businesses and individuals are incredibly rational, always looking for ways to reduce costs or enhance utility. Based on the rational actions of diverse individual agents, the model replicates the gross fluxes of workers between different states (unemployment, employment, and idleness). In his improved version, Lewkovicz (2010) integrates the role of social networks in job search, studies age discrimination in the labor market against older workers and investigates the impact of the introduction of a new type of employment contract on the unemployment rate in this group of workers. Goudet (2015) investigates the impact of the removal of fixed-term contracts on French unemployment. Finally, Goudet (2015) examines the effect of the elimination of fixed-term contracts on unemployment in France in the most recent update of WorkSim.

The EURACE model (Deissenberg et al. 2008) is a software platform that uses agents to design European economic policy. It is constituted of a huge number of autonomous agents that interact in a complex economic context. The Eurace model considers capital, various goods, credit, finance, and various labor markets in a spatial framework. Dawid et al. (2008, 2009, 2012) investigated the regional allocation of funding for human capital investment in the presence of labor market frictions and the role of labor market integration policies as part of the Eurace project, while Teglio et al. (2012) investigated the impact of bank capital adequacy regulations on GDP growth, unemployment rate, and aggregate capital stock.

Neugart published various studies on the labor market. He initially simulated the matching process between enterprises and workers and investigated the emergent matching function (Neugart 2004). He also developed a model to measure the total impact of labor market policy. The model is composed of a diverse group of people and firms from various industries. Workers' talents vary, and companies in each sector have different skill requirements. Workers can learn skills that will enable them to work in industries other than their current one (Neugart 2004). Neugart and Gemkow also constructed a model based on labor market agents that described an endogenous social network and studied its implications on hiring and inequalities (Gemkow and Neugart 2011).

Richiardi et al. (2006) developed a labor market model including job search and attempts to reproduce stylized facts such as the Beveridge, Wage, and Okun curves. Workers in the model differ in terms of productivity and entrepreneurial proclivity. Companies, on the other hand, have the possibility for incremental expansion and make decisions regarding the number of vacancies and wages they offer.

Similarly, to adapt important pure economic models, Boulahbel-Bachari et al. (2017, 2018, 2019) and Ait Mokhtar and El Saadi (2022) successfully adjusted Todaro's (1969) and Krugman's (1991) models, respectively. The first work sought to describe employees' mobility between two different locations and heterogeneous workers. The second one attempted to explain labor mobility at the company level and introduced the impact of charging capability of companies.

Other works, such as those by Moreno Baruffini (2013) and Mitkova (2016), were concerned with job searching.

This was a brief review of the literature on labor market modeling using agent-based models. The proposed labor market model is described in the section that follows.

3 Proposed Labor Market Model

The suggested labor market model describes the socioeconomic properties of agents and their behaviors using mathematical equations and simulates these behaviors using an agent-based model.

This model is inspired by the theory of human capital by taking into account individual characteristics in the job search and by the dual theory (Harris and Todaro 1970) in the description of two labor markets and the existence of worker migration due to wage differentials.

The key hypothesis of our model is that people want to obtain a good job to shield themselves from potential hazards such as health problems and unsafe working conditions.

The model simulates the labor market in two regions: one with a relatively low rate of decent job creation, a big income disparity, and a weak formal sector, and another with more economic potential activities. These features broadly characterize the two regions, rural and urban.

We also suppose that people in the first region can look for work in the second region if the expected wage and likelihood of finding work in this region are higher than in the first.

The agent-based model proposed considers two types of agents:

- The Individual Agent, that represents the key element of the model. It can be either an employee, employer, unemployed or inactive. He is described by a set of socioeconomic characteristics such as: age, employment status, level of education and level of qualification.
- The Agent Region includes businesses and firms. The Agent Region is described by the unemployment rate, labor market tension, probability of finding a job and wages paid. Creation and destruction of jobs occur inside the Agent Region. The model does not explicitly simulate companies but considers that jobs are randomly distributed according to a job creation rate specific to each region.

The model simulates two important processes: the migration process and the job search process. At the end of the two stages, an update of the demographic and socioeconomic parameters is performed (births, deaths, marriages, divorces, entry and exit on the labor market).

3.1 The Migratory Process

The determination of migrants is based on the calculation of a discriminant function of socio-economic variables and the expected gain of migration.

The discriminant function (Fj (xi)) is weighted by the tension of the labor market (MT) and a realization (u) of the uniform variable on [0, 1] allowing to consider the possibilities of employment and to integrate a hazard in the choice of individuals. This function is compared to a Threshold to differentiate high potential individuals for migration from others.

The expected gain from migration and the probability of finding a formal job $\boldsymbol{W^e_{i,D,(O)}}$ determine who will look for a job on the first region (non-migrants) and who will move to the second region to do it (migrants).[1]

Migration Decision Process Pseudocode

```
Take an individual "j" unemployed from Origin and aged
[15, 59]
   Calculation of the propensity to migrate Fj
    if    Fj > Threshold x MT x u, then
          Potentiel_Migrant = True
   Calculation of migration's gains for potential migrants
    If W^e_{i,D} > W^e_{i,O}, then
        Migrant = True
All agents « j » of the Origin region are treated?
   No: Return to the first point;
   Yes: Go to step job search
```

[1] See Boulahbel-Bachari and El Saadi (2018, 2019) for more details.

3.2 The Job Search Process

The proposed model describes the labor market and the different inputs and outputs of this market. Thus, we consider several statuses for the individual through which it can transit. An agent can be unemployed, employed, an employer, inactive, or a student. He cannot be considered unemployed until he reaches the age of sixteen. He is no longer regarded a member of the active population after 60 years.

Figure 1 shows the transitions that can occur during the life cycle of a worker.

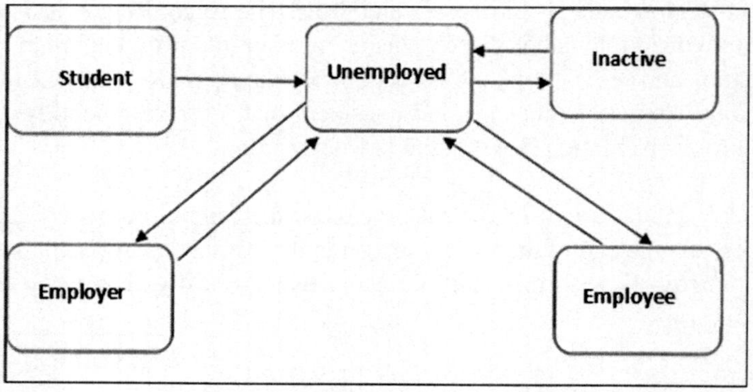

Fig. 1. Different states of the agent individual

All unemployed individuals (aged between 16 and 59) are involved in the job search, whether they are native or migrant. Those who are qualified can apply for qualified and unqualified positions unless those who are not qualified will have to settle for unskilled jobs. The qualification of individuals depends on their level of education.

In the proposed model, we assume that the use of social networks enhances the chances of getting a job, which increases with the amount of time spent looking for a job. The more time a person spends looking for work, the larger his social network grows and the more likely he is to be recruited. We further suppose that the positive relationship between search duration and hiring probability reverses after a certain period, due to the job seeker's discouragement and the depreciation of his qualifications.

Candidates with higher probabilities will be hired first. Those who cannot be recruited will keep searching.

At time t, the probability of being recruited for individual "i" in Region "j" is defined as follows:

$$\pi_i^j(t) = \pi_i^j(t-1) + \left(1 - \pi_i^j(t-1)\right) \times P_i^j(t) \tag{1}$$

With:

- $\pi_i^j(t)$ the probability of being recruited for individual i in Region j at time t;

- $P_i^j(t)$ the selection probability for the individual i in Region j at time t, which is defined as follows:

$$P_i^j(t) = \begin{cases} \frac{NQPos_j}{A_j} \\ \frac{Pos_j}{A_j} \end{cases} \tag{2}$$

With:

- $NQPos_j$ the number of unskilled vacancies in Region j;
- Pos_j the total number of vacancies in Region j;
- A_j represents job seekers in Region j.
- In each region and each year, a number of jobs are released as a result of retirements, others are created, and some may be destroyed.
- Individuals cannot remain indefinitely unemployed; they can be discouraged after a few attempts and will give up looking for a job and become inactive.

Pseudo Code of the Job Search

```
Agents "i" of Region j, aged between [15, 59], unemployed or
informal workers
 Calculation of π^j_i (t);
 Ranking agents in descending order according to π^j_i (t);
 Assigning Agents to vacancies;
 For the remaining job seekers:
   If Unemployed = True, then
    Searching_time = Searching_time + 1
    If Searching_time = Searching_time_max then,
      Inactive = True
All agents « i » in Region j, aged between [15, 59], unemployed
workers are treated?
  No: Return to the first point and randomly choose an agent;
 Yes: Go to next step "updates of socioeconomic and demographic
parameters".
```

4 Simulations and Results

The model was tested on data representing two regions with different levels of development that can be assimilated to rural and urban regions in a developing country. We assume that rural workers can migrate to the urban region to search for a job. We will call the rural-type region the region of origin and the urban-type the destination region relative to the flow direction.

The distribution of wages in these two regions is different: the region of origin has low wages and widely dispersed due to the preponderance of precarious jobs. The Destination Region has higher wages and lower dispersion and a larger share of skilled jobs.

The data used describes a young population with 62% that can be potentially active and approximately 38% of the population is occupied. For the unemployed, 42.7% of them are graduates against 28.9% for the occupied. Employees represent the largest proportion of the employed persons (67% of the employed individuals). The urban-type region has a slightly higher unemployment rate than the other region but remains attractive because of its potential in terms of job opportunities relative to the other region.

The number of individuals in each region is around 9900 individuals (Table 1).

We introduce in the model an aspect of migration that is not often considered: the self-selective aspect of migration. Thus, individuals with distinct sociodemographic characteristics and identical wage differentials will have different propensities to migrate. Individuals with limited work experience and no children will have the highest propensities to migrate unlike those with children attending school, for example.

Table 1. Regional characteristics

Origin region		Destination region	Total
Population	9950	9900	19,850
Potential active population	62.2	61.9	62.1
Employment	15.8	17.5	16.6
Unemployment rate	12.9	16.2	14.7

We assume that the birth and death rates are identical in both regions (2.478% and 0.441% respectively).

4.1 Scenarios Tested

Two scenarios are tested to analyze the evolution of the labor market: the first one shall consider internal migration of workers while the second one operates without it.

At first, we will analyse the evolution of the population and the potential active population. Then, we will analyse the evolution of employment and unemployment in the two regions.

Population evolution:

The population of the region of origin varies according to whether or not the migration of workers is taken into account.

During the first 20 years, the difference is not very important. After this period, the gap become larger. At the end of the considered period (50 years), the population of the region of origin reaches 11,526 individuals when migration is considered and 17,754 individuals in the absence of migration (Fig. 2).

For the evolution of the destination population, a discrepancy is also observed between the two situations with and without migration even if it remains less important than the difference in the evolution of the population of the region of origin (Fig. 3).

The population of the destination region is larger in the presence of migrant workers and the gap between the two developments in this population is increasing at the end

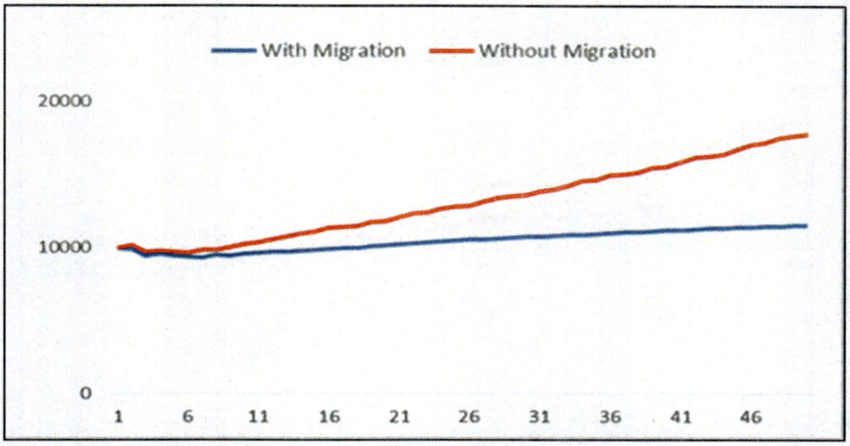

Fig. 2. The evolution of the origin region's population

of the considered period. At the end of this period, this population, in the presence of a labor migration, exceeds 41,000 individuals while in the absence of migrant workers, it slightly exceeds the 30,000 people.

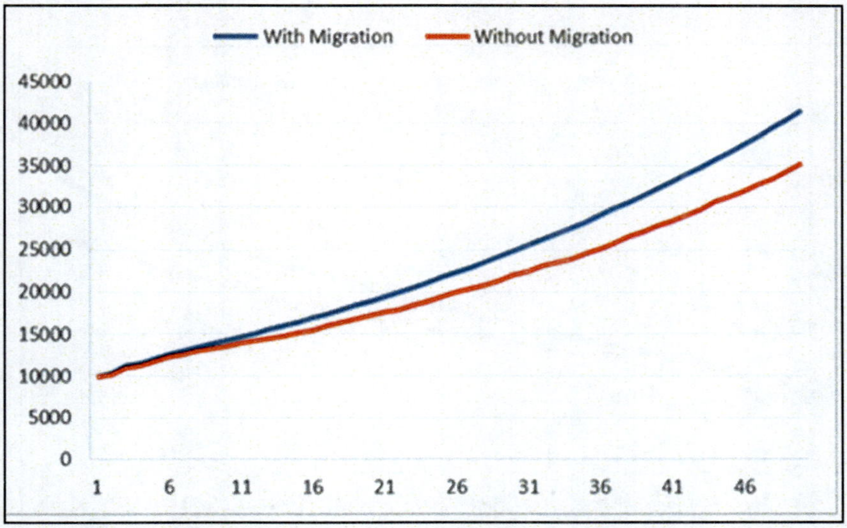

Fig. 3. The evolution of the destination region's population

Evolution of potentially active populations (population aged between 15 and 60):

Regarding the evolution of the potentially active population in the region of origin, after a small decrease during the first ten years, the value was found to increase slightly to exceed 6000 individuals and stagnates around this value for the rest of the period.

In the presence of a labor migration, the potential active population decreased over the entire period at a faster rhythm to reach values below 3000 persons (Fig. 4).

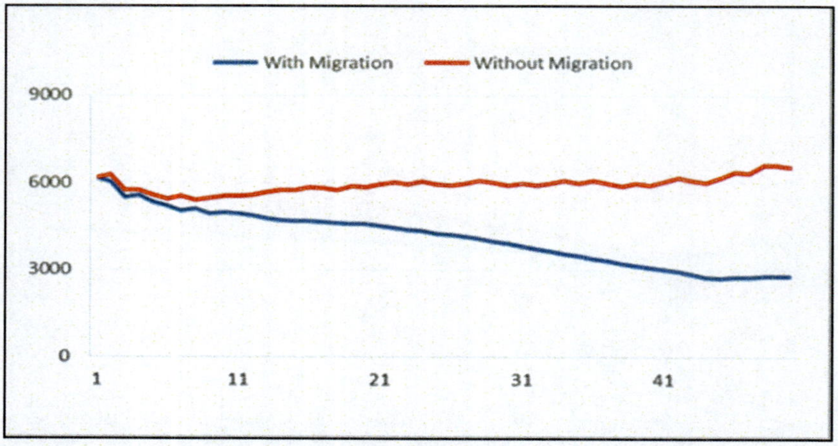

Fig. 4. Potentially active population evolution in the origin region

Conversely, the potentially active population in the destination region increases in both cases (presence or absence of migrant workers). This increase was found to be greater in the presence of migrant workers (20,485 people compared to 16,723 people at the end of the considered period) (Fig. 5).

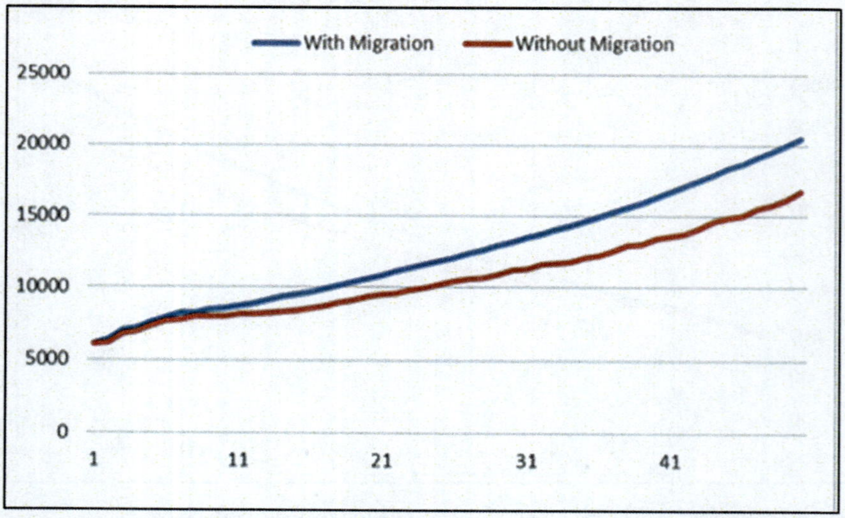

Fig. 5. Potentially active population evolution in the destination region

In the long term, the effect of migration becomes more and more palpable on the evolutions of populations. Since young people are the most concerned by migration, their departure does not allow the region of origin to deploy in favor of the region of destination. It is important to remember that both regions have, at the start of the simulations, almost the same population size and identical birth and death rates.

Evolution of employment:

In the region of origin, when applying a job creation rate of 3%, the employment was found to increase in the absence of migration and decrease in its presence until it cancels out at the end of the period.

Workers from the origin region move to the destination region which presents a higher job creation rate (10%) and therefore better job opportunities impoverishes the region of origin and does not allow it to develop (Fig. 6).

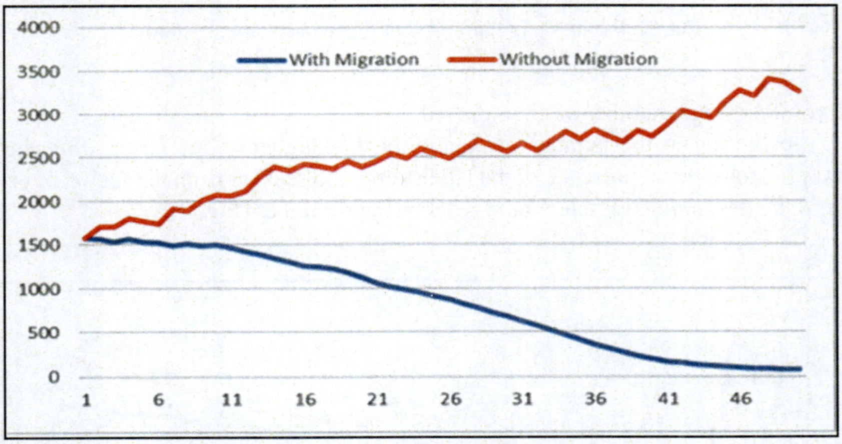

Fig. 6. Evolution of employment in the origin region

For the Destination region, we apply a job creation rate of 10% in both situations. Employment is increasing in both cases with higher values in the presence of migrant workers.

Thus, these workers participate in the deployment of employment in the Destination area (Fig. 7).

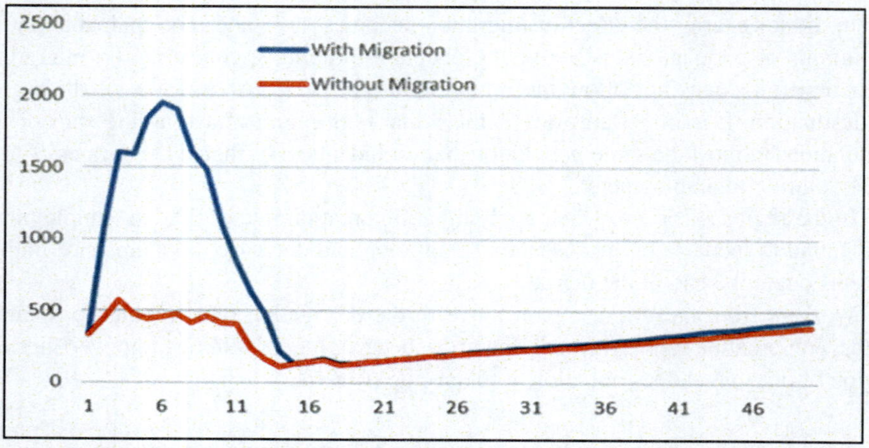

Fig. 7. Evolution of employment in the destination region

Evolution of the number of unemployed:

The evolution in the number of unemployed is higher in the absence of worker migration. However, in presence of the latter, some job seekers from the region of origin move to the destination region, which reduces their number (Fig. 8).

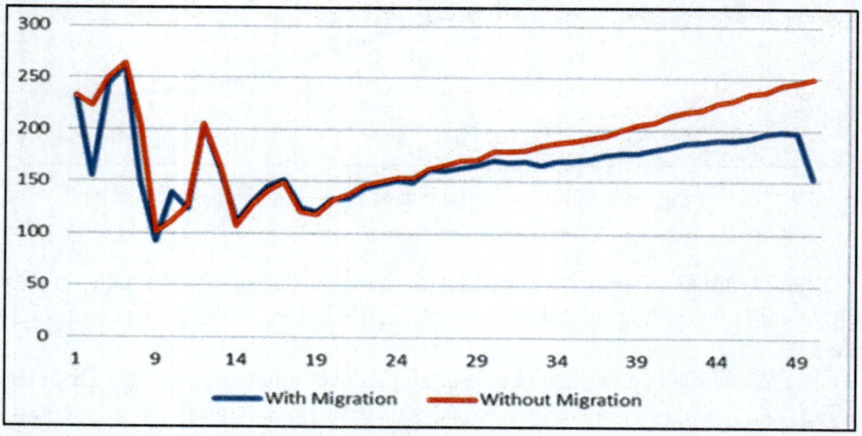

Fig. 8. Evolution of unemployment in the origin region

For the evolution of the number of unemployed in the region of destination, we find that the number of job seekers is higher when we take into account the migration of workers, but after a certain period, their number is almost identical in absence of migrant workers. This is due to the high job creation rate of 10%, which allows this flow of migrants to be absorbed without extending the unemployed population (Fig. 9).

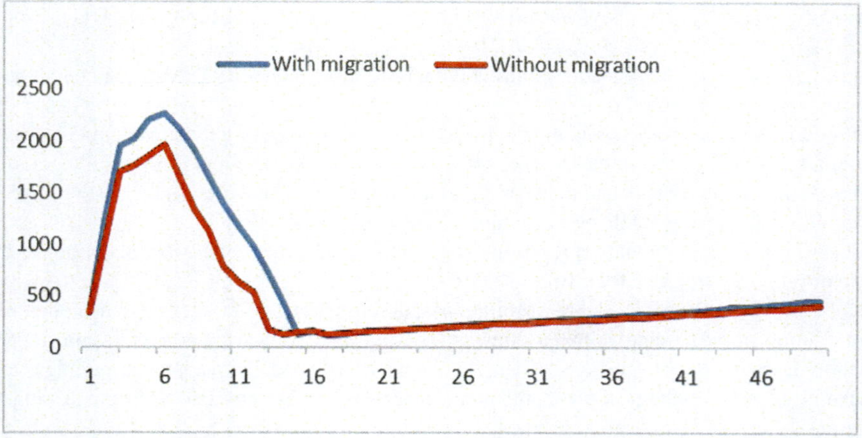

Fig. 9. Evolution of unemployment in the destination region

This section highlights the important role played by the migrant workers in the development or precariousness of regions.

5 Conclusion

The work presented in this article illustrates the importance of considering labor migration in the analysis of the labor market that is unfortunately not considered in the literature. The scarcity of data is certainly one of the main causes of this absence. The use of multi-agent systems in the study of the labor market has allowed us to take them into account. Their use has also allowed us to integrate the aspect of heterogeneity of individuals in terms of characteristics and behavior.

The incorporation of the probabilities of finding a job by qualification as well as the individual salaries makes possible the description of the behaviors of the agents according to their profile.

However, the model has some limitations that need to be improved. The integration of a matching function for the job search will allow agents to better target job positions according to their level of qualification. Also, the study of the influence of social networks will give a more realistic description of the job search. These extensions could support the design of a more robust and comprehensive integrated model that can facilitate the development and evaluation of public employment policies for sustainable rural development and controlled urbanization of cities.

References

Ait Amokhtar, S., El Saadi, N.: On the industrial clustering: a view from an agent-based version of Krugman model. In: Congress on Intelligent Systems. Springer, Singapore (2022)

Azariadis, C.: Implicit contracts and underemployment equilibria. J. Polit. Econ. **83**(6), 1183–1202 (1975)

Baily, M.N.: Wages and employment under uncertain demand. Rev. Econ. Stud. **41**(1), 37–50 (1974)

Ballot, G.: Modeling the labor market as an evolving institution: model ARTEMIS. J. Econ. Behav. Organ. **49**(1), 51–77 (2002)

Baruffini, M.: An agent-based simulation of the Swiss labour market: an alternative for policy evaluation. In: 53rd Congress of the European Regional Science Association: Regional Integration: Europe, the Mediterranean and the World Economy, pp. 27–31. Palermo, Italy (2013)

Becker, G.S.: Human capital theory. Columbia, New York (1964)

Bergmann, B.R.: Occupational segregation, wages and profits when employers discriminate by race or sex. East. Econ. J. **1**(2), 103–110 (1974)

Boulahbel-Bachari, S., El Saadi, N.: Migration and self-selection: an agent-based model. In: Proceedings of the Computational Methods in Systems and Software. Springer, Cham (2019)

Boulahbel-Bachari, S., El Saadi, N., Bah, A.: Agent based modelling approach of migration dynamics. In: Proceedings of the Computational Methods in Systems and Software. Springer, Cham (2017)

Boulahbel-Bachari, S., El Saadi, N.: Labor migration, from an economic model to an agent based model. In: Loster, T., Pavelka, T. (eds.) The 11th International Days of Statistics and Economics. Conference Proceedings. Melandrium, Prague (2017)

Boulahbel-Bachari, S., El Saadi, N., Bah, A.: Agent based modelling approach of migration dynamics. In: Advances in Intelligent Systems and Computing, vol. 662, pp. 338–349. Springer (2018)

Bulow, J.I., Summers, L.H.: A theory of dual labor markets with application to industrial policy, discrimination, and Keynesian unemployment. J. Labor Econ. **4**(3, Part 1), 376–414 (1986)

Deissenberg, C., Van Der Hoog, S., Dawid, H.: EURACE: a massively parallel agent-based model of the European economy. Appl. Math. Comput. **204**(2), 541–552 (2008)

Dawid, H., Gemkow, S., Harting, P., Van der Hoog, S., Neugart, M.: The eurace@ unibi model: an agent-based macroeconomic model for economic policy analysis (2012)

Dawid, H., Gemkow, S., Harting, P., Neugart, M.: On the effects of skill upgrading in the presence of spatial labor market frictions: an agent-based analysis of spatial policy design. J. Artif. Soc. Soc. Simul. **12**(4), 5 (2009)

Dawid, H., Gemkow, S., Harting, P., Kabus, K., Neugart, M., Wersching, K.: Skills, innovation, and growth: an agent-based policy analysis. Jahrbücher für Nationalökonomie und Statistik **228**(2–3), 251–275 (2008)

Dosi, G., Fagiolo, G., Napoletano, M., Roventini, A.: Income distribution, credit and fiscal policies in an agent-based Keynesian model. J. Econ. Dyn. Control **37**(8), 1598–1625 (2013)

Eliasson, G.: Competition and market processes in a simulation model of the Swedish economy. Am. Econ. Rev. **67**(1), 277–281 (1977)

Gemkow, S., Neugart, M.: Referral hiring, endogenous social networks, and inequality: an agent-based analysis. J. Evol. Econ. **21**(4), 703–719 (2011)

Goudet, O.: La modélisation multi-agent du marché du travail français. PhD diss., University of Paris VI (2015)

Haldane, A.G., Turrell, A.E.: An interdisciplinary model for macroeconomics. Oxf. Rev. Econ. Policy **34**(1–2), 219–251 (2018)

Harris, J.R., Todaro, M.P.: Migration, unemployment and development: a two-sector analysis. Am. Econ. Rev. **60**(1), 126–142 (1970)

Krugman, P.: Increasing returns and economic geography. J. Polit. Econ. **99**(3), 483–499 (1991)

Leibenstein, H.: Allocative efficiency vs. "X-efficiency". Am. Econ. Rev. **56**(3), 392–415 (1966)

Lewkovicz, Z.: Modélisation et simulation du marché du travail avec un système multi-agents, Thèse de doctorat en informatique, Université Pierre et Marie Curie. PhD thesis (2010)

Mitkova, M.: Labour market model with on-the-job search, an agent-based approach (2017)

Neugart, M.: Complicated dynamics in a flow model of the labor market. J Econ Behav. Organ. **53**(2), 193–213 (2004)

Neugart, M., Richiardi, M.: Agent-based models of the labor market (2018)

North, M., Conzelmann, G., Koritarov, V., Macal, C., Thimmapuram, P., Veselka, T.: E-laboratories: agent-based modeling of electricity markets (No. ANL/DIS/CP-107570). Argonne National Lab., IL (US). 2018 (2002)

Perrot, A., Zylberberg, A.: Salaire d'efficience et dualisme du marché du travail. Revue économique 5–20 (1989)

Piore, M.J., Doeringer, P.: Internal labour markets and manpower adjustment. New York DC Heath and Company (1971)

Richiardi, M.G., Leombruni, R., Saam, N.J., Sonnessa, M.: A common protocol for agent-based social simulation. J. Artif. Soc. Soc. Simul. **9**(1) (2006)

Stigler, G.J.: Information in the labour market. J. Polit. Econ. **70**(5, Part 2), 94–105 (1962)

Stigler, G.J., Friedland, C.: The citation practices of doctorates in economics. J. Polit. Econ. **83**(3), 477–507 (1975)

Teglio, A., Raberto, M., Cincotti, S.: The impact of banks' capital adequacy regulation on the economic system: an agent-based approach. Adv. Complex Syst. **15**(supp02), 1250040 (2012)

Todaro, M.P.: A model of labour migration and urban unemployment in less developed countries. Am. Econ. Rev. **59**(1), 138–148 (1969)

United Nations. Economic Commission for Africa. The state of African cities 2008: a framework for addressing urban challenges in Africa. UN-HABITAT (2008)

Zylberberg, A.: Salaires et emploi dans une économie de partage. Économie & prévision. **92**(1), 153–160 (1990)

Predicting Threat Degree for Onset of Type 2 Diabetes Mellitus Based on Machine Learning Methods

Gyuzel Shakhmametova[1](✉), Nikita Vakkazov[1], and Sofya Klimets[2]

[1] Computer Science and Robotics Department, Ufa State Aviation Technical University, Ufa, Russia
g.shakhmametova@gmail.com
[2] Faculty of General Medicine, Bashkir State Medical University, Ufa, Russia

Abstract. This article discusses the use of machine learning methods to predict the degree of threat for onset of type 2 diabetes mellitus in patients aged 25 years. Type 2 diabetes mellitus is a disease that complicates the course of other concomitant diseases, and predicting the threat of its occurrence is an important element in forming the trajectory of diagnosis and treatment of the patient. A binary classification model based on the logistic regression method has been proposed and developed. The developed approach identifies the threat of diabetes mellitus onset using indicators such as age, sex, body mass index (BMI) and glycated hemoglobin (HbA1c). The paper describes how to build a model, and how to generate and prepare data for binary logistic regression. To implement the described approaches, the Python programming language, the Jupyter Lab development environment and scikit-learn, scipy, pandas, and numpy packages were used. Performance analysis showed accuracy of the proposed model as 0.98 on test data. The developed software can be used as a separate application and be built as a module into the clinical decision support system #COMESYSO1120.

Keywords: Machine learning · Predicting · Logistic regression · Type 2 diabetes mellitus

1 Introduction

The introduction of IT technologies in healthcare facilitates the ability to collect, store and process huge amounts of medical data which can be presented in the form of numbers, text, images, videos, or sound recordings. In the vast majority of cases, the data are heterogeneous and poorly structured [1]. Currently existing machine learning methods provide an effective set of tools for the analysis and processing of biomedical data. Utilizing machine learning methods allows solving a number of problems related to diagnosis of diseases, monitoring and prediction of patient's condition, research and development of new drugs, and other problems [2].

R. Silhavy et al. (Eds.): CoMeSySo 2022, LNNS 596, pp. 770–779, 2023.
https://doi.org/10.1007/978-3-031-21435-6_66

Subsequent paragraphs, however, are indented. Despite the rapid development of video and sound analysis, a significant part of medical data is stored, as a rule, in tabular form [3]. Data is provided as a set of rows and columns that contain numeric or text information. In this form, it is convenient to store various information about the patient, about the results of medical research, about the medical history of a particular patient, etc. Data stored in tabular form are used in endocrinology to record the results of studies of various patient indicators and further analyze these indicators to detect the dynamics of disease development [4]. Endocrinological monitoring is very important for controlling diabetes mellitus, as the disease can very quickly get out of control and lead to death. The WHO global report [5] notes that the number of people with diabetes has increased by a factor of 4 since 1980. And by 2030, diabetes could be the 7th leading cause of death worldwide. Analysis of endocrinological monitoring results makes it possible not only to control the course of diabetes mellitus, but also to detect the threat of its appearance. To detect the degree of threat for the onset of this disease, it is proposed to use machine learning and data analysis methods.

This article describes an algorithm for predicting the degree of threat for the type 2 diabetes mellitus based on logistic regression. For model training, a sample consisting of 71,833 records was used, containing information on all patients with diagnosed type 2 diabetes mellitus in a single region of the Russian Federation, and including a synthesized data sample containing corresponding indicators of healthy people. The synthesized sample was generated using a developed algorithm based on data analysis of patients with type 2 diabetes mellitus. The second part of the article presents existing solutions in the field of predicting the degree of threat of diabetes mellitus. In the third part, the algorithm for predicting the degree of threat of type 2 diabetes mellitus is discussed in detail. The fourth part presents the results of testing and analysis of the effectiveness of the developed algorithm.

2 Problem Definition and Suggested Decision

2.1 State of Art

Presently, there is a large variety of software products for monitoring the condition of a patient with diabetes. This list includes:

- electronic diaries and journals;
- programs for monitoring measuring devices;
- calorie programs;
- electronic consultants.

Let us review some of them:

Diasend [6] is an online data management system to download data from various glucometers and insulin pumps. This is a web application, so its advantage is ease of access in case of working Internet. Its disadvantage is lack of intelligent analysis.

Glucose Buddy—Diabetes Helper 2.0 [7] is a mobile app that stores information about glucose levels, carbohydrates consumed, insulin, necessary medications and activity logs. There are also built-in reminders about taking medications and warnings. On the plus side, it allows adding product information using barcodes. Its main disadvantage is absence of Intelligent Data Processing System.

Health Tracker [8] is a program that helps the user track and display any health-related measurements. Its advantages are Windows and Mac platforms and simple user interface. Disadvantages include paid distribution and lack of data analysis functions.

Diabetes Pilot [9] makes it possible to print reports, save data for use in other programs or e-mails as well as edit and analyze data on your computer. Its advantages are a large list of parameters available for monitoring. Disadvantages include no version for Android.

GNU Gluco Control [10] is intended for type 1 diabetes. It supports storage of insulin, HA and food data. Its advantage is being an open source. However, this project rarely updated, which is its major flaw.

The main disadvantage of all ready-made solutions is the orientation towards the already existing disease of diabetes mellitus in the patient and the lack of functionality to predict the degree of threat of its occurrence, which today is an urgent and sought-after option.

2.2 Date Structure and Synthesis

The full sample of data for all patients with diabetes mellitus contains 154,970 records and has the following structure:

1. Patient Code.
2. Date of birth.
3. Sex.
4. Year of diagnosis.
5. Type of diabetes mellitus.
6. Class diagram.
7. Year of beginning of insulin therapy.
8. HbA1c, % (Last Visit).
9. Weight, kg (Last visit).
10. Height, cm (Last visit).
11. BMI, kg/m^2. (Last visit).
12. Others.

To construct a model for predicting the degree of threat of type 2 diabetes mellitus, records of patients over 25 years old and with type 2 diabetes mellitus disease were selected. This sub-sample included 36,833 records, each of which was a set of features for a unique patient. The following fields were selected as characteristics of experts:

1. Sex.
2. Age.
3. Body mass index (BMI).
4. Glycated hemoglobin (HbA1c), %.

This dataset contains 36,833 records of people with already developed diabetes mellitus. To teach the model, it is necessary to obtain similar data from healthy people. To do this, 35,000 records were synthesized with corresponding measures of healthy people.

Data generation is performed according to the following rules:

1. ***Sex*** $\in \{0, 1\}$. 50% share of one class.
2. ***Age*** $\in [25, 90]$ and has a normal distribution with $\mu = 60$, $\sigma = 30$. Noise generated from normal distribution is added to the data $\mu = 0$, $\sigma = 0.2$ (Fig. 1).

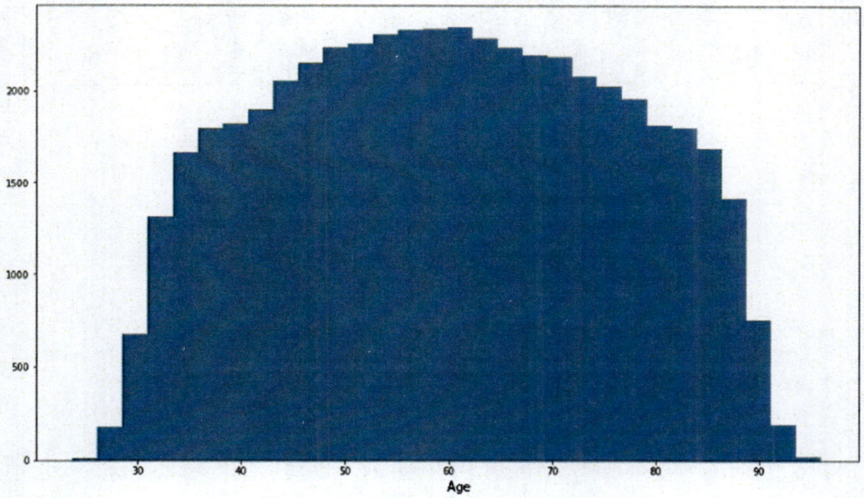

Fig. 1. Characteristic distribution ***Age***

3. ***Body mass index (BMI)*** $\in [19.5, 25.9]$ and has a normal distribution with $\mu = 21$, $\sigma = 2.5$, Noise generated from normal distribution is added to the data $\mu = 0$, $\sigma =$ 0.5 (Fig. 2).

Glycated hemoglobin (HbA1c), %, $\in [3.6, 5.6]$ and has a normal distribution with $\mu =$ 4.6, $\sigma = 0.5$. Noise generated from normal distribution is added to the data $\mu = 0$, $\sigma =$ 0.2 (Fig. 3).

Data analysis and generation were performed in the Jupyter Lab environment [11] using the Python 3.9 programming language and Scipy [12], Matplotlib [13] libraries.

```
lower = 3.6
upper = 5.6
mu = 4.6
sigma = 0.5
```

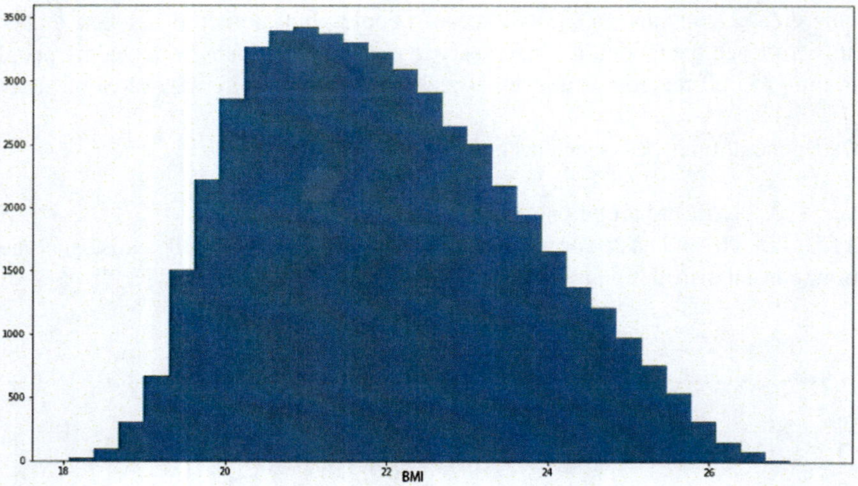

Fig. 2. ***BMI*** characteristic distribution

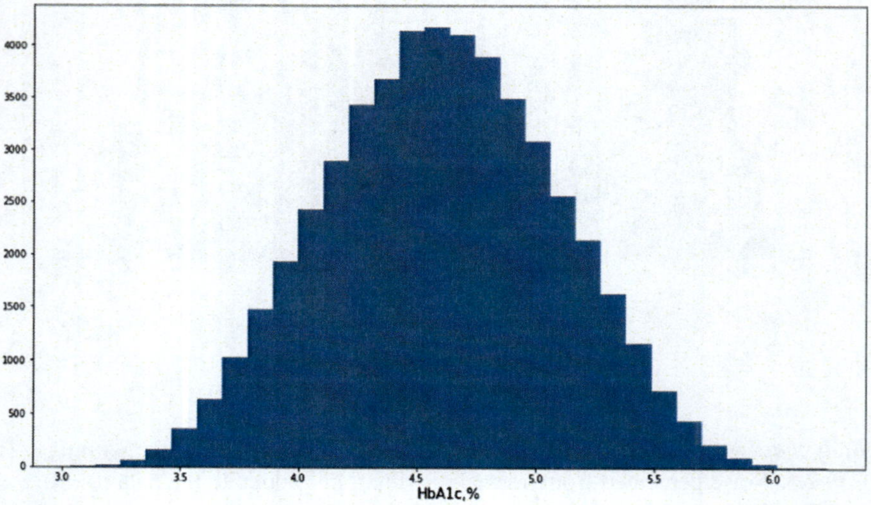

Fig. 3. Characteristic distribution HbA1c

```
hba1c_norm = scipy.status.truncnorm.rvc((lower-mu)/sigma,
                               (upper-mu)/sigma, loc=mu,
                               scale=sigma,
                               size=N)
hba1c = hba1c_norm + hba1c_noise
```

[Computer program code. Example of characteristic generation HbA1c]

The RVS method from the Scipy library stats.truncnorm module was used to generate normally distributed data. To generate noise, the normal method from the random library module Numpy was used. At the end, the sample and noise were added up to obtain the final generation result.

2.3 Construction of Logistic Regression Model

The task of predicting the degree of threat of type 2 diabetes mellitus can be reduced to the task of binary classification with the determination of the probability of belonging to each class. To solve this problem, a mathematical model based on the logistic regression method was chosen.

Logistic regression [14] is a method of constructing a linear classifier that allows evaluating the posterior probabilities of belonging to classes of objects.

Logistic regression equation:

$$\sigma(Z) = \frac{1}{1 + e^{-\hat{y}}} \tag{1}$$

$$\hat{y} = \sum_{i=0}^{n} W_i x_i \tag{2}$$

Error functionality:

$$L(\hat{y}, y) = -\left(y \log \hat{y} + (1 - y) \log\left(1 - \hat{y}\right)\right) \tag{3}$$

The work uses binary logistic regression. The result of this algorithm is the probability of belonging to each class (class 0 - there is no diabetes, class 1 - there is diabetes). Classification threshold is 0.5.

To build the model, the Python programming language and the Sklearn [15] linear_mode module with the implementation of logistic regression were used. The random state parameter is set for the reproducibility of the experiment while multi class = ovr indicates that the model will be binary.

```
clf = LogisticRegression(random_state=0, multi_class='ovr')
            .fit(X_train, y_train)
```

[Computer program code. Model object creation example and training run]

To assess the degree of threat, it is necessary to obtain the probability of the patient belonging to one of two classes. To do this, use the predict_proba method, which returns two numbers between 0 and 1. The first is the probability of belonging to class 0 (the patient does not have type 2 diabetes), and the second is the probability of belonging to class 1 (the patient has type 2 diabetes).

3 Results

Patient data from both the main sample and the synthesized set were used to test the model. The Sklearn.model_selection module was used for correct data splitting. Testing was carried out using the Sklearn.linear_model module. Metrics were calculated using the Sklearn.metrics module. Jupyter Lab was used.

Testing was performed on a sample of 30,786 records. Of these, 15,000 records belonged to class 0 (no diabetes of the 2nd degree) and 15,786 records belonged to class 1 (that is, diabetes of the 2nd degree). Figure 4 shows the error matrix.

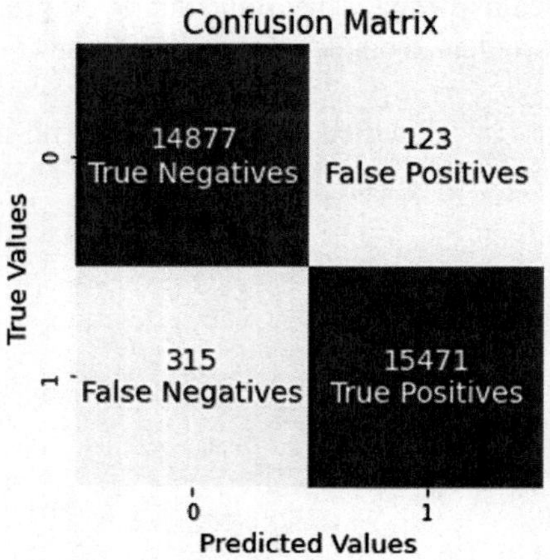

Fig. 4. Error matrix on test sample

Analysis of the error matrix shows that 30,348 records were classified correctly (of which 14,877 had no diabetes and 15,471 had diabetes), and 438 records were classified incorrectly (of which 123 records, the model classified the presence of diabetes and 315 labeled the absence of diabetes).

Calculation of metrics:

$$P = \frac{15471}{15471 + 123} = 0.99 \quad (4)$$

$$R = \frac{15471}{15471 + 315} = 0.98 \quad (5)$$

$$F = 2 \cdot \frac{0.99 \cdot 0.98}{0.99 + 0.98} = 0.99 \quad (6)$$

The metrics turned out to be almost the same. For the result, take the value of F measure.

$$\text{Error of the 1st kind} = \frac{123}{30,786} = 0.004 \quad (7)$$

$$\text{Error of the 2nd kind} = \frac{315}{30,786} = 0.01 \quad (8)$$

The efficiency of the algorithm was tested on 257 real new records. Figure 5 illustrates a fragment of 10 records (patients).

	Sex	Age	BMI	HbA1c, %	target
0	0.0	64	42.820000	9.240000	1.0
1	0.0	65	30.220000	10.000000	1.0
2	1.0	46	43.070000	6.660000	1.0
3	1.0	71	27.660000	7.250000	1.0
4	0.0	66	28.910000	7.100000	1.0
5	1.0	76	24.212306	3.821603	0.0
6	1.0	39	20.565085	4.161126	0.0
7	1.0	36	24.700218	4.976281	0.0
8	1.0	54	20.580443	4.592623	0.0
9	1.0	72	25.070596	4.712172	0.0

Fig. 5. Fragment of patient data and healthy people

4 Discussions

Each record is represented by the characteristics described earlier and the target field, which contains a class label (0—healthy, 1—sick). Figure 6 shows the classification errors.

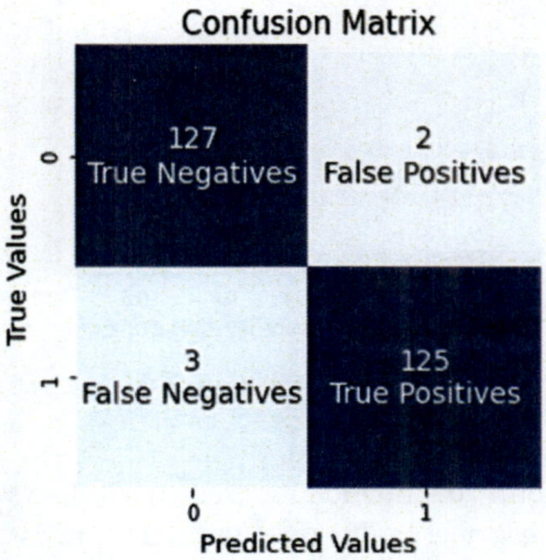

Fig. 6. Error matrix on a sample of 257 entries

Analysis of the error matrix shows that 252 records were classified correctly (of which 127 did not have diabetes and 125 had diabetes), and 5 real new records were classified incorrectly (of which 2 records, the model classified the presence of diabetes, and 3 marked the absence of diabetes).

Calculation of metrics:

$$P = \frac{125}{125 + 2} = 0.98 \tag{9}$$

$$R = \frac{125}{125 + 3} = 0,98 \tag{10}$$

$$F = 2 \cdot \frac{0.98 \cdot 0.98}{0.98 + 0.98} = 0.98 \tag{11}$$

Next, we calculate errors of the 1st and 2nd kind:

$$\text{Error of the 1st kind} = \frac{2}{257} = 0.008 \tag{12}$$

$$\text{Error of the 2nd kind} = \frac{3}{257} = 0.012 \tag{13}$$

Thus, the proposed model with 98% accuracy classified all the examples provided.
In addition to classification, the degree of threat for onset of diabetes may also be predicted. Figure 7 shows the probability distribution of such forecasts.

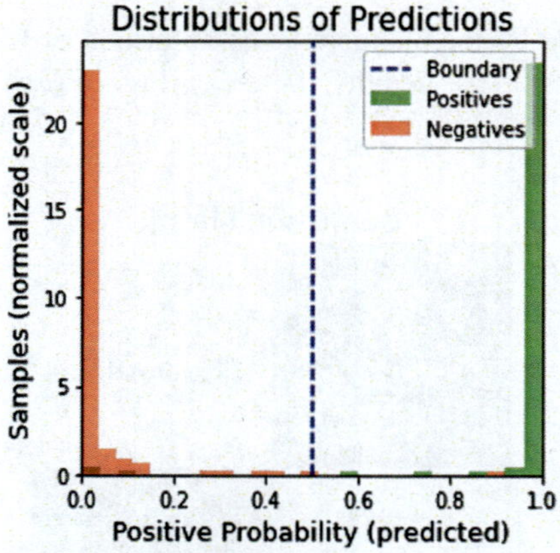

Fig. 7. Probability distribution of the forecast

Analysis of the probability distribution for predictions shows that most of them are close to 0 or 1. This suggests that in this case, the algorithm assigns a particular class label with high confidence.

5 Conclusions

The result of the study is the development of a model to assess the degree of threat for onset of type 2 diabetes mellitus in patients over 25 years of age based on machine learning methods. Model development involves 2 stages, i.e., preparation of real and synthesized data for further model training, and training in binary logistic regression. Statistical and data processing methods are used to implement the data preparation phase. The software is implemented in the Python 3.9 programming language; Pandas 1.1.3,

Numpy 1.19.5, Scipy 1.6.2 packages were used for data processing, and Scikit-learn 0.23.2 packages were used for machine learning algorithms. The test results showed a high classification accuracy of 99%. The developed software is cross-platform and can function both as a separate application and as an embedded module in the clinical decision support system for endocrinologists.

At the moment, the developed software functions as a separate application, but in the future it will be built as a module into the clinical decision support system for the prevention and treatment of bronchopulmonary diseases.

Acknowledgements. The reported study was funded by RSF, project number 22-19-00471.

References

1. Abiteboul, S.: Querying semi-structured data. In: Afrati, F., Kolaitis, P. (eds.) Database Theory—ICDT '97. Lecture Notes in Computer Science, vol. 1186, pp. 1–18. Springer, Heidelberg (1997). https://doi.org/10.1007/3-540-62222-5_33
2. Muthalaly, R.G., Evans, R.M.: Applications of machine learning in cardiac electrophysiology. Arrhythm Electrophysiol. Rev. **9**, 71–77 (2020)
3. Mostafa, F., Hasan, E., Williamson, M., Khan, H.: Statistical machine learning approaches to liver disease prediction. Livers **1**, 23 (2021)
4. Bland, M., Peacock, J.: Statistical questions in evidence-based medicine (2000)
5. World Health Organization. https://www.who.int/ (date of the request: 15.08.22)
6. Diasend. https://diasend.com//us. Last Accessed 15 Jul 2022
7. Glucose Buddy. https://www.glucosebuddy.com (date of the request: 15.08.22)
8. Health tracker. http://www.blackcatsystems.com/software/healthtracker.html. Last Accessed 15 Jul 2022
9. Diabetes Pilot. http://www.diabetespilot.com/desktop. Last Accessed 15 Jul 2022
10. GNU Gluco control. http://ggc.sourceforge.net/. Last Accessed 15 Jul 2022
11. Jupyter documentation. https://jupyter.org/. Last Accessed 15 Jul 2022
12. Scipy documentation. https://docs.scipy.org/doc/scipy/. Last Accessed 15 Jul 2022
13. Matplotlib documentation. https://matplotlib.org/stable/index.html. Last Accessed 15 Jul 2022
14. Hosmer, D.W. Jr., Lemeshow, S., Sturdivant R.X.: Applied Logistic Regression, 3rd edn
15. Scikit-learn documentation. https://scikit-learn.org/stable/index.html. Last Accessed 15 Jul 2022

A Creating of Processes and Threads in MS Windows

Martin Sysel(✉)

FAI, Tomas Bata University, Nad Stráněmi 4511, 760 05 Zlín, Czech Republic
Sysel@utb.cz

Abstract. This article discusses the differences between processes and threads in the MS Windows operating system. First, the mechanism of creating a new process and thread is explained, and then several measurements are performed under different conditions. It focuses on the time required to create a process or thread; several measurement methods are described. Finally, the advantages and disadvantages of processes and threads are summarized with recommendations.

Keywords: MS Windows · Process · Thread · Time measurement

1 Introduction

The Traditional concept of processes can be divided into two parts: processes which dealing resource ownership (in the simplest term, is an executing program) and threads which are called a unit of execution (a stream of instructions) [1, 2] and is the basic unit to which the operating system allocates processor time [3]. Normally when a program starts up and becomes a process, it starts with a single thread, e.g., `main` function in the C language. It can be created extra threads from any of its threads, so one or more threads run in the context of the process.

2 Threads Versus Processes

A question is whether processes or threads are better to use. The answer is: it depends. A process or thread creating time is not the most important reason. The only one advantage of a process is its own address space. If process crashes or has a buffer overrun, it does not affect any other processes, whereas if a thread crashes, it influences other threads in the process. An unshared address space could be beneficial for performance on a multiprocessor system because synchronizing shared memory between different processors is time expensive [10].

R. Silhavy et al. (Eds.): CoMeSySo 2022, LNNS 596, pp. 780–785, 2023.
https://doi.org/10.1007/978-3-031-21435-6_67

Threads share memory of the process with other threads of the same process, inter-thread communication is faster, context switching can be less expensive because it exists chance to use cache memory content [9]. Synchronization is faster, but it depends on application programmer, not on the kernel, so there is a much greater risk of race condition.

2.1 Processes

An application consists of one or more processes. Each process provides the resources needed to execute a program. A process has a virtual address space, executable code, open handles to system objects, a security context, a unique process identifier, environment variables, a priority class, minimum and maximum working set sizes, and at least one thread of execution. A job object allows groups of processes to be managed as a unit. Job objects are namable, securable, sharable objects that control attributes of the processes associated with them [3].

Windows applications launch either classic or modern applications, the Executive includes support for additional kinds of processes that must be started by bypassing the Windows API, such as native processes, minimal processes, or Pico processes [4]. The Windows API provides several functions for creating processes. The simplest function is `CreateProcess`, which attempts to create a process with the same access token as the creating process. Creating a modern application process requires more than just calling `CreateProcess` with the correct executable path. There are some required command-line arguments [4].

Creating a Process

Creating a Windows process consists of several stages carried out in three parts of the operating system: the Windows client-side library Kernel32.dll, the Windows executive, and the Windows subsystem process. The following describes the main stages of creating a process with the `CreateProcess` function. Many steps of are related to the setup of the process virtual address space [4]:

1. Validate parameters; convert Windows subsystem flags and options to their native counterparts; parse, validate, and convert the attribute list to its native counterpart.
2. Open the image file (.exe) to be executed inside the process.
3. Create the Windows executive process object.
4. Create the initial thread (stack, context, and Windows executive thread object).
5. Perform post-creation, Windows subsystem—specific process initialization.
6. Start execution of the initial thread.
7. In the context of the new process and thread, complete the initialization of the address space (for example, load required DLLs) and begin execution of the program's entry point.

2.2 Threads

A thread exists within a process and can be scheduled for execution. Applications can schedule their own threads using User-mode scheduling (UMS). The ability to switch between threads in user mode is more efficient than thread pools for short-duration work items that require few system calls [3]. There are still fibers that run in the thread context and are manually scheduled by the application. Each thread can schedule multiple fibers.

All threads of a process share its virtual address space and system resources. In addition, each thread maintains exception handlers, a scheduling priority, thread local storage, a unique thread identifier, and a set of structures used by system to save the thread context until it is scheduled. The thread context includes the thread's set of machine registers, the kernel stack, a thread environment block, and a user stack in the address space of the thread's process. An application can use the thread pool to reduce the number of application threads and provide management of the worker threads [3].

Creating a Thread

The simplest creation function in user mode is `CreateThread`, which creates a thread in the current process (in the context of some running thread, e.g., main function). After a request to create a thread, the process manager allocates space for the thread object and calls the kernel to initialize the thread control block. The following steps are taken in Kernel32.dll to create a Windows thread [4]:

1. The function converts the Windows API parameters to native flags and builds a native structure describing object parameters.
2. It builds an attribute list with two entries: client ID and Thread Environment Block (TEB) address.
3. It determines whether the thread is created in the calling process, or another process indicated by the handle passed in.
4. It calls `NtCreateThreadEx` (in Ntdll) to make the transition to the executive in kernel mode and continues inside a function with the same name and arguments.
5. `NtCreateThreadEx` creates and initializes the user-mode thread context and then calls `PspCreateThread` to create a suspended executive thread object. Then the function returns.
6. The thread is now resumed so that it can be scheduled for execution.
7. The thread handle and the thread ID are returned to the caller.

3 A Time Measurement

It exists many methods but there are often not enough accurate or portable to other platforms because they depend on operating system or compiler version. It is necessary to define difference between the CPU time and wall time before measure any time-interval. CPU time refers to the time the CPU processed instructions. In the CPU time is not included time spent by waiting for other things to complete. The end user is mainly interested in wall time, it is simply the total time elapsed during the measurement [5]. Windows provides APIs that can be used to acquire high-resolution time stamps, or measure time intervals. Unfortunately, `GetProcessTimes` and `GetThreadTimes` functions are not applicable for this measurement due to low precision (measures the number of quanta, a typical scheduler quantum is about 15 ms).

The primary API for native code is QueryPerformanceCounter (QPC) when time stamps with a resolution of 1 μs or better are necessary. QPC is independent of, and isn't synchronized to, any external time reference. QPC is typically the best method to use to time-stamp events and measure small time intervals that occur on the same system or virtual machine [6].

Another suitable method for short time-interval measurement is a function timespec_get() which set the current time in the struct pointed to by the argument. The struct has two members, tv_sec and tv_nsec. The tv_sec value is set to the whole number of seconds and tv_nsec to the integral number of nanoseconds, rounded to the resolution of the system clock [7].

The <chrono> library is probably the most portable way to measure wall time which works on Linux and Windows operating systems. It is necessary to note that it requires supported C++ 11 or later versions compiler [5, 8]. The clock with the highest resolution available is high_resolution_clock and provide nanoseconds resolution. In the current version of the Microsoft C++ implementation is <chrono> based on QueryPerformanceCounter().

4 Measuring the Creation Time of Processes and Threads

All time measurement has been processed on the PC with CPU AMD Ryzen 7 1700X Eight-Core 3.40 GHz, 16 GB RAM; SSD in M.2 PCIe/NVMe slot. Operating system Windows 10 version 21H2.

All three of the above time measurement methods were used and gave comparable results. Firstly, the execution time of the CreateThread function in the parent thread is measured. When the CreateThread function is used for the first time in the loop than this measured time is more than 60% greater (It takes about 50 μs) than the following calling CreateThread function (It takes about 28 μs).

Executing the CreateThread function and continuing in the child thread takes more time, the reason is wait states and starting this thread on another processor core. The variance of the measured values is very large, the average value is about 130 μs, the minimum achieved value is 60 μs, on the other hand the maximum is 1800 μs. Unfortunately, the results provided are not as consistent as were measured in the case of Linux [10], the behavior is not predictable. This may have various reasons (probably wait states or even the scheduler) and will be further researched.

The allocated huge virtual address space in the process (4 GB allocated by int * p= new int[1e9];) does not influence executing the CreateThread function in the parent thread or continuing in the child thread. The values are comparable to previously measured ones. When the measured times are compared with the values measured in Linux [10], it can be stated that the behavior are comparable to each other (assuming that the extreme measured values in the case of Windows are not considered). Creation thread in Linux is slightly faster.

Furthermore, the execution time of the CreateProcess function is measured. When the CreateProcess function is used for the first time in the loop than this measured time is significantly greater (about 7700 μs) than the following calling CreateProcess function (about 1550 μs). Measurement provides repeatedly consistent results.

Measuring startup time of a new process (the time to execute the `CreateProcess` function in the parent process to start the child process is measured) gives consistent results too. When the process creation is used for the first time in the loop than this measured time is about 60% greater (about 16 ms) than the following process creations (about 10 ms).

Process creation in Windows works on a different mechanism than in Linux. Therefore, it can be assumed that the size of the virtual address space of the parent process will not affect the speed of starting a new process in Windows. This assumption has been confirmed.

4.1 Time Measurement Summary

Table 1 shows measured times with descriptions. The value in rounded brackets indicates the value at the function first call. Values in the square brackets indicates the range of measured values.

Table 1. Time measurement summary.

Measured time [μs]	Description
28 (50)	Execution time of the `CreateThread` function and continuing in the parent thread
130 [60–1800]	Execution time of the `CreateThread` function and continuing in the child thread
1550 (7700)	Execution time of the `CreateProcess` function and continuing in the parent process
about 10,000 (about 16,000)	Execution time of the `CreateProcess` function and continuing in the child process

5 Conclusions

The Windows operating system uses a different process creation mechanism than Linux, there is no duplication of the parent process, but create the new Windows executive process object. The size of the virtual address space of the parent process do not affect the speed of starting a new process in Windows. If a small process in Linux creates a new process by using the `clone()` system call, it is significantly faster than creating a new process in Windows.

Creating a new thread is less time consuming than creating a new process, unfortunately in Windows the behavior is not consistent and predictable, this will be further researched. Processes are safer and more secure than threads because each process runs in its own virtual address space, but processes have a higher overhead; it depends on the size of the virtual address space (unlike threads). Inter-process communication is harder and slower than inter-thread communication. While maintaining all the rules of secure programming, it is preferable to use threads rather than processes.

References

1. Stallings, W.: Operating Systems: Internals and Design Principles, 8th ed. Pearson Education Limited, UK (2014)
2. Silberschatz, A., Galvin, P.B., Gagne, G.: Operating System Concepts, 9th edn. Wiley, Hoboken, NJ (2013)
3. Microsoft: processes and threads. Microsoft Docs. https://docs.microsoft.com/en-us/windows/win32/procthread/processes-and-threads. Last accessed 2022/06/07
4. Yosifovich, P., Ionescu, A., Russinovich, M.E., Solomon, D.A.: Windows Internals, Part 1, System Architecture, Processes, Threads, Memory Management, and More, 7th edn. Microsoft Press, Redmond (2017)
5. Ferreira, C.R.: 8 ways to measure execution time in C/C++. Level up Coding. https://levelup.gitconnected.com/8-ways-to-measure-execution-time-in-c-c-48634458d0f9. Accessed 2020/07/20
6. Microsoft: Acquiring high-resolution time stamps. Microsoft Docs. https://docs.microsoft.com/en-us/windows/win32/sysinfo/acquiring-high-resolution-time-stamps. Accessed 2022/06/12
7. Microsoft: timespec_get(), _timespec32_get, _timespec64_get. Microsoft Docs. https://docs.microsoft.com/en-us/cpp/c-runtime-library/reference/timespec-get-timespec32-get-timespec64-get1. Accessed 2022/06/12
8. Microsoft: <chrono>. Microsoft Docs. https://docs.microsoft.com/en-us/cpp/standard-library/chrono. Accessed 2022/06/10
9. Farrell, J., Buttlar, D., Nichols, B.: PThreads Programming. O'Reilly online learning. https://www.oreilly.com/library/view/pthreads-programming/9781449364724/ch01.html. Accessed 2020/06/25
10. Sysel, M.: A comparison of processes and threads creation. In: Silhavy, R., Silhavy, P., Prokopova, Z. (eds) Software Engineering Perspectives in Intelligent Systems. CoMeSySo. Advances in Intelligent Systems and Computing, vol. 1294. Springer, Cham (2020). https://doi.org/10.1007/978-3-030-63322-6_85

Agriculture, Hunting, Forestry, and Fishing Industry of Kazakhstan Economy: Input-Output Analysis

Seyit Kerimkhulle(✉), Zhangeldi Aitkozha, Ainur Saliyeva, Zhandos Kerimkulov, Alibek Adalbek, and Roman Taberkhan

L.N. Gumilyov Eurasian National University, Nur-Sultan, Kazakhstan
kerimkhulle@gmail.com, rn_82@bk.ru

Abstract. In this research considers of the problems of (+) over- and/or (−) under-evaluation of aggregated flows in monetary terms of accounts and budgets of the agents of two sectoral economy, agriculture, hunting, forestry, and fishing industry and the rest of national economics. To achieve the goal of the study, the input-output table statistical data obtained from the reports of the Asian Development Bank for 2007–2020 for the Republic of Kazakhstan at current prices in of U.S. billions of dollars. Here was determined the time phases of divergence, convergence, and equilibrium of the values of such macroeconomic indicators as interindustry Produce/selling and Purchase/buying, Gross value added and Final demand, Imports and Exports. The results obtained based on this systematic research on input-output analysis can be used in decision-making in macroeconomic policy in the face of external challenges and uncertainty.

Keywords: System research · Input-output analysis · Economic statistics · Evaluation · Divergence · Convergence · Equilibrium · Kazakhstan

1 Introduction

The problems of input-output analysis and related supply chain analysis in the sectors of the national economy, in particular the sectors of agriculture, hunting, forestry, and fishing Industry have been studied in many scientific papers. The following articles [1–5] consider production processes in the market supply chain. It is known that the supply chain can be analyzed as a network of production processes that can be localized in a certain geographical area and beyond. Each process can be defined as a system with input streams resulting in output streams.

R. Silhavy et al. (Eds.): CoMeSySo 2022, LNNS 596, pp. 786–797, 2023.
https://doi.org/10.1007/978-3-031-21435-6_68

The input-output approach based on production processes is considered in the work [3]. It here is used proposes an input-output approach based on production processes to develop specific models that investigate flows among production processes either of a whole (global) supply chain or of a part of the supply chain localized within a given area (local). We also note that developed models can help improve design and management of supply chains also in terms of sustainable development of a local area.

And the position of the Thai economy in the entire world market supply chain is studied in [4]. Data on the country's economy show that growing inflows of intermediate, engineering, and foreign direct investment (FDI) coincide with manufacturing exports. To reflect these connections, the World Wide Web Database (WIOD) has been expanded to include data on Thailand's domestic and international trade statistics. Meanwhile, to obtain the main structure of the international production network, Leontiev's inverse and direct factor calculation methods and structural path analysis (SPA) were applied to the newly created WIOD table. As a result, according to the obtained calculations, it is shown that the electronics manufacturing sector has the highest value of international inverse and direct multipliers among the manufacturing activities of Thailand. This highlights the importance of the transmission mechanism that spreads through the Thai economy and the international production network, as well as the interconnected supply chain.

Also, the impact of bilateral trade relations between Vietnam and China is studied in [5]. It is known that China is a country with the world's leading economic potential, and Vietnam has a trade deficit with China. This situation is getting more complicated every year. The study was based on the 2015 national output-output tables of the two countries, as well as statistical data from the Vietnam Household Living Standards Survey. During the study, 36 sectors of bilateral trade between Vietnam and China were identified. The two-way output-input model is used to analyze how final demand and resource use in production in these countries stimulate output and value added in another country. As a result, it is determined that China benefits more from Vietnam's production and consumption than Vietnam. It is noted that Vietnam's cross-industry structure does not encourage domestic production due to the absence of secondary products as raw materials in the production process.

2 Methods, Models, Data

The input-output analysis of agriculture, hunting, forestry, and fishing industry with the rest of Kazakhstan economy is carried out based on system research as a system for distributing flows output products between agents of purchases and a system for distributing of input products between selling agents in monetary terms.

Assume that by products selling sector $\mathbf{x}' = \left[x_1', \ldots, x_n'\right]$—the value of total outlays of one sector are distributed through purchases in monetary terms to other sectors of the economy as $\boldsymbol{\varepsilon}'_v = \left[\varepsilon_{v1}', \ldots, \varepsilon_{vn}'\right]$—payments of employment resources as total value-added expenditures by each sector according and $\boldsymbol{\varepsilon}'_m = \left[\varepsilon_{m1}', \ldots, \varepsilon_{mn}'\right]$—the payment for imports to the system of equations [2]:

$$\mathbf{x}' = \mathbf{i}'\mathbf{Z} + \boldsymbol{\varepsilon}'_v + \boldsymbol{\varepsilon}'_m, \tag{1}$$

and the product of purchases sector $\mathbf{x} = \begin{bmatrix} x_1 \\ \dots \\ x_n \end{bmatrix}$—the total outputs of one sector are distributed through selling in monetary terms to other sectors of the economy as $\mathbf{f}_d = \begin{bmatrix} f_{1d} \\ \dots \\ f_{nd} \end{bmatrix}$—the final demand and the $\mathbf{f}_e = \begin{bmatrix} f_{1e} \\ \dots \\ f_{ne} \end{bmatrix}$—the payment for exports to the system of equations [2]:

$$\mathbf{x} = \mathbf{Zi} + \mathbf{f}_d + \mathbf{f}_e, \tag{2}$$

where $\mathbf{i}' = [1, \dots, 1]$—vector $1 \times n$ row consisting of n elements of 1; $\mathbf{Z} = \begin{bmatrix} z_{11} & \cdots & z_{1n} \\ \vdots & \ddots & \vdots \\ z_{n1} & \cdots & z_{nn} \end{bmatrix}$—interindustry flows for a particular period in monetary terms.

So the sum ε'_{vj}—the values added of input flows that purchased of products by the sector $j = \overline{1, n}$ and ε'_{mj}—the values of imports flows that purchased of products by the $j = \overline{1, n}$'s sector of sales flows will be equal to sum f_{id}—the final demand for sector $i = \overline{1, n}$'s products that received from sales flows and f_{ie}—the export products $i = \overline{1n}$ to of foreign purchases flows [2]:

$$(\boldsymbol{\varepsilon}'_v + \boldsymbol{\varepsilon}'_m)\mathbf{i} = \mathbf{i}'(\mathbf{f}_d + \mathbf{f}_e), \tag{3}$$

and the sum x'_j—the total outlays for sector $j = \overline{1, n}$'s products of sales flows will be equal to the sum x_i—the total outputs for sector $i = \overline{1, n}$'s products of purchases flows [2]:

$$\mathbf{x}'\mathbf{i} = \mathbf{i}'\mathbf{x}. \tag{4}$$

Further, we were created an information database compiled from source of ADB Report [6] to solve the system of models (1)–(4) and evaluate solutions for (+) over- and/or (−) under-evaluation of the equilibrium state of accounts and budgets to agents of economic system. Indeed, the source of Asian Development Bank Report [6] has data of input-output table for 2007–2020, however, in Table 1, due to the limited amount of work, only 2020 is given. Also, we used the theory and methodology of system analysis, solutions of a system of algebraic equations, comparative analysis, assessment of the behavioral characteristics of economic agents from such sources of scientific literature as [7–18].

3 Results

Assume that Kazakhstan economic statistics consists of the distribution system Agriculture, hunting, forestry, and fishing industry and Rest of economics flows, then we identified the following disbalances and (+) over- and/or (−) under-evaluations.

Table 1. Input-output table of interindustry flows of agriculture, hunting, forestry, and fishing goods of Kazakhstan statistics, U.S. billion dollars, 2020

	Agriculture industry (z_{i1})	Rest of economics (z_{i2})	Inter industry purchase/buying ($\sum_j z_{ij}$)	Final demand (f_{di})	Rest of world: exports (f_{ei})	Output (x_i)
Agriculture industry (z_{1j})	2.138	1.488	3.626	11.632	1.616	16.874
Rest of economics (z_{2j})	3.431	81.389	84.820	122.698	51.775	259.294
Interindustry produce/selling ($\sum_i z_{ij}$)	5.569	82.877	88.447	134.330	53.391	276.168
Gross value added (ε'_{vj})	9.187	153.082	162.269		187.721	
Rest of world: imports (ε'_{mj})	2.118	23.334	25.452	187.721		
Input (x'_j)	16.874	259.294	276.168			

where z_{ij}—the represent interindustry sales flows from sector $i = \overline{1,2}$ to all buyers, in particular to sector $j = \overline{1,2}$; f_{di}—the final demand for sector $i = \overline{1,2}$'s product that received from sales flows; f_{ei}—the export sales flows product $i = \overline{1,2}$ to foreign purchases; x_i—the total output (production) stock of h sector that received from sales; ε'_{vj}—the values added of input flows that purchased of product by the sector $j = \overline{1,2}$; ε'_{mj}—the values of imports flows that purchased of product by the $j = \overline{1,2}$'s sector; x'_j—the total outlays (input) stock that purchased of sector $j = \overline{1,2}$
Compiled from source ADB Report [6]

3.1 Disbalance and (+) Over- and/or (−) Under-Evaluation in Agriculture Industry Flows

According to the Kazakhstan economic statistics the agriculture, hunting, forestry, and fishing industry flows to have the following phases of interindustry disbalances:

- the Produce/selling have the converging (−) under-evaluation phase and the Purchase/buying have the converging (+) over-evaluation phase on 2007–2012; the Produce/selling and Purchase/buying have the equilibrium phase on 2012–2017; the Produce/selling have the divergent (+) over-evaluation phase and the Purchase/buying have the divergent (−) under-evaluation phase on 2017–2020 (see Table 2, column (i)–(iii) and see Fig. 1);
- the Final demand have the converging (−) under-evaluation phase and the Gross value added have the converging (+) over-evaluation phase on 2007–2012; the Final demand and Gross value added have the equilibrium phase on 2012–2017; the Final demand have the divergent (+) over-evaluation phase and the Gross value added have the

divergent (−) under-evaluation phase on 2017–2020 (see Table 2, column (iv)–(vi) and see Fig. 2);

- the Imports have the converging (−) under-evaluation phase and the Exports have the converging (+) over-evaluation phase on 2007–2012; the Imports and Exports have the equilibrium phase on 2012–2017; the Imports have the divergent (+) over-evaluation phase, and the Exports have the divergent (−) under-evaluation phase on 2017–2020 (see Table 2, column (vii)–(ix) and see Fig. 3).

3.2 Disbalance and (+) Over- and/or (−) Under-Evaluation in Rest of Economics Flows

According to the Kazakhstan economic statistics the rest of economics flows to have the following phases of interindustry disbalances:

- the Produce/selling have the converging (+) over-evaluation phase and the Purchase/buying have the converging (−) under-evaluation phase on 2007–2012; the Produce/selling and Purchase/buying have the equilibrium phase on 2012–2017; the Produce/selling have the divergent (−) under-evaluation phase and the Purchase/buying have the divergent (+) over-evaluation phase on 2017–2020 (see Table 3, column (i)–(iii) and see Fig. 4);
- the Final demand have the divergent (−) under-evaluation phase and the Gross value added have the divergent (+) over-evaluation phase on 2007–2012; the Final demand have the converging (−) under-evaluation phase and the Gross value added have the converging (+) over-evaluation phase on 2017–2020 (see Table 3, column (iv)–(vi) and see Fig. 5);
- the Imports have the divergent (−) under-evaluation phase and the Exports have the divergent (+) over-evaluation phase on 2007–2012; the Imports have the converging (−) under-evaluation phase, and the Exports have the converging (+) over-evaluation phase on 2012–2020 (see Table 3, column (vii)–(ix) and see Fig. 6).

4 Discussions

The input-output analysis of agriculture, hunting, forestry, and fishing with the rest of Kazakhstan economy was carried out based on system research as a system for distributing flows output products between agents of purchases and a system for distributing of input products between selling agents in monetary terms. Then to ensure the equilibrium growth of the economic system, it is necessary to identify and evaluate the imbalances in the corresponding accounts and budgets of the agents of the sectors of the national economy of the country.

Indeed, a new methodology to measure and decompose global DMU efficiency into efficiency of inputs (or outputs) was established in [19]. Generally, managers should be able to correctly assess the effectiveness of Global DMU. And due to the lack of detailed information about the costs and results necessary for decision-making, efficiency indicators do not have a full opportunity to manage resources. The article proposes a method for measuring the global efficiency of a DMU along with the efficiency of its input (or output) components. This, in turn, helps directive agencies to evaluate, improve and reorient their strategies and implement programs aimed at improving best practices and reducing costs.

Also, a neoclassical model that studies and explains the dynamics of the Chinese economy is presented in the article [20]. The study examines different perspectives on China's rapid economic growth between 2010 and 2017. In particular, the authors consider neoclassical models that study and explain the dynamics of the Chinese economy through the accumulation and development of domestic production factors. From the point of view of active monetary stimulus, it allowed us to conclude about the significant contribution of domestic demand to China's economic growth. Thus, the novelty of the study is ensured by the fact that Thirlwall's Law does not apply to modern China. It is concluded that the stimulation of the economy in China does not lead to a decrease in the positive balance of the trade balance at the expense of monetary and financial sovereignty, industrial competitiveness, and innovative economic development.

Contradictions that arise during the analysis of the growth rate of the economy are considered in the work [21]. These mainly arise when external or personal factors are not considered during planning. The article analyzes the components of conventional feedback indices (BLs) and type I for-ward multipliers (t.I-Ms) to provide a comparison and understanding of the contradictions that appear in their derivative industry ratings. As a result of the research, it is shown that BL indices tend to grow, and t.I-Ms are suitable for growth planning in the medium and long term. Data Envelopment Analysis (DEA) method is one of the methods of analyzing the efficiency of the enterprise. This is a non-parametric method of evaluating a group of operating objects, which allows identifying efficient and inefficient groups of enterprises.

5 Conclusion

Thus, in this research considers of the problems of (+) over- and/or (−) under-evaluation of aggregated flows in monetary terms of accounts and budgets of the agents of two sectoral economy, agriculture, hunting, forestry, and fishing industry and the rest of national economics. To achieve the goal of the study, the input-output table statistical data obtained from the reports of the Asian Development Bank for 2007–2020 for the Republic of Kazakhstan at current prices in of U.S. billions of dollars. Here was determined the time phases of divergence, convergence, and equilibrium of the values of such macroeconomic indicators as interindustry Produce/selling and Purchase/buying, Gross value added and Final demand, Imports and Exports. The results obtained based on this systematic research on input-output analysis can used in decision-making in macroeconomic policy in the face of external challenges and uncertainty.

Acknowledgments. This research was funded by the Science Committee of the Ministry of Education and Science of the Republic of Kazakhstan (Grant No. AP09259435).

Appendix

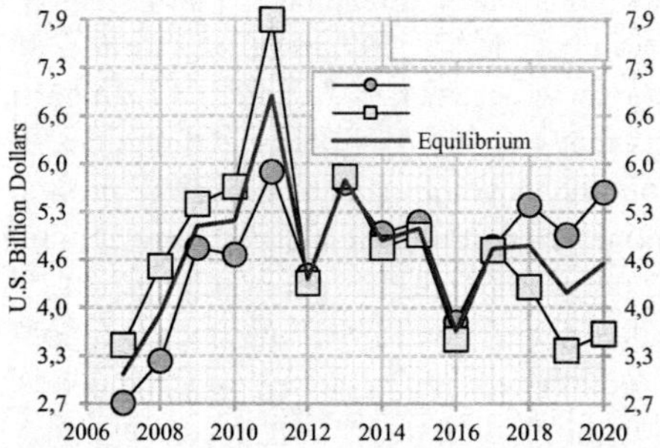

Fig. 1. Disbalance and/or equilibrium in agriculture industry: produce/selling and purchase/buying

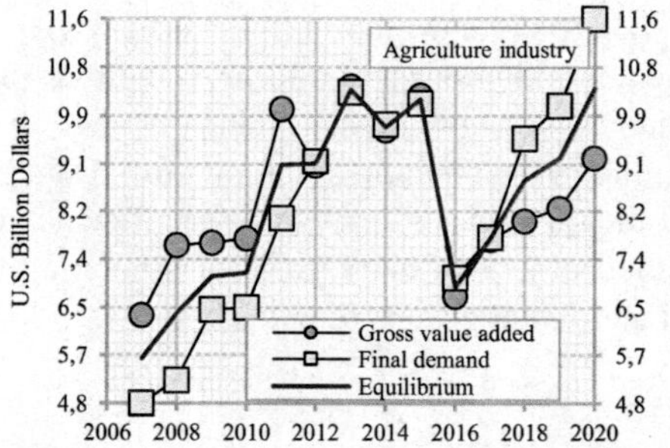

Fig. 2. Disbalance and/or equilibrium in agriculture industry: gross value added and final demand

Table 2. Disbalance and (+) over- and/or (−) under-evaluation in agriculture industry flows

Years	Interindustry produce/selling and purchase/buying +/– evaluation			Gross value added and final demand +/– evaluation			Rest of world economics: imports and exports +/– evaluation		
	(i)	(ii)	(iii)	(iv)	(v)	(vi)	(vii)	(viii)	(ix)
2007	0.795	–0.398	0.398	−1.563	0.781	–0.781	0.768	–0.384	0.384
2008	1.305	–0.652	0.652	–2.397	1.199	−1.199	1.092	–0.546	0.546
2009	0.608	–0.304	0.304	−1.166	0.583	–0.583	0.558	–0.279	0.279
2010	0.935	–0.467	0.467	−1.211	0.606	–0.606	0.276	–0.138	0.138
2011	2.085	–1.042	1.042	−1.950	0.975	–0.975	–0.135	0.068	–0.068
2012	–0.107	0.053	–0.053	0.083	–0.042	0.042	0.024	–0.012	0.012
2013	0.088	–0.044	0.044	–0.098	0.049	–0.049	0.010	–0.005	0.005
2014	–0.198	0.099	–0.099	0.103	–0.051	0.051	0.096	–0.048	0.048
2015	–0.169	0.085	–0.085	–0.138	0.069	–0.069	0.307	–0.154	0.154
2016 ara>	–0.229	0.114	–0.114	0.322	–0.161	0.161	–0.093	0.046	–0.046
2017	–0.041	0.021	–0.021	0.032	–0.016	0.016	0.010	–0.005	0.005
2018	–1.120	0.560	–0.560	1.491	–0.746	0.746	–0.371	0.186	–0.186
2019	–1.585	0.793	–0.793	–1.585	0.793	–0.793	8.274	10.109	9.191
2020	–1.943	0.972	–0.972	–1.943	0.972	–0.972	9.187	11.632	10.410

where (i) is Disbalance in interindustry Produce/selling and Purchase/buying on Agriculture industry; (ii) is +/– evaluation of Produce/selling; (iii) is +/– evaluation of Purchase/buying; (iv) is Disbalance in Gross value added and Final demand on Agriculture industry; (v) is +/– evaluation of Gross value added; (vi) is +/– evaluation of Final demand; (vii) is Disbalance in Imports and Exports on Agriculture industry; (viii) is +/– evaluation of Imports; (ix) is +/– evaluation of Exports

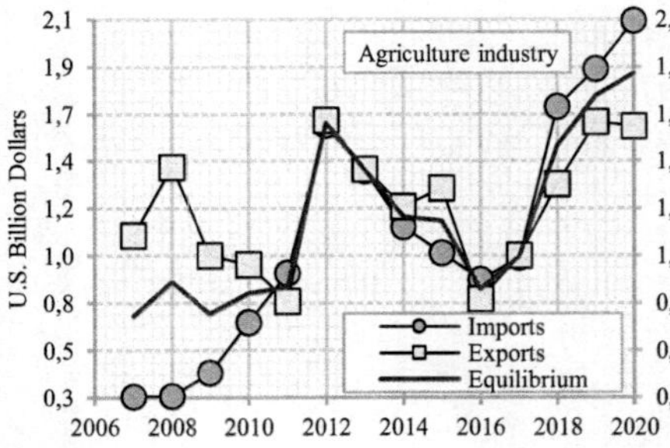

Fig. 3. Disbalance and/or equilibrium in agriculture industry: rest of world: imports and exports

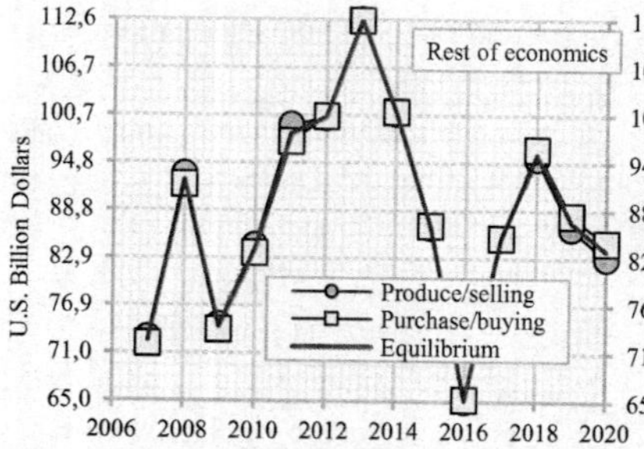

Fig. 4. Disbalance and/or equilibrium in rest of economics: produce/selling and purchase/buying

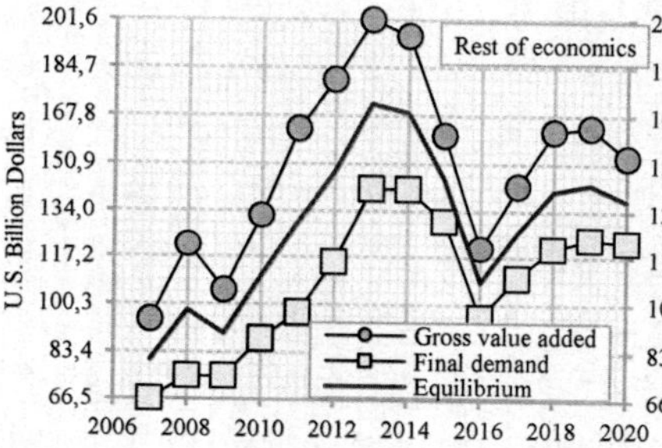

Fig. 5. Disbalance and/or equilibrium in rest of economics: gross value added and final demand

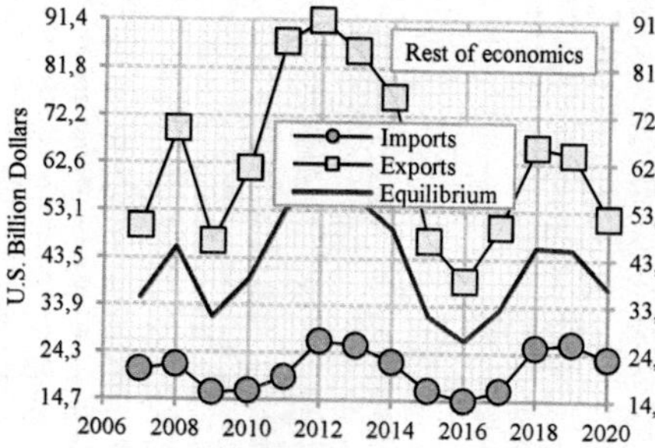

Fig. 6. Disbalance and/or equilibrium in rest of economics: rest of world: imports and exports

Table 3. Disbalance and (+) over- and/or (–) under- evaluation in rest of economics flows

Years	Interindustry produce/selling and purchase/buying +/– evaluation			Gross value added and final demand +/– evaluation			Rest of world economics: imports and exports +/– evaluation		
	(i)	(ii)	(iii)	(iv)	(v)	(vi)	(vii)	(viii)	(ix)
2007	−0.795	0.398	−0.398	−28.286	14.143	−14.143	29.081	−14.540	14.540
2008	−1.305	0.652	−0.652	−46.528	23.264	−23.264	47.833	−23.916	23.916
2009	−0.608	0.304	−0.304	−30.080	15.040	−15.040	30.688	−15.344	15.344
2010	−0.935	0.467	−0.467	−44.132	22.066	−22.066	45.066	−22.533	22.533
2011	−2.085	1.042	−1.042	−65.550	32.775	−32.775	67.635	−33.817	33.817
2012	0.107	−0.053	0.053	−64.858	32.429	−32.429	64.751	−32.375	32.375
2013	−0.088	0.044	−0.044	−59.400	29.700	−29.700	59.488	−29.744	29.744
2014	0.198	−0.099	0.099	−53.771	26.886	−26.886	53.573	−26.787	26.787
2015	0.169	−0.085	0.085	−30.398	15.199	−15.199	30.229	−15.114	15.114
2016	0.229	−0.114	0.114	−24.435	12.217	−12.217	24.206	−12.103	12.103
2017	0.041	−0.021	0.021	−33.174	16.587	−16.587	33.133	−16.566	16.566
2018	1.120	−0.560	0.560	−41.756	20.878	−20.878	40.635	−20.318	20.318
2019	1.585	−0.793	0.793	−39.824	19.912	−19.912	38.239	−19.120	19.120
2020	1.943	−0.972	0.972	−30.384	15.192	−15.192	28.441	−14.220	14.220

where (i) is Disbalance in interindustry Produce/selling and Purchase/buying on Rest of economics; (ii) is +/− evaluation of Produce/selling; (iii) is +/− evaluation of Purchase/buying; (iv) is Disbalance in Gross value added and Final demand on Rest of economics; (v) is +/− evaluation of Gross value added; (vi) – +/− evaluation of Final demand; (vii) is Disbalance in Imports and Exports on Rest of economics; (viii) is +/− evaluation of Imports; (ix) is +/− evaluation of Exports

References

1. Leontief, W.: Quantitative input and output relations in the economic systems of the United States. Rev. Econ. Stat. **18**(3), 105–125 (1936)
2. Miller, R.E., Blair, P.D.: Input-Output Analysis: Foundations and Extensions, 2nd ed., pp. 1–750. Cambridge University Press, Cambridge (2009)
3. Albino, V., Izzo, C., Kühtz, S.: Input-output models for the analysis of a local/global supply chain. Int. J. Prod. Econ. **78**(2) (2002)
4. Puttanapong, N.: Tracing Thailand's linkages to global supply chain: applications of world input-output database (WIOD) and structural path analysis. Int. J. Appl. Bus. Econ. Res. **14**(1), 411–438 (2016)
5. Nguyen, Q.T., Trinh, B., Ngo, T.L., Tran, M.D.: Analysis of bilateral input-output trading between Vietnam and China. J. Asian Financ. Econ. Bus. **7**(6), 157–172 (2020)
6. Asian Development Bank, Kazakhstan: Input-output economic indicators. www.data.adb.org. Last accessed 2022/08/08
7. Sterman, J.D.: Business Dynamics: Systems Thinking and Modeling for a Complex World. McGraw-Hill Inc. (2000)
8. Cao, Y., Liu, Z., Zeng, Y., Huang, X., Liu, D., Han, J.: Coordinated development of energy ecosystem based on carbon neutrality and carbon peak targets. In: 2021 International Conference on Power System Technology (POWERCON), pp. 461–465 (2021)
9. Khalilova, M.K., Davydov, V.A., Niyazbekova, S.U.: P2P lending as a new model of digital bank. Stud. Big Data **110**, 101–107 (2022)
10. Burkaltseva, D., et al.: Methodological foundations of the risk of the stock markets of developed and developing countries in the conditions of the crisis. J. Risk Financ. Manag. **15**(1), 3 (2022)
11. Gnap, J., Konečný, V., Slávik, R., Beňová, D.: Possible impacts of regulating the weekly rest of road freight drivers on logistics in EU countries. Nase More **65**(4), 259–265 (2018)
12. Low, S., Ullah, F., Shirowzhan, S., Sepasgozar, S.M.E., Lee, C.L.: Smart digital marketing capabilities for sustainable property development: a case of Malaysia. Sustainability **12**(13), 5402 (2020)
13. Balabanova, A., Petrova, S., Fomenko, V., Keschyan, N.: Labor potential of youth for the development of ecology in the digital economy. In: E3S Web of Conferences, vol. 258, p. 06048 (2021)
14. Balabanova, A., Keschyan, N., Borisova, T., Hachemizova, E.: Using digital platforms for environmental management. In: E3S Web of Conferences, vol. 244, p. 07006 (2021)
15. Stone, M.: Developments in B to B and B to C marketing and sales automation systems. J. Bus. Bus. Mark. **28**(2), 203–222 (2021)
16. Kadyraliev, A., et al.: The impact of cargo turnover on the ecology of Kyrgyzstan. IOP Conf. Ser. Earth Environ. Sci. **937**(2), 022086 (2021)
17. Yerzhanov, A., et al.: Building information security skills among young transport professionals. Transp. Res. Procedia **63**, 1481–1488 (2022)
18. Butsenko, E.: Development of information system for monitoring the rate of digital currency for investing. In: Lecture Notes in Networks and Systems, vol. 502, pp. 176–185
19. Mosbah, E., Zaibet, L., Dharmapala, P.S.: A new methodology to measure efficiencies of inputs (outputs) of decision-making units in data envelopment analysis with application to agriculture. Socioecon. Plann. Sci. **72**, 100857 (2020)

20. Bykov, A.A., Tolkachev, S.A., Parkhimenka, U.A., Shablinskaya, T.V.: China's economic growth in 2010–2017: analysis from the perspective of the input-output model and modern monetary theory. Financ. Theor. Pract. **25**(2), 166–184 (2021)
21. Kolokontes, A.D., Kontogeorgos, A., Loizou, E., Chatzitheodoridis, F.: Decomposition analysis for the comparison, and the comprehension of conventional input-output impacts' indicators: an empirical paradigm. Sci. Ann. Econ. Bus. **67**(2), 193–217 (2020)

Interdisciplinary Research and Development Projects' Application Domains Classification

A. D. Uliev[1], V. L. Rozaliev[1](✉), O. K. Kasymov[1], Y. V. Kulikova[1], and S. V. Pronichkin[2]

[1] Volgograd State Technical University, Volgograd, Russia
vlrozaliev@mail.ru

[2] Federal Research Center "Computer Science and Control" of Russian Academy of Sciences, Moscow, Russia

Abstract. Supply and demand for interdisciplinary research is closely integrated into the market structure of the economic system and responds to the factors that determine its functioning. From the standpoint of the formation of innovative development strategies, it should be noted the need for a comprehensive combination of interdisciplinary projects in accordance with various classification features of application domains. As a methodological basis for classification, it is proposed to use algorithms for verbal decision analysis. By analyzing information about the topology of the global scientific collaboration network, the decision maker can discover all possible ways to implement interdisciplinary projects. Software has been developed to balance the load at each stage of the implementation of research and development.

Keywords: System analysis · Digitalization · Application domain · Research and development

1 Introduction

Currently, global instability is manifested in the fall of the main macroeconomic indicators, the growth of inflationary processes, the fading of investment activity, the decrease in purchasing power, currency fluctuations and the closeness of a number of traditional markets for national enterprises. All this raises the question of finding such competitive strategies that will fully implement the concept of crisis-resistant management, first of all, at the systemic level of the digital economy—innovative enterprises [1, 2].

R. Silhavy et al. (Eds.): CoMeSySo 2022, LNNS 596, pp. 798–804, 2023.
https://doi.org/10.1007/978-3-031-21435-6_69

An evidence-based long-term action plan is needed to achieve strategic goals and advantages over competitors, which determines how national research support systems can achieve a stable and advantageous position against competitive forces operating in the target market, i.e. to provide conditions for sustainable functioning on it. At the same time, the target market for scientific research and development is closely integrated into the market structure of the economic system and reacts to the factors that determine its functioning [3, 4].

Innovative enterprises are faced with the task of a gradual transition to the use of such competitive strategies that will lay precisely the long-term conditions for crisis-resistant development. This requires the construction of an appropriate scheme of actions for the development and implementation in the practice of managing application domains of interdisciplinary research and development projects.

2 Materials and Methods

The classification of application domains of interdisciplinary research and development projects is based on a clear understanding of the essence of competitiveness, which should be understood as a comprehensive comparative characteristic of an innovative enterprise, which determines its position among many similar business entities and lies in the ability to use various types of scientific and technical resources and competencies are more efficient than competitors, depending on the conditions of the functioning environment. Such an understanding of competitiveness indicates the need for innovative enterprises to maintain and increase their market positions in the long term, taking into account the need to determine the factors and conditions for the implementation of interdisciplinary research and development projects.

At the same time, the issue of developing competitive strategies based on a comprehensive analysis and evaluation of both portfolios of interdisciplinary research and development projects and crisis phenomena is being updated [5, 6]. In this aspect, the use of the entire set of methods for assessing competitiveness is justified, especially given the possibility of combining them.

Methods based on the construction of multi-level structures deserve special attention:

- innovative market potential/interdisciplinary projects' application domain competitiveness;
- industry potential/research areas competitiveness;
- market development stage/innovative enterprise competitive position;
- product life cycle stage/research and development competitive position.

It is also advisable to apply calculation methods, such as analysis of the comparative advantages of competing enterprises; methods based on the theory of effective competition; integrated indicators.

These evaluation methods are superimposed on the diagnosis of crisis phenomena, since the analysis of influencers itself contains parameters for classifying the application domains of interdisciplinary research and development projects. The assessment of competitiveness and crisis phenomena has a common methodological basis [7, 8].

It is the classification of application domains of interdisciplinary research and development projects, factors of influence on the activities of an innovative enterprise that makes it possible to assess the competitive environment and the parameters of the functioning of the knowledge economy. It provides an opportunity to determine the presence or absence of crisis phenomena, actualizes the issue of the type of crisis and determines the range of possible alternative directions for preventing, overcoming or eliminating its consequences based on the formation or use of competitive strategies [9, 10].

An important aspect of assessing competitiveness and crisis phenomena is a clear understanding of the goals and levels of activity of an innovative enterprise, their structuring, which lay the foundation for the target direction and evaluation of interdisciplinary research and development projects. It is proposed to consider four main levels of the attitude of an innovative enterprise to competitiveness: neutral, situational imitation, strategic and adaptive-strategic. Each of them shows the readiness of the enterprise to introduce real competitive strategies for interdisciplinary research and development projects focused on crisis management. This affects the methods for assessing and diagnosing application domains of interdisciplinary projects, as well as algorithms for substantiating strategies, the ability to maintain market positions in times of crisis [11, 12].

Particular attention should be paid to methodological and organizational support, criteria, methods of control and resulting indicators, which will be the basis for its implementation. In the process of forming competitive strategies, one should clearly realize one's strength in the process of influencing the factors of the innovation environment as the basis for supporting and building up competitive positions. Innovation environment factors can be classified into managed and unmanaged [13, 14]. It is necessary to focus the strategy on the maximum possible consideration of the influence of the unmanaged and adaptation to them and flexible management of the managed, taking into account the already assessed state of the innovative enterprise.

From the standpoint of the choice or formation of strategies, it should be noted the need for their complex combination in accordance with various classification features of the application domains of interdisciplinary projects and focusing on the stage of the life cycle of an innovative enterprise.

As a methodological basis for classifying application domains of interdisciplinary research and development projects, it is proposed to use algorithms for verbal decision analysis. It is proposed to use three groups of verbal criteria for estimating interdisciplinary projects:

- $C_1 = \{c_1^1, c_1^2, c_1^3\}$—perspective;
- $C_2 = \{c_2^1, c_2^2, c_2^3\}$—resource intensity;
- $C_3 = \{c_3^1, c_3^2, c_3^3\}$—feasibility.

For each application domain class D_i of interdisciplinary research and development projects, the most informative tuple is found:

$$\text{If } (c_1^3, c_2^1, c_3^2) \in D_1 \text{ then } (c_1^2, c_2^1, c_3^2), (c_1^1, c_2^1, c_3^2), (c_1^3, c_2^1, c_3^1), (c_1^2, c_2^1, c_3^1) \in D_1;$$
$$\text{If } (c_1^3, c_2^1, c_3^2) \in D_2 \text{ then } (c_1^3, c_2^2, c_3^2), (c_1^3, c_2^3, c_3^2), (c_1^3, c_2^1, c_3^3), (c_1^3, c_2^2, c_3^3) \in D_2.$$

An index of informativeness of the combination of multi-criteria assessments is calculated and the consistency of the preferences of the decision maker is checked. Next, the construction of the upper and lower boundaries of the classification of application domains of interdisciplinary research and development projects is carried out:

$$M^L - d_1^1 d_2^1 d_3^1 d_4^1 d_5^1 > d_3^2 \approx d_4^2 \succ d_3^3 \succ d_1^2 \approx d_2^2 \succ d_4^3 \succ d_5^2 \succ d_1^3 \succ d_2^3 \approx d_5^3;$$

$$M^H - d_1^1 \succ d_3^1 \succ d_5^1 \succ d_4^1 \succ d_2^1 \succ d_3^2 \succ d_4^2 \succ d_1^2 \succ d_2^2 \succ d_5^2 > d_1^3 d_2^3 d_3^3 d_4^3 d_5^3.$$

To build decisive classification rules, a number of successive steps are required, the leading ones among which are: a comprehensive assessment of competitiveness and crisis phenomena; an analysis of available resources and scientific competencies; diagnosing potential opportunities; separating the impact factors into manageable/partially manageable/unmanageable; development of control systems and parameters for determining the effectiveness of management decisions. In addition, an innovative enterprise should introduce into its activities the key principles for the development and selection of competitive strategies, actually putting them into practice. It is necessary to fully determine competitive advantages, evaluate the most appropriate competitive strategy and clearly implement it in direct scientific and technical activities.

3 Results

The objective economic category of the interdisciplinary research and development projects' effectiveness characterizes profit as the financial result of an innovative enterprise. Profitability is a synthetic indicator that reflects many aspects of the activity of an innovative enterprise for a certain period. Increasing the level of profitability of interdisciplinary research and development projects means strengthening the financial condition, therefore, increasing the funds used for material incentives for employees of an innovative enterprise. For the management of an innovative enterprise, the indicator of profitability of interdisciplinary research and development projects is information about the results, tactics and strategies used, as well as the feasibility of its adjustment [15, 16].

The calculation of profitability based on the classification of application domains of interdisciplinary projects makes it possible to more correctly take into account the objective differences that exist in technical equipment and other subjective factors of research and development. With regard to interdisciplinary projects in the chemical and technological field, increasing the productivity of scientific work is considered as a fundamental factor in reducing the cost of scientific research and development. This is due to the fact that the share of wages here reaches 45%, and the coefficient of use of scientific equipment does not exceed 0.52.

The result of increasing profitability in the chemical and technological sphere should be the scientific and technical management of material and energy resources, the costs of which account for more than 62% of the total costs of innovative enterprises. To ensure load balancing between several research teams of interdisciplinary projects, the decision maker needs to solve two problems—how to distribute scientific and technical subtasks and by what criteria to divide the load on research teams.

4 Discussion

By analyzing information about the topology of the global scientific collaboration network, the decision maker can discover all possible ways to implement interdisciplinary research and development projects. Taking advantage of the global vision of the network of scientific cooperation, it is possible to assess the state of the load of each performer of an interdisciplinary project. To support the management of the implementation of interdisciplinary projects at each stage of research and development, load balancing software has been developed. In order to choose the least labor-intensive way to implement an interdisciplinary project in real time, the balancer processes the collected information about the state of the scientific cooperation network. When the controller has performed load balancing and application domain classification, the load balancer returns the least labor-intensive implementation of interdisciplinary research and development projects.

When finding multiple ways to implement an interdisciplinary project, the controller will pass many options for classifying application domains for load balancing. The load balancer calculates the integrated load for each mode and selects one of them. A less loaded configuration of research and development performers is selected in order to get feedback to the controller for the implementation of interdisciplinary projects.

The controller receives the selected method from the load balancer and generates decision rules to assign co-executors of the interdisciplinary research and development project.

The systematic error of the proposed scientific and methodological approaches to the classification of application domains of interdisciplinary projects is the smallest among all the others, which means that the coefficient of use of the competencies of all participants in an interdisciplinary project and the target result is achieved faster.

The proposed balancing system, using the global vision of the network of scientific cooperation, solves the problem of effective classification of application domains of interdisciplinary research and development projects. The system is based on an analysis of the state of participants in the national research support system. The developed software analyzes the scientific and technical potential use coefficient. Experimental results show that the proposed strategy provides higher scientific performance and achieves a 15% reduction in the latency of the scientific collaboration network.

5 Conclusion

When managing the implementation of interdisciplinary research and development projects, the activation of network human capital is of paramount importance. All characteristics of production and management efficiency must be optimized together. All factors affecting the profitability of interdisciplinary projects of innovative enterprises can be divided into groups: material and technical, which include the composition and technical level of scientific research and development, the degree of automation, the intensity of scientific and technological processes; organizational, characterized by the level of concentration and specialization of scientific research, the degree of its continuity, rhythm, balance, the perfection of management systems, organization and maintenance, the conditions of scientific work; economic factors, the system of material interest of the scientific team in achieving the highest results, the effectiveness of planned, financial and other economic measures; education and scientific qualifications, the effectiveness of forms of encouragement. The key factors are economic factors that most fully cover all elements of the classification of application domains of interdisciplinary projects, namely the methods and subject of research and development. Profitability management of an innovative enterprise is a unifying factor for the whole complex of tasks for organizing, planning and managing the implementation of interdisciplinary research and development projects.

Acknowledgments. The article is prepared with the financial support of the Russian Science Foundation, project № 18-78-10047-P.

References

1. Wang, J., Wang, B., Dong, X.: How does the digital economy improve high-quality energy development? The case of China. Technol. Forecast. Soc. Change **26**, 121960 (2022)
2. Chaudhuri, A., Subramanian, N., Dora, M.: Circular economy and digital capabilities of SMEs for providing value to customers: combined resource-based view and ambidexterity perspective. J. Bus. Res. **142**, 32–44 (2022)
3. Sheng, C., Zhang, D., Huang, Y.: Research on risk mechanism of China's carbon financial market development from the perspective of ecological civilization. J. Comput. Appl. Math. **381**, 112990 (2021)
4. Connor, M., Rafferty, M., Sheikh, A.: Equity compensation and the sensitivity of research and development to financial market frictions. J. Bank. Financ. **37**, 2510–2519 (2013)
5. Gomez, J., Aradros, B., Salazar, I.: How to beat early movers: the role of competitive strategy and industry dynamism on followers' performance in the telecommunications industry. Long Range Plann. **55**, 102244 (2022)
6. Adomako, S., Tran, M.: Sustainable environmental strategy, firm competitiveness, and financial performance: evidence from the mining industry. Resour. Policy **75**, 102515 (2022)
7. Hermundsdottir, F., Aspelund, A.: Competitive sustainable manufacturing—sustainability strategies, environmental and social innovations, and their effects on firm performance. J. Clean. Prod. **370**, 133474 (2022)
8. Sultana, S., Akter, S., Kyriazis, E.: How data-driven innovation capability is shaping the future of market agility and competitive performance? Technol. Forecast. Soc. Change. **174**, 121260 (2022)
9. Liu, Z., Du, Y.: Open knowledge disclosure and technical standard competition in transition economies: a legitimacy perspective. Technol. Soc. **70**, 102001 (2022)
10. Zhao, S., Liu, X., Shenkar, O.: Knowledge management of emerging economy multinationals. J. World Bus. **57**, 101255 (2022)
11. Portillo, A., Gonzalez, M., Perez, J.: The role of innovation in the relationship between digitalisation and economic and financial performance. A company-level research. Eur. Res. Manag. Bus. Econ. **28**, 100190 (2022)
12. Fuller, J., Hutter, K., Tekic, Z.: How AI revolutionizes innovation management—perceptions and implementation preferences of AI-based innovators. Technol. Forecast. Soc. Change **178**, 121598 (2022)
13. Zhao, X., Yi, C., Guo, M.: Business environment distance and innovation performance of EMNEs: the mediating effect of R&D internationalization. J. Innov. Knowl. **7**, 100241 (2022)

14. Lin, B., Ma, R.: Green technology innovations, urban innovation environment and CO_2 emission reduction in China: fresh evidence from a partially linear functional-coefficient panel model. Technol. Forecast. Soc. Change **176**, 121434 (2022)
15. Lee, I., Kim, S., Green, S.: Social enterprises and market performance: the moderating roles of innovativeness, sectoral alignment, and geographic localization. J. Bus. Res. **132**, 491–506 (2021)
16. Wang, L., Zeng, T., Li, C.: Behavior decision of top management team and enterprise green technology innovation. J. Clean. Prod. **367**, 133120 (2022)

Non-fungible Tokens: Value for Digital Arts and Pricing Factors

Vladislav Rutskiy[1,2], Sudhir Allam[3], Sergei Elkin[4,5], Zafarjon Beknazarov[6], Ekaterina Semina[1,2], Oleg Ikonnikov[1,7], Roman Kuzmich[1], and Roman Tsarev[8,9,10](✉)

[1] Siberian Federal University, Krasnoyarsk, Russia
[2] National Research University Higher School of Economics (HSE University), Moscow, Russia
[3] University of Cumberlands, Portland, OR, USA
[4] PT Taplink Digital Global, Kuta, Indonesia
[5] Institute of Economics of the Ural Branch of the Russian Academy of Sciences, Yekaterinburg, Russia
[6] Tashkent Institute of Finance, Tashkent, Uzbekistan
[7] Reshetnev Siberian State University of Science and Technology, Krasnoyarsk, Russia
[8] MIREA—Russian Technological University (RTU MIREA), Moscow, Russia
rtsarev@acadst.org
[9] V.A. Trapeznikov Institute of Control Sciences of Russian Academy of Sciences, Moscow, Russia
[10] International Academy of Science and Technologies, Moscow, Russia

Abstract. The paper examines the pricing factors for non-fungible tokens (NFT) as a new class of blockchain-based cryptocurrency assets originally associated with unique digital art. One of the key challenges of the digital art industry is the problem of monetization and popularization of digital art objects. NFT can be a solution to the problem of financing digital art and at the same time a mechanism for realizing specific rights of a wide audience of token collectors. The authors of the paper carry out econometric modelling of the pricing factors of NFT art objects on the example of the electronic avatars collection "Unicly CryptoPunks", the "wrapped" (tokenized) version of "CryptoPunks" which is one of the first and the most popular NFT art collections with the highest monetization. The results of the analysis show that the dynamics of the collection's quotes is only partially related to the market factors of the cryptocurrency industry. The collection's historical quotes and the factors behind the development of the NFT industry are found to be more important. Thus, crypto market actors may appreciate the added value of NFT associated with digital art. The results of the study show the possibilities of NFT art to manage the risks of investors in cryptocurrency assets and to monetize digital art with the help of NFT. #COMESYSO1120.

Keywords: Digital art · Creative industries · Blockchain · Cryptocurrency · Non-fungible tokens · NFT · CryptoPunks · Unicly CryptoPunks

R. Silhavy et al. (Eds.): CoMeSySo 2022, LNNS 596, pp. 805–816, 2023.
https://doi.org/10.1007/978-3-031-21435-6_70

1 Introduction

The digital art industry is one of the most innovative and dynamic sectors of the creative economy. However, one of the main challenges in this industry is the problem of monetizing and popularizing digital art objects that can be easily copied and distributed without considering the interests of their creators and copyright holders. This is especially true for digital art of obscure creators at an early stage of the audience awareness or for types of digital art that are difficult to monetize (for example, street art).

The so-called non-fungible tokens (NFT) could give the solution to these problems. It is a new class of cryptocurrency assets based on blockchain technology (a distributed ledger for storing and exchanging data using cryptography), originally associated with unique pieces of digital art. NFTs are based on blockchain networks with support for smart contracts and are capable of realizing the various entitlements of token holders to digital art.

In this study, we aim to identify and model NFT pricing factors. This research topic is quite relevant since it refers to an innovative, little-studied, but dynamically developing mechanism for the creation and circulation of digital art objects. First of all, we describe the trends that determine the development of the digital art industry and the history of the blockchain technology application to monetize digital art in the form of NFT. On this basis, we are modelling possible factors of NFT art pricing. We use the example of the electronic avatars collection "Unicly CryptoPunks", the "wrapped" (tokenized) version of "CryptoPunks" which is one of the first and the most popular NFT art collections with the highest monetization in the world.

2 NFT Art: Concept and Factors of Pricing

The digital technology of the decentralized ledger (blockchain) was created in 2009 [1]. The technology was originally associated with Bitcoin (the first cryptocurrency), emerging as a reaction to the growing distrust of centralized financial institutions [2, 3]. Over the years, technology has offered various alternative tools for solving traditional problems of the financial sector of the economy [4–6]. The range of types and properties of digital assets is quite wide—from traditional cryptocurrencies that have no declared intrinsic value and are used as a means of payment and subject of speculative trading, to complex digital assets with many properties regulated by smart contracts. One of the modern trends in the field of cryptocurrencies and blockchain assets is the "tokenization" of various sectors of the economy. The tokenization of the "creative economy" around the world means linking real and digital items to digital tokens on the blockchain.

One of the relatively new types of cryptocurrency assets are the so-called digital "non-fungible tokens" (NFT). These assets revolutionize the creation, authentication, commercialization, exchange and storage of digital content for its creators (for example, artists, musicians, etc.) and holders [7]. These are "unique" digital assets that do not have their material form but can be freely bought and sold on the blockchain network for cryptocurrency. Ideas for tokens with such properties appeared back in 2013 [8]. However, the NFT only gained widespread public attention in March 2021 when a collage of digital paintings by the artist Beeple was sold in NFT form for $ 69.3 million.

Public attention to NFTs skyrocketed in 2020–2021, when the crypto-asset market showed record sales of individual NFTs. But so far, still little is known about the general structure and evolution of this very dynamic market. NFTs sales across various blockchain networks reached nearly $ 2.5 billion in the first half of 2021, while sales in 2020 were only around $ 95 million. Sports and collectible NFTs became the most popular categories in NFT circulation in early 2021. As cryptocurrency assets spread on the market, individual tokens acquired very high value. This raises the question of the presence, first of all, of speculative factors of demand for NFTs, at least for the audience in certain market segments.

The academic literature places much emphasis on the technical or legal aspects of NFT technology [9, 10]. According to forecasts by the Gartner Hype Cycle for Emerging Technologies [11], NFTs are almost at the peak of their initial hype started by NFT art. According to Gartner, this is one of the key technologies to watch out for. For example, NFT could significantly change the future marketing processes for many brands seeking to generate consumer interest and purchasing behaviour [12]. NFTs can help bypass existing platforms like Amazon and Instagram as they allow marketers to monetize and distribute NFTs without intermediaries, directly to consumer audiences. The possible areas of NFT application appear to go beyond technical, legal or artistic issues and will have a profound impact on marketing in the future.

NFTs are tied to unique real or intangible objects using a software algorithm (smart contract) within the blockchain network. They can implement various functions and realize the rights of token holders regarding the digital entity (item) with which the token is associated. Most often, NFTs represent objects such as art, collectibles and in-game items that can be purchased with cryptocurrency or fungible tokens on the corresponding blockchain network [13]. Issuing and trading NFTs is possible on any blockchain protocol with support for smart contracts. The most popular blockchain protocols for issuing and trading NFTs are Ethereum (as well as level 2 Ethereum networks: Immutable X, Polygon), Bitcoin Cash, Cardano, Flow, GoChain, Solana, Tezos, Binance smart chain, Polkadot. Since its inception in 2017, NFTs have spread to almost all areas of the decentralized blockchain industry and have become highly demanded assets in the cryptocurrency asset market in 2020–2021 due to the ability to "attach" them to almost any real or digital subject.

Due to smart contract abilities, the rights of the NFT holder are related to assets on the Internet or in the real world. NFTs enable any blockchain network user to find out the current state of a particular token. If a user has also access to the cryptocurrency wallet on which the NFT is stored, then he or she becomes the owner of the NFT. He or she could not only view but also change the NFT data, saving all changes in the blockchain network data. If you integrate such digital files as images or videos, in blockchain-tokens, their replication can give ownership of such digital artworks [14]. Due to these abilities, NFTs become an object of experiments for many representatives of creative industries who are trying to enter new markets for their products, primarily in the fashion industry, cinema, animation, design, advertising services, music and other industries based on intellectual activity and creation of digital content.

The use of NFT in creative production implies the following principles of the cryptocurrency economy:

1. Creators of digital content (for example, artists, musicians, video makers, bloggers, developers of computer games, etc.) can make money without intermediaries, receiving funds in cryptocurrency directly from the audience of fans.
2. NFTs and their smart contracts allow the addition of detailed attributes to a digital entity, such as the identity of the owner, metadata, secure file links—all these attributes potentially help content creators such as artists to retain and protect the copyright of digital work.

At the moment it is still difficult to talk about determining some kind of universal "fair" price for NFTs, mainly due to the novelty of this class of assets. If we talk about physical works of art, such as paintings, basketball cards, or other real collectible objects, then we can assume that the NFTs attached to them have more explicit intrinsic value, since they are associated with objects in the real world. However, for a significant part of NFTs that are created exclusively in the digital environment, there is no definite answer to the question of the advisability of acquiring NFTs and their fair valuation.

If we consider the characteristics of the tokens, both well-known creators of NFTs and the platforms intended for issuing and trading NFTs distinguish three key factors that affect the valuation of tokens: rarity, usefulness and "tangibility".

Rarity. The rarer and more difficult to obtain a particular token, the higher the likelihood that its price will be high. Good examples of rare NFTs are tokens issued by renowned digital art creators, popular celebrities, or brands. Also, NFTs could be often highly rated if they give some important property to a character in a computer game. This is typical for many Play-to-Earn NFT games. Such tokens attract people because of their intrinsic value, because owning an NFT means owning a certain item that has value in the game. Such a right can be confirmed, and therefore transferred within the blockchain network a game is built on. This makes the rare tokens somewhat premium and gives the token holder a sense of being different from other users of the blockchain network. The first "cryptokitties" (characters from the "Cryptokitties" game implemented on the Ethereum blockchain) are one of the best-known examples of the rarity effect in the valuation of NFTs. Another example is a collage of digital art objects "Everydays—The First 5000 Days" by the renowned digital artist Beeple.

Utility. The assessment of NFT is positively influenced by the possibility of its real application in the physical world or digital environment. Non-fungible tokens can be more than just collectibles. The examples of utility NFTs are are game cards representing virtual land plots, avatars, objects, weapons, character skills necessary for the successful completion of computer games. The more popular the project where the item will be used, the higher the NFT token valuation. As the decentralized gaming community develops on some blockchain network, users are willing to pay an ever higher price for unique NFT cards. Examples of such NFTs include football tickets, virtual land plots in Decentraland, collectible cards for card games (for example, the Geralt of Rivia NFT for the Witcher Universe Gwent Card Game).

Tangibility. At the moment, on the NFT market, there are also tokens tied to real-world objects that can be "touched". In this case, the cryptographic token is a confirmation of the property's immutability right. Almost any object can be used as an item for NFT. Moreover, such an item does not have to be rare. It may simply bring pleasure or benefits in life. The NFT valuation in this case depends on the level of satisfaction of the token holder. Examples of such NFTs are digital copies of famous real paintings sold at auctions organized by public museums of State Hermitage Museum in Russia (in cooperation with the blockchain platform Binance NFT) and the British Museum in the UK (in cooperation with the blockchain platform La Collection) [15]. At those auctions, anyone could buy a desired NFT copy of the painting, which were released in different editions with the valuation depending on the rarity of the token programmed by a smart contract in the blockchain network. Moreover, in both cases, according to the terms of the auction, half of the NFT circulation issued for the selected paintings was sold at the auction, and the other half, along with the original paintings, remained at the disposal of museums to confirm the value of the issued NFT art items. Some experts argue that from an investment point of view, such "tangible" NFT are most suitable for short-term trading in the crypto market. Many kinds of these "tangible" NFT may have a limited duration of value, in particular, this applies to NFT tickets to events. However, there are also "tangible" NFT that can add value over the long term. For example, tokens issued for limited editions of clothing, books, or NFT copies of original items that are publicly destroyed in order to transfer the value of the destroyed original to its NFT copy.

Investments in NFT and their use are likely to become more widespread. However, the regulation of blockchain technology and its adoption in various industries is likely to take some time. In the meantime, in some countries, smart contracts on the blockchain are not allowed, and in some countries, there are trends of tightening the legislation [16]. For the spread of successful NFT practices, regulation is needed to ensure fair conditions for the use of smart contracts related to NFT for any participant in the NFT industries.

The most important tenet of NFT in the creative economy is the ability of the creator to create digital content and sell it without intermediaries. That is, the concept of selling digital content changes from a demand-driven to a supply-driven model, when a digital content creator, upon reaching a certain level of fame, may not look for intermediaries to sell his works. Collectors and customers have the opportunity to find a creator on their own, for example, on a digital platform intended for issuing and trading NFT.

3 Modeling the Pricing Factors of NFT Art Collection "Unicly CryptoPunks"

In this study, on the one hand, we set the goal to study the factors of NFT pricing considering the factors of cryptocurrency assets pricing identified in the literature previously. On the other hand, we are not going to investigate NFT market as a whole. We would like to identify possible pricing factors for a particular collection of NFT art. We have chosen the most popular and highly monetized collection of NFT art as an object of our research, namely the "CryptoPunks" collection of electronic avatars.

In the quantitative part of this study, we conduct an econometric analysis of the relationship between prices and sales volumes of instruments on the NFT collection "Unicly CryptoPunks" (variants of the dependent variable of the regression model) and several possible NFT pricing factors (independent variables). The main research objective was to test the hypothesis that, in addition to the pricing factors of cryptocurrencies and traditional financial assets, the set of NFT pricing factors include factors that make the NFTs as an asset class in its own right. As for the method of quantitative studies this paper applies an autoregressive integrated moving average model (ARIMA) to estimate the NFT pricing factors. Firstly, ARIMA is one of the most common models used in the analysis of assets in traditional financial markets, which, in our opinion, can also be applied to the analysis of cryptocurrency quotes [17]. Secondly, to the best of our knowledge, no attempt has yet been made to use ARIMA to model NFT pricing factors.

We choose possible pricing factors from recent studies. The influence of cryptocurrencies and traditional financial markets (for example, stock indices) played an important role in our investigation of NFT pricing factors. As such, we include various data sets of traditional assets pricing factors and cryptocurrencies factors to study the results of simulating NFT pricing. Furthermore, we study the relationship of the price quotes of the "Unicly CryptoPunks" collection with the factors of the development of the NFT industry to test if some common NFT (including the "CryptoPunks" collection) demonstrate lower price volatility than the leading cryptocurrencies during stressful periods of development of cryptocurrency markets. In other words, would art-based NFT be able to diversify the risks of investors in crypto assets?

For an overview of variables studied, see Table 1. Most of the time series values for variables were obtained from data on the websites of blockchain industry tracker (coi nmarketcap.com), NFT industry tracker (nonfungible.com), and stock exchanges and financial news agencies (nasdaq.com, bloomberg.com).

Due to the limited period of some specific indicators of the NFT industry of interest to this paper, we have chosen daily values of the time series to build variants of the model. Data for two periods were used: (a) from October 23, 2020 to October 23, 2021 (250 observations), (b) from March 8, 2021 to October 23, 2021 (161 observations). The period limitation was due to the availability of data on the index and trading volume indicators for infrastructure projects in the NFT sphere: "NFTI Price" and "NFTI Total Volume".

Following the methodology for building ARIMA (p, d, q), we tested different versions of the model with different values of the parameters p (time series variables) and q (time series variables in stationary form) for the time series differences provided by the optimal order d (order of difference, at which the stationarity of the time series is achieved).

Subsequently, we selected those variants of the models that are statistically most suitable. We performed all calculations using the Gretl 1.9.92 software package.

Before building a model, we checked the studied variables for stationarity, since the estimation of a non-stationary series can lead to distortion of the results or the appearance of the so-called "fictitious regression", the estimates of which do not allow making adequate conclusions about the relationship between the variables. In case the results of the Dickey-Fuller test and the ACF and PACF charts show that the original time series are not stationary, we calculate successive differences until stationarity is obtained. The presence of ARCH processes implies the dependence of the conditional variance on the past variance values. Accordingly, the assumption of the absence of such relationships is violated, it could lead to some distortion of the results and forecasts. Typically, such outliers in the price quotes of financial assets are explained by special periods of high volatility (usually short), which are not unusual for ARCH processes.

Table 1. Description of endogenous and exogenous variables of ARIMA models.

Indicator	Unit	Period	Comments
Endogenous (dependent) variables			
Unicly CryptoPunks collection price (P)	USD	23.10.2020–23.10.2021	Selling prices of the Unicly CryptoPunks, an instrument for the "wrapped" NFT-collection "CryptoPunks"
Unicly CryptoPunks collection primary sales (PS)	USD		Primary sales volume of Unicly CryptoPunks for the "wrapped" NFT collection "CryptoPunks"
Unicly CryptoPunks collection secondary sales (SS)	USD		Secondary sales volume of Unicly CryptoPunks for the "wrapped" NFT collection "CryptoPunks"
Exogenous (explanatory) variables			
S&P 500 index (P)	USD	23.10.2020–23.10.2021	The S&P 500 Index is built on the quotes of the 500 largest-cap companies traded on the US stock exchanges. It is one of the key indicators of traditional financial markets
NASDAQ index (P)	USD		The NASDAQ-100 Index is built on the stock quotes of the 100 largest non-financial corporations on the NASDAQ platform, reflecting the dynamics of the value of companies in the sectors of industry, technology, retail, telecommunications, biotechnology, healthcare, transportation, media and services. The index reflects the conjuncture in the US high-tech markets and the reaction of the business community to political and economic events in the country that affect the high-tech markets. The USA is one of the countries with the most massive investments in cryptocurrencies

(continued)

Table 1. (*continued*)

Indicator	Unit	Period	Comments
Bitcoin price (P)	USD		An innovative private payment network and a new kind of decentralized (private) "digital" money. Bitcoin serves as a key indicator of the cryptocurrency assets market. The first cryptocurrency in terms of capitalization
Bitcoin total volume (TV)	USD		
Ethereum price (P)	USD		Cryptocurrency and platform for creating decentralized blockchain services based on smart contracts. This cryptocurrency is actively used on largest platforms for issuing and trading NFTs. The second-largest cryptocurrency in terms of capitalization
Ethereum total volume (TV)	USD		
Binance coin price (P)	USD		A cryptocurrency on the Binance Smart Chain launched by Binance, one of the world's largest centralized cryptocurrency exchanges. The currency is used to pay for trading commissions, listing, exchange fees, and launching new projects. The third-largest cryptocurrency by capitalization
Binance coin total volume (TV)	USD		
Unicly CryptoPunks collection active market wallets (MW)	Units		The number of active wallets that own part of the Unicly CryptoPunks collection. Shows the number of investors in the market
Unicly CryptoPunks collection number of sales (NS)	Units		The number of deals concluded with Unicly CryptoPunks. Using this indicator, you can track the speculative sales of the Unicly CryptoPunks collection
NFTI price (P)	USD	08.03.2021–23.10.2021	NFT cryptocurrency industry index. The purpose of the index is to track the dynamics of price quotes for a set of 8 key tokens for infrastructure projects in the NFT sphere: Polygon (Matic), Enjin, Decentraland, Sand, Axie Infinity, Aavegotchi, Rarible, Meme
NFTI total volume (TV)	USD		

After developing and testing different variants of 6 ARIMA models for two types of samples, we found that a large number of investigated variables are insignificant (with linear, square-logarithmic and semi-logarithmic values). The smaller the order of p or q, the lower the test statistics and the lower the level of significance of other parameters. Of all the studied and modelled ARIMA (p, d, q) models, we chose those with the best statistics and minimum values of the parameters p and q. After analyzing the intermediate models, insignificant variables were excluded from the regressions. The resulting models are presented in Table 2.

Table 2. Final ARIMA models after excluding non-significant variables.

No.	Indicators	R-squared
Model 1	d_UPUNK_P = 238471d_UPUNK_MW – 66964.5d_UPUNK_NS + 0.267917phi_1 – 1.81595theta_1 + 0.815954theta_2	0.859414
Model 2	d_UPUNK_P = 193119d_UPUNK_MW + 669900d_l_NFTI_TV – 0.440670phi_1 – 0.335437phi_2 – 0.99999theta_1	0.892722
Model 3	d_UPUNK_PS = 18595.3d_UPUNK_MW + 328159d_l_BTC_TV – 291271d_l_ETH_TV + 146733d_l_BNB_TV – 5384.19d_UPUNK_NS – 0.133583phi_1 – 0.348445phi_2 – 1.55580theta_1 + 0.555802theta_2	0.794435
Model 4	d_UPUNK_PS = 1418.8d_UPUNK_MW + 53668.4d_l_NFTI_TV – 4.94064e + 06d_l_NASDAQ_P + 413191d_l_BTC_TV – 430940d_l_ETH_TV – 0.119577phi_1 – 0.362009phi_2 – 1.59749theta_1 + 0.597489theta_2	0.822543
Model 5	d_UPUNK_SS = 225840d_UPUNK_MW – 68705.5d_UPUNK_NS – 0.550529phi_1 + 0.300021phi_2 – 0.998956theta_1 – 0.697946theta_2 + 0.696903theta_3	0.859066
Model 6	d_UPUNK_SS = 176507d_UPUNK_MW + 497491d_l_NFTI_V + 0.251308phi_1 – 1.79436theta_1 + 0.794361theta_2	0.896870

All final models were obtained with fairly high quality in terms of the value of the coefficient of determination (R-squared). As you can see, the historical price quotes of the "wrapped" collection of "CryptoPunks" are among the factors influencing the valuation of the collection, namely its price or sales volume. The prices of the leading cryptocurrencies Bitcoin and Ethereum are of less importance. At the same time, there is an unexpected negative relationship between the collection and Ethereum, but there is a positive relationship between the collection and Bitcoin and Binance coin. This, perhaps, shows that NFT is a dynamically developing segment of the blockchain industry. We have not found a significant relationship between the value of the "wrapped" collection of "CryptoPunks" and the volume of transactions in the Bitcoin and Ethereum markets. Probably, this is due to the liquidity flows from the traditional cryptocurrency assets sector to the dynamically developing NFT sector in the period studied. The models did not confirm the expected positive relationship between the value of the "wrapped" collection and the Ethereum price quotes (although the collection is released and circulates precisely on the Ethereum blockchain network). However, we have found a positive relationship between the value of the "wrapped" collection of "CryptoPunks" and the quotes or sales volumes of both Bitcoin and of the third-largest cryptocurrency, Binance Coin. This is probably due to some statistical fluctuations of simulation and due to the fact that internal factors of the collection have greater impact on the collection valuation than the traditional cryptocurrency factors.

The collection's historical quotes and the factors of the development of the NFT industry are the most important for the value of the "wrapped" CryptoPunks collection. In particular, the higher the distribution of the collection among the holders of cryptocurrency wallets, the higher the cost of NFTs. We did not find a positive relationship between the NFTI (index of NFT industry infrastructure projects) and the prices of the "wrapped" collection "CryptoPunks". The relationship between the NFTI and the secondary sales volumes of the wrapped CryptoPunks collection is positive. Also in most of the models that contain the "NFTI" factor, there is no relationship between the value of the "wrapped" collection of "CryptoPunks" and the quotes of leading cryptocurrencies. Apparently, the factor of traditional crypto assets is already considered in the infrastructure index "NFTI".

It is important to mention that the factors of traditional financial markets were statistically insignificant for the most of the simulation cases. The quotes of the Nasdaq-100 index turned out to be a significant factor in only one of the 6 considered models, for the dependent variable of the primary sales of the "wrapped" collection "CryptoPunks".

It should be noted that, perhaps, for a more accurate result and analysis in subsequent studies of NFT pricing factors, it is necessary to develop an improved model that would take into account the presence of conditional autoregressive heteroscedasticity, for example, the ARCH/GARCH model.

4 Conclusions

Art-based NFTs are a relatively new phenomenon in the crypto economics and digital art. So, the question arises whether NFT are a new asset class on the merge of these two industries. In our paper we studied trends of digital arts and built a regression model to study what influences the price of NFTs on the example of the electronic avatars collection "Unicly CryptoPunks", the "wrapped" (tokenized) version of "CryptoPunks" which is one of the first and the most popular NFT art collections with the highest monetization in the world. The major interest was to study if the NFT prices are an outcome of larger forces at work—like the price of the underlying cryptocurrency (Ethereum)—or if digital art develops its quotation independently. This would be indicative for the emergence of an own asset class.

The digital art industry in the form of non-fungible tokens as digital assets that allow tokenizing a wide class of real and virtual assets in blockchain networks, is developing at a very fast pace in the world. NFT seem to belong to a new asset class and the respective markets are still in their infancy, making it difficult to measure the fair value of NFT. Nevertheless, there are many experiments in various areas of NFT application, primarily in the field of digital art, network computer games, fashion, sports, marketing of cryptocurrency projects in the blockchain industry.

The simulation results of our study showed that the sales volumes of "wrapped" assets derived from the NFT collection in study ("Unicly CryptoPunks"), as a rule, do not depend on the price quotes of traditional financial assets and only partially depend on the price quotes of the leading cryptocurrencies Bitcoin and Ethereum. Perhaps this is due to the transfer of the added value from traditional cryptocurrencies into the dynamically developing NFT market. The historical quotes of the collection and the factors of the NFT industry development are more important for the value of the NFT collection. In particular, we mean the distribution of the collection among holders of cryptocurrency assets and the price quotes of the NFTI, an index of NFT industry infrastructure projects. Perhaps the results obtained reflect the market assessment of the specific intrinsic value of the NFT compared to classic crypto assets. This internal valuation, for example, can be associated with such functional properties of many NFTs as rarity and ease of issuance and transfer in the blockchain network, as well as opportunities for realizing the rights of token holders with the help of smart contracts.

Blockchain technology in the form of NFT can find good use in organizing the creation and circulation of digital art and financing various activities in the creative industries related to the creation and widespread distribution and monetization of unique content. In general, the possibilities and limitations of the use of NFT in digital art and other areas of creative industries require more detailed research, both quantitative and qualitative.

References

1. Rutskiy, V., Javed, S., Azizam, S.H., Chudopal, N., Zhigalov, K., Kuzmich, R., Pupkov, A., Tsarev, R.: The price determinants of bitcoin as a new digital form of money. Adv. Intell. Syst. Comput. **1294**, 784–792 (2020). http://doi.org/10.1007/978-3-030-63322-6_66
2. Patashkova, Y., Niyazbekova, S., Kerimkhulle, S., Serikova, M., Troyanskaya, M.: Dynamics of bitcoin trading on the Binance cryptocurrency exchange. Econ. Ann. XXI **187**(1–2), 177–188 (2021). https://doi.org/10.21003/EA.V187-17
3. Rutskiy, V., Javed, S., Rayk, V., Kulakova, N., Derindag, O.F., Chikizov, A., Kukhar, I., Knyazkov, A., Tsarev, R.: Market acceptance of the blockchain technology on the example of bitcoin. In: Lecture Notes in Networks and Systems, vol. 229, pp. 306–315 (2021). https://doi.org/10.1007/978-3-030-77445-5_28
4. Derindag, O.F., Yarygina, I.Z., Tsarev, R.: International trade and blockchain technologies: implications for practice and policy. IOP Conf. Ser. Earth Environ. Sci. **421**(2), 022051 (2020). https://doi.org/10.1088/1755-1315/421/2/022051
5. Niyazbekova, S., Moldashbayeva, L., Kerimkhulle, S., Dzholdoshev, N., Dzholdosheva, T., Serikova, M.: "Green" bonds—a tool for financing green projects in countries. In: E3S Web of Conferences, vol. 244, p. 10060 (2021). http://doi.org/10.1051/e3sconf/202124410060
6. Omarkhanova, Z.M., Niyazbekova, S.U., Varzin, V.V., Kerimkhulle, S.Y., Nurekenova, E.S.: Financial provision of the agro-industrial complex of Kazakhstan: problems and solutions. In: Popkova, E.G., Sergi, B.S. (eds.) Sustainable Agriculture. Environmental Footprints and Eco-design of Products and Processes, pp. 27–32. Springer, Singapore (2022). http://doi.org/10.1007/978-981-19-1125-5_4
7. Malhotra, A., O'Neill, H., Stowell, P.: Thinking strategically about blockchain adoption and risk mitigation. Bus. Horiz. **65**(2), 159–171 (2022). https://doi.org/10.1016/j.bushor.2021.02.033

8. Chen, Y.: Blockchain tokens and the potential democratization of entrepreneurship and innovation. Bus. Horiz. **61**(4), 567–575 (2018). https://doi.org/10.1016/j.bushor.2018.03.006
9. Arcenegui, J., Arjona, R., Román, R., Baturone, I.: Secure combination of IoT and blockchain by physically binding IoT devices to smart non-fungible tokens using PUFs. Sensors **21**(9), 3119 (2021). https://doi.org/10.3390/s21093119
10. Fairfield, J.: Tokenized: the law of non-fungible tokens and unique digital property. Indiana Law J. **97**, 1261 (2022)
11. Gartner: Gartner identifies key emerging technologies spurring innovation through trust, growth and change (2021). https://www.gartner.com/en/newsroom/press-releases/2021-08-23-gartner-identifies-key-emerging-technologies-spurring-innovation-through-trust-growth-and-change. Last accessed 2022/08/20
12. Kaczynski, S., Kominers, S.D.: How NFTs create value. https://hbr.org/2021/11/how-nfts-create-value. Last accessed 2022/08/20
13. Dowling, M.: Is non-fungible token pricing driven by cryptocurrencies? Financ. Res. Lett. **44**, 102097 (2022). https://doi.org/10.1016/j.frl.2021.102097
14. Nadini, M., Alessandretti, L., Di Giacinto, F., Martino, M., Aiello, L.M., Baronchelli, A.: Mapping the NFT revolution: market trends, trade networks, and visual features. Sci. Rep. **11**(1), 1–11 (2021). https://doi.org/10.1038/s41598-021-00053-8
15. Valeonti, F., Bikakis, A., Terras, M., Speed, C., Hudson-Smith, A., Chalkias, K.: Crypto collectibles, museum funding and OpenGLAM: challenges, opportunities and the potential of Non-Fungible Tokens (NFTs). Appl. Sci. **11**(21), 9931 (2021). https://doi.org/10.3390/app11219931
16. Morkunas, V.J., Paschen, J., Boon, E.: How blockchain technologies impact your business model. Bus. Horiz. **62**(3), 295–306 (2019). https://doi.org/10.1016/j.bushor.2019.01.009
17. Ariyo, A.A., Adewumi, A.O., Ayo, C.K.: Stock price prediction using the ARIMA model. In: Proceedings of the 16th International Conference on Computer Modelling and Simulation, UKSim-AMSS, pp. 106–112. IEEE, Cambridge, UK (2014). http://doi.org/10.1109/UKSim.2014.67

Intelligent Methods for Restructuring Digital Science-Intensive Solutions of Network Innovation Institutions

A. V. Zubkov[1], S. V. Pronichkin[2], O. K. Kasymov[1], V. V. Gilka[1], and V. L. Rozaliev[1](✉)

[1] Volgograd State Technical University, Volgograd, Russia
vlrozaliev@mail.ru

[2] Federal Research Center "Computer Science and Control" of Russian Academy of Sciences, Moscow, Russia

Abstract. The success of any method of restructuring the digital science-intensive solutions of network innovation institutions depends on the correct choice of representative indicators of the state of innovative enterprises, the forms of links between them, as well as the volume and composition of the data sample under which the model is built. Few authors in their research use exactly the built complex of models, most scientific works are limited to individual coefficients or discriminant indicators. For a systematic analysis of the restructuring of digital knowledge-intensive solutions, it is proposed to jointly use intelligent predicative and normative models. A decision support system has been developed for the analysis of complex scientific and technical systems based on dominance verbal scales.

Keywords: Decision support system · Innovations · Digital economy · Science-intensive solutions

1 Introduction

Low efficiency, lack of incentives for entrepreneurial activity, structural imbalances, inflation, low level of competitiveness in global markets—all this is an incomplete list of problems caused by global instability. Anti-crisis management of innovative enterprises in countries with developed economies is an element of a unified system for managing the national scientific and technical potential and is an integral system of measures aimed at protecting the national system of supporting scientific research from crisis phenomena. The process of anti-crisis management in network innovation institutions is carried out continuously: starting from the creation of a start-up and throughout all subsequent life cycles of an innovative enterprise.

R. Silhavy et al. (Eds.): CoMeSySo 2022, LNNS 596, pp. 817–823, 2023.
https://doi.org/10.1007/978-3-031-21435-6_71

Anti-crisis management should begin from the moment the idea of creating a knowledge-intensive enterprise arises. It is at this moment that the initiator of entrepreneurial activity must determine the potential negative and positive factors that may affect the functioning of an innovative enterprise in the chosen direction. It is necessary to take into account all possible threats that, for example, can lead an innovative enterprise to bankruptcy.

To ensure sustainable development and prevent bankruptcy, a model of anti-crisis management is needed. The model should include effective structures and mechanisms for restructuring digital knowledge-intensive solutions of network innovation institutions for anti-crisis management in the context of global instability.

2 Materials and Methods

The restructuring model of digital knowledge-intensive solutions of network innovation institutions is a management system for an innovative enterprise aimed at localizing or eliminating unfavorable situations for sustainable development by means of artificial intelligence and economic and mathematical modeling. Justifying the structure of intellectual models, it is necessary to single out two components:

- anti-crisis management;
- anti-crisis regulation.

An integral element of the intellectual model of anti-crisis management of an innovative enterprise is its restructuring. The restructuring of digital high-tech solutions is the implementation of a set of organizational, financial, economic, legal, technical measures aimed at increasing the investment attractiveness of high-tech solutions, increasing the output of competitive products, and increasing the scientific and technical efficiency of production. Restructuring is carried out in compliance with the requirements of regulatory legal acts, as well as taking into account the decisive rules of efficiency. It is important to take into account the implementation of a protectionist policy to improve viable companies or enterprises belonging to strategically important sectors of the national economy [1, 2].

The procedure for restructuring digital science-intensive solutions of network innovation institutions provides for:

- making a decision on restructuring;
- creation of an expert commission on restructuring;
- development and implementation of a restructuring plan.

The use of quantitative econometric models for the restructuring of digital knowledge-intensive solutions has a number of methodological defects. Therefore, it is considered most appropriate to use models of verbal analysis of decisions to assess the technical and economic state of an innovative enterprise, since the vast majority of indicators are known at the current time and can be easily modeled for the following periods (such indicators as loan obligations, fixed and variable costs, production funds, wage funds are projected for future periods). Strategic decisions should be made precisely on the basis of a verbal analysis of the state of an innovative enterprise, modeling the desired technical and economic state and the parameters necessary to achieve it.

The group for the restructuring of digital knowledge-intensive solutions can include economically "healthy" enterprises that have fallen into a difficult financial situation due to the insolvency of their customers. The possibility of a delay in the fulfillment of current contractual obligations by partners leads to the risk of losses associated with violation of payments and profit-making by the enterprise. By analogy with restructuring, reengineering distinguishes two significantly different types of activities:

- crisis reengineering, which is a crisis management tool;
- development reengineering—a tool for proactive management.

Assessment of the technical and economic state and diagnosis of the possibility of a financial crisis is an integral task of managing an innovative enterprise in modern conditions. One of the common methods for assessing the technical and economic state of an enterprise and the likelihood of its activity stagnation is the use of discriminant models [3, 4]. The technical and economic situation at an innovative enterprise is determined by a single criterion obtained by synthesizing a number of coefficients weighted according to their degree of significance.

In general, the advantage of discriminant models is that the value of the weight coefficients for research can be selected based on the state of the innovative enterprise, the state of the market for supply and demand for new scientific knowledge [5, 6]. Discriminant analysis as a method of modeling the assessment of the propensity of an innovative enterprise to bankruptcy has a number of disadvantages. Such methods are significantly limited by the linear dependencies of the effective indicator on a number of factors, and the accuracy depends not only on the correct choice of factors, but also on the correct and reasonable calculation of weight coefficients for a particular country or industry. On the other hand, for a system analysis of the restructuring of digital knowledge-intensive solutions, the most suitable is the joint use of predicative and normative models.

Predicative models are models of a predictable nature. They are used to predict the income of an innovative enterprise and the future state of the national research support system [7, 8].

Normative models make it possible to compare the actual results of the activity of innovative enterprises with the expected ones, calculated according to the budget. These models are mainly used for internal technical and economic analysis [9, 10].

The most common method for assessing the technical and economic state of an innovative enterprise and the probability of a financial crisis is linear discriminant models that differ in the number and composition of the factors used. Among other methods, neural networks stand out, which, based on the analysis of the training sample, can assess the technical and economic state of an innovative enterprise and the effectiveness of restructuring with a high degree of probability [11, 12].

It should be noted that few authors in their studies use a set of models of technical and economic analysis, most scientific works are limited to individual coefficients or discriminant financial indicators.

A complex intellectual model has been developed that makes it possible to predict the effectiveness of restructuring in a certain segment of the digital science-intensive solutions market, which is extremely important in a systematic analysis of the activity of an innovative enterprise.

First, a subset of variants for restructuring digital high-tech solutions of network innovation institutions is formed. Boundary solutions $A^L = (a_1^L, a_2^L, \ldots, a_M^L)$ and $A^H = (a_1^H, a_2^H, \ldots, a_M^H)$ are given.

Further, various variants for technical and economic development are compared according to the following risk criteria for the restructuring of innovative enterprises:

- $K_1 = \{k_1^1, k_1^2, k_1^3\}$—restructuring of an absent debtor;
- $K_2 = \{k_2^1, k_2^2, k_2^3\}$—general restructuring procedure;
- $K_3 = \{k_3^1, k_3^2, k_3^3\}$—debtor restructuring, liquidated by the owner;
- $K_4 = \{k_4^1, k_4^2, k_4^3\}$—sanitation of the debtor by his director.

For each criterion K_i, a pairwise comparison of variants A_j for restructuring digital science-intensive solutions of network innovation institutions is carried out:

$$\begin{aligned}&\left(\ldots, a_i^1, \ldots, a_j^1, \ldots\right) \succ \left(\ldots, a_i^1, \ldots, a_j^2, \ldots\right) \succ \left(\ldots, a_i^2, \ldots, a_j^1, \ldots\right)\\ &\succ \left(\ldots, a_i^1, \ldots, a_j^3, \ldots\right) \succ \left(\ldots, a_i^3, \ldots, a_j^1, \ldots\right)\\ &\succ \ldots \succ \left(\ldots, a_i^L, \ldots, a_j^H, \ldots\right) \approx \left(\ldots, a_i^H, \ldots, a_j^L, \ldots\right).\end{aligned}$$

Then a unified scale of verbal assessments of dominance is calculated for all risk criteria for the restructuring of innovative enterprises:

$$\begin{aligned}&S_{12}^L - a_1^1 a_2^1 > a_1^2 \approx a_2^2 \succ a_1^3 \succ a_2^3;\\ &S_{13}^L - a_1^1 a_3^1 > a_3^2 \succ a_3^3 \succ a_1^2 \succ a_1^3;\\ &S_{14}^L - a_1^1 a_4^1 > a_4^2 \succ a_1^2 \succ a_4^3 \succ a_1^3;\\ &S_{15}^L - a_1^1 a_5^1 > a_1^2 \succ a_5^3 \succ a_1^3 \succ a_5^3;\\ &S_{23}^L - a_2^1 a_3^1 > a_3^2 \succ a_3^3 \succ a_2^2 \succ a_2^3;\\ &S_{24}^L - a_2^1 a_4^1 > a_4^2 \succ a_2^2 \succ a_4^3 \succ a_2^3;\\ &S_{25}^L - a_2^1 a_5^1 > a_2^2 \succ a_5^2 \succ a_5^3 \approx a_2^3;\\ &S_{34}^L - a_3^1 a_4^1 > a_3^2 \approx a_4^2 \succ a_3^3 \succ a_4^3;\\ &S_{35}^L - a_3^1 a_5^1 > a_3^2 \succ a_3^3 \succ a_5^2 \succ a_5^3;\\ &S_{45}^L - a_4^1 a_5^1 > a_4^2 \succ a_4^3 \succ a_5^2 \succ a_5^3.\end{aligned}$$

The success of applying complex technical and economic methods for restructuring digital science-intensive solutions of network innovation institutions depends on the correct choice of representative indicators of the state of innovative enterprises, the forms of connections between them, as well as the volume and composition of the data sample, in which an intellectual model is built.

3 Results

As an object of study, digital high-tech solutions of innovative enterprises in the electrical field were considered. The classification and system analysis of the factors, according to the degree of their influence on the errors of measuring current transformers, has been carried out. The main goal of restructuring digital science-intensive solutions of innovative enterprises in the electrical field is to increase the accuracy of measurements and the reliability of data transmission by an automated electricity metering system. The structure of the system for collecting and processing data on electricity consumption should ensure the accuracy and simultaneity of obtaining data on electricity consumption [13, 14].

The metrological characteristics of the automated electricity metering system are determined by the characteristics of the measuring instruments and the parameters of the technical means that are part of the system and affect the results and measurement errors of the electricity parameters [15, 16].

The structure of digital science-intensive solutions for collecting and processing data on electricity consumption is built as a three-level structure and includes:

- level of metering points;
- level of accounting objects;
- local level.

The level of metering points includes all measuring circuits used in the metering points of electric energy of the system of network innovation institutions. The measuring circuit includes:

- measuring current and voltage transformers;
- electricity meters;
- measuring circuits providing connection of meters to measuring transformers.

The level of accounting objects is built on the basis of functional electronic meters, communication devices for transmitting data from meters to the terminal via digital communication channels. The local level of accounting, built on the basis of a modem, has a data transfer interface for working as part of an automatic system for monitoring and accounting for electricity of an electric power supply organization.

4 Discussion

One of the effective ways to improve the measurement accuracy is to introduce corrections into the measured values that exclude systematic errors in measuring current transformers. It is proposed to take into account these corrections when programming electronic meters at each metering point. This will reduce the total error of the measuring channel to values that meet the requirements of a decision maker.

As the study showed, the measurement channel error is affected by the transmission of an analog signal from the measuring current transformer to the meter. A systematic analysis of alternative options for restructuring digital high-tech solutions of network innovation institutions in the electrical field was carried out:

- data transmission via power network;
- data transfer via serial interfaces;
- data transmission via computer network technology;
- data transfer via GSM network.

Laying additional communication cables requires additional financial costs and coordination with various local organizations. As the dominant variant for the restructuring of digital science-intensive solutions, the technology of data transmission using the GSM network was chosen.

Information about current transformer errors that need to know to determine the corrective characteristics of digital science-intensive solutions can be obtained by the following means:

- experimental determination of errors of current transformers under real load;
- calculation of the current transformer error under the condition of a real secondary load according to individual metrological characteristics based on the results of their metrological certification;
- calculation of current transformers errors according to its parameters, including the magnetization characteristic, under the condition of a real secondary load.

It is also proposed to transfer measuring current transformers from the category of analog devices to the digital category by installing a microprocessor-based analog-to-digital converter in close proximity to the current transformer.

5 Conclusion

The systems for collecting and processing data on electricity consumption, operated by energy supply companies of network innovation institutions, are considered. The multicriteria decision-making problem of choosing the best digital science-intensive solutions in the electrical field has been solved. The advantages of an automated system for collecting and processing data on electricity consumption are emphasized, which makes it possible to automatically collect and process information on the consumption of active and reactive electricity, and transmit data at information collection nodes. The system also allows automated control of the technical and economic characteristics of the operation of electrical networks and the technical condition of electricity metering devices. Thus, the costs and human resources for collecting and processing data are reduced. Reducing the error of the automated electricity metering system will lead to an increase in the accuracy of accounting for electricity supplied to consumers, which will provide a significant economic effect.

Acknowledgments. The article is prepared with the financial support of the Russian Science Foundation, project № 18-78-10047-P.

References

1. Metiu, N.: Anticipation effects of protectionist U.S. trade policies. J. Int. Econ. **113**, 103536 (2021)
2. Blanchard, E., Willmann, G.: Unequal gains, prolonged pain: a model of protectionist overshooting and escalation. J. Int. Econ. **135**, 103559 (2021)
3. Zhou, Y., Yu, L., Liu, X.: An enterprise default discriminant model based on optimal misjudgment loss ratio. Expert Syst. Appl. **205**, 117641 (2022)
4. Wang, J., Zhang, X., Fu, B.: Prediction model for blast-induced structural damage based on an optimized Bayes discriminant analysis model. Appl. Math. Model. **110**, 338–366 (2022)
5. Guindalini, C., Verreynne, M., Kastelle, T.: Taking scientific inventions to market: mapping the academic entrepreneurship ecosystem. Technol. Forecast. Soc. Change **173**, 121144 (2021)
6. Viti, M., Lowe, R., McKnight, U.: Knowledge gaps and future research needs for assessing the non-market benefits of nature-based solutions and nature-based solution-like strategies. Sci. Total. Environ. **841**, 156636 (2022)
7. Schleinkofer, U., Moz, D., Lang, A.: Knowledge acquisition in product planning of frugal manufacturing systems for emerging markets. Procedia CIRP **81**, 246–251 (2019)
8. Yi, L., Wang, Y., Yin, Y.: Knowledge spillover, knowledge management capabilities, and innovation among returnee entrepreneurial firms in emerging markets: Does entrepreneurial ecosystem matter? J. Bus. Res. **130**, 283–294 (2021)
9. An, P., Li, H., Chen, F.: The evolution analysis of listed companies co-holding non-listed financial companies based on two-mode heterogeneous networks. Phys. A Stat. Mech. Appl. **484**, 558–568 (2017)
10. Aigbedo, H.: An empirical analysis of the effect of financial performance on environmental performance of companies in global supply chains. J. Clean. Prod. **278**, 121741 (2021)
11. Feng, Q., Chen, H., Jiang, R.: Analysis of early warning of corporate financial risk via deep learning artificial neural network. Microprocess. Microsyst. **87**, 104387 (2021)
12. Jin, X., Liu, Q., Long, H.: Impact of cost–benefit analysis on financial benefit evaluation of investment projects under back propagation neural network. J. Comput. Appl. Math. **384**, 113172 (2021)
13. Sujjaviriyasup, T.: A new class of MODWT-SVM-DE hybrid model emphasizing on simplification structure in data pre-processing: a case study of annual electricity consumptions. Appl. Soft Comput. **54**, 150–163 (2017)
14. Amayri, M., Silva, C., Ploix, S.: Flexibility characterization of residential electricity consumption: a machine learning approach. Sustain. Energy Grids Netw. **32**, 100801 (2022)
15. Lazzari, F., Mor, G., Solsona, F.: User behavior models to forecast electricity consumption of residential customers based on smart metering data. Energy Rep. **8**, 3680–3691 (2022)
16. Tiwari, A., Pindoriya, N.: Automated demand response in smart distribution grid: a review on metering infrastructure, communication technology and optimization models. Electr. Power Syst. Res. **206**, 107835 (2022)

Decision Support System for Multi-criteria Information-Logical Analysis of Scientific and Technical Projects of Innovative Enterprises

A. S. Kuznetsova[1], Y. V. Kulikova[1], V. L. Rozaliev[1](✉), S. V. Pronichkin[2], and O. K. Kasymov[1]

[1] Volgograd State Technical University, Volgograd, Russia
vlrozaliev@mail.ru
[2] Federal Research Center "Computer Science and Control" of Russian Academy of Sciences, Moscow, Russia

Abstract. The search for effective approaches to the analysis of scientific and technical projects, taking into account the conditions for the development of a modern competitive environment, is one of the main tasks of managing innovative enterprises. The determining factor in the effective functioning of the organizational and economic mechanism of scientific research and development is the successful operation of all its constituent elements. A decision support system has been developed that implements a conceptual model of the organizational and economic mechanism for evaluating scientific and technical projects. For multi-criteria information-logical analysis of scientific and technical projects, it is proposed to use technical and organizational subsystems on multisets of binary relations.

Keywords: Scientific research · Decision making · System approach · Knowledge economy

1 Introduction

In the context of global transformation, new commercial opportunities are opening up for the subjects of scientific and technical activity, but this at the same time leads to increased competition between them. Innovative enterprises are forced to find new approaches to managing competitiveness and require intellectual methodological support for the evaluation of scientific and technical projects. Important aspects of the formation and development of the theoretical, methodological and practical base for ensuring competitiveness in the context of global transformation have not received adequate coverage. The issues of creating an effective mechanism for ensuring sustainable competitive positions of innovative enterprises, identifying the impact of marketing orientation on competitiveness, and especially assessing the competitiveness of research and development, taking into account changes in the dynamic macroeconomic environment, need further research.

R. Silhavy et al. (Eds.): CoMeSySo 2022, LNNS 596, pp. 824–830, 2023.
https://doi.org/10.1007/978-3-031-21435-6_72

In conditions of global instability, an important factor is the efficient use of energy resources through the introduction of science-intensive energy-saving technologies. This leads to the study of the process of managing scientific and technical projects at innovative enterprises, which is a necessary condition for increasing the level of profitability, and also affects the improvement of the quality of life of the population. At the same time, the problem of organizing the process of choosing scientific and technical projects at innovative enterprises should be considered comprehensively, taking into account the entire set of conditions that affect this process. Relevant research is the formation of organizational and economic mechanisms for a multi-criteria analysis of scientific and technical projects in the field of energy saving, the construction of a conceptual model using the principles and criteria for evaluating the effectiveness of scientific research and development.

2 Materials and Methods

One of the main tasks of managing innovative enterprises is to assess the level of competitiveness of scientific research and development, which in turn requires the search for effective approaches to the analysis of scientific and technical projects that take into account the conditions for the development of the modern competitive environment [1, 2]. It should be noted that the competitiveness of an innovative enterprise is the ability to create, produce and sell science-intensive goods and services, the price and non-price qualities of which are more attractive than those of similar science-intensive products of competitors.

There are a number of methods for assessing the competitiveness of scientific and technical projects, taking into account both price and non-price factors. Assessment of competitiveness in the market of supply and demand for scientific knowledge should be based on a systematic analysis of the technological, production, financial and marketing capabilities of an innovative enterprise. Such an assessment is the last stage of marketing research and is designed to find out the potential of an innovative enterprise and the measures that should be taken to ensure the competitiveness of scientific and technical projects in a competitive market.

A sufficiently large number of methods for assessing the competitiveness of scientific and technical projects of an innovative enterprise have been developed:

- profile method;
- metric method;
- structural-functional method;
- integral method;
- rating method.

The above methods can be used in the process of analysis both separately and together with others, it depends on the goals of the analysis, as well as the specifics of the activities of the innovative enterprises under study. The advantages of existing approaches are the simplicity of calculation and the availability of operational information about the quality of a science-intensive product, which allows to quickly respond to changes in the market environment. As a drawback, a quantitative approach to assessing the qualitative characteristics of scientific research and development can be noted.

The determining factor in the effective functioning of the organizational and economic mechanism for the implementation of scientific research and development is the successful operation of all the constituent elements of the national scientific and technical system. The economic mechanism includes planning and stimulating the activities of an innovative enterprise. Any innovative enterprise operates under market conditions, so it is necessary to take into account the characteristic features of knowledge-intensive production in society and the dynamics of productive forces. Thus, for the multi-criteria information-logical analysis of scientific and technical projects, it is proposed to use two subsystems of relations: organizational-economic and technical-economic.

The proposed scientific and methodological approaches combine organizational and economic goals, incentives and management decisions aimed at achieving the set goals, using various methods and management tools. As an object of research, innovative enterprises in the field of energy saving are considered.

The organizational and economic mechanism of energy saving is the synthesis and implementation of organizational and economic goals, a motivational mechanism and managerial decisions aimed at energy saving using management tools and methods, the formation of which will help reduce energy dependence, increase competitiveness and increase the efficiency of using innovative technologies. Energy saving is one of the components of energy management and the practical side of its activities [3, 4].

The formation of the organizational and economic mechanism of energy saving at an innovative enterprise is carried out in various directions:

- energy and economic survey of the enterprise;
- definition of energy saving measures;
- implementation of the planned volume of energy saving measures.

Monitoring of energy consumption and control over compliance with energy saving measures will ensure an increase in the level of energy efficiency at an innovative enterprise, with subsequent integration at the regional and state levels [5]. The organizational and economic mechanism of energy saving at an innovative enterprise is a system of interrelated economic and organizational elements that contribute to the economical use of energy resources, the introduction of energy-saving measures, taking into account innovative achievements. The set of organically connected technical and economic means is aimed at stimulating energy-saving processes, investment support for energy saving at all levels of managing energy saving processes [6, 7].

Improving energy efficiency is impossible without the formation of an effective organizational and economic mechanism for energy saving at the level of an innovative enterprise. Energy saving is a kind of resource-saving activity aimed at reducing electricity consumption [8, 9]. Each direction of energy-saving activity is ultimately aimed at conserving resources by reducing energy losses or replacing one energy resource with another. To implement the concept of energy saving at an innovative enterprise, a system of qualitative criteria has been developed for the information and logical analysis of binary relations of scientific and technical projects:

- $C_1 = \{c_1^1, c_1^2, c_1^3, c_1^4\}$ - substantiation of the goals and main tasks of scientific research;
- $C_2 = \{c_2^1, c_2^2, c_2^3, c_2^4\}$ - determination of the principles on which the organizational and economic mechanism of scientific research is based;
- $C_3 = \{c_3^1, c_3^2, c_3^3, c_3^4\}$ - formation of stages of implementation of a scientific and technical project;
- $C_4 = \{c_4^1, c_4^2, c_4^3, c_4^4\}$ - substantiation of the principles of the effectiveness of the implementation of a scientific and technical project;
- $C_5 = \{c_5^1, c_5^2, c_5^3, c_5^4\}$ - development of methods and tools for monitoring the effectiveness of the implementation of a scientific and technical project.

Each of the projects P_i is associated with a multiset M_i based on the set of all criterion estimates of the hyperscale with dimension $n = 5$:

$$B = \bigcup_{j=1}^{n} C_j = \left\{c_1^1, c_1^2, c_1^3, c_1^4; \ldots; c_j^1, c_j^2, c_j^3, c_j^4; \ldots; c_n^1, c_n^2, c_n^3, c_n^4\right\}.$$

The multiset M_i of multiplicities k_i of variants for implementing research and development has the following form:

$$M_i = \left\{k_i\left(c_1^1\right) \circ c_1^1, \ldots, k_i\left(c_1^4\right) \circ c_1^4; \ldots; k_i\left(c_n^1\right) \circ c_n^1, \ldots, k_i\left(c_n^4\right) \circ c_n^4\right\}.$$

Taking into account the developed efficiency criteria and the principle of optimality (1), a conceptual model of the organizational and economic mechanism for the implementation of a scientific and technical project is built:

$$E_i = \left(\sum_{j=1}^{n}\left(M_j^{max} - M_{ij}\right)^2\right)^{\frac{1}{2}} \tag{1}$$

The implementation of the principles of the organizational and economic mechanism of energy saving of an innovative enterprise should be carried out through a number of stages of the implementation of scientific and technical projects:

- involvement in energy saving activities of all stakeholders;
- creation of an inspection to supervise the use of funds from the energy saving fund;
- development and implementation of quality management in the production of science-intensive products and services;
- implementation of energy saving measures;
- adjusting the parameters of the organizational and economic mechanism.

In the decision support system for managing the technical and economic activities of an innovative enterprise, one of the main tasks is to conduct a systematic analysis of promising scientific and technical projects that can indicate areas that an enterprise should develop in order to compete in the supply and demand market for new scientific knowledge.

3 Results

A systematic analysis of scientific and technical projects of innovative enterprises in the electrical field of energy saving was carried out. The measurement of short time intervals is the most urgent task in the electrical field. In particular, repeating short time intervals are an informative parameter for measuring systems in the electric power industry [10, 11].

When solving scientific and technical problems, time intervals in most cases are given in the form of electrical signals, which can be single or repeated. To measure them, the method of direct comparison with the period of quantizing pulses is usually used. The main disadvantage of this method is the limitation of the minimum duration of the measured time intervals by the allowable quantization error, which depends on the maximum possible frequency of quantizing pulses.

Another related problem is the limitation of the use of modern microcontroller element base, which usually operates at much lower clock frequencies. For most microcontrollers, quartz clock generators with a low output signal frequency are used. Therefore, the metrological characteristics of quartz oscillators are known and stable, which allows them to be used to measure time intervals. There are a significant number of microcontrollers operating at much higher clock frequencies due to the use of built-in phase-locked loop [12, 13]. However, the metrological characteristics of the increased clock frequency obtained in this way are unstable due to the instability of the phase-locked loop system, which does not allow it to be used to measure time intervals with normalized metrological characteristics.

4 Discussion

Most of the existing meters for short time intervals have high metrological characteristics [14, 15]. High metrological characteristics are due to the use of a high-speed element base, which leads to a significant increase in the cost of implementing scientific and technical projects of innovative enterprises in the field of energy saving.

As the study showed, to measure short time intervals, for example, a quantization error of 0.3% and the use of an element base of wide consumption and low cost, it is necessary to use special measurement methods.

One of the possible ways to solve this problem is to use the statistical method of measurement in the time interval when using low-cost microcontrollers for wide consumption. The essence of the statistical method is to determine the duration of the pulses by estimating the probability of the coincidence of the pulses that specify the measured time interval and the quantizing pulses.

This method makes it possible to measure time intervals of much shorter periods of quantizing pulses, which makes it possible to use microcontrollers with low speed.

To implement scientific and technical projects in the electrical field based on the statistical method, it is necessary to ensure that the following conditions are met:

- repeatability of measured time intervals;
- uncorrelated sequence of pulses that set the measured intervals and quantizing pulses;
- significantly shorter duration of quantizing pulses in comparison with the measured ones.

When using this method, an instrumental error arises due to the limited duration of the quantizing pulse. Due to the replacement of probability by its estimate, a methodological error arises, which depends on the number of measured time intervals (sample size). The measured time interval is set by the duration of the positive half-wave of the harmonic signal or the duration of the incoming pulse of a rectangular signal to the pulse shaper, built on the basis of a high-speed comparator. The research results showed that the additive error of the shaper does not exceed 0.11%. A high-speed counter counts the pulses corresponding to the coincidence of the pulses of the measured interval and the quantizing ones. To ensure the uncorrelatedness of the measured intervals and quantizing pulses, the frequency of quantizing pulses changes randomly according to a uniform distribution law within 5% of the nominal value. The signal from the generator is fed to a short pulse shaper controlled by a microcontroller based on delay lines and a logic element $\oplus$. With its help, quantizing pulses are formed. The choice of the required duration is carried out using the switch. The developed software makes it possible to study the methodological error of the statistical method by changing the number of measured intervals (sample size), as well as the instrumental error by changing the duration of quantizing pulses. For structure and research purposes, the meter has 10 test points that can be selected by the user and connected to a dual-channel oscilloscope.

5 Conclusion

The implementation of energy saving should be carried out in accordance with the energy saving strategy and its formation at the regional level. Software has been developed that implements a conceptual model of the organizational and economic mechanism of multi-criteria information-logical analysis of scientific and technical projects to ensure energy saving. A significant part of the functional blocks is implemented in software, which allows to change the algorithm of work only by software methods, display debugging information and programmatically correct errors. To conduct research of a statistical meter, a circuit was assembled, which included a pulse generator and a digital oscilloscope. The quantizing pulse generator is implemented by software on the basis of a separate microcontroller. The performed calculations show that the use of the statistical method makes it possible to obtain the lower limit for measuring time intervals with a quantization error of 0.5%. The results of the multi-criteria information-logical analysis of scientific and technical projects in the electric power industry show that a further direction of research is to study ways to improve the accuracy of a statistical time interval meter by optimizing the sample size and duration of quantizing pulses.

Acknowledgments. The article is prepared with the financial support of the Russian Science Foundation, project № 18-78-10047-P.

References

1. Brandl, F., Roider, N., Reinhart, G.: Selecting practices in complex technical planning projects: a pathway for tailoring agile project management into the manufacturing industry. CIRP J. Manuf. Sci. Technol. **33**, 293–305 (2021)
2. Cunha, P., Ribeiro, P.: Definition of a technique for characterizing the expected benefits of a project. Procedia Comput. Sci. **196**, 1007–1012 (2022)
3. Thirunavukkarasu, G., Seyedmahmoudian, M., Stojcevski, A.: Role of optimization techniques in microgrid energy management systems - a review. Energy Strategy Rev. **43**, 100899 (2022)
4. Hegde, B., Ahmed, Q., Rizzoni, G.: Energy saving analysis in electrified powertrain using look-ahead energy management scheme. Appl. Energy **325**, 119823 (2022)
5. Cheng, L., Liu, Y., Yang, Y.: Does technology conglomeration promote innovative outcomes of new energy vehicle enterprises? The moderating effect of divisive faultlines. J. Clean. Prod. **324**, 129232 (2021)
6. Li, F., Zhang, J., Li, X.: Research on supporting developing countries to achieve green development transition: based on the perspective of renewable energy and foreign direct investment. J. Clean. Prod. **372**, 133726 (2022)
7. Zheng, C., Deng, F., Yang, Z.: The impact of China's western development strategy on energy conservation and emission reduction. Environ. Impact Assess. Rev. **94**, 106743 (2022)
8. Hao, Y., Li, Y., Wu, H.: Digitalization and electricity consumption: does internet development contribute to the reduction in electricity intensity in China? Energy Policy **164**, 112912 (2022)
9. Chaudhari, A., Mulay, P.: Algorithmic analysis of intelligent electricity meter data for reduction of energy consumption and carbon emission. Electr. J. **32**, 106674 (2019)
10. Meng, M., Pang, T.: Operational efficiency analysis of China's electric power industry using a dynamic network slack-based measure model. Energy **251**, 123898 (2022)
11. Cai, L., Duan, J., Wang, L.: Pathways for electric power industry to achieve carbon emissions peak and carbon neutrality based on LEAP model: a case study of state-owned power generation enterprise in China. Comput. Ind. Eng. **170**, 108334 (2022)
12. Castaneda, C., Chiu, R., Mancilla, D.: Electronic locking devices based on microcontrollers and chaotic maps using Model-Matching Control. Microprocess. Microsyst. **86**, 104338 (2021)
13. Kumar, A., Patel, N., Gupta, V.: Design, analysis and implementation of electronically interfaced photovoltaic system using ARM Cortex-M4 microcontroller. Comput. Electr. Eng. **98**, 107701 (2022)
14. Wang, W., Chen, D., Lu, Z.: Fast lock-in amplifier electrochemical impedance spectroscopy for big capacity lead-acid battery. J. Energy Storage **40**, 102693 (2021)
15. Martinez, A., Gonzalez, C., Chong, A.: Low-cost, microcontroller-based phase shift measurement system for a wireless power transfer prototype. HardwareX **11**, e00311 (2022)

An Expert System for the Priorities Formation in Sustainable Development Fundamental Research

A. V. Zubkov[1], I. B. Mamai[2], S. V. Pronichkin[3](✉), and A. V. Kholstov[4]

[1] Volgograd State Technical University, Volgograd, Russia
[2] Federal Centre of Theoretical and Applied Sociology of Russian Academy of Sciences, Moscow, Russia
[3] Federal Research Center "Computer Science and Control" of Russian Academy of Sciences, Moscow, Russia
pronichkin@mail.ru
[4] N.N. Semenov Federal Research Center for Chemical Physics Russian Academy of Sciences, Moscow, Russia

Abstract. Determining the priorities for sustainable development involves critical problems identifying, the conditions creating necessary for successfully solving such problems, and identifying a set of methods and tools to ensure that these conditions are met. The existing quantitative approaches to the priorities definition do not consider the problems of resource provision of scientific priorities. The developed expert system makes it possible to use effectively multilevel information-logical structures for the systematic analysis of scientific priorities according to many criteria. It allows ensuring the transition from research goals to methods of achieving these goals, the choice of specific means and the development of the necessary research methodology.

Keywords: System analysis · Decision making · Semantic analysis · Scientific research

1 Introduction

The place of science in the life of modern society, the influence of its achievements on the rate of scientific and technological progress have led to the close attention of management specialists and politicians to the sustainable development scientific priorities determination. The processes occurring in science, factors affecting the effectiveness of the implementation of scientific directions, the social significance of scientific results for determining scientific priorities have become the subject of detailed and comprehensive study.

Identifying sustainable development priorities is a process whose purpose is to highlight critical problems, create the conditions necessary for successfully solving those problems, and identify a set of methods and tools to ensure that these conditions are met. The set of relevant activities is interdependent and forms a system that constantly

R. Silhavy et al. (Eds.): CoMeSySo 2022, LNNS 596, pp. 831–839, 2023.
https://doi.org/10.1007/978-3-031-21435-6_73

needs to be adjusted and updated due to changes in external conditions, for example, the environment quality deterioration (air, water, land), the natural resources reduction, as well as the growth of risks associated with inequality in life quality. The scientific research purpose in the sustainable development field is to obtain new knowledge, to identify the properties and patterns of social and economic processes in order to develop the best management decisions. Various manifestations of these properties are studied using methods and means specific for each area of sustainable development, the choice of which is due to the completeness and accuracy of the tasks of the research being carried out.

2 Materials and Methods

At present, no holistic picture of the key factors in the development of applied areas and the creation of promising and breakthrough technologies based on the results of fundamental research in the field of sustainable development has been proposed. A stable system of criteria for assessing the content and results of fundamental research has not yet developed. For example, it is proposed to evaluate fundamental research on the basis of economic indicators [1–3]. Scientific research is considered as the basis for innovative projects. Thus, as a criterion for the effectiveness of fundamental research, it is proposed to use the net present value, adjusted for the probability of success of a scientific project. On the contrary, in [4] it is argued that it is extremely difficult to assess the effectiveness of costs for fundamental research using traditional economic methods. We can only talk about the effectiveness of fundamental research, but the development of unified approaches to assessing performance is in its infancy. Existing approaches are limited only by the number of published articles, their citation and the impact factor of journals.

The paper [5] can be noted as an example of the application of the quantitative approach to defining scientific priorities. The paper proposes a method for automatically identifying areas of scientific research based on an assessment of the thematic proximity of scientific publications based on scientometric indicators. Databases of patent services and scientific journals are used to identify promising areas. As a result, it turns out that the definition of promising areas of research depends on the structure of search sections based on keywords [6, 7]. Such structures, built automatically on the basis of past scientific publications, may not correspond to real changes and prospects of sustainable development, and, accordingly, promising research directions will not be adequately formulated. It is important to note that in order to apply such approaches, it is necessary to be very careful about the formation of the reference base of the study, to include mainly foresight articles, since the results obtained "yesterday" are reflected in publications and patents. But "today" the boundaries of scientific knowledge have already expanded and by the time the scientific results are published, they may already be "outdated." In addition, some areas are developing faster than others. Every day, new knowledge that has not yet been formalized appears and one cannot do without the opinion of experts.

It is important to note that the existing approaches to the definition of priorities do not consider the problems of resource provision of scientific priorities for sustainable development. There is no simple formula linking the direction importance indicators and the required funding, since each scientific problem has its own level of "readiness" for the application of funds.

It is obvious that the existing formal automatic approaches can be easily transferred to any area of research [8, 9], but the more universal the method for determining scientific priorities, the less applicable it is for solving specific practical problems.

Basic research is aimed at understanding the underlying principles of phenomena and observable facts without specific applications [10, 11]. Applied research, on the other hand, is aimed at obtaining new knowledge necessary to determine the means by which a specific need can be satisfied [12, 13]. For basic research, it is very difficult to determine reliable quantitative relationships between elements of future research, especially between costs and results, and here the uncertainty in the timing and results of research is extremely high. Fundamental research, in contrast to applied research, is characterized by the difficulty of establishing unambiguous links between the volume of research and its result. Therefore, in the fundamental sciences, planning is inapplicable in the sense that it is understood in applied research, where it is possible to set clear deadlines for the development completion.

3 Results

Determining the priorities of scientific research will be effective only if it is carried out as a systemic process, where several operations are carried out simultaneously, including:

- the selection of key science and technology policy goals of sustainable development;
- assessment of the correspondence of scientific research directions to the set goals according to a variety of quality criteria;
- an analysis of the actual state of research in the selected areas, identifying a set of methods and means, i.e. matching resource requirements with actual opportunities.

The proposed modeling option combines both an analysis of the possible consequences (results) of the implementation of scientific directions for achieving goals, and an assessment of the possibilities of their material and technical support. Prioritization is based on the use of the following body of accumulated knowledge about logical connections between various elements of the scientific and technical planning process:

- the relationship between the study of specific problems of sustainable development through experiments and those conclusions and generalizations that are a consequence of understanding the results of these experiments in the system of existing theoretical concepts;
- the relationship between the observed features of dysfunctions of various classes, as well as the methods and means used to mitigate them.

Let us consider the information-logical relationship between the key elements of multi-level information-logical modeling. The scientific problem of specific research for sustainable development is, as a rule, some aspects of solving particular problems (either to test intuitive guesses, or to confirm theoretical predictions). The following generalization of various studies results, which is a deepening of knowledge about the observed phenomena and processes, is already a particular goal. The aim of the set of sciences on sustainable development is to obtain new knowledge about the basic laws of the structure, functioning and development of socio-economic systems.

Identifying scientific problems and defining goals is the first element of multi-level information-logical modeling. Further, in order to achieve the goal, it is necessary to study specific dysfunctions belonging to different classes, i.e. research is developing in various directions. The study of the characteristics of a variety of specific dysfunctions can be considered as "directions for achieving the goal". Directions can be decomposed into their component tasks. Finally, success in the study of specific dysfunctions essentially depends on the chosen research method and the choice of means of implementing the planned directions, the availability of a material and technical base, staffing, etc. Accordingly, the place of these elements in the process of multi-level information-logical modeling is determined as follows: methods and means are "a way to achieve the set goal".

Thus, the essence of multilevel information-logical modeling can be defined as an analysis of possible logical connections between scientific problems, directions and means, and the totality of these connections is established on the basis of a multidimensional expert analysis, and the assessment is based on a multi-criteria expertise. As a result, it is possible to assess the prospects for the knowledge increasing and improving on the most important problems of sustainable development based on the study of various dysfunctions, to clarify the amount of resource provision necessary for the most effective tasks implementation.

The analysis of the links also opens up new possibilities for retrospective studies, i.e. making it possible to assess whether scientific policy was relevant, on the one hand, and technically sound, on the other, in certain periods of development and to examine the impact of the level of research logistics on the degree to which the objectives have been achieved. It is possible to identify new opportunities for the use of scientific results of past years.

Multilevel information-logical modeling is focused precisely on identifying the entire variety of connections and relationships that take place in the current system of scientific knowledge. To carry out such an analysis, it is necessary to single out all the elements of research - various scientific problems, possible directions and necessary ways to solve problems, and then compare them element by element - in order to identify their mutual dependence, influence, etc. It is necessary to develop rules for searching and selecting these elements, as well as the relationships between them, i.e. reproduce the processes of functioning of research components of the study of sustainable development.

4 Discussion

The construction of multi-level information-logical structures for determining priorities is possible only on the basis of identifying sets of elements of all types (goals, objectives, methods and means) and collecting information about these elements and their interrelation.

Since fundamental research of both humanitarian and natural science prevails in the sciences of sustainable development, information about the elements of information-logical structures is of a qualitative nature and can be obtained only with the help of qualified experts.

In view of the need to have as complete information as possible about the elements of multi-level information-logical structures, they should be assessed according to many criteria reflecting various aspects of their complexity. The system of criteria is designed in such a way as to accurately and independently identify the various factors affecting the achievement of goals. As a mathematical model for representing the set of expert assessments of sustainability S, it is proposed to use the multiplicities of multisets:

$$S = \frac{\left(\sum_{i=1}^{N}\left(k_i^+ - k_i\right)^2\right)^{\frac{1}{2}}}{\left(\sum_{i=1}^{N}\left(k_i^- - k_i\right)^2\right)^{\frac{1}{2}} + \left(\sum_{i=1}^{N}\left(k_i^+ - k_i\right)^2\right)^{\frac{1}{2}}}, \tag{1}$$

where k_i is the multiplicity of the multiset by the criterion i; k_i^+ is the maximum value of the multiplicity of the multiset by the criterion i; k_i^- is the minimum value of the multiplicity of the multiset by the criterion i.

In practical tasks of determining priorities of sustainable development, the number of elements under consideration and their relationships is quite large, therefore, a comprehensive analysis of the obtained expert assessments is possible only with the help of software. For this purpose, a software has been developed for the construction and analysis of multi-level information-logical structures for determining priorities. The software block diagram is shown in Fig. 1.

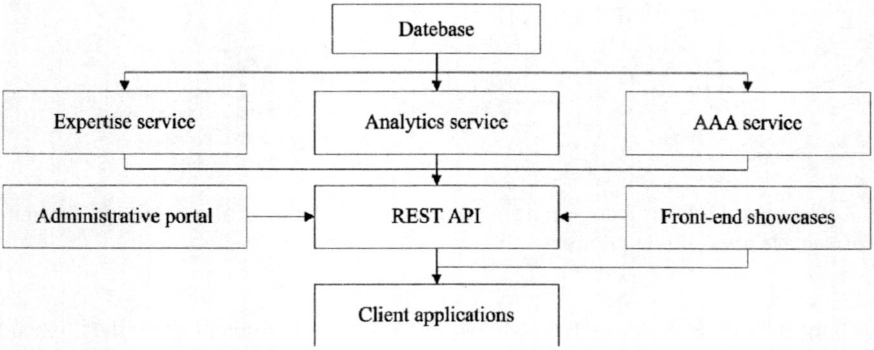

Fig. 1. Block diagram of the expert system for determining the priorities of fundamental research in the sciences of sustainable development

The API server is used by all components of the expert system responsible for the preparation, assessment and interpretation of the examination results. MySQL is used as a database for logic and workflow. Mongodb is used as a database of blocks of expertise and analytics. The server for transformation, preparation and indexing of content is made in python. The index server is based on elastic search and is designed to store indexed data in full-text search. In terms of loading mechanism, for the greatest flexibility, the API is fully compatible with the API AWS S3. Thus, a convenient integration of existing ones, including third-party applications, with the API in question is provided. The content download server is implemented on the basis of the Minio project with a balancing proxy based on nginx.

The developed software ensures the solution of scientific and scientific-technical problems of determining scientific priorities, taking into account the mutual influence of all elements of multi-level information-logical structures. Such software can become part of an integrated information support system for regional government bodies.

The software for modeling multi-level information-logical structures for determining priorities of sustainable development was developed in accordance with the requirement to ensure the possibility of transition from research objectives formulated in a general form to methods for achieving these goals, i.e. to the purposeful allocation of resources, to the choice of specific means, to the development of the necessary research methods, etc. At the same time, the ways of achieving the set goals should be determined, i.e. specific directions in which scientific research should be developed. The program also makes it possible to provide a reverse transition: from methods through directions to scientific problems. A conceptual framework for modeling multi-level information logic structures is shown in Fig. 2.

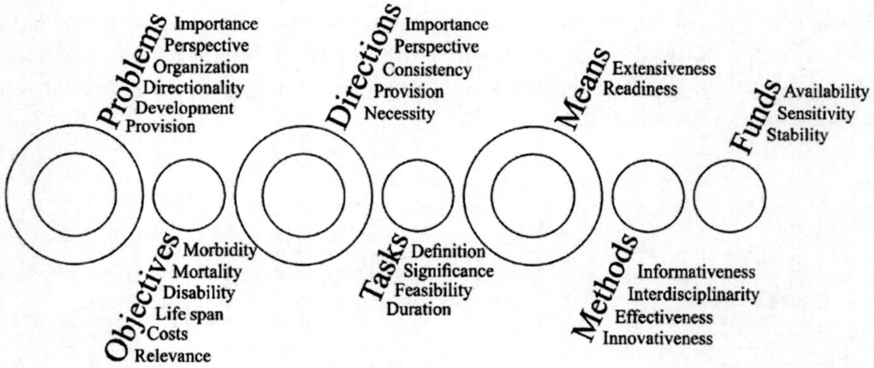

Fig. 2. Conceptual scheme for modeling multi-level information-logical structures for determining sustainable development priorities

These requirements are met as follows. At the research problem level, the impact of the research problem on sustainable development is assessed, and the status and future prospects of research on the problem are identified. The following aspects of scientific problems are analyzed fundamental and applied importance; prospects; organization and focus of scientific research; state and development trends of theoretical and applied works; resource endowment. A joint assessment of fundamental and applied importance will allow achieving both short-term improvements and long-term fundamental discoveries that, over time, will lead to the development of the ability to understand the causes of dysfunctions. All scientific problems can be compared with each other for each of these indicators.

At the level of directions for solving scientific problems, the following characteristics are analyzed: the possible impact of work in this direction on the achievement of goals and the types of research required for this implementation within the direction; state, prospects and availability of research resources; interconnection of directions and methods.

The importance of a direction for any problem is understood as the degree of influence of the results obtained in this direction on progress in solving this problem, i.e. the contribution of this direction is estimated. The results can contribute to a qualitative leap in solving the problem as a whole, or only its individual aspects. Since various characteristics of the problems themselves are assessed independently by experts, and the selection of the most important problems is carried out in accordance with a given scientific and technical policy, the search for directions is due to the already made choice of problems and therefore is much less subjective.

Prospects for the development of each direction are assessed in the light of what kind of research is most needed at the present time to achieve the set goals - either the collection of all kinds of data, or the interpretation of already available information, or the dysfunctions study with fundamentally new methods.

When assessing the state of research in this area, the degree of coordination of work, staffing, needs for material and technical base are considered, the level of theoretical and applied work is revealed. All directions can be compared with each other according to these indicators.

Assessments of the directions contribution in problems solving, the prospects for the development of directions and the relationship of directions, as well as methods, provide mutual transitions between the level of research goals and the levels of methods and means of achieving these goals.

At the level of the way research problems are solved, it becomes clear: coverage, i.e. how widely each method is used in research in various fields; what is its information content, how reliable are the results obtained with its help. Readiness is assessed, i.e. the effectiveness and elaboration of the method, its dependence on the perfection of technical means, as well as on the type and quality of the materials used.

The needs of this area in material and technical support are determined taking into account the applied research methods. However, the analysis and comparison of the methods themselves is carried out on the assumption that the necessary funds are available. In this way, the influence of the quality of the material and technical base on the implementation of the method is divided.

To select the means of achieving goals, each type of material and technical support is assessed according to key parameters and compared with the world level for the same type of support, and in one case, the best of the currently widely used samples is assessed, and in the second, the level of perfection of experimental samples of these types of provision is assessed, which allows to some extent represent the immediate prospects in the field of innovative technologies.

The above connections between the elements of multi-level information and logical structures allow us to move from purely qualitative assessments to the definition of very specific material and technical requirements for research, make it possible, based on the set goals of scientific policy, to determine the necessary means to achieve the goals, including staffing.

5 Conclusion

Assessment of the necessary resources and the development of scientific and technical policy for sustainable development should be carried out using evidence-based approaches and modern methods of decision support. Using the developed software, it is possible to make an informed choice of sustainable development priority areas in such a way as to ensure that as many problems as possible are contributed to the solution of as many problems as possible at the expense of limited resources. The best, respectively, will be those directions that make the maximum contribution to solving the greatest number of problems at the expense of less funds. Using multi-criteria assessments of directions, it is necessary to select those directions that satisfy this principle of optimality. Then, for the selected areas, it is necessary to evaluate the ways of their implementation, the necessary methods and means, thus defining the range of areas, the effective development of which is permissible within the framework of the available funds. The developed scientific and methodological approaches open up the opportunity to formulate a well-thought-out and grounded scientific and technical policy, corresponding to the current level and trends of sustainable development.

Acknowledgments. The article is prepared with the financial support of the Russian Science Foundation, project № 18-78-10047.

References

1. Onofri, L., Paulo, A., Nunes, L.: Economic valuation for policy support in the context of ecosystem-based adaptation to climate change: an indicator, integrated based approach. Heliyon **6**, e04650 (2020)
2. Dziallas, M., Blind, K.: Innovation indicators throughout the innovation process: an extensive literature analysis. Technovation **80**, 3–29 (2018)
3. Coluccia, D., Dabic, M., Solimene, S.: R&D innovation indicator and its effects on the market. An empirical assessment from a financial perspective. J. Bus. Res. 119, 259–271 (2020)
4. Rousseau, S., Catalano, G., Daraio, C.: Can we estimate a monetary value of scientific publications? Res. Policy **50**, 104116 (2021)
5. Laborde, A., Habit, E., Kemp, P.: Strategic methodology to set priorities for sustainable hydropower development in a biodiversity hotspot. Sci. Total Environ. **714**, 136735 (2020)
6. Rodriguez, R., Svensson, G., Wood, G.: Sustainability trends in public hospitals: efforts and priorities. Eval. Program. Plann. **78**, 101742 (2020)
7. Rons, N.: Bibliometric approximation of a scientific specialty by combining key sources, title words, authors and references. J. Informetr. **12**, 113–132 (2018)
8. Mohadab, M., Bouikhalene, B., Safi, S.: Bibliometric method for mapping the state of the art of scientific production in Covid-19. Chaos Solitons Fractals **139**, 110052 (2020)
9. Gao, H., Ding, X., Wu, S.: Exploring the domain of open innovation: bibliometric and content analyses. J. Clean. Prod. **275**, 122580 (2020)

10. Ke, Q.: Technological impact of biomedical research: the role of basicness and novelty. Res. Policy **49**, 104071 (2020)
11. Beck, E.: Access and benefit sharing - the perspective of basic research. Phytomedicine **53**, 302–307 (2019)
12. Fukuda, K.: Effects of trade liberalization on growth and welfare through basic and applied researches. J. Macroecon. **62**, 103058 (2018)
13. Balajee, A., Hande, M.: History and evolution of cytogenetic techniques: current and future applications in basic and clinical research. Mutat. Res. Genet. Toxicol. Environ. Mutagen. **836**, 3–12 (2018)

Applying Verbal Decision Analysis for Public Research and Technology Advancement

D. V. Polevoi[1], S. V. Pronichkin[2](✉), and Y. S. Chernyshova[1]

[1] Federal Center of Theoretical and Applied Sociology of Russian Academy of Sciences, Moscow, Russia

[2] Federal Research Center "Computer Science and Control" of Russian Academy of Sciences, Moscow, Russia

pronichkin@mail.ru

Abstract. Ensuring sustainable development of scientific organizations is impossible without coordinating development strategies at all levels of management, taking into account the totality of factors and environmental conditions. The possibility of using probabilistic modeling tools in making management decisions to ensure sustainable development of scientific organizations is considered. A methodology for modeling the management decisions implementation results is proposed. An algorithm for adapting the decision trees structure to new data arriving in real time has been implemented.

Keywords: Management decisions · Scientific organization · Verbal analysis · Decision making

1 Introduction

An important task of increasing the scientific organizations efficiency is the introduction of analytical decision support systems. At present, the level of information support for the scientific organizations activities is rather low [1, 2]. Usually, the computerization of documenting quantitative scientific results is carried out without their analytical processing. The lack of information systems of a wide profile leads to the impossibility of using any intelligent decision support systems in scientific organizations.

Formation of a unified information system of scientific organizations with broad analytical

support for user activities is an urgent task. This means that it is necessary to create a unified knowledge base with a unified format for presenting scientific data, design software interfaces for working with the knowledge base, and also develop a computer decision support system using information from this knowledge base. With this approach, conditions are automatically created for obtaining information in a standardized representation, which makes it possible to apply the necessary analytical processing [3, 4].

Consequently, there are prerequisites for using formalized decision support systems in the scientific organizations activities (for example, using algorithms for solving optimization problems, forecasting, planning, analyzing the scientific research state [5, 6]).

R. Silhavy et al. (Eds.): CoMeSySo 2022, LNNS 596, pp. 840–846, 2023.
https://doi.org/10.1007/978-3-031-21435-6_74

In the future, the accumulated data and algorithmic procedures can become the basis for the knowledge base of a more powerful intellectual system - an expert system with a wide range of tasks.

The scientific organizations activities computerization relies on the core of the information system in the form of a knowledge base and corresponding algorithms for data processing in order to model and predict sustainable development [7, 8].

2 Materials and Methods

The process of accumulating knowledge is divided into obtaining knowledge from an expert, organizing knowledge that ensures the system effective operation, and presenting knowledge in a form understandable to a decision-maker [9, 10]. The process of intellectualization of the analytical decision support system means providing the user with a fundamental opportunity not only to receive information based on data processing, but to use this information, as well as the accumulated experience and knowledge in the work process.

Bayesian networks are one of the highly effective tools for mathematical modeling, forecasting and decision support [11, 12]. They arose at the junction of probability theory and mathematical statistics and have a number of advantages over other modeling methods. In particular, the ability to take into account verbal and numerical variables in one model, and the number of variables can reach several hundred, the availability of alternative methods for generating probabilistic inference and the correct presentation of cause-and-effect relationships.

Bayesian networks are composed of many nodes and a collection of directed edges that connect these nodes to each other. Links define causal relationships in the domain. The reliability of a statement is represented using probability. The concept of Bayesian networks is to update probabilities when additional information arrives. Each node of the network can receive information, since the method of updating the probabilities is invariant with respect to the information propagation direction along the edges of the network. Thus, the Bayesian network, as the basis of the expert system, significantly expands the possibilities of analysis and decision-making, since it allows making both direct and reverse logical conclusions at the same time. In addition, the simultaneous introduction of information about the states of several nodes does not change the network processing algorithm, which makes it possible to exclude situations of logical contradiction [13, 14]. Bayesian networks in expert systems have such significant advantages as the inference algorithms computational interpretation possibility, the flexibility of the information dissemination process and adaptation to new data.

The adaptation of the constructed network to the new data is performed in the following sequence:

- implementation of the procedure for correcting the model structural part (removing arcs that do not correspond to the data set and adding new arcs);
- procedure for correcting the model probabilistic part.

Since at the initial stage of training a Bayesian network, the probabilistic component of the model is represented by tables of conditional probability distributions obtained directly from the frequency analysis of the appearance of variable values in observations, therefore, changes in the procedure for correcting the model probabilistic part are immediately determined. To obtain a positive effect from adaptation, it is necessary to compensate for the losses from extracting an arc and from adding a new arc. The arc is estimated by calculating the value of the local quality functional. Accordingly, the applicant for the addition must determine the input arcs configuration that has the greatest value of the local quality functional of the expert system.

3 Results

The expert system must implement the strategic analysis process in consultation with the user, providing, on the basis of generalized expert experience, its assessment of strategic alternatives for ensuring sustainable development.

The implementation of such a system on the basis of Bayesian networks means that it is necessary to build a network by highlighting the incoming nodes (probabilities of user-defined states) and the resulting nodes in it. Further, by assigning certain values to the states of the input nodes, it is necessary to obtain the resulting nodes states probabilities.

An expert system based on a domain model in the form of a Bayesian network can also be used to determine the optimal development strategy for a scientific organization.

The goals of modeling when developing a strategy for the development of a scientific organization are:

- assessment of the current management strategy;
- determination of possible methods and means of managing the organization development;
- comparative analysis of the certain management decisions implementation consequences;
- selection of the best development strategy.

To achieve the set goals, it is necessary to determine the criteria that determine the effectiveness degree of the implemented strategy. In addition, it is necessary to determine the factors influencing the final result. The main criteria that determine the level of competitiveness and prospects for the development of a scientific organization are the following: the scientific personnel value; the organization development level; the scientific results commercialization level. The scientific personnel value describes the level of motivation and qualifications. This criterion makes it possible to assess the quality of the labor resources of a scientific organization.

The level of development is an indicator of the scientific and technical structure of the organization. This criterion determines the effectiveness of the internal organization and the scientific activity excellence degree.

The scientific results commercialization level is an integral indicator of the organization's activities. It allows to assess the relevance and feasibility of scientific results based on qualitative and quantitative indicators.

The following indicators are used as factors influencing the value of the organization's performance criteria: scientific results; fixed assets; research costs. The key factor of efficiency is the organization's labor resources - the number of employees, labor productivity and average wages of employees. These factors affect all criteria for the development of a scientific organization.

In the terminology of Bayesian networks, the above indicators and integral criteria constitute the nodes of the network. Taking into account the cause-and-effect relationships between these factors, a Bayesian network is being built. Further, for each node entering the network, it is necessary to determine the set of possible states. Each node can take only two verbal states - "increasing" or "decreasing". Conditional options for the development of a scientific organization are expert assessments on the basis of which the expert system is built, and the adequacy of the constructed model depends specifically on their accuracy.

Conditional options for the development of a scientific organization, created on the basis of expert information, should ensure the implementation of logical relationships characteristic of the subject area. This means that the following qualitative conditions must be met: a decrease in the average wage increases the number of people dismissed; an increase in average wages leads to an increase in the scientific results volume; an increase in labor productivity leads to an increase in the scientific results volume; all types of costs increase the scientific activity cost; an increase in the scientific activity cost reduces the scientific results commercialization rates; an increase in labor productivity leads to an increase in the scientific results commercialization rates.

With an increase in the number of employees and a simultaneous increase in labor productivity, the scientific results volume will increase, and the probability of this growth, according to experts, is very high. With a simultaneous decrease in the number of employees and an increase in their labor productivity, the expert assessment is complicated, and the probabilities of growth and decrease of this indicator are assumed to be equal.

4 Discussion

The first step in determining the optimal strategy for sustainable development is the scientific organization current state analysis. The root nodes of the decision tree for the development of a scientific organization are considered as objects of management, that is, indicators, changing the values (states probabilities) of which allows to obtain the desired result. Leaf nodes of the tree are interpreted as indicators of modeling results. The probabilities of the descendants of these nodes obtained as a result of modeling are the results of modeling the development strategy of a scientific organization. Since in the process of analyzing causal relationships between network nodes, experts take into account the current state of a scientific organization, then the current strategy is modeled by stationary states of root nodes.

An expert assessment of the current strategy gives the initial values of the development criteria, with which the results of modeling other strategies can be compared. The next step in the analysis is to model the implementation of other strategic decisions. To make a managerial decision, it is necessary to find an integral criterion for the development of a scientific organization for each strategy, the best value of which will determine the best strategy.

Let the simulation consider C criteria and L strategies for the development of a scientific organization. As a result of modeling each strategy for verbal assessments of each expert on the Bayesian network, the values of the multiplicities of the probabilities of increasing the criteria $k_{cl},\ c = 1 \ldots C,\ l = 1 \ldots L$ are obtained and the initial values of k_{c0} for the current state of the scientific organization are known. Then, for each strategy, a weighted indicator of the increase in the criteria for the development of a scientific organization is calculated:

$$k_l = \left(\sum_{c=1}^{C} w_c^2 \left(k_{cl}^* - k_{cl}\right)^2 \right)^{\frac{1}{2}}, \quad l = 0 \ldots L, \tag{1}$$

where w_c is a weighting factor that determines the importance of increasing the cth criterion for the development of a scientific organization. The weighted indicator of the multiplicity of multisets [15] of verbal assessments k_l can be considered as the probability of an increase in some integral criterion for the sustainable development of a scientific organization, that is, k_l is a value that characterizes the probability of a general improvement in the level of development of a scientific organization when implementing the lth strategy in comparison with the ideal situation k_{cl}^*.

An integral criterion for the development of a scientific organization can be found as the ratio of the weighted indicator $k_l,\ l = 1 \ldots L$ to the indicator of the current state k_{c0}:

$$S_l = \frac{k_l}{\left(\sum_{c=1}^{C} w_c^2 \left(k_{cl}^* - k_{c0}\right)^2 \right)^{\frac{1}{2}}}, \quad l = 1 \ldots L. \tag{2}$$

Then, as the optimal strategy z^*, one should take one for which the integral development criterion has the maximum value:

$$z^* = \arg\max_l S_l. \tag{3}$$

The strategic importance of the considered criteria allows one to determine the cause-and-effect relationships between the network nodes, thereby defining the edges of the network and their direction. Determination of conditional probabilities is based on information received from experts. Decision tree nodes are qualitative and relate to expectations in future time periods.

A method for predicting the indicators values of sustainable development of a scientific organization for the coming period has been developed. With these predictive values, the Bayesian network nodes states can be determined. The required series of data, which will be processed by statistical methods, can be obtained by executing the corresponding queries to the knowledge base.

It is necessary to make a forecast of the series value at the moment of time $t + 1$. The resulting predicted value is denoted by $\hat{x}_{t+1}$. This value can be considered as the value of a random variable. Then, with an error that can be neglected for solving practical problems, the fair equality $\left(\hat{x}_{t+1} \in [B_{min}; B_{max}]\right) \cong 1$, where B_{min} and B_{max} are determined according to the six sigma rule [16, 17].

The following equation is assigned to each state of a node in a Bayesian network:

$$y_i = \frac{(I - i)B_{min} + (i - 1)B_{max}}{I - 1}, \quad i = 1 \ldots I. \tag{4}$$

A network node is associated with a measure, and all of its states are sorted in ascending order of that measure. Then the node will take the state i, if $\hat{x}_{t+1} = y_i$, and in the general case the probability of each state can be determined based on the distance $\hat{x}_{t+1}$ to each y_i, $i = 1 \ldots I$:

$$k_i = \begin{cases} 0, if \ \hat{x}_{t+1} \leq y_1 \\ \frac{1}{\hat{x}_{t+1} - y_i}, if \ B_{min} < \hat{x}_{t+1} < B_{max} \\ 1, if \ \hat{x}_{t+1} \geq y_I \end{cases}. \quad (5)$$

The above equations determine how the developed expert system integrates statistical data from the knowledge base into an intelligent decision support system.

5 Conclusion

The Bayesian network is a powerful probabilistic tool for modeling processes of various natures that allows to develop decision-making strategies in the face of different nature uncertainties. On the example of the scientific organizations activities, the possibility of solving the problems of choosing a strategic development plan has been demonstrated. To solve the main problem, it is necessary to create a Bayesian network on a set of related events with known a priori probabilities, that is, to build an acyclic graph characterized by a set of probabilistic parameters; to investigate the characteristics of the resulting network with the goal of further application to the analysis and assessment of the state of a scientific organization, as well as determine the strategies for its further development. The concept of modeling the scientific organizations activities and identifying a set of strategic goals of its development is described; the factors influencing the achievement of these goals are highlighted; a model of the activity of a scientific organization in the form of a Bayesian network has been built. A procedure for constructing Bayesian networks based on statistical data and expert assessments for predicting the sustainable development of scientific organizations is proposed.

Acknowledgments. The article is prepared with the financial support of the Russian Science Foundation, project № 19-78-10055.

References

1. Son, H., Chung, Y., Hwang, H.: Do technology entrepreneurship and external relationships always promote technology transfer? Evidence from Korean public research organizations. Technovation **82**, 1–15 (2019)
2. Pearce, J.: Ensuring technical competency for management of research-focused organizations. J. High Technol. Manag. Res. **29**, 172–180 (2018)
3. Vancauwenbergh, S.: Governance of research information and classifications, key assets to interoperability of CRIS systems in inter-organizational contexts. Procedia Comput. Sci. **106**, 335–342 (2017)
4. Castro, P., Dvorak, J., Simons, E.: OpenAIRE compatibility for CRIS systems: recent progress. Procedia Comput. Sci. **146**, 182–189 (2019)

5. Biesenbender, S., Petersohn, S., Thiedig, C.: Using Current Research Information Systems (CRIS) to showcase national and institutional research (potential): research information systems in the context of Open Science. Procedia Comput. Sci. **146**, 142–155 (2019)
6. Guillaumet, A., Garcia, F., Cuadron, O.: Analyzing a CRIS: from data to insight in university research. Procedia Comput. Sci. **146**, 230–240 (2019)
7. Sokhanvar, S., Matthews, J., Yarlagadda, P.: Importance of knowledge management processes in a project-based organization: a case study of research enterprise. Procedia Engineering. **97**, 1825–1830 (2014)
8. Leitne, K., Warden, C.: Managing and reporting knowledge-based resources and processes in research organisations: specifics, lessons learned and perspectives. Manag. Account. Res. **15**, 33–51 (2004)
9. Martinez, J., Gamero, J., Delgado, L.: Empirical analysis of organizational archetypes based on generation and adoption of knowledge and technologies. Technovation **96**, 102145 (2020)
10. Nicolaescu, S., Florea, A., Zanetti, P.: Human capital evaluation in knowledge-based organizations based on big data analytics. Future Gener. Comput. Syst. **111**, 654–667 (2020)
11. Sun, B., Zhou, Y., Zhang, W.: A new PC-PSO algorithm for Bayesian network structure learning with structure priors. Expert Syst. Appl. **184**, 115237 (2021)
12. Liu, Y., Wang, L., Mammadov, M.: Learning semi-lazy Bayesian network classifier under the c.i.i.d assumption. Knowl. Based Syst. **208**, 106422 (2020)
13. Chen, G., Ge, Z.: Robust Bayesian networks for low-quality data modeling and process monitoring applications. Control. Eng. Pract. **97**, 104344 (2020)
14. Atoui, M., Cohen, A., Kobi, A.: A single Bayesian network classifier for monitoring with unknown classes. Eng. Appl. Artif. Intell. **85**, 681–690 (2019)
15. Pronichkin, S., Chernyshova, Y., Kholstov, A., Mamai, I.: Multiplicity based algorithms for processing group multi-criteria expert assessments. J. Phys. Conf. Ser. **2060**, 012010 (2021)
16. Kaswan, M., Rathi, R.: Analysis and modeling the enablers of Green Lean Six Sigma implementation using interpretive structural modeling. J. Clean. Prod. **231**, 1182–1191 (2019)
17. Wang, C., Chen, K.: New process yield index of asymmetric tolerances for bootstrap method and six sigma approach. Int. J. Prod. Econ. **219**, 216–223 (2019)

Finding Research Community Structures Based on Semantic Relations of Concepts

Y. S. Chernyshova[1], S. V. Pronichkin[2(✉)], and A. V. Gayer[1]

[1] Federal Center of Theoretical and Applied Sociology of Russian Academy of Sciences, Moscow, Russia
[2] Federal Research Center "Computer Science and Control" of Russian Academy of Sciences, Moscow, Russia
pronichkin@mail.ru

Abstract. Sustainable development as an alternative to extensive growth is a global problem, the state of which determines the civilization progress. Scientific and methodological approaches to the analysis of sustainable development projects of the main subsystems of the socio-economic system (social, industrial, ecological) have been developed. Algorithms have been developed for finding research community structures based on semantic relations of concepts. The combined use of artificial neural networks and genetic algorithms made it possible to combine the flexibility of network configuration and the adaptability of the evolutionary approach for modeling more complex connections.

Keywords: Semantic relations · Systems analysis · Knowledge engineering · Conceptual analysis

1 Introduction

To ensure sustainable development on time and within certain resources, innovative projects must be carefully planned and well managed. In practice, mistakes in the selection of innovative projects, risk analysis and conceptual planning lead to the following consequences: limited resources are used for clearly ineffective operations; the research community risks are increasing [1, 2].

Errors in planning and implementation of joint innovation projects have the following consequences: excess of the initial cost and failure to meet the deadlines for implementation; the launch of fixed assets is delayed, resulting in a failure to meet the sustainable development goals; information systems projects run out of schedule and over budget, which negatively impacts management, overall costs and the performance of the research community.

To make informed decisions, it is necessary to formulate criteria, the planned values of the main indicators of a joint innovative project to ensure sustainable development. Usually, the actual values as a result of the implementation of a joint innovation project deviate from the planned [3, 4]. Large-scale deviations can lead to large losses. Therefore, for the research community considering several options for implementing a joint

R. Silhavy et al. (Eds.): CoMeSySo 2022, LNNS 596, pp. 847–852, 2023.
https://doi.org/10.1007/978-3-031-21435-6_75

innovation project, it is important to have data on the predicted values of the project implementation risk, the deviation of costs and the project duration from the planned ones, obtained at the stage of pre-project planning, before starting the implementation of the joint innovation project.

2 Materials and Methods

The problem of forecasting deviations of costs and duration of the implementation of a joint innovative project from those planned at the stage of pre-project planning is relevant. The values of deviations obtained at this stage, and therefore the values of the actual costs and duration of implementation, allow to effectively manage the project at an early stage. This will allow not only to reduce the most critical factors that can lead to deviation, but also to effectively manage resources.

Management of a joint innovative project and resources of the main subsystems of the socio-economic system (social, industrial, environmental), in view of the obtained predicted values, allows to directly reduce the deviation of costs, avoid losses due to missed deadlines, idle time or untimely provision of resources, which also increases costs [5, 6].

Effective management of joint innovative projects is a multi-criteria decision support task. The existing methods of multivariate optimization strongly depend on the choice of the initial search point and impose additional restrictions on the objective control function properties [7, 8].

Effective management of joint innovative projects for sustainable development is possible on the basis of a neural network approach. The properties of neural networks are highly dependent on their architecture.

A network containing one hidden layer builds multiple sigmoid slopes. Their number corresponds to the number of neurons in the hidden layer, and the original neuron combines a common surface from them. There are areas on the surface that correspond to the levels of belonging of the input data to a certain class. If the network contains two hidden layers, the surface will have a more complex structure and may contain more vertices. Using a network of this type, it is possible to reproduce a surface of any complexity, which makes it possible to effectively solve most classification problems.

In the process of training in the multilayer perceptron, the weights of interneuronal connections and the bias parameters of the adders of neurons are adjusted. Accordingly, the response surfaces of each of the neurons rotate in the desired direction and change the slope, adjusting in the best way for the classification of the input data.

The functioning of the radial-basis neural network is based on a different principle of the output signal formation. This principle works by analogy with the functioning of a specific type of neurons, in which the output signal extends only to a certain limited area of space. In a radial-basis neural network, artificial neurons are characterized by activation functions that radially change around a selected center and take nonzero values only in a certain vicinity of this center.

Radial base networks have a number of advantages over the multilayer perceptron: there is no need to study the optimal number of hidden layers (since only one is always used) and a linear combination of the outputs of the neurons of the hidden layer in the original layer. Due to these advantages, the radially basis network learns quickly. On the other hand, to model complex initial functional dependencies, a radially basis network requires more elements than a perceptron. Therefore, the choice of the network type when solving the problem of finding the research community structures remains a problem, the solution of which requires experimental calculations.

In addition to choosing the network type, the problem of determining the optimal network configuration arises, which also requires a number of experiments. First, some initial configuration is selected. For example, a multilayer perceptron with one hidden layer and the number of neurons, at which the number of network parameters (weights of interneuronal connections between all neurons of each pair of adjacent layers and parameters of displacement of adders of all neurons) will not exceed a third of the number of input vectors. Next, a series of experiments are carried out with different configurations. In addition, several experiments should be performed for each individual configuration to ensure that the training process does not result in a local minimum of simulation error.

3 Results

The adequacy criterion of the neural network for each new configuration can be the percentage of correctly classified examples from the test set. If, as a result of the experiment, the perceptron does not demonstrate satisfactory classification accuracy, it is necessary to change the configuration by increasing the number of neurons or the number of hidden layers. If there is a retraining of the network, then the number of neurons in the hidden layers or the number of the most hidden layers should be reduced. Similar experiments are carried out to find the optimal configuration of the radially basic network [9, 10].

A lot of algorithms and methods that are used to find a solution evolutionary principles are combined under the general name - evolutionary algorithms, one of the main types of which is the genetic algorithm [11, 12].

From the point of view of artificial information processing systems, genetic search is a specific method for solving the optimization problem, while such an iterative search adapts to the characteristics of the objective function: new chromosomes that appear in the process of crossing test ever wider regions of the search space and are predominantly located in the optimum region, and relatively rare mutations prevent the degeneration of the gene pool, which is equivalent to a rare but continuous search for the optimum in other areas of the search space [13, 14].

Genetic algorithms are used to tune the structure of the neural network, while the network architecture is encoded in the chromosome [15, 16]. In this case, gradient algorithms can be used for training. The fitness of each individual representing the structure of the network is assessed depending on the learning outcomes. In this case, the neuro-evolutionary approach makes it possible to automate the process of choosing a neural network architecture with the possibility of considering irregular architectures, which can lead to a better result.

In the process of searching for the most effective architecture for assessing the fitness of a particular individual, it is possible to train networks of the corresponding architecture based on standard methods. To reduce the influence of random factors on the neural network structure assessment (the learning results are sensitive to the initial conditions and the learning algorithm parameters values), it is proposed to carry out several independent learning operations, and the best result can be used as an assessment of the neural network topology - the fitness of an individual.

The neuro-evolutionary approach optimizes the weights of neural networks, the values of which are encoded in chromosomes [17, 18]. To represent the neural networks weights, the chromosome of each individual was represented by a multiset of parameters.

An information criterion has been developed, which allows to evaluate, using the slope tangent, how much the actual result of a joint innovation project differs in relation to the predicted result of ensuring sustainable development. Moreover, if the sign of the slope of the predicted deviation does not correspond to the slope of the actual deviation, then the result of their ratio is not taken into account, which makes it possible to evaluate not the optimized value of the result, but its information content for the decision-maker.

4 Discussion

Algorithms have been developed for predicting the deviation of the actual results of an innovative sustainable development project from those planned at the stage of pre-project planning based on neural networks and an evolutionary approach to finding research community structures based on the semantic relations of the concepts:

- to form training and test samples in the form of $(\{e_i, t_i\}, z_i)$, where e_i is the efficiency of the ith innovative project; t_i is the risk of the ith innovative project (calculated on the basis of the method for predicting the risk of implementing innovative projects at the stage of pre-project planning based on ensembles of neural networks with clustering); z_i is the deviation of the actual cost (duration) of the ith innovative project from the planned one;
- highlight concepts based on all available data using self-organization maps based on semantic relations;
- for each concept, define association rules:

$$R_i^a = \sup(e_i \to z_i) + \sup(t_i \to z_i); \tag{1}$$

- then the interest of each of the identified rules is assessed:

$$R_i^e = \frac{\sup(e_i \to z_i)}{\sup(e_i)} \tag{2}$$

$$R_i^t = \frac{\sup(t_i \to z_i)}{\sup(t_i)} \tag{3}$$

- select from the training sample a subsample, which includes projects related to the ith concept;
- on the basis of a training subsample using the genetic algorithm, select the optimal architecture and calculate the weight of the connections of the cascade neural network;
- using the cascading neural network for each project for a test sample related to concept i, , determine the predicted volume of actual costs (duration) of an innovative project.

The choice of architecture and setting the weights of the connections of the cascading neural network can be implemented in two steps. It is proposed to select an architecture using the genetic algorithm based on gradient learning methods, and then configure the neural network for the selected architecture using the genetic algorithm.

The neuro-evolutionary approach proposed for adjusting the architecture and weights of connections made it possible to obtain a prediction result, the relative deviation of which from the actual one exceeds almost three times the corresponding result based on conventional regression models. The developed finding algorithms for the structures of the research community based on the semantic relations of concepts make it possible to model more complex connections due to the presence of connections not only between adjacent layers of the network.

The use of two informative features (the project efficiency value and risk) made it possible to improve the forecasting results.

Thus, the proposed method made it possible to improve the forecasting results based on the automated tuning of the architecture, the weights of the neural network connections, the use of performance indicators (social, industrial, environmental) and risk, as well as concepts for the specialization of the neural network.

5 Conclusion

The study of the problems of predicting the actual results of an innovative sustainable development project at the stage of pre-project planning based on indicators of costs and duration of implementation is carried out. The effectiveness of the application of models based on neural networks of different architectures has been studied. It is proposed to use a neuro-evolutionary approach to select the architecture and weights of neural network connections for forecasting based on the available training sample. This made it possible not only to automate this process, but also to achieve a significant improvement in the average results of models based on neural networks. The use of two informative features (efficiency and risk) and specialization of neural networks based on semantic relations of concepts instead of one informative feature also made it possible to improve the results. The systems criterion is proposed, on the basis of which the informativeness of the results of forecasting the structures of the research community is assessed. The scientific novelty lies in the fact that a new method is proposed for predicting the deviation of the actual results of an innovative sustainable development project from the planned indicators at the stage of pre-project planning, which made it possible to increase the accuracy of forecasting results in comparison with existing approaches.

Acknowledgments. The article is prepared with the financial support of the Russian Science Foundation, project № 19-78-10055.

References

1. Wu, W., Liu, Y., Yu, B.: How innovative knowledge assets and firm transparency affect sustainability-friendly practices. J. Clean. Prod. **229**, 32–43 (2019)
2. Battaglia, M., Annesi, N., Sargolini, M.: Future perspectives of sustainable development: an innovative planning approach to inner areas. Experience of an Italian alpine region. Futures **114**, 102468 (2019)
3. Overturf, E., Ravasio, N., Ubiali, D.: Towards a more sustainable circular bioeconomy. Innovative approaches to rice residue valorization: the RiceRes case study. Bioresource Technol. Rep. **11**, 100427 (2020)
4. Ma, Y., Liu, Y.: Turning food waste to energy and resources towards a great environmental and economic sustainability: an innovative integrated biological approach. Biotechnol. Adv. **37**, 107414 (2019)
5. Roth, T., Spieth, P.: The influence of resistance to change on evaluating an innovation project's innovativeness and risk: a sensemaking perspective. J. Bus. Res. **101**, 83–92 (2019)
6. Rathore, H., Jakhar, S., Madhumitha, E.: Examining the mediating role of innovative capabilities in the interplay between lean processes and sustainable performance. Int. J. Prod. Econ. **219**, 497–508 (2018)
7. Li, M., Cao, X., Shang, R.: Sustainable management of agricultural water and land resources under changing climate and socio-economic conditions: a multi-dimensional optimization approach. Agric. Water Manag. **259**, 107235 (2021)
8. Lai, X., Hao, J., Yue, D.: Diversity-preserving quantum particle swarm optimization for the multidimensional knapsack problem. Expert Syst. Appl. **149**, 113310 (2020)
9. Lucas, P., Alves, M., Guimaraes, F.: Reference evapotranspiration time series forecasting with ensemble of convolutional neural networks. Comput. Electron. Agric. **177**, 105700 (2020)
10. Lella, E., Vessio, G.: Ensembling complex network 'perspectives' for mild cognitive impairment detection with artificial neural networks. Pattern Recognit. Lett. **136**, 168–174 (2020)
11. Mamai, I., Savelyev, B., Pronichkin, S., Kholstov, A.: Using evolutionary algorithms to determine the environmental projects effectiveness. J. Phys. Conf. Ser. **2060**, 012009 (2021)
12. Polap, D.: An adaptive genetic algorithm as a supporting mechanism for microscopy image analysis in a cascade of convolution neural networks. Appl. Soft Comput. **97**, 106824 (2020)
13. Chen, R., Yang, B., Wang, S.: A self-learning genetic algorithm based on reinforcement learning for flexible job-shop scheduling problem. Comput. Ind. Eng. **149**, 106778 (2020)
14. Alhaddad, W., Halabi, Y., Yu, Z.: Optimum design method for simplified model of outrigger and ladder systems in tall buildings using genetic algorithm. Structures **28**, 2467–2487 (2020)
15. Yang, J.: Indoor space compositions based on genetic algorithms to optimize neural networks. Phys. Commun. **42**, 101167 (2020)
16. Chen, Z., Huang, A., Qiang, X.: Improved neural networks based on genetic algorithm for pulse recognition. Comput. Biol. Chem. **88**, 107315 (2020)
17. Yan, C., Li, M., Qi, M.: Improved adaptive genetic algorithm for the vehicle Insurance Fraud Identification Model based on a BP neural network. Theor. Comput. Sci. **817**, 12–23 (2019)
18. Esfeh, H., Safikhani, H., Motahar, S.: Multi-objective optimization of cooling and heating loads in residential buildings integrated with phase change materials using the artificial neural network and genetic algorithm. J. Energy Storage **32**, 101772 (2020)

Author Index

R. Silhavy et al. (Eds.): CoMeSySo 2022, LNNS 596, pp. 853–855, 2023.
https://doi.org/10.1007/978-3-031-21435-6